▭	Metals
▭	Metalloids
▭	Nonmetals

10 VIII	11 IB	12 IIB	13 IIIA	14 IVA	15 VA	16 VIA	17 VIIA	18 VIIIA
								helium 2 **He** 4.00
			boron 5 **B** 10.81	carbon 6 **C** 12.01	nitrogen 7 **N** 14.01	oxygen 8 **O** 16.00	fluorine 9 **F** 19.00	neon 10 **Ne** 20.18
			aluminum 13 **Al** 26.98	silicon 14 **Si** 28.09	phosphorus 15 **P** 30.97	sulfur 16 **S** 32.07	chlorine 17 **Cl** 35.45	argon 18 **Ar** 39.95
nickel 28 **Ni** 58.69	copper 29 **Cu** 63.55	zinc 30 **Zn** 65.39	gallium 31 **Ga** 69.72	germanium 32 **Ge** 72.61	arsenic 33 **As** 74.92	selenium 34 **Se** 78.96	bromine 35 **Br** 79.90	krypton 36 **Kr** 83.80
palladium 46 **Pd** 106.42	silver 47 **Ag** 107.87	cadmium 48 **Cd** 112.41	indium 49 **In** 114.82	tin 50 **Sn** 118.71	antimony 51 **Sb** 121.75	tellurium 52 **Te** 127.60	iodine 53 **I** 126.90	xenon 54 **Xe** 131.29
platinum 78 **Pt** 195.08	gold 79 **Au** 196.97	mercury 80 **Hg** 200.59	thallium 81 **Tl** 204.38	lead 82 **Pb** 207.2	bismuth 83 **Bi** 208.98	polonium 84 **Po** (209)	astatine 85 **At** (210)	radon 86 **Rn** (222)
ununnilium 110 **Uun** (269)	unununium 111 **Uuu** (272)	ununbium 112 **Uub** (277)						

gadolinium 64 **Gd** 157.25	terbium 65 **Tb** 158.93	dysprosium 66 **Dy** 162.50	holmium 67 **Ho** 164.93	erbium 68 **Er** 167.26	thulium 69 **Tm** 168.93	ytterbium 70 **Yb** 173.04	lutetium 71 **Lu** 174.97
curium 96 **Cm** (247)	berkelium 97 **Bk** (247)	californium 98 **Cf** (251)	einsteinium 99 **Es** (252)	fermium 100 **Fm** (257)	mendelevium 101 **Md** (258)	nobelium 102 **No** (259)	lawrencium 103 **Lr** (260)

Chemistry Fundamentals

An Environmental Perspective

SECOND EDITION

Phyllis Buell
James Girard
American University

JONES AND BARTLETT PUBLISHERS
Sudbury, Massachusetts
BOSTON TORONTO LONDON SINGAPORE

World Headquarters
Jones and Bartlett Publishers
40 Tall Pine Drive
Sudbury, MA 01776
978-443-5000
info@jbpub.com
www.jbpub.com

Jones and Bartlett Publishers Canada
2406 Nikanna Road
Mississauga, ON L5C 2W6
CANADA

Jones and Bartlett Publishers International
Barb House, Barb Mews
London W6 7PA
UK

Dedication

Dedicated to the Memory of William C. Buell IV and Mary Catherine Girard

Production Credits
Chief Executive Officer: Clayton Jones
Chief Operating Officer: Don W. Jones, Jr.
Executive V.P. & Publisher: Robert W. Holland, Jr.
V.P., Design and Production: Anne Spencer
V.P., Sales and Marketing: William Kane
V.P., Manufacturing and Inventory Control: Therese Bräuer
Executive Editor: Stephen L. Weaver
Senior Developmental Editor: Dean W. DeChambeau
Senior Production Editor: Louis C. Bruno, Jr.
Senior Marketing Manager: Nathan Schultz
IT Manager: Nicole Healey
Cover Design: Anne Spencer
Cover Images: © PhotoDisc, Inc.
Production Service: Jane Hoover/Lifland et al., Bookmakers
Composition: Monotype Composition Company, Inc.
Text Artwork: Gayle F. Hayes
Printing and Binding: Courier Companies
Cover Printing: Courier Companies

Credits appear on pages 641–642, which constitute a continuation of the copyright page.

The first edition of this book was published by Prentice Hall, Inc.

Library of Congress Cataloging-in-Publication Data unavailable at time of printing.

Printed in the United States of America
06 05 04 03 02 10 9 8 7 6 5 4 3 2 1

Brief Contents

Contents

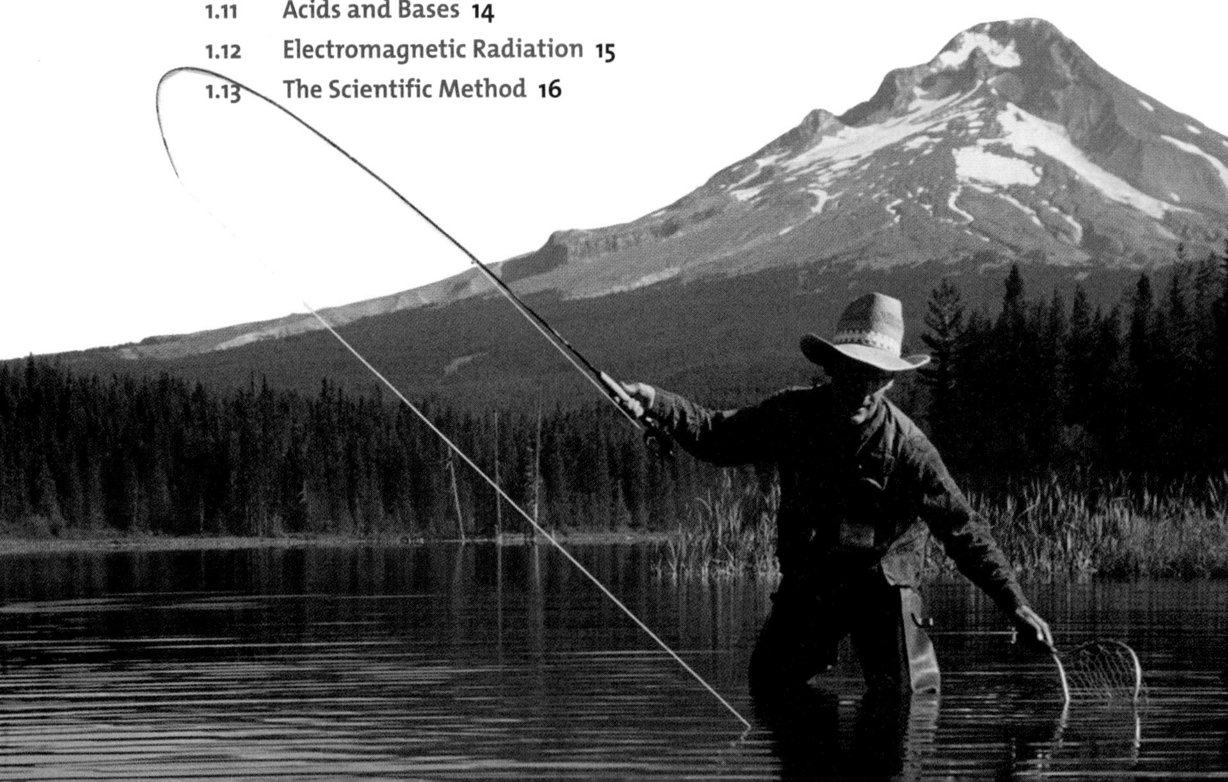

Chapter 2 Planet Earth: Rocks, Life, and Energy 23

Chapter 3 Atoms and Atomic Structure 55

Chapter 4 Atoms and Chemical Bonding 82

Chapter 5 Nuclear Chemistry: The Risks and Benefits of Nuclear Radiation 116

Chapter 6 Chemical Reactions 146

Chapter 7 Reactions in Solution: Acids and Bases and Oxidation-Reduction Reactions 176

Chapter 10 Water Resources and Water Pollution 280

Chapter 11 The Air We Breathe 316

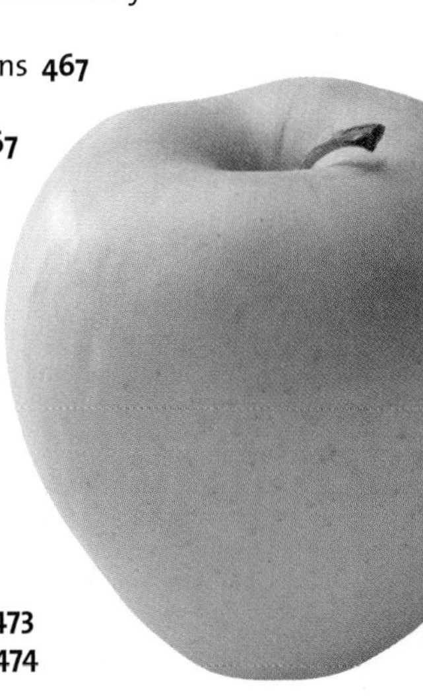

Chapter 16 Agricultural Chemicals: Feeding the Earth's People 482

Chapter 17 Household Chemicals 510

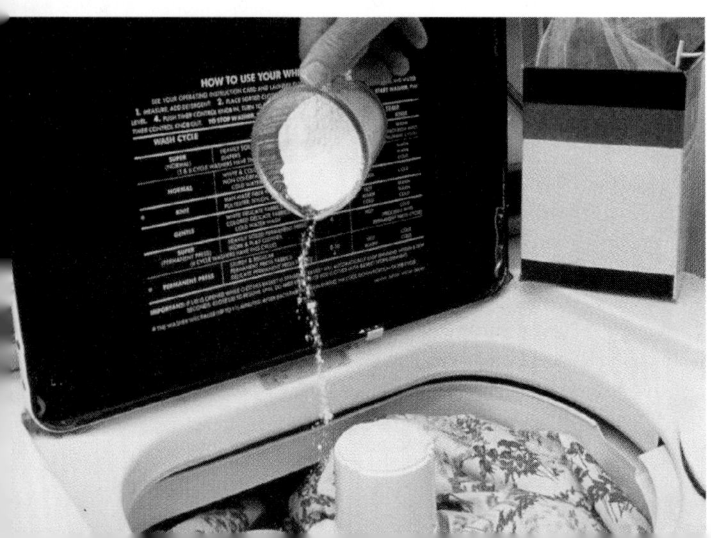

Chapter 18 Toxicology 534

Chapter 19 The Disposal of Dangerous Wastes 558

Preface

At present, there is worldwide concern that many human activities are endangering—perhaps permanently—the quality of the environment and that the time for action to address these problems is running out. The public is becoming increasingly aware of the environmental damage caused by pesticides, toxic wastes, chlorofluorocarbons, nuclear radiation, oil spills, the greenhouse effect, and other human inputs. Environmental organizations such as the Sierra Club, the National Wildlife Federation, and Friends of the Earth are gaining support—especially on college campuses—and are becoming a major influence in the political arena. Articles on environmental issues appear daily in the newspapers, and Congress is introducing legislation to combat threats to the environment.

Paralleling this widespread concern for the environment is the realization that the majority of U.S. citizens, including those with a college degree, are virtually science-illiterate and are not, therefore, well-equipped to make informed decisions on environmental issues. As a response to this crisis in science education, many universities and colleges, American University included, are altering their general curriculums to require science courses for all entering students— nonscience majors as well as science majors. To meet the new science requirements, more and more chemistry departments are offering courses with an environmental perspective, hoping to capture the interest of students who otherwise would not choose a chemistry course.

Objectives

We designed *Chemistry Fundamentals: An Environmental Perspective* as a liberal arts chemistry text for nonscience majors who have little or no previous knowledge of chemistry. Our primary objective in this text is to enable students to make informed judgments on crucial issues that are of current concern worldwide while providing a basic understanding of chemical principles and practices. We aim to guide students to the knowledge that humans live in a chemical environment and that chemistry affects every aspect of life. The text emphasizes that all living and nonliving parts of our environment are made up of chemicals, and that all the natural processes occurring continuously in the environment involve chemical reactions. Once they have grasped this notion of interdependence, students begin to see that without some understanding of chemistry, it is impossible to fully understand environmental issues such as ozone depletion, global warming, air and water pollution, and the hazards of radioactivity.

Organization

The organization and approach of this text differ in several ways from other chemistry books intended for nonscience majors. This book places an environmental perspective within the proven framework of basic chemistry, assuming little or no scientific background.

The opening chapter introduces basic chemical principles and familiarizes students with the language of chemistry. The next chapter provides an early orientation to the earth and its ecosystems. This early coverage of the dynamic nature of the earth and its natural cycles not only establishes the importance of maintaining a sustainable natural world but also gives students a firm foundation for the further study of chemical principles.

The middle portion of the text focuses primarily on chemical processes and principles, with applications to the environment and other areas interwoven throughout. Chapters 3–7 offer a solid introduction to the core chemical principles normally included in chemistry texts for nonscience majors. Topics include the atom, electron configuration, bonding, nuclear reactions, the mole, acids and bases, and oxidation-reduction reactions.

Chapter 8, which introduces students to organic chemistry, is followed by a chapter devoted to synthetic polymers. Wherever possible in these two chapters, the chemistry is discussed in the context of natural processes, building a bridge to understanding the chemistry-related environmental issues examined in the chapters that follow.

The next four chapters (Chapters 10–13) show students the relevance of the basic chemistry just covered by focusing on specific applications that pertain to the environment. Chapter 10 covers water—its properties, its importance to life on earth, and the dangers of polluting and misusing it. The focus in Chapter 11 turns to the earth's atmosphere, with discussions of pollution, ozone depletion, global warming, and the greenhouse effect. Chapters 12 and 13 are devoted to energy, addressing the chemistry of fossil fuels and their use as our major energy source. Emphasized are the dangers of depleting these nonrenewable resources and the need to explore other energy sources such as nuclear and wind power and geothermal and solar energy.

Chapter 14, which introduces students to the more advanced subject of biochemistry, is a useful introduction to the coverage of food and nutrition in Chapters 15 and 16. These two chapters include insightful discussions of the risks and benefits of food additives, synthetic fertilizers, and pesticides, with respect to both human health and environmental well-being.

Chemicals used in the home environment every day are discussed in Chapter 17. The last two chapters examine toxic and hazardous chemicals and their effects on human health, with the final chapter, Chapter 19, concentrating on the laws governing the proper disposal of hazardous and radioactive chemicals. For example, the Clean Air Act the Clean Water Act, EPA regulations, and Superfund are all considered.

Chapter Elements

Chapter Objectives Each chapter opener has a list of the concepts and subject matter students should understand after reading the chapter. This statement of learning objectives provides goals that students should strive for and serves as a useful guide for reviewing each chapter.
Introduction The chapter text begins with an introduction that explains the importance of the subject matter to our understanding of the environment and outlines the material that will be covered.
Examples and Practice Exercises Illustrative worked examples, each one accompanied by a challenging practice exercise, are included in most chapters, particularly those covering basic chemical principles.
Explorations The two-page Explorations essays near the ends of chapters explore ways in which chemistry affects our dynamic world. They profile people such as Marie Curie and Linus Pauling, advances such as the pacemaker and the hybrid car, and events such as the eruption of Krakatoa and environmental terrorism during the Persian Gulf War.
Chapter Summary Each chapter ends with a summary of the main topics covered in the chapter.
Key Terms Lists of key terms introduced in each chapter are included at chapter's end to help reinforce the most important information.
Questions and Problems Each chapter includes a large selection (40–50) of problems and questions, with answers to all even-numbered ones given in an appendix. Quantitative, review, and discussion-type questions are included.

Course Use

Chemistry Fundamentals: An Environmental Perspective offers the flexibility to tailor a course to suit both instructors' preferences and the needs of particular audiences. The full text may be used for a comprehensive two-semester course, or the book may be broken down in several ways for a one-semester course. One option for a one-semester course is to use the first seven chapters followed by choices from the remaining chapters on more advanced chemistry and environmental applications according to the instructor's preferences. For a more traditional course, Chapter 2 may be omitted, and following Chapters 3–7, the course can be rounded out with organic chemistry (Chapter 8) and biochemistry (Chapter 15), and choices from the remaining chapters. If students already have a strong chemistry background, Chapters 3–7 may be omitted, and the course can be devoted to the environmental issues discussed in Chapter 2 and in later chapters.

Ancillary Materials

Jones and Bartlett Publishers offers traditional print and interactive multimedia supplements to assist instructors and aid students in mastering chemistry. Additional information and review copies of any of the following items are available through your Jones and Bartlett Sales Representative.

For the Instructor

Instructor's ToolKit CD-ROM This CD-ROM contains full-color illustrations from the text for computer projection, an electronic test bank, test-generating software, and

Microsoft PowerPoint™ Lecture Outline Slides that can be edited to suit your lectures. The image bank illustrations can be printed on acetates to create your own transparencies or used in the Lecture Outline Slides. Also included are the complete text files, written by James Girard, for the lecture outlines, sample syllabi, and answers to all of the end-of-chapter questions and problems. The test bank contains approximately 1500 questions in a variety of formats.

The CD-ROM also provides the complete Instructor's Manual for the Laboratory Manual. The Instructor's Manual contains directions for preparing the reagents and other materials used in each experiment. It also provides useful comments regarding the experiments, suggestions for proper disposal of used and hazardous materials, and additional sources for reagents and equipment.

Color Transparencies One hundred full-color acetates provide clear and effective lecture illustrations of important diagrams from the text.

For the Student

Laboratory Manual to accompany Chemistry Fundamentals: An Environmental Perspective, Second Edition Written by Phyllis Buell and James Girard, the laboratory manual is a well-tested collection of twenty experiments that parallel the topics in Chemistry Fundamentals. The experiments focus on important chemical principles while assuming a mininum of laboratory skills.

Study Guide to accompany Chemistry Fundamentals: An Environmental Perspective, Second Edition Designed to help the students prepare for the course, the study guide contains learning objectives, chapter overviews, study tips, an expanded glossary for each chapter's key terms, and practice tests.

Acknowledgements

We would like to thank our colleague at American University, Professor Albert Cheh, who gave us helpful advice with the biochemistry chapter.

We would also like to thank the many people at Jones and Bartlett who worked with us to produce this book. Special thanks go to development editor, Dick Morel, who gave us many valuable ideas for improving our manuscript.

We are very much indebted to all those at Lifland et al., Bookmakers who expertly converted our manuscript into its final form. Photo researcher Gail Magin did a wonderful job of locating photos. Our special thanks go to production coordinator Jane Hoover, who with great patience and editing skill guided us through the production of this book.

We and our team at Jones and Bartlett have labored long and hard to make this book as error-free as possible. If, despite all our efforts, a reader discovers an error or has suggestions for ways to improve this book, we would be delighted to hear from her or him.

Phyllis Buell
James Girard

Chapter 1

What Is Chemistry?

An Introduction to the Central Science

Chapter Objectives

In this chapter, you should gain an understanding of:

- How technology involves the application of basic scientific discoveries
- The difference between physical and chemical changes
- The different states of matter
- How all matter can be divided into pure substances and mixtures
- The symbols used to represent the elements
- The names chemists use for elements, compounds, and subatomic particles
- The application of the scientific method

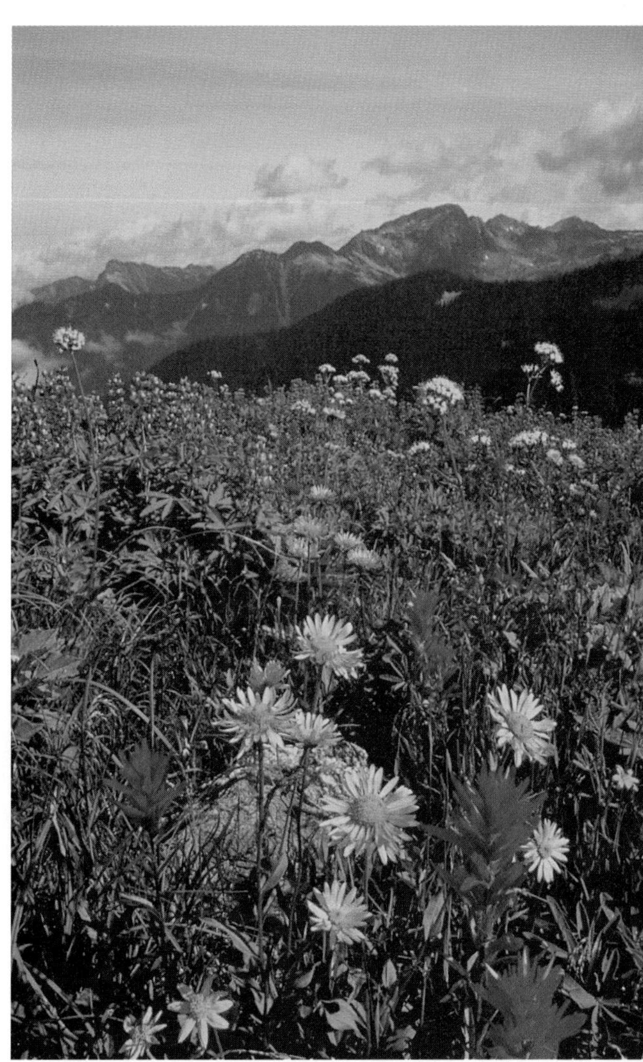

Everything on earth—including mountains, flowers, clouds, and the air we breathe—is made of chemicals.

ONLINE FEATURES

www.jbpub.com/chemistry

- Chemistry and the Environment
- eLearning
- Chemistry in the News
- Special Topics
- Research & Reference Links
- Ask the Authors

CHEMISTRY IS THE SCIENCE that is concerned with the composition of matter and with the changes that matter undergoes. We practice chemistry all the time in our daily lives: when we cook, do the laundry, take medicine, fertilize the lawn, paint the house, or strike a match. In all these activities, substances interact, and chemical changes occur. In our bodies, as we breathe, walk, and digest food, chemical reactions occur constantly. The environmental problems we face—such as disposing of toxic wastes, controlling smog, and removing asbestos—are all essentially chemical problems.

In this chapter, we look at some of the basic properties of matter and see how matter is classified, and we identify the basic units of all matter: elements, compounds, atoms, and ions. We examine the two major types of changes—physical and chemical—that matter can undergo. We consider the language of chemistry: the symbols and names that represent and identify chemical substances, and the equations that describe and quantify the chemical reactions that substances undergo. We also consider the radiant energy that reaches us from the sun and sustains all life on earth.

Chemistry is an experimental science. At the end of this chapter, we discuss the scientific method—the method of gaining knowledge that is fundamental to the advancement of chemistry.

1.1 Science, Technology, and Chemistry

What Is Science?

Science is the study of all aspects of the world around us. It seeks to understand the order in the natural world and to explain in a logical way how and why things happen as they do. Because science covers such an enormous body of knowledge and because it can be studied in so many different ways, it is subdivided into a number of disciplines. Some of the major ones are astronomy, botany, chemistry, geology, physics, and zoology. There are no definite boundaries between the various scientific disciplines. Figure 1.1 shows how chemistry overlaps with several of the other disciplines.

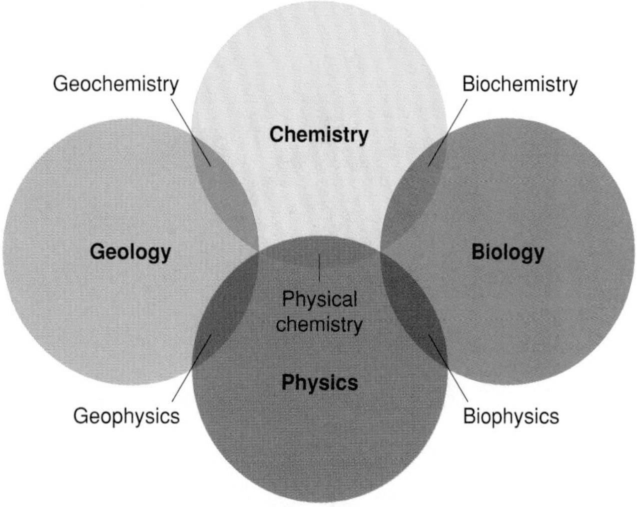

FIGURE 1.1

Chemistry overlaps several other scientific disciplines.

What Is Technology?

Technology is often defined as *applied* science. It is the application of knowledge to the development of new materials and new processes, primarily those that are likely to have a practical use. Technology involves the modification of natural materials to generate products that improve the quality of life or satisfy society's wants.

Today, nearly all technological advances are based on scientific knowledge, but most of the early developments were achieved accidentally or by trial and error. For example, people knew how to make alcohol from grain and glass from a mixture of limestone, sand, and soda ash (sodium carbonate) long before they had any understanding of the scientific basis for these processes.

Technological progress has brought both benefits and problems. Advances in chemistry during the 20th century include penicillin, artificial fertilizers, plastics, the microchip, the pesticide DDT, Freon (used as a refrigerant), and nuclear weapons, not all of which have been of lasting benefit to humans (**Fig. 1.2**). When DDT and Freon were first introduced, only their many advantages were apparent. It was not until later that their harmful effects on the environment were recognized. For each new technological development, the costs must be carefully balanced against the benefits.

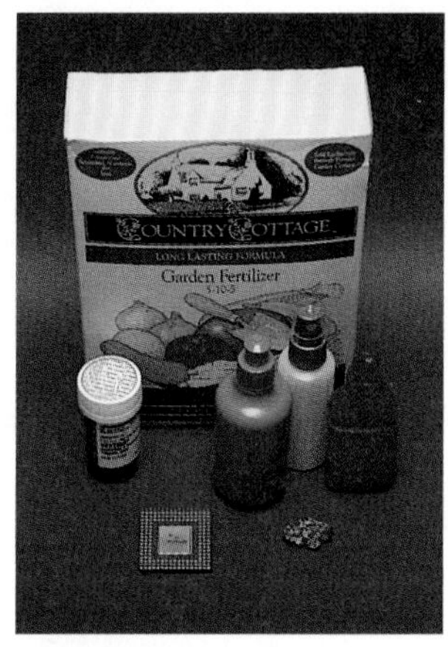

FIGURE 1.2

Each of these products—fertilizer, penicillin, plastic containers, microchip, and uranium ore—has benefits and risks.

What Is Chemistry?

Chemistry is the study of the composition and structure of matter and the changes that matter undergoes. **Matter** is anything that occupies space and has mass. Mass is a measure of the *quantity* of matter. (The *weight* of an object is related to the pull of gravity on the object and is not the same as its mass. The difference between mass and weight is explained in Appendix B.)

Because everything in the universe is composed of matter, chemistry is the study of our material world. Chemistry touches our lives and influences our activities in so many ways that it is often called the **central science.**

1.2 Properties of Matter

The distinguishing characteristics that we use to identify different samples of matter are called **properties.** We can recognize our own car among hundreds of other cars in a parking lot by characteristics such as make, model, color, dents in the body, and the belongings left on the back seat. In much the same way, we can recognize different substances by their characteristics, or properties. Properties of substances can be grouped into two main categories: physical and chemical.

Our environment is the planet earth, a small sphere in the vastness of space. Everything in this environment is made of chemicals.

Physical Properties

The **physical properties** of a substance are those characteristics that can be observed without changing the substance into another substance. Color, odor, taste, hardness, density, solubility, melting point (the temperature at which a substance melts), and boiling point (the temperature at which a substance boils or vaporizes) are all physical properties. For example, some physical properties of pure copper are as follows. Copper is a bright, shiny metal. It is malleable (can be beaten into thin sheets) and ductile (can be drawn into fine wire), and it is a good conductor of electricity (**Fig. 1.3**). It melts at 1083°C (1981°F) and boils at 2567°C (4653°F); its density is 8.92 grams per milliliter (g/mL) (units of measurement are explained in Appendix B). No matter what its source, pure copper always has these properties.

Each chemical substance has a unique set of properties that distinguishes it from every other substance.

To determine the melting point of a substance, we must change it from a solid to a liquid. When we do so, we are not changing the composition of the substance. For example, to determine the melting point of ice, we must change the ice to water. In this process, the appearance changes, but there is no change in composition.

Chemical Properties

The **chemical properties** of a substance are those characteristics that can be observed when the substance undergoes a change in its composition. One chemical property of water is that it undergoes a process called *electrolysis:* If an electrical current is passed through a container of liquid water (H_2O), a change in composition occurs, and the water is broken down to yield the gases hydrogen (H_2) and oxygen (O_2) (**Fig. 1.4**).

a.

b.

FIGURE 1.3

Some physical properties of copper: (a) Copper is ductile and can be drawn into wires that can be used to conduct an electrical current. (b) It is malleable and can be beaten into sheets, from which items such as kettles are made.

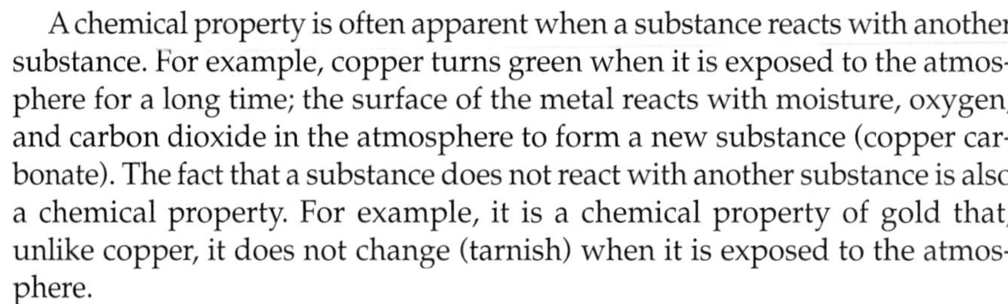

A chemical property is often apparent when a substance reacts with another substance. For example, copper turns green when it is exposed to the atmosphere for a long time; the surface of the metal reacts with moisture, oxygen, and carbon dioxide in the atmosphere to form a new substance (copper carbonate). The fact that a substance does not react with another substance is also a chemical property. For example, it is a chemical property of gold that, unlike copper, it does not change (tarnish) when it is exposed to the atmosphere.

Some physical and chemical properties of water, iron, and gold are shown in Table 1.1.

1.3 Changes in Matter

Changes in matter occur all the time in the natural world: Snow melts, leaves on trees change color in the fall, dead creatures decompose, and iron rusts. And men and women have been purposely changing natural substances into new products since well before the beginning of recorded history. Early changes brought about by humans for their own advantage include the conversion of natural clays into pottery and limestone rock into building materials.

The raw materials from which today's manufactured goods are made must undergo many chemical changes. For example, chemical changes convert iron ore into the steel that is used to make car bodies, and they convert simple chemicals into Dacron, nylon, and other synthetic fibers used in the manufacture of clothing and carpeting. Controlled chemical change is thus a major factor in the attainment of our present standard of living. Changes in matter can be divided into physical changes, chemical changes, and nuclear changes.

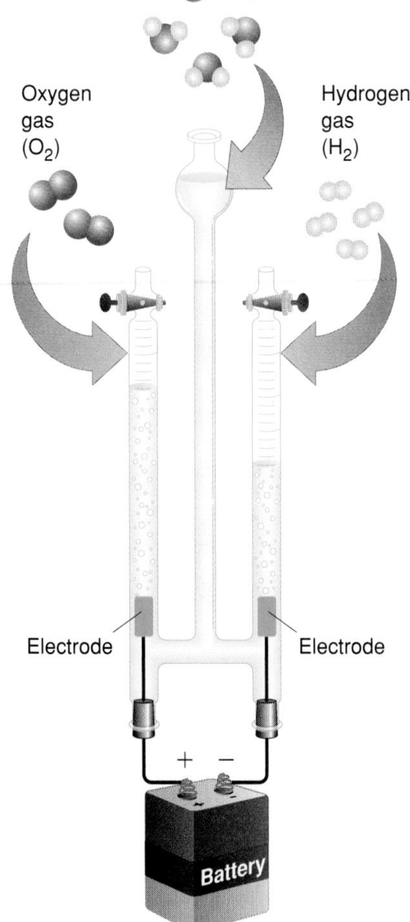

FIGURE 1.4

A chemical change: The breakdown of water (H_2O) into the gases hydrogen (H_2) and oxygen (O_2) by the process of electrolysis. (Hydrogen and oxygen are produced in a volume ratio of 2:1.)

Table 1.1 Some Physical and Chemical Properties of Water, Iron, and Gold

	Physical Properties			
Substance	**Color**	**Melting Point (°C)**	**Boiling Point (°C)**	**Chemical Properties**
Water (liquid)	Colorless	0	100	Breaks down to hydrogen and oxygen in the process of electrolysis
Iron (solid)	Gray	1535	3000	Reacts with moisture and oxygen in the air to form rust, an iron oxide
Gold (solid)	Yellow	1065	2808	Dissolves in a mixture of hydrochloric and nitric acids to form a yellow solution

Physical Changes

A **physical change** in a substance is one that does not alter the composition of the substance. The commonest type of physical change is a change in physical state—for example, a change from solid to liquid. The freezing of water, the melting of ice, the evaporation of water, and the condensation of water vapor are all examples of physical changes (Fig. 1.5).

No new substance is ever formed when a physical change occurs.

$$\text{ice} \underset{\text{freezing}}{\overset{\text{melting}}{\rightleftharpoons}} \text{liquid water} \underset{\text{condensation}}{\overset{\text{evaporation}}{\rightleftharpoons}} \text{water vapor}$$

Most of the materials we use in our daily lives, such as medicines, plastics, and synthetic fibers, are produced by controlled chemical changes.

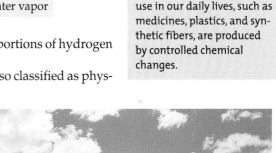

None of these changes involves a change in composition. The proportions of hydrogen and oxygen in ice, liquid water, and water vapor are identical.

Certain changes in the physical appearance of substances are also classified as physical changes (Fig. 1.6). Chopping up cabbage in a food processor brings about a physical change, as does dissolving salt in water to form a salt solution. The salt does not change into another substance when it dissolves in water; if we evaporate the water, we can recover the salt unchanged.

FIGURE 1.5

A physical change does not alter the composition of a substance. Ice, water, and water vapor are the same substance.

Chemical Changes

A **chemical change,** or **chemical reaction,** occurs when the composition of a substance is changed. In the process, the substance is changed into one or more other substances having physical and chemical properties different from those of the original substance. For example, when iron rusts, a red-brown compound, quite different from the original metal, is formed. The decomposition of water into hydrogen and oxygen by electrolysis is another example of a chemical change. Examples of chemical reactions in the natural world are the combustion of wood in a forest fire, the digestion of food in an animal's stomach, and the dissolution of limestone rocks by rain.

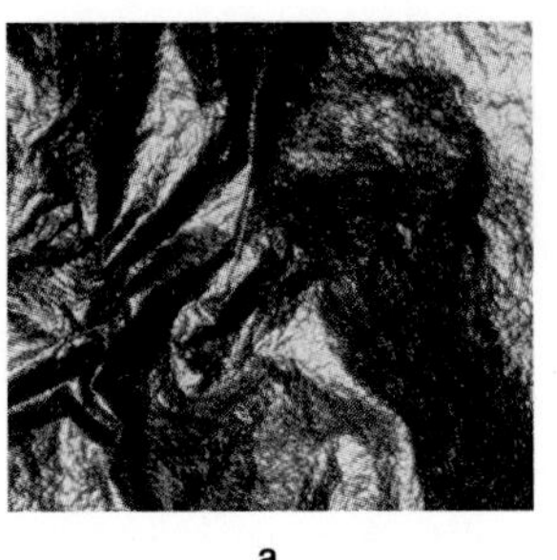
a.

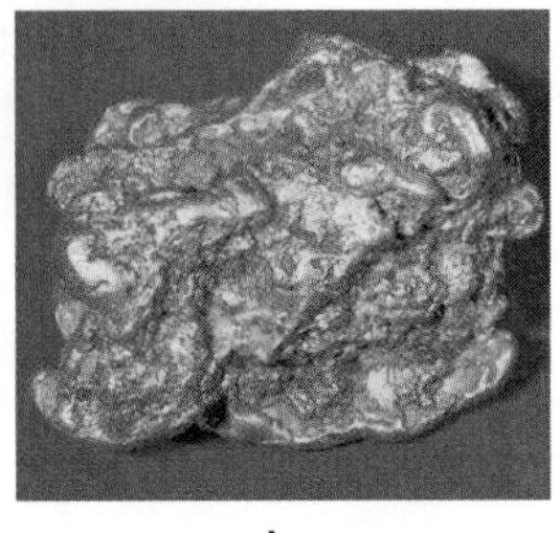
b.

c.

FIGURE 1.6

Changes in the outward appearance of a substance are classified as physical changes. In the three forms of gold shown, the properties of gold remain the same.

EXAMPLE 1.1 Differentiating between physical and chemical changes

Identify each of the following as a physical or chemical change: **a.** hard boiling an egg; **b.** silver becoming tarnished; **c.** boiling water.

Solution

a. Chemical change. New substances are formed as the egg hardens.

b. Chemical change. The silver and the tarnish have different compositions.

c. Physical change. The water changes from a liquid to a gas (a different physical state), but its composition is not changed.

Practice Exercise 1.1

Which of the following is a physical change and which a chemical change?
a. fruit ripening **b.** dissolving sugar in tea **c.** chopping up an onion

Nuclear Changes

The splitting of uranium atoms in an atomic bomb and the generating of elements in stars are examples of nuclear reactions.

Any pure substance is composed of basic units called *elements.* In any chemical change, the pure substances involved are changed, but the elements that make up the substances are not changed; they are rearranged to form new substances. In a **nuclear change,** or **nuclear reaction,** however, one or more elements may be changed into another element. Nuclear reactions, which are sometimes associated with the production of enormous amounts of energy, will be discussed in Chapter 5.

Energy and Changes in Matter

Chemical energy is stored in each molecule of a substance; it can be released during a chemical reaction.

Energy is the capacity to do work. Energy exists in many forms, including light, heat, and electricity. Energy is not made of matter, but energy and matter are closely related. Energy is associated with all changes in matter; in nuclear reactions, some matter is actually converted into energy. Heat energy is required to change water into steam, and green plants need light energy from the sun to convert water and carbon dioxide into sugars in the process of photosynthesis.

Radiant energy from the sun will be discussed later in this chapter. The flow of energy through the natural environment will be studied in Chapter 2, and the energy use of our society will be discussed in Chapters 12 and 13.

1.4 States of Matter

Matter can be classified according to its three familiar states: solid, liquid, and gas **(Fig. 1.7)**.

Solids such as ice cubes, pebbles, books, and most household objects occupy a fixed volume and retain their shape no matter where they are located. The particles that make up the structure of a solid are held tightly together in fixed positions; thus their motion is limited.

Liquids have a fixed volume but not a fixed shape. They take the shape of the container in which they are placed. The particles that make up a liquid are held together less tightly than are those of solids, and liquids have considerable freedom of motion. If milk is poured into a glass, the milk takes the shape of the glass up to the point at which it overflows or no more milk is added.

Gases maintain neither volume nor shape. The forces holding their particles together are relatively weak, and the particles move freely and completely fill whatever vessel contains them.

Whether a substance exists as a solid, a liquid, or a gas at any particular time depends on its temperature, the surrounding pressure (the force per unit area exerted on it), and the strength of the forces holding its component particles together. Water, the commonest liquid in our environment, is a very unusual substance in that it exists naturally on earth in all three states: solid ice, liquid water, and gaseous water vapor. Many substances, however, can be changed from one state to another only by means of extremes of temperature or pressure or both. Iron, for example, is a solid at normal temperatures and must be heated to 1535°C (2795°F) to change it to a liquid and to 3000°C (5432°F) to change it to a vapor (refer to Table 1.1).

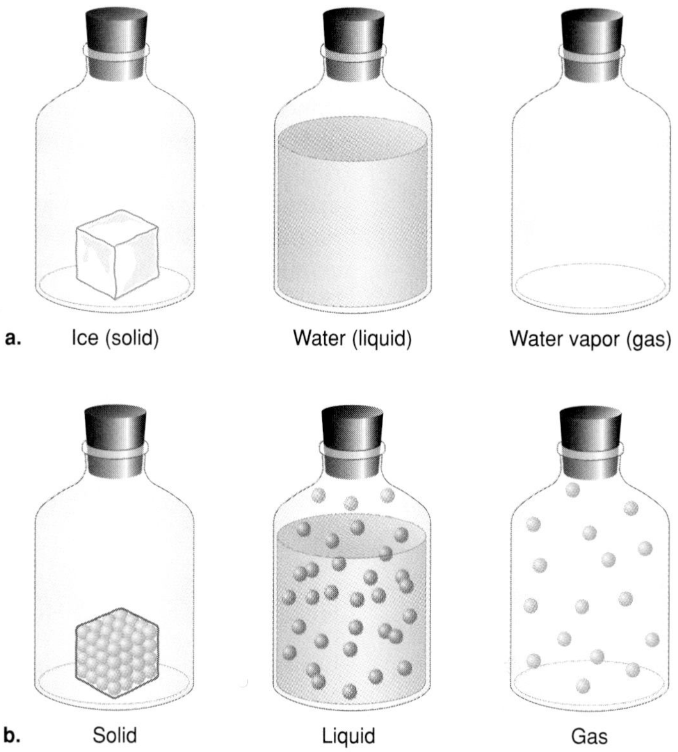

a. Ice (solid) Water (liquid) Water vapor (gas)

b. Solid Liquid Gas

FIGURE 1.7

The three physical states of matter: (a) The shape and volume of a solid, a liquid, and a gas are influenced by the vessels containing them. (b) The arrangement of particles in the three states (diagrammatic and not to scale).

Plasmas are a fourth state of matter. Plasmas do not exist naturally in the earth's environment, but the plasma state is probably the most common state of matter in the universe as a whole. A plasma is similar to a gas except that the particles that make up a plasma are charged and extremely energetic. The atmosphere of the stars and much of the matter in outer space exist in the plasma state, and our solar system was formed out of a swirling cloud of plasma.

Gases have no definite volume and are easily compressed into steel cylinders for storage and transportation.

1.5 Pure Substances and Mixtures

All matter can be divided into two classes: pure substances and mixtures. A **pure substance** always has a definite and constant composition and, regardless of its source, always has the same properties under a certain set of conditions. Water is a pure substance; no

Substance is a general term used to describe any type of matter. *Pure substance* is used to denote a substance with a constant composition and a specific set of properties.

matter what the source, water is always 11% by mass hydrogen and 89% by mass oxygen, and at sea level it always melts at 0°C (32°F) and boils at 100°C (212°F). Gold is another pure substance; pure gold is always 100% gold. Other common examples of pure substances are salt, sugar, aluminum, and nitrogen gas.

Mixtures are combinations of two or more pure substances in which each substance retains its own identity. The components of a mixture are physically mixed but are not chemically combined. If we make a mixture of salt and ordinary sand, we can still distinguish the colorless salt crystals from the larger light-brown sand particles. Mixing changes neither the salt nor the sand in any way. A mixture can have variable composition. In our example, we can obviously alter the proportions of salt and sand and make any number of mixtures of varying composition.

Another characteristic of mixtures is that the components can be separated by physical means. With the salt–sand mixture, it would be possible, though tedious, to separate the components by picking out all the sand particles with a pair of tweezers. Alternatively, we could add water to the mixture to dissolve the salt, then remove the sand from the salt solution using a fine strainer, and finally evaporate the water to recover the salt.

Heterogeneous and Homogeneous Mixtures

Examples of mixtures in the natural world include rocks, soil, milk, blood, seawater, and air. In many mixtures (rocks, soil, and salt–sand mixtures, for example), the different components are visibly distinguishable from one another **(Fig. 1.8)**. Such mixtures are called **heterogeneous mixtures.** Other mixtures (clear seawater, samples of air, and water containing dissolved sugar) are uniform in appearance; the separate ingredients in the mixtures cannot be seen. Mixtures such as these have uniform composition and uniform properties throughout the sample and are called **homogeneous mixtures,** or **solutions.** The composition of a solution can vary from one sample to another. A given sugar solution, for example, is uniformly of the same sweetness throughout, but the sweetness of another sugar solution may be different.

a.

b.

c.

FIGURE 1.8

In a heterogeneous mixture, the components are clearly visible, as in (a) chocolate chip cookies, (b) a piece of quartz containing a vein of gold, and (c) oil and vinegar.

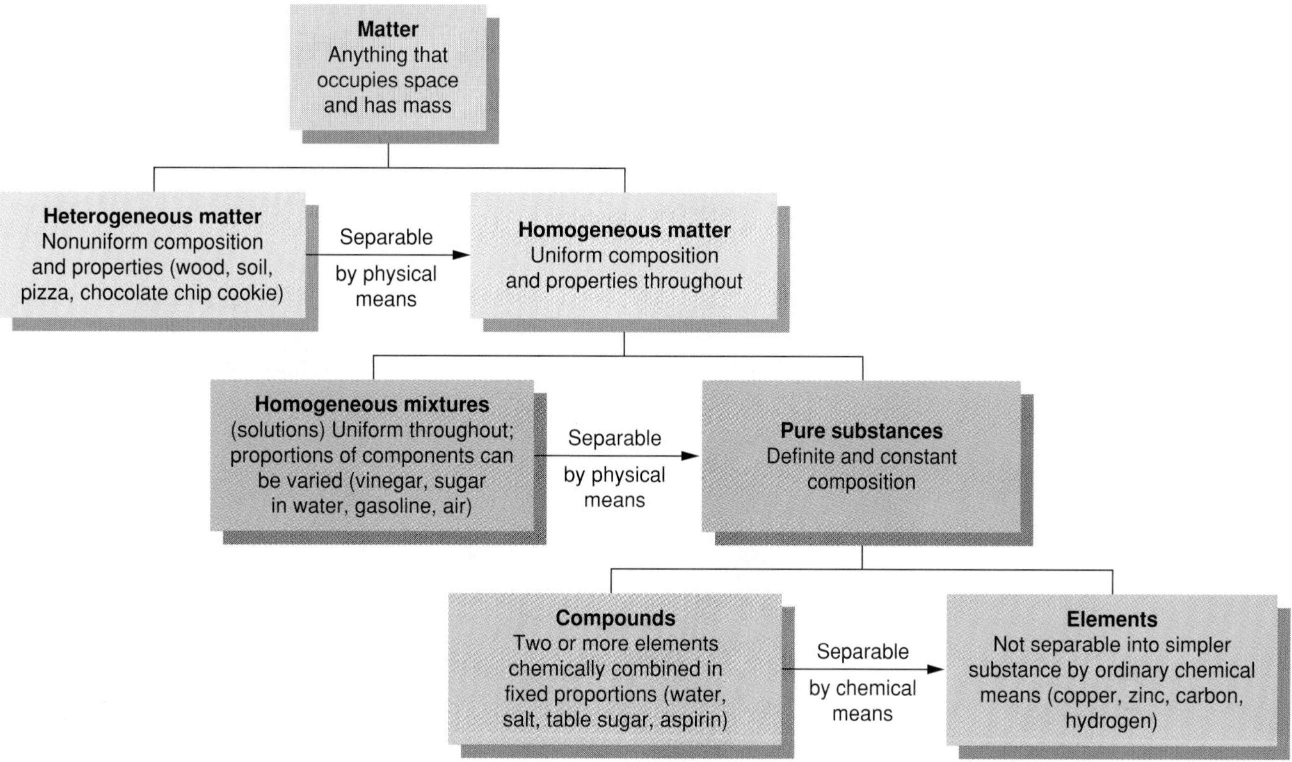

FIGURE 1.9

A classification of matter.

1.6 Elements and Compounds

Pure substances can be subdivided into elements and compounds. **Elements** are the building blocks from which all matter is constructed. Elements cannot be decomposed into simpler substances by ordinary chemical means. **Compounds** are pure substances that are composed of two or more elements combined chemically in fixed proportions. A classification of matter based on composition and properties is outlined in **Figure 1.9**.

Compounds can be decomposed by chemical means to yield the elements from which they are constructed. The properties of a particular compound are always the same under a given set of conditions but are different from the properties of the elements from which it is composed **(Fig. 1.10)**.

FIGURE 1.10

The properties of a compound are very different from the properties of the elements from which it is composed. For example, chlorine gas (in the glass container) combines with sodium metal to form sodium chloride (table salt).

Examples of simple compounds (together with the elements of which they are composed) are water (hydrogen and oxygen), carbon dioxide (carbon and oxygen), salt (sodium and chlorine), and sulfuric acid (hydrogen, sulfur, and oxygen). In all these compounds, two or more elements are chemically combined in fixed proportions.

There are 92 naturally occurring elements and 17 others that have been created by scientists in nuclear reactions (Chapter 5). Every compound that exists is made up of some combination of these 109 elements. **Figure 1.11** shows the distribution in the universe, the earth as a whole, the earth's crust, and the human body of the most abundant of the elements. The universe is composed almost

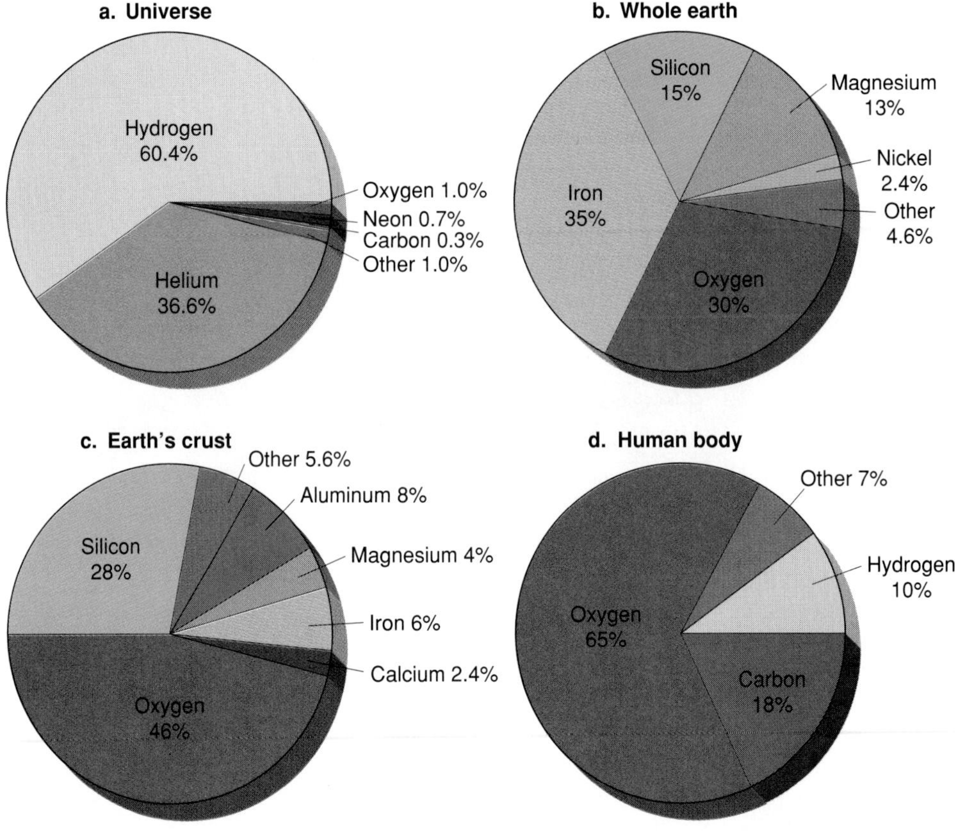

FIGURE 1.11

The relative abundance (as percent by mass) of the major elements in (a) the universe, (b) the earth as a whole, (c) the earth's crust (including the oceans and the atmosphere), and (d) the human body. Note the significant differences in composition that distinguish the earth's crust (the part of the planet with which we are in contact), the earth as a whole, and the universe.

More than 6 billion different compounds are known.

entirely of hydrogen and helium; on the earth, iron and oxygen predominate. Just 5 of the 92 naturally occurring elements make up over 90% of all the matter in the earth's crust. The three elements oxygen, carbon, and hydrogen make up over 90% of our bodies. A complete list of the elements can be found inside the back cover of this book.

Symbols

Chemists use symbols consisting of one or two letters to represent the elements. The first letter of an element's symbol is always capitalized, and the second letter is always lowercased. Co is the symbol for the element cobalt, whereas CO denotes the compound carbon monoxide. **Table 1.2** lists some common elements with their symbols.

The first (or in some cases, only) letter of a symbol is usually the first letter of the element's name (e.g., O, oxygen; Mg, magnesium); the second letter is often the second letter in the element's name (e.g., Br, bromine; Si, silicon). When names of elements share the same second letter, another letter from the name of one of them is chosen (e.g., Ca, calcium; Cd, cadmium). Some of the elements shown in Table 1.2—mainly those that have been known for hundreds of years—have symbols that are derived from their Latin names **(Table 1.3)**.

Table 1.2 Common Elements and Their Symbols

Element	Symbol	Element	Symbol
Aluminum	Al	Iron	Fe
Antimony	Sb	Lead	Pb
Argon	Ar	Lithium	Li
Arsenic	As	Magnesium	Mg
Barium	Ba	Mercury	Hg
Beryllium	Be	Neon	Ne
Bismuth	Bi	Nickel	Ni
Boron	B	Nitrogen	N
Bromine	Br	Oxygen	O
Cadmium	Cd	Phosphorus	P
Calcium	Ca	Platinum	Pt
Carbon	C	Potassium	K
Chlorine	Cl	Radium	Ra
Chromium	Cr	Selenium	Se
Cobalt	Co	Silicon	Si
Copper	Cu	Silver	Ag
Fluorine	F	Sodium	Na
Gold	Au	Sulfur	S
Helium	He	Tin	Sn
Hydrogen	H	Uranium	U
Iodine	I	Zinc	Zn

Table 1.3 Elements That Have Symbols Derived from Their Latin Names

Element	Latin Name	Symbol
Antimony	*Stibium*	Sb
Copper	*Cuprum*	Cu
Gold	*Aurum*	Au
Iron	*Ferrum*	Fe
Lead	*Plumbum*	Pb
Mercury	*Hydrargyrum*	Hg
Potassium	*Kalium*	K
Silver	*Argentum*	Ag
Sodium	*Natrium*	Na
Tin	*Stannum*	Sn

EXAMPLE 1.2 Recognizing symbols for elements and compounds

Which of the following are elements and which compounds?

Be BN CO Cr

Solution Be and Cr, which represent beryllium and chromium, are elements. BN is a compound of B (boron) and N (nitrogen). CO is a compound of C (carbon) and O (oxygen).

Practice Exercise 1.2

Identify the element(s) and compound(s).

Si KI He HI

1.7 Atoms and Molecules

Atoms

Elements are composed of atoms. Copper is an element, and even a very small piece of pure copper wire is made up of billions and billions of copper atoms. An **atom** can be defined as the smallest unit of an element that can take part in a chemical change. An atom of any particular element differs from an atom of any other element (Chapter 3).

Molecules

Very few elements exist in nature as isolated atoms. Most matter is composed of **molecules.** A molecule is a combination of two or more atoms. It can be defined as the smallest unit of a pure substance (element or compound) that can exist and still retain the physical and chemical properties of the substance. Molecules are the particles that undergo chemical changes in a reaction. They are composed of atoms held together by forces known as chemical bonds (Chapter 4).

Molecules may be composed of two or more *identical* atoms or two or more *different* atoms. For example, two atoms of oxygen bond together to form a molecule of ordinary oxygen gas: O_2. A molecule of water consists of two atoms of hydrogen bonded to a single atom of oxygen: H_2O. A molecule of glucose comprises six atoms of carbon, twelve atoms of hydrogen, and six atoms of oxygen: $C_6H_{12}O_6$. Proteins and many other biological molecules are made up of several thousand atoms bonded together in unique ways.

1.8 Subatomic Particles and Ions

Atoms are composed of three kinds of subatomic particles: **protons, electrons,** and **neutrons.** A proton carries a one-unit positive charge, an electron carries a one-unit negative charge, and a neutron is uncharged. In an atom of a neutral (uncharged) element, the number of protons is balanced by an equal number of electrons. An element is defined by the number of protons it contains. We will discuss the structure of the atom in detail in Chapter 3.

In certain circumstances, atoms can lose or gain one or more electrons. When electrons (negatively charged) are removed from or added to neutral atoms (or molecules), charged particles called **ions** are formed. A positively charged particle is called a **cation;** a negatively charged particle is called an **anion.** For example, a sodium atom (Na) can lose an electron to become the cation Na^+, and a chlorine atom (Cl) can gain an electron to become the anion Cl^-. Sodium chloride (NaCl), common table salt, is a compound formed from Na^+ and Cl^- ions.

Ionic solids such as NaCl do not exist as individual molecules. As we shall see in Chapter 4, they are more accurately described in terms of formula units.

1.9 Chemical Formulas

A chemical formula is the shorthand chemists use to represent the composition of a pure substance.

The composition of a compound can be represented by a **chemical formula.** The formula indicates the number of atoms of each element that are present in a molecule of the compound. The chemical formulas of some familiar compounds are given in Table 1.4.

The numerical subscripts to the right of the symbols for the elements indicate the number of atoms of each element present in a molecule of the compound. If there is no subscript following the element (as for the O in H_2O), it means that only one atom of that element is present.

It might seem that the formula for glucose ($C_6H_{12}O_6$) could be simplified to CH_2O. In this simpler formula, the C, H, and O atoms are in the same ratios to each other (1:2:1) as in $C_6H_{12}O_6$. However, for glucose to have its characteristic properties, there must be 6 atoms of carbon, 6 atoms of oxygen, and 12 atoms of hydrogen bonded together in a unique way.

The manner in which atoms are joined in a molecule can be depicted in a number of ways (**Fig. 1.12**). A molecular formula shows the number of atoms of each element that are present. A structural formula shows which atoms are joined to which other atoms in the molecule. Ball-and-stick and space-filling models give more information. They show the relative sizes of the atoms and their geometric arrangement. A space-filling model shows the relative distances between the atoms and is the most informative representation of a molecule.

Table 1.4 Some Compounds and Their Formulas

Compound	Formula
Aluminum sulfate	$Al_2(SO_4)_3$
Ammonia	NH_3
Ammonium phosphate	$(NH_4)_3PO_4$
Calcium carbonate	$CaCO_3$
Carbon dioxide	CO_2
Carbon monoxide	CO
Glucose	$C_6H_{12}O_6$
Kaolinite	$Al_2Si_2O_5(OH)_4$
Methane	CH_4
Silica	SiO_2
Water	H_2O

Some of the chemical formulas in Table 1.4 include parentheses. These formulas indicate that the grouping of atoms within the parentheses is repeated more than once in the molecule. For example, aluminum sulfate, which is used in water purification, includes three sulfate, SO_4, units. Ammonium phosphate, a fertilizer, includes three ammonium, NH_4, units. Kaolinite, a clay, includes four hydroxide, OH, units. Multiatom

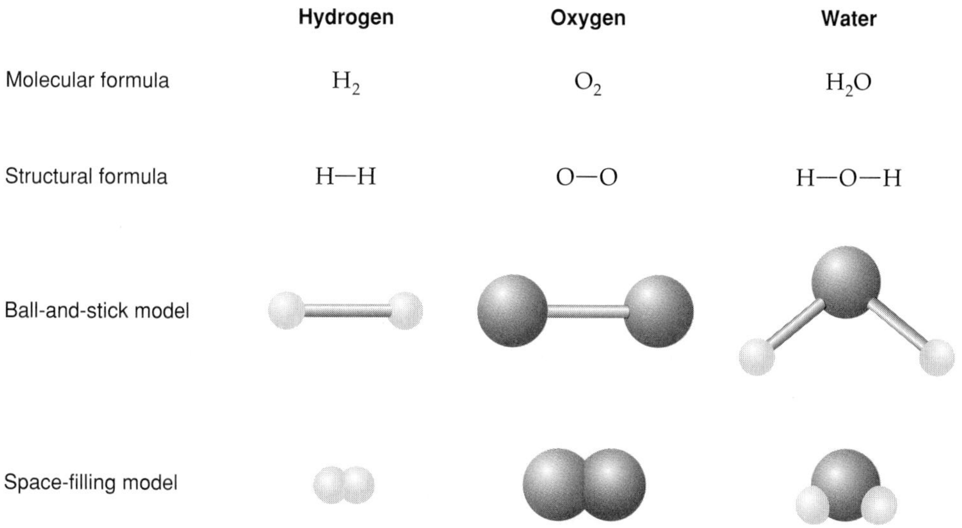

	Hydrogen	Oxygen	Water
Molecular formula	H_2	O_2	H_2O
Structural formula	H—H	O—O	H—O—H
Ball-and-stick model			
Space-filling model			

FIGURE **1.12**

Structural formulas, ball-and-stick models, and space-filling models showing how hydrogen and oxygen atoms join together to form hydrogen (H_2), oxygen (O_2), and water (H_2O) molecules. Structural formulas show which atoms are joined together; ball-and-stick and space-filling models show the relative sizes of the atoms and their geometric arrangements; space filling models also show the relative distances between the atoms and give the most accurate representations of the actual molecules.

units such as NO_3, SO_4, PO_4, OH, and NH_4 remain intact in chemical reactions. They will be considered in more detail in Chapter 4.

1.10 Chemical Equations

The changes that occur during chemical reactions can be described by means of **chemical equations.** For example, when coal—which is composed mainly of carbon—is burned, it combines with oxygen gas in the atmosphere to produce the gas carbon dioxide. This information can be summarized concisely in the following chemical equation:

$$\underset{\text{carbon}}{C} + \underset{\text{oxygen}}{O_2} \longrightarrow \underset{\text{carbon dioxide}}{CO_2}$$

The symbols for the elements are not only abbreviations of the names of the elements; each symbol represents one atom of the element. The above equation tells us that one atom of carbon reacts with one molecule of oxygen (two oxygen atoms bonded together) to produce one molecule of carbon dioxide (one carbon and two oxygen atoms bonded together).

Coal often contains some sulfur, and when coal burns, the sulfur reacts with atmospheric oxygen to produce sulfur trioxide (SO_3), a toxic gas that is a major source of air pollution.

$$2\ S + 3\ O_2 \longrightarrow 2\ SO_3$$

The numbers in front of the formulas in this equation ($\underline{2}$ S, $\underline{3}$ O_2, $\underline{2}$ SO_3) are called **coefficients.** They indicate the minimum numbers of atoms and molecules that take part in the reaction. In the above example, the coefficients tell us that two atoms of sulfur react with three molecules of oxygen to produce two molecules of sulfur trioxide. If there is no number in front of a formula—as for the formulas in the equation for the burning of coal—it is understood that one atom or one molecule is involved. In other words, chemical equations don't use 1 as a coefficient.

Sulfur trioxide released into the atmosphere as a result of the burning of sulfur-containing coal reacts with moisture in the atmosphere to form sulfuric acid (H_2SO_4). The acid falls to earth in acid rain.

$$SO_3 + H_2O \longrightarrow H_2SO_4$$

Acid rain is harmful in many ways. It slowly dissolves monuments and statues made of limestone, and it may make lake water too acid for fish to survive.

1.11 Acids and Bases

Sulfuric acid and two other common acids, hydrochloric acid (HCl) and nitric acid (HNO_3), are strong acids. However, most acids we encounter are weak acids that are not harmful to the environment. Common weak acids include citric acid found in citrus fruits (e.g., oranges and grapefruits), vinegar, and ascorbic acid (vitamin C).

Bases are compounds that can neutralize the effects of acids. For example, the base lime (calcium oxide, CaO) is used to neutralize acid soil. Other common bases are sodium hydroxide (NaOH), a strong base found in drain cleaners, and ammonia (NH_3), which is a common ingredient in many household cleaners. Acids and bases will be discussed in Chapter 7.

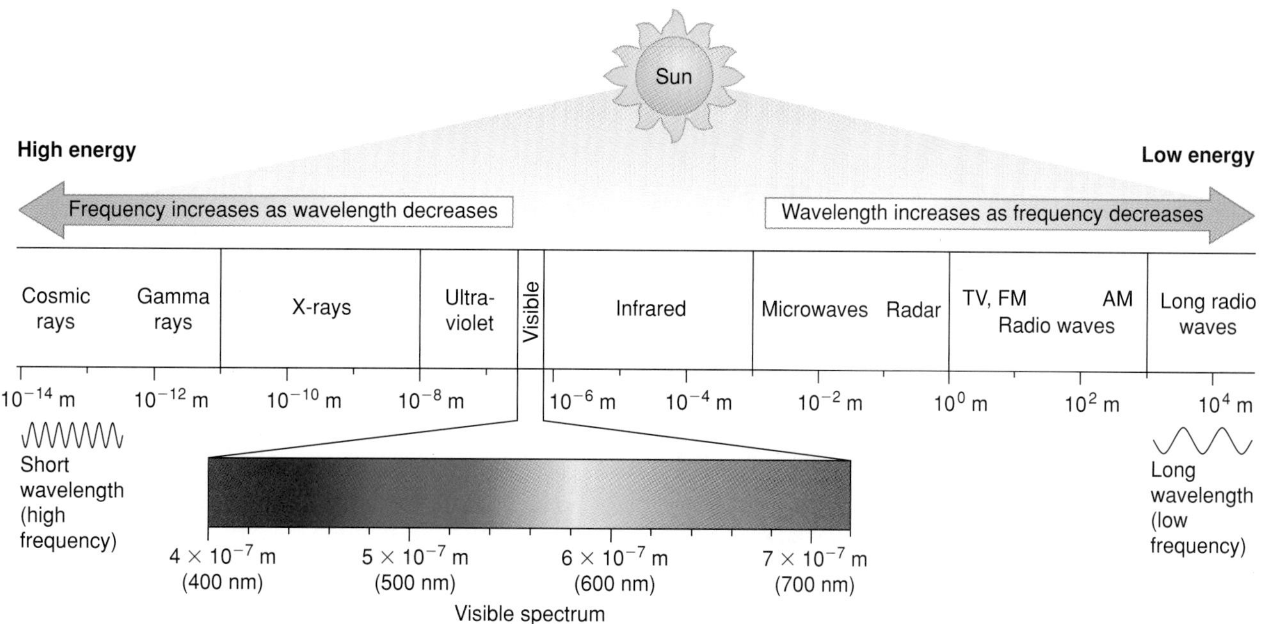

FIGURE 1.13

The electromagnetic spectrum. As the wavelength of electromagnetic radiation increases from cosmic rays to radio waves, the energy and frequency of the waves decrease. The visible wavelengths make up a very small part of the electromagnetic spectrum. [One nanometer (nm) is equal to 10^{-9} meters (m). Exponential notation, like 10^{-9}, is explained in Appendix A.]

1.12 Electromagnetic Radiation

The energy source that sustains all life on earth is the sun. It provides the light energy that plants need for photosynthesis and the heat that warms the earth. It controls our climate, drives our weather systems, and regulates the life cycles of all plant and animal species. The radiant energy, or **electromagnetic radiation (EMR),** that the sun continually transmits through space reaches the earth in many forms. The most familiar forms are light and radiant heat; other forms include cosmic rays, X-rays, ultraviolet (UV) radiation, microwaves, and radio waves **(Fig. 1.13)**.

EMR travels in waves that can be compared to the waves that ripple outward on the surface of a pond when a stone is dropped into the water. Electromagnetic waves differ from water ripples, however, in that they travel outward in all directions. All electromagnetic waves, whether they are light waves, microwaves, radio waves, or any other type, travel at the same rate. This rate is at a maximum (300,000 kilometers per second, or 186,000 miles per second) when the waves travel in a vacuum; electromagnetic waves are slowed very slightly when they travel through air or any medium where they encounter atoms or molecules.

> Interactions with gas molecules and other particles prevent about half of the EMR transmitted by the sun from reaching the earth's surface (Chapter 11).

For convenience, the electromagnetic spectrum is divided at arbitrary intervals into a number of different types of radiation. In reality, the spectrum is continuous, with the **wavelength** of the radiation gradually increasing from cosmic rays to radio waves, as shown in Figure 1.13. The wavelength (λ) is the distance between adjacent wave crests **(Fig. 1.14)**, and it is usually measured in nanometers (nm). The number of crests that pass a fixed point in 1 second is termed the **frequency** (ν). A frequency of

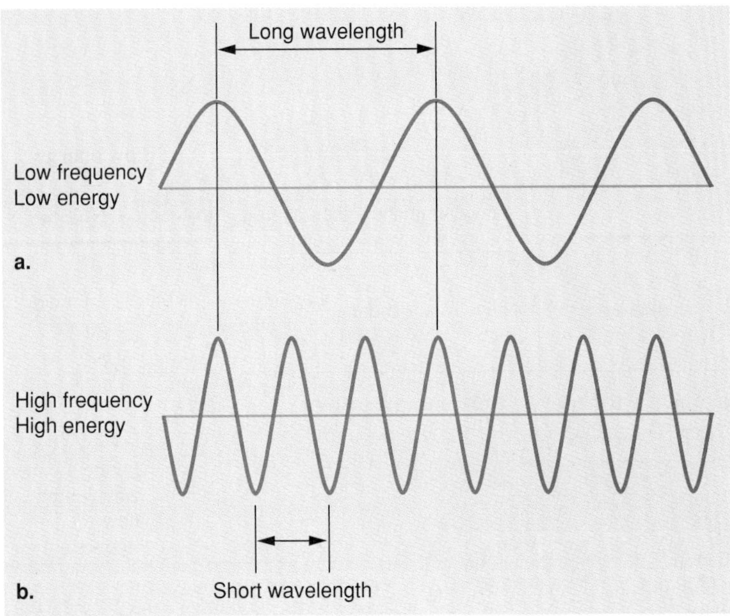

FIGURE 1.14

Wavelength and frequency of electromagnetic radiation. The frequency of the high-frequency (high-energy) radiation shown in (b) is three times that of the low-frequency (low-energy) radiation shown in (a).

1 cycle (passage of one complete wave) per second is equal to 1 hertz (Hz). Wavelength and frequency are related as shown in the following equation:

$$\nu = \frac{c}{\lambda} \quad \text{(where } c = \text{the speed of light in a vacuum)}$$

The various forms of radiation seem very different, but they are all manifestations of the same phenomenon; they differ from each other only in energy, wavelength, and frequency. The shorter the wavelength (and the higher the frequency) of the radiation, the greater the energy it transmits. High-energy radiation is very damaging to living tissues. The harmful effects of gamma rays (which are emitted by radioactive elements) and X rays will be discussed in Chapter 5. We are protected from the harmful effects of UV radiation from the sun primarily by the ozone layer in the upper atmosphere. The potentially dangerous results of depleting the ozone layer will be covered in Chapter 11.

Visible light, the kind we see, makes up only a small part of the entire electromagnetic spectrum. Its wavelength ranges from approximately 400 nm (violet) to approximately 700 nm (red). Infrared radiation (which has a wavelength greater than 700 nm) is invisible but can be felt as heat when, for example, it is emitted by a stove burner or electric heater. Microwave radiation is used for radio communication, microwave cooking, and weather tracking (radar). Long-wavelength radio waves are emitted by large antennae at broadcasting stations.

1.13 The Scientific Method

If a scientist wishes to find the answer to a specific problem, he or she will apply the **scientific method,** an orderly step-by-step procedure involving observation and measurement, speculation, experimentation, and deduction. To illustrate this method, we will examine the historic experiment conducted by James Lind in 1747 to find the cause of scurvy, a terrible disease that afflicted sailors on long sea voyages. Scurvy caused vomiting, weight loss, bleeding gums, tooth loss, and very slow healing of wounds. More seamen died from scurvy in wartime than were killed in battle.

James Lind, who was a Scottish naval surgeon and physician, had observed thousands of cases of scurvy on board ships, and, as he notes in his account of his experiment, had also studied the papers that had been published about the condition. On the basis of his own observations and those described by others, Lind suggested that scurvy might be related to diet. A suggestion like this, which is speculation and has not been proved, is termed a **hypothesis.** Lind described the experiments he carried out to test his hypothesis:

> On the 20th May, 1741, I took twelve patients in the scurvy on board the *Salisbury* at sea. Their cases were as similar as I could have them. They all in general had putrid gums, the spots and lassitude, with weakness of their knees. They lay together in one place, being a proper apartment for the sick in the forehold; and had one diet in common to all, viz., water gruel sweetened with sugar in the morning; fresh mutton broth often times for dinner; at other times puddings, boiled biscuit with sugar etc.; and for supper, barley, raisins, rice and currants, sago and wine, or the like. Two of these were ordered each a quart of cyder a day. Two others took twenty five gutts [drops] of elixir vitriol three times a day upon an empty stomach, using a gargle strongly acidulated with it for their mouths. Two others took two spoonsfuls of vinegar three times a day upon an empty stomach, having their gruels and their other food well acidulated with it, as also the gargle for the mouth. Two of the worst patients, with the tendons in the ham rigid (a symptom none [of] the rest had) were put under a course of sea water. Of this they drank half a pint every day and sometimes more or less as it operated by way of gentle physic. Two others had each two oranges and one lemon given them every day. These they eat with greediness at different times upon an empty stomach. They continued but six days under this course, having consumed the quantity that could be spared. The two remaining patients took the bigness of a nutmeg three times a day of an electuary [medicated paste] recommended by an hospital surgeon made of garlic, mustard seed, *rad. raphan.*, balsam of Peru and gum myrrh, using for common drink barley water well acidulated with tamarinds, by a decoction of wich [sic], with the addition of *cremor tartar*, they were gently purged three or four times during the course.
>
> The consequence was that the most sudden and visible good effects were perceived from the use of the oranges and lemons; one of those who had taken them being at the end of six days fit four [sic] duty. The spots were not indeed at that time quite off his body, nor his gums sound, but without any other medicine than a gargarism or elixir of vitriol he became quite healthy before we came into Plymouth, which was on the 16th June. The other was the best recovered of any in his condition, and being now deemed pretty well was appointed nurse to the rest of the sick. . . .
>
> As I shall have occasion elsewhere to take notice of the effects of other medicines in this disease, I shall here only observe that the result of all my experiments was that oranges and lemons were the most effectual remedies for this distemper at sea. I am apt to think oranges preferable to lemons. . . .

After reading Lind's descriptions of the treatments he prescribed for the unfortunate sailors, we can only conclude that, with the exception of oranges and lemons, the treatments must have added considerably to the patients' misery. But the experiments did indicate that adding citrus fruits to the diet rapidly reversed the effects of scurvy.

As a result of Lind's work, all ships in the British Navy were eventually required to add lemon juice to the sailors' diet. Because of this practice, British sailors to this day are often called "limeys," even though lemons are usually the fruit used to prevent scurvy.

The scientific method can be summarized as follows (**Fig. 1.15**):

1. *Observation.* As a result of observation, a question is raised. In our example, Lind observed that sailors on long sea voyages suffered from scurvy, and he wanted to know why. In science, many of the often complex questions that are raised originate in the laboratory. Observations are not confined solely to what can be seen, heard, smelled, touched, or tasted with the unaided senses; they also include information gained from analytical instruments.

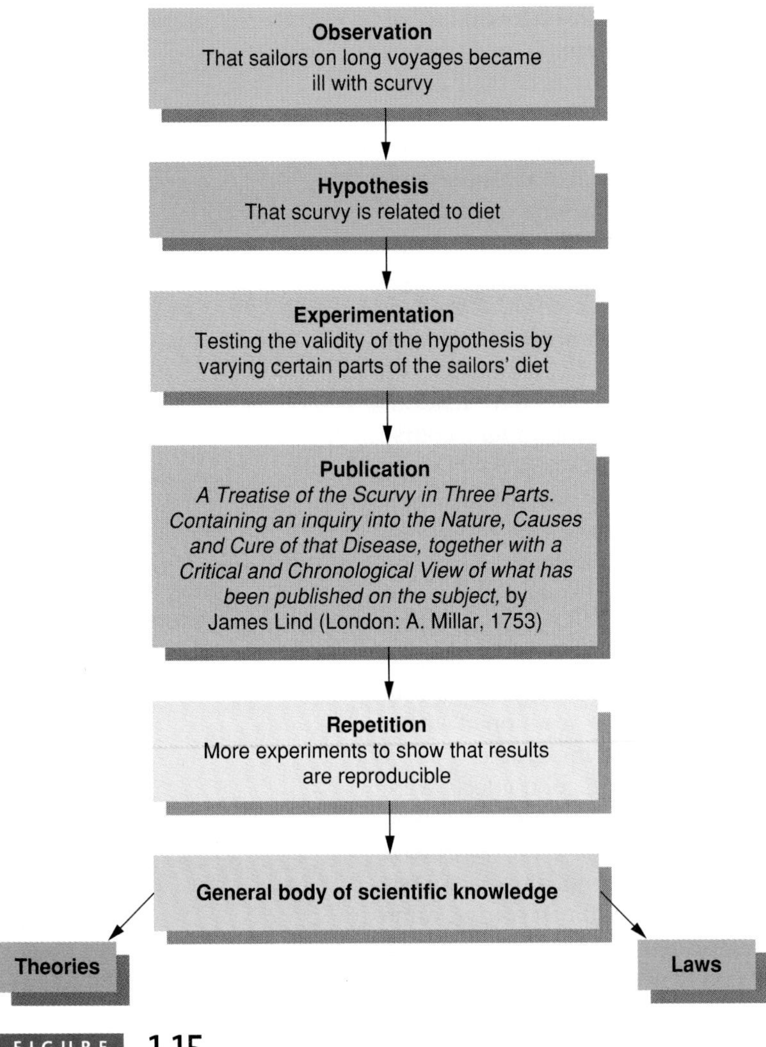

Observation
That sailors on long voyages became ill with scurvy
Hypothesis
That scurvy is related to diet
Experimentation
Testing the validity of the hypothesis by varying certain parts of the sailors' diet
Publication
A Treatise of the Scurvy in Three Parts. Containing an inquiry into the Nature, Causes and Cure of that Disease, together with a Critical and Chronological View of what has been published on the subject, by James Lind (London: A. Millar, 1753)
Repetition
More experiments to show that results are reproducible
General body of scientific knowledge
Theories
Laws

FIGURE 1.15
The scientific method.

2. *Hypothesis.* Based on the relevant information that is available, an educated guess is made to explain the initial observations. This guess is a hypothesis. Lind's hypothesis was that scurvy is related to diet.
3. *Experimentation.* A series of experiments is devised to test the validity of the hypothesis. The experiments must be well designed and carefully controlled, and the exact conditions under which they are conducted must be recorded. Lind designed his experiments carefully. He limited the variables in his study as far as possible by making sure that the cases of scurvy were similar, and that all the patients had the same accommodations and a common general diet. To show that his results were reproducible, he administered each of the different treatments to two individuals, not just to one.
4. *Revision of original hypothesis.* If the results of the experiments are inconclusive or if they prove that the original hypothesis is incorrect, a new hypothesis must be formulated, and another set of experiments must be devised and conducted. In Lind's case, this step was unnecessary because his experiments confirmed his original hypothesis.
5. *Publication and repetition.* Once it has been established that the experimental data uphold the hypothesis, the data should be published. Other scientists can then repeat the experiments to determine if, using similar conditions, they are able obtain the same results. Lind published his results in 1753, and since then, it has been confirmed numerous times that citrus fruits can both prevent and cure scurvy.

A **theory** brings together the results of many experiments and is used to build what scientists call a **model,** a representation of some aspect of physical reality. Unlike a hypothesis, a theory allows us to make predictions. Theories, just like hypotheses, can be tested and rejected. But more often a theory evolves into a more refined theory and a better model. You have been introduced to atoms in this chapter. Later we will see how atomic theory evolved as experiments revealed more and more about the structure of atoms. Although some atomic models were rejected, the basic theory that matter is composed of atoms remained intact.

Later in this text, we will discuss laws. A **law** is a statement that summarizes a large body of scientific data. A law describes a general property of nature and, like a theory, allows us to make predictions. One very important law in chemistry is the law of conservation of mass, which states that *there is no detectable change in mass in an ordinary chemical reaction.* It has been shown numerous times that when a scientist carries out a chemical reaction in a closed, leak-proof container, the mass of the container plus its contents before the start of the reaction is always equal to the mass of the container plus its contents after the reaction has occurred. No one has ever observed either a gain or a loss of mass under these conditions. Once a law has been established, scientists seek to explain why the law is true. As we shall see later, the law of conservation of mass follows directly from atomic theory.

Chapter Summary

1. Chemistry is the study of the composition of matter and the changes that matter undergoes.
2. A substance has both physical and chemical properties and can undergo both physical and chemical changes.
3. There are three states of matter: solid, liquid, and gas.
4. A pure substance has a definite and constant composition.
5. Mixtures are combinations of two or more pure substances.
6. Pure substances can be divided into elements and compounds. Elements cannot be decomposed into simpler substances by ordinary chemical means; compounds are pure substances composed of two or more elements.
7. Elements are represented by symbols consisting of one or two letters.
8. An atom is the smallest unit of an element that can take part in a chemical change.
9. A molecule is a combination of two or more atoms. It is the smallest unit of a pure substance (element or compound) that can exist and still retain the physical and chemical properties of the substance.
10. Atoms are composed of protons (which have a one-unit positive charge), electrons (which have a one-unit negative charge), and neutrons (no charge).
11. Atoms and molecules can lose or gain electrons to form cations (positively charged) or anions (negatively charged), respectively.
12. A chemical formula indicates the number of atoms of each element that are present in a molecule of a compound.
13. A chemical equation summarizes the changes that occur in a chemical reaction.
14. Electromagnetic radiation (EMR) from the sun includes X-rays, ultraviolet (UV) radiation, visible light, infrared (IR) radiation, and radio waves.
15. Wavelength (λ) and frequency (ν) of radiation are related as shown in the equation:

    $\nu = \frac{c}{\lambda}$ (where c = the speed of light)
16. Short-wavelength (high-frequency) radiation has more energy than longer-wavelength (lower-frequency) radiation.

17. The scientific method includes several steps: observation, hypothesis, experimentation, revision of original hypothesis, publication, and repetition.
18. A theory brings together a body of knowledge that is used to predict a basic truth about a scientific problem.
19. A law is a statement that summarizes a large body of scientific data.

Key Terms

anion (p. 12)
atom (p. 12)
cation (p. 12)
central science (p. 3)
chemical change (p. 6)
chemical equation (p. 14)
chemical formula (p. 12)
chemical properties (p. 4)
chemical reaction (p. 6)
chemistry (p. 3)
coefficient (p. 14)
compound (p. 9)
electromagnetic radiation (EMR) (p. 15)
electron (p. 12)
element (p. 9)
energy (p. 6)
frequency (p. 15)
gas (p. 7)
heterogeneous mixture (p. 8)
homogeneous mixture (p. 8)
hypothesis (p. 16)
ion (p. 12)
law (p. 18)
liquid (p. 7)
matter (p. 3)
mixture (p. 8)
model (p. 18)
molecule (p. 12)
neutron (p. 12)
nuclear change (p. 6)
nuclear reaction (p. 6)
physical change (p. 5)
physical properties (p. 3)
plasma (p. 7)
properties (p. 3)
proton (p. 12)
pure substance (p. 7)
science (p. 2)
scientific method (p. 16)
solid (p. 7)
solution (p. 8)
technology (p. 2)
theory (p. 18)
wavelength (p. 15)

Questions and Problems

1. Classify the following as physical or chemical changes:
   a. dissolving sugar in water
   b. setting gasoline on fire
   c. pulverizing (smashing) a rock
   d. a bronze statue turning green after exposure to wind and rain
   e. discharging carbon dioxide from a smokestack into the air
   f. milk souring
2. Dry ice (solid CO_2) evaporates to form gaseous CO_2 by a process called *sublimation*. Is sublimation a physical or chemical change?
3. Define and give an example of each of the following:
   a. pure substance b. mixture c. solution
4. Classify the following as either homogeneous or heterogeneous mixtures:
   a. sand and water b. sand and salt
   c. salt dissolved in water d. apple juice
   e. orange juice f. coffee with sugar
5. How would you separate the components of the following mixtures?
   a. sand and water b. sugar and water
   c. oil and water d. salt and sand
6. Which of the following are heterogeneous mixtures?
   a. iced tea that is sweetened with sugar
   b. apple juice
   c. vegetable soup
7. Which of the following are homogeneous mixtures?
   a. distilled water
   b. paint
   c. chocolate chip cookie
8. Classify the following as compounds or elements:
   a. sodium, Na b. carbon dioxide, CO_2

c. arsenic, As
d. helium, He
e. sulfur dioxide, SO_2
f. aluminum, Al
g. water, H_2O
h. octane, C_8H_{18}
i. zinc, Zn
j. nitrogen dioxide, NO_2
k. sulfuric acid, H_2SO_4
l. mercury, Hg
m. calcium carbonate, $CaCO_3$
n. iron, Fe

9. Write the chemical symbol for each of the following elements:
   a. uranium
   b. sulfur
   c. oxygen
   d. carbon
   e. calcium
   f. potassium
   g. radon
   h. nitrogen
   i. nickel
   j. lead
   k. gold
   l. chromium
   m. cadmium
   n. silver

10. Write the names of the following elements:
   a. Mg
   b. Cu
   c. Co
   d. C
   e. B
   f. Ni
   g. P
   h. Si
   i. S
   j. Al
   k. Br
   l. I
   m. Hg
   n. O

11. Classify the following as anions, cations, or neutral atoms:
   a. Cl^-
   b. Na^+
   c. He
   d. Ca^{2+}
   e. F^-
   f. C
   g. O^{2-}
   h. Fe^{3+}
   i. Si
   j. K^+
   k. I^-
   l. Ne

12. Classify the following as anions, cations, or neutral molecules:
   a. CO_2
   b. $PO_4{}^{3-}$
   c. Fe^{2+}
   d. $CO_3{}^{2-}$
   e. NO
   f. Co^{2+}
   g. SO_2
   h. $NH_4{}^+$
   i. $C_6H_{12}O_6$
   j. $NO_3{}^-$

13. About 90% of the human body is composed of only three elements. Name them.

14. Which element is the second most abundant in the crust of the earth?

15. How many oxygen atoms are contained in each of the following molecules?
   a. NaOH
   b. $MgSO_4$
   c. $NaHCO_3$
   d. $Al(NO_3)_3$

16. Describe in words the chemical reactions that are summarized below:
   a. $C + O_2 \longrightarrow CO_2$
   b. $4\ Fe + 3\ O_2 \longrightarrow 2\ Fe_2O_3$
   c. $2\ Na + Cl_2 \longrightarrow 2\ NaCl$

17. Consider the following reaction:

$$P_4O_{10} + 6\ H_2O \longrightarrow 4\ H_3PO_4$$

   a. How many reacting atoms are in this equation?
   b. How many reacting molecules are in this equation?
   c. How many molecules are produced in this reaction?
   d. How many atoms are produced in this reaction?

18. Consider the following reaction:

$$N_2 + 3\ H_2 \longrightarrow 2\ NH_3$$

   a. How many elements are in this equation?
   b. How many reacting atoms are in this equation?
   c. How many reacting molecules are in this equation?
   d. How many molecules are produced in this reaction?
   e. How many atoms are produced in this reaction?

19. Which of the following are physical changes, and which are chemical changes?
   a. cooking a hamburger on a charcoal grill
   b. adding sugar to tea
   c. frying an egg
   d. freezing ice cubes
   e. baking bread

20. How do the gas, liquid, and solid forms of a particular material differ in physical properties?

21. List two chemical reactions you carried out today.

22. Define each of the following and give the appropriate units for each. How are the three measurements related to each other?
   a. wavelength
   b. frequency
   c. speed of light, *c*

23. The visible spectrum to which the human eye is sensitive goes from red through orange, yellow, green, blue, and indigo to violet. Is an orange beam of light of higher or lower energy than a blue beam?

24. Define each of the following:

a. electromagnetic radiation

b. energy

c. nanometer (nm)

d. wavelength

25. Sunscreen creams absorb high-energy ultraviolet (UV) radiation and thus protect the skin from damage.

a. What is the wavelength of UV radiation?

b. Is UV radiation of shorter or longer wavelength than visible light?

c. Is UV radiation of higher or lower energy than infrared (IR) radiation?

26. X-rays are useful for diagnostic medical evaluation. The Surgeon General has warned that excessive X-ray exposure may pose a risk to health. Referring to Figure 1.13:

a. Explain why X-rays pose a risk.

b. Are X-rays more dangerous than microwaves?

c. TV and radio waves permeate our environment. Are they dangerous?

d. Cosmic rays from the sun bombard our planet. Should we be concerned about the effects of cosmic rays?

e. Tanning salons use UV lights to darken (tan) the skin. Why is UV light used? Is there a risk?

27. The earth is warmed by sunlight during the day. On a clear and cloudless winter night, the temperature will drop lower than it would if clouds were present. Explain.

28. Can you think of anything you have done today that did not involve a chemical reaction?

29. Answer the following questions about electromagnetic radiation:

a. Which form of electromagnetic radiation is used to cook food?

b. Which form of electromagnetic radiation causes sunburn?

c. Which form of electromagnetic radiation can be used to tell if an arm is broken?

d. Which form of electromagnetic radiation is used to expose an image on film?

30. Name two of your senses that can be used to detect some forms of electromagnetic radiation.

31. One of your friends tells you that scientists have performed experiments and have formed a hypothesis that a person's ability to solve math problems is directly related to the size of his or her shoes. Can you tell where the scientists did these experiments? Do you think the scientists would be able to repeat the experiment in your class?

Answers to Practice Exercises

1.1 **a.** chemical change
b. physical change
c. physical change

1.2 Si and He are elements. KI and HI are compounds.

Chapter 2

Planet Earth: Rocks, Life, and Energy

Chapter Objectives

In this chapter, you should gain an understanding of:

- How the earth, the oceans, and the atmosphere were formed
- The earth's mineral resources, and how they are used by society
- Important metal elements that are extracted from ores
- The origin of life on earth
- The producers and consumers of energy in the natural environment
- The flow of energy through ecosystems
- The nutrient cycles that provide living organisms with the chemicals they need for life

On earth, energy and materials flow through all ecosystems. Energy from the sun reaches all living creatures through plants.

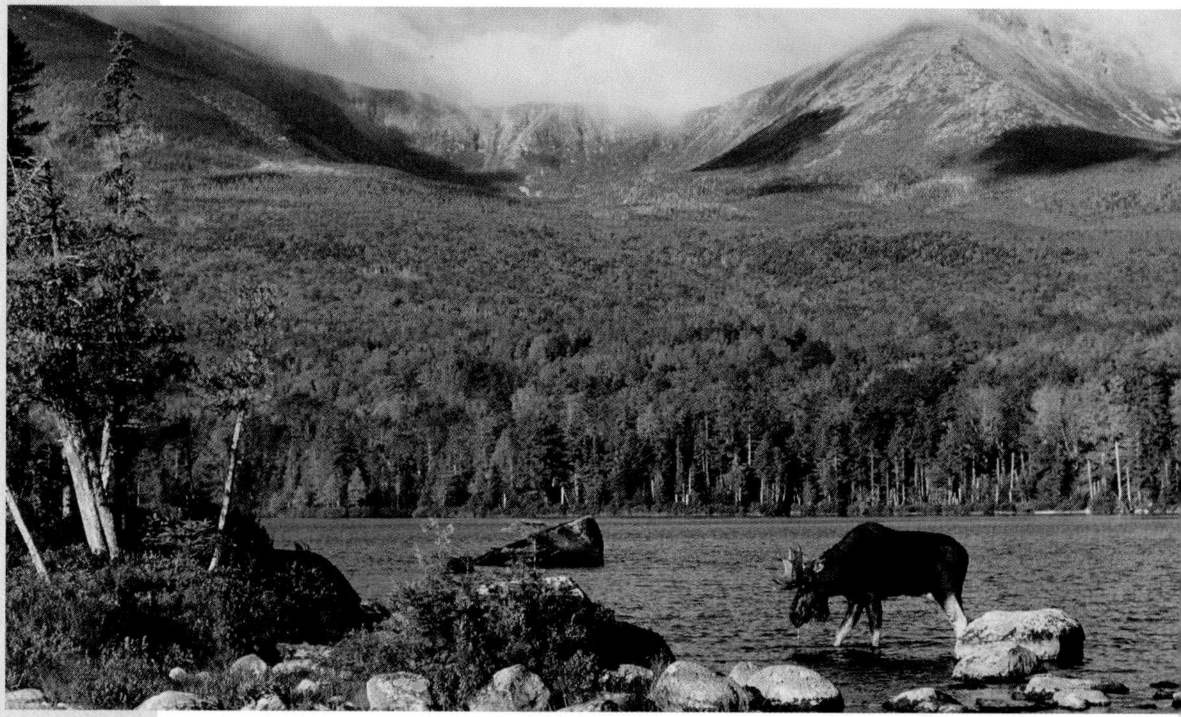

ONLINE FEATURES

www.jbpub.com/chemistry

- Chemistry and the Environment
- eLearning
- Chemistry in the News
- Special Topics
- Research & Reference Links
- Ask the Authors

TO UNDERSTAND HOW OUR ENVIRONMENT WORKS, we must first look back billions of years to find out how the earth was born and how it evolved into the life-supporting planet we inhabit today.

In this chapter, we consider the formation of the universe, including the origin of the galaxies, the stars, and our own planet earth. We look at how the oceans, the atmosphere, and the rocky surface on which we live were formed, and we examine the earth's mineral resources and discuss the ways in which society uses them. We see how life developed on earth, how all living organisms interact with their physical surroundings and with each other, how all these interactions are intertwined, and how the whole system is fueled by a continuing flow of energy through all its parts.

2.1 The Formation of the Universe

If we gaze up at the sky on a clear night, away from the lights of any city, we see myriads of stars. All the stars we see are just a part of our galaxy, the **Milky Way.** This pinwheel-shaped body, which is made up of clouds of gas and cosmic dust, and billions and billions of stars, includes our solar system—the sun and its nine orbiting planets. What we see is only a minute fraction of the whole **universe.** Beyond the Milky Way, extending into space for distances beyond our comprehension, are countless other galaxies. It was probably only when we humans first ventured out into space in the 1960s that we began to appreciate the smallness and insignificance of our planet in relation to the universe as a whole. The first photographs of the earth taken from the moon showed us our planet suspended in the black vastness of space (Fig. 2.1).

FIGURE 2.1

The earth seen from the surface of the moon.

According to the most recent research, the universe began between 12 billion and 13.5 billion years ago. There are still differences of opinion, but many scientists believe that all the matter in the universe was once compressed into an infinitesimally small and infinitely dense mass that exploded with tremendous force. This explosion of unimaginable proportions—appropriately called the **big bang**—generated enormous amounts of light, heat, and energy and released the cosmic matter from which the galaxies and stars were eventually formed. The universe began expanding in all directions and, according to most astronomers, has been expanding ever since.

In 1992, the Hubble Space Telescope detected new evidence that the universe will not collapse back on itself billions of years from now but will probably go on expanding forever.

Galaxies and Stars

As the universe expanded, it cooled very, very slowly and cosmic matter gradually condensed to form the first galaxies. Atoms of hydrogen—the simplest and lightest of all the elements—formed in the swirling clouds of condensing matter. Over billions of years, the galaxies gave birth to the early stars, and these stars generated sufficient heat to cause hydrogen atoms to fuse (join) together to form atoms of helium, the second lightest of the elements. The energy released during these fusion reactions initiated further fusion reactions, in which all 90 of the remaining naturally occurring elements found on earth were formed. In the universe as a whole, 90% of all atoms are hydrogen, 9% are helium, and the remaining 1% are atoms of all the other elements. Scientists believe that subsequent explosions of the early stars scattered the elements and that our sun was born from the debris of one of these explosions. The sun, which to us appears so very bright, is an average-sized star located toward the edge of the Milky Way.

The Planets in Our Solar System

Scientists still do not know with any certainty how the planets in our **solar system** developed (a solar system is a group of planets that revolve around a star), but it is generally believed that they began to form about 5 billion years ago from hot, mainly gaseous matter rotating about the sun. With time, the matter slowly cooled, and solid particles condensed out from the gases. The particles gradually coalesced into clumps of matter. Larger clumps had stronger gravity and gradually drew in and retained additional particles, eventually forming the nine planets that revolve around the sun: Mercury, Venus, Earth, Mars, Jupiter, Saturn, Uranus, Neptune, and Pluto **(Fig. 2.2)**.

The clumps of matter formed by the coalescing particles are called *planetesimals.*

The four planets closest to the sun—Mercury, Venus, Earth, and Mars—are called *terrestrial planets* and are small and dense. The more distant *giant planets*—Jupiter, Saturn, Uranus, and Neptune—are much larger and of lower density than the terrestrial planets.

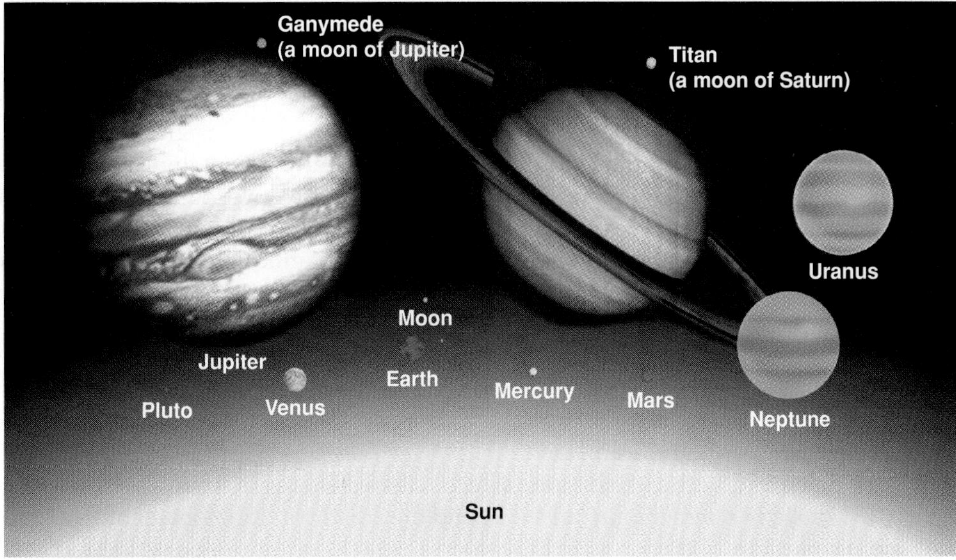

FIGURE 2.2

The relative sizes of the planets in our solar system.

The earth and the other terrestrial planets formed close to the sun and were so hot that lighter, easily evaporated materials could not condense and were swept away. Only substances with extremely high boiling points, such as metals and minerals, condensed on these planets. Mercury, the planet closest to the sun and therefore the hottest, is composed mainly of the metal iron. On the earth, which formed at a somewhat lower temperature, silicates and other metals besides iron were able to condense. (Silicates are minerals formed from the elements silicon, oxygen, and a variety of metals.) The larger planets, with their greater mass and thus stronger gravitational pull, retained gases—mostly hydrogen and helium—in the atmospheres surrounding them. Some important features of the planets, as they exist today, are listed in **Table 2.1**.

The Sun

The sun is the ultimate source of energy for life on earth. It makes up 99.9% of the mass of the solar system, and its diameter is approximately 110 times as great as that of the earth. Scientists estimate that temperatures near the center of this immense rotating sphere of extremely hot gases reach almost 15 million degrees Celsius (27 million degrees Fahrenheit). Fusion reactions occur at these incredibly high temperatures, continually releasing tremendous amounts of energy in the form of heat and light. These fusion reactions have allowed the sun to shine brightly for billions of years and will allow it to continue to shine for billions more.

2.2 Differentiation of the Earth into Layers

Exactly how the earth evolved to its present state is not known, but earth scientists believe that when the earth was first formed some 4.7 billion years ago, it was homogeneous in composition—a dense rocky sphere with no water on its surface and no atmosphere. Then, over time, the interior of the sphere gradually grew hotter, and the earth became differentiated into layers, each layer having a different chemical composition. This crucial period in the development of the earth led to the formation of its magnetic field, atmosphere, oceans, and continents—and ultimately to life.

Why the Earth Heated Up

Three factors are believed to have caused the earth to heat up. First, the cosmic particles that collided and clumped together to form the earth were drawn inward by the pull of gravity. As more particles collided with the developing planet, heat was released. Some of this heat was retained within the earth; this heat gradually built up as more and more material accumulated.

Second, as the earth grew, material in the center was compressed by the weight of new material that struck the surface and was retained. Some of the energy expended in compression was converted to heat and caused a further rise in the temperature within the earth.

The third, and very significant, factor in the warming of the earth was the decay of radioactive elements within the interior, which released energy in the form of heat. The atoms in radioactive elements are unstable, and they disintegrate spontaneously, emitting atomic particles and energy. In this process, which continues today, the radioactive elements are converted into atoms of other elements (this topic will be discussed in Chapter 5). Only a very small percentage of naturally occurring elements have atoms that disintegrate in this way, and the heat generated with each disintegration is

Table 2.1 Important Features of the Planets in Our Solar System

Planet	Diameter		Mass	Density	Gravity	Time for One Rotation on Axis	Time for One Revolution around Sun	Distance from Sun		Composition of
	(km)	(mi)	(earth = 1)	(water = 1)	(earth = 1)	(earth days or hours)	(earth years)	(million km)	(million mi)	Atmosphere
Terrestrial										
Mercury	4,835	3,004	0.055	5.69	0.38	59 days	0.24	57.7	36.8	None
Venus	12,194	7,577	0.815	5.16	0.89	243 days	0.62	107	66.9	CO_2
Earth	12,756	7,926	1.00	5.52	1.00	1.00 days	1.0	149	92.6	N_2, O_2
Mars	6,760	4,200	0.108	3.89	0.38	1.03 days	1.9	226	141	CO_2, N_2, Ar
Giant										
Jupiter	141,600	87,986	318	1.25	2.64	9.83 hours	12	775	482	H_2, He
Saturn	120,800	75,061	95.1	0.62	1.17	10.23 hours	29	1421	883	H_2, He
Uranus	47,100	29,266	14.5	1.60	1.03	23.00 hours	84	2861	1777	H_2, He, CH_4
Neptune	44,600	27,713	17.0	2.21	1.50	22.00 hours	165	4485	2787	H_2, He, CH_4

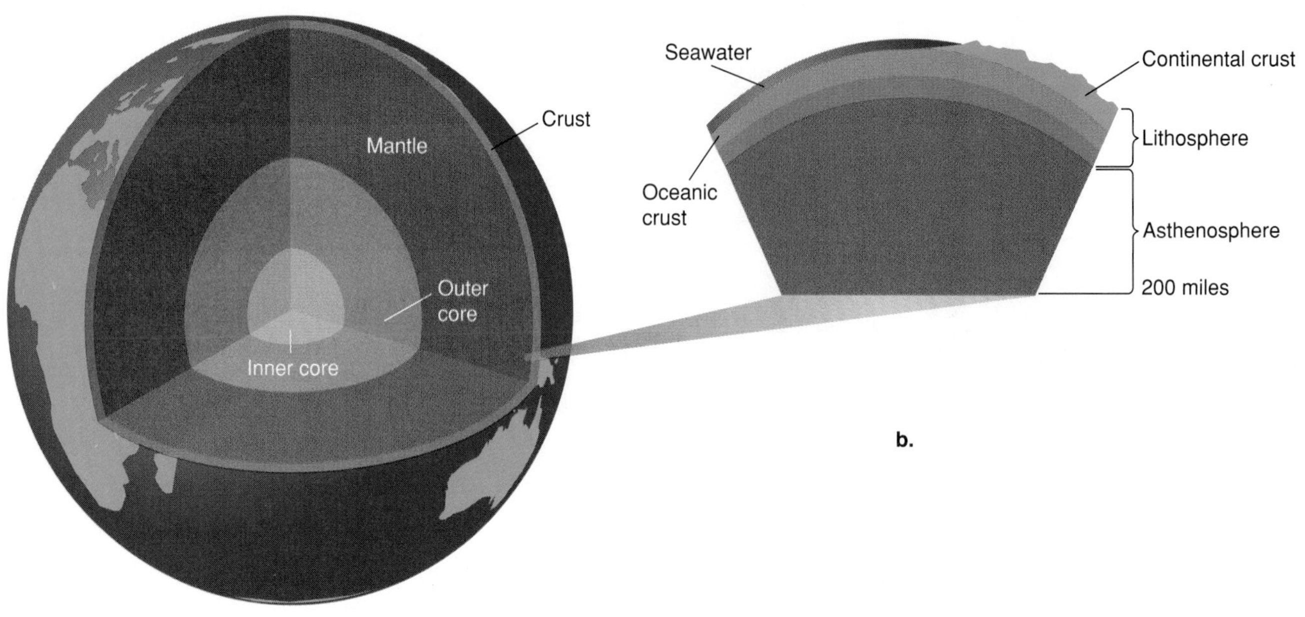

FIGURE 2.3

The structure of the earth: (a) The earth is differentiated into three distinct layers called the core, the mantle, and the crust. (b) The lithosphere, which comprises the continental and oceanic crust together with the solid upper part of the mantle, rests on the partially molten asthenosphere.

extremely small. Nonetheless, earth scientists have calculated that the retention of this heat within the earth over billions of years (together with the heat released as new material accumulated and was compressed) would have been sufficient to raise the temperature of the material at the center of the earth to the point where it became molten.

It seems probable that this critical temperature was reached about 1 billion years after the earth was born. Metallic iron, which melts at 1535°C (2795°F) and makes up over 30% of the mass of the earth, began to melt. The heavy molten iron, together with some molten nickel, sank to the center of the earth. As the molten iron sank, less dense material was displaced and rose toward the surface. As a result, the earth ceased to be homogeneous and eventually became differentiated into three distinct layers: the core, the mantle, and the crust **(Fig. 2.3)**.

The Core

The earth's **core,** which extends 3500 kilometers (2200 miles) from the planet's center, is believed to be composed of iron and small amounts of nickel. The metals are thought to be in solid form in the inner core and molten in the surrounding outer core (Fig. 2.3a). Because the core is inaccessible to us, there is no way to prove that it consists primarily of iron, but there is considerable indirect evidence to support this view. For example, analysis of light emitted by the sun and stars has revealed that iron is the most abundant metal in the universe, and most of the meteorites that have landed on the earth from outer space are composed of iron. Furthermore, analysis of waves generated by earthquakes has shown that the core is very dense, and iron is the densest metal found in any quantity on earth.

The Mantle

The earth's **mantle,** which lies between the core and the crust, is approximately 2900 kilometers (1800 miles) thick (Fig. 2.3). The relatively thin upper part of the mantle is solid and rigid, but the layer below it—called the **asthenosphere**—although essentially solid, is able to flow extremely slowly, like a very thick viscous liquid. In the deep mantle, below the asthenosphere, the rock is believed to be rigid.

The Crust

Above the mantle is the **crust,** which forms the thin outer skin of the earth (Fig. 2.3b). The crust is thicker beneath the continents than beneath the oceans. Its thickness ranges from 6 kilometers (4 miles) under the oceans to 70 kilometers (45 miles) under mountainous regions. Although the crust makes up a very small part of the earth as a whole, we gather from it practically all the resources that sustain our way of life.

Together, the crust and the solid upper part of the mantle make up the relatively cool and rigid **lithosphere,** which floats on the hotter, partially molten asthenosphere. The boundary between the lithosphere and the asthenosphere is not caused by a difference in the chemical composition of their rocks but by a change in the physical properties of the rocks that occurs as temperature and pressure increase with depth.

Lithosphere is derived from *lithos*, the Greek word for "stone."

Relative Abundance of the Elements in the Earth

By mass, the four most abundant elements in the earth are iron, oxygen, silicon, and magnesium, which together account for about 93% of the earth's mass **(Fig. 2.4a)**. Nickel, sulfur, calcium, and aluminum make up another 6.5%. The remaining 0.5% or so of the earth's mass is made up of the other 84 naturally occurring elements.

Primarily because most of the iron sank to the center of the earth during the period of differentiation, the relative abundance of the elements in the crust differs greatly from that in the earth as a whole **(Fig. 2.4b)**. Seventy-four percent of the crust consists of oxygen and silicon, while aluminum, iron, magnesium, calcium, potassium, and sodium together account for 25%.

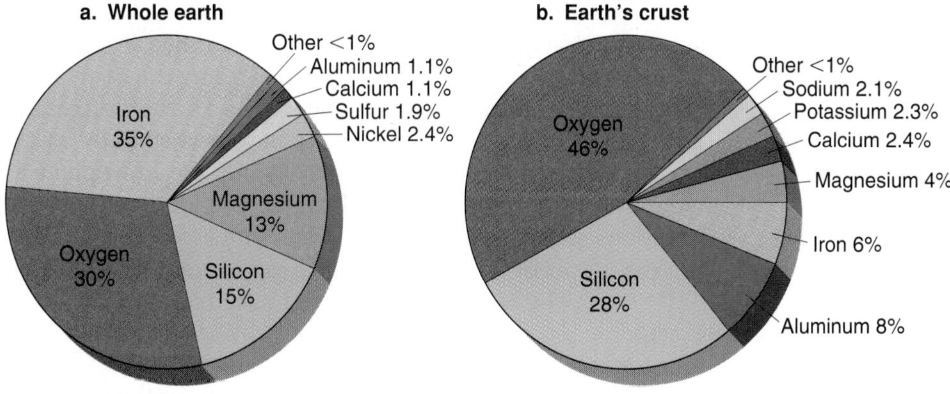

FIGURE 2.4

The relative abundance (by mass) of elements in the whole earth and in the earth's crust. Because of the differentiation that occurred early in the earth's history, the percentage of iron in the crust (b) is lower than that in the whole earth (a), and the percentages of aluminum, silicon, and oxygen (the elements that combine to form silicates) are higher.

Table 2.2 The Relative Abundance of the Economically Valuable Elements in the Earth's Crust

Name	Chemical Symbol	Abundance in Crust (% by mass)
Aluminum	Al	8.00
Iron	Fe	5.80
Magnesium	Mg	2.77
Potassium	K	1.68
Titanium	Ti	0.86
Hydrogen	H	0.14
Phosphorus	P	0.101
Fluorine	F	0.0460
Sulfur	S	0.030
Chlorine	Cl	0.019
Chromium	Cr	0.0096
Zinc	Zn	0.0082
Nickel	Ni	0.0072
Copper	Cu	0.0058
Cobalt	Co	0.0028
Lead	Pb	0.00010
Arsenic	As	0.00020
Tin	Sn	0.00015
Uranium	U	0.00016
Tungsten	W	0.00010
Silver	Ag	0.000008
Mercury	Hg	0.000002
Platinum	Pt	0.0000005
Gold	Au	0.0000002

Source: From F. Press and R. Siever, *Earth*, 3rd ed. (New York: W. H. Freeman, 1982), p. 553.

It might have been expected that as the earth became differentiated into layers, the elements would have been distributed strictly according to mass, with the heavier elements falling to the earth's center and the lighter ones rising to the surface. This distribution did not occur, however, because some elements combined with other elements to form compounds, and the melting points and densities of the compounds (rather than those of the elements from which they were formed) primarily determined how the elements were distributed in the earth. For example, silicon, oxygen, and various metals combined to form silicates, which are relatively light compounds that melt at relatively low temperatures. When the earth's interior was hot, these silicates rose to the surface. They are the most abundant minerals in the earth's crust.

As a result of the chemical changes that occurred during the period of differentiation, the distribution of the elements on the earth is very uneven. The relative abundance in the earth's crust of the economically valuable elements is shown in **Table 2.2**. Of these, only four—aluminum, iron, magnesium, and potassium—are present in amounts greater than 1% of the total mass of the crust. It is fortunate for us that as the result of geological processes that have been occurring for millions of years, the less abundant (but valuable) elements such as gold and silver are concentrated in specific regions of the world. If these elements had been distributed evenly throughout the earth's crust, their concentrations would be too low to make their extraction technically or economically feasible.

2.3 Formation of the Oceans and the Atmosphere

It is generally accepted that there was no water on the earth's surface for billions of years after the planet formed. Then, as the interior of the earth heated up, minerals below the earth's surface became molten. The molten material rose to the surface, and oxygen (O) and hydrogen (H) atoms that were chemically bound to certain minerals escaped explosively into the atmosphere as clouds of water (H_2O) vapor. In these tremendous volcanic eruptions, which were widespread and numerous, carbon dioxide (CO_2), nitrogen (N_2), and other gases were also released from the planet's interior (**Fig. 2.5**). The lighter gases escaped into space, but the heavier ones, including water vapor and carbon dioxide, were held by gravity and formed a thick blanket of clouds surrounding the earth. In time, as the earth's surface cooled, the water vapor condensed, and the clouds released their moisture. For the first time, rain fell on the earth. During the next several million years, volcanoes continued to erupt, and the oceans filled with water as more rain fell.

EXPLORATIONS describes a tremendous volcanic eruption of modern times (see pp. 48–49).

The earth's first atmosphere was quite different from today's. Volcanic eruptions were still common long after the earth's surface had cooled to the point where water vapor began to condense to form the oceans. Evidence suggests that, in addition to water vapor

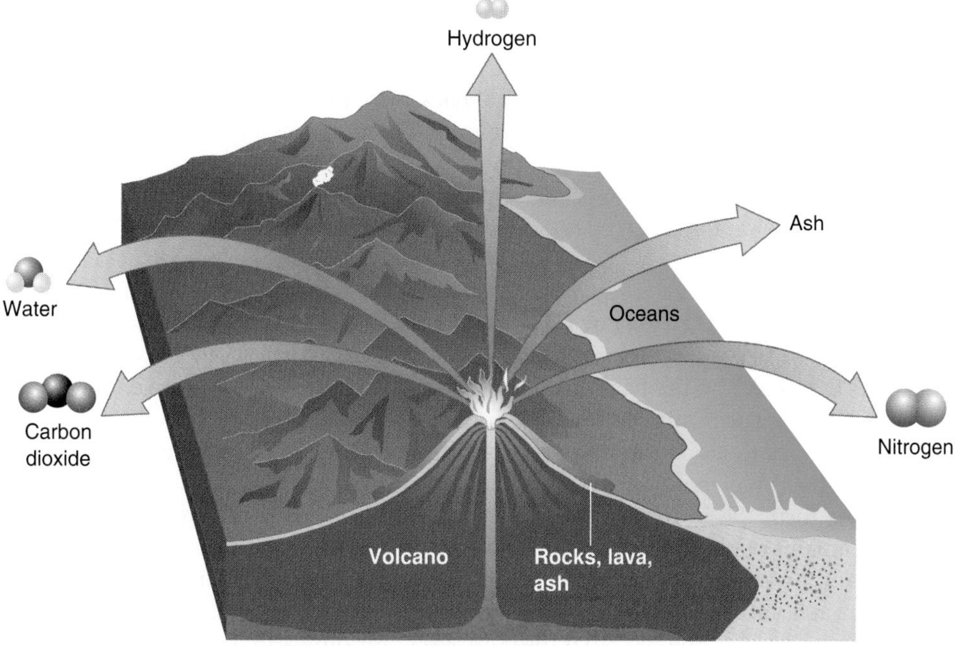

FIGURE 2.5

In volcanic eruptions, huge quantities of rocks, lava, ash, and gases (mainly water vapor, carbon dioxide, and nitrogen) are ejected. Early in the earth's history, when volcanic eruptions were frequent and widespread, water vapor released to the atmosphere condensed and fell to earth as rain, and the oceans were formed; carbon dioxide and nitrogen became the main constituents of the early atmosphere.

and carbon dioxide, the enormous volumes of gases emitted were mostly nitrogen, with smaller amounts of carbon monoxide, hydrogen, and hydrogen chloride—the same gases that are emitted by erupting volcanoes today. Hydrogen gas, being very light, was lost into space, but the other gases were held near the surface by the earth's gravitational pull.

After millions of years of volcanic activity, the atmosphere was rich in nitrogen and carbon dioxide but completely devoid of oxygen. Today, the earth's atmosphere is still rich in nitrogen (78%), but only 0.04% of the atmosphere is carbon dioxide, while oxygen accounts for 21%. There were two main ways in which the excess carbon dioxide was removed: First, when rain began to fall on the earth, very large quantities of carbon dioxide dissolved in the oceans that were formed, and much of it combined with calcium in the water to form limestone (calcium carbonate). Second, some 3 billion years ago, the first primitive blue-green algae, or **cyanobacteria,** developed in shallow waters. Like the more advanced plants on earth today, these organisms used energy from the sun to convert carbon dioxide and water into simple carbohydrates (compounds of carbon, hydrogen, and oxygen that are the main food for plants) and oxygen by the process of photosynthesis. Oxygen escaped from the water and, for the first time, entered the atmosphere. Life had created the conditions for its own success.

Small amounts of oxygen gas may have been formed by water vapor breaking down into its elements. Any oxygen formed in this way would have combined with other compounds and elements (for example, with iron to form iron oxides) and been removed from the atmosphere.

As cyanobacteria multiplied, more and more carbon dioxide was removed from the atmosphere and replaced with oxygen. Eventually, when photosynthesis had been going on for millions of years, the earth's atmosphere attained its present composition. Although photosynthesis was important in reducing the carbon dioxide content of the

Combustion of coal provides the energy needed to convert water to steam. The steam is used to turn the turbines that generate electrical power.

atmosphere, dissolution in the oceans followed by the formation of limestone was the major factor in its removal.

Ever since the beginning of the Industrial Revolution, humans have been pouring unprecedented amounts of carbon dioxide back into the atmosphere by burning carbon-containing fuels such as coal and petroleum. There is growing concern that the continuing increase in atmospheric carbon dioxide may lead to a rise in the earth's temperature, which could have serious consequences for our planet. This problem will be discussed in Chapter 11.

2.4 Rocks and Minerals

The rocks that make up the hard surface of the earth, the lithosphere, are composed of one or more substances called **minerals.** A mineral is a naturally occurring, usually crystalline, substance that has a definite composition or a restricted range of composition; it may be either an element or a compound.

Minerals form the **inorganic** part of the earth's crust; materials derived from the decayed remains of plants and animals make up the **organic** part. The terms "inorganic" and "organic" were introduced in the 18th century to distinguish between compounds derived from nonliving matter and compounds derived from plant and animal sources. At that time, chemists believed that the complex compounds that make up living matter—such as carbohydrates and proteins—could be produced only by living organisms. Once it was discovered that these compounds, which all contain carbon, could be synthesized in the laboratory, the definition of organic compounds was broadened to include all nonmineral compounds of carbon.

Some minerals have been known and used since ancient times. There is evidence that very early in human history, flint and obsidian (a volcanic glass) were shaped to make weapons and primitive knives, and clay was formed into pottery vessels and bricks. Gold, silver, copper, and brightly colored minerals such as jade and amethyst were fashioned into jewelry and other objects, and pigments were made from red and black iron oxides (compounds in which oxygen is bonded to iron).

Over 2500 distinct minerals have been identified, but only a few of them are distributed widely over the earth's surface. Many of the more valuable minerals are found in only a few limited regions of the world, where they became concentrated as a result of the upheaval and subsidence of crust materials and other rock-forming processes that have gone on for millions of years. The study of the composition of rocks has been an important factor in helping to explain how the earth was formed.

2.5 Rocks as Natural Resources

Rocks and minerals are natural resources. A natural resource is anything taken from the physical environment to meet the needs of society. Such resources may be **renewable** or **nonrenewable.** Resources such as soil, natural vegetation, fresh water, and wildlife are all renewable; if not depleted too rapidly, they are replaced in natural recycling processes. Rocks and minerals—as well as oil, natural gas, and coal—are nonrenewable. They are present in the earth in fixed amounts and are not replaced as they are used.

Rocks and minerals are quarried and used widely, often in modified form, in the construction and chemical industries and for making ceramics and many other products. In the next section, we will consider the chemical makeup and some of the uses of the most widespread of all minerals, the silicates.

2.6 Silicates

The basic unit of all **silicates** is the SiO_4 tetrahedron, in which a silicon atom at the center is bonded (joined) to four oxygen atoms located at the corners of a four-sided pyramid (Fig. 2.6). Because the radius of a silicon atom is only about one-third that of an oxygen atom, the silicon atom fits comfortably between the four oxygen atoms. Other properties related to atomic structure, which will be discussed in Chapter 4, contribute to making this a very stable arrangement.

Because each oxygen atom can bond to two silicon atoms, adjacent SiO_4 units can be linked through shared oxygen atoms. This linking allows long chains, sheets, or three-dimensional networks of SiO_4 units to be formed.

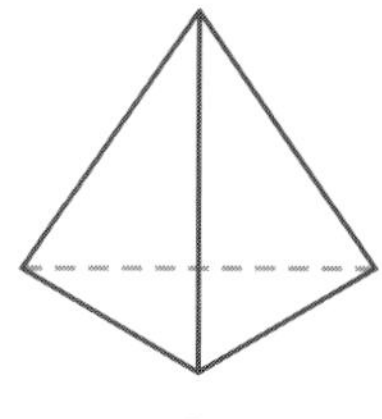

a.

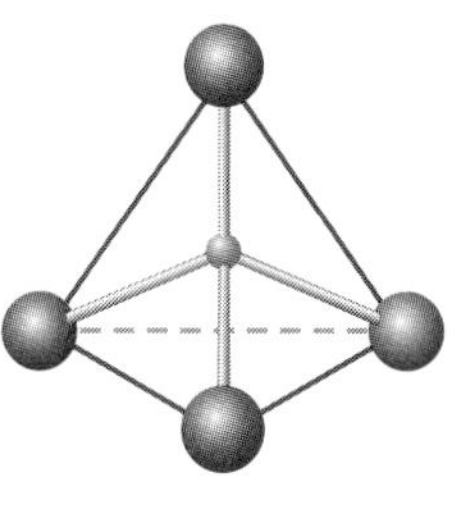

b.

FIGURE 2.6

(a) A tetrahedron. (b) The tetrahedral arrangement of the SiO_4 unit in silicates. The silicon atom at the center is joined to four oxygen atoms located at the corners of the tetrahedron.

Three-Dimensional Networks and Sheets

The mineral quartz (Fig. 2.7a) is composed entirely of SiO_4 units joined together to form a three-dimensional network (Fig. 2.7b).

In micas, the SiO_4 units form sheetlike arrays in which each tetrahedron is joined to three others (Fig. 2.8a). Because of the planar arrangement of the SiO_4 tetrahedra, micas are easily cleaved into thin sheets, as shown in Fig. 2.8b. A common mica is muscovite, a pearly white mineral that was used to make window panes in medieval Europe, before glass became readily available. Other micas include talc, a soft mineral used to make talcum powder, and kaolinite, a clay mineral.

Long Chains

Asbestos is a general term for a number of natural silicates formed from double chains of SiO_4 tetrahedra (Fig. 2.9). *Chrysotile,* the most abundant type of asbestos, takes the form of curly fibers. The much rarer *crocidolite* crystallizes as sharp, narrow needles.

The simplest formula for quartz is SiO_2. Although quartz is made up of SiO_4 units, the ratio of Si atoms to O atoms in its structure is 1:2, not 1:4, because all the O atoms are shared by adjacent tetrahedra.

a.

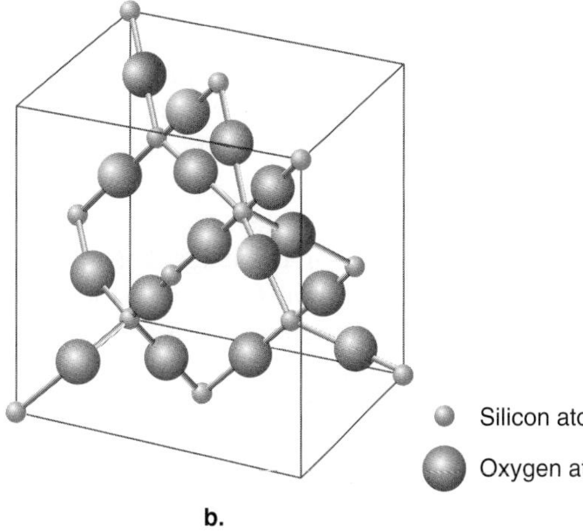

b.

FIGURE 2.7

(a) A specimen of quartz. (b) In quartz, silicon and oxygen atoms join to form three-dimensional networks.

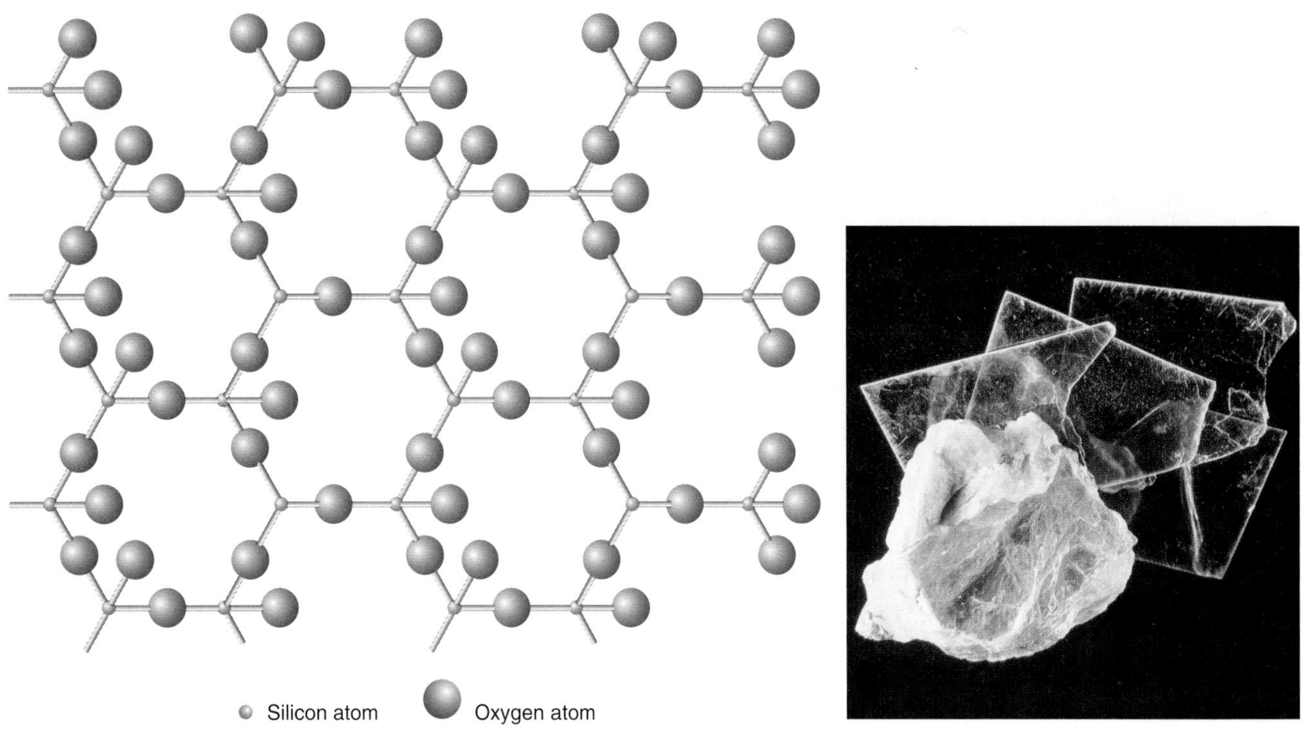

FIGURE 2.8

(a) In micas, silicon and oxygen atoms join to form sheets. (b) A specimen of mica showing how it can be separated into thin sheets.

Asbestos is a very versatile material. It is strong, flexible, and resistant to corrosion, and it serves as an excellent thermal insulator. Until the 1970s, when it became evident that it posed a serious health hazard, asbestos was widely used to insulate steam pipes and other heating units and to make protective clothing for firefighters, welders, and other people exposed to high temperatures. It was also used in brake linings, roofing materials, hair dryers, and other products.

It is well established that workers exposed to large amounts of fine airborne asbestos fibers over a considerable period of time develop serious lung disorders. These disorders include asbestosis, a disease that makes breathing very difficult, and lung cancer. Thousands of shipyard workers contracted crippling, often fatal, lung diseases as a result of exposure to asbestos in the 1940s. The risk of cancer for asbestos workers who smoke is very much higher than for those who do not smoke. This relationship between asbestos inhalation and smoking is an example of **synergism**—the working together of two factors to produce an effect greater than the sum of their individual effects.

Because of the health risks, the EPA banned the use of asbestos for insulation and fireproofing in 1974, and many communities began removing asbestos from public buildings, particularly from schools. In 1989, the agency proposed a ban on all other uses of asbestos.

The process of asbestos removal is not only extremely costly, it is also likely to release fibers into the air and expose removal workers to unusually high levels of airborne fibers. The EPA now recommends that, unless it is crumbling, asbestos be left in place, encapsulated in a plastic coating.

a.

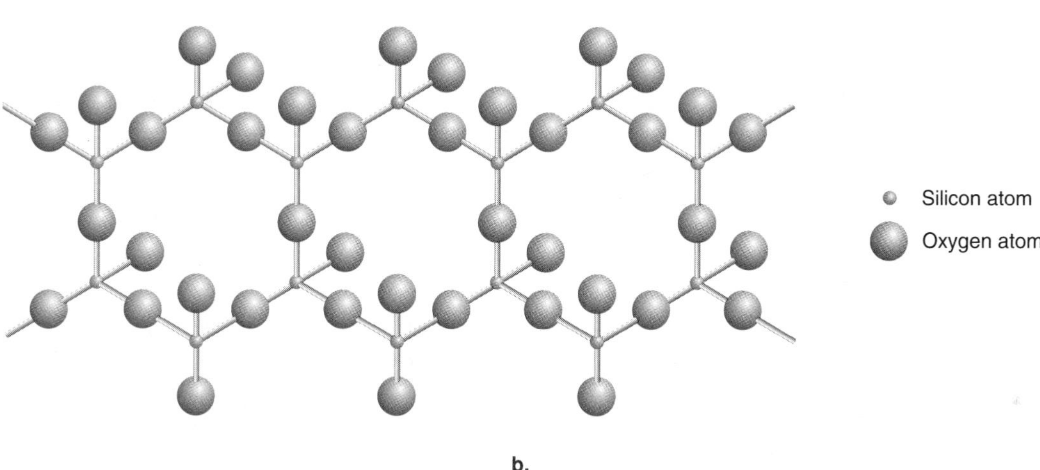

b.

Silicon atom

Oxygen atom

FIGURE 2.9

(a) Asbestos fibers. (b) In asbestos, silicon and oxygen atoms join to form long chains. Two parallel rows of linked SiO_4 tetrahedra join through oxygen atoms to form double chains.

Uses of Silicates

Since well before recorded history, men and women have used natural silicates in one form or another in their daily lives. Silicates were first used to fashion primitive weapons and cutting tools; later, as new skills were learned, these minerals were modified and transformed into pottery utensils. Today, silicates are starting materials for a wide range of products, including bricks, china, glass, cement, and other materials that have a wide range of specialized properties.

2.7 Ores and Metals

Minerals in which a particular metallic element occurs in sufficiently high concentration to make mining and extracting it economically feasible are termed **ores.** Silicates, although very abundant, are seldom used as ores because extraction of their metallic elements involves high costs and technological difficulties. Instead, most metals are extracted from sulfide, oxide, carbonate, chloride, and phosphate ores that are found concentrated in a relatively small number of regions of the world. We will consider briefly the sources and uses of a few of the metals that are essential for our modern way of life.

Iron

Large deposits of iron ores are located in Minnesota, Sweden, France, Venezuela, and the former Soviet Union.

An alloy is a mixture of two or more metals.

Iron, the fourth most abundant element in the earth's crust (refer to Fig. 2.4), is the metal used in greatest quantity by industrialized nations. The ores from which it is extracted usually contain a mixture of two iron oxides: hematite (Fe_2O_3) and magnetite (Fe_3O_4). Nearly all the iron extracted from ores is used to manufacture steel, an alloy of iron with a small amount of carbon. The percentage of carbon determines the properties of steel. Low-carbon steel (less than 0.25% carbon) is relatively soft and suitable for making cans and wire. High-carbon steel (up to 1.5% carbon) is very hard and strong and is used for making tools and surgical instruments. Steels with a variety of properties and uses are made by alloying iron with small amounts of other metals.

Although world reserves of iron are still large—and the United States has an abundant supply within its borders—the demand for iron ore and steel continues to rise, and known sources of ores must eventually run out.

Aluminum

Aluminum is the second most highly utilized metal in industrialized nations and the third most abundant element in the earth's crust (refer to Fig. 2.4). Nearly all naturally occurring aluminum is a component of complex silicates, but at present there is no economically viable way to extract it from these silicates. Instead, the source of practically all aluminum is bauxite, an ore rich in aluminum oxide that is found in quantity in only a few places in the world. Very large deposits of bauxite are found in Jamaica and Australia.

Aluminum is a light, strong metal used primarily for making beverage cans. In the building industry, it is used to make doors, windows, and siding; it is ideal for cooking utensils and many household appliances. Because it is a good conductor of electricity, aluminum is used extensively for high-voltage transmission lines. When alloyed with magnesium, it forms a light but strong material vital to the manufacture of airplane bodies.

Aluminum corrodes less easily than iron does; this feature is an important advantage for certain purposes such as the construction of homes, but it also means that aluminum cans are very slow to degrade. Tin cans (which are made of steel and coated with tin) eventually break down completely, but discarded aluminum cans remain in the environment for a very long time.

Aluminum oxides that include traces of certain metal impurities are valuable as gemstones. Rubies are crystalline aluminum oxide that is colored red with traces of chromium. Sapphires, which occur in various shades of yellow, green, and blue, owe their colors to traces of nickel, magnesium, cobalt, iron, or titanium.

Copper

Another valuable and extensively used metal is copper, which today is obtained from low-grade copper sulfide ores. Because the copper content of these ores is 1% or less, the cost of obtaining the pure metal is high. Valuable by-products of copper production are gold and silver, which are frequently present in very small quantities in the original ores. Increasingly, pure copper is being obtained by recycling copper-containing materials.

Copper is an excellent conductor of electricity and is used extensively for electrical wiring. It is also used for plumbing fixtures and as a constituent of alloys. Brass is an alloy of copper and zinc, and bronze is an alloy of copper and tin. Copper and copper alloys are also used for coinage.

Strategic Metals

Because metal ores are very unevenly distributed in the earth's crust, many countries must depend on imports for their supplies. Within the United States, for instance, there are no deposits of ores of many **strategic metals,** metals that are essential for industry and defense. Large reserves of chromium (Cr), manganese (Mn), and the platinum (Pt) group of metals (platinum, palladium, rhodium, iridium, osmium, and ruthenium), which are needed for the manufacture of specialty steels, heat-resistant alloys, industrial catalysts, and parts for automobiles and aircraft, are found only in South Africa and the former Soviet Union.

2.8 Mineral Reserves

The world's human population continues to grow at an ever-increasing rate, and with it grows the demand not only for food but also for material goods. To meet these material needs, metals and minerals (nonrenewable resources) are being consumed at a tremendous rate, one that is bound to rise as the developing nations become increasingly industrialized.

The industrialized nations currently consume a disproportionate amount of the earth's mineral reserves. The North American continent, for example, has less than 10% of the world's population, but it consumes almost 75% of the world's production of aluminum. The same disproportionate usage rate holds true for many other metals. At the present rate of consumption, supplies of many important metals will be severely depleted by the middle of the 21st century.

It is possible that new mineral deposits will be discovered in the future. However, since most of the earth's surface has already been thoroughly explored by geologists, it is unlikely that significant quantities of ores will be found. One source of minerals that has not yet been fully explored is the ocean floor. Areas where new floor is being formed have been shown to be rich sources of manganese and polymetallic sulfides. As deep-sea mining technology develops, sites on the ocean floor may well provide a much needed source of metals.

One of the most important ways to conserve mineral supplies is to recover metals by recycling. More and more communities are collecting aluminum beverage cans (and also glass containers and paper) for recycling, which saves energy as well as conserving natural resources. Approximately half as much energy is required to make new aluminum cans from old cans as is needed to make them from bauxite.

2.9 The Origin of Life on Earth

So far in this chapter, we have considered only the nonliving part of the earth. We will now turn our attention to the living creatures that inhabit the planet.

It is generally agreed that life on earth began between 3.5 and 4 billion years ago, but exactly how it began will probably never be fully understood. The early atmosphere is thought to have consisted mainly of carbon dioxide (CO_2) and nitrogen (N_2), with smaller amounts of ammonia (NH_3) and methane (CH_4), and these gases would have dissolved to some extent in the early oceans. Some scientists believe life began in tidal pools or lagoons, where evaporation would have concentrated the dissolved chemicals, making it possible for them to combine to form simple amino acids (Chapter 14), organic compounds that contain carbon (C), oxygen (O), nitrogen (N), and hydrogen (H). Amino acids, which are basic building blocks of living tissues, might then have joined together to form simple proteins, and further reactions in the chemical "soup" of the ocean pools or lagoons could have produced other compounds essential for life. The essential compounds produced in this way might then have gradually clumped together to form larger masses. Membranes formed around these masses, separating them from the surrounding environment. The organic matter gradually acquired the characteristics of living cells.

In the 1950s, the American chemist Stanley Miller demonstrated that if a mixture of water (H_2O), ammonia (NH_3), hydrogen (H_2), and methane (CH_4) was subjected to electrical sparks, amino acids were formed.

Other researchers believe that life is more likely to have begun near volcanic vents on the ocean floor, where there was heat and protection from destructive UV radiation. Still others believe that the first living organisms did not arise on earth but came from outer space in meteorites or interplanetary dust.

No matter how the first one-celled organisms were formed, they developed in an environment devoid of oxygen. These early anaerobic bacteria (bacteria that require an oxygen-free environment) flourished until the development of oxygen-producing cyanobacteria. Oxygen was lethal to the anaerobic bacteria, and, except in a few specialized locations, they gradually died out.

2.10 The Uniqueness of the Earth

The earth is unique. It is the only planet in our solar system that developed an environment capable of supporting life as we know it. The position of the earth relative to the sun made possible the formation of the atmosphere and the oceans, which together maintain the temperature on the earth's surface within a very narrow range—a range that extends approximately from the freezing point of water (0°C, 32°F) to the boiling point of water (100°C, 212°F). If the earth had formed a little closer to the sun, it would have been too hot to support life as we know it; if it were a little farther away, it would have been too cold.

The size of our planet is another important factor. If the earth were much smaller, the pull of gravity would be too weak to hold the atmosphere around the earth. Without an atmosphere, we would be exposed to life-destroying amounts of UV radiation from the sun. If the earth were much larger, the atmosphere would be thicker and would contain more kinds of gases, many of them poisonous.

2.11 The Environment

Water, carbon dioxide, oxygen, and minerals are abiotic factors.

When we speak of the **environment,** we are referring to all the factors, both living and nonliving, that in any way affect living organisms on earth. The living, or **biotic,** factors include plants, animals, fungi, and bacteria. The nonliving, or **abiotic,** factors include physical and chemical components such as temperature, rainfall, nutrient supplies, and sunlight.

2.12 Ecosystems

For purposes of study and for the sake of simplicity, it is useful to subdivide the environment, which comprises the entire earth, into small functional units called **ecosys-**

tems. An ecosystem consists of all the different organisms living within a finite geographic region and their nonliving surroundings. It may be a forest, a desert, a grassland, a marsh, or just a pond or a field **(Fig. 2.10)**. Interrelationships between the organisms and the surroundings are such that an ecosystem is usually self-contained and self-sustaining.

FIGURE 2.10

An ecosystem is a group of plants and animals interacting with one another and their surroundings. It may cover a small area, such as a pond.

Producers and Consumers

Ecosystems are sustained by the energy that flows through them. The biotic part of any ecosystem can be divided into producers of energy and consumers of energy. Green plants and cyanobacteria (blue-green algae) are the **producers;** they are able to manufacture all their own food. By means of **photosynthesis,** they absorb light energy from the sun and use it to convert water (H_2O) and carbon dioxide (CO_2) from the air into the simple carbohydrate glucose ($C_6H_{12}O_6$). At the same time, oxygen gas (O_2) is released to the atmosphere. By further reactions between glucose and chemicals obtained from water and soil, plants manufacture all the complex materials they need **(Fig. 2.11)**. The plant world includes trees, bushes, flowers, grasses, mosses, and algae.

Consumers are unable to harness energy from the sun to manufacture their own food; they must consume plants or other creatures to obtain the nutrients and energy they need. Consumers can be divided into four main groups according to their food source: herbivores, carnivores, omnivores, and decomposers.

Herbivores feed directly on producers. Examples of herbivores are deer, cows, mice, and grasshoppers. **Carnivores** eat other animals and include spiders, frogs, hawks, and all cats (lions, tigers, and domestic cats). The animals that are eaten by carnivores may be herbivores, carnivores, or omnivores. **Omnivores** are creatures, including rats, raccoons, bears, and most humans, that feed on both plants and animals.

Decomposers feed on **detritus,** the freshly dead or partly decomposed remains of plants and animals. Decomposers include bacteria, fungi, earthworms, and many insects. Decomposers perform the very useful task of breaking down complex organic compounds in dead plants and animals into simpler chemicals and returning them to the soil for reuse by producers. In this way, many nutrients are endlessly recycled through an ecosystem.

2.13 The Flow of Energy through Ecosystems

All the activities that go on in an ecosystem require energy. Without a constant flow of energy from the sun to producers and then to consumers, an ecosystem would not be able to maintain itself. Before you can understand these energy relationships, you need to have some appreciation of what is meant by the term *energy*. We all know that it takes enormous amounts of energy to maintain our industrialized society. Energy is needed to run automobiles, heat and cool buildings, supply light, and grow food. We receive energy from the food we eat and use it up when we perform any activity. Energy exists in many forms, including light, heat, electrical energy, nuclear energy, and chemical energy. The ultimate source of energy for our planet is the sun. But what exactly is energy?

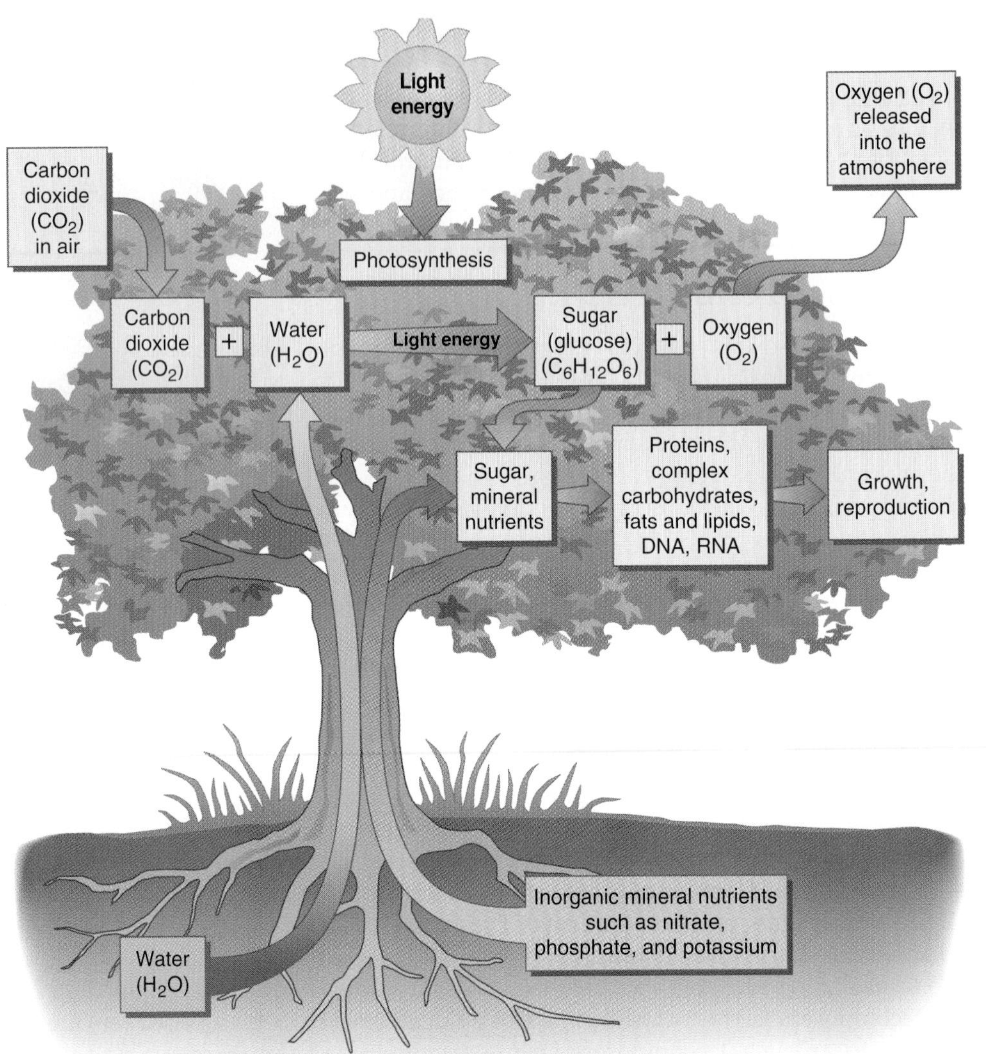

FIGURE 2.11

Green plants like this tree are producers. In the process of photosynthesis, they use light energy from the sun to convert carbon dioxide and water to glucose and oxygen. The oxygen is released to the atmosphere; the glucose, together with mineral nutrients from the soil, is used to produce the complex organic compounds that make up plant tissues.

What Is Energy?

Energy is usually defined as the ability to do work or bring about change (see Chapter 1). **Work** is done whenever any form of matter is moved over a distance. Everything that goes on in the universe involves work, in which one form of energy is transformed into one or more other forms of energy. Energy is the capacity to make something happen.

All forms of energy can be classified as either kinetic or potential. **Kinetic energy** is energy of motion. A moving car, wind, swiftly flowing water, and a falling rock all have kinetic energy and are capable of doing work. Wind, for example, can turn a windmill,

a.

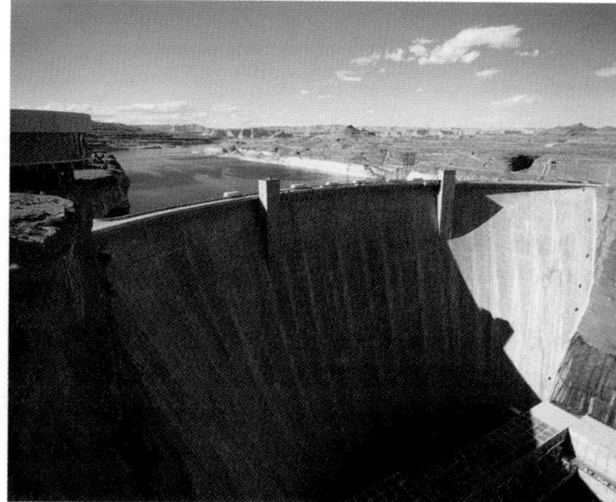

b.

FIGURE 2.12

(a) The kinetic energy of water rushing from a dam can be harnessed to do useful work. (b) The potential energy of the water in the lake behind a dam is converted to kinetic energy as it is released from the dam.

and water rushing from a dam **(Fig. 2.12a)** can turn a turbine to produce electricity. **Potential energy** is stored energy, energy that is converted to kinetic energy when it is released. Water held behind a dam **(Fig. 2.12b)** or a rock poised at the edge of a cliff has potential energy due to its position. When the water is released or the rock falls off the cliff, the potential energy is converted to kinetic energy. Potential energy is also stored in chemical compounds, such as those present in food and gasoline. When food is digested, chemical bonds are broken, and energy that the body needs to function is released. Similarly, combustion of gasoline in a car engine releases energy to set the car in motion.

Energy Transformations

In the universe, energy transformations occur continuously. Stars convert nuclear energy into light and heat, plants convert light energy from the sun into chemical energy in the bonds within sugar molecules, and animals convert the chemical energy in sugars into energy of motion. None of these transformations is 100% efficient; in every case, a part of the energy is converted to some useless form of energy, usually heat. But, in any transformation, no new energy is created and no energy is destroyed. As stated in the **first law of thermodynamics** (also known as the law of conservation of energy): *Energy can be neither created nor destroyed; it can only be transformed from one form to another.*

The efficiency of any energy-transfer process is defined as the percentage of the total energy that is transformed into some useful form of energy. For example, the efficiency of an incandescent electric lightbulb is very low. Only about 5% of the electrical input is converted into light energy; the remainder, as anyone who has touched a lighted bulb knows, is converted to heat **(Fig. 2.13a)**. If the lightbulb is in its normal surroundings in a home or office, there is no way for the heat energy to do useful work or be converted into some useful form of energy—it is essentially lost.

$$\% \text{ efficiency} = \frac{\text{usable energy}}{\text{total energy}} \times 100$$

a.

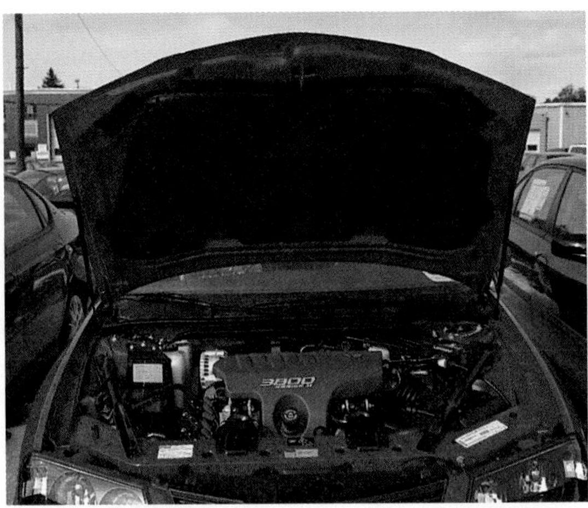

b.

FIGURE 2.13

(a) When electrical energy passes through the filament of a light bulb, only about 5% of it is converted to light; the rest is lost as heat. (b) An automobile engine converts about 10% of the chemical energy in gasoline to mechanical energy that can be used to drive the vehicle; 90% is lost as heat.

The loss of useful energy is summed up in the **second law of thermodynamics:** *In every energy transformation, some energy is always lost in the form of heat energy that thereafter is unavailable to do useful work.* This statement means that all systems tend to run "downhill"; in other words, whenever any work is done, high-quality energy is converted into lower-quality energy. All the life-sustaining processes that go on in the human body and in all other living organisms follow this pattern of energy flow. (The first and second laws of thermodynamics are considered more rigorously in Chapter 6, in which the related topic of entropy is introduced.)

Although energy is never destroyed, the fact that energy is lost as heat in all transformations means that, unlike many material resources, energy cannot be recycled. This fact has important implications for our society, which is dependent on so many energy-inefficient machines. The gasoline motor **(Fig. 2.13b)**, for instance, is only 10% efficient.

Food Chains and Trophic Levels

Green plants (producers) are the only organisms that can take energy from the sun and use the process of photosynthesis to store some of that energy in chemical bonds in sugars, starches, and other large molecules. When an animal eats a plant, the sugars and other chemical substances in the plant are broken down in chemical reactions in the animal's body (see Chapter 15). Bonds that had connected the atoms in the plant's molecules are broken, and energy is released. The energy is used to power the many activities that enable the animal to grow and survive.

Any ecosystem has innumerable feeding pathways, or **food chains,** through which energy flows. In one typical food chain, grasshoppers eat green leaves, frogs eat grasshoppers, and fish eat frogs **(Fig. 2.14a)**. Each step in the chain is called a **trophic level.** In the example just mentioned, there are four trophic levels: green leaves (plants) at the first trophic level, grasshoppers (herbivores) at the second trophic level, frogs (carnivores eating herbivores) at the third trophic level, and fish (carnivores eating carnivores) at the fourth trophic level, at the top of this food chain. A human eating a carrot would be at the second trophic level; a human eating beef that has been raised on corn would be at the third trophic level. Organisms at any one trophic level are dependent on the organisms at the level below them for their energy needs. Ultimately, all animals—including humans—are dependent on producers for their existence.

Omnivores function at more than one trophic level. A bear eating berries is at a lower trophic level than a bear eating fish.

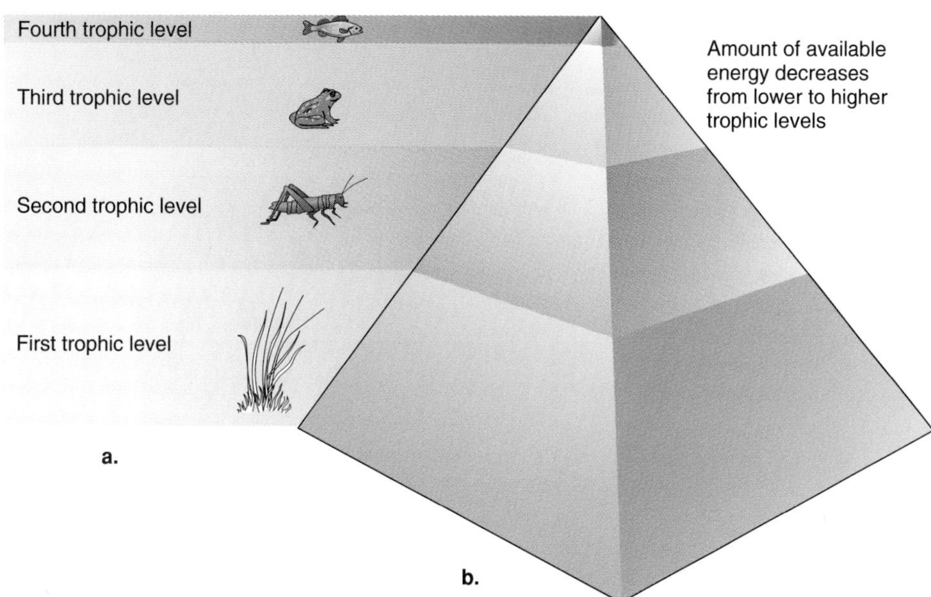

FIGURE 2.14

(a) A typical food chain with four trophic levels. (b) As energy passes to a higher trophic level in a food chain, approximately 90% of the useful energy is lost. High trophic levels contain less energy and fewer organisms than lower levels.

Energy and Biomass

At each trophic level in a food chain, energy is used by the organisms at that level to maintain their own life processes. Inevitably, as a result of the second law of thermodynamics, some energy is lost to the surroundings as heat. It is estimated that in going from one trophic level to the next, about 90% of the energy that was present at the lower level is lost **(Fig. 2.14b)**. Thus, in any ecosystem, the energy at the second trophic layer (herbivores) is only about 10% of the energy at the first trophic level (producers). The energy at the third trophic level (carnivores) is a mere 1% of that at the first trophic level (producers). This progession has an important implication for humans. It means that it is much more efficient to eat grain than to eat beef that has been fed on grain **(Fig. 2.15)**. It also means that to support a given mass of herbivores requires a mass of producers that is 10 times as large. To support a given mass of carnivores requires a mass of producers that is 100 times as large. Because of these mass requirements, food chains rarely go beyond four trophic levels.

2.14 Nutrient Cycles

To survive, a community of plants and animals in an ecosystem requires a constant supply of both energy and **nutrients.** The energy that sustains the system is not recycled. It flows endlessly from producers to consumers, entering as light from the sun and leaving as waste heat that cannot be reused. Nutrients, however, are continually recycled and reused. When living organisms die, their tissues are broken down and vital chemicals are returned to the soil, water, and atmosphere.

Analysis of tissues from living organisms shows that more than 95% of the mass of the tissues is made up from just 6 of the earth's 92 naturally occurring elements: carbon (C), hydrogen (H), oxygen (O), nitrogen (N), sulfur (S), and phosphorus (P). These six elements are the main building blocks for the manufacture of carbohydrates, pro-

FIGURE 2.15

Energy is used much more efficiently if humans eat plants (first trophic level) instead of meat (second trophic level). A given area of farmland can support many more people if the crops are fed directly to them rather than to livestock that they then eat.

teins, and fats. These compounds, along with water, make up almost the entire mass of all living organisms. Plants are composed primarily of carbohydrates, while animals are composed primarily of proteins.

Small amounts of other elements are also required in order for plants and animals to survive and thrive. Iron (Fe), magnesium (Mg), and calcium (Ca) together make up most of the remaining 5% of the mass of living organisms. In many animals, iron is bound to hemoglobin, the protein in the blood that supplies oxygen to all parts of the body. In green plants, magnesium is bound to chlorophyll, the protein that absorbs light from the sun and is a vital part of the process of photosynthesis. Animals with skeletons need calcium as well as phosphorus to make bones and cartilage.

Trace amounts of approximately 16 other elements are also required. Copper (Cu) and zinc (Zn), for example, are essential components of certain enzymes, specialized proteins that facilitate many vital chemical reactions within animals' bodies.

Plants obtain the essential elements from the soil and the atmosphere. Animals obtain them from their food: plants and other animals. Let's consider the cycles by which supplies of carbon, nitrogen, and oxygen are constantly renewed.

The Carbon Cycle

The **carbon cycle** is illustrated in **Fig. 2.16**. The major sources of carbon for our planet are the carbon dioxide gas in the atmosphere and the carbon dioxide dissolved in the oceans. Enormous quantities of carbon are also present in rocks, tied up in carbonates such as limestone, but this source recycles so slowly that it is not available to plants and animals for their daily needs.

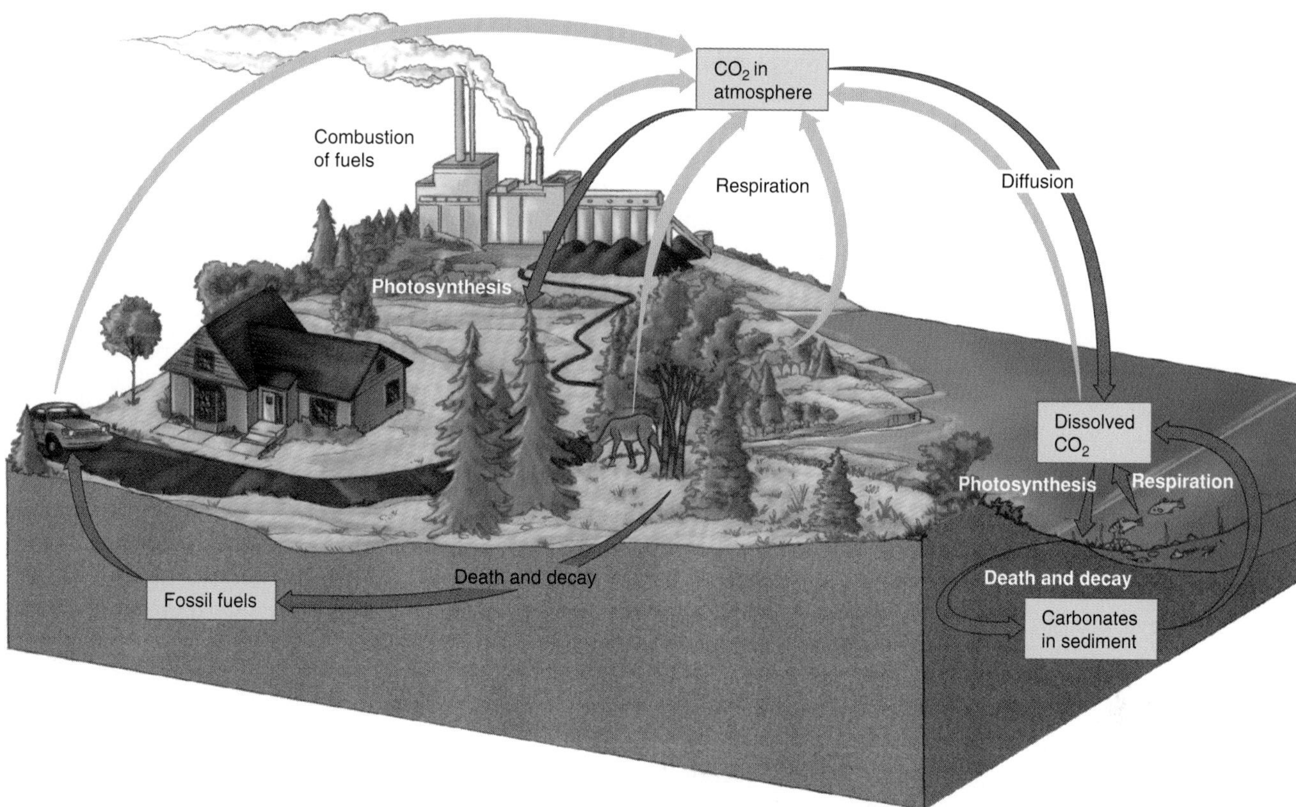

FIGURE 2.16

The carbon cycle: Atmospheric carbon dioxide is consumed by green plants in photosynthesis. Respiration in animals and plants and combustion of fossil fuels return carbon dioxide to the atmosphere. Carbon also cycles through water; dissolved carbon dioxide reacts with minerals and water to form carbonates, which are deposited in sediments.

Atmospheric carbon dioxide, although it makes up only 0.037% of the atmosphere by volume, is the starting material on which all living organisms depend. In the process of photosynthesis, carbon dioxide is taken into the leaves of green plants, where it combines with water to form sugar (glucose) and oxygen, as shown in the following equation.

> Specialized structures in plant cells, called *chloroplasts,* capture radiant energy from the sun and convert it into chemical energy.

$$\underset{}{\text{solar energy}} + \underset{\text{carbon dioxide}}{6\ CO_2} + \underset{\text{water}}{6\ H_2O} \xrightarrow{\text{chlorophyll}} \underset{\text{glucose}}{C_6H_{12}O_6} + \underset{\text{oxygen}}{6\ O_2}$$

Plants on land obtain the needed water from the soil; aquatic plants obtain it from their surroundings. The water is absorbed through a plant's roots and then transported to its leaves. Part of the sugar that is formed is stored in the leaves, and part is converted into the complex carbohydrates and other large molecules that make up plant tissues. The oxygen is released to the atmosphere.

Photosynthesis is a very complex process that is still not fully understood. It involves many chemical reactions in which the green pigment chlorophyll plays an important role. We will consider only the overall result of photosynthesis as summarized by the above equation.

> Photosynthesis consists of two series of reactions: one that occurs only in the presence of light and another that can occur in the dark.

When an animal such as a deer or rabbit eats a green plant, the carbohydrates in the plant are digested and broken down into simple sugars, including glucose. Glucose is absorbed into the bloodstream and carried to the cells of the animal's body, where, in the process of **respiration,** it reacts with oxygen in the blood. Carbon dioxide and water are formed, and energy is released.

$$C_6H_{12}O_6 + O_2 \longrightarrow 6\ CO_2 + 6\ H_2O + \text{energy}$$

Part of the energy released in respiration is used to power the many activities that go on in living cells, and part is lost as heat.

Notice that the overall reaction for respiration is the same as the overall reaction for photosynthesis written backward. However, although cellular respiration is essentially the reverse of photosynthesis, the complex intermediate steps that are involved in the two processes are very different.

Plants engage in both photosynthesis and respiration. During the day, photosynthesis is the dominant process. At night, when there is no sunlight, respiration is dominant. Decomposers also play a part in the carbon cycle. They feed on the dead remains of plants and animals and, via respiration, release carbon dioxide and water to the atmosphere.

About 300 million years ago, huge quantities of dead and decaying plant and animal remains became buried deeply under sediments before they could be completely broken down. Over time, these remains were compressed, and chemical reactions gradually transformed them into the **fossil fuels:** coal, oil, and natural gas. When these fuels are burned to release the chemical energy stored within their molecules' bonds, oxygen from the atmosphere is used to convert their carbon into carbon dioxide.

$$C + O_2 \longrightarrow CO_2 + \text{energy}$$

The combustion of the fossil fuels that powers our industrialized society therefore forms an integral part of the carbon cycle. Humans also intervene in the carbon cycle when they cut down more trees than they replace, thereby decreasing the amount of carbon dioxide that otherwise would be taken from the atmosphere and converted to nutrients. The climatic implications of these two activities, both of which tend to increase carbon dioxide concentration in the atmosphere, will be studied in Chapter 11.

Another important part of the carbon cycle is the continual exchange of carbon dioxide between the atmosphere and the oceans, a process that is important in maintaining the carbon dioxide concentration of the atmosphere at a constant level. The amount of carbon dioxide that dissolves in the oceans depends mainly on the temperature of the ocean water and the relative concentrations of carbon dioxide in the atmosphere and in the water. When the temperature falls or when carbon dioxide concentration in the water becomes relatively low, more atmospheric carbon dioxide dissolves. A very small portion of the dissolved carbon dioxide reacts with chemicals in the water, such as calcium, to form carbonates. The carbonates, primarily limestone ($CaCO_3$), are insoluble (do not dissolve) in water and settle on the ocean floor.

Rock formation and weathering are other aspects of the carbon cycle. Sedimentary rocks such as limestone and dolomite [$CaMg(CO_3)_2$] were formed millions of years ago from the skeletal remains of coral and other marine creatures, which were rich in calcium carbonates. When chemically weathered by rain with a slight natural acidity, limestone rocks very gradually dissolve, releasing carbon dioxide into the atmosphere. This recycling of carbon through rock is a very slow process.

The Nitrogen Cycle

The **nitrogen cycle** is illustrated in **Fig. 2.17**. Nitrogen is an essential component of proteins and of the genetic material that makes up DNA (deoxyribonucleic acid), and a constant supply is vital for all living organisms. Although 78% of the earth's atmosphere is composed of nitrogen gas (N_2), plants and animals cannot use this nitrogen directly.

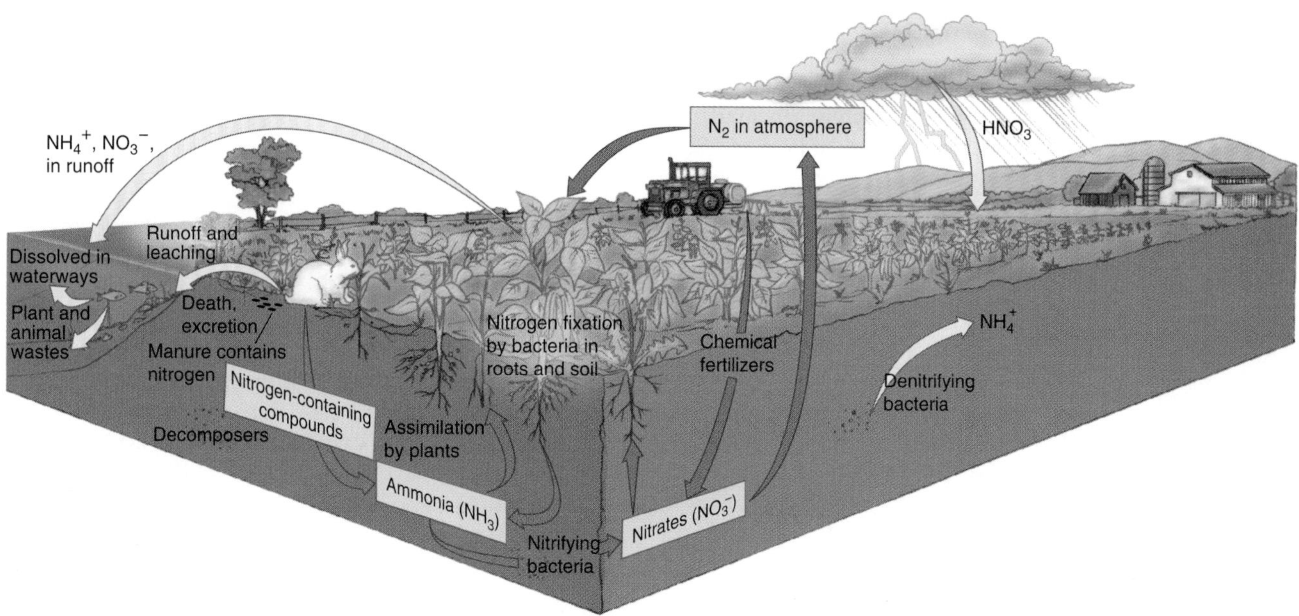

FIGURE **2.17**

The nitrogen cycle: In nitrogen fixation, specialized bacteria convert atmospheric nitrogen to ammonia and nitrates, which plants absorb through their roots. Some nitrogen gas is fixed by lightning. Animals obtain the nitrogen they need to make tissues from plants. The wastes animals produce and their dead bodies return nitrogen to the soil in forms plants can use. In denitrification, bacteria convert nitrates in soil back to nitrogen gas.

Atmospheric nitrogen must be converted to other nitrogen compounds before it can be absorbed through the roots of plants. This change is achieved by **nitrogen fixation,** a process carried out by specialized bacteria that have the ability to transform atmospheric nitrogen into ammonia (NH_3). Some nitrogen-fixing bacteria live in soil. Others live in nodules on the roots of leguminous plants such as peas, beans, clover, and alfalfa (see Fig. 2.17). The ammonia produced in root nodules is converted into a variety of nitrogen compounds that are then transported through the plant as needed.

Another means by which atmospheric nitrogen is converted to a usable form is lightning. The electric discharges in lightning cause nitrogen and oxygen in the atmosphere to combine and form oxides of nitrogen, which in turn react with water in the atmosphere to form nitric acid (HNO_3). The nitric acid, which reaches the earth's surface dissolved in rainwater, reacts with substances in soil and water to form nitrates (NO_3^-) which are directly absorbed through plant roots. Compared to biological fixation, lightning accounts for only a small fraction of the usable nitrogen in soil.

Although some trees and grasses can absorb ammonia produced by nitrogen-fixing bacteria directly from the soil, most plants can only use nitrogen that is in the form of nitrates. The transformation of ammonia into nitrates is carried out by specialized soil bacteria in a process known as **nitrification.**

Plants convert ammonia or nitrates that they get from soil or root nodules into proteins and other essential nitrogen-containing compounds. Animals get their essential nitrogen supplies by eating plants. When plants and animals die and decompose, the nitrogen-containing compounds in their tissues are broken down by decomposers; ammonia is eventually formed and returned to the soil. Nitrogen is also returned to the

EXPLORATIONS

The Eruption of Krakatoa

On August 26, 1883, the small island of Krakatoa, which is located in the Sunda Strait between the larger islands of Java and Sumatra, exploded in one of the biggest and most catastrophic volcanic eruptions in recorded history. The explosion was heard over nearly 8% of the earth's surface, thousands of people died, and the aftereffects were apparent worldwide for many months.

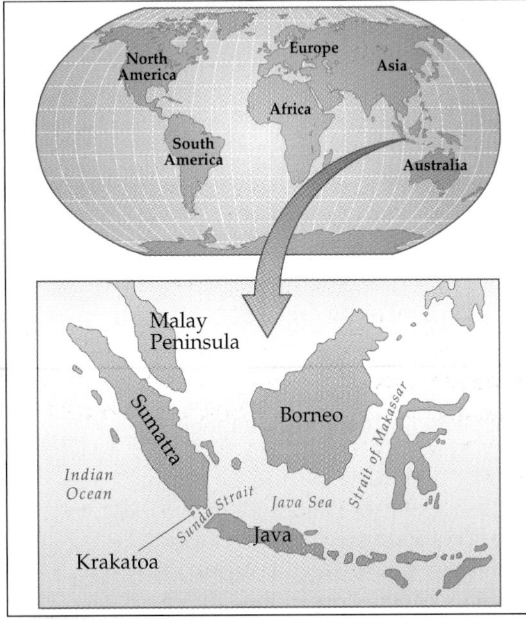

The location of Krakatoa.

The island of Krakatoa is situated in one of the most unstable regions of the world. Its volcano is just one of 500—of which more than 100 are still active—that form a 1600-kilometer (1000-mile) arc across Java and Sumatra, the two main islands of modern Indonesia. The people on these densely populated islands have always lived with the menace of volcanoes. The dangers are great, but the volcanic ash makes the soil some of the richest in the world. The Krakatoa volcano was considered to be extinct, and the Dutch authorities who ruled Indonesia at the time of the eruption had no idea that the small island was a time bomb waiting to explode.

Causes of the Explosive Eruption

Why do some volcanoes erupt explosively, while others—like the Hawaiian volcanoes—emit quietly flowing lava that does no harm except to objects in its path? It depends on the silica (SiO_2) and water content of the molten magma that is ejected. When molten magma rises toward the earth's surface during an eruption, the water in the magma turns to steam, expanding 1000 times as it does so. The force of the expanding steam and other hot gases propels the magma up through the volcano's chimney. Low-silica magma produces a liquid lava from which gases can escape easily. If the water content of this lava is low, it flows gently; if high, it surges up in lava fountains. High-silica magma forms a viscous lava from which gases escape with difficulty. The thick magma often solidifies and forms a plug in the chimney. As gases try to escape, pressure builds up, until eventually the plug is blown out with explosive force, somewhat like a cork from a champagne bottle. The frothy, gas-saturated, high-silica magma is ejected as pumice, a lightweight material that is full of small gas-filled cavities and floats on water. If more had been understood about volcanoes in 1883, the emission of pumice in the early stages of the eruption would have been recognized as a clue that Krakatoa had the potential to be extremely dangerous.

Chronology of Events

The first signs that Krakatoa was awakening came on May 20, when rumblings and detonations were heard on shore and a dark column of ash appeared above the island. Pumice and ash fell on ships passing close by, and the forests of Krakatoa were seen to be on fire. Activity continued intermittently for 3 months, no one realizing that subterranean fires beneath Krakatoa were being stoked and much more violent disturbances were to follow.

Then on Sunday, August 26, at 1 p.m., Krakatoa erupted in one of the most awesome displays of nature's might ever witnessed. Ear-splitting blasts of increasing violence followed one another in quick succession, and an enormous black mushroom cloud billowed out above the Sunda Strait. In towns and villages along the coasts of Java and Sumatra, the inhabitants were stifled by sulfurous fumes and could barely see through the falling ash. To the accompaniment of strong winds and violent thunderstorms, tremendous roars continued through the night and into the next morning. The sea receded and then advanced. Boats were at one moment stranded on the beach and minutes later smashed in pieces by inrushing waves, which at times were 3 meters (10 feet) high.

The captain of a British ship, the *Charles Bal,* which sailed within 16 kilometers (10 miles) of Krakatoa during the night of August 26, told of blinding lightning and a hail of hot pumice stone. Holes were burned in the ship's sails and in the clothes of the sailors as they tried to clear pumice from the decks.

Shortly before 10 a.m. on August 27, after a terrifying night, it suddenly became ominously quiet, the explosions died down, and the frightened people on shore began to hope that the eruption had ended. In fact, the worst was still to come. At exactly 10:02 a.m., almost the entire island of Krakatoa was obliterated in a titanic explosion greater than any that had gone before. An immense column of fiery rocks and ash rose into the atmosphere, and total darkness fell on Jakarta 160 kilometers (100 miles) away. A gigantic sea wave was unleashed and sped outward toward the shores of Sumatra and Java. A great black wall of water, in places 40 kilometers (130 feet) high, swept inland for distances up to 10 kilometers (6 miles), destroying everything and everyone in its path. Not a house was left standing, and thousands of people were swept out to sea by the receding waters. For nearly 2 hours, a succession of mountainous waves followed the first one, and then, just before noon, Krakatoa roared for the last time. An area almost as large as the island of Manhattan had vanished, over 36,000 people were dead, and 165 towns and villages had been destroyed.

The sound from the most violent explosion was heard almost 5000 kilometers (3000 miles) away—farther than any sound has ever carried before or since.

A Scientist Investigates

Soon after the eruption, the Dutch geologist R. D. M. Verbeek visited what remained of the island and established the sequence of events that had led to its final violent collapse. He was puzzled that among the materials blasted from the volcano, he could find none of the rock that had once formed the cone—all of the materials were either pumice or ash from the magma chamber. To explain this mystery, Verbeek concluded that 19 hours of continual eruption had emptied a large, shallow magma chamber that lay under the volcano. Once it was empty, there was nothing left to support the rock walls, and the volcanic cone and most of the surrounding island crashed with a deafening roar into the chamber. The sea rushed in and then exploded violently outward in a tremendous column of steam and debris. The collapse set off the killer waves.

The sound from the most violent explosion was heard almost 5000 kilometers (3000 miles) away—farther than any sound has ever carried before or since. A pressure wave swept around the earth seven times, its passage recorded on instruments at scattered locations around the world. An increase in atmospheric pressure was first noted in New York 14 hours after the explosion. As immense ocean waves sped outward, they raised sea levels many feet along coastlines worldwide. Six successive surges at 30-minute intervals were recorded as far away as the English Channel. A layer of pumice—which analysis showed to be very high in silica (72%)—spread into the Indian Ocean from the Sunda Strait, which for a time was impassable to shipping. Corpses and debris were found embedded in the pumice for many months. Ash from the cloud, which rose to an estimated height of 80 kilometers (50 miles), fell on ships as far as 1600 kilometers (1000 miles) away. Fine dust reached the stratosphere and, carried by the winds, circled the earth many times, creating spectacular sunsets and sunrises for several years.

A circular depression under the water, known as a *caldera,* was all that was left after most of Krakatoa vanished. But Krakatoa was not dead. Beginning with steam clouds in 1927, intermittent eruptions from within the caldera gradually built up a new island, named Anak Krakatoa (child of Krakatoa), which emerged from the sea in 1952. A new cone has appeared, and the island continues to grow. Perhaps several hundred years from now it will once again give a demonstration of the mighty forces that are at work beneath the surface of the earth.

References R. Furneaux, *Krakatoa* (Englewood Cliffs, NJ: Prentice Hall, 1964); *Planet Earth: Volcano* (Chicago, IL: Time/Life, 1982).

soil in animal wastes. Both urine and feces have a high content of nitrogen-containing compounds. In this way, nitrogen is continually cycled through food chains.

Not all the ammonia and nitrates that are formed in soil by the processes just described become available for plants. Both ammonia and nitrates are very soluble in water. As rainwater percolates downward through the ground, these compounds are leached out of topsoil. They are often carried away in runoff into nearby streams, rivers, and lakes, where they are recycled through aquatic food chains. Another process that removes nitrates from the soil is **denitrification,** in which bacteria carry out a series of reactions that convert nitrates back to nitrogen gas.

These sodium nitrate deposits are known as Chile saltpeter.

In a few locations around the world, nitrates have accumulated in large mineral deposits. In Chile, for example, as mountain streams originating in the Andes Mountains flowed across dry, hot desert toward the sea over thousands of years, much of the water evaporated, leaving behind huge deposits of sodium nitrate.

In the natural environment, a balance is maintained between the amount of nitrogen removed from the atmosphere and the amount returned. However, because most soils contain an insufficient amount of nitrogen for maximum plant growth, farmers frequently apply synthetic inorganic fertilizers containing ammonia and nitrates. As a result of runoff from fertilized farmland, the extra nitrogen-containing compounds reaching rivers and lakes may upset the natural balance, sometimes with damaging consequences for the environment. (We will discuss the problem of water pollution in more detail in Chapter 10.)

An alternative to the use of synthetic fertilizers for replenishing nitrogen in the soil is to plant a nitrogen-fixing crop, such as clover, and plow it back into the soil. Another method is to spread manure and allow the natural soil bacteria to degrade it and release the nitrogen-containing compounds that plants can absorb.

The Oxygen Cycle

Oxygen is all around us. As oxygen gas (O_2), it makes up 21% of the atmosphere, and it is a component of all the important organic compounds in living organisms. Oxygen is a very reactive element that combines readily with many other elements. It is a component of carbon dioxide (CO_2), nitrate (NO_3^-), and phosphate (PO_4^{3-}) and thus is an integral part of the recycling of carbon, nitrogen, and phosphorus. Oxygen is also a constituent of most rocks and minerals, including silicates, limestone ($CaCO_3$), and iron ores (Fe_2O_3, Fe_3O_4). The **oxygen cycle** is very complex and is interconnected with many other cycles. Thus, here we will briefly consider only some of the more important pathways in the cycle.

Photosynthesis and respiration are the basis of both the carbon cycle and the oxygen cycle. Oxygen is released during photosynthesis, and it is consumed during respiration. The oxygen and carbon cycles are also interconnected when coal, wood, or any other organic material is burned. During burning, oxygen is consumed and carbon dioxide is released.

Water, another vital part of our environment that is constantly being recycled, will be discussed in Chapter 10.

Another part of the oxygen cycle is the constant exchange of oxygen between the atmosphere and bodies of water, especially oceans. Oxygen dissolved at the surface of water is carried to deeper levels by currents. Dissolved oxygen is essential for fish and other aquatic life.

Nature's Cycles in Balance

Life on earth depends on a continual supply of energy from the sun and a continual recycling of materials. As you have learned in this chapter, the numerous and complex processes that sustain life are interlinked and interdependent. Because all aspects of an ecosystem are so completely interwoven, human intervention in any part of the system can easily lead to widespread disturbances of the natural balance in the environment.

Chapter Summary

1. The universe is believed to have been formed about 13 billion years ago by the big bang.
2. The sun is the ultimate source of energy for life on earth. It makes up 99.9% of the mass of the solar system.
3. The earth is divided into three layers: the core, the mantle, and the crust.
4. The core is believed to be composed mainly of molten iron at a temperature of about 1535°C (2795°F); the mantle and crust are composed mainly of silicates.
5. The lithosphere is made up of the crust and the solid upper part of the mantle.
6. Iron, oxygen, silicon, and magnesium—the four most abundant elements in the earth—make up over 90% of its mass.
7. The earth's oceans and atmosphere were formed as a result of volcanic eruptions.
8. Oxygen first entered the atmosphere when primitive cyanobacteria began using energy from the sun in the process called photosynthesis.
9. Silicates are made up of tetrahedral SiO_4 units, which can join to form chains, sheets, and three-dimensional arrays.
10. Valuable metals such as iron, aluminum, and copper are extracted from ores present in the earth's crust.
11. The earth's mineral reserves are expected to be severely depleted by the middle of the 21st century.
12. Life on earth is believed to have begun between 3.5 and 4 billion years ago.
13. Living creatures can be divided into producers (which manufacture their own food) and consumers (which do not).
14. Consumers can be subdivided into carnivores, omnivores, herbivores, and decomposers.
15. Energy is classified as either potential or kinetic.
16. The first law of thermodynamics states that energy can be neither created nor destroyed; the second law of thermodynamics states that in every energy transformation, some energy is lost in the form of heat.
17. A typical food chain has three or four trophic levels.
18. Going from one trophic level to the one above it involves the loss of about 90% of the energy present in the lower level.
19. Nutrients in an ecosystem are continually recycled and reused, but the energy that sustains the system is not recycled. Important nutrient cycles are those involving carbon, nitrogen, and oxygen.

Key Terms

abiotic (p. 38)
asbestos (p. 33)
asthenosphere (p. 29)
big bang (p. 24)
biotic (p. 38)
carbon cycle (p. 44)
carnivore (p. 39)
consumer (p. 39)
core (p. 28)
crust (p. 29)
cyanobacteria (p. 31)
decomposer (p. 39)
denitrification (p. 50)
detritus (p. 39)
ecosystem (p. 38)
environment (p. 38)
first law of thermodynamics (p. 41)
food chain (p. 42)
fossil fuels (p. 46)
herbivore (p. 39)
inorganic (p. 32)
kinetic energy (p. 40)
lithosphere (p. 29)
mantle (p. 29)
Milky Way (p. 24)
minerals (p. 32)
nitrification (p. 47)
nitrogen cycle (p. 46)
nitrogen fixation (p. 47)
nonrenewable resource (p. 32)
nutrient (p. 43)
omnivore (p. 39)
ore (p. 35)
organic (p. 32)
oxygen cycle (p. 50)
photosynthesis (p. 39)
potential energy (p. 41)
producer (p. 39)
renewable resource (p. 32)
respiration (p. 46)
second law of thermodynamics (p. 42)
silicates (p. 33)
solar system (p. 25)
strategic metal (p. 37)
synergism (p. 34)
trophic level (p. 42)
universe (p. 24)
work (p. 40)

Questions and Problems

1. Draw a diagram of the solar system that shows the following:
   a. relative size of each planet
   b. distances between the planets
   c. size and position of the sun relative to the planets
2. List:
   a. the terrestrial planets
   b. the giant planets
   c. What are the main differences between the two groups of planets?
3. What percentage of the mass of the solar system is accounted for by the mass of the sun?
   a. less than 1%
   b. approximately 10%
   c. nearly 30%
   d. more than 99%
4. Why does the large number of hydrogen atoms in the universe suggest that other elements were built from hydrogen rather than by larger elements breaking up into smaller ones?
5. Make a diagram showing a cross section of the earth and label the three main layers into which the earth can be divided. What element predominates at the center of the earth?
6. By weight, which are the four most abundant elements in the whole earth? Explain why elements are not uniformly distributed throughout the earth.
7. How did water form on the earth's surface?
8. Why does earth's sister planet, Venus, not support life?
9. Describe the chemistry carried out by primitive cyanobacteria (blue-green algae). Name this chemical process.
10. The early earth is thought to have been devoid of oxygen. How did oxygen first appear in the earth's atmosphere?
11. What are anaerobic bacteria?
12. What is the basic chemical unit that is present in all silicates? How are the atoms in this unit arranged? Illustrate your answer with a diagram.
13. The basic units in the following silicates are arranged as three-dimensional arrays, as two-dimensional sheets, or as chains. For each mineral, indicate which arrangement is present:
    a. talc
    b. chrysotile
    c. quartz

14. There is great debate over the hazards posed by asbestos.

a. What properties make asbestos a useful material?

b. Give three examples of the industrial uses of asbestos.

c. "Friable" asbestos is defined as that which can be easily crumbled into a powder. Should we be more concerned about friable asbestos than asbestos that is encapsulated in a product (such as an asbestos floor tile)?

15. Describe the health hazards associated with asbestos for smokers and nonsmokers.

16. Name the mineral from which most of the world's aluminum is extracted. From what countries does the United States obtain its supply of this mineral?

17. If aluminum and steel cans are thrown into the city dump, which will degrade first? Briefly explain your answer.

18. Making an aluminum can from recycled aluminum rather than starting with aluminum ore produces an energy savings of about:

a. 10% **b.** 50% **c.** 75%

19. Give three uses for the metal copper.

20. Give three uses for the metal aluminum.

21. What is the purpose of adding carbon to steel?

22. What is meant by a strategic metal?

23. Significant quantities of strategic metals are found in:

a. South America

b. Africa

c. Europe

d. Australia

24. The danger of asbestos in schools and office buildings is a matter of debate. Some argue that any exposure is to be carefully avoided, while others say that there is danger only from high concentrations of airborne asbestos fibers. What do you think should be done? Should asbestos already in school buildings be removed or coated over with a plastic sealant? Remember that removal will cause some of the fibers to become airborne during the removal process and that if the coating technique is used, the underlying asbestos may be disturbed during sealing.

25. Discuss each of the following claims:

a. New discoveries of mineral deposits will provide all we need in the future.

b. Our children will have sufficient mineral resources because they will be able to mine lower-grade ore deposits.

c. The ocean will supply all our mineral needs in the future.

d. Recycling scarce mineral resources will guarantee a continued ample supply of those minerals.

26. What is meant by *biotic* and *abiotic?* Give examples of each.

27. Define the word *ecosystem.* Give an example of a terrestrial and an aquatic ecosystem.

28. Living organisms can be divided into producers and consumers. Describe how:

a. producers make their food

b. consumers get the food they need

29. Consumers can be divided into groups according to their food source. Name the main groups and give two examples of each.

30. Organisms that consume the remains of dead plants and animals are necessary for a balanced ecosystem. Explain their purpose.

31. Define what is meant by *energy.* Name four forms of energy.

32. State the first law of thermodynamics. Give an example of the conversion of one form of energy to another.

33. State the second law of thermodynamics. Give an example that illustrates this law.

34. Of what type is the unusable energy that is released during almost all energy transformations?

35. Make a diagram that shows a four-level food chain. In which direction does energy flow?

36. If 10,000 units of energy are available to organisms at the first trophic level of a food chain, how many units of energy will be available to organisms that occupy the third trophic level in the food chain?

37. Why do most food chains contain only three or four trophic levels?

38. Suppose a bear has a choice between trout and blueberries for dinner.
    a. At which trophic level is the bear if it eats the trout?
    b. At which trophic level is the bear if it eats the blueberries?
    c. Which food represents the more efficient use of energy?
39. Why are people with limited food resources often herbivores?
40. Make a diagram of a simple food chain. Give your food chain a name.
41. Each time energy flows from a trophic level to the one above it, approximately what percentage of useful energy is lost?
42. What four elements are most abundant in the bodies of plants and animals? Name three other elements that are essential for life.
43. Make a diagram that illustrates one of the following nutrient cycles:
    a. carbon cycle b. nitrogen cycle
44. What is the main source of carbon for living things? Describe the process by which plants use this source of carbon to make their food.
45. When a cow grazes, the grass is a fuel for the animal. Explain how the carbon cycle is involved.
46. What part do decomposers play in the carbon cycle?
47. What part does each of the following play in the carbon cycle?
    a. the oceans b. limestone c. crude oil
48. What is the name of the class of compounds formed in addition to oxygen as products in the photosynthesis reaction?
49. Name the process by which the nitrogen of the atmosphere is converted into a form that can be used by plants.
    a. Where are the bacteria that bring about this process found?
    b. What chemical compound is produced by these bacteria?
50. Lightning can convert atmospheric nitrogen to a form that plants can use. Explain the process.
51. Why don't nitrates occur as large deposits in many places in the crust of the earth?
52. List two ways streams and rivers receive excessive amounts of nitrogen-containing compounds because of the activities of people.
53. What is meant by *nitrification?* How is the nitrogen in animal waste recycled by nature?
54. Describe two ways by which a farmer might replenish the nitrogen in her fields, without using inorganic fertilizers.
55. Two parts of the oxygen cycle are intertwined with the carbon cycle. Name these and briefly explain them.
56. As Americans have become more aware of the delicate balance required to maintain our ecosystems, we have begun to bring environmental concerns into decision-making processes. The possible extinction of the snail darter, a small fish, almost stopped the development of a large dam in Tennessee. More recently, the northern spotted owl's habitat has been threatened by the logging of mature timber in the Pacific Northwest.
    a. Should the possible extinction of one species of animal be of such concern that it causes development to be halted?
    b. Should alternative habitats for endangered species (artificial, if necessary) be created so that development may proceed?

Chapter 3

Atoms and Atomic Structure

Chapter Objectives

In this chapter, you should gain an understanding of:

- The basic laws that govern the way substances react and elements combine to form compounds
- The experimental evidence that led to the discovery of the composition of the atom
- The nature and properties of electrons, protons, and neutrons, the fundamental particles of which atoms are composed
- The relationship between numbers of electrons, protons, and neutrons and the mass of an atom
- The difference between isotopes of the same element
- The experimental basis for the arrangement of elements in the periodic table
- Similarities between elements in the same group in the periodic table
- The trends that occur in successive periods of the periodic table

A scanning electron micrograph of Velcro, one of the many modern materials scientists have produced by combining atoms in various ways.

ONLINE FEATURES

www.jbpub.com/chemistry

- Chemistry and the Environment
- eLearning
- Chemistry in the News
- Special Topics
- Research & Reference Links
- Ask the Authors

THE IDEA THAT TINY, INDIVISIBLE PARTICLES called *atoms* are the building blocks of all matter is an ancient one, but it was not until the early part of the 18th century that there was any convincing experimental evidence to support this theory. By the middle of the 20th century, scientists realized that atoms, although fundamental particles, are not the ultimate, indivisible units of matter, as had been previously supposed. Atoms of all the elements that form our world are themselves built from three fundamental subatomic particles: electrons, protons, and neutrons. The identity of each element is determined by the internal structure of its atoms. The atomic theory of matter is the foundation on which modern chemistry is built.

As more and more elements were discovered during the latter part of the 19th century, it became evident that certain groups of them had very similar properties. Scientists began to look for a systematic way to organize the elements on the basis of the observed similarities. This search led to the development of the periodic table of the elements, the table that is found in all chemistry textbooks today.

In this chapter, we examine the experimental evidence that led to the development of the atomic theory and see how the theory evolved to include new knowledge of the structure of the atom. We also study the periodic table, the invaluable reference tool that summarizes and correlates a great quantity of information about the elements from which every single thing—both living and nonliving—in our environment is made.

The Greeks believed that the natural world was constructed from four elements: earth, air, fire, and water.

3.1 The Atomic Nature of Matter

The proposal that matter is composed of tiny particles that cannot be subdivided was first made in the 5th century B.C. by the Greek philosopher Leucippus and his pupil Democritus. They argued that if one divided a piece of a material such as iron into ever smaller and smaller pieces, one would eventually have many invisibly small particles of iron that could not be subdivided further and still preserve the properties of iron. Democritus gave these ultimate particles the name *atomos* (which literally means "uncuttable" in Greek). He believed that all matter was made up of various arrangements of atoms and that all atoms were composed of the same basic material. He asserted that the differences between materials—for example, between iron and gold—were due to differences in size and shape of the atoms that made up the materials. There was no way to verify or refute Democritus's hypothesis, and many influential philosophers of the time, including Plato (427–347 B.C.) and Aristotle (384–322 B.C.), rejected it and persisted in their belief that matter was continuous and could be endlessly subdivided.

Dalton was color-blind and was the first to describe this condition.

Although the idea that matter was made up of atoms was discussed from time to time during the following centuries, it was not until 1803 that the concept was seriously revived by John Dalton (1766–1844), an English schoolteacher and amateur meteorologist (**Fig. 3.1**). By that time, Dalton was able to present considerable experimental evidence to support the atomic theory.

FIGURE 3.1

John Dalton (1766–1844) is regarded as the father of chemical theory in recognition of the importance of his atomic theory to the development of chemical knowledge.

3.2 Dalton's Atomic Theory

The main points in **Dalton's atomic theory** are as follows:

1. All matter is composed of tiny, indivisible particles called **atoms**.
2. All atoms of a particular element are identical, but the atoms of one element differ from the atoms of any other element.

3. Atoms of different elements combine with each other in certain whole-number proportions to form compounds.
4. In a chemical reaction, atoms are rearranged to form new compounds; they are not created, destroyed, or changed into atoms of any other element.

Dalton was not able to verify all these statements during his lifetime, but all were in agreement with the data available to him.

FIGURE 3.2
Antoine-Laurent Lavoisier (1743–1794) was one of the first scientists to recognize the importance of accurate weight measurements in experiments.

Lavoisier supported his scientific work with the profits from money invested in a private company authorized by Louis XVI to collect taxes. This activity, not his noble birth, led to his death sentence during the French Revolution.

3·3 Evidence Supporting the Atomic Theory

Evidence to support the atomic theory was provided by three fundamental laws of chemistry that were established by Dalton and other scientists of his time. These three laws are: (1) the law of conservation of mass, (2) the law of definite proportions, and (3) the law of multiple proportions.

The Law of Conservation of Mass

One of the first scientists to carry out systematic quantitative chemical experiments and draw logical conclusions from them was the Frenchman Antoine Laurent Lavoisier (1743–1794) **(Fig. 3.2)**. Lavoisier, often called "the father of modern chemistry," described his experiments in the first chemistry textbook. He established chemistry as a quantitative science by showing the importance of accurately weighing the quantities of chemicals that were used up and produced in experiments. It is one of the tragedies of history that this brilliant man was guillotined at the age of 50, during the excesses of the French Revolution.

Some of Lavoisier's most important experiments were concerned with combustion, or burning. He observed, for example, that when coal was burned in a closed container, the mass of the container plus its contents *before* combustion equaled the mass of the container plus its contents *after* combustion, even though the substances originally in the container had been changed **(Fig. 3.3)**. Lavoisier was able to show that when the coal burned, it gained something from the air in the container, and that the amount of mass gained by the coal was equal to the amount of mass lost from the air. He named the substance lost from the air *oxygen*.

Lavoisier summarized the results of these and many other experiments in the form of a fundamental law known as the **law of conservation of mass,** which states that *in a chemical reaction, matter is neither created nor destroyed*. In modern terms, the law is more accurately written as follows: *There is no detectable change in mass during an ordinary chemical reaction.*

The law of conservation of mass is consistent with Dalton's atomic theory. If reacting substances are made up of atoms (statement 1), and if the atoms in the different elements in these substances are unique (statement 2) and can be neither created

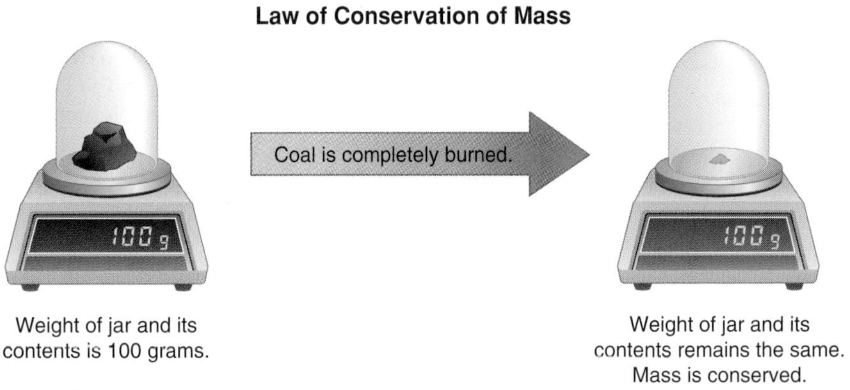

Law of Conservation of Mass

100 g

Coal is completely burned.

100 g

Weight of jar and its contents is 100 grams.

Weight of jar and its contents remains the same. Mass is conserved.

FIGURE 3.3
When a substance such as coal is burned in a closed container, the weight of the container and its contents before burning will equal the weight of the container and its contents after burning.

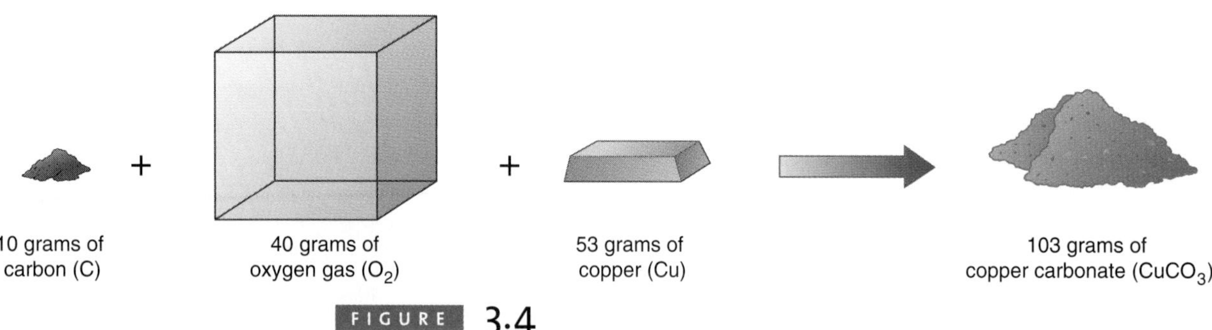

FIGURE 3.4

No matter how it is prepared, pure copper carbonate always has the same composition.

EXPLORATIONS describes the life of Joseph Priestley, whose work was vital to that of Lavoisier (see pp. 74–75).

During the French Revolution, Proust escaped to Madrid, where he continued his scientific work until his laboratory was destroyed during Napoleon's invasion of Spain.

nor destroyed but only rearranged (statement 4), then it follows that the total mass of the products must equal the total mass of the reacting substances.

The Law of Definite Proportions

Another Frenchman, Joseph Proust (1754–1826), a contemporary of Lavoisier, is responsible for establishing the second fundamental law that supports the atomic theory. Proust showed that the compound copper carbonate ($CuCO_3$), whether obtained from natural sources or synthesized in the laboratory, always had the same composition. It always contained the same three elements—copper (Cu), carbon (C), and oxygen (O)—in the same proportions by mass, namely, 53 parts of copper to 40 parts of oxygen to 10 parts of carbon **(Fig. 3.4)**. These findings, together with careful analysis of many other compounds, led Proust to formulate the **law of definite proportions** (sometimes called the *law of constant composition*), which can be stated as follows: *Different samples of any pure compound contain the same elements in the same proportions by mass.*

Proust's findings can also be explained in terms of Dalton's atomic theory. According to the theory, all atoms of copper are alike, all atoms of carbon are alike, and all atoms of oxygen are alike, but the three types of atoms differ from each other (statement 2). Therefore, atoms of the three elements must have different masses. Thus, if a fixed number of copper atoms combine with a fixed number of carbon atoms and a fixed number of oxygen atoms to form copper carbonate, it follows that a pure sample of this compound always has these three elements in the same proportions by mass.

EXAMPLE 3.1 Using the Law of Definite Proportions

Three parts by mass of hydrogen combine with 14 parts by mass of nitrogen to form the gas ammonia. How many grams of hydrogen will combine with 70 g of nitrogen? (See Appendix C for an explanation of problem solving by dimensional analysis.)

Solution Given that 3 g of hydrogen combined with 14 g of nitrogen, the ratio of grams (mass) of hydrogen to grams (mass) of nitrogen is 3/14. Therefore,

$$70\ \cancel{\text{g nitrogen}} \times \frac{3\text{ g hydrogen}}{14\ \cancel{\text{g nitrogen}}} = 15\text{ g hydrogen}$$

That is, 15 g of hydrogen will combine with 70 g of nitrogen.

Practice Exercise 3.1

When decomposed, the gas methane yields 3 parts by mass of carbon for every 1 part by mass of hydrogen. What mass of hydrogen will be obtained if 320 g of methane is decomposed?

The Law of Multiple Proportions

Based on his own work and that of others, Dalton realized that two elements often combined to form more than one compound. The compounds were distinct from each other, and each obeyed the law of definite proportions. For example, two different compounds containing only carbon and oxygen were known. One was the poisonous gas carbon monoxide, and the other was carbon dioxide, a product of respiration and the burning of wood. Dalton analyzed both compounds and determined that 3 parts by mass of carbon combined with 4 parts by mass of oxygen to form carbon monoxide, but that 3 parts by mass of carbon combined with 8 parts by mass of oxygen to form carbon dioxide. For equal quantities of carbon (3 parts), the ratio of oxygen in the two compounds was 4 to 8 (or 1 to 2). After analyzing other sets of compounds formed from two elements, Dalton summarized his conclusions in the **law of multiple proportions:** *The masses of one element that can combine chemically with a fixed mass of another element are in a ratio of small whole numbers.*

For example, the two compounds of carbon and oxygen can be explained in terms of atomic theory as follows: In carbon monoxide (CO), one atom of oxygen is combined with one atom of carbon, while in carbon dioxide (CO_2), two atoms of oxygen are combined with one atom of carbon (**Fig. 3.5**).

Carbon monoxide (CO)

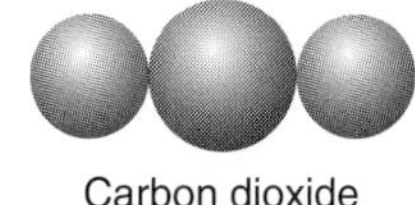

Carbon dioxide (CO_2)

FIGURE 3.5 The law of multiple proportions is explained in terms of atomic theory. In carbon monoxide (CO), one atom of carbon is combined with one atom of oxygen; in carbon dioxide (CO_2), one atom of carbon is combined with two atoms of oxygen.

EXAMPLE 3.2 Using the Law of Multiple Proportions

Nitrogen forms three compounds with oxygen: nitrous oxide (N_2O), nitric oxide (NO), and nitrogen dioxide (NO_2). Represent these compounds in terms of the two kinds of atoms they contain, and show how the law of multiple proportions applies.

Solution

	N_2O	NO	NO_2
Atoms of N (●) and O (●)	●●●	●●	●●●
No. of O atoms combined with a single N atom	1/2	1	2

It is not possible to have *half* an O atom. Therefore, the simplest ratio is given by the number of O atoms that combine with 2 N atoms:

1	2	4

Practice Exercise 3.2

Nitrogen reacts with hydrogen to form two compounds: ammonia (NH_3), a fertilizer, and hydrazine (N_2H_4), a rocket fuel. Represent these compounds in terms of the two kinds of atoms they contain, and show how the law of multiple proportions applies (N = ●, H = ○).

Dalton's atomic theory, although challenged, was widely accepted because its assumptions were supported by quantitative measurements and were consistent with a large body of experimental data. Although it later became evident that atoms are not the indivisible particles that Dalton had envisioned, his basic conclusions are still valid. His atomic theory was a major breakthrough in the development of chemistry and remains a remarkable feat of scientific deduction.

3.4 The Structure of the Atom: The Experimental Evidence

By the end of the 19th century, it was evident that atoms were much more complex than Dalton had imagined. New work showed that an atom was not a single uniform particle but was composed of a number of smaller particles. In this section, we will review the experimental evidence that led to the discovery of the three fundamental types of particles that make up all atoms: electrons, protons, and neutrons.

The Electrical Nature of Matter

During the 19th century and the early part of the 20th century, work by scientists in several fields indicated that matter had an electrical component. As experimental observations were gradually pieced together, it became evident that atoms must be composed of electrically charged particles.

The English chemists Humphry Davy (1778–1829) and Michael Faraday (1791–1867) were among the first to demonstrate the electrical nature of matter. They showed that the passage of an electric current through certain molten compounds or through water containing dissolved salts resulted in chemical changes in the compounds and the salts in solution. Faraday concluded that the electric current must be carried through the molten compounds and the solutions by charged atoms.

More information about the electrical nature of matter came in the latter part of the 19th century from experiments with **gas-discharge tubes**, also known as **cathode-ray tubes**, the forerunners of today's TV picture tubes, neon signs, and fluorescent lights. A simplified version of a cathode-ray tube is shown in **Figure 3.6a**. It consists of a glass tube with a metal **electrode** sealed into each end and a sidearm that can be attached to a vacuum pump. When the tube is in use, the electrodes are connected to a source of electric power, and the vacuum pump removes gas from the tube. As gas in the tube is pumped out, pressure inside the tube decreases.

In experiments using a variety of gases in the tube, scientists observed that when most of the gas had been pumped out, an electrical discharge was created between the electrodes, and the small amount of remaining gas began to glow **(Fig. 3.6b)**. If a screen coated with a fluorescent material such as zinc sulfide (ZnS) was placed between the electrodes, pinpoint flashes of light were emitted randomly from the side of the screen facing the cathode, thus proving conclusively that the electrical discharge was coming from the **cathode** (negative electrode) and flowing to the **anode** (positive electrode). For this reason, the beam of current was said to be composed of *cathode rays*. This demonstration also indicated that the rays were not made up of waves, as many believed, but were composed of minute particles, each of which produced a flash of light when it hit the screen. The particles later became known as *electrons*.

Fluorescent substances absorb electromagnetic radiation (EMR) of one wavelength (often invisible UV radiation) and then emit EMR of a longer wavelength (often visible light of a particular color).

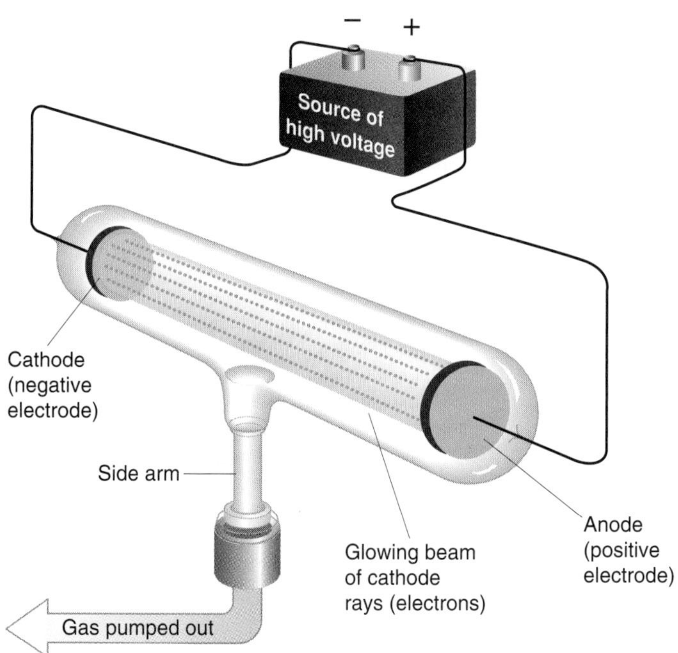

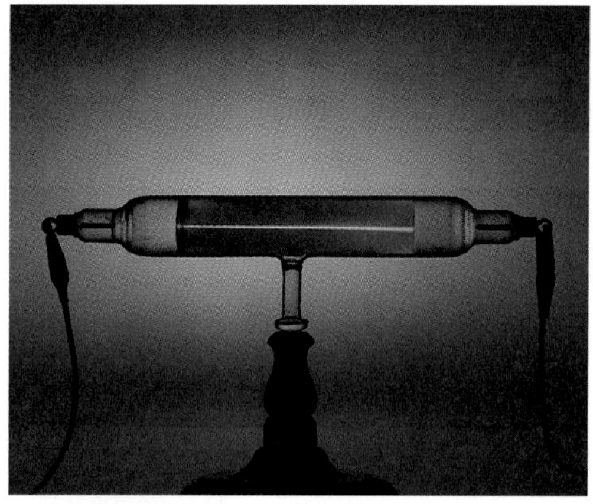

FIGURE 3.6

When a gas-discharge tube, or cathode-ray tube, is partially evacuated and an electrical current is applied, the gas remaining in the tube begins to glow. The glowing beam is composed of cathode rays (electrons).

Electrons

The fundamental nature of cathode-ray particles was established by the English physicist J. J. Thomson (1856–1940). In 1897, in one of his most important experiments with a gas-discharge tube, Thomson showed that, in an electric field, a beam of cathode rays was deflected (or bent) toward the positively charged plate **(Fig. 3.7a)**. Because like charges repel and unlike charges attract, this deflection proved that cathode rays were negatively charged. The direction in which the rays were deflected when a magnetic field was applied further confirmed the negative charge **(Fig. 3.7b and c)**.

Seven of the men who worked under Thomson in the Cavendish Laboratory at Cambridge University (including Ernest Rutherford) went on to win Nobel Prizes.

Thomson also showed that, regardless of the nature of the gas in the discharge tube or the material composing the electrodes, cathode rays always had the same properties. His experimental findings led him to conclude that cathode rays must be composed of very energetic, negatively charged particles, or **electrons,** which are a fundamental part of all matter. For this work, Thomson was awarded the Nobel Prize in physics in 1906.

Thomson was not able to determine either the weight of an electron or its charge, but he was able to determine the ratio of its charge to its mass. In 1909, the American physicist Robert A. Millikan (1868–1953) determined the charge on the electron. From this value and Thomson's value of the ratio of charge to mass, the mass of an electron was calculated to be 9.1×10^{-28} gram (negative exponents are explained in Appendix A), an extremely small number. Millikan received the Nobel Prize in physics in 1923.

When an electric charge is passed through a gas-discharge tube, electrons in the atoms of the gas are energized and escape, and they move toward the anode (+). The atoms, stripped of electrons, acquire a positive charge and move toward the cathode (−) as cathode rays.

Once electrons had been characterized, it was clear that it had been these particles that carried the current through molten compounds and solutions in the earlier experiments by Davy and Faraday (described at the beginning of this section).

Canal Rays

Another scientist who made important discoveries while experimenting with gas-discharge tubes was the German physicist Eugen Goldstein (1850–1930). In 1886, using a gas-discharge tube that had a perforated cathode **(Fig. 3.8)**, Goldstein discovered that, in addition to cathode rays that flowed from cathode to anode, positively charged rays

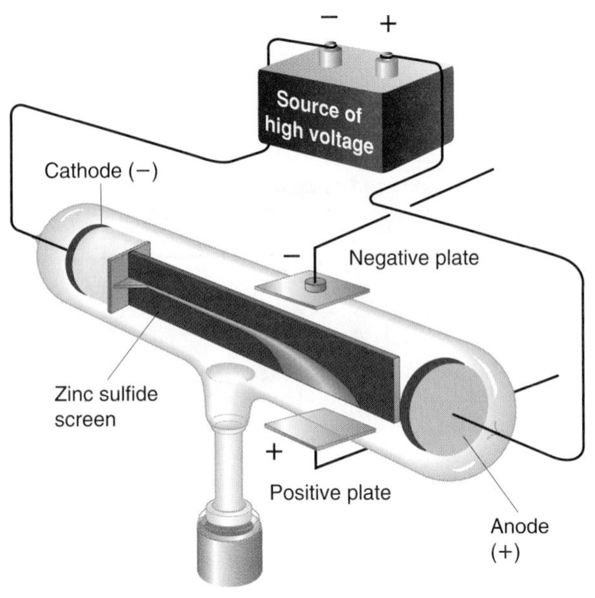

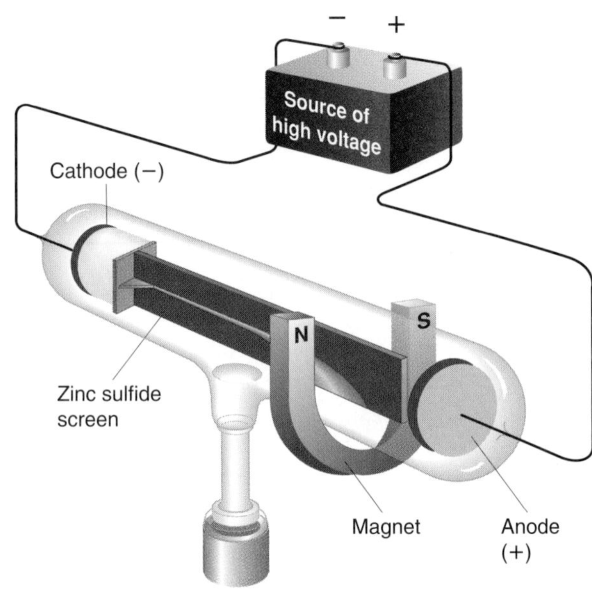

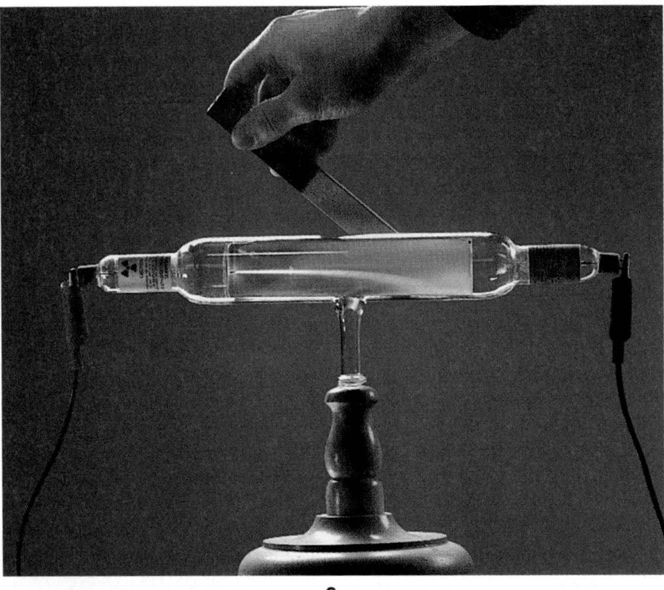

FIGURE 3.7

The direction in which cathode rays were deflected in (a) an electric field and (b) a magnetic field indicated that the rays were negatively charged. (c) The beam from a cathode-ray tube is deflected in a magnetic field.

flowed in the opposite direction through the holes (or canals) in the cathode. He called these rays **canal rays.**

Later research revealed that unlike electrons, which are all identical in mass, canal-ray particles are of varying mass. The lightest canal-ray particles were obtained when the gas in the gas-discharge tube was hydrogen. Even these particles, however, were very much heavier than electrons. Scientists had to wait until the discovery of the nucleus of the atom at the beginning of the 20th century before the full significance of canal rays was understood.

3.5 The Discovery of the Nucleus of the Atom

Based largely on the results of gas-discharge tube experiments, Thomson concluded that an atom must be a positively charged sphere of almost uniform density in which negatively charged electrons are embedded much like "raisins in a plum pudding" (or, to use a more American analogy, like blueberries in a blueberry muffin) **(Fig. 3.9)**. The positive and negative charges in the atom must be balanced so that the charge on the atom as a whole is zero. This view was generally accepted until 1911, when the New Zealand physicist Ernest Rutherford (1871–1937), who conducted most of his research in England and Canada, reported the results of an experiment that showed conclusively that Thomson's model of the atom was incorrect.

Gold was chosen because it can be beaten into extremely thin sheets.

The classic experiment was carried out by Ernest Marsden, an undergraduate, and Johannes Geiger, a German physicist, who were working in Rutherford's laboratory. They bombarded a very thin sheet of gold foil with alpha particles, which are very energetic, positively charged particles emitted at high speed by certain radioactive elements (radioactivity and alpha particles will be discussed in Chapter 5). The alpha-emitting radioactive source was placed in a lead-lined box that had a small hole in it. Alpha particles, which cannot penetrate lead, passed through the hole in the box in a narrow beam that was aimed at the extremely thin sheet of gold foil **(Fig. 3.10)**. Almost completely surrounding the metal foil was a fluorescent zinc sulfide screen that recorded the impact of alpha particles on it as flashes of light. In this way, the paths followed by the alpha particles could be observed.

Assuming that the mass and positive charge of each gold atom were uniformly distributed throughout a sphere, as envisioned by Thomson (refer to Fig. 3.11), Rutherford anticipated that this diffuse charge would have little effect on the fast-moving, energetic alpha particles. He expected the alpha particles to pass right through the thin gold foil without being deflected to any extent. He was, therefore, astonished to find that although most of the alpha particles (99.9%) did pass straight through the foil, as expected, a small but significant number were quite sharply deflected from their original paths. A very few were deflected directly back along the path they had traveled from the source. As Rutherford himself said, "It was about as credible as if you had fired a 15-inch shell at a piece of tissue paper and it came back and hit you."

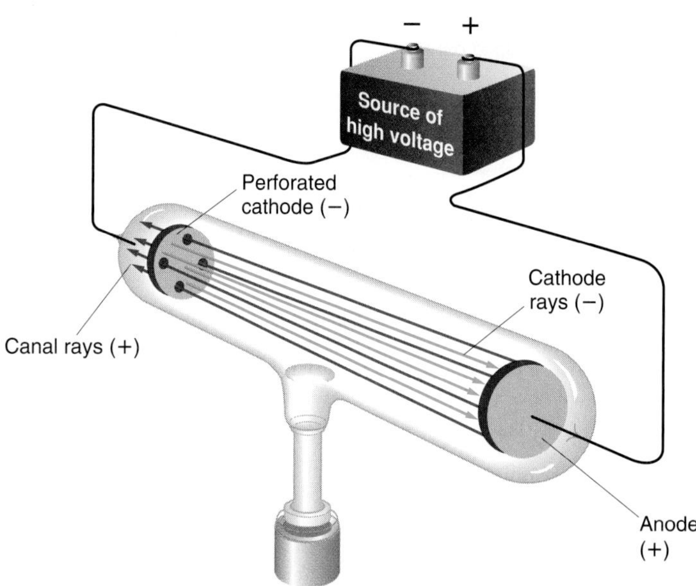

FIGURE 3.8

Using a gas-discharge tube with a perforated cathode, Goldstein showed that positively charged particles (canal rays) flowed from anode to cathode and through the holes in the cathode.

Rutherford interpreted the results as follows: (1) Because almost all the alpha particles passed through the foil without being deflected, the volume occupied by an atom must be mostly empty space, and (2) each of the alpha particles that was so dramatically deflected from its original path must have encountered a dense, similarly charged particle that repelled it and thus caused it to change direction. The closer an alpha particle came to one of these dense positively charged particles, the more sharply it was deflected. Rutherford concluded that at the center of an atom, there must be a minute, very dense **nucleus** that accounts for almost all of the mass of the atom and contains all the positive charge. Electrons must be distributed in the space surrounding the nucleus; they would be too light to have any effect on the fast-moving, much heavier (approximately 8000 times heavier) alpha particles **(Fig. 3.11)**.

The revolutionary model of the atom proposed by Rutherford has been verified many times by later experiments. Rutherford was awarded the Nobel Prize in chemistry in 1908 for his work on radioactivity, but his greatest contribution to science was undoubtedly his nuclear theory of the atom.

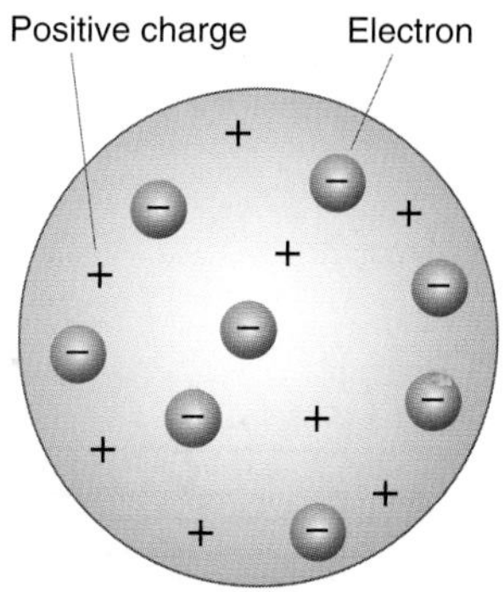

FIGURE 3.9

Thomson's plum pudding model of the atom (which can also be thought of as the blueberry muffin model). Negatively charged electrons are embedded in a positively charged sphere of uniform density. The negative and positive charges are balanced, yielding an electrically neutral model.

The Structure of the Nucleus of the Atom

Once the existence of the nucleus of the atom had been established, the next step was to determine its exact composition. Rutherford suggested that the positively charged particle in the atomic nucleus was identical to the positively charged particles produced in Goldstein's canal-ray experiments when the gas in the discharge tube was hydrogen. He named these particles **protons.**

Rutherford deduced that atoms of hydrogen, the lightest of all the elements, contain one proton and one electron. In Goldstein's canal-ray experiments, the high-energy elec-

Radioactivity is a spontaneous process in which certain unstable elements emit penetrating radiation.

trons streaming from cathode to anode knocked the single electron from some of the hydrogen atoms, leaving behind positively charged protons. Atoms of elements heavier than hydrogen contain increasingly larger numbers of protons. The heavier, positively charged particles observed in Goldstein's canal rays when a gas other than hydrogen was in the discharge tube were atoms whose nuclei contained more than one proton and from which an electron had been removed. Rutherford later obtained confirmation of his theory that protons are fundamental particles present in all atoms by demonstrating that when nitrogen atoms are bombarded with alpha particles, protons are ejected (see Chapter 5).

By 1914, it was possible to determine the number of protons in the nucleus of an atom, and Rutherford showed that, except for hydrogen, only about one-half of the nuclear mass of the atoms of the lighter elements—such as boron, carbon, nitrogen, and oxygen—could be accounted for by the number of protons present. He therefore concluded that in addition to protons, the nuclei of atoms must contain electrically neutral (uncharged) particles with mass approximately equal to the mass of the protons. It was not until 1932 that the English scientist James Chadwick (1891–1974) confirmed the existence of this third basic unit of the atom, the **neutron.** Chadwick showed that when atoms of beryllium were bombarded with alpha particles, uncharged particles with mass almost identical to that of the proton were emitted (see Chapter 5).

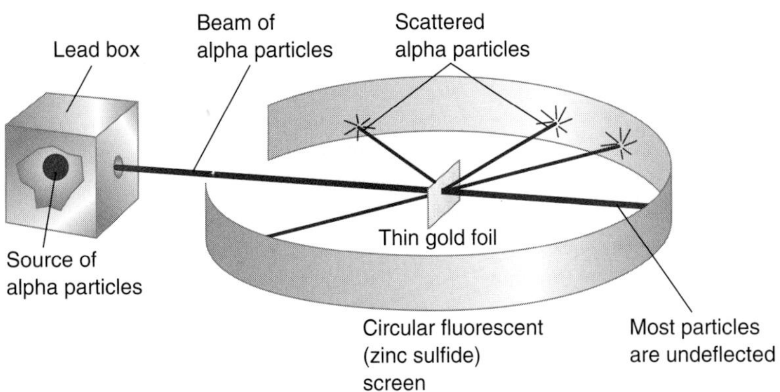

FIGURE 3.10

Rutherford's gold-foil experiment: Most of the alpha particles aimed at the gold foil passed straight through it, but a few were deflected at large angles.

H. G. Moseley, who was killed in World War I at age 27, discovered the relationship between the X-ray spectrum of an element and the number of charges in its nucleus. From this relationship, the number of protons in an atom of the element could be determined.

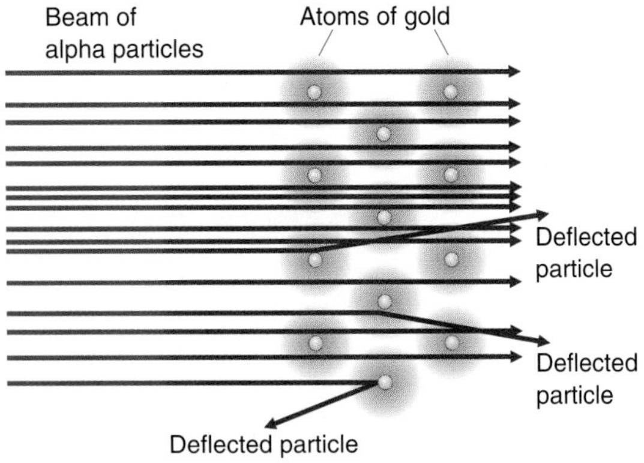

FIGURE 3.11

Rutherford's interpretation of the results of his gold-foil experiment: Most of an atom is empty space. The positive charge and mass of an atom are concentrated in a tiny nucleus at the center of the space. Most of the positively charged alpha particles passed straight through the empty space of gold atoms, without being deflected. The very few that came close to a nucleus were repelled by the like charge and deflected.

3.6 The Subatomic Particles: Their Properties and Arrangement in the Atom

In Chapter 2, we noted that it is difficult to believe that everything found on earth—whether living or nonliving, natural or synthesized by humans—is made up from fewer than 92 naturally occurring elements. Now we have to accept the even more startling fact that everything on earth is constructed from just three basic **subatomic particles:** electrons, protons, and neutrons. **Table 3.1** summarizes what we know about these three fundamental types of particles.

Electrons

An electron, abbreviated e^-, is a negatively charged (-1) particle. It is the smallest of the subatomic particles. Compared to the mass of a proton or a neutron, its mass is negligible and is taken to be zero for most purposes.

Protons

A proton, abbreviated p, is a positively charged $(+1)$ particle. The charge on a proton is opposite, but exactly equal to, the charge

Table 3.1 Characteristics of the Three Basic Subatomic Particles

Name of Particle	Symbol	Location in Atom	Relative Electrical Charge	Mass (g)	Approximate Relative Mass (amu)*
Electron	e^-	Surrounding the nucleus	−1	9.1×10^{-28}	0
Proton	p	Nucleus	+1	1.7×10^{-24}	1
Neutron	n	Nucleus	0	1.7×10^{-24}	1

*Relative mass in amu is explained in Section 3.8.

on an electron. A proton, although an extremely small particle, is approximately 1800 times heavier than an electron.

Neutrons

A neutron, abbreviated n, is a neutral particle; that is, it has no electrical charge. The mass of a neutron is very slightly greater than that of a proton, but for almost all purposes, it can be taken to be equal to that of a proton.

The Arrangement of Subatomic Particles in an Atom

An atom has two distinct regions: the center, or nucleus, and an area surrounding the nucleus, which contains the electrons **(Fig. 3.12)**. The very small, very dense nucleus is made up of protons and neutrons packed closely together. Because it contains protons, the nucleus is positively charged. A nucleus is so dense that 1 cubic centimeter (cm^3) of its material would have a mass equal to 90 million metric tons. Almost all (99.9%) of the mass of an atom is concentrated in the nucleus.

Moving around the nucleus are the electrons. The region occupied by the electrons, often termed the **electron cloud,** is mostly empty space. Compared with the size of the nucleus, this region is very large. If we could expand a nucleus so that it was the size of the period at the end of this sentence, then the space occupied by the electrons would be approximately as large as a sphere with a diameter of 50 yards (46 meters).

An atom, as a whole, has no charge because the number of positively charged protons in the nucleus of an atom is balanced by an equal number of negatively charged electrons in its electron cloud. As we shall see in later chapters, this relationship between protons and electrons in an atom explains many of the physical and chemical properties of the elements.

Nucleus (protons and neutrons) (positive charge)
Electrons
Electron cloud (area around nucleus in which the negatively charged electrons travel)

FIGURE 3.12
An atom consists of a very dense, compact, and positively charged nucleus made up of protons and neutrons, which is completely surrounded by an almost massless cloud of negatively charged electrons.

The Size of an Atom

Although the dimensions of atoms cannot be measured directly because they are so small, it is possible, based on measurements made on large collections of atoms, to calculate the diameters and masses of individual atoms. It has been determined that the diameter of an atom of an element ranges from 1×10^{-8} centimeter for hydrogen, the smallest atom, to 5×10^{-8} centimeter for the largest atoms. This means that if 100 million hydrogen atoms, each with a diameter of 1×10^{-8} centimeter, were lined up side by side, they would extend a distance of only 1 centimeter **(Fig. 3.13)**.

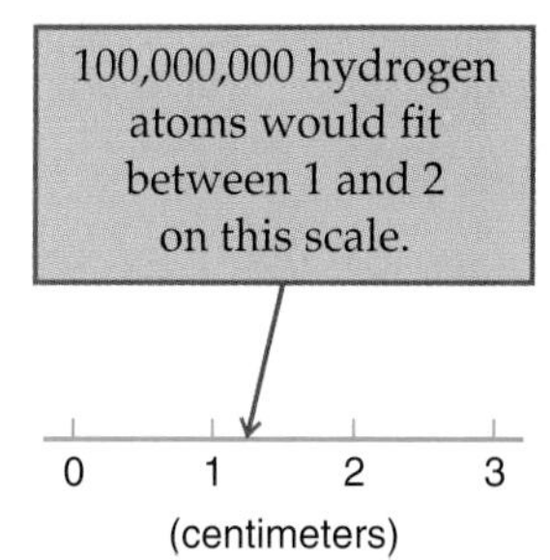
100,000,000 hydrogen atoms would fit between 1 and 2 on this scale.
0
1
2
3
(centimeters)

FIGURE 3.13
If 100 million hydrogen (H) atoms, each with a diameter of 1×10^{-8} cm, were lined up in a row, the line would extend only about 1 cm.

One of the reasons that Democritus's and Dalton's ideas about the atomic nature of matter were not immediately accepted was that, to many philosophers and scientists, it seemed irrelevant to speculate about particles that could not be seen. If atoms

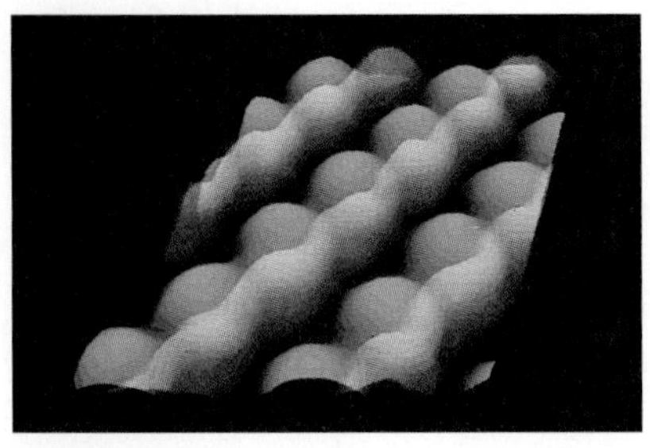

FIGURE 3.14

Scanning tunneling electron microscopes have allowed scientists to confirm the existence of atoms. Shown here is gallium arsenide with gallium (Ga) atoms in red and arsenic (As) atoms in blue.

could not be seen, even with the most powerful microscopes available at the time, how could anyone be certain that they existed? Recently, electron microscopes with magnification factors of several millions have made it possible to view images of atoms (**Fig. 3.14**), and their existence can no longer be disputed.

3.7 Atomic Numbers, Mass Numbers, and Isotopes

We have seen that all atoms are made up of electrons, protons, and neutrons. Now we will see how these three basic units are put together to form atoms of the different elements.

Atomic Number

All atoms of a particular element have the same number of protons in their nuclei. This number, which determines the identity of an element, is the **atomic number** for that particular element.

Because atoms are electrically neutral, the number of protons in the nucleus of an atom must be balanced by an equal number of electrons in the space surrounding the nucleus:

number of protons = number of electrons = atomic number

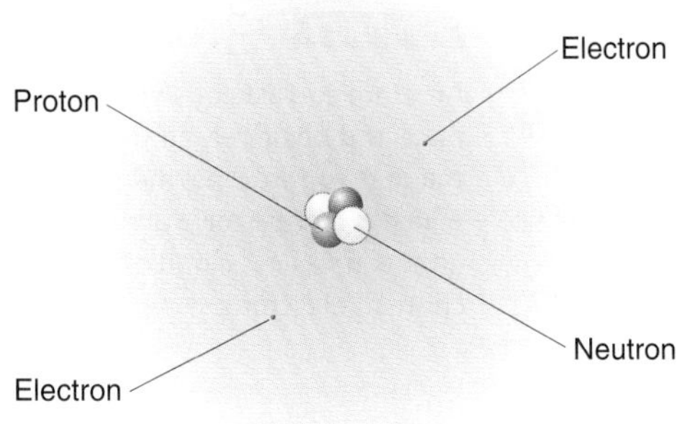

Atomic number = number of protons = number of electrons = 2
Mass number = number of protons + number of neutrons = 4

FIGURE 3.15

A helium (He) atom has 2 electrons, 2 protons, and 2 neutrons.

Mass Number

The sum of the number of protons and the number of neutrons in the nucleus of an atom is termed the **mass number** of the element (**Fig. 3.15**):

number of protons + number of neutrons = mass number

The mass of a proton is taken to be 1, and since the mass of a neutron is for all practical purposes identical to that of a proton, the mass of a neutron is also 1.

The arrangement of protons, neutrons, and electrons in an atom of the element helium (He) is illustrated in Figure 3.15. **Table 3.2** shows how atomic numbers and mass numbers describe the structure of the atoms of four common elements: hydrogen (H), nitrogen (N), phosphorus (P), and radon (Rn).

Table 3.2 The Relationship between Atomic Number and Mass Number for Four Common Elements

Element	Symbol	Number of of Protons	Number of Electrons	Number of Neutrons	Atomic Number	Mass Number
Hydrogen	H	1	1	0	1	(1 + 0) = 1
Nitrogen	N	7	7	7	7	(7 + 7) = 14
Phosphorus	P	15	15	16	15	(15 + 16) = 31
Radon	Rn	86	86	136	86	(86 + 136) = 222

EXAMPLE 3.3 Working with Atomic Numbers and Mass Numbers

An atom of iron (Fe) has 26 electrons. The mass number of Fe is 56.

a. What is the atomic number of Fe?

b. How many protons and neutrons does the nucleus of an Fe atom contain?

Solution

a.
$$\text{number of protons} = \text{number of electrons} = \text{atomic number}$$
$$\text{number of protons} = 26 = \text{atomic number}$$

Therefore, for iron, the atomic number is 26.

b. From part a, the number of protons is equal to the number of electrons, 26. Using the equation for mass number gives us

$$\text{number of neutrons} = \text{mass number} - \text{number of protons}$$
$$\text{number of neutrons} = 56 - 26 = 30$$

Therefore, the number of neutrons is 30.

Practice Exercise 3.3

Copper (Cu) has a mass number of 63 and an atomic number of 29. How many electrons, protons, and neutrons are contained in a Cu atom?

Isotopes

The number of protons (and the number of electrons) in each atom of a given element is always constant, but the number of neutrons may vary. Atoms of an element that have the same numbers of protons but different numbers of neutrons are called **isotopes**. It therefore follows that isotopes of a particular element all have the same atomic number but different mass numbers.

Research has shown that most naturally occurring elements are mixtures of isotopes in which one isotope usually predominates. For example, helium (atomic number 2) is a mixture of two isotopes. Just a few atoms in every thousand have one neutron; the remainder have two neutrons **(Fig. 3.16)**. Oxygen (atomic number 8) is a mixture of three isotopes. Over 99.7% of oxygen atoms have 8 neutrons; the remaining approximately 0.3% have either 9 or 10 neutrons.

The following notation is used to specify a particular isotope of an element:

$${}^{A}_{Z}\text{X} \qquad \begin{aligned} \text{X} &= \text{symbol of the element} \\ A &= \text{mass number} \\ Z &= \text{atomic number} \end{aligned}$$

With this notation, the two isotopes of helium (He) are written

$${}^{3}_{2}\text{He} \qquad {}^{4}_{2}\text{He}$$

The three isotopes of oxygen (O) are written

$${}^{16}_{8}\text{O} \qquad {}^{17}_{8}\text{O} \qquad {}^{18}_{8}\text{O}$$

Numerous studies have established that, no matter what the source of a particular element, the relative abundance of

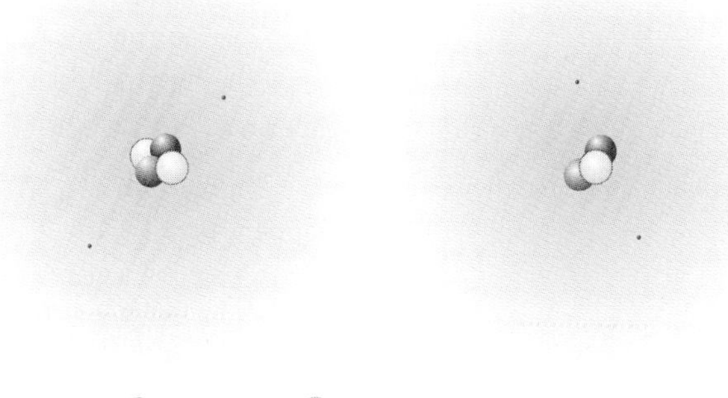

FIGURE 3.16

There are two isotopes of helium: Most helium atoms have two neutrons, but a very few have only one.

its isotopes is very nearly constant. For example, samples of oxygen obtained from air, from the decomposition of water, and from calcium carbonate ($CaCO_3$) contain practically the same percentages of the three isotopes of oxygen.

Table 3.3 lists the isotopes of a number of naturally occurring elements, with their atomic numbers, numbers of protons, numbers of neutrons, and percent natural abundance (the atomic mass, given in the last column, will be explained in the next section).

Table 3.3 Isotopes of Selected Common Elements

Element	Isotope	Atomic Number	Number of Protons	Number of Neutrons	Natural Abundance (%)	Atomic Mass (amu)
Hydrogen	$^{1}_{1}H$	1	1	0	99.98	1.01
	$^{2}_{1}H$	1	1	1	0.02	2.01
Helium	$^{3}_{2}He$	2	2	1	0.0001	3.02
	$^{4}_{2}He$	2	2	2	100.00	4.00
Lithium	$^{6}_{3}Li$	3	3	3	7.42	6.01
	$^{7}_{3}Li$	3	3	4	92.58	7.02
Beryllium*	$^{9}_{4}Be$	4	4	5	100.00	9.01
Boron	$^{10}_{5}B$	5	5	5	19.6	10.01
	$^{11}_{5}B$	5	5	6	80.4	11.01
Carbon	$^{12}_{6}C$	6	6	6	98.89	12.00†
	$^{13}_{6}C$	6	6	7	1.11	13.00
	$^{14}_{6}C$	6	6	8	trace	14.00
Nitrogen	$^{14}_{7}N$	7	7	7	99.63	14.00
	$^{15}_{7}N$	7	7	8	0.37	15.00
Oxygen	$^{16}_{8}O$	8	8	8	99.76	15.99
	$^{17}_{8}O$	8	8	9	0.04	17.00
	$^{18}_{8}O$	8	8	10	0.20	18.00
Fluorine*	$^{19}_{9}F$	9	9	10	100.00	19.00
Sulfur	$^{32}_{16}S$	16	16	16	95.00	31.97
	$^{33}_{16}S$	16	16	17	0.76	32.97
	$^{34}_{16}S$	16	16	18	4.22	33.97
	$^{36}_{16}S$	16	16	20	0.01	35.97
Chlorine	$^{35}_{17}Cl$	17	17	18	75.53	34.97
	$^{37}_{17}Cl$	17	17	20	24.47	36.97
Copper	$^{63}_{29}Cu$	29	29	34	69.09	62.93
	$^{65}_{29}Cu$	29	29	36	30.91	64.93
Uranium	$^{234}_{92}U$	92	92	142	0.01	234.04
	$^{235}_{92}U$	92	92	143	0.72	235.04
	$^{238}_{92}U$	92	92	146	99.27	238.05

*Beryllium (Be) and fluorine (F) are 2 of the 20 elements that have only one naturally occurring isotope.
†By international agreement, the mass of $^{12}_{6}C$ is exactly 12.0000 amu.

EXAMPLE 3.4 Identifying Isotopes

There are three isotopes of neon which have 10, 11, and 12 neutrons, respectively. Write the complete notation for each isotope, showing the symbol, atomic number, and mass number.

Solution

1. Refer to the list of elements on the inside back cover of this book to find neon. The symbol for neon is Ne. Its atomic number is 10.
2. We know the following:

$$\text{mass number} = \text{number of protons} + \text{number of neutrons}$$

$$\text{number of protons} = \text{atomic number}$$

Therefore, the number of protons is 10.

3. For the isotope with 10 neutrons:

$$\text{mass number} = 10 + 10 = 20$$

4. The notation for this isotope is

$${}^{20}_{10}\text{Ne}$$

Similarly, the notations for the isotopes with 11 and 12 neutrons, are respectively,

$${}^{21}_{10}\text{Ne} \qquad {}^{22}_{10}\text{Ne}$$

Practice Exercise 3.4

Three of the isotopes of mercury have 118, 120, and 124 neutrons, respectively. Write the complete notation for each isotope. Refer to the list of elements on the inside back cover of this book to find mercury.

EXAMPLE 3.5

Which of the following **a.** represent isotopes of the same element; **b.** have the same mass numbers; **c.** have the same number of neutrons?

$$(1)\ {}^{12}_{6}\text{X} \quad (2)\ {}^{14}_{6}\text{X} \quad (3)\ {}^{14}_{7}\text{X} \quad (4)\ {}^{15}_{7}\text{X} \quad (5)\ {}^{11}_{5}\text{X}$$

Identify the elements by referring to the atomic numbers listed in Table 3.3.

Solution

a. Notations 1 and 2 represent isotopes of carbon (atomic number 6) with 6 (12 − 6) and 8 (14 − 6) neutrons, respectively. Notations 3 and 4 represent isotopes of nitrogen (atomic number 7) with 7 (14 − 7) and 8 (15 − 7) neutrons, respectively.

b. Isotopes 2 and 3 have the same mass number (14).

c. Isotopes 1 and 5 (an isotope of boron, B, atomic number 5) both have 6 neutrons (12 − 6 = 6 and 11 − 5 = 6, respectively).

Practice Exercise 3.5

a. Which of the following represent isotopes of the same element? **b.** Which have the same mass number? **c.** Which have the same number of neutrons?

$$(1)\ {}^{32}_{16}\text{X} \quad (2)\ {}^{33}_{16}\text{X} \quad (3)\ {}^{36}_{16}\text{X} \quad (4)\ {}^{32}_{15}\text{X} \quad (5)\ {}^{30}_{14}\text{X}$$

In each case, give the name(s) of the element(s).

In view of the existence of isotopes, Dalton's original atomic theory had to be modified to include the fact that the atoms of most elements are not all identical; they may differ in the number of neutrons in the nucleus. As we shall see later in this chapter, it is the number and arrangement of the electrons in an atom—not the number of neutrons—that determine the chemical properties of an element. Isotopes of a particular element share the same chemical properties; they differ from each other in mass only.

3.8 The Atomic Mass Unit and Atomic Masses (Atomic Weights)

The atomic mass of an element can be determined with great accuracy, but because atoms are so very small, it is not convenient to express the mass of an individual atom in grams. For example, the weight of one of the heaviest atoms known—the $^{238}_{92}U$ isotope of uranium—is only 3.95×10^{-22} gram, an amount that is impossible to work with in the laboratory. To avoid this difficulty, scientists established an arbitrary unit of mass, the **atomic mass unit (amu).** This unit expresses the mass of an atom *relative* to the mass of an atom of a different element. By international agreement, the mass of the most common isotope of carbon, $^{12}_{6}C$, is taken to be exactly 12 amu. The masses of the atoms of all other elements are determined relative to that value (see Table 3.3).

When describing chemical reactions in quantitative terms, one rarely needs to specify the masses of the individual isotopes that are taking part in the reaction because all the isotopes of an element behave in the same way in a chemical reaction. Instead, the *average* mass of the various isotopes that make up an element is used. This average mass, which takes into account the relative abundance of the different isotopes, is termed the **atomic mass,** or **atomic weight** of the element. Chlorine, for example, which is 75.53% $^{35}_{17}Cl$ and 24.47% $^{37}_{17}Cl$, has an atomic mass of 35.45 amu. The atomic masses of the known elements are listed inside the back cover of this book.

3.9 The Periodic Table: Mendeleev's Contribution

By 1865, 63 elements were known, and the atomic masses of many of them had been determined. It was also known that groups of them shared similar properties, and scientists began to look for ways to arrange the elements systematically to reflect the similarities.

The first really useful classification of the elements, in the form of a **periodic table,** was published in 1869 by the Russian chemist Dmitri Ivanovich Mendeleev (1834–1907) (Fig. 3.17). Mendeleev had noticed that if the 63 elements then known were arranged in order of increasing atomic mass, those sharing similar properties appeared at regular, or periodic, intervals. He constructed his table by first positioning hydrogen by itself and then arranging the remaining elements in order of increasing atomic mass in a series of horizontal rows such that the elements with similar properties became stacked in the same columns. The arrangement of elements in the first two rows of Mendeleev's table was as follows:

Li	Be	B	C	N	O	F
Na	Mg	Al	Si	P	S	Cl

FIGURE 3.17 Dmitri Mendeleev (1834–1907), Russia's most famous chemist, was born in Siberia, the youngest of 17 children. A popular teacher, he was for many years a professor of chemistry at the University of St. Petersburg.

Each element in the first row is very similar to the element immediately below it: Lithium (Li) closely resembles sodium (Na), beryllium (Be) resembles magnesium (Mg), and so on, until we reach fluorine (F), which resembles chlorine (Cl).

A striking feature of Mendeleev's table was that, where necessary, he left spaces in it so that only elements with similar properties would occupy any given column, and he correctly predicted that new elements would be discovered to fit into the spaces he had left.

Table 3.4 Comparison of Properties of Eka-silicon Predicted by Mendeleev (in 1871) with Observed Properties of Germanium (discovered in 1886)

Property	Eka-silicon (X)	Germanium (Ge)
Color	Gray	Gray
Atomic mass	72	72.59
Density (g/cm^3)	5.5	5.35
Formula of oxide	XO_2	GeO_2
Density of oxide (g/cm^3)	4.7	4.70
Formula of chloride	XCl_4	$GeCl_4$
Melting point of chloride (°C)	below 100	84

Another important feature of Mendeleev's table was that, on the basis of the position of a missing element, he was able to predict its properties with great accuracy. Furthermore, his prediction helped to accelerate the discovery of the element. For example, knowing the properties of neighboring elements, Mendeleev correctly predicted the properties of germanium (Table 3.4), which he called *eka-silicon*. Mendeleev's table not only brought order to the study of chemistry, it stimulated the search for new elements.

Eka means "one" or "first" in Sanskrit. *Eka-silicon* means "one place from silicon."

3.10 The Modern Periodic Table

The modern periodic table shown in Figure 3.18 and on the inside front cover of this textbook evolved directly from Mendeleev's original table. The main differences between the current version and the original version are (1) the modern table includes 112 elements instead of 63, and (2) the elements are now arranged in order by atomic *number,* instead of atomic mass. The atomic number, as we shall see in the next chapter, explains the periodic repetition of similar properties more accurately than does the atomic mass. It also gives each element its own identifying number.

Each element in the periodic table is represented by its symbol; included with the symbol are the element's atomic number and atomic mass. The first 92 elements—hydrogen through uranium—occur naturally; the remainder have been synthesized in laboratory-controlled nuclear reactions (see Chapter 5).

Periods and Groups

The periodic table is divided into horizontal rows called **periods** and vertical columns called **groups** (Fig. 3.19). Although in many ways it is simpler to use the new international system of numbering the groups 1 through 18, most textbooks still use the older system, in which Roman numerals identify the groups and the letters A and B distinguish families within a group.

Certain sets of elements are given special names. The eight A groups of elements (Groups IA through VIIIA) are called the **representative elements,** and the B groups of elements that lie between the two blocks of representative elements are called the **transition elements.** The table is kept compact by listing two groups of transition elements, the lanthanides (elements 58 to 71) and the actinides (elements 90 to 103), separately at the bottom of the table.

For historical reasons, some groups of representative elements are often referred to by common names. For example, the Group IA elements are known as the **alkali**

PERIODIC TABLE OF THE ELEMENTS

Key	
Element	hydrogen
Atomic Number	1
Symbol	**H**
*Atomic Mass	1.01

Period	1 IA	2 IIA	3 IIIB	4 IVB	5 VB	6 VIB	7 VIIB	8 VIII	9 VIII	10 VIII	11 IB	12 IIB	13 IIIA	14 IVA	15 VA	16 VIA	17 VIIA	18 VIIIA
																		helium 2 **He** 4.00
2	lithium 3 **Li** 6.94	beryllium 4 **Be** 9.01											boron 5 **B** 10.81	carbon 6 **C** 12.01	nitrogen 7 **N** 14.01	oxygen 8 **O** 16.00	fluorine 9 **F** 19.00	neon 10 **Ne** 20.18
3	sodium 11 **Na** 22.99	magnesium 12 **Mg** 24.31											aluminum 13 **Al** 26.98	silicon 14 **Si** 28.09	phosphorus 15 **P** 30.97	sulfur 16 **S** 32.07	chlorine 17 **Cl** 35.45	argon 18 **Ar** 39.95
4	potassium 19 **K** 39.10	calcium 20 **Ca** 40.08	scandium 21 **Sc** 44.96	titanium 22 **Ti** 47.88	vanadium 23 **V** 50.94	chromium 24 **Cr** 52.00	manganese 25 **Mn** 54.94	iron 26 **Fe** 55.85	cobalt 27 **Co** 58.93	nickel 28 **Ni** 58.69	copper 29 **Cu** 63.55	zinc 30 **Zn** 65.39	gallium 31 **Ga** 69.72	germanium 32 **Ge** 72.61	arsenic 33 **As** 74.92	selenium 34 **Se** 78.96	bromine 35 **Br** 79.90	krypton 36 **Kr** 83.80
5	rubidium 37 **Rb** 85.47	strontium 38 **Sr** 87.62	yttrium 39 **Y** 88.91	zirconium 40 **Zr** 91.22	niobium 41 **Nb** 92.91	molybdenum 42 **Mo** 95.94	technetium 43 **Tc** (99)	ruthenium 44 **Ru** 101.07	rhodium 45 **Rh** 102.91	palladium 46 **Pd** 106.42	silver 47 **Ag** 107.87	cadmium 48 **Cd** 112.41	indium 49 **In** 114.82	tin 50 **Sn** 118.71	antimony 51 **Sb** 121.75	tellurium 52 **Te** 127.60	iodine 53 **I** 126.90	xenon 54 **Xe** 131.29
6	cesium 55 **Cs** 132.91	barium 56 **Ba** 137.33	lanthanum 57 **La** 138.91	hafnium 72 **Hf** 178.49	tantalum 73 **Ta** 180.95	tungsten 74 **W** 183.85	rhenium 75 **Re** 186.21	osmium 76 **Os** 190.2	iridium 77 **Ir** 192.22	platinum 78 **Pt** 195.08	gold 79 **Au** 196.97	mercury 80 **Hg** 200.59	thallium 81 **Tl** 204.38	lead 82 **Pb** 207.2	bismuth 83 **Bi** 208.98	polonium 84 **Po** (209)	astatine 85 **At** (210)	radon 86 **Rn** (222)
7	francium 87 **Fr** (223)	radium 88 **Ra** (226)	actinium 89 **Ac** (227)	rutherfordium 104 **Rf** (261)	dubnium 105 **Db** (262)	seaborgium 106 **Sg** (263)	bohrium 107 **Bh** (262)	hassium 108 **Hs** (265)	meitnerium 109 **Mt** (266)	ununnilium 110 **Uun** (269)	unununium 111 **Uuu** (272)	ununbium 112 **Uub** (277)						

Series														
Lanthanide Series	cerium 58 **Ce** 140.12	praseodymium 59 **Pr** 140.91	neodymium 60 **Nd** 144.24	promethium 61 **Pm** (147)	samarium 62 **Sm** 150.36	europium 63 **Eu** 151.97	gadolinium 64 **Gd** 157.25	terbium 65 **Tb** 158.93	dysprosium 66 **Dy** 162.50	holmium 67 **Ho** 164.93	erbium 68 **Er** 167.26	thulium 69 **Tm** 168.93	ytterbium 70 **Yb** 173.04	lutetium 71 **Lu** 174.97
Actinide Series	thorium 90 **Th** 232.04	protactinium 91 **Pa** (231)	uranium 92 **U** 238.03	neptunium 93 **Np** (237)	plutonium 94 **Pu** (244)	americium 95 **Am** (243)	curium 96 **Cm** (247)	berkelium 97 **Bk** (247)	californium 98 **Cf** (251)	einsteinium 99 **Es** (252)	fermium 100 **Fm** (257)	mendelevium 101 **Md** (258)	nobelium 102 **No** (259)	lawrencium 103 **Lr** (260)

*Note: For radioactive elements, the mass number of an important isotope is shown in parenthesis; for thorium and uranium, the atomic mass of the naturally occurring radioisotopes is given.

FIGURE 3.18

The periodic table. Note that hydrogen, the lightest element, is not included in the main body of the table.

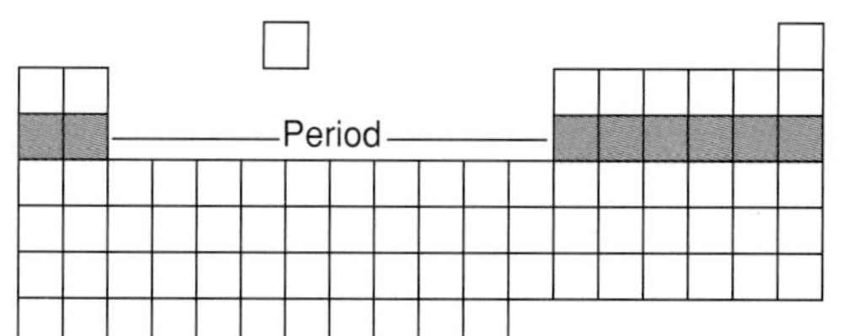

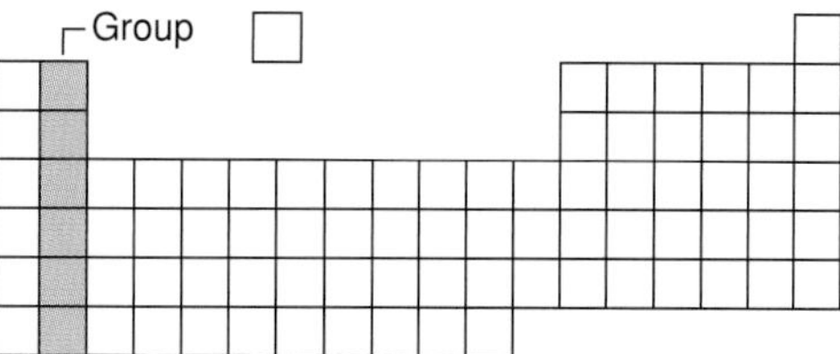

FIGURE 3.19

The periodic table is divided into periods (horizontal rows) and groups (vertical columns).

metals; Group IIA is referred to as the **alkaline earth metals;** Group VIIA elements are the **halogens;** and Group VIIIA elements are called the **noble gases.**

Elements within a group, particularly if they are representative elements, have similar physical and chemical properties. For example, the alkali metals—lithium (Li), sodium (Na), potassium (K), rubidium (Rb), and cesium (Cs)—are all soft, shiny, silvery metals **(Fig. 3.20)** with low densities and low melting points. They are very reactive and combine readily with many other elements, including oxygen, sulfur, phosphorus, and the halogens. Because they react vigorously with water, they have to be stored under oil. The halogens—fluorine (F), chlorine (Cl), bromine (Br), and iodine (I)—are all so reactive that they do not occur as elements in nature. However, they do combine readily with many other elements to form a very large number of compounds that are widely distributed.

For many years after their discovery, no compounds of the Group VIIIA elements were known, and these elements were called *the inert gases.* Following the discovery of compounds of krypton and xenon, the group was renamed *the noble gases.*

EXAMPLE 3.6 Using the Periodic Table

Refer to the periodic table to determine which two elements in the following list are most alike in their physical and chemical properties.

sodium (Na) carbon (C) chlorine (Cl) sulfur (S) bromine (Br)

Solution Chlorine (Cl) and bromine (Br) are most alike because they are in the same group (VIIA) of the periodic table.

Practice Exercise 3.6

Which of the following elements would you expect to show similar physical and chemical properties?

magnesium (Mg) potassium (K) aluminum (Al) fluorine (F) calcium (Ca)

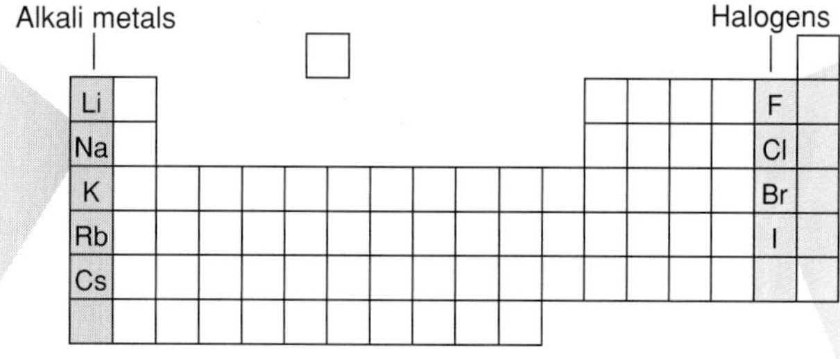

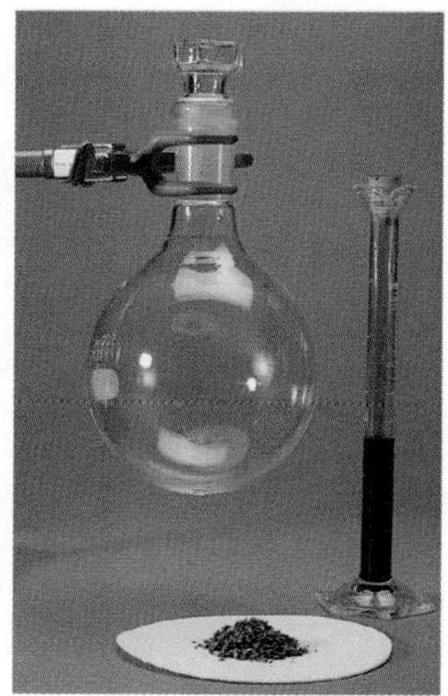

FIGURE 3.20

Elements within the same group in the periodic table have similar properties. Examples are the alkali metals (Group IA) and the halogens (Group VIIA).

EXPLORATIONS

Joseph Priestley—Radical Thinker and Inspired Dabbler in Chemistry

Antoine-Laurent Lavoisier is rightly remembered and revered as the father of chemistry, but, as he himself acknowledged, he owed a great deal to his English contemporary, the clergyman Joseph Priestley. Lavoisier named oxygen and correctly explained its relationship to combustion and respiration, but it was Priestley's experiments that led him to his conclusions.

Joseph Priestley

Priestley is scarcely remembered today outside the world of science, but in his time he was well known as a philosopher, theologian, and politician and counted among his friends Adam Smith, Edmund Burke, Benjamin Franklin, John Adams, and Thomas Jefferson.

His Early Career

Born in England in 1733 in a small Yorkshire village, Priestley was destined for the ministry from an early age. Because he was a Dissenter, someone who did not adhere to the established beliefs and practices of the Church of England, Priestley was barred by English law from attendance at either Oxford or Cambridge University. Instead, in 1752, he enrolled in one of the Dissenting Academies, which, in contrast to the conservative Oxford and Cambridge, were hotbeds of liberal thought and free inquiry. In this stimulating atmosphere, Priestley acquired many of the unorthodox religious beliefs and liberal ideas that were later to bring him into conflicts.

Priestley's life as a clergyman was difficult not only because his theology often offended his congregation but also because he stammered. His speech impediment had one good result: It took him to London in search of a cure. Although he never obtained more than temporary relief, his visits to the capital brought him in contact with persons whose views and intellects matched his own.

Priestley had always been interested in the study of nature, but it was not until he met Benjamin Franklin in London that he began to devote himself seriously to the study of science. Encouraged by Franklin, he published his first scientific work, "History of Electricity," in 1767. His interest in science was further sparked by a fortunate circumstance: While a minister in Leeds, he lived next door to a brewery. He conducted many experiments with the "fixed air" (carbon dioxide) produced in the fermentation process and made a notable contribution to human enjoyment—he discovered a practical way of "impregnating water with fixed air" to produce "soda water" (carbonated water). This useful discovery led to his being asked to accompany Captain Cook on his second voyage to the South Seas. A supply of soda water was carried on the ship in the mistaken belief that it would prevent scurvy, the curse of all long sea voyages. However, when Priestley's unorthodox religious opinions became known, it was thought wiser not to tempt Providence by having a Dissenter on board, and so the invitation was withdrawn.

By 1773, Priestley's ministerial duties were such that he had more time to indulge his increasing interest in chemistry. The fact that he had no prior knowledge of this subject did not deter him, and in the course of the next 7 years, he added five new gases to the list of only three—"air," carbon dioxide, and hydrogen—that were then recognized. Priestley discovered "nitrous air" (nitric oxide, NO), "red nitrous vapor" (nitrogen dioxide, NO_2), "diminished nitrous air" (nitrous oxide, N_2O), "marine acid air" (hydrogen chloride, HCl), and "alkaline air" (ammonia, NH_3). One of the main reasons for his success was his ability to design his own ingenious laboratory apparatus.

The Phlogiston Theory

In the 18th century, chemistry was not far removed from alchemy and was still dominated by the *phlo-*

giston theory, which held that any substance capable of being burned contained a fundamental component called *phlogiston* that escaped in the act of burning. When metals such as calcium or lead were strongly heated in air, for example, their changed appearance was said to be due to loss of phlogiston. The indisputable fact that the metal actually gained weight when heated was ingeniously explained by giving phlogiston a negative weight! A flame was said to move upward because it possessed "a quality of levity."

At this time, Priestley and other scientists realized that there was a connection between combustion and respiration. That a candle burning in a limited supply of air went out and that an animal placed in a closed container died were explained by assuming, in both cases, that phlogiston was transferred to the air. Both burning and life ceased when the air in the confined space became saturated with phlogiston. Priestley, in the course of many experiments, showed that during combustion or respiration in a confined space, the volume of air was reduced by about one-fifth; he also observed that a growing plant restored the air that had been used up by either a burning candle or a respiring animal (in this case, a mouse). His most important combustion experiment was performed in 1774. He heated red "calcinatus per se" (mercury oxide, HgO) and obtained a colorless gas. He observed that a candle burned in this gas with a "remarkably bright flame." Priestley, a staunch believer in the phlogiston theory, called his gas "dephlogisticated air."

Priestley was better at devising experiments than at interpreting them, and he did not appreciate the significance of his discovery of this new "air." However, in that same year in Paris, he had his only meeting with Lavoisier—a meeting that was very significant for the future of chemistry. Priestley told Lavoisier about his new "air," and Lavoisier, who was a much more serious and careful scientist than Priestley, recognized the significance of the discovery. After repeating Priestley's experiment, and doing many other related experiments, he published his conclusions in 1779. Lavoisier realized that Priestley's dephlogisticated air was a component of the atmosphere. He gave it the name *oxygen* and correctly deduced its role in respiration and combustion. Lavoisier's conclusions completely discredited the phlogiston theory, but despite all the opposing evidence, Priestley upheld the theory to the end of his life.

Revolution and Emigration

At its beginning in 1789, the French Revolution was generally hailed as the dawning of an era of peace and brotherhood. One of its first acts, the establishment of religious liberty, was warmly welcomed by the English Dissenters, who were still denied full rights of citizenship in their own country. Priestley, already unpopular because of his attacks on orthodox theology, made further enemies by openly and ardently supporting the French Revolution, which, as it progressed, appeared to many to threaten both the English crown and the established Church. In 1791, in Priestley's hometown of Birmingham, a second anniversary celebration of the start of the revolution became the excuse for a mob to attack his home. Priestley and his family escaped unharmed, but his laboratory and library were completely destroyed. Continuing hostility eventually forced Priestley to flee to America, where he was warmly welcomed by his old friend Benjamin Franklin. He was offered the post of professor of chemistry at the University of Pennsylvania but declined it, preferring to devote himself to preaching, theology, and experimenting in his new laboratory. Priestley and his family settled in a small town in central Pennsylvania, and, out of contact with scientific research and clinging to outdated theories, he made no further useful contributions to chemistry. He died in 1804.

Continuing hostility eventually forced Priestley to flee to America, where he was warmly welcomed by his old friend Benjamin Franklin.

Priestley was a true pioneer. A gentle, pious man, he devoted his life to the advancement of individual liberty, confident in the belief that political and religious freedom and the application of science would lead to human progress. He hoped to be remembered as a theologian and would no doubt be surprised to know that 200 years after his death he is revered instead for his contribution to chemistry.

References: A. Holt, *A Life of Joseph Priestley* (London: Oxford University Press, 1931); K. S. Davis, *The Cautionary Scientists (Priestley, Lavoisier, and the Founding of Modern Chemistry)* (New York: G. P. Putnam Sons, 1966); W. R. Aykroyd, *The Three Philosophers: Lavoisier, Priestley and Cavendish* (Westport, CT: Greenwood Press, 1935)

Table 3.5 Some Properties of the Alkali Metals (Group IA or 1) and the Halogens (Group VIIA or 17)

Group	Element	Symbol	Appearance	Boiling Point (°C)	Melting Point (°C)	Reactivity
Alkali Metals	Lithium	Li	Shiny, silvery, soft metals	1336	181	Reactive (increasing reactivity from Li to Cs)
	Sodium	Na		883	98	
	Potassium	K		758	63	
	Rubidium	Rb		700	39	
	Cesium	Cs	Increasing softness from Li to Cs	670	29	Very reactive
Halogens	Fluorine	F	Pale yellow gas	−188	−219	Very reactive (decreasing reactivity from F to I)
	Chlorine	Cl	Yellow-green gas	−34	−101	
	Bromine	Br	Red-brown liquid	59	−7	
	Iodine	I	Purple solid	184	114	Reactive

In addition to sharing similarities, the properties of the elements in each group vary in quite regular ways from top to bottom within the group. For instance, melting points and boiling points decrease from lithium to cesium in the alkali metals and increase from fluorine to iodine in the halogens (**Table 3.5**). The increase in melting points and boiling points in the halogen group results in a change in state from gas (F and Cl) to liquid (Br) to solid (I) (Fig. 3.20). Chemical reactivity also shows a definite trend within the groups. As we proceed down the alkali metal group, the elements become increasingly reactive from lithium to cesium; down the halogen group from fluorine to iodine, they become less reactive. Similar trends are found in other groups in the periodic table.

Metals and Nonmetals

Elements can be classed into two main types: metals and nonmetals. **Metals** have many distinctive properties. They are good conductors of heat and electricity, and most of them have a characteristic lustrous (shiny) appearance. Metals are ductile (i.e., they can be drawn out into a fine wire) and malleable (i.e., they can be rolled out into a thin sheet). All metals are solids at room temperature, except mercury (Hg), which is a liquid. Familiar metals are sodium (Na), aluminum (Al), calcium (Ca), chromium (Cr), iron (Fe), copper (Cu), silver (Ag), tin (Sn), platinum (Pt), and gold (Au).

Nonmetals, unlike metals, usually do not conduct heat or electricity to any significant extent. They have little or no luster and are neither ductile nor malleable. At room temperature, many nonmetals—including hydrogen (H), nitrogen (N), oxygen (O), chlorine (Cl), and the noble gases—exist as gases. One nonmetal—bromine (Br)—is a liquid. Solid nonmetals include carbon (C), phosphorus (P), sulfur (S), and iodine (I).

In the periodic table displayed inside the front cover of this book, metals and nonmetals are shown in different colors. The change from metal to nonmetal properties is not abrupt but gradual, and the elements shaded light gold yellow in the table have properties that lie between those of metals and nonmetals. These elements, which are called **semimetals,** or **metalloids,** include boron (B), which conducts electricity well only at a high temperature, and the **semiconductors**—silicon (Si), germanium (Ge), and arsenic (As)—which conduct electricity better than nonmetals but not as well as metals such as copper and silver. The special conducting ability of the semiconductors, particularly silicon, accounts for their use in computer chips and electronic calculators.

Chapter Summary

1. The idea that matter is composed of indivisible particles was suggested by the ancient Greeks but was not proved until 1803, by Dalton.
2. Dalton's atomic theory states that:
   a. All matter is composed of tiny, indivisible particles called atoms.
   b. All atoms of a particular element are identical, but the atoms of one element differ from the atoms of any other element.
   c. Atoms of different elements combine with each other in certain whole-number proportions to form compounds.
   d. In a chemical reaction, atoms are rearranged to form new compounds; they are not created, destroyed, or changed into atoms of any other element.
3. Three fundamental laws provide evidence to support the atomic theory: the law of conservation of mass, the law of definite proportions, and the law of multiple proportions.
4. The law of conservation of mass was based largely on the work of Lavoisier. It states that in a chemical reaction, matter is neither created nor destroyed.
5. The law of definite proportions was established by Proust. It states that different samples of any pure compound contain the same elements in the same proportions by mass.
6. The law of multiple proportions was established by Dalton. It states that the masses of one element that can combine chemically with a fixed mass of another element are in a ratio of small whole numbers.
7. Experiments with cathode-ray tubes showed that atoms contain negative particles, which were named electrons.
8. The work of Thomson and Millikan established that an electron has practically no mass.
9. Using a gas-discharge tube, Goldstein showed that atoms contain positive particles, which he named cathode rays. Later, it was shown that when the gas in the tube was hydrogen, these particles were protons.
10. Rutherford's gold-foil experiment established that there is a minute, dense, positively charged nucleus at the center of every atom.
11. The nucleus of an atom is composed of protons, each of which has a $+1$ charge, and neutrons, each of which has almost the same mass as a proton but no charge. The space surrounding the nucleus is occupied by electrons, each of which has a -1 charge and negligible mass.
12. An atom of an element has no charge because the number of positively charged protons it has is balanced by an equal number of negatively charged electrons.
13. The number of protons in an atom of a particular element identifies the element and is called its atomic number.
14. The sum of the number of protons and the number of neutrons in the nucleus of an atom is termed its mass number.
15. Atoms of an element that have the same number of protons but different numbers of neutrons are called isotopes.

16. The atomic mass unit (amu) of a particular element expresses the mass of an atom of that element relative to the mass of the most common isotope of carbon, ${}^{12}_{6}C$, which is taken to be exactly 12 amu.

17. The first useful classification of the elements in the form of a periodic table was made by Mendeleev. He arranged the known elements in order of increasing atomic mass in horizontal rows in such a way that elements with similar properties were placed in the same columns.

18. In the modern periodic table, the elements are arranged in order of atomic number. The table is divided into horizontal rows called periods and vertical columns called groups. Two groups of transition elements, the lanthanides and the actinides, are listed separately.

19. Elements in the same group—particularly if they are representative elements—have similar physical and chemical properties.

20. The properties of elements in a group vary in a regular way from top to bottom of the column.

21. Depending on their properties, elements can be classified as metals, nonmetals, or semimetals. Semimetals have properties between those of metals and those of nonmetals.

Key Terms

alkali metals (p. 71)
alkaline earth metals (p. 73)
anode (p. 60)
atom (p. 56)
atomic mass (weight) (p. 70)
atomic mass unit (amu) (p. 70)
atomic number (p. 66)
canal rays (p. 62)
cathode (p. 60)
cathode-ray tube (p. 60)
Dalton's atomic theory (p. 56)
electrode (p. 60)
electron (p. 61)
electron cloud (p. 65)
gas-discharge tube (p. 60)
group (p. 71)
halogens (p. 73)
isotope (p. 67)
law of conservation of mass (p. 57)
law of definite proportions (p. 58)
law of multiple proportions (p. 59)
mass number (p. 66)
metalloids (p. 76)
metals (p. 76)
neutron (p. 64)
noble gases (p. 73)
nonmetals (p. 76)
nucleus (p. 63)
period (p. 71)
periodic table (p. 70)
proton (p. 63)
representative elements (p. 71)
semiconductors (p. 76)
semimetals (p. 76)
subatomic particles (p. 64)
transition elements (p. 71)

Questions and Problems

1. Many cities are having tremendous problems disposing of their garbage. Landfills are full, and there is no room for more trash. Many suggest burning (incinerating) the combustible portion. Others suggest recycling reusable material or composting vegetable waste. How does the law of conservation of mass apply to this problem, and what does it suggest as possible solutions?

2. Medieval alchemists tried to change lead into gold by many different methods. Why didn't they succeed?

3. When a 5-kg piece of wood is burned, only 0.025 kg of solid ash is left. Does this result violate the law of conservation of mass?

4. What does the law of conservation of mass prove?

5. How does Dalton's atomic theory explain each of the following?
   a. the law of conservation of mass
   b. the law of definite proportions
   c. the law of multiple proportions

6. What experimental facts indicate that all atoms contain:

a. at least one negatively charged electron?

b. at least one positively charged proton?

7. Rutherford used the results of Marsden and Geiger's experiment to conclude that the nucleus contains most of the mass of the atom and that it has a positive charge. Describe the experimental results and show how they confirmed Rutherford's hypothesis (model).

8. How does Rutherford's model differ from Dalton's model?

9. Cholesterol is a compound suspected of causing hardening of the arteries. The formula of cholesterol is $C_{27}H_{46}O$.

a. A sample of cholesterol is isolated from the arterial lining of a 60-year-old American male. What is the formula of this compound?

b. Cholesterol is extracted from a chicken's egg. What is the formula of this compound?

10. Octane, a major component of gasoline, always contains 84% carbon and 16% hydrogen. What law do these facts illustrate?

11. State, in your own words, the law of multiple proportions.

12. Sulfur and oxygen form two compounds. By weight, one of the compounds is 40% sulfur and 60% oxygen; the other is 50% oxygen and 50% sulfur. What law is illustrated?

13. Which of the three laws described in this chapter is confirmed or violated by each of the following statements?

a. Lavoisier found that when mercury oxide (HgO) decomposed, the mass of mercury and oxygen formed equaled the mass of HgO decomposed.

b. The formula of sulfuric acid, an acid formed from some air pollutants, is $H_{1.2}SO_{4.9}$.

c. The atomic ratio of oxygen to nitrogen is twice as large in one compound containing the two elements as it is in another.

d. The formula of water found in China is H_4O, but water found commonly in the United States has the formula H_2O.

14. Twelve parts by mass of carbon combine with 32 parts by mass of oxygen to form a gas. How many grams of carbon will combine with 16 grams of oxygen to form the same gas?

15. When a gas decomposes, it releases 4 parts by mass of carbon for every one part by mass of hydrogen. What mass of hydrogen will be obtained in the decomposition of 100 g of this gas?

16. If a compound contains only carbon and hydrogen in a ratio of one carbon atom to 2.67 hydrogen atoms, what is its formula?

17. Twelve parts by mass of carbon will combine with 16 parts by mass of oxygen to form a gas. How many grams of the gas can be formed using 12 g of carbon and a sufficient amount of oxygen?

18. What are the distinguishing characteristics of each of the following?

a. protons

b. neutrons

c. electrons

19. Indicate whether the following statements are true or false:

a. An electron and a proton repel each other.

b. A proton and a proton attract each other.

c. An electron and an electron repel each other.

d. A proton and a neutron repel each other.

20. **a.** Cathode rays are composed of __________.

b. A neutral atom has an equal number of __________ and __________.

c. The mass of a proton is __________ amu.

21. What happens if the number of protons in the nucleus of an atom stays the same, but the number of neutrons changes?

22. What happens if the number of neutrons in the nucleus of an atom stays the same, but the number of protons changes?

23. Atoms contain protons which are positively charged, and electrons, which are negatively charged. Why doesn't the atom, as a whole, have a charge?

24. **a.** What is the difference between atomic number and mass number?

b. What is the difference between mass number and atomic mass?

c. What units are used for atomic mass and atomic number?

25. Define the following terms:
  a. isotope of an element
  b. atomic number of an element
  c. mass number of an element.

26. Sodium chloride containing the sodium isotope $^{24}_{11}Na$ is used to trace blood clots.
  a. How many protons are in its nucleus?
  b. How many neutrons are in its nucleus?
  c. How many electrons are in a sodium atom?

27. The element carbon has an atomic number of 6 and an atomic mass of 12. What do these numbers tell you about the numbers of protons, neutrons, and electrons in a neutral carbon atom?

28. How many protons, neutrons, and electrons are present in each of the following atoms? (Obtain the atomic numbers from the periodic table on the inside front cover of this book.)
  a. ^{208}Pb **b.** ^{204}Pb **c.** ^{37}Cl **d.** ^{27}Al **e.** ^{31}P

29. If a neutral atom has an atomic number of 9 and an atomic mass of 19, how many electrons does it have?

30. If an element has an atomic number of 15 and an atomic mass of 31, how many protons and neutrons does each atom of the element contain?

31. Complete the table below:

Atom	Atomic No.	Mass No.	No. of Protons	No. of Neutrons	No. of Electrons
Zn	30	64	____	____	____
Eu	____	153	63	____	____
U	____	____	____	143	92
Pd	____	____	46	60	____

32. Complete the following table using the list of elements on the inside back cover of this book.

Atom	No. of Protons	No. of Neutrons	No. of Electrons	Symbol
____	10	11	____	____
Barium		82	____	____
____	21	24	____	____
____	15	16	____	____

33. If magnesium has two isotopes, one that weighs 24 amu and another that weighs 25, and one out of every three atoms of magnesium has a mass of 25, then what is the atomic mass for naturally occurring magnesium?

34. Explain how the three isotopes of neon, ^{21}Ne, ^{22}Ne, and ^{23}Ne, differ.

35. Copper has two isotopes, ^{63}Cu and ^{65}Cu. The lighter isotope has a natural abundance of 70.5% and the heavier, 29.5%. What is the atomic mass of naturally occurring copper?

36. Arrange the following atoms in order of increasing mass (amu):
  a. N **b.** Zn **c.** Cl **d.** Xe **e.** Hg

37. Chlorine has two natural isotopes: ^{35}Cl and ^{37}Cl. Hydrogen reacts with chlorine to form hydrochloric acid, HCl.
  a. Would a given mass of hydrogen react with different masses of each chlorine isotope, if the reactions were carried out separately? Explain.
  b. Does this outcome conflict with the law of definite proportions?

38. List any two physical properties of an element that can be predicted using the periodic table.

39. Describe what is meant by the terms *group* and *period* in relation to the periodic table.

40. Classify the following elements as metals or nonmetals:
  a. Mg **b.** Si **c.** Rn **d.** Ti **e.** Ge **f.** Eu
  g. Au **h.** B **i.** Am **j.** Bi **k.** At **l.** Br

41. Are the elements in the periodic table arranged by increasing atomic mass, increasing atomic number, or alphabetically?

42. Give the chemical symbols for two elements that have chemical properties very similar to those of magnesium.

43. Isotopes of carbon always have different numbers of ________.

44. An element with chemical properties similar to those of Ne is ________.

45. Write the chemical symbols for two elements that are halogens.

46. Semiconductors are made from elements that are called ________.

47. Give two distinguishing characteristics of metals and nonmetals.

48. Which of the following sets of elements are all in the same group in the periodic table?

a. Fe, Ru, Os **b.** Rh, Pd, Ag **c.** Sn, As, S
d. Se, Te, Po **e.** N, P, O **f.** C, Si, Ge
g. Rb, Sn **h.** Mg, Ca

49. List: **a.** the noble gases; **b.** the alkali metals; **c.** the halogens; **d.** the alkaline earth metals.

50. Using the periodic table, give the number of protons and neutrons in the nucleus of each of the following atoms:

a. ^{15}N **b.** ^{3}H **c.** ^{207}Pb **d.** ^{151}Eu **e.** ^{107}Ag **f.** ^{109}Ag

51. Identify each of the following elements by referring to the periodic table:

a. an element that has chemical properties similar to those of sulfur
b. the halogen in the third period
c. the alkaline earth metal in the second period
d. the alkali metal in the fourth period
e. the noble gas in the seventh period

52. The element with atomic number 22 forms crystals that melt at 1668°C, and the liquid boils at 3313°C. The crystals are hard, conduct heat and electricity, can be drawn into thin wires, and emit electrons when exposed to light. On the basis of these properties, classify the element as a metal or nonmetal. Which element is it?

53. If element 18 is an inert gas, in what groups would you expect to find elements 17 and 19?

Answers to Practice Exercises

3.1 80 g of hydrogen

3.2

		H/N
ammonia	●●●●	3/1
hydrazine	●●●●●●	2/1

3.3 29 electrons, 29 protons, 34 neutrons

3.4 $^{198}_{80}Hg$ $^{200}_{80}Hg$ $^{204}_{80}Hg$

3.5 **a.** Notations 1, 2, and 3 represent isotopes of sulfur.

b. Isotopes 1 (sulfur) and 4 (phosphorus) have the same mass number (16).

c. Isotopes 2 and 4 have the same number of neutrons (17); 1 and 5 (silicon) also have the same number of neutrons (16).

3.6 magnesium (Mg) and calcium (Ca), which are both in Group IIA

Chapter 4

Atoms and Chemical Bonding

Chapter Objectives

In this chapter, you should gain an understanding of:

- The arrangement of electrons in the atoms of the different elements
- The relationship between electron configuration and position in the periodic table
- The octet rule and how it is used to explain the formation of compounds
- How to draw Lewis electron-dot structures
- How ionic, covalent, and polar covalent bonds are formed
- Why some molecules are polar
- How forces of attraction operate between molecules

The brilliant colors seen in fireworks displays are the result of the electron configurations of the atoms of certain elements. When compounds containing these elements are heated, they produce the colorful emissions we see.

ONLINE FEATURES

www.jbpub.com/chemistry

- Chemistry and the Environment
- eLearning
- Chemistry in the News
- Special Topics
- Research & Reference Links
- Ask the Authors

THE NATURALLY OCCURRING ELEMENTS are the building blocks of which everything on earth is constituted. Of these elements, only the noble gases exist as free atoms. Atoms of all other elements join (or bond) with one another to form the billions of substances that make up both the living and the nonliving parts of our world. The forces that hold atoms together are called **chemical bonds.**

Why do atoms have a tendency to bond with other atoms to form distinctive units, units that may range in size from two atoms to thousands of atoms? The answer lies in the **electron configuration** of each individual atom—that is, the way the electrons are arranged around the nucleus of the atom. By forming bonds, individual atoms achieve a more stable configuration.

In this chapter, we first examine the way electrons are arranged in the atoms of different elements and how these arrangements determine the properties of the elements and explain their positions in the periodic table. We then study the various types of chemical bonds that hold atoms together and learn how electron configurations account for the various attractive forces responsible for the formation of the numerous substances that exist in the world. We concentrate on bonds formed between atoms of representative elements (see Chapter 3), those elements that make up the majority of naturally occurring compounds.

4.1 The Electron Configuration of Atoms

Until the structure of the atom was elucidated, the theoretical basis for the orderly arrangement of the elements in the periodic table was not understood. Scientists could not explain why the properties of certain elements are repeated in a periodic manner. Why, for example, are lithium (Li), sodium (Na), and potassium (K) in Group IA so similar to each other but so different from the halogens (fluorine, chlorine, bromine, and iodine) in Group VIIA and from the unreactive noble gases (helium, neon, argon, and krypton) in Group VIIIA?

By 1911, it was known that an atom consists of a nucleus of protons and neutrons surrounded by a cloud of rapidly moving electrons. However, the way in which the electrons were arranged about the nucleus remained a mystery. Scientists knew that, reading from left to right across a row of the periodic table, each successive element had one more electron in its atoms than did the element before it. From this, they deduced that electron configuration must be the key to understanding the chemical behavior of the elements.

In 1913, the Danish physicist Niels Bohr (1885–1962) **(Fig. 4.1)** proposed the first useful model to explain the electron configuration of the atom. His concept was based on well-established experimental evidence. Before you can understand Bohr's model, we need to consider the information available to him concerning the nature of light and the emission of light by different elements.

FIGURE 4.1

In the 1920s and 1930s, Neils Bohr (1885–1962) headed the prestigious Institute of Theoretical Physics in Copenhagen, Denmark. To avoid imprisonment by the Nazis, who occupied Denmark in 1940, Bohr escaped to Sweden and then to the United States, where he worked with other noted physicists on the development of the atomic bomb.

Continuous and Line Spectra

When white light from an incandescent lightbulb is passed through a glass prism, the light is separated into a **continuous spectrum** of all the visible colors: violet, blue, green, yellow, orange, and red. The colors merge smoothly into one another in an unbroken band **(Fig. 4.2a)**. The reason we see a rainbow in the sky when the sun reappears after a shower is that the falling raindrops act as prisms and disperse the sun-

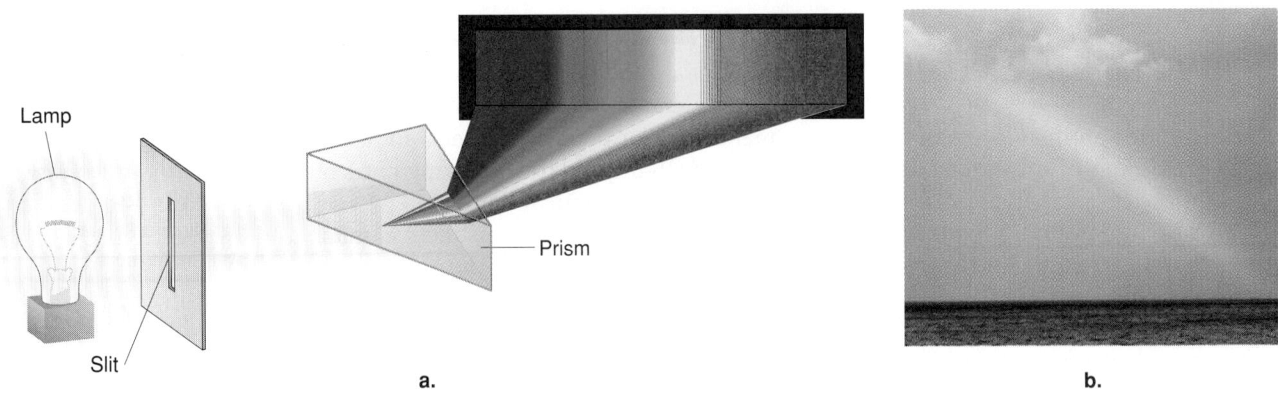

FIGURE 4.2

(a) A continuous spectrum is formed when a narrow beam of white light (sunlight or light from an incandescent bulb) is passed through a glass prism, which separates the light into its component colors. Adjacent colors merge into one another in an unbroken band. (b) A rainbow is a continuous spectrum. When sunlight passes through raindrops, the drops act as prisms, and each wavelength of light is bent at a different angle, forming the rainbow.

light. The different colors produced by a prism or by raindrops represent light with different amounts of radiant energy, with the shorter-wavelength violet end of the visible spectrum having more energy than the longer-wavelength red end (Chapter 1). A rainbow is a continuous spectrum **(Fig. 4.2b)**.

If, instead of white light, the light from a gas-discharge tube containing hydrogen, or another element, is passed through a prism and focused onto a photographic film, a series of lines separated by black spaces is seen **(Fig. 4.3a)**. This type of spectrum is called a *discontinuous spectrum*, or a **line spectrum.** The pattern of the lines produced is unique for each element and can be used to identify the element. **Figure 4.3b** shows the visible portions of the characteristic line spectra of hydrogen (H), sodium (Na), and

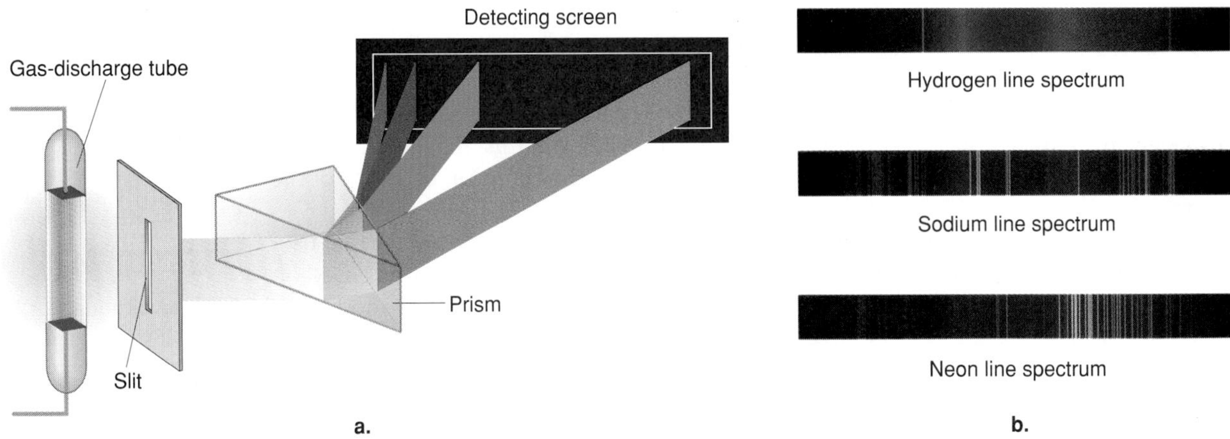

FIGURE 4.3

(a) A line spectrum (or discontinuous spectrum) is formed when light emitted from a gas-discharge tube containing hydrogen is passed through a prism. Lines in the visible region are colored. (b) Line spectra of visible colors emitted by hydrogen, sodium, and neon. Each element has its own characteristic line spectrum, which differs from that of any other element. (A continuous spectrum is shown for comparison.)

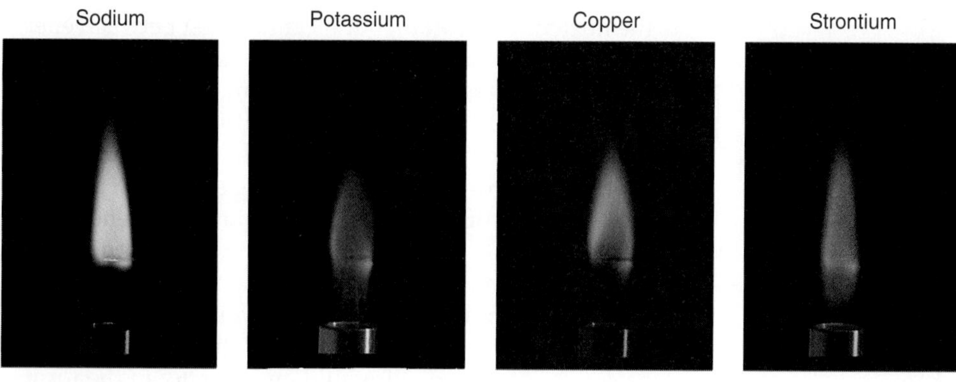

FIGURE 4.4

When heated in a flame, compounds of certain elements impart characteristic colors to the flame. The color identifies the element.

The flame test can help identify an unknown compound. For example, if a compound emits a brilliant red color when heated in a flame, it is a strontium salt.

neon (Ne). Other lines that cannot be seen occur in the ultraviolet and infrared parts of the spectra. By observing the line spectra of light coming from stars and planets, scientists have been able to identify elements in these distant bodies.

Line spectra are also produced if the light from a flame in which a metallic compound is heated is passed through a prism. In some cases, an element in the compound imparts a characteristic color to the flame itself. For example, sodium compounds give a persistent bright yellow flame, potassium compounds give lavender, copper compounds give blue-green, and strontium compounds give red **(Fig. 4.4)**. The presence of these and other compounds is responsible for the colors seen in fireworks displays.

According to Bohr, electrons can exist only in certain specified energy states.

The Bohr Model of the Atom

As one requirement, a successful model of the atom had to be able to explain and predict the line spectra of different elements. Bohr's proposed model appeared to meet these requirements. He visualized the electrons in an atom as orbiting around the nucleus in much the same way as the planets revolve around the sun. According to his model, each orbit was associated with a definite **energy level.**

The analogy of a bookcase can be used to clarify Bohr's model **(Fig. 4.5)**. A book can be placed on any shelf in a bookcase, but it cannot be placed between the shelves. In a similar way, Bohr suggested, an electron can be in one energy level or another around

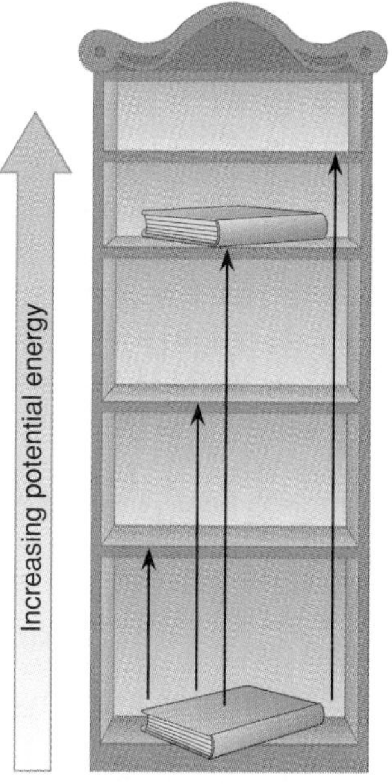

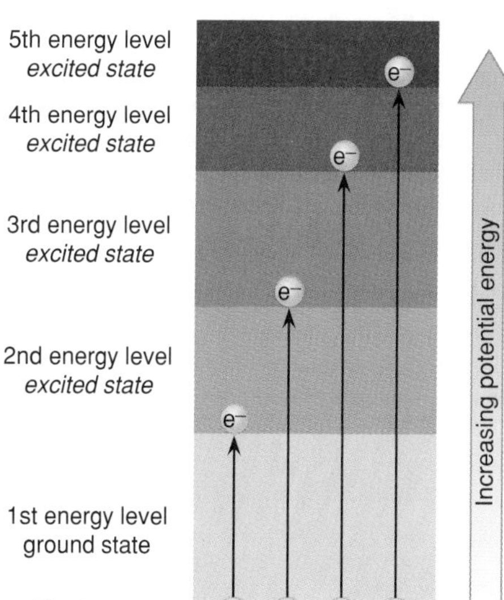

FIGURE 4.5

A bookcase provides an analogy to Bohr's model of the arrangement of electrons in an atom. Just as books can rest on any shelf in a bookcase but cannot rest between shelves, electrons are restricted to certain permitted orbits, or energy levels around the nucleus. If a book is moved to a higher shelf, it acquires potential energy. Similarly, if an electron is excited so that it moves to a higher energy level, its energy increases by a definite amount called a quantum.

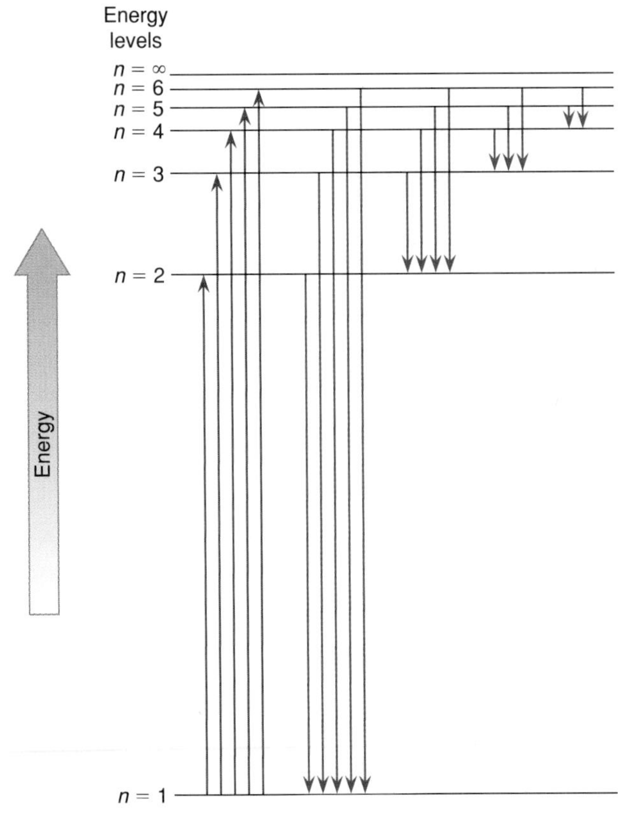

FIGURE 4.6

Upward-pointing arrows (red) show transitions of excited-state electrons from allowed energy levels to higher ones. When the electrons return to their orginal energy levels (downward-pointing arrows, blue), they produce line spectra. Some of the lines are in the visible part of the spectrum; others are in the ultraviolet and infrared regions.

the nucleus of an atom, but it cannot exist between energy levels. Further, a book can be moved from a lower shelf to a higher one, but a certain amount of energy must be expended to do this. Similarly, if sufficient energy is supplied to an electron, the electron can move from a lower energy level to a higher one. Bohr suggested that a definite, fixed amount of energy, called a **quantum** of energy, was needed to excite an electron so that it would jump from one energy level to a higher one. In going to the higher energy level, the electron—like the book—would acquire potential energy (energy of position). When the electron dropped back to a lower level, it would emit energy in the form of light—an experimental fact that could be recorded as a line spectrum.

Bohr explained the line spectrum of hydrogen as follows: When electricity was passed through a gas-discharge tube containing hydrogen, the hydrogen atoms became energized. The single electron in each individual atom jumped from its original energy level—the one closest to the nucleus—to a higher energy level, an **excited state**. Depending on their degree of excitation, different electrons jumped to different energy levels. Some reached the second energy level; other, more highly excited ones reached the fifth or sixth energy levels **(Fig. 4.6)**. The characteristic line spectrum of hydrogen is created when the excited-state electrons drop back down to lower energy levels in a series of "jumps," eventually returning to the lowest level, the **ground state.** In each transition—from level 5 to level 1, from level 3 to level 2, from level 6 to level 3, and so on—a definite amount of light energy is released. Because the energy of light is related to the wavelength of the light, which in the visible range is related to the color of the light, those transitions that produce energy in the form of visible light are seen as lines in the spectrum (refer to Fig. 4.3). On the basis of this model, Bohr calculated the expected positions of the lines for hydrogen. His calculated values agreed almost exactly with the observed values, and his model was generally accepted. Bohr was awarded the Nobel Prize in physics in 1922.

Short-wavelength violet light has more energy than longer-wavelength red light.

Building Atoms with the Bohr Model

Bohr visualized the single electron in the hydrogen atom as being at a particular energy level, moving in a circular path at a fixed distance from the nucleus. Electrons in the atoms of heavier elements were seen as circling around the nucleus in a series of more distant paths, at higher energy levels, according to certain rules. Bohr deduced that only a fixed number of electrons could be accommodated in any one energy level, and he calculated that this number was given by the formula $2n^2$, where n is equal to the number of the energy level. For the first energy level—the lowest level, or ground state—n equals 1, and the maximum number of electrons allowed is $2(1)^2$, or 2. For the second energy level ($n = 2$), the maximum number of electrons is $2(2)^2$, or 8. Values for the first four energy levels are given in **Table 4.1**.

Table 4.1 The Maximum Number of Electrons That Can Occupy a Particular Energy Level According to the Bohr Model

Energy level (n)	1	2	3	4
Maximum number of electrons ($2n^2$)	2	8	18	32

Let's see how, according to the Bohr model, electrons are arranged in the first few elements that follow hydrogen in the periodic table. Helium (He, atomic number 2) has two electrons, both of which can be accommodated in the first energy level. The first energy level is then filled. Lithium (Li, atomic number 3) has three electrons. The first two can go in the first energy level, but the third one must be placed in the second energy level. For the remaining seven elements in the second period—beryllium (Be, atomic number 4) through neon (Ne, atomic number 10)—two electrons are placed in the first energy level, and the remaining electrons are placed in the second energy level. With neon, the second energy level is filled. Thus, for the next element, sodium (Na, atomic number 11), the additional electron must be placed in the third energy level. The arrangement of electrons in the atoms of the first 20 elements in the periodic table, according to Bohr, is shown in the column headed "Bohr Model" in Table 4.2.

EXAMPLE 4.1 Using the Bohr model to determine electron configuration

How are the electrons arranged in a nitrogen atom, according to the Bohr model? Refer to the periodic table to find nitrogen, which is in Group VA.

Solution The atomic number of nitrogen is 7. Therefore, nitrogen has seven electrons. The first energy level can accommodate two of these electrons. The remaining five are placed in the second energy level, which can accommodate eight electrons. Representing this arrangement diagrammatically gives:

N) $2e^-$) $5e^-$

Practice Exercise 4.1

Use the Bohr model to show the arrangement of electrons in a sulfur atom. Sulfur is in Group V1A of the periodic table.

The Bohr model worked perfectly for hydrogen and worked well in many ways for the rest of the first 20 elements in the periodic table, but unfortunately it was a flawed model. It could not accurately predict the line spectrum of any element other than hydrogen, and it could not adequately explain the electron configurations of the transition elements. Therefore, it had to be discarded.

FIGURE 4.7

In 1933, Austrian scientist Erwin Schrödinger (1881–1961) was awarded the Nobel Prize in physics. Like many of his European colleagues, he was forced to flee his country when it was occupied by the Nazis, in 1938. Schrödinger had many interests besides theoretical physics. A poet and sculptor, he had a wide knowledge and love of literature and art.

The Wave-Mechanical Model of the Atom

In 1926, the Austrian physicist Erwin Schrödinger (1881–1961) (Fig. 4.7) developed the currently accepted **wave-mechanical model** of the atom. In this model, electrons are treated as both particles and waves, and complex mathematical equations are used to describe their arrangement and motion within atoms. We will consider a simplified form of Schrödinger's model.

Schrödinger used the very complex mathematical theory of quantum mechanics to develop mathematical equations that describe the motion of electrons.

Table 4.2 Electron Arrangements of the First 20 Elements in the Periodic Table

Element	Atomic Number	Bohr Model	Wave Mechanical Model
Hydrogen (H)	1	$1e^-$	$1s^1$
Helium (He)	2	$2e^-$	$1s^2$
Lithium (Li)	3	$2e^-$ $1e^-$	$1s^2 2s^1$
Beryllium (Be)	4	$2e^-$ $2e^-$	$1s^2 2s^2$
Boron (B)	5	$2e^-$ $3e^-$	$1s^2 2s^2 2p^1$
Carbon (C)	6	$2e^-$ $4e^-$	$1s^2 2s^2 2p^2$
Nitrogen (N)	7	$2e^-$ $5e^-$	$1s^2 2s^2 2p^3$
Oxygen (O)	8	$2e^-$ $6e^-$	$1s^2 2s^2 2p^4$
Fluorine (F)	9	$2e^-$ $7e^-$	$1s^2 2s^2 2p^5$
Neon (Ne)	10	$2e^-$ $8e^-$	$1s^2 2s^2 2p^6$
Sodium (Na)	11	$2e^-$ $8e^-$ $1e^-$	$1s^2 2s^2 2p^6 3s^1$
Magnesium (Mg)	12	$2e^-$ $8e^-$ $2e^-$	$1s^2 2s^2 2p^6 3s^2$
Aluminum (Al)	13	$2e^-$ $8e^-$ $3e^-$	$1s^2 2s^2 2p^6 3s^2 3p^1$
Silicon (Si)	14	$2e^-$ $8e^-$ $4e^-$	$1s^2 2s^2 2p^6 3s^2 3p^2$
Phosphorus (P)	15	$2e^-$ $8e^-$ $5e^-$	$1s^2 2s^2 2p^6 3s^2 3p^3$
Sulfur (S)	16	$2e^-$ $8e^-$ $6e^-$	$1s^2 2s^2 2p^6 3s^2 3p^4$
Chlorine (Cl)	17	$2e^-$ $8e^-$ $7e^-$	$1s^2 2s^2 2p^6 3s^2 3p^5$
Argon (Ar)	18	$2e^-$ $8e^-$ $8e^-$	$1s^2 2s^2 2p^6 3s^2 3p^6$
Potassium (K)	19	$2e^-$ $8e^-$ $8e^-$ $1e^-$	$1s^2 2s^2 2p^6 3s^2 3p^6 4s^1$
Calcium (Ca)	20	$2e^-$ $8e^-$ $8e^-$ $2e^-$	$1s^2 2s^2 2p^6 3s^2 3p^6 4s^2$

The wave-mechanical model of Schrödinger retains Bohr's idea that electrons are limited to definite energy levels, but within each **principal energy level,** it distinguishes sublevels called **orbitals.** The orbitals are identified by the letters *s*, *p*, *d*, and *f*. Each successive principal energy level, going outward from the nucleus, has one more orbital than the one before. Each orbital is described by the *number* of the energy level and the *letter* of the orbital, as shown in **Table 4.3**.

Table 4.3 Orbitals in the First Four Energy Levels in the Wave-Mechanical Model of the Atom

Principal Energy Level	Number of Orbitals	Orbitals
1	1	1*s*
2	2	2*s*2*p*
3	3	3*s*3*p*3*d*
4	4	4*s*4*p*4*d*4*f*

Table 4.4 The Maximum Numbers of Electrons Permitted in Each Orbital and in Each Principal Energy Level

Energy Level (n)	Maximum Number of Electrons in Each Orbital				Maximum Number of Electrons in Each Energy Level ($2n^2$)
	s	*p*	*d*	*f*	
1	2				2
2	2	6			$2 + 6 = 8$
3	2	6	10		$2 + 6 + 10 = 18$
4	2	6	10	14	$2 + 6 + 10 + 14 = 32$

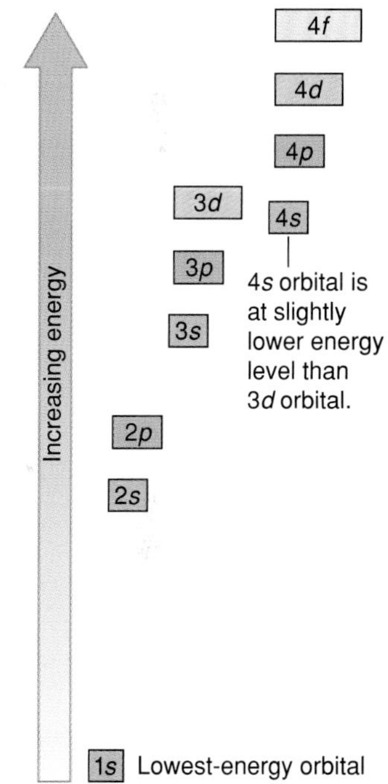

FIGURE 4.8

Energy levels of orbitals. Electrons are added to orbitals in order of increasing energy. Because the 4*s* orbital is of lower energy than the 3*d* orbital, it is filled first.

The maximum number of electrons that can be accommodated in each principal energy level of the wave-mechanical model is the same as in the Bohr model (see Table 4.1), but the electrons are divided among the different orbitals in each energy level. Each orbital, like each energy level, can hold only a certain number of electrons. The maximum permitted number of electrons for each orbital is shown in Table 4.4. The sum of the electrons in all the orbitals of an energy level equals the maximum number of electrons permitted according to Bohr's model.

Building Atoms with the Wave-Mechanical Model

Building atoms using the wave-mechanical model is the same as building atoms using the Bohr model, except that the electrons must be placed in the appropriate orbital in each principal energy level. Superscripts are used to indicate the number of electrons in each orbital. Hydrogen (H), with one electron, is designated $1s^1$; helium (He) with two electrons, is $1s^2$; lithium (Li), with three electrons, is $1s^22s^1$. The electron configurations of the first 20 elements according to the wave-mechanical model are shown in the last column in Table 4.2. With argon (Ar, atomic number 18), the 3*p* orbital is filled. You will notice that for the next two elements—potassium (K, atomic number 19) and calcium (Ca, atomic number 20)—the additional electrons are located in the 4*s* orbital and not in the 3*d* orbital, as might be expected. This location is used because the 4*s* orbital actually has slightly lower energy than the 3*d* orbital (Figure 4.8), and orbitals are always filled in order of increasing energy. The 3*d* orbital fills as electrons are added for the elements in the first row of transition elements—scandium (Sc, atomic number 21) through zinc (Zn, atomic number 30).

The order in which orbitals are filled does not have to be memorized; it can be readily determined by reference to an Aufbau diagram (Fig. 4.9). The order in which orbitals are filled is found by following the diagonal arrows from top to bottom, starting with the top left arrow: 1*s*, 2*s*, 2*p*, 3*s*, 3*p*, 4*s*, 3*d*, 4*p*, 5*s*, 4*d*, 5*p*, 6*s*, and so on.

The German word *Aufbau* means "building up."

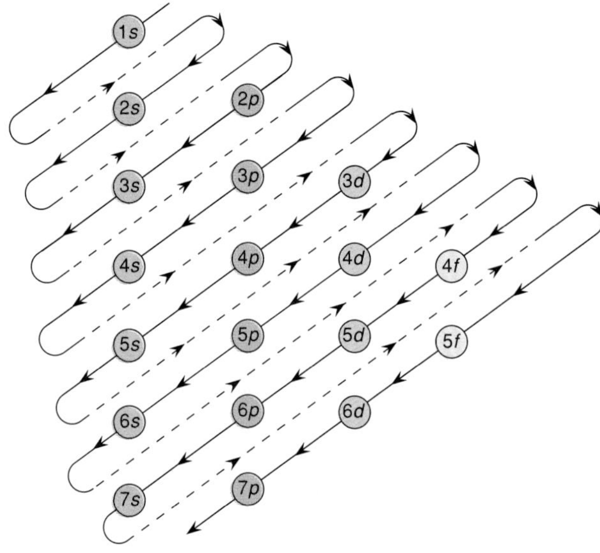

FIGURE 4.9

An Aufbau diagram can be used to determine the order in which orbitals are filled. Follow the arrows starting at 1s.

Writing electron configurations can seem difficult and confusing. It need not be so if you follow the steps given in Example 4.2.

EXAMPLE 4.2 Writing electron configurations

Write the electron configuration of sulfur (S, Group VIA, atomic number 16).

Solution

1. Each S atom has 16 electrons.
2. Refer to Figure 4.9 to determine the *order* in which the orbitals, are filled.
3. Refer to Table 4.4 to determine how many electrons can be placed in each orbital.

Order of filling orbitals	1*s*	2*s*	2*p*	3*s*	3*p*	4*s*	3*d*
Max. no. of electrons per orbital	2	2	6	2	6	2	10

The number of electrons that have been placed when orbitals 1*s* through 3*s* have been filled is

$$2 + 2 + 6 + 2 = 12$$

Four electrons remain to be placed. The next orbital to be filled, 3*p*, can hold up to six electrons. The four remaining electrons are therefore placed in the 3*p* orbital. The electron configuration of sulfur (atomic number 16) is

$$\text{S} \quad 1s^2 2s^2 2p^6 3s^2 3p^4$$

Note that the sum of the superscripts equals 16.

Practice Exercise 4.2

Write the electron configurations for:
a. sodium (Na, Group IA) **b.** argon (Ar, Group VIIIA)
c. zinc (Zn, Group IIB)

The Positions of Electrons in Orbitals

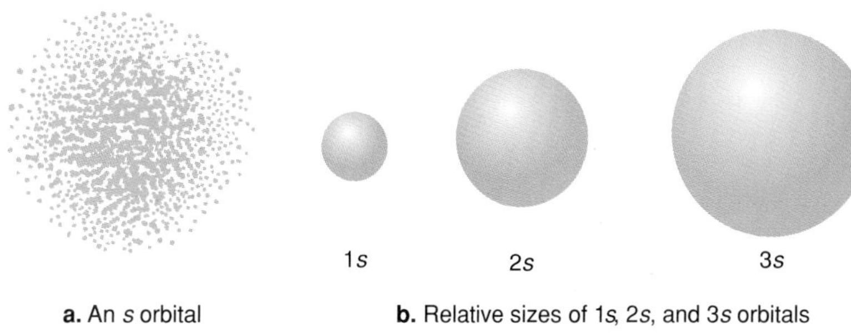

a. An *s* orbital

b. Relative sizes of 1*s*, 2*s*, and 3*s* orbitals

FIGURE 4.10

(a) Each *s* orbital is spherical. The area near the nucleus where the dots are most concentrated is the region in the sphere where the probability of finding an electron is greatest. (b) The relative sizes of the 1*s*, 2*s*, and 3*s* orbitals.

The Bohr model visualized electrons circling the nucleus in well-defined pathways, or energy levels. If this model were correct, then it should be possible to predict the exact location of an electron in its energy level at a given time, similar to predicting the position of a planet moving in space around the sun at any time. In fact, predicting the exact position of an electron cannot be done. Schrödinger, using sophisticated mathematical methods, concluded that, at any given time, it is possible to predict only the **probability** (or likelihood) of finding an electron in a certain volume of space around the nucleus.

We can use another analogy to explain this concept of probability: Assume that a bird feeder is set up in a yard and that there is one bird feeding from it. During the course of an hour, the bird will eat at the feeder, fly off—often to nearby bushes and sometimes to more distant trees—and return to the feeder many times. If the bird is hungry, it will spend most of its time at, or close to, the feeder. By observing and recording the bird's position at intervals over the course of a half-hour, for example, we can determine the area of the yard in which the bird spends 90% of its time. In other words, we can determine the 90% probability of finding the bird in a definite area of the yard at a particular time. In a somewhat similar way, Schrödinger was able to calculate the volume of space in which an electron could be expected to be found 90% of the time.

For an electron in the hydrogen 1*s* orbital, the volume in which it can probably be found 90% of the time is in the shape of a sphere. The probability of finding the electron at a particular location within the sphere is greatest in the region close to the nucleus—that is, where the dots in **Figure 4.10a** are most concentrated. The sphere represents the 1*s* orbital. The 2*s*, 3*s*, and higher-numbered s orbitals are also spherical; they can be represented by increasingly larger spheres, as shown in **Figure 4.10b**.

A *p* orbital is visualized as divided into three suborbitals, each shaped like a dumbbell. The three dumbbells are aligned in three spatial orientations designated *x*, *y*, and *z*, as shown in **Figure 4.11a**. Only two electrons can be present in any one of these *p* suborbitals, all of which are at the same energy level. As was the case for the *s* orbital, the probability of finding an electron at a particular location within a dumbbell-shaped *p* suborbital is greatest in the regions closest to the nucleus **(Fig. 4.11b)**.

The electrons in a *p* orbital of an atom are spread out as far as possible among the p_x, p_y, and p_z suborbitals. For example, carbon's two *p* electrons are located in different suborbitals—one in $2p_x$ and the other in $2p_y$ ($2p_x^1$, $2p_y^1$). Nitrogen's three *p* electrons are placed one in each of the $2p_x$, $2p_y$, and $2p_z$ suborbitals ($2p_x^1$, $2p_y^1$, $2p_z^1$). Only when we reach the next element, oxygen, which has four 2*p* electrons, are two electrons located in the same *p* suborbital ($2p_x^2$, $2p_y^1$, $2p_z^1$).

The *d* and *f* orbitals, like the *p* orbitals, are divided into suborbitals, each one holding two electrons. The shapes of the *d* and *f* suborbitals—the regions around the nucleus that define where the electrons are most likely to be located—are complex and need not be considered here.

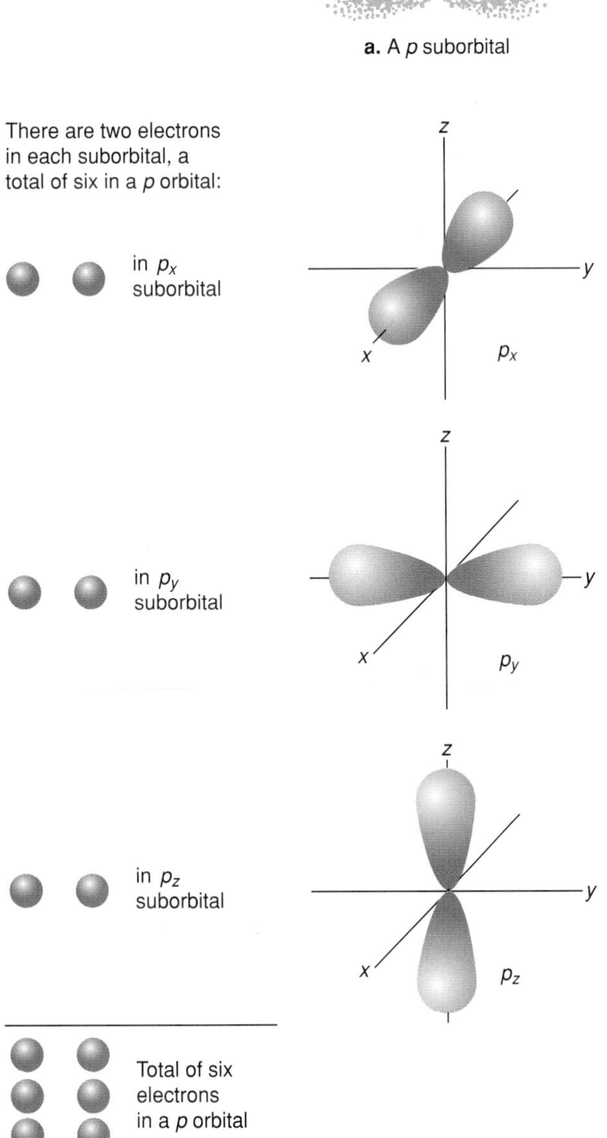

FIGURE 4.11

(a) All the *p* suborbitals are dumbbell-shaped. Each *p* suborbital can hold a maximum of 2 electrons. The areas near the nucleus where the dots are most concentrated represent the regions in the dumbbell where the probability of finding an electron is greatest. (b) The subscripts *x*, *y*, and *z* indicate the axes along which the *p* suborbitals are oriented.

4.2 Electron Configuration and the Periodic Table

Once the electron configurations of the atoms of the different elements were understood, it became evident that electron configuration was the basis for the arrangement of the elements in the periodic table. When the elements are arranged in order of increasing atomic number, elements with similar properties recur at periodic intervals because similar electron configurations recur at periodic intervals.

The correlation between the periodic table and the filling of orbitals is shown in **Figure 4.12**. The alkali and alkali earth metals (Groups IA and IIA) are formed as one and two electrons are added to the *s* orbitals. The other representative elements (Groups IIIA–VIIIA) are formed as the *p* orbitals are filled. The main block of transition elements is formed as the *d* orbitals are filled, and the two rows of inner transition elements (the lanthanides and actinides) are formed as the *f* orbitals are filled.

To help you better understand this correlation between chemical properties and electron configurations, let's look at the first few members of three groups of elements: (1) the alkali metals of Group IA, (2) the halogens of Group VIIA, and (3) the noble gases of Group VIIIA.

The alkali metals, which are very reactive, all have one electron in their highest energy level (shaded blue):

Li	$1s^2 2s^1$
Na	$1s^2 2s^2 2p^6 3s^1$
K	$1s^2 2s^2 2p^6 3s^2 3p^6 4s^1$

The halogens, which are also very reactive, all have seven electrons in their highest energy level and are therefore one electron short of filling those orbitals:

F	$1s^2 2s^2 2p^5$
Cl	$1s^2 2s^2 2p^6 3s^2 3p^5$
Br	$1s^2 2s^2 2p^6 3s^2 3p^6 4s^2 3d^{10} 4p^5$

The noble gases, which are all very inert and undergo only a few chemical reactions, have filled orbitals. Except for helium, which has a total of only two electrons, they all have eight electrons in their highest energy level.

He	$1s^2$
Ne	$1s^2 2s^2 2p^6$
Ar	$1s^2 2s^2 2p^6 3s^2 3p^6$
Kr	$1s^2 2s^2 2p^6 3s^2 3p^6 4s^2 3d^{10} 4p^6$

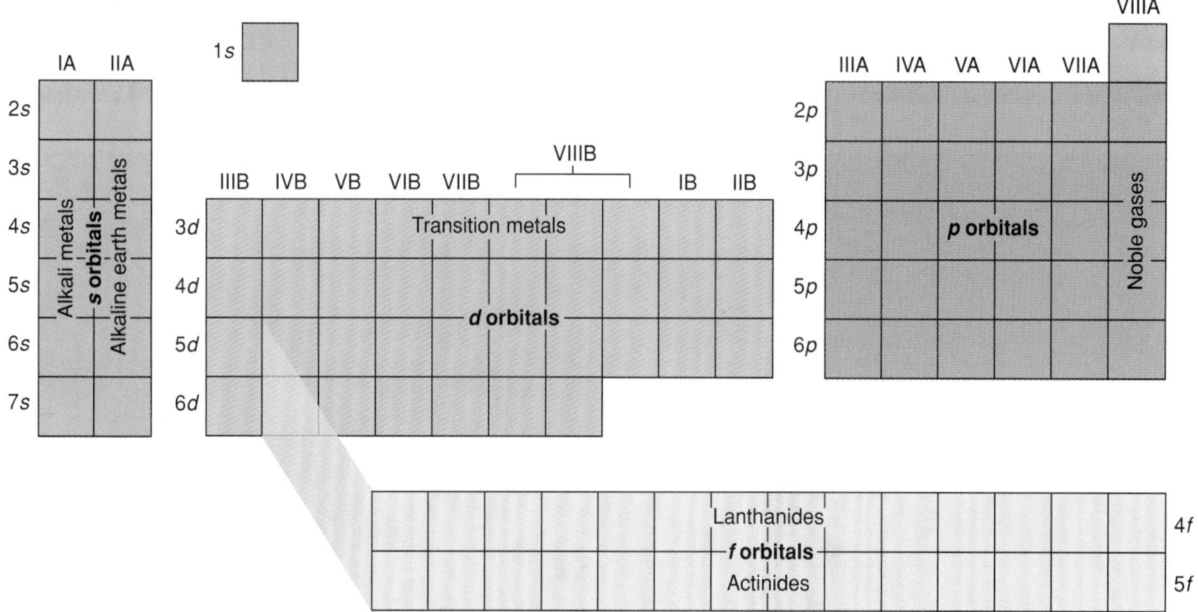

FIGURE 4.12

Correlation between electron configuration and periodic table. There are 2, 6, 10, and 14 elements, respectively, in each period in the *s*, *p*, *d*, and *f* areas of the periodic table. These numbers are the maximum numbers of electrons that each type of orbital can hold.

The above examples illustrate the repeating patterns in electron configurations. They also show us that elements with *filled* orbitals at their highest energy levels (noble gases) are very inert and unreactive, whereas elements that have *one more* electron than a filled highest energy level (alkali metals) or *one less* (halogens) are very reactive. As we shall see in the next section, the most important factor in determining how atoms bond is the number of electrons in the highest energy level.

4.3 Electron Configuration and Bonding

Valence Electrons

The electrons in an atom's *highest-numbered* energy level are known as **valence electrons.** As shown above, for potassium (K) the highest energy level is 4, and there is one electron in this level. A potassium atom, therefore, has one valence electron. The highest energy level for chlorine is 3, and it includes both a 3*s* and a 3*p* orbital. The electrons in both these orbitals are valence electrons. A chlorine atom, therefore, has seven (2 + 5) valence electrons. For bromine, the highest energy level is 4. Only electrons in this level are valence electrons; the 3*d* electrons are not counted as valence electrons. Therefore, bromine—like chlorine—has seven valence electrons in its atoms.

The number of valence electrons is the same for each element in any particular group of the periodic table.

The concept of valence electrons is simplified when each valence electron is represented by a dot, as suggested by G. N. Lewis in 1916. **Table 4.5** shows **electron-dot symbols** (or **Lewis symbols**) and electron configurations for the first 20 elements in the periodic table. The arrangement of the dots around the element symbol is arbitrary, but dots are never paired unless more than four are present (except in the case of helium, which has only 2 electrons). You will notice that the number of valence electrons (dots)

Table 4.5 Electron-Dot Symbols and Electron Configurations (with Valence Electrons Shaded) for the First 18 Elements

IA	IIA	IIIA	IVA	VA	VIA	VIIA	Noble Gases VIIIA
H•							He:
$1s^1$							$1s^2$
Li•	•Be•	•B•	•C•	:N•	:O•	:F:	:Ne:
$1s^2 2s^1$	$1s^2 2s^2$	$1s^2 2s^2 2p^1$	$1s^2 2s^2 2p^2$	$1s^2 2s^2 2p^3$	$1s^2 2s^2 2p^4$	$1s^2 2s^2 2p^5$	$1s^2 2s^2 2p^6$
Na•	•Mg•	•Al•	•Si•	:P•	:S•	:Cl:	:Ar:
$1s^2 2s^2 2p^6 3s^1$	$1s^2 2s^2 2p^6 3s^2$	$1s^2 2s^2 2p^6 3s^2 3p^1$	$1s^2 2s^2 2p^6 3s^2 3p^2$	$1s^2 2s^2 2p^6 3s^2 3p^3$	$1s^2 2s^2 2p^6 3s^2 3p^4$	$1s^2 2s^2 2p^6 3s^2 3p^5$	$1s^2 2s^2 2p^6 3s^2 3p^6$

is the same as the group number (given as a Roman numeral) in the periodic table. Thus, Group IA elements—lithium (Li), sodium (Na)—have one (I) valence electron; Group IIA elements—beryllium (Be), magnesium (Mg)—have two (II); Group VIIA elements—fluorine (F), chlorine (Cl)—have seven (VII); and so on.

EXAMPLE 4.3 Writing electron-dot symbols

Write electron-dot symbols for aluminum (Al, Group IIIA), nitrogen (N, Group VA), and bromine (Br, Group VIIA).

Solution The number of dots in the symbol is equal to the group number.

Al (Group IIIA)	N (Group VA)	Br (Group VIIA)
3 dots	5 dots	7 dots
•Al• (one dot above)	•N: (one dot above, one below)	•Br: (two dots above, two below)

Practice Exercise 4.3

Write electron-dot symbols for magnesium (Mg, Group IIA), carbon (C, Group IVA), sulfur (S, Group VIA), sodium (Na, Group IA), argon (Ar, Group VIIIA), and chlorine (Cl, Group VIIA).

You will also notice in Table 4.5 that (1) the maximum number of valence electrons for any of the elements listed is eight, and (2) with the exception of helium, which has only two electrons, all the noble gas (Group VIIIA) elements have the maximum number of eight valence electrons. The noble gases are the most unreactive of all elements and rarely form compounds. In contrast to the noble gases, elements that have either one (Group IA) or seven valence electrons (Group VIIA) are extremely reactive and readily form compounds. The correlation between the noble gas electron configuration and stability led Lewis to conclude that having eight valence electrons in some way conferred great stability and prevented the noble gases from combining with other elements. He further concluded that other elements combined with one another in order to acquire a noble gas electron configuration. On the basis of these conclusions, Lewis formulated the octet rule—or rule of eight.

The Octet Rule

The **octet rule** can be stated as follows: *In forming compounds, atoms gain, lose, or share one or more valence electrons in such a way that they achieve the electron configuration of the nearest noble gas in the periodic table.* Although there are many exceptions to the octet rule, it provides a useful framework for explaining the many basic concepts of bonding that are described in the following sections.

4.4 Ionic Bonds: Donating and Accepting Electrons

When common salt (NaCl) is formed, sodium atoms transfer their single valence electrons to chlorine atoms. The sodium atom, in losing a negatively charged electron, becomes positively charged and achieves the stable electron configuration of neon (Ne) (see Table 4.5). The chlorine atom, in gaining an electron, becomes negatively charged and achieves the electron configuration of argon (Ar) (see Table 4.5). We can write the *transfer* of electrons as follows:

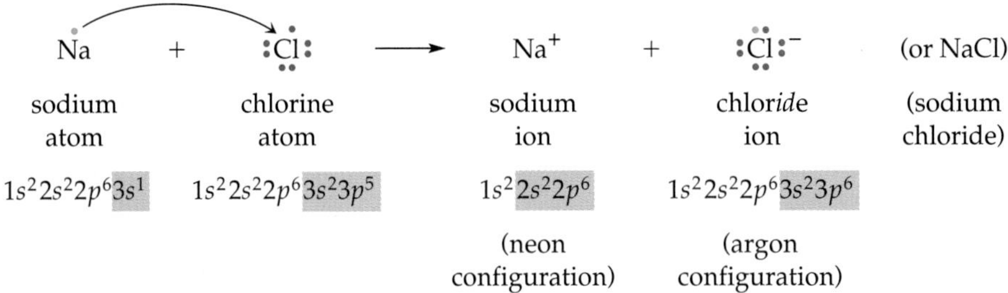

Although in the above equation, and in subsequent equations, electrons are given more than one symbol (in this case, • and •) in order to identify their source, it is very important to understand that all electrons are identical, regardless of origin. For example, the electron added to a chlorine atom to form a chloride ion is indistinguishable from the other electrons already present in the chlorine atom.

The charged atoms of sodium and chlorine are called **ions.** Positively charged ions are termed **cations** and negatively charged ions are termed **anions.**

It must be remembered that although the electron configurations of Na^+ and neon (Ne) are the same, as are those of Cl^- and argon (Ar), the ions and atoms are *not* identical. The number of protons does not change during electron transfer, and therefore the identity of the element does not change. The Na^+ ion has 11 protons and is charged; the Ne atom has 10 protons and is uncharged. Similarly, Cl^- has 17 protons and is charged, while Ar has 18 protons and is uncharged.

Because of their opposite charges, sodium ions (Na^+) and chloride ions (Cl^-) are attracted to each other. The electrostatic forces that hold the ions together are called **ionic**

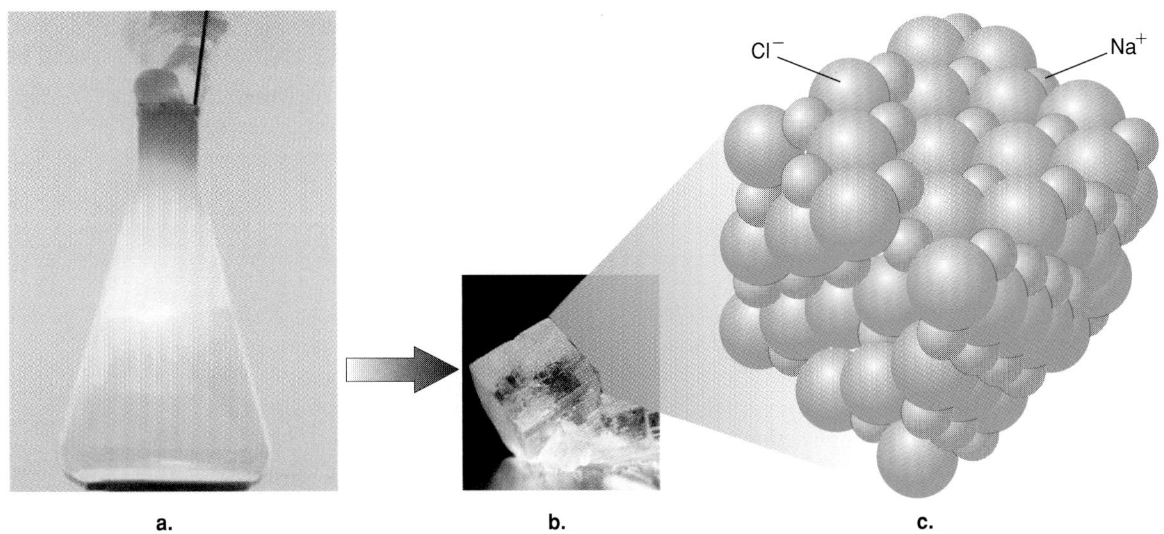

FIGURE 4.13

(a) Solid sodium metal reacts vigorously with chlorine gas to produce solid sodium chloride. (b) A crystal of sodium chloride. (c) The closely packed alternating arrangement of sodium (Na^+) and chloride (Cl^-) ions explains the cube-shaped crystals of sodium chloride (NaCl).

bonds, and the compounds that are formed, such as sodium chloride, are called ionic compounds.

Sodium and chlorine, which are both very reactive elements, react to form solid sodium chloride (NaCl) (Fig. 4.13a). Crystals of sodium chloride (table salt) are shaped like cubes (Fig. 4.13b). The crystals are made up of billions and billions of Na^+ and Cl^- ions that, because of electrostatic attractions between the oppositely charged ions, are arranged in an orderly fashion, with Na^+ ions and Cl^- ions alternating. Figure 4.13c shows how the ions are packed together in a salt crystal. The ratio of Na^+ ions to Cl^- ions in the crystal is 1:1, and the simplest formula for the ionic compound is NaCl.

A *crystal* can be defined as a solid whose internal arrangement of ions, atoms, or molecules is repeated regularly in any direction throughout its interior.

The octet rule predicts the number of atoms of one element that will react with an atom (or atoms) of another element. Let us see how magnesium (Mg), a Group IIA element, bonds with chlorine (Cl), a Group VIIA element. To achieve the stable neon electron configuration, a Mg atom must lose two electrons (see Table 4.5). We have just seen that a Cl atom needs to gain one electron to achieve a stable configuration. Therefore, to form a compound, a Mg atom transfers one electron to each of two Cl atoms, as shown below. In losing two electrons, Mg acquires a 2+ charge. The ionic compound formed ($MgCl_2$) has no net charge; the 2+ charge on Mg is balanced by the two 1− charges on Cl.

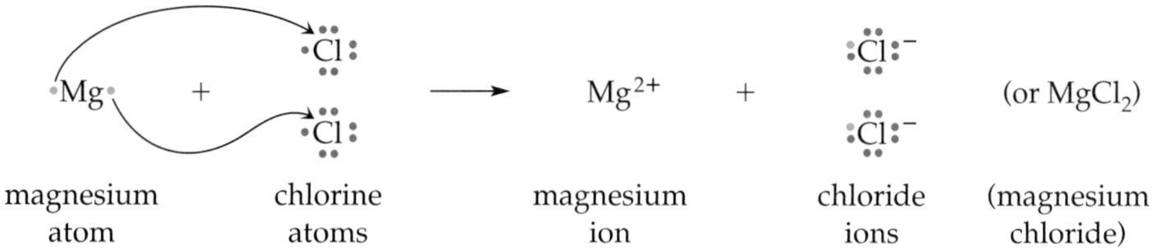

$$\cdot Mg \cdot \; + \; \begin{matrix}\cdot\ddot{\underset{..}{Cl}}: \\ \cdot\ddot{\underset{..}{Cl}}:\end{matrix} \longrightarrow Mg^{2+} \; + \; \begin{matrix}:\ddot{\underset{..}{Cl}}:^- \\ :\ddot{\underset{..}{Cl}}:^-\end{matrix} \quad (\text{or } MgCl_2)$$

magnesium atom — chlorine atoms — magnesium ion — chloride ions — (magnesium chloride)

The numbers of valence electrons in Mg and Cl atoms explain why the formula of magnesium chloride is $MgCl_2$ and not, for example, MgCl or Mg_2Cl.

EXAMPLE 4.4 Determining the formulas of ionic compounds

What is the formula of the ionic compound formed by potassium (K, Group IA) and oxygen (O, Group VIA)?

Solution

1. Refer to Table 4.5 to find the electron-dot symbols of potassium and oxygen. A K atom needs to *lose* one electron to achieve the argon electron configuration; an O atom needs to *gain* two electrons to achieve the neon electron configuration.

$$\begin{matrix}K\cdot \\ K\cdot\end{matrix} \; + \; :\dot{\underset{\cdot}{O}}: \longrightarrow \begin{matrix}K^+ \\ K^+\end{matrix} \; + \; :\ddot{\underset{..}{O}}:^{2-} \quad (K_2O)$$

potassium atoms — oxygen atom — potassium ions — oxide ion — (potassium oxide)

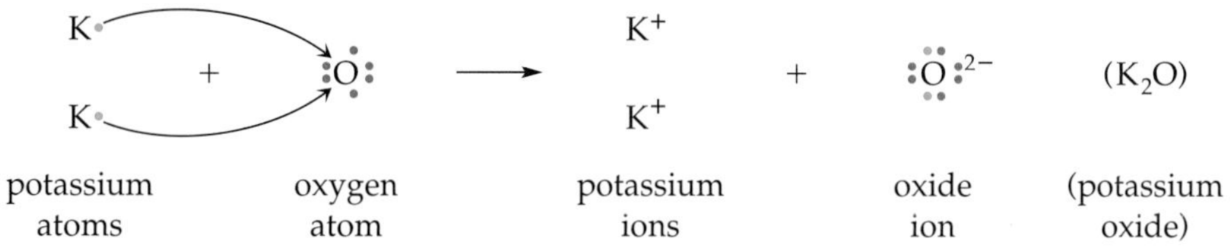

2. Since the oxygen atom gains two electrons, it ends up with a 2− charge. Each potassium atom ends up with a 1+ charge because each loses one electron.
3. The formula of potassium oxide is K_2O.

Practice Exercise 4.4

What is the formula of the ionic compound formed by each pair of elements?
a. lithium (Group IA) and sulfur (Group VIA)
b. calcium (Group IIA) and fluorine (Group VIIA)
c. magnesium (Group IIA) and oxygen (Group VIA)

Positive and negative ions always form together in a complementary process. For ions to form, at least two atoms must be present: one to donate an electron, and another to accept the donated electron. For example, a sodium atom will not lose an electron to form Na^+ unless a chlorine (or some other) atom that can accept the electron is present. As we saw in the examples given above, ionic compounds (NaCl, $MgCl_2$, and K_2O) are neutral; in each case, the total positive charge is balanced by the total negative charge.

Ionic compounds are formed primarily when metals on the left-hand side of the periodic table donate electrons to nonmetals (excluding the noble gases) on the right-hand side of the table.

Metals in Groups IA, IIA, and IIIA give up one, two, and three electrons to form ions with 1+, 2+, and 3+ charges. Examples are

Li^+	(Group IA)
Ca^{2+}	(Group IIA)
Al^{3+}	(Group IIIA)

Nonmetals in Groups VA, VIA, and VIIA gain three, two, and one electron(s) to form ions with 3−, 2−, and 1− charges. Examples are

N^{3-}	(Group VA)
S^{2-}	(Group VIA)
F^-	(Group VIIA)

Among the representative elements, the tendency to form ions is greatest for elements in Group IA, at the far left of the table, and elements in Group VIIA, at the far right. This tendency decreases going toward the center of the table and is at a minimum in Group IVA.

Transition Element Ions

All transition elements are metals.

So far, we have discussed only ionic compounds formed by representative elements. The formation of these compounds can be predicted by the octet rule. Transition metals, however, and some representative metals in the lower part of the area of the periodic table in which *p* orbitals are being filled (see Fig. 4.12), form ionic compounds in less predictable ways. For example, it is a characteristic of transition metals that they form more than one type of ion. Depending on conditions, iron forms Fe^{2+} or Fe^{3+} ions, and copper forms Cu^+ or Cu^{2+} ions.

Examples of common ions and their positions in the periodic table are given in Figure 4.14.

Naming Two-Element Ionic Compounds

Ionic compounds containing just two elements are named as follows: The name of the metallic element is given, followed by, as a separate word, the name stem of the non-

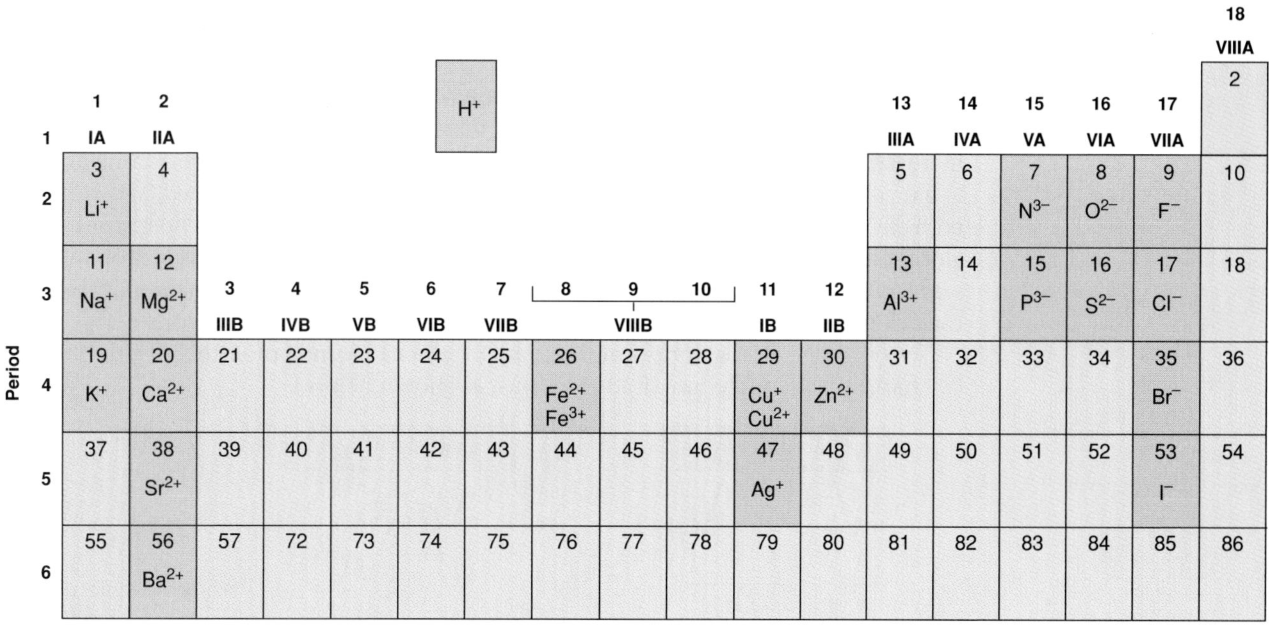

H^+

Period	1 IA	2 IIA	3 IIIB	4 IVB	5 VB	6 VIB	7 VIIB	8 VIIIB	9 VIIIB	10 VIIIB	11 IB	12 IIB	13 IIIA	14 IVA	15 VA	16 VIA	17 VIIA	18 VIIIA
1																		2
2	3 Li^+	4											5	6	7 N^{3-}	8 O^{2-}	9 F^-	10
3	11 Na^+	12 Mg^{2+}											13 Al^{3+}	14	15 P^{3-}	16 S^{2-}	17 Cl^-	18
4	19 K^+	20 Ca^{2+}	21	22	23	24	25	26 Fe^{2+} Fe^{3+}	27	28	29 Cu^+ Cu^{2+}	30 Zn^{2+}	31	32	33	34	35 Br^-	36
5	37	38 Sr^{2+}	39	40	41	42	43	44	45	46	47 Ag^+	48	49	50	51	52	53 I^-	54
6	55	56 Ba^{2+}	57	72	73	74	75	76	77	78	79	80	81	82	83	84	85	86

FIGURE 4.14

Common ions and their positions in the periodic table.

metallic element with the suffix *-ide,* as shown in **Table 4.6**. Thus, the compound CaO is named calcium oxide, AlF_3 is aluminum fluoride, Li_2S is lithium sulfide, and Na_3N is sodium nitride.

The ionic compounds of metallic elements that form more than one ion are named by writing within parentheses and immediately following the name of the metal a Roman numeral corresponding to the magnitude of the positive charge. Thus, the chloride of Fe^{2+} ($FeCl_2$) is named iron(II) chloride, and the chloride of Fe^{3+} ($FeCl_3$) is iron(III) chloride. Similarly, CuO and Cu_2O are named copper(II) oxide and copper(I) oxide, respectively. The appropriate charge on the metal ion in the compound can always be calculated from the known charge on the nonmetal ion. For example, the oxygen ion has a 2− charge. The copper ion in CuO must, therefore, have a 2+ charge to balance oxygen's negative charge. Similarly, each copper ion in Cu_2O must have a 1+ charge.

Table 4.6 Names of Common Nonmetallic Ions

Element	Symbol	Name of Anion	Symbol
Fluorine	F	Fluoride	F^-
Chlorine	Cl	Chloride	Cl^-
Bromine	Br	Bromide	Br^-
Iodine	I	Iodide	I^-
Oxygen	O	Oxide	O^{2-}
Sulfur	S	Sulfide	S^{2-}
Nitrogen	N	Nitride	N^{3-}
Phosphorus	P	Phosphide	P^{3-}

4.5 Covalent Bonds: Sharing Electrons

In two-element ionic compounds, the atoms involved in forming the ionic bonds are metal atoms and nonmetal atoms. As we have seen, when metal atoms approach nonmetal atoms, there is a strong tendency for the metal atom to lose electrons and for the nonmetal atom to gain electrons in such a way that positive and negative ions are formed, and atoms of both elements attain a stable configuration with an octet of electrons. When two *identical* atoms come together, however, their tendencies, if any, to lose and gain electrons are the same. Neither atom is more likely than the other to transfer an electron. Identical atoms can achieve the stable noble gas electron configuration by *sharing* electrons.

Two chlorine atoms, for example, come together to form a molecule of chlorine gas by sharing their single unpaired electrons, as shown below.

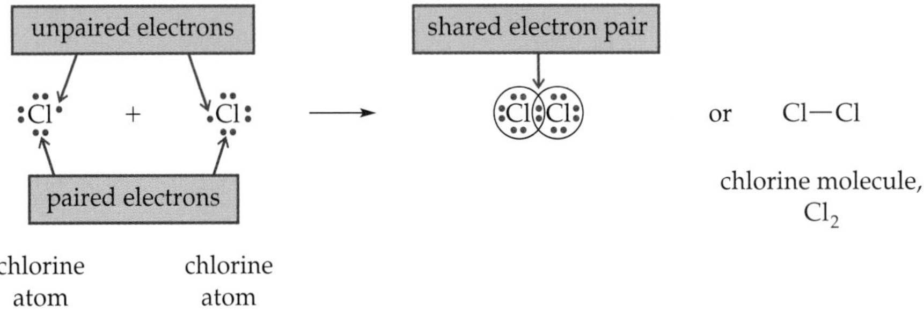

The chlorine molecule has no charge because there is no net gain or loss of electrons. The paired electrons in each chlorine atom that are not involved in bonding are called *unshared electrons*, or **nonbonding electrons.** The chlorine molecule can be represented as Cl—Cl. The dash represents the shared electrons, and the unshared electron pairs are not indicated.

A *molecule* can be defined as a unit of a pure substance in which the atoms are held together by covalent bonds.

Free chlorine atoms are very unstable, but by combining with other chlorine atoms, they achieve the stable argon electron configuration. In accordance with the octet rule, each chlorine atom in the chlorine molecule has eight electrons around it. Bonds formed in this way, by sharing one or more pairs of electrons, are called **covalent bonds.**

Hydrogen molecules, like chlorine molecules, are formed by a sharing of electrons.

H· + ·H ⟶ (H:H) or H—H

hydrogen atom, hydrogen atom, hydrogen molecule, H_2

By sharing their single electrons, the hydrogen atoms achieve the stable two-electron helium configuration.

A more realistic picture of the covalent bond between two hydrogen atoms is obtained if we remember that the single electron in each hydrogen atom is located within a spherical 1*s* orbital. When two hydrogen atoms come close together, their 1*s* orbitals merge and overlap as shown in **Figure 4.15**.

Besides chlorine and hydrogen, all other elements whose names end in -ine or -gen form diatomic (two-atom) molecules by sharing electrons. The other halogens (fluorine, bromine, and iodine) and the common gases nitrogen and oxygen, all form diatomic molecules: F_2, Br_2, I_2, N_2, and O_2.

Covalent bonds also form between atoms of *different* elements, usually between atoms of nonmetals. For example, the nonmetal carbon and the nonmetal chlorine combine by means of covalent bonds to form carbon tetrachloride, CCl_4. Carbon (Group IVA) needs four electrons to complete its octet; chlorine needs one. The number of cova-

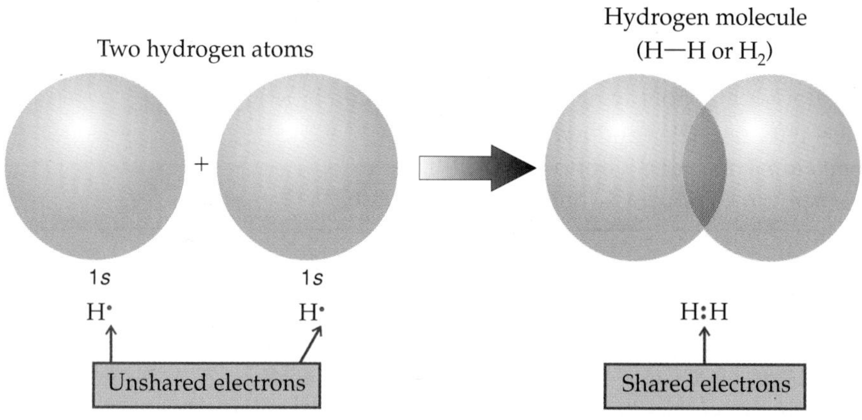

FIGURE 4.15
Overlap of 1s orbitals in the hydrogen molecule. The two shared electrons are distributed within the overlapping orbitals. The probability of finding an electron at any particular moment is greatest in the space between the nuclei.

lent bonds that an atom forms equals the number of electrons it needs to attain the noble gas electron configuration. Carbon needs four electrons and will therefore form four bonds; each chlorine atom needs one electron and will form one bond

4 :C̈l· + ·Ċ· ⟶ (electron-dot structure) or Cl—C(Cl)(Cl)—Cl

chlorine atoms — carbon atom — carbon tetrachloride molecule, CCl_4

In the CCl_4 molecule, the C atom and each of the four Cl atoms have eight electrons surrounding them. Thus, each atom has achieved the stable noble gas configuration.

EXAMPLE 4.5 Writing electron-dot structures

Write an electron-dot structure for the compound that results when nitrogen (N) and hydrogen (H) react to form covalent bonds.

Solution

1. Determine the numbers of electrons that nitrogen and hydrogen need to attain a stable noble gas configuration.
2. Nitrogen (Group VA) has five valence electrons (see Table 4.5) and needs three more to complete its octet. A nitrogen atom will, therefore, form three covalent bonds.
3. Hydrogen has one valence electron and needs one more to attain the two-electron helium configuration. A hydrogen atom will, therefore, form one covalent bond.
4. Thus, three H atoms will bond covalently with one N atom to form NH_3, as shown. In each bond, the N atom shares an electron with one of the three H atoms.

·N̈· + 3 H· ⟶ (electron-dot structure) or H—N(H)—H

nitrogen atom — hydrogen atoms — ammonia, NH_3

Practice Exercise 4.5

Write the electron-dot structure and the formula for the molecule that is formed when atoms of the following pairs of elements combine:

a. silicon (Group IVA) and hydrogen
b. two atoms of iodine (Group VIIA)
c. phosphorus (Group VA) and chlorine (Group VIIA)

Multiple Covalent Bonds

The sharing of electrons is not limited to sharing one pair of electrons. Nitrogen, as we saw in Example 4.5, has five valence electrons and needs three more electrons to complete its octet. In forming N_2, the two N atoms cannot attain the stable eight-electron configuration by sharing a single pair of electrons.

:Ṅ̤· + ·Ṅ̤: ⟶ :Ṅ̤:Ṅ̤: (incorrect)

Each N atom in the above arrangement has only six electrons around it. However, the desired octet can be attained if *three* pairs of electrons are shared.

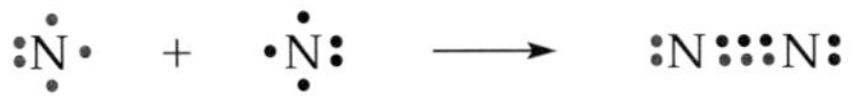

:Ṅ̤· + ·Ṅ̤: ⟶ :N⋮⋮N: or N≡N

The bond formed between the two N atoms is termed a **triple bond (≡).** Each N atom is surrounded by an octet of electrons.

In a similar way, a **double bond (=)** is formed when two atoms share *two* pairs of electrons. This occurs in carbon dioxide, CO_2. The C atom needs four electrons to complete its octet. It will, therefore, form four bonds. Each O atom needs two electrons, and each will form two bonds. Thus, in forming CO_2, the C atom is joined to the two O atoms by double bonds.

2 :Ö· + ·Ċ· ⟶ :Ö::C::Ö: or O=C=O

Although the octet rule cannot explain the formation of all covalent compounds, it accounts for a great many of them. As we will discuss in more detail in Chapter 8, it also explains the enormous number and variety of compounds formed by carbon.

EXAMPLE 4.6 Writing electron-dot structures

Write the electron-dot structure that explains the formation of ethylene, C_2H_4, from carbon (Group IVA) and hydrogen atoms.

Solution

1. Draw the electron-dot symbols for carbon and hydrogen:

·Ċ· H·

2. Each H atom needs one more electron to attain the stable two-electron helium configuration. Each hydrogen will form one bond by sharing its electron with carbon.

3. Each C atom needs four more electrons to attain an octet. Each carbon will form four bonds.

4. If the C atoms share one electron each to form a carbon-carbon single bond, and each of the four H atoms shares one electron with a C atom as shown below, then the total number of electrons around each carbon is only seven. This does not satisfy the octet rule.

$$\begin{array}{c} \text{H}\;\text{H} \\ \text{H}:\ddot{\text{C}}:\ddot{\text{C}}:\text{H} \\ \cdot\;\;\cdot \end{array} \quad \text{(incorrect)}$$

5. The only way the C atoms can attain an octet is if each of them shares two electrons with the other, forming a double bond.

$$2\ \cdot\dot{\underset{\cdot}{\text{C}}}\cdot \;+\; 4\ \text{H}\cdot \longrightarrow \begin{array}{c}\text{H}\quad\;\text{H} \\ \text{C}::\text{C} \\ \text{H}\quad\;\text{H}\end{array} \quad \text{or} \quad \begin{array}{c}\text{H}\quad\quad\text{H} \\ \text{C}=\text{C} \\ \text{H}\quad\quad\text{H}\end{array}$$

ethylene

Practice Exercise 4.6

Write electron-dot structures for each of the following (carbon is in Group IVA, oxygen in Group VIA, and nitrogen in Group VA):

a. carbon monoxide, CO **b.** nitrous oxide, N_2O (nitrogen-nitrogen-oxygen)

4.6 Polar Covalent Bonds

In molecules like H_2 and Cl_2, where both atoms are identical, the electron pair forming the bond is shared equally between the two atoms. The bond is a nonpolar covalent bond. When two *different* atoms are covalently bonded, however, the bonding electrons are not shared equally but are shifted more toward one atom than the other. This occurs because one of the atoms attracts electrons more strongly to itself than the other.

Electronegativity

The ability to attract bonding electrons can be expressed in terms of electronegativity, a concept that was introduced by Linus Pauling (1901–1994) in 1934. **Electronegativity** can be defined as the relative tendency of an atom in a molecule to attract a shared pair of electrons in a bond to itself. Electronegativity values for representative elements are shown in Figure 4.16. These values are relative and are based on an arbitrary value of 4.0 for the most electronegative element, fluorine (F). The higher the electronegativity value for an element, the greater is the ability of an atom of that element to attract electrons to itself.

EXPLORATIONS considers the long and colorful career of Linus Pauling (see pp. 108–109).

With one or two exceptions, electronegativity increases from left to right across the periodic table and decreases from top to bottom down the table. The most electronegative element (fluorine) is located in the top right-hand corner of the periodic table. Other elements with high electronegativity values (3.0 or greater) are nitrogen and oxygen, located just to the left of fluorine, and chlorine, which is located just below it.

Let's consider the hydrogen chloride molecule, HCl. To achieve a stable noble gas electron configuration, the H atom and the Cl atom both need one more electron; therefore, they share an electron as shown below. But Cl (electronegativity, 3.0) attracts electrons more strongly than does H (electronegativity, 2.2), and therefore the shared pair of electrons is drawn toward the Cl atom. As a result, the chlorine end of the molecule

Electronegativity increases →

Electronegativity decreases ↓

1 IA	2 IIA		13 IIIA	14 IVA	15 VA	16 VIA	17 VIIA	18 VIIIA
1 H 2.2								2 He —
3 Li 1.0	4 Be 1.6		5 B 1.8	6 C 2.5	7 N 3.0	8 O 3.4	9 F 4.0	10 Ne —
11 Na 0.9	12 Mg 1.3		13 Al 1.6	14 Si 1.9	15 P 2.2	16 S 2.6	17 Cl 3.2	18 Ar —
19 K 0.8	20 Ca 1.0		31 Ga 1.8	32 Ge 2.0	33 As 2.2	34 Se 2.6	35 Br 3.0	36 Kr —
37 Rb 0.8	38 Sr 0.9	Transition elements	49 In 1.8	50 Sn 1.8	51 Sb 2.0	52 Te 2.1	53 I 2.7	54 Xe —
55 Cs 0.8	56 Ba 0.9		81 Tl 2.0	82 Pb 2.3	83 Bi 2.0	84 Po 2.0	85 At 2.2	86 Rn —
87 Fr 0.8	88 Ra 0.9							

FIGURE 4.16

Electronegativity values for the representative elements. In general, electronegativity decreases from top to bottom in a group and increases from left to right across a period. Fluorine (electronegativity = 4) is the most electronegative element.

acquires a *partial* negative charge, and the hydrogen end acquires a corresponding partial positive charge. The result is a **polar molecule.**

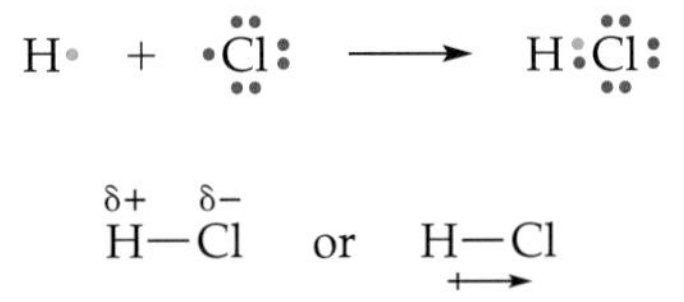

$$H\cdot + \cdot\ddot{\underset{\cdot\cdot}{Cl}}: \longrightarrow H:\ddot{\underset{\cdot\cdot}{Cl}}:$$

$$\overset{\delta+}{H}-\overset{\delta-}{Cl} \quad \text{or} \quad \underset{\longleftrightarrow}{H-Cl}$$

Partial charges are indicated by symbols using the Greek letter delta with a plus or minus sign, δ^+ and δ^-, or by an arrow with a plus sign (+) on its tail. In magnitude, partial charges are always less than the 1− and 1+ charges of electrons and protons. A molecule that has an unequal distribution of charge has a **dipole,** and bonds formed by unequal sharing of electron pairs are called **polar covalent bonds.**

The greater the difference in electronegativity between two atoms forming a bond, the more polar the bond is. In general, if the electronegativity difference between bonding atoms is greater than 1.7, the bond formed is ionic; if the difference lies between 0 and 1.7, the bond is polar covalent; and if the difference is zero, the bond is covalent.

4.7 Polyatomic Ions

In each of the compounds having more than one bond that we have discussed so far, the bonds have been of the same kind. For example, in $MgCl_2$, the two bonds between the Cl atoms and the Mg atom are both ionic bonds. In CCl_4, the four bonds joining the Cl atoms to the C atom are all polar covalent bonds. In many common substances, however, atoms are held together in part by covalent bonds and in part by ionic bonds. In these substances, atoms of more than one element bond together covalently to form **polyatomic ions** (*polyatomic* means "many-atom"); the polyatomic ions form ionic bonds with ions of opposite charge.

Common polyatomic ions are listed in Table 4.7. All except the ammonium ion are negatively charged. Polyatomic ions are stable and generally maintain their identity during chemical reactions. They are always associated with an ion (or ions) of equal but opposite charge.

Polyatomic ions are found in many familiar commercial products and in the minerals that make up the earth's crust. Compounds containing phosphate ions (PO_4^{3-}), and nitrate ions (NO_3^-) are major constituents of fertilizers—and are also potential water pollutants (Chapter 10). Calcium carbonate ($CaCO_3$) is the main ingredient in limestone and marble. Sodium bicarbonate ($NaHCO_3$) is a common antacid and plays a vital role in maintaining acid-base balance in the body. Sodium hydroxide (NaOH), an important industrial chemical, is among the ten chemicals produced in greatest quantity in the United States. Silicates, which contain the SiO_4^{4-} ion, are the most abundant minerals on earth and are the main constituents of rock, sand, and clay.

The formation of covalent and ionic bonds in sodium hydroxide (NaOH) is readily understood if we write the electron-dot symbols for each of the atoms in the formula—Na, O, and H—and then determine how electrons must be donated, accepted, or shared in order for each atom to acquire a stable noble gas electron configuration.

$$\text{Na}\cdot + :\ddot{\text{O}}\cdot + \text{H}\cdot \longrightarrow :\ddot{\underset{\cdot\cdot}{\text{O}}}:\text{H}^- + \text{Na}^+$$

hydroxide ion sodium ion
(sodium hydroxide, NaOH)

Stable configurations are achieved if the H and O atoms form a covalent bond by sharing an electron pair and the Na atom transfers its single valence electron to the O atom. As a result of the electron transfer, the OH group acquires a negative charge and the Na atom acquires a positive charge. An ionic bond is formed between the OH^- ion and the Na^+ ion. Because both the O atom and the H atom in OH^- have a noble gas electron configuration, the hydroxide ion is very stable.

Calcium carbonate ($CaCO_3$) and sodium phosphate (Na_3PO_4) are formed in a similar way.

$$\text{Ca} + \cdot\dot{\text{C}}\cdot + 3\ :\ddot{\text{O}}\cdot \longrightarrow \text{Ca}^{2+} + [\text{O}_3\text{C}]^{2-}$$

calcium ion carbonate ion
(calcium carbonate, $CaCO_3$)

A crystal of calcium carbonate and a space-filling model of the CO_3^- ion that is present in the crystal are shown in **Figure 4.17**.

$$3\ \text{Na} + \cdot\ddot{\text{P}}\cdot + 4\ :\ddot{\text{O}}\cdot \longrightarrow 3\ \text{Na}^+ + [\text{PO}_4]^{3-}$$

sodium ions phosphate ion
(sodium phosphate, Na_3PO_4)

Table 4.7 Some Common Polyatomic Ions

Ammonium	NH_4^+
Hydroxide	$OH-$
Nitrate	NO_3^-
Nitrite	NO_2^-
Bicarbonate	HCO_3^-
Acetate	$CH_3CO_2^-$
Cyanide	CN^-
Carbonate	CO_3^{2-}
Sulfate	SO_4^{2-}
Sulfite	SO_3^{2-}
Chromate	CrO_4^{2-}
Phosphate	PO_4^{3-}

The bonds within the CO_3^{2-} and PO_4^{3-} ions are covalent bonds. The bonds between the Ca^{2+} ion and the CO_3^{2-} ion and between the Na^+ ions and the PO_4^{3-} ion are ionic bonds.

a.

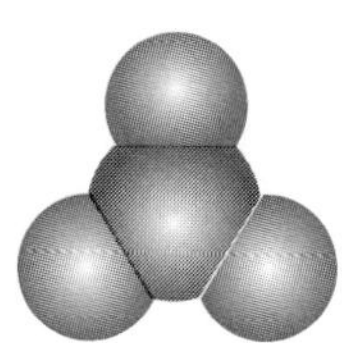

Carbonate ion (CO_3^{2-})

b.

FIGURE 4.17

(a) Calcium carbonate crystals consist of Ca^{2+} ions, and polyatomic CO_3^{2-} ions. (b) Space-filling model of the carbonate ion.

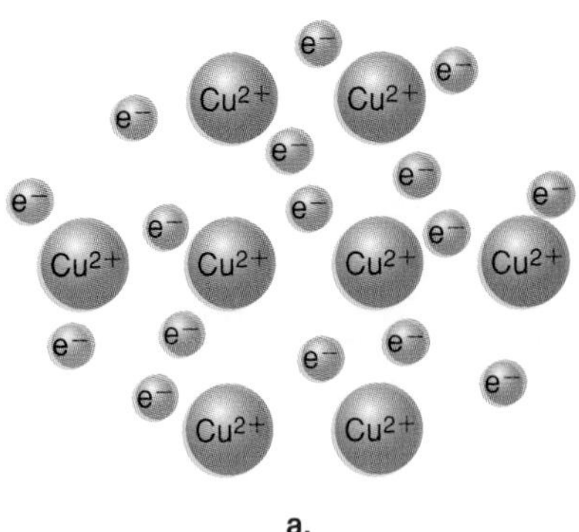

a.

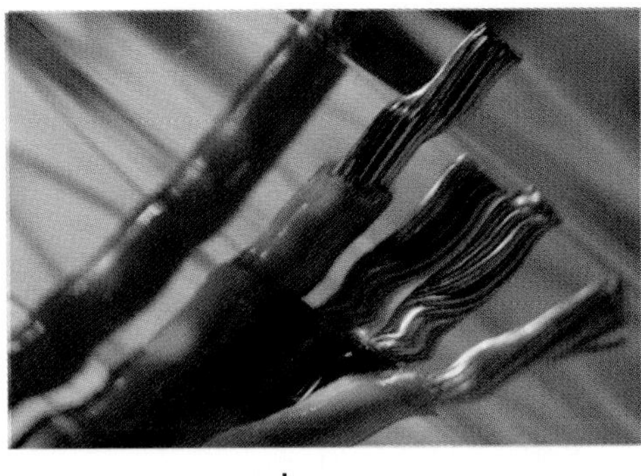
b.

FIGURE 4.18

(a) The electron sea model of metal bonding applied to copper. Valence electrons are uniformly distributed throughout the metal; they are free to wander and are not associated with particular copper nuclei. (b) Because electrons are free to move about, metals like copper conduct electricity.

4.8 Metallic Bonding

The bonds, or forces, that hold atoms together in metals are unlike any of the bonds we have described so far and are not yet completely understood. As we noted in Chapter 3, metals have many characteristic properties that distinguish them from other substances. They are, for example, good conductors of electricity and heat. Any model of **metallic bonding** must account for these and other distinctive properties.

According to one model, a metallic element in its solid state consists of a regular lattice of positively charged metal ions surrounded by a "sea" of valence electrons. The electrons are not associated with particular positive ions but wander through the lattice, as shown in Figure 4.18a for the metal copper. Because they can move freely, the electrons are able to conduct electricity and heat through the metal (Figure 4.18b).

4.9 Polar Molecules

Hydrogen chloride (HCl) was used earlier as an example of a polar molecule. As we saw, the H and Cl atoms in this molecule are joined by a single bond, and because of the unequal distribution of charges, the chlorine end of the molecule is slightly more negative than the hydrogen end. Many molecules, unlike HCl, are held together by more than one polar covalent bond. These molecules, as a whole, may be polar or nonpolar depending on the spatial arrangement of the bonds in the molecules.

Carbon dioxide is an example of a *nonpolar* molecule that is held together by two *polar* bonds. It is a linear molecule in which the two O atoms form polar covalent double bonds with the single C atom. Oxygen is more electronegative than carbon, and therefore the oxygen ends of the molecule have partial negative charges relative to the central carbon atom.

$$\overset{\longleftarrow\!\!+}{\text{O}{=}}\text{C}\overset{+\!\!\longrightarrow}{{=}\text{O}}$$

carbon dioxide,
a linear nonpolar molecule

Despite the polarity of the bonds, the carbon dioxide molecule as a whole is nonpolar because the partial negative charges on the two oxygen atoms cancel each other out. The pull on one pair of bonding electrons, pulling them toward one oxygen atom, is exactly compensated by the pull on the second pair of bonding electrons, pulling them in the opposite direction toward the other oxygen atom.

In water, the situation is different. Water is an angular, or bent, polar molecule. The partial charges associated with its two pairs of bonding electrons do not cancel each other out. The two bond polarities are of equal magnitude but are not oriented in opposite directions. Thus, the water molecule as a whole is polar.

$$2\ H\cdot \quad + \quad \cdot\ddot{\underset{\cdot\cdot}{O}}\cdot \quad \longrightarrow \quad H\!:\!\ddot{\underset{\cdot\cdot}{O}}\!:\!H \quad \text{or} \quad H\!-\!O\!-\!H$$

water,
an angular polar molecule

London forces are named after the German-American physicist who first explained them.

4.10 Intermolecular Forces

So far we have discussed only *intra*molecular forces, the forces that hold atoms together *within* molecules. In addition, **intermolecular forces** act *between* molecules and draw molecules toward one another. Although much weaker than ionic bonds or covalent bonds, intermolecular forces have a strong influence on the properties of a substance, particularly on its melting point, boiling point, and solubility in different solvents.

London forces and dipole-dipole interactions between polar molecules are also known as *van der Waals forces*, after the Dutch physicist who first explained how they affect the behavior of gases and liquids.

Intermolecular forces result from electrostatic attraction between dipoles—that is, attraction between the positive end of one molecule and the negative end of another. In order of increasing strength, the three main types of dipole-dipole interactions are (1) London forces, (2) dipole-dipole interactions between polar molecules, and (3) hydrogen bonds.

London Forces

London forces exist because of temporary spontaneous shifts in electron distribution that occur within atoms. For example, in nonpolar molecules such as H_2 and Cl_2, the electrons are, on average, as close to one nucleus in the molecule as the other, and the molecule is uniformly neutral. However, at any instant, the electrons, which are in constant motion, may be slightly more concentrated on one side of the molecule than on the other. Momentarily, one side of the molecule becomes very slightly negatively charged compared to the opposite side. An instantaneous dipole is formed and induces a similar dipole in an adjacent molecule (**Figure 4.19**). A momentary and weak electrostatic attractive force, called a **London force,** is established between the two dipoles.

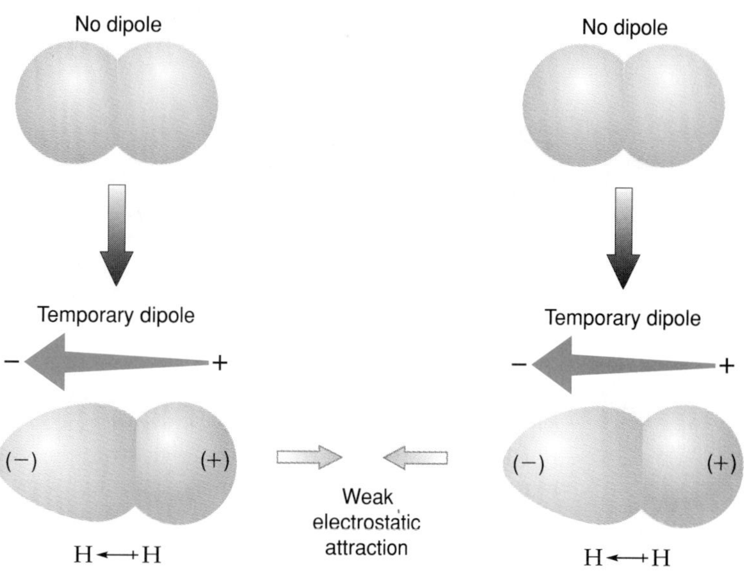

FIGURE 4.19

Temporary dipoles on two adjacent hydrogen molecules create a momentary attractive electrostatic force (a London force) between them.

Linus Pauling—The Unconventional Genius from California

Linus Pauling ranks as one of the greatest scientists of the 20th century. A brilliant, charismatic man of boundless energy, he had an uncanny genius for solving chemical problems intuitively. In a career spanning 70 years, he made important contributions in the fields of chemistry, mineralogy, physics, and biology, and he received the Nobel Prize for chemistry in 1954 in recognition of his work on chemical bonding and protein structure. His interests were not limited to science. In the 1950s, he was an outspoken and courageous advocate for banning atmospheric testing of nuclear weapons. For these activities he was both stigmatized and praised: The U.S. government labeled him a communist and traitor; the Nobel Prize committee awarded him the 1963 Peace Prize.

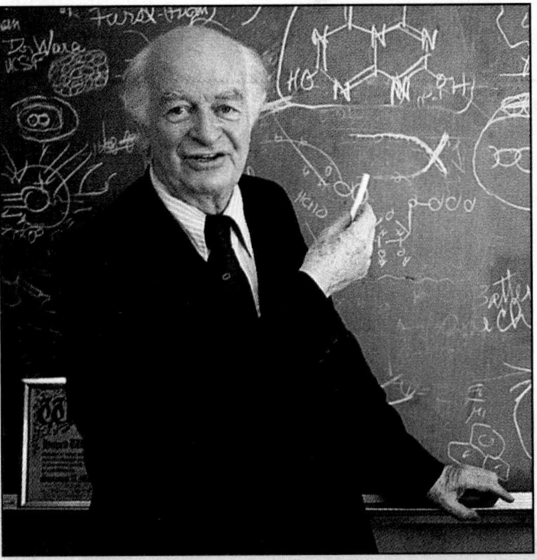

Linus Pauling

Linus Pauling was born in a small Oregon town in 1901. He grew up in great poverty in Portland, where his father, a druggist, struggled to make a living until his untimely death in 1910. Pauling was a precocious, intensely curious child with a great thirst for knowledge. From an early age, he went his own way with little regard for conventional behavior, a trait he perhaps inherited from his colorful maternal forebears. A great uncle was a spiritualist who lived in the hope that a dead Indian with whom he kept in touch would lead him to a long-lost gold mine. His aunt, Stella "Fingers" Darling, was renowned for her skill in opening safes.

During his school years, Pauling delivered milk and worked as a photographer's assistant and motion picture projectionist to help pay the family's bills. Despite his strenuous schedule and the chaotic conditions in his mother's rowdy boarding house, he excelled in his studies. He left high school at 16 but was not given a diploma because he had refused to take a civics course, telling the principal he could learn all he needed by reading on his own. A diploma was presented to him after he won his first Nobel Prize.

An Early Interest in Chemistry

Pauling began experimenting in chemistry when he was very young. As a schoolboy, he sprinkled explosive chemicals on the trolley tracks and watched them blow up when the cars went over them. And he experimented in his basement with chemicals "borrowed" from a laboratory at an abandoned smelter belonging to the Oregon Iron and Steel Company.

In 1917, over the strong objections of his mother, who wanted him to help support his two younger sisters, Pauling enrolled in the engineering school at tuition-free Oregon Agricultural College, where his exceptional talents were immediately recognized. By keeping his expenses to a minimum and working at a number of odd jobs, he managed to send money home. However, when his mother became seriously ill at the end of his sophomore year, he was forced to leave school and obtain full-time work. He was able to return a year later to complete his degree, helped by a paid teaching position. One of the students in the class he taught was Ava Helen Miller, who in 1923 became his wife. To Ava Helen's annoyance, Pauling, fearful of being accused of favoritism, always gave her lower grades than she deserved.

In 1922, again in opposition to his mother's wishes, Pauling accepted a fellowship to do graduate work at the California Institute of Technology (Caltech). In his research into the structure of minerals using X-ray crystallography, Pauling demonstrated the talent that was to characterize his career—a remarkable ability to predict structure and then adapt his "guess" to the experimental data. He received his doctorate in 1925, the same year in which his first child, Linus, Jr., was born. Two other sons and a daughter were to follow during the next 12 years.

Pauling recognized that the revolutionary concept of quantum mechanics then sweeping through Europe had profound implications for chemical bonding, the subject that had become his primary interest. He therefore accepted a one-year Guggenheim Fellowship to study in Munich, Germany with the eminent physicist Arnold Sommerfeld. His stay in Europe was productive, and soon after his return to Caltech, he published a paper on the application of quantum mechanics to chemical bonding. This paper later formed the basis of his classic book *The Nature of the Chemical Bond*.

Working on the Structure of Proteins

In the mid-1930s, Pauling turned his attention to the conflicting theories concerning the structure of proteins, the giant biological molecules that form the structural material of the human body. Pauling became convinced that the amino acid molecules, the building blocks from which proteins are composed, must be joined together in long coiled chains held together by hydrogen bonds. However, it was not until several years later that he was able to make a model to fit his theory. One day in 1948, sick in bed and bored, he drew a chain of amino acid molecules on paper and began folding and twisting the paper into a spiral resembling a coiled spring. He instantly realized that with this configuration (which he named an alpha-helix), hydrogen bonds could form between opposing amino acids at each turn of the helix. It was for this work and his work on chemical bonding that he received the 1954 Nobel Prize in chemistry. His confirmation of the helical structure of proteins led directly to James Watson and Francis Crick's discovery, in 1953, that a double helix was the basic structure of DNA.

Pauling became convinced that the amino acid molecules, the building blocks from which proteins are composed, must be joined together in long coiled chains held together by hydrogen bonds.

Activism and Nobel Peace Prize

Pauling engaged in military research during World War II (1939–1945). After the war, he became increasingly concerned about the dangers of radioactive fallout, and he led a crusade to ban atmospheric testing of nuclear weapons. These activities, coupled with his outspoken left-of-center political opinions, were viewed by the U.S. government as an effort to undermine its ability to counter a threat from the Soviet Union, and Pauling's loyalty was questioned. Although no proof was ever found, he was labeled a communist, and on two occasions he was denied a passport to travel abroad. The world community viewed his activities in a different light, and in 1963 he was awarded the Noble Peace Prize.

Pauling's involvement in political demonstrations was an embarassment to Caltech, and in 1963, after 40 years at the institute and with his best scientific work behind him, he resigned. After he moved to Stanford University in 1965, he become increasingly interested in the role of nutrients in human disorders and began his controversial work on Vitamin C.

He first advocated large doses of Vitamin C as a cure for the common cold and later, in association with the Scots physician Ewan Cameron, as a treatment for cancer. Conclusive proof of the therapeutic value of the vitamin was elusive, and Pauling's work was severely criticized by his peers. Angered by their reaction, Pauling went directly to the public and, despite the disapproval of the medical profession, began promoting Vitamin C in popular magazine articles and on TV talk shows. (It is a sad irony that his beloved wife, who had been taking massive doses of Vitamin C for several years, died of stomach cancer in 1981.)

After his retirement from Stanford University in 1973, Pauling established the Linus Pauling Institute of Science and Medicine in Menlo Park, California, where he remained active until his death in 1994.

During his long career, Pauling was often criticized for his tendency to jump to conclusions before obtaining sufficient evidence to support his theories, and his unconventional behavior and love of the limelight did not always endear him to his colleagues. But, undoubtedly, he will be remembered and honored as one of the most influential and productive scientists of the 20th century.

References: A. Serafini, *Linus Pauling: A Man and His Science* (New York: Paragon House, 1989).

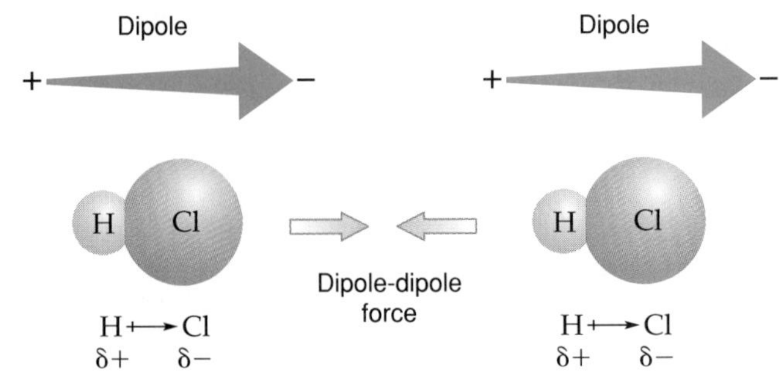

FIGURE 4.20

Electrostatic forces called dipole-dipole interactions exist between polar molecules such as HCl. The positive end of one molecule tends to orient itself adjacent to the negative end of another molecule.

London forces, which are approximately a thousand times weaker than single covalent bonds, exist between all chemical species: ions, atoms, and molecules. Even though these forces are so weak, by drawing atoms and molecules toward each other, they cause nonpolar substances such as the noble gases and the halogens to condense into liquids and freeze into solids at very low temperatures.

Dipole-Dipole Interactions between Polar Molecules

Dipole-dipole interactions occur between polar molecules, such as HCl, that have permanent dipoles. The negative end of one dipole (δ−) is attracted toward the positive end (δ+) of another dipole, and the molecules tend to orient themselves as shown in **Figure 4.20**. Dipole-dipole interactions between polar molecules are slightly stronger than London forces but are still much weaker than covalent bonds or the electrostatic forces that hold ions together in ionic compounds.

Hydrogen Bonds

Particularly strong dipole-dipole interactions exist between polar molecules that contain hydrogen atoms bonded to one of the following very electronegative elements: fluorine, oxygen, or nitrogen. These dipole-dipole interactions, which are approximately a tenth as strong as covalent bonds, are called **hydrogen bonds.** They typically occur in hydrogen fluoride (HF), water (H_2O), and ammonia (NH_3). Because fluorine, oxygen, and nitrogen are very electronegative compared to hydrogen (refer to Fig. 4.16), the F—H, O—H, and N—H bonds are very polar, and the H atoms are strongly attracted to the F, O, and N atoms in adjoining molecules **(Fig. 4.21)**.

Hydrogen bonds also occur between alcohol molecules (molecules that have an —O—H group) and between important biological molecules that have —N—H groups, including proteins, enzymes, DNA, and RNA. Hydrogen bonds have a profound effect on the properties of the substances that contain them. In DNA, the numerous hydrogen bonds are responsible for the structural stability of the genetic material. As we shall see in Chapter 10, hydrogen bonds account for many of the unusual, often unique, properties of water that make life on earth possible.

Hydrogen bonds exist *within* molecules as well as *between* molecules. The three-dimensional structures of many proteins are determined by the formation of hydrogen bonds between different parts of their large molecules. The vital role that hydrogen bonding plays in many biological processes will be discussed in more detail in Chapter 14.

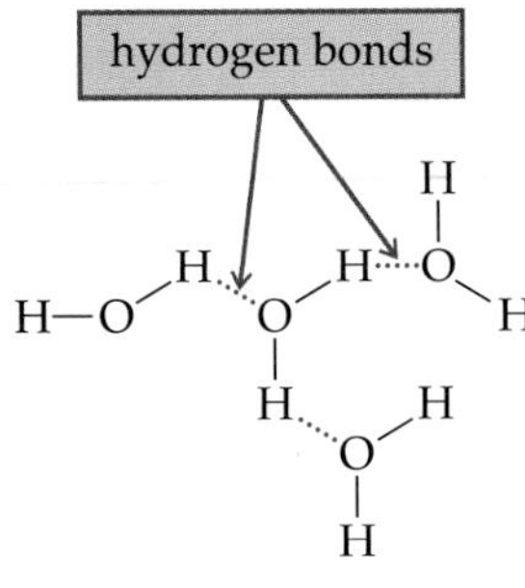

water (H_2O)

ammonia (NH_3)

hydrogen fluoride (HF)

FIGURE 4.21

Hydrogen bonding in water, ammonia, and hydrogen fluoride. Hydrogen bonds are shown as dotted lines.

Chapter Summary

1. When white light is passed through a glass prism, a continuous spectrum containing the colors of the rainbow is formed.
2. When light from a gas-discharge tube containing a particular element is passed through a prism, a discontinuous, or line, spectrum is formed. The pattern of lines in the spectrum is unique for each element.
3. According to Bohr's model of the atom, the electrons in an atom orbit around the nucleus in definite energy levels. A fixed amount of energy called a quantum is needed to raise an electron to the next higher energy level. Characteristic line spectra are created when excited electrons drop back to lower levels.
4. The maximum number of electrons that can be accommodated at a particular energy level is given by $2n^2$, where n equals the number of the energy level.
5. Bohr's model was replaced by Schrödinger's wave-mechanical model, which includes sublevels called orbitals that are identified by the letters *s*, *p*, *d*, and *f*. Going outward from the nucleus, each principal energy level has one more orbital than the one before.
6. The *s* orbitals are spherical; *p* orbitals are dumbell-shaped and are arranged in three spatial orientations designated x, y, and z.
7. The electron configurations of their atoms determine the arrangement of the elements in order of increasing atomic number in the period table. The correlation between segments of the periodic table and the filling of orbitals is as follows:
   a. alkali and alkali earth metals—*s* orbitals
   b. other representative elements—*p* orbitals
   c. main block of transition elements—*d* orbitals
   d. lanthanides and actinides—*f* orbitals
8. Elements that have filled orbitals at their highest energy levels (noble gases) are very unreactive; elements that have one more electron than a filled highest level (alkali metals) or one less (halogens) are very reactive.
9. The electrons in an atom's highest-numbered energy level are called valence electrons.
10. The octet rule states that in forming compounds, atoms gain, lose, or share one or more valence electrons in such a way that they achieve the electron configuration of the nearest noble gas in the periodic table.
11. An ionic bond is formed when a metal atom donates one or more electrons to a nonmetal atom.
12. Covalent bonds are formed when two identical atoms or two unlike atoms share one or more pairs of electrons.
13. Electronegativity increases from left to right across the periodic table and decreases from top to bottom down the table.
14. Polar covalent bonds are formed when the electrons in a bond between two different atoms are not shared equally because of differences in the atoms' electronegativity.

15. Atoms in polyatomic ions are held together by both ionic and covalent bonds.
16. Metals conduct electricity because their electrons are not associated with particular positive ions but wander throughout the metal lattice.
17. Molecules that are held together by more than one polar covalent bond may be polar or nonpolar as a whole, depending on the spatial arrangement of their bonds.
18. Intermolecular forces are (in order of increasing strength) London forces, dipole-dipole interactions between polar molecules, and hydrogen bonds.

Key Terms

anion (p. 96)
cation (p. 96)
chemical bond (p. 83)
continuous spectrum (p. 83)
covalent bond (p. 100)
dipole (p. 104)
dipole-dipole interactions (p. 108)
double bond (p. 102)
electron-dot symbols (p. 93)
electronegativity (p. 103)
electron configuration (p. 83)
energy level (p. 85)
excited state (p. 86)
ground state (p. 86)
hydrogen bonds (p. 108)
intermolecular forces (p. 107)
ion (p. 96)
ionic bond (p. 96)
Lewis symbols (p. 93)
line spectrum (p. 84)
London force (p. 107)
metallic bonding (p. 106)
nonbonding electrons (p. 100)
octet rule (p. 95)
orbitals (p. 88)
polar covalent bond (p. 104)
polar molecule (p. 104)
polyatomic ions (p. 104)
principal energy level (p. 88)
probability (p. 91)
quantum (p. 86)
triple bond (p. 102)
valence electrons (p. 93)
wave-mechanical model (p. 87)

Questions and Problems

1. **a.** When sunlight is passed through a prism, a _____ spectrum is produced.
   b. When electrons lose energy and fall from an excited state to a lower energy state, they produce _____ spectra.
2. Although a hydrogen atom has only one electron, the line spectrum of hydrogen has many lines. How does the Bohr model of the atom explain this?
3. Using the periodic table and Figure 4.9 (Aufbau diagram), write the electron configuration for:
   a. silicon **b.** krypton **c.** zinc
4. Using the periodic table and Figure 4.9 (Aufbau diagram), write the electron configuration for:
   a. arsenic **b.** sodium **c.** bromine
5. The atomic numbers of phosphorus, carbon, and potassium are 15, 6, and 19, respectively. Predict:
   a. the number of electrons in each energy level for each element
   b. the electron configuration of each element
6. Fill in the blanks:
   a. The fifth energy level can hold a maximum of _____ electrons.
   b. The designations used for the first four types of orbitals are _____.
   c. The maximum number of orbitals in each of the first four principal energy levels is ____.
   d. The maximum number of electrons that can occupy a *d* orbital is ___.
7. Fill in the numerical values that correctly complete each of the following statements.
   a. A 2*p* orbital can hold a maximum of ____ electrons.
   b. A 4*f* orbital can hold a maximum of ____ electrons.
   c. A 3*d* orbital can hold a maximum of ____ electrons.
   d. A 2*s* orbital can hold a maximum of ____ electrons.

8. Name the element with the lowest atomic number that has:

a. a filled 3*d* orbital

b. a filled 3*s* orbital

c. one 4*s* electron

d. three 3*p* electrons

9. What is wrong with each of the following attempts to write an electron configuration?

a. $1s^21p^62s^22p^6$ b. $1s^22s^22p^63s^23p^63d^{10}$

10. Does each of the following correctly represent the electron configuration of an element? Explain your answer.

a. $1s^22s^23s^2$ b. $1s^22s^22p^62d^9$

11. How many valence electrons do atoms of each of the following elements have?

a. Be b. F

c. Na d. K

e. Se f. Sn

12. Each of the following electron-dot structures represents a period 2 element. Identify each.

a. X• b. :Ẍ• (with dots above and below) c. :Ẍ• (with one dot below) d. •Ẋ•

13. Would you expect each of the following atoms to gain or lose electrons when forming ions? What is the most likely ion each will form? Explain your answer.

a. Na b. Sr c. Ba d. I

14. The following atoms can form ions. Explain why, in doing so, they either lose or gain electrons.

a. Al b. S c. B d. Cs

15. Give the octet rule in your own words.

16. How many valence electrons do the most unreactive elements have?

17. From the electron configurations listed below, write the electron-dot symbols for the elements.

a. $1s^22s^22p^5$ b. $1s^22s^22p^63s^1$ c. $1s^22s^22p^63s^23p^3$

18. What elements are represented by the following electron configurations?

a. $1s^22s^22p^63s^23p^64s^1$ b. $1s^22s^22p^63s^23p^64s^2$

19. A chloride ion results when a chlorine atom gains an electron. What other atom has the same electron configuration as the chloride ion?

20. Write the electron-dot symbols for the following elements:

a. barium b. chlorine

c. aluminum d. sulfur

21. The ions Na^+ and Mg^{2+} occur in chemical compounds, but the ions Na^{2+} and Mg^{3+} do not. Explain.

22. Find cesium, strontium, and bromine in the periodic table. Write a formula for the ionic compound formed from:

a. cesium and bromine

b. strontium and bromine

23. What is the formula of the ionic compound formed by sodium and sulfur?

24. Classify the bonds in the following compounds as ionic or covalent.

a. NaF b. MgS c. MgO d. $AlCl_3$

25. Write the formula for an ionic compound formed from the oxide ion (O^{2-}) and each of the following ions:

a. magnesium ion b. lithium ion

c. beryllium ion

26. Write the formula of the ionic compound containing each of the following and the chloride ion.

a. sodium ion b. calcium ion c. aluminum ion

27. Write the formulas for the ionic compounds formed from the following pairs of elements. Name each compound.

a. sodium and sulfur

b. potassium and oxygen

c. chlorine and lithium

28. Write the formulas of the ionic compounds that can be formed from the following pairs of elements. Name the compounds.

a. beryllium and fluorine

b. aluminum and phosphorus

c. bromine and magnesium

29. Describe each kind of bond in your own words:

a. ionic bond b. covalent bond

30. Explain what is meant by the following:

a. polar covalent bond b. polar molecule

31. Water is a polar molecule. How is its shape related to its polarity?

32. The forces of attraction between molecules are
    a. magnetic forces
    b. electrostatic forces
    c. elastic forces

33. The weakest forces of attraction between molecules are called _____.

34. What is the difference between an intramolecular force and an intermolecular force?

35. Using electron-dot symbols, show the bonding in each of the following diatomic molecules:
    a. O_2 b. N_2 c. H_2

36. One of the major components of marsh gas is methane, CH_4, which is produced by the decay of plant and animal matter. Methane is also produced in animals' intestinal tracts as a by-product of bacterial metabolism. Write the Lewis electron-dot structure for CH_4.

37. Draw electron-dot structures to illustrate the covalent bonding in each of the following molecules:
    a. Br_2 b. BrCl c. HBr

38. What is the formula of a compound formed by the combination of one carbon atom with fluorine atoms?

39. Draw electron-dot structures to show the covalent bonding in the following compounds:
    a. NCl_3 b. OF_2 c. PH_3

40. What is the difference between a single covalent bond, a double covalent bond, and a triple covalent bond?

41. Draw the Lewis electron-dot structure for C_2H_4.

42. Draw the Lewis electron-dot structures for the following molecules, each of which contains at least one double or triple bond. (The arrangement of the atoms in each molecule is shown with each formula.)

    a. N_2F_2 F N N F

    b. C_3H_4

```
H       H
  C C C
H       H
```

43. Each of the following molecules contains at least one double or triple bond. Draw the electron-dot structure of each. (The arrangement of the atoms is shown with each molecular formula.)

    a. C_2H_3N

```
  H
H C C N
  H
```

    b. C_2N_2 N C C N

44. Write formulas for the following compounds:
    a. calcium fluoride b. carbon tetrachloride
    c. magnesium bromide d. nitrogen trichloride

45. Write the formulas for the following compounds.
    a. silicon dioxide b. sodium hydroxide
    c. sulfur hexafluoride d. cesium bromide

46. What type of bond is formed between two atoms with the same electronegativity?

47. In which portion of the periodic table are the most electronegative elements located?

48. In each of the following pairs of elements, indicate which element is more electronegative.
    a. H and F b. Be and N c. N and O

49. Which element in the following pairs of elements is the more electronegative?
    a. Cl and Br b. Na and Mg c. O and S

50. Arrange each of the following sets of bonds in order of increasing polarity.
    a. H—Cl, H—O, H—F
    b. N—O, P—O, Al—O
    c. H—Cl, Br—Br, B—N
    d. P—N, S—O, Be—F

51. Write formulas (including charges) for the following polyatomic ions:
    a. nitrate ion b. hydroxide ion c. sulfate ion

52. What polyatomic ion is found in both limestone and marble?

53. Write the formulas for the following polyatomic ions. Include the charge.
    a. ammonium ion b. phosphate ion
    c. carbonate ion

54. Write the chemical formula for calcium phosphate.

55. Describe a metallic bond.

56. Potassium perchlorate has the formula $KClO_4$. What formula would you expect for lithium perbromate and sodium periodate?

Answers to Practice Exercises

4.1 S) 2e⁻) 8e⁻) 6e⁻)

4.2 a. $1s^22s^22p^63s^1$
b. $1s^22s^22p^63s^23p^6$
c. $1s^22s^22p^63s^23p^64s^23d^{10}$

4.3 Refer to Table 4.5 for the answers.

4.4 a. Li_2S b. CaF_2 c. MgO

4.5 a. H:Si:H (with H above and below Si) SiH_4
b. :I:I: I_2
c. :Cl:P:Cl: (with :Cl: below P) PCl_3

4.6 a. :C:::O: or C≡O
b. N::N::O or N=N=O

Nuclear Chemistry

The Risks and Benefits of Nuclear Radiation

Chapter Objectives

In this chapter, you should gain an understanding of:

- The three types of radiation emitted from atomic nuclei
- Nuclear reactions and how to write balanced nuclear equations
- The detection and measurement of radiation
- The relative harmfulness of each type of radiation to living things
- Beneficial uses of radioisotopes
- The process of nuclear fission

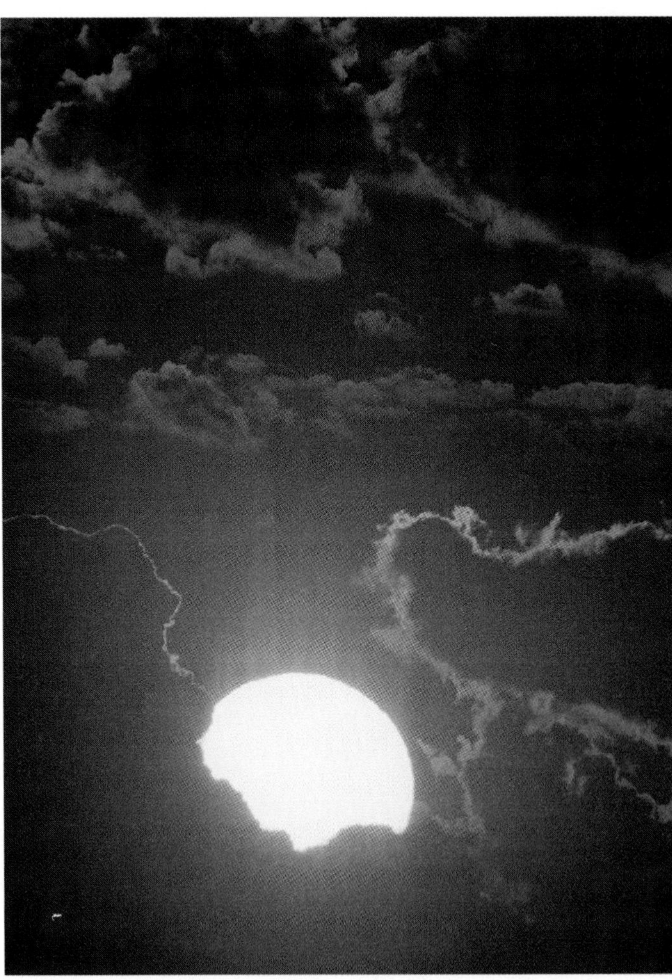

Nuclear fusion reactions in the sun release the energy that powers all life on earth.

ONLINE FEATURES

www.jbpub.com/chemistry

- Chemistry and the Environment
- eLearning
- Chemistry in the News
- Special Topics
- Research & Reference Links
- Ask the Authors

IN PREVIOUS CHAPTERS, WE EXAMINED the electron configurations of elements and the way electrons are involved in bonding atoms together to form compounds. In this chapter, we focus on the nuclei of unstable atoms that are the source of nuclear energy and nuclear radiation.

The explosion of the first atomic bomb in 1945 demonstrated the enormous destructive power of the atomic nucleus and the dangers of nuclear radiation. With the end of the Cold War and the collapse of the Soviet Union in 1992, the threat of nuclear war has diminished, but many concerns remain. Radioactive wastes produced by nuclear weapons plants and the nuclear power industry threaten the environment, and safe disposal of these wastes is a major problem. There is also fear that another catastrophic accident, like the one at Chernobyl in the Ukraine in 1985, could occur at one of Eastern Europe's aging nuclear power plants.

Nuclear radiation can destroy tissue cells and cause death, and it is particularly frightening because it cannot be seen, smelled, or felt. But nuclear radiation can also be enormously beneficial. In medicine, it is extremely useful for diagnosis and for cancer therapy. It has applications in biological research, agriculture, archeology, and industry, and many countries depend on nuclear power for most of their energy needs. Nuclear radiation is all around us in the environment. Some of it is artificially produced, but most occurs naturally. You need to understand nuclear radiation so that you can better assess both the perils and the benefits.

In this chapter, we examine the reactions within the atomic nucleus that produce nuclear radiation. We discuss how nuclear radiation was discovered and how and why it occurs. We study the properties of the different types of nuclear radiation and find out how nuclear reactions can be used to make synthetic isotopes and elements, which are unknown in nature. The events that led to the construction and detonation of the first nuclear bomb will be described. We examine both the harmful effects and the many benefits of nuclear radiation.

5.1 The Atomic Nucleus

In this chapter, we focus on the *nucleus* of the atom, the tiny dense core at the center of every atom. The nucleus of an atom is made up of protons and neutrons packed tightly together. We saw in Chapter 3 that the number of protons in the atoms of a given element—the element's *atomic number*—is always constant, but that the number of neutrons may vary. The sum of the number of protons and the number of neutrons in the nucleus is the *mass number.* Atoms of an element that have the same number of protons but differing numbers of neutrons are called *isotopes*. For example, the symbols $^{235}_{92}U$ and $^{238}_{92}U$ identify two different isotopes of uranium: $^{235}_{92}U$ has 92 protons and 143 (235 − 92) neutrons; $^{238}_{92}U$ has 92 protons and 146 (238 − 92) neutrons.

number of protons + number of neutrons = mass number

It is rarely necessary to consider the different isotopes of elements participating in ordinary chemical reactions, because chemical reactions involve the rearrangement of electrons and do not affect the nucleus. However, for **nuclear reactions**—reactions in which changes occur in the nucleus of the atom—it is essential to identify the specific isotopes taking part.

Nuclear Stability

The nuclei of the atoms of most—but not all—naturally occurring elements are very stable, despite the electrical repulsions between the positively charged protons that tend

Over 1100 unstable radioisotopes are known. About 65 occur naturally; the rest have been produced in laboratory-controlled nuclear reactions.

to pull them apart. This stability is achieved by a powerful, although not well understood, localized force within each nucleus that overcomes the repulsions and holds the nucleus firmly together. Nuclei with certain ratios of neutrons to protons and nuclei with very large total numbers of protons and neutrons, however, are not stable. They spontaneously emit high-energy radiation as a means of achieving greater stability. These unstable isotopes are called **radioisotopes** because they are *radioactive.* About 25 naturally occurring elements, including all whose atomic number is greater than 83, have one or more radioisotopes.

5.2 The Discovery of Radioactivity

X-rays, which pass through most body tissues except bone, are used to diagnose broken bones. An image of a bone produced on a photographic plate placed behind the body part being X-rayed will show if there is a break in the bone.

As we saw in Chapter 3, work with cathode-ray tubes led to the discovery of electrons and protons. It also led to another important discovery, X-rays. In 1895, the German physicist Wilhelm Roentgen (1845–1923) observed that when cathode rays struck certain metals and glass, a new type of ray was emitted. These rays behaved quite differently from cathode rays; they were not deflected by electric or magnetic fields, and they could penetrate deeply into matter, even passing through walls. Roentgen named these extraordinary new rays **X-rays.** They were later identified as a high-energy, short-wavelength form of electromagnetic radiation (Chapter 1).

Roentgen's discovery prompted many scientists, including the French physicist Henri Becquerel (1852–1908), to study these new rays. Becquerel was working with substances that become luminous after exposure to sunlight, and then continue to glow after being placed in the dark. He wondered if this phenomenon, which is called *fluorescence,* was related to X-rays. While working with a fluorescent uranium ore, he made an unexpected discovery. He inadvertently placed a piece of the ore on top of an unexposed photographic plate inside a desk drawer. Some days later, he removed the plate from the drawer and was surprised to find that, although the plate had been wrapped in dark paper to protect it from accidental exposure to light, it showed an image of the piece of uranium ore he had placed on it. Becquerel concluded that the uranium ore must have spontaneously emitted some form of radiation that had penetrated the protective wrapping and exposed the photographic plate. He had discovered another type of radiation.

EXPLORATIONS describes the life and ground-breaking career of Marie Curie in detail (see pp. 140–141).

Following this discovery, Becquerel's student, the Polish-born chemist Marie Sklodowska Curie (1867–1934) and her husband, French physicist Pierre Curie (1859–1906), began a systematic search for other substances that would spontaneously emit radiation. They found that all uranium ores, regardless of type and source, emitted radiation, and that this property, which Marie Curie called **radioactivity,** was a characteristic of the element uranium (U, atomic number 92). After a great deal of arduous work, the Curies isolated two new radioactive elements from uranium ore: polonium (Po, atomic number 84), which is 400 times more radioactive than uranium, and radium (Ra, atomic number 88), which is more than 1 million times more radioactive than uranium. Several tons of uranium ore had to be processed to obtain just 0.1 g of pure radium. In 1903, Becquerel and the Curies were jointly awarded the Nobel Prize in physics for their discovery of radioactivity.

Marie Curie was the first person to win two Nobel Prizes. She won her second prize in 1911 for the discovery of polonium and radium.

5.3 The Nature of Natural Radioactivity

Soon after Becquerel's discovery that uranium ores were radioactive, Ernest Rutherford (Chapter 3) determined that the radiation emitted by naturally occurring radioactive materials is of three distinct types. Rutherford passed a beam of such radiation between electrically charged plates onto a photographic plate, as shown in **Figure 5.1**. He found

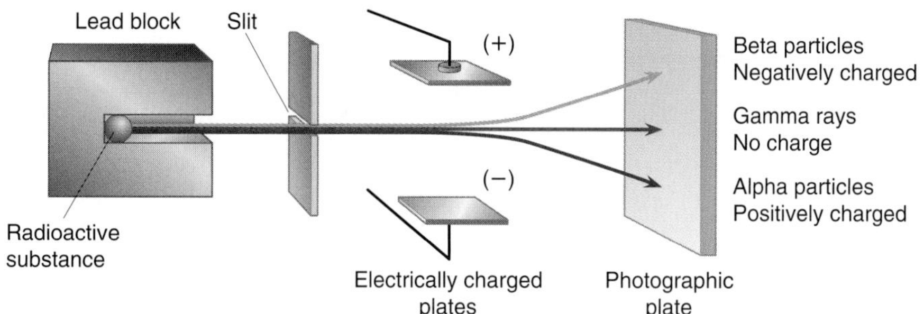

FIGURE 5.1

In an electric field, radiation from a radioactive source splits into three components: alpha particles, beta particles, and gamma rays. The positively charged alpha particles are attracted to the negatively charged plate; the negatively charged beta particles are attracted to the positively charged plate; and the uncharged gamma rays are not deflected from their original path.

Rutherford showed that alpha particles combine with electrons to form atoms of helium. He therefore concluded that an alpha particle must consist of the positively charged nucleus of a helium atom.

that the beam of radiation split into three components. One component was attracted to the negatively charged plate and was therefore positively charged; a second component was attracted to the positively charged plate and was therefore negatively charged; the third component was not deflected from its original path and could be assumed to have no charge. Rutherford named these components alpha (α), beta (β), and gamma (γ) rays, after the first three letters in the Greek alphabet. It was not until some 30 years later that these three kinds of radiation were completely characterized.

Alpha rays are, in fact, made up of particles, each consisting of two protons and two neutrons. An **alpha particle,** therefore, has a mass of 4 amu and a 2+ charge and is identical to a helium nucleus (a helium atom minus its two electrons). Alpha particles are represented as ${}^{4}_{2}\alpha$, or ${}^{4}_{2}He$.

Beta rays are also made up of particles. A **beta particle** is identical to an electron and therefore has negligible mass and a charge of 1−. It is usually represented as ${}^{0}_{-1}\beta$ or ${}^{0}_{-1}e$. Although a beta particle is identical to an electron, it does not come from the electron cloud surrounding an atomic nucleus, as might be expected. It is produced *inside* the atomic nucleus and then ejected. (We will discuss this process further in the next section.)

Gamma rays, unlike alpha and beta rays, are not made up of particles and therefore have no mass. Similar to X-rays, they are a form of high-energy electromagnetic radiation with a very short wavelength and are usually represented as ${}^{0}_{0}\gamma$. All three types of radiation have sufficient energy to break chemical bonds and thus disrupt living organisms.

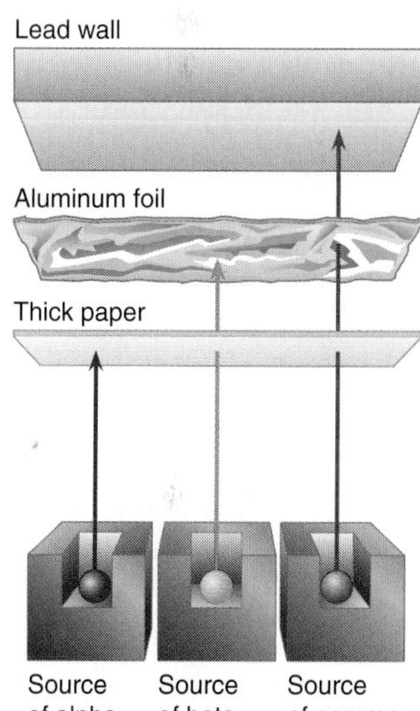

FIGURE 5.2

The penetrating abilities of alpha particles, beta particles, and gamma rays differ: Alpha particles are the least penetrating and are stopped by a thick sheet of paper or the outer layer of the skin. Beta particles pass through paper but are stopped by aluminum foil or a block of wood. Gamma rays can be stopped only by a lead wall several centimeters thick or a concrete wall several meters thick.

The Penetrating Power and Speed of the Types of Radiation

Alpha particles, beta particles, and gamma rays are emitted from radioactive nuclei at different speeds and have different penetrating powers. Alpha particles are the slowest. They are emitted at speeds approximately equal to a tenth of the speed of light and can be stopped by a sheet of paper or by the outer layer of a person's skin **(Fig. 5.2)**. Beta particles are emitted at speeds almost equal to the speed of light, and, because of their greater velocity and smaller size, their penetrating power is approximately 100 times greater than that of alpha particles. Beta particles can pass through paper and several millimeters of skin but are stopped by aluminum foil. Gamma rays are released from nuclei at the speed of light and are even more penetrating than X-rays. They pass easily into the human body and can be stopped only by several centimeters of lead or

Table 5.1 Properties of the Three Types of Radiation Emitted by Radioactive Elements

Name	Symbol	Identity	Charg	Mass (amu)	Velocity	Penetrating Power
Alpha	${}^{4}_{2}\alpha$, ${}^{4}_{2}He$	Helium nucleus	2+	4	1/10 the speed of light	Low, stopped by paper
Beta	${}^{0}_{-1}\beta$	Electron	1−	0	Close to the speed of light	Moderate, stopped by aluminum foil
Gamma	${}^{0}_{0}\gamma$	High-energy electromagnetic radiation	0	0	Speed of light (3×10^{10} cm/s)	High, stopped by several centimeters of lead

several meters of concrete. Properties of the three types of natural radiation are summarized in **Table 5.1**.

5.4 Nuclear Reactions

Both gamma-rays and X-rays are forms of electromagnetic radiation (see Chapter 1). Gamma rays have a shorter wavelength and a higher energy than X-rays.

When a radioactive isotope of an element emits an alpha or a beta particle, a nuclear reaction occurs, and the nucleus of that isotope is changed. The changes that occur during a nuclear reaction can be represented by a nuclear equation.

As examples, let's write nuclear equations to explain the radioactivity that Becquerel and the Curies observed in their studies of uranium ores. The element uranium has several radioactive isotopes, including ${}^{238}_{92}U$ (also written uranium-238), which spontaneously emit alpha particles. The notation ${}^{238}_{92}U$ indicates that the atomic number of uranium is 92 and the mass number of this particular isotope is 238. The isotope ${}^{238}_{92}U$ therefore has 92 protons and 146 (238 − 92) neutrons in each nucleus.

When a uranium-238 nucleus emits an alpha particle, ${}^{4}_{2}\alpha$, it loses 4 atomic mass units (2 protons and 2 neutrons). The resulting atomic nucleus therefore has an atomic mass number of 234 (238 − 4) and, because it has lost 2 protons, an atomic number of 90 (92 − 2). An atom with an atomic number of 90 is no longer an atom of uranium. If you refer to the periodic table, you will see that the element with atomic number 90 is thorium (Th). Thus, the spontaneous emission of an alpha particle from a uranium atom results in the formation of a completely different element. This **transmutation** of uranium to thorium is represented by the following nuclear equation:

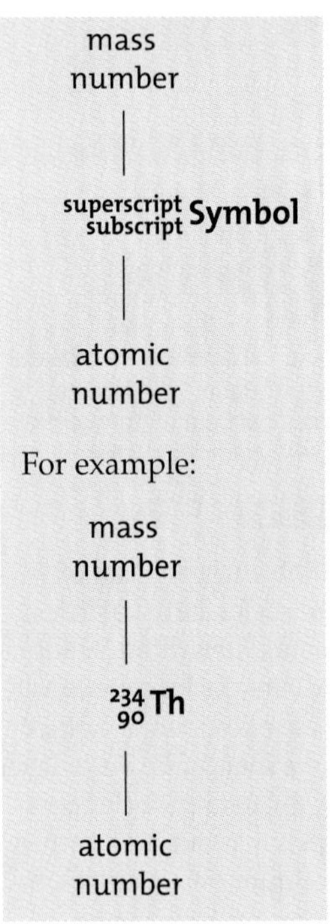

mass number = 238 — sum of mass numbers = 234 + 4 = 238

$$ {}^{238}_{92}U \longrightarrow {}^{234}_{90}Th + {}^{4}_{2}\alpha $$

atomic number = 92 — sum of atomic numbers = 90 + 2 = 92

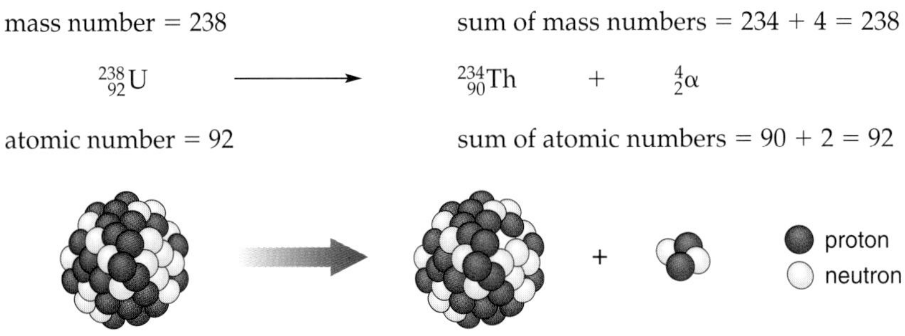

For a nuclear equation to be properly balanced, the mass number (superscript) on the left side of the equation must equal the sum of the mass numbers (superscripts) on the right side of the equation. Similarly, the atomic number (subscript) on the left must equal the sum of the atomic numbers (subscripts) on the right.

Thorium-234, the main product in the above equation, is always found in uranium-238 ore deposits. (The Curies isolated it from a uranium ore before anyone could explain exactly how it had been formed.) It, too, is radioactive but emits beta particles instead of alpha particles. It may seem improbable that a negatively charged beta particle ($_{-1}^{0}\beta$) identical to an electron can be released from a nucleus made up of positively charged protons ($_{1}^{1}p$) and uncharged neutrons ($_{0}^{1}n$), but this does occur. The accepted explanation is that a neutron in the nucleus changes into a proton and a beta particle according to the following equation:

$$_{0}^{1}n \longrightarrow {}_{1}^{1}p + {}_{-1}^{0}\beta + \text{energy}$$

Notice that this equation is balanced: The sums of the superscripts on both sides of the equation are equal ($1 = 1 + 0$); the subscripts are similarly balanced ($0 = 1 - 1$). Once formed, the beta particle is ejected from the nucleus, leaving the nucleus with an additional proton but one less neutron. Because the mass of a proton equals that of a neutron and the mass of a beta particle is essentially zero, there is no net change in the mass of the nucleus. But as a result of exchanging a neutron for a proton, the nucleus gains a 1+ charge, and a new element is formed.

Here is the balanced nuclear equation showing the emission of a beta particle from an atom of thorium-234 and the formation of a new element:

mass number (superscript) = 234 sum of superscripts = 234 + 0 = 234

$$_{90}^{234}Th \longrightarrow {}_{91}^{234}Pa + {}_{-1}^{0}\beta$$

atomic number (subscript) = 90 sum of subscripts = 91 − 1 = 90

proton
neutron
electron

Because the atomic number has changed, a new element has been formed. This time the atomic number has increased by 1. If you refer to the periodic table, you can confirm that the element with atomic number 91 is protactinium (Pa).

After emitting an alpha or a beta particle, the nuclei of naturally occurring radioisotopes still have excess energy, which they release by emitting gamma rays. Since gamma rays are waves, not particles, and have no mass and no charge, the identity of the emitting element does not change when a gamma ray is emitted. Gamma rays are not therefore included in nuclear equations. The changes in mass number and atomic number that occur when unstable nuclei emit alpha or beta particles are shown in **Table 5.2**.

When an alpha-particle ($_{2}^{4}\alpha$) is emitted, the atomic number decreases by 2 and the mass number decreases by 4.

When a beta-particle ($_{-1}^{0}\beta$) is emitted, the atomic number increases by 1; there is no change in the mass number.

New elements are never produced in a chemical reaction, but new elements are often produced in a nuclear reaction.

Table 5.2 Changes in Mass Number and Atomic Number That Occur When Radioactive Elements Decay

Type of Emission*	Symbol	Mass Number (protons plus neutrons)	Charge	New Element: Change in Mass Number	New Element: Change in Atomic Number
Alpha	$_{2}^{4}\alpha$, $_{2}^{4}He$	4	2+	Decreased by 4	Decreased by 2
Beta	$_{-1}^{0}\beta$	0	1−	No change	Increased by 1

*Gamma rays are not included because they have no mass and no charge. Therefore, when they are emitted, no new element is formed.

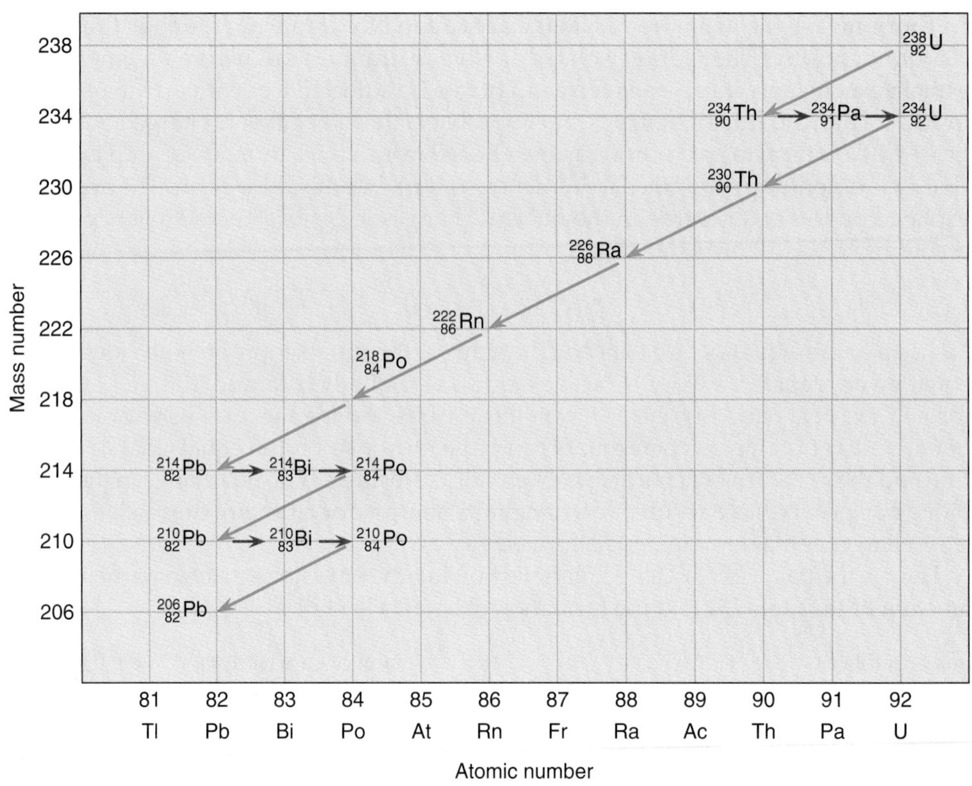

FIGURE 5.3

The uranium decay series: $^{238}_{92}U$ spontaneously decays in a series of steps until the stable element $^{206}_{82}Pb$ is formed. Each time an alpha particle is emitted, the mass number decreases by 4 and the atomic number decreases by 2 (blue arrows). Each time a beta particle is emitted, the mass number remains the same but the atomic number increases by 1 (red arrows). Note that the gas radon (Rn), which accounts for most of our everyday exposure to radiation, is formed in the series as a result of the disintegration of $^{226}_{88}Ra$.

Radioactive Decay Series

The transformation of one element into another element as a result of radioactive emissions is termed **radioactive decay.** The spontaneous decay of uranium-238 to form thorium-234 and that of thorium-234 to form proactinium-234 are the first two steps in the *uranium decay series*, which continues through a total of 14 steps to finally yield the stable isotope lead-206 **(Fig. 5.3).** Of particular interest in this decay series is radon-222 ($^{222}_{86}Rn$), the radioactive gas that (as you will learn later in this chapter) accounts for most of the potentially harmful radiation in the environment.

Besides the uranium decay series, two other naturally occurring decay series exist: the *thorium series*, which starts with thorium-232 and ends with lead-208; and the *actinium series*, which starts with uranium-235 and ends with lead-207. All the naturally occurring radioisotopes with high atomic numbers belong to one of these three decay series.

5.5 Artificial Transmutations

The transmutation reactions we have considered so far occur naturally as a result of the spontaneous emission of alpha or beta particles from unstable nuclei. **Artificial**

transmutations of one element into another element can be brought about by bombarding certain stable nuclei with alpha particles, neutrons, or other subatomic particles.

The first artificial transmutation was achieved by Rutherford in 1919. He bombarded stable nitrogen atoms with alpha particles and produced stable oxygen-17 atoms and protons ($^{1}_{1}p$). This experiment confirmed that, as predicted, protons are fundamental particles that are present in the nuclei of all atoms (Chapter 3). The nuclear equation for this reaction is as follows:

$$^{14}_{7}N \quad + \quad ^{4}_{2}\alpha \quad \longrightarrow \quad ^{17}_{8}O \quad + \quad ^{1}_{1}p$$

Notice that this equation is properly balanced: The sum of the superscripts on the left of the equation ($14 + 4 = 18$) equals the sum of the superscripts on the right ($17 + 1 = 18$). The subscripts are also balanced ($7 + 2 = 8 + 1$).

During the 1920s and 1930s, many other artificial transmutations were carried out. One of the most important was done by James Chadwick, who bombarded beryllium with alpha particles and obtained carbon atoms and uncharged particles whose mass was almost identical to the mass of a proton. This reaction, which is shown below, confirmed the existence of the neutron (Chapter 3).

$$^{9}_{4}Be \quad + \quad ^{4}_{2}\alpha \quad \longrightarrow \quad ^{12}_{6}C \quad + \quad ^{1}_{0}n$$

New Isotopes and New Elements

The isotopes of carbon and oxygen, $^{12}_{6}C$ and $^{17}_{8}O$, that were produced in the bombardment reactions carried out by Rutherford and Chadwick are stable and occur naturally. Bombardment reactions can also produce isotopes that are unstable and unknown in nature.

The first radioactive isotopes unknown in nature were made by Irene Curie (the daughter of Marie and Pierre Curie) and Frederic Joliot (her husband) in 1932. In one of their experiments, they bombarded $^{27}_{13}Al$ with alpha particles and obtained an unstable isotope of phosphorus:

$$^{27}_{13}Al \quad + \quad ^{4}_{2}\alpha \quad \longrightarrow \quad ^{30}_{15}P \quad + \quad ^{1}_{0}n$$

They found that the particles emitted by the radioactive phosphorus were not the expected alpha or beta particles or gamma rays. These new particles had the same mass as an electron or beta particle but had a positive charge. Called **positrons,** these particles are represented by the symbol $^{\ 0}_{+1}e$ or $^{\ 0}_{+1}\beta$. The reaction showing their emission can be written as follows:

$$^{30}_{15}P \quad \longrightarrow \quad ^{30}_{14}Si \quad + \quad ^{\ 0}_{+1}e$$

proton
neutron
positron

Like the emission of a beta particle, the emission of a positron from a nucleus seems improbable. It is explained as the conversion, within the nucleus, of a proton to a neutron.

$$^{1}_{1}p \quad \longrightarrow \quad ^{1}_{0}n \quad + \quad ^{\ 0}_{+1}e$$

Notice that the equation is properly balanced: The sums of the superscripts on both sides of the equation are equal ($1 = 1 + 0$); similarly, the subscripts are balanced ($1 = 0 + 1$). In 1935, the Joliot-Curies received a Nobel Prize in chemistry for their work.

Linear accelerators and cyclotrons are often referred to as "atom smashers." The National Accelerator Laboratory is located near Chicago in Batavia, Illinois.

Most of the early bombardment reactions were carried out with slow-moving alpha particles obtained from naturally occurring radioactive materials. It was soon realized that many more nuclear changes could be achieved if nuclei could be bombarded with faster-moving particles. In the 1940s, special instruments called *linear accelerators* (or *cyclotrons*) were developed. These devices could generate fast-moving charged particles, including alpha and beta particles, protons, and the nuclei of light elements. By directing these charged particles at target nuclei at high speed and with great force, thousands of radioisotopes unknown in nature were produced. All the presently known elements that are not naturally occurring (generally those with atomic numbers greater than 92) were formed in bombardment reactions using accelerated particles. Elements with atomic numbers near or above 100 are so unstable and so short-lived that most are no more than scientific curiosities.

Although scientists still cannot transmute base metals, such as lead, into gold—the long-sought goal of the alchemists of the Middle Ages—they are able to use artificial transmutation to produce many very useful isotopes, including all those used in medicine.

EXAMPLE 5.1 Writing balanced nuclear equations

Write a balanced nuclear equation for the following reaction:
Hafnium-181 (Group IVB) emits a beta particle.

Solution

1. Look up hafnium (Hf) in the periodic table, and find its atomic number (72).
2. Write the equation with the reactants and products that you know:

$$^{181}_{72}\text{Hf} \longrightarrow {}^{0}_{-1}\beta + ?$$

3. Determine the isotope that is formed when hafnium emits a beta particle. Beta particles have almost no mass and a negative charge (1−). Therefore, the isotope formed will have the same mass as hafnium (181) and an atomic number that has increased by 1 (from 72 to 73). Look in the periodic table for an element with atomic number 73. It is tantalum (Ta).
4. Complete the nuclear equation:

$$^{181}_{72}\text{Hf} \longrightarrow {}^{0}_{-1}\beta + {}^{181}_{73}\text{Ta}$$

5. To make sure the equation is correct, check that (a) the sum of the superscripts on the left side of the equation equals the sum of the superscripts on the right side and (b) the sum of the subscripts on the left side of the equation equals the sum of the subscripts on the right side:

$$\text{superscripts: } 181 = 0 + 181$$

$$\text{subscripts: } 72 = -1 + 73$$

The equation is correct.

Practice Exercise 5.1

Write a balanced nuclear equation for the following reaction: Zirconium-93 (Group IVB) emits a beta particle.

EXAMPLE 5.2

Write a balanced nuclear equation for the following: Radium-226 (Group IIA) decays to a radon (Group VIIIA) isotope.

Solution

1. Look up radium (Ra) in the periodic table, and find its atomic number (88).
2. Look up radon (Rn) in the periodic table, and find its atomic number (86).
3. Write the equation with the reactants and products that you know:

$$^{226}_{88}\text{Ra} \longrightarrow ? + \,^{?}_{86}\text{Rn}$$

4. The atomic number has decreased by 2 (from 88 to 86). Therefore, Ra must have lost an alpha particle ($^{4}_{2}\alpha$).
5. Write the complete equation:

$$^{226}_{88}\text{Ra} \longrightarrow \,^{4}_{2}\alpha + \,^{222}_{86}\text{Rn}$$

6. The radon isotope must be Rn-222 to agree with the requirements that (a) the sum of the superscripts of the reactants equals the sum of the superscripts of the products ($226 = 4 + 222$) and (b) the sum of the subscripts of the reactants equals the sum of the subscripts of the products ($88 = 2 + 86$).

Practice Exercise 5.2

Write a balanced nuclear equation for the following: Hydrogen-1 bombards chlorine-35 (Group VIIA), which emits an alpha particle.

5.6 The Half-Life of Radioisotopes

Different radioisotopes—whether naturally occurring or artificially produced—decay at characteristic rates. The more unstable the isotope is, the more rapidly it will emit alpha or beta particles and change into a different element. The rate at which a particular radioisotope decays is expressed in terms of its **half-life,** *the time required for one-half of any given quantity of the isotope to decay.* There is no way of knowing when a particular nucleus will decay, but after one half-life, half the nuclei in the original sample will have done so.

Half-lives range from billionths of a second to billions of years. For example, the half-life of boron-9 is only 8×10^{-19} second, that of thorium-234 is 24 days, and that of uranium-238 is 4.5 billion years. Half-livesof a number of radioisotopes are given in **Table 5.3**.

Table 5.3 Half-lives of Some Radioactive Isotopes

Radioactive Isotope	Half-life
Oxygen-13	8.7×10^{-3} seconds
Bromine-80	17.6 minutes
Iodine-132	2.4 hours
Technetium-99	6.0 hours
Radon-222	3.8 days
Barium-140	12.8 days
Hydrogen-3 (tritium)	12.3 years
Strontium-90	28.1 years
Radium-226	1620 years
Carbon-14	5730 years
Plutonium-239	24,400 years
Beryllium-10	4.5 million years
Potassium-40	1.3 billion years
Uranium-238	4.5 billion years

We can construct a decay curve **(Fig. 5.4)** to show graphically the quantity of a given radioisotope that will remain after a given length of time. Assume that we start with 16 grams (g) of $^{32}_{15}\text{P}$, a radioisotope that has a half-life of 14 days and decays by emitting a beta particle to form $^{32}_{16}\text{S}$. After one half-life (14 days), one-half of the original 16 g of $^{32}_{15}\text{P}$ will have decayed and been converted to $^{32}_{16}\text{S}$; 8 g of $^{32}_{15}\text{P}$ will therefore remain (and 8 g of $^{32}_{16}\text{S}$ will have been formed). At the end of two half-lives (28 days), one-half of the 8 g of $^{32}_{15}\text{P}$ that was present at the end of one half-life will have decayed; 4 g will remain. At the end of three half-lives (42 days), 2 g will remain, and

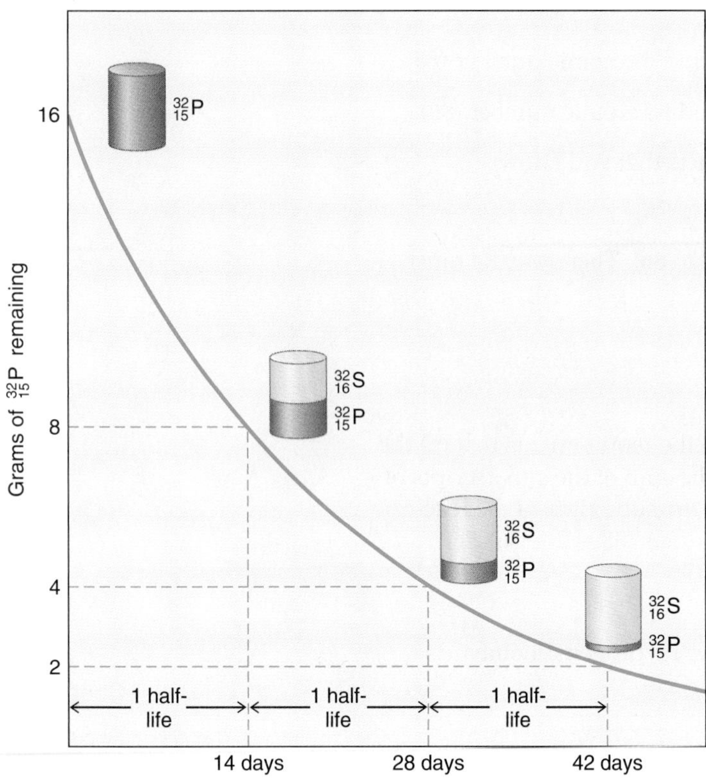

FIGURE 5.4

Decay curve for a 16-g sample of $^{32}_{15}P$ (half-life, 14 days) to $^{32}_{16}S$ by emission of beta particles. After one half-life, one-half of the original 16 g of $^{32}_{15}P$ will have been converted to $^{32}_{16}S$; that is, 8 g will remain. After another 14 days, half of that 8 g will remain. At the end of 42 days (three half-lives), 2 g of $^{32}_{15}P$ will remain and 14 g of $^{32}_{16}S$ will have been formed, and so on.

so on. After six half-lives (84 days), just 0.25 g of the original amount of $^{32}_{15}P$ will remain; the rest of the sample will be $^{32}_{16}S$. Even after many half-lives, a minute fraction of the original radioisotope will remain.

Many of the radioisotopes in the wastes generated in the production of nuclear weapons and the operation of nuclear power plants have long half-lives. This is a matter of concern. If long-lived radioisotopes escape into the environment, they persist for many years, and there is the possibility that they will accumulate in food chains. At least 10 half-lives must elapse before a radioisotope has decayed to the point at which it is no longer considered to be a radiation hazard (Chapter 19).

5.7 The Harmful Effects of Radiation on Humans

When radioactivity was discovered at the beginning of the 20th century, its harmful effects were not recognized. Marie Curie, who worked for many years with radioactive materials, suffered from anemia and died of leukemia; she was probably one of the first victims of radiation poisoning. Other early sufferers were women who, in the 1920s, worked in factories painting radium on watch dials to make them glow in the dark (Fig. 5.5). The women frequently licked their brushes to obtain a fine point and often developed cancer of the lips; as a result of ingesting the radioactive material, many of them developed bone cancer or leukemia and died at an early age.

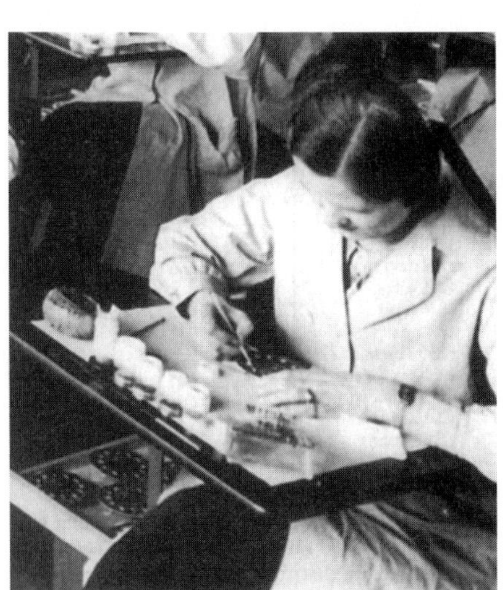

FIGURE 5.5

Women painted radium on watch dials to make them glow in the dark.

Why Is Radiation Harmful?

Radiation emitted by radioisotopes is harmful because it has sufficient energy to knock electrons from atoms and thus form positively charged ions. For this reason, it is called **ionizing radiation.** Some of these ions are highly reactive and, by disrupting the normal workings of cells in living tissues, can produce abnormalities in the genetic material DNA (Chapter 14) and increase the risk of cancer.

Factors Influencing Radiation Damage

The degree of damage caused by ionizing radiation depends on many factors, including (1) the type and penetrating power of the radiation, (2) the location of the source of the radiation (inside or outside the body), (3) the type of tissue exposed, and (4) the amount and frequency of exposure.

Alpha particles from a source outside the body constitute the least dangerous type of ionizing radiation because alpha particles cannot penetrate the skin and enter the body. However, if an alpha emitter is ingested, for example, in contaminated food, or is inhaled by breathing air containing radon gas, damage to internal body tissues can be severe. Once inside the body, alpha particles are more damaging than either beta particles or gamma rays because they are more effective in forming ions in the surrounding tissues. They travel only short distances through tissues and rapidly transfer the bulk of their energy to a small area. Beta particles and gamma rays travel farther than alpha particles and transfer their energy over a wider area of tissue. As a result, the radiation received per unit area is lower for beta particles and gamma rays, and the ability to form ions is reduced. The greater the penetrating power of a particular source of radiation, the weaker is its ionizing power.

Beta particles, unlike alpha particles, can penetrate the outer layers of skin and clothing and produce severe burns. They can also cause skin cancer and cataracts. Inside the body, beta particles are less disruptive to individual cells than are alpha particles; however, outside the body, they are more damaging than alpha particles. Gamma rays, because of their great penetrating power (see Fig. 5.2), are more dangerous than alpha or beta particles outside the body; they are less dangerous than the other two inside the body.

X-rays are a fourth type of ionizing radiation. Their penetrating power, and thus their ability to cause tissue damage, lies between that of beta particles and gamma rays.

Body tissues vary widely in their sensitivity to ionizing radiation. Rapidly dividing cells are very vulnerable to radiation. Such cells are found in bone marrow, the lining of the gastrointestinal tract, the reproductive organs, the spleen, and the lymph glands. Embryonic tissue is particularly easily damaged. Therefore, unless there are compelling medical reasons, pregnant women should avoid exposure to all radiation, including X-rays. Cancer cells, which divide very rapidly, are more easily killed by radiation than are healthy cells; this fact explains the success of radiation treatment for some types of cancer.

Red blood cells are formed in bone marrow.

Detection of Radiation

In order to measure radiation exposure, it is necessary to have a means of detecting it. The commonest instrument for detecting and measuring radioactivity is the **Geiger counter** (Fig. 5.6), which is essentially a modified cathode-ray tube (Chapter 3). Argon gas is contained in a metal cylinder, which acts as the cathode; a wire anode runs down the axis of the tube. Radiation from a radioactive source—for example, contaminated soil— enters the tube through a window of thin mica; when it does so, it causes ionization of the argon gas. As a result of the ionization, pulses of electric current flow between the electrodes, and these pulses are amplified and converted into a series of clicks that are counted automatically.

Units of Radiation

Nuclear disintegrations are measured in **curies** (Ci); 1 Ci is equal to 3.7×10^{10} disintegrations per second. A curie represents a very high dose of radiation. Natural background radiation amounts to only about 2 disintegrations per second. The damage caused by radiation depends not only on the number of disintegrations per second but also on the radiation's energy and penetrating power. Another unit of radioactivity, the **rad,** measures the amount of energy released in tissue (or another medium) when it is struck by radiation. A single diagnostic X-ray is equivalent to 1 rad. The rad is being replaced by a new international unit, the **gray** (1 gray = 100 rad).

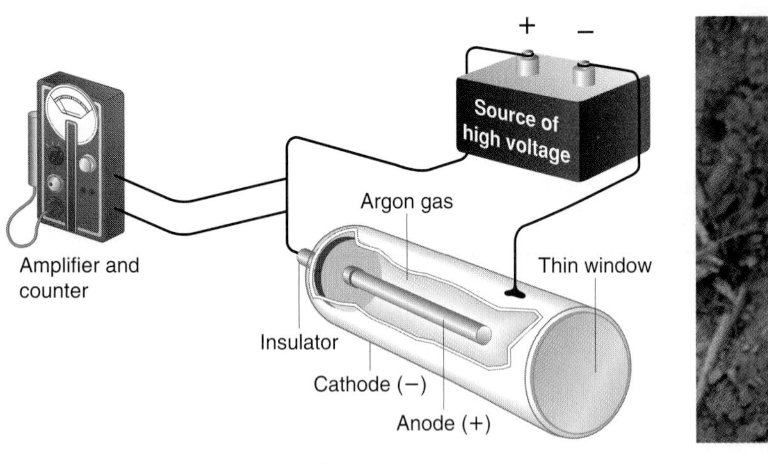

FIGURE 5.6

(a) In a Geiger counter, the metal cylinder, which is filled with a gas (usually argon), acts as the cathode; the metal wire projecting into the cylinder acts as the anode. The window in the cylinder is permeable to alpha, beta, and gamma radiation. When radiation enters the cylinder, the gas is ionized and small pulses of electric current flow between the wire and the metal cylinder. The electrical pulses are amplified and counted. The number of pulses per unit of time is a measure of the amount of radiation. (b) Checking rock for radioactivity.

A more useful unit for measuring radiation is the **rem,** which takes into account the potential damage to living tissues caused by the different types of ionizing radiation. For X-rays, gamma rays, and beta particles, 1 rad is essentially equivalent to 1 rem, but for alpha particles, because of their greater ionizing ability, 1 rad is equivalent to 10 to 20 rems. The new international unit to replace the rem is the **sievert** (1 sievert = 100 rem).

How Much Radiation Is Harmful?

Scientists still do not know for certain how much radiation is harmful. The study of Japanese survivors of the World War II bombings of Hiroshima and Nagasaki and of people exposed to fallout from nuclear power plant accidents—primarily those at Three Mile Island near Harrisburg, Pennsylvania in 1979 and at Chernobyl in 1986—has revealed the probable effects of high short-term doses of radiation given to the whole body. But there is no agreement on how the effects of these high short-term doses can be used to predict the effects of low doses. A summary of the effects of high doses is given in **Table 5.4**.

Some scientists believe that because the body has the ability to repair radiation damage, many small doses of radiation over a long period of time produce no lasting effects. Others believe that there is no safe level of radiation exposure. Recently, a study of survivors of the atomic bomb blasts in Japan found that although large numbers of those who were closest to the explosions died of cancers, those with limited exposure to the radiation from the bombs were actually living longer than Japanese people who had been far enough from the bombs to avoid exposure. Another study has shown that U.S. soldiers who participated in the nuclear weapons testing in the early 1960s have no more health problems than veterans who were not involved in the tests. At the present time, there is no agreement on what constitutes a "safe" annual dose of radiation.

Table 5.4 Effects on Humans of Short-Term Whole-Body Exposure to Various Doses of Radiation

Dose (rem)	Effects
<50	Effects inconsistent and difficult to demonstrate.
50–250	Fatigue, nausea, decreased production of white cells and platelets in blood; increased probability of leukemia.
250–500	Same as for 50–250 rem, but more severe; vomiting, diarrhea, damage to intestinal lining; very susceptible to infections because of low white cell count; hemorrhaging because of impaired clotting mechanism; 50% die within months.
500–1000	Damage to cardiovascular system, intestinal tract, and brain; death within weeks.
1000–10,000	Same as for 500–1000 rem, but more severe; coma; death within hours at 10,000 rem.
100,000	Immediate death.

5.8 Everyday Exposure to Radiation

It is impossible to avoid exposure to every source of radiation in daily life. Low-level radiation is all around us in the environment. It has been estimated that the *average* exposure of a member of the U.S. population to ionizing radiation amounts to approximately 360 millirem (mrem) per year (note that a millirem is 0.001 rem, the unit used in Table 5.4). By far, the largest contribution—82%—to this average exposure comes from natural sources. The remaining 18% percent comes primarily from medical procedures, consumer products, and occupational activities **(Table 5.5)**. Medical personnel and other people who work near radioactive substances wear badges containing film that is sensitive to radiation **(Fig. 5.7)**. Any radiation affects the film in the same way that light affects a photographic film. The amount of darkening of the film gives a measure of the person's exposure.

Natural Sources of Radiation

The average dose from all natural sources of radiation is about 300 mrem/year. It is now recognized that the major natural source is **radon,** a gas that, on average, contributes 200 mrem to the total annual exposure (see Table 5.5). Radon-222 (half-life of 3.8 days), an alpha emitter, is a naturally occurring product of the decay of uranium-238 (refer to Fig. 5.3), a radioisotope of uranium that is present in widely varying concentrations in most soils and rocks. Radon-222 is an odorless, tasteless, inert gas, which, as it escapes from soil and rocks, enters the surrounding water and air. The Environmental Protection Agency (EPA) has conducted a nationwide survey of radon in U.S. homes to determine where problems are likely to exist **(Fig. 5.8)**. In areas of the country where crustal rocks and soil have high concentrations of uranium-238, radon can seep into homes through basements and, in well-sealed houses, may reach potentially dangerous levels. Inhalation of radon gas can increase the risk of cancer. This effect is due less to the radon itself than to its alpha- and beta-emitting decay products (see Fig. 5.3), which, when the gas is inhaled, become deposited in the respiratory tract.

Table 5.5 Average Annual Exposure (1990) to Radiation for U. S. Residents

Source of Radiation	Dose (mrem)	Percent of Total Dose
Natural		
Radon gas	200	55
Cosmic rays	27	8
Terrestrial (radiation from rocks and soil other than radon)	28	8
Inside the body (naturally occurring radioisotopes in food and water)	39	11
Total natural	294	82
Artificial		
Medical		
X-rays	39	11
Nuclear medicine	14	4
Consumer products (building materials, water)	10	3
Other		
Occupational (underground miners, X-ray technicians, nuclear plant workers)	<1	<0.03
Nuclear fuel cycle	<1	<0.03
Fallout from nuclear weapons testing	<1	<0.03
Miscellaneous	<1	<0.03
Total artificial	64	18
Total natural plus artificial	358	100

Source: National Council on Radiation Protection and Measurement (NCRP87b), Washington DC: National Academy Press, 1990.

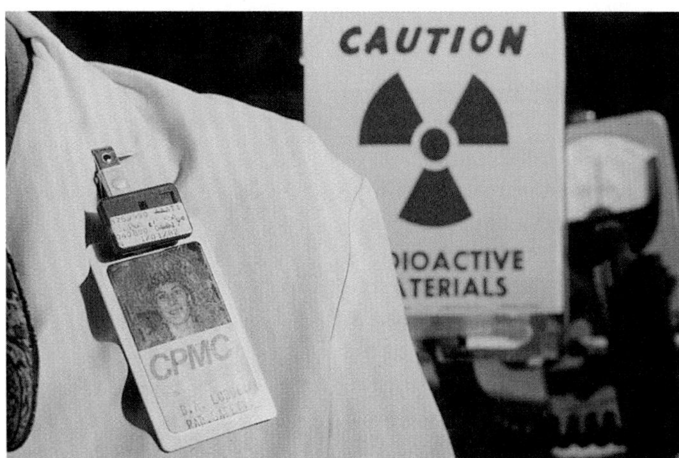

FIGURE 5·7

Lapel badges containing film that darkens on exposure to radiation are used to monitor the extent of workers' exposure.

Cosmic rays—a form of short-wavelength electromagnetic radiation that reaches the earth's surface from outer space—account for 8% of natural radiation, or an average dose of 27 mrem/year. Because the dose increases with altitude, residents of mile-high Denver receive about 50 mrem more radiation each year from this source than do people living at sea level in New Orleans.

Terrestrial radiation in rocks and soil, originating from radioisotopes other than radon, accounts for another 8% of natural radiation (28 mrem/year). Naturally occurring carbon-14 and traces of potassium-40, thorium-223, and uranium-238 are present in food, water, and air and enter the body when ingested or inhaled. This inside-the-body source makes up about 11% of natural radiation (39 mrem/year).

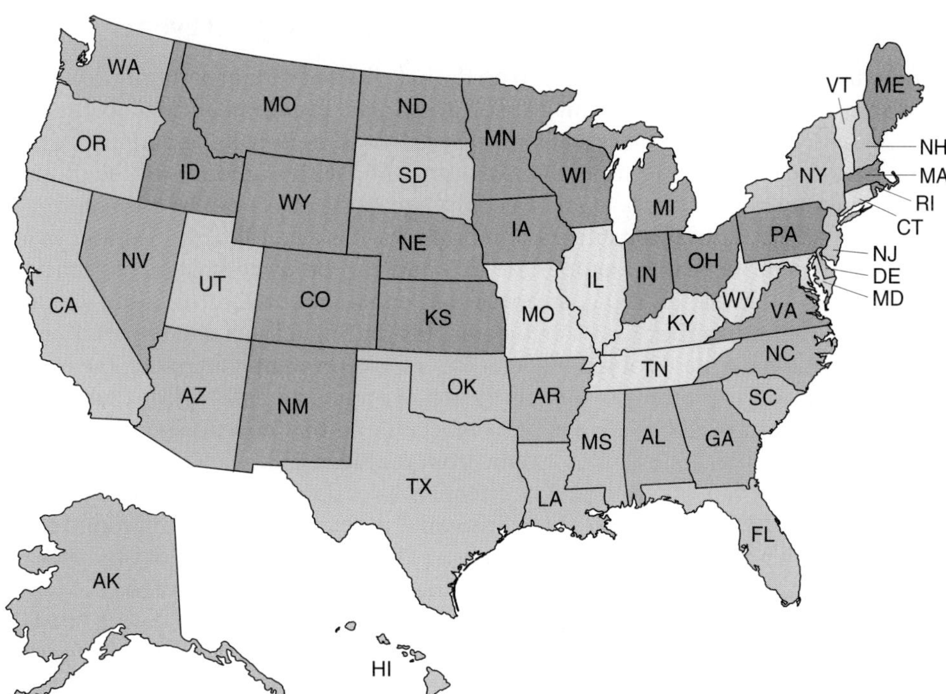

FIGURE 5.8

Results of an Environmental Protection Agency 1993 survey of radon exposure in U.S. residential dwellings. Radon levels below 4 pCi/L are considered to be average, and generally no action is recommmended. If the level is much greater than 4 pCi/L, action to reduce the radon exposure is recommended. (pCi/L = picocuries per liter of air; 1 pCi = 1×10^{-12} Ci)

Radiation from Human Activities

The average annual dose of radiation arising from everyday human activities amounts to about 64 mrem (see Table 5.5). Diagnostic X-rays account for 39 mrem/year, and other medical procedures and treatments add 14 mrem/year (together accounting for 15% of the total average annual dose). Consumer products add 10 mrem/year, with most of this radiation coming from radon dissolved in well water, and in building materials—stone, brick, and concrete—which frequently contain uranium ores.

Smokers receive an additional dose of radiation from polonium-210, a naturally occurring alpha emitter present in tobacco. Workers who are potentially at risk of higher exposure to radiation than the average for the U.S. population include underground miners, radiologists, X-ray technicians, and nuclear power plant workers. The average additional dose for these workers is less than 1 mrem/year. Although residents downwind from nuclear weapons testing sites were exposed to radioactive fallout in the past, nuclear weapons production and nuclear power plant operations contribute negligible amounts of radiation today. However, nuclear wastes from both of these activities pose a serious hazard (Chapter 19).

5.9 Uses of Radioisotopes

Although it may present potential health hazards, radiation from radioisotopes has useful applications in fields as diverse as archeology, medical diagnosis and treatment, agriculture, scientific research, and the power industry.

Determining the Dates of Archeological and Geological Events

An ingenious and reliable method based on the half-lives of certain naturally occurring radioisotopes such as carbon-14 can be used to estimate the age of ancient objects **(Fig. 5.9)**. A radioisotope decays at a constant rate that is defined by its half-life, and as it decays, it changes into a different isotope. By measuring the relative amounts of the original and the product radioisotopes in an object, the age of the object can be estimated.

Uranium-238, a natural constituent of most rocks, has a half-life of 4.5 billion years. It decays to form the stable isotope lead-206 and can be used to date very ancient rocks. To take a simple example, suppose that equal amounts of uranium-238 and lead-206 are found in a rock we wish to date. This finding tells us that the uranium in the rock must have gone through one half-life (4.5 billion years); therefore, the rock is 4.5 billion years old. In coming to this conclusion, we make the assumption that the rate of decay of uranium-238 has always been constant and that no lead-206 was initially present in the sample; that is, all of it came from uranium-238.

In 1988, carbon dating showed that the linen cloth known as the Shroud of Turin was made after 1350 A.D. This proved that the image on the cloth could not have been imprinted by the body of Jesus while it was wrapped in the shroud—as many believed—because Jesus died in about 30 A.D.

Using the decay rate of uranium-238, scientists have dated the oldest rocks on earth at between 3 and 3.5 billion years. Meteorites and moon rocks have been found to be approximately 4.5 billion years old. On the basis of this information, geologists conclude that the planets, including the earth, were formed 4.5–5.0 billion years ago.

To obtain an accurate date, the half-life of the radioisotope used for dating must be fairly close to the age of the object being studied. Uranium-238 is appropriate for dating rocks that are billions of years old, but for objects that are several thousand years old, carbon-14, which has a half-life of 5730 years, is more suitable. Other radioisotopes may also be used, depending on the age and nature of the material being studied.

Carbon-14 is formed naturally at a fairly constant rate in the earth's upper atmosphere when neutrons ejected from stable nuclei by cosmic rays interact with ordinary nitrogen nuclei:

$$^{14}_{7}\text{N} + ^{1}_{0}\text{n} \longrightarrow ^{14}_{6}\text{C} + ^{1}_{1}\text{p}$$

The radioactive carbon-14 reacts with oxygen in the atmosphere to form radioactive carbon dioxide ($^{14}_{6}CO_2$). Both radioactive carbon dioxide and ordinary carbon dioxide ($^{12}_{6}CO_2$ are incorporated into plants through photosynthesis and into animals through normal food chains and respiration. While an organism is alive, the ratio of carbon-14 to carbon-12 in its tissues remains constant, but when the organism dies and incorporation of carbon dioxide ceases, the ratio begins to change. The carbon-14 in the tissues slowly disappears as it emits beta particles and becomes nitrogen-14, but the nonradioactive carbon-12 does not change. The ratio of carbon-14 to carbon-12 decreases.

$$^{14}_{6}\text{C} \longrightarrow ^{14}_{7}\text{N} + ^{0}_{-1}\beta$$

The carbon-14 emits beta particles at a constant rate given by its half-life. Thus, measuring the carbon-14 content in the plant or animal artifact gives the age of the artifact.

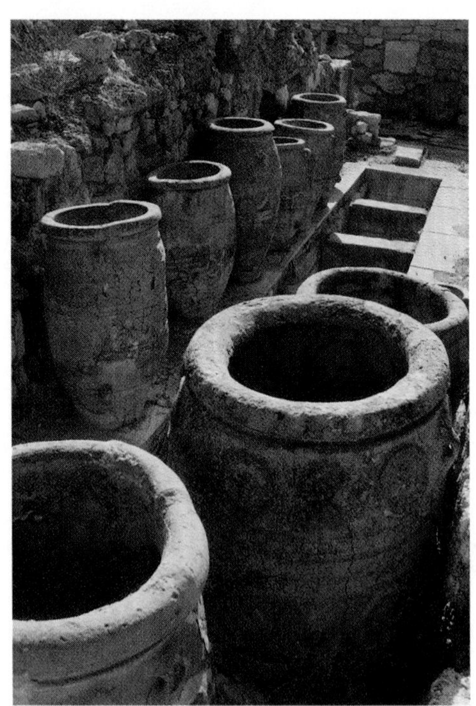

FIGURE 5.9

Carbon-14 dating of the charcoal found in these ancient pots was used to estimate how old they are.

Carbon-14 dating, also known as **radiocarbon dating,** is used to date ancient artifacts such as cloth, wood, leather, and bone, which were derived from once-living matter and thus contain carbon. The method assumes that the flow of carbon-14 into the environment has always been constant. Studies of the annual growth rings in trees have shown this assumption to be justified over the past 7000 years. There were some fluctuations, however, in earlier times. Even so, other dating methods generally agree very well with carbon-14 dating.

EXAMPLE 5.3 Using the half-lives of radioisotopes

The carbon-14 content of a piece of ancient wood is one quarter that of new wood. What is the age of the ancient wood? The half-life of carbon-14 is 5730 years.

Solution

1. First, represent the carbon-14 content by a diagram like that in Figure 5.4. After one half-life, the content decreases by one-half. After a second half-life, the content is halved again, and one-quarter of the original content remains.
2. The carbon-14 in the ancient wood is old enough to have gone through two half-lives:

$$\text{age of wood} = 2 \text{ half-lives}$$

The wood is therefore 2×5730, or 11,460 years old.

Practice Exercise 5.3

The carbon-14 content of an ancient piece of linen cloth is one-eighth that of linen cloth produced today. Given that the half-life of carbon-14 is 5730 years, determine the age of the ancient cloth.

EXAMPLE 5.4

If you have 100 g of radioactive bromine-80 (half-life of approximately 20 minutes), how many grams of the isotope will remain after 1 hour?

Solution After 20 minutes (one half-life), one half of the original sample of bromine-80, or 50 g, will remain. After another 20 minutes, 25 g will remain. After another 20 minutes (that is, after a total of 1 hour has elapsed) 12.5 g of bromine-80 will remain.

Practice Exercise 5.4

The half-life of tritium (hydrogen-3) is 12.3 years. If 0.005 g of tritium is released from a nuclear power plant during an accident, what mass of this isotope remains after **a.** 12.3 years; **b.** 49.2 years?

Medical Diagnosis and Treatment

Radioisotopes are used in **nuclear medicine** for both diagnosis and therapy. In diagnosis, a particular radioisotope, or a compound containing that radioisotope, is ingested or injected into the body. The radioisotope emits its characteristic radiation, which is converted to an electrical signal, amplified, and then processed by a computer to produce a visual image of the site being studied.

Iodine-131, a beta and gamma emitter, is useful for diagnosing thyroid disease. Nearly all the iodine that we normally ingest in food and water becomes concentrated

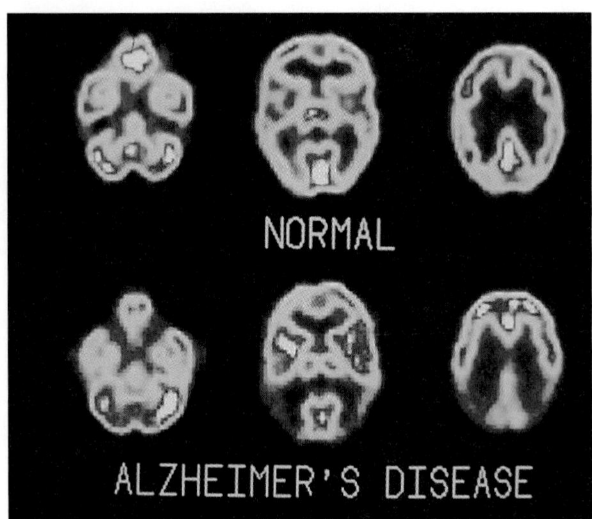

FIGURE 5.10

A PET scan showing brain activity is generated using positron emission.

in the thyroid gland. If a patient drinks a solution containing iodine-131, the radioisotope behaves like nonradioactive iodine and accumulates in the thyroid gland. By monitoring the radioactive emissions in the area of the gland, a physician can determine whether it is functioning normally. A visual image shows whether uptake of iodine-131 is greater or less than for a normal person, or if the thyroid is cancerous.

Positron emission tomography (PET) makes use of positron emitters such as carbon-11 and fluorine-18, both of which have short half-lives, to pinpoint areas of abnormal activity in the brain, such as abnormal glucose uptake or abnormal blood flow **(Fig. 5.10)**. When a compound containing one of these radiosotopes is injected, the radioisotope decays by emitting a positron. The positron collides almost immediately with an electron, which results in mutual annihilation and the emission of energy in the form of two gamma rays.

$$^{11}_{6}\mathrm{C} \longrightarrow {}^{11}_{5}\beta + {}^{0}_{+1}\mathrm{e}$$

$$^{0}_{+1}\mathrm{e} + {}^{0}_{-1}\mathrm{e} \longrightarrow 2\gamma$$

The two gamma rays are emitted in exactly opposite directions and are detected by monitors placed 180° apart.

There is obviously some risk attached to introducing radioactive materials into the body. To keep the risk of tissue damage at a minimum, alpha emitters are never used for diagnostic purposes. Beta emitters or, even better, gamma emitters with very short half-lives are used, and the dose is kept as small as possible. Besides iodine-131, carbon-11, and fluorine-18, radioisotopes used in diagnosis include gadolinium-153, which is used to evaluate bone mineralization, and sodium-24, which can detect constriction and obstruction in blood vessels.

Like other rapidly dividing cells, cancer cells are very sensitive to radiation. Radiation therapy aims to destroy only cancer cells, but inevitably some rapidly growing healthy cells—particularly intestinal and blood cells—are also killed. As a result, patients receiving radiation therapy usually experience nausea and vomiting, and their white blood cell counts fall. Aiming the radiation, which may amount to 150–200 rems per treatment, as precisely and narrowly as possible at the site of the cancer minimizes damage to normal cells. Thyroid cancer can be treated by ingesting iodine-131; the radioisotope concentrates in the thyroid, where it bombards the cancer cells with radiation and destroys them. For other forms of cancer, penetrating radiation from an outside source is generally used. The most common forms are gamma rays from cobalt-60 or cesium-137 and X-rays. **Table 5.6** lists a number of radioisotopes and their applications in nuclear medicine.

Applications in the Home

Although most of us may not be aware of it, most low-cost smoke detectors **(Fig. 5.11)** are ionization devices that contain the radioisotope americium-241, which is an alpha emitter:

$$^{241}_{95}\mathrm{Am} \longrightarrow {}^{4}_{2}\alpha + {}^{237}_{93}\mathrm{Np}$$

Table 5.6 Applications of Radioisotopes in Medicine

Radioisotope	Symbol	Radiation Emitted	Application
Carbon-11	$^{11}_{6}C$	Positron, gamma	Brain imaging
Chromium-51	$^{51}_{24}Cr$	Gamma	Determination of blood flow through heart and of lifetime of red blood cells
Cobalt-57	$^{57}_{27}Co$	Gamma	Detection of defects in uptake of vitamin B-12
Cobalt-60	$^{60}_{27}Co$	Beta, gamma	Treatment of cancer
Cesium-137	$^{137}_{55}Cs$	Gamma	Treatment of cancer
Gadolinium-153	$^{153}_{64}Gd$	Beta, gamma	Determination of bone density
Iodine-131	$^{131}_{53}I$	Beta, gamma	Determination of activity of thyroid gland; treatment of thyroid cancer
Iron-59	$^{59}_{26}Fe$	Beta	Assessment of iron metabolism in blood
Sodium-24	$^{24}_{11}Na$	Beta, gamma	Detection of constrictions and obstructions in the circulatory system
Technetium-99m*	$^{99m}_{43}Tc$	Gamma	Obtaining images of organs, e.g., heart, lungs, liver, kidney
Tritium	$^{3}_{1}H$	Beta	Determination of total body water

*Technetium-99m, with a half-life of about 6 hours, is one of the most useful radioisotopes in medicine. The m indicates that the isotope is metastable; by emitting gamma rays, it disintegrates to a more stable form of the same isotope, technetium-99.

The alpha particles that are released hit air that is flowing through the smoke detector, and oxygen and nitrogen molecules in the air become ionized as electrons are stripped from them. The electrons that have been stripped from the air molecules are attracted to a positive electrode in the smoke detector, setting up a constant flow of electrons called a *standing current*. The electron circuitry of the smoke detector senses this flow, and, as

a.

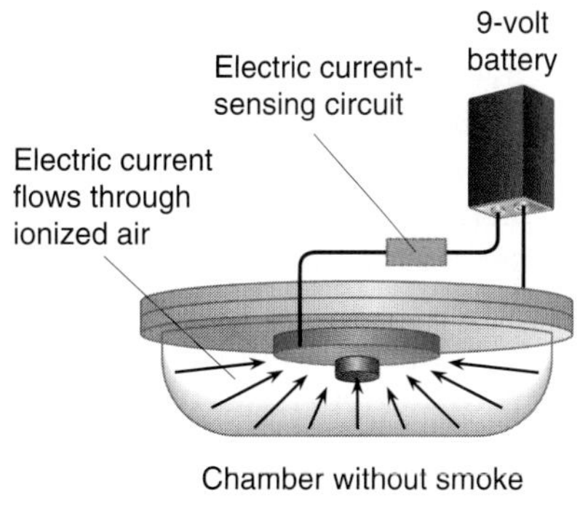

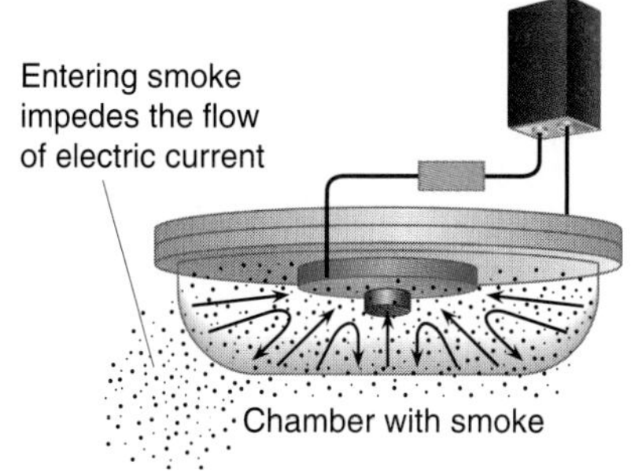

b.

FIGURE 5.11

(a) The inside of a typical smoke detector. (b) Alpha particles from the radioisotope americium-241 ionize the air inside the detector, which sets up a standing current. If smoke enters the detector, the flow of electricity is interrupted, causing the alarm to go off.

long as the current is flowing, the circuit keeps the alarm shut off. In the event of a fire, smoke, which consists primarily of small, black particles of carbon, enters the detector. The electron flow is disrupted by these particles, and the flow of electrons to the positive electrode decreases. The electronic circuit senses the drop in the standing current and trips the alarm.

There is no danger of exposure to radiation from a smoke detector because the alpha particles emitted by the americium-241 cannot pass through the smoke detector's cover. The few particles that exit through the openings in the cover combine with electrons stripped from oxygen and nitrogen molecules in the air to form helium, which is harmless.

Americium-241 has a half-life of 432 years, and so the sample of the radioisotope far outlasts the smoke detector. The 9-V battery that is required to sound the alarm lasts for at least a year because its power is only consumed when the alarm sounds.

According to a recent study by the National Fire Protection Association, more than 80% of fire fatalities and more than 80% of fire injuries occur in homes that have no smoke detectors.

a.

b.

FIGURE 5.12

Irradiation prevents the growth of mold on strawberries. (a) Irradiated strawberries kept at 4°C are still fresh after 15 days. (b) Strawberries that are not irradiated become moldy.

Applications in Agriculture, Industry, and Scientific Research

Because all isotopes of a given element, whether radioactive or not, behave almost identically in chemical reactions, a radioisotope of an element can be used to label that element for identification. For example, if phosphate fertilizer labeled with radioactive phosphorus-32 is fed to a plant, the labeled phosphate will be taken up through the plant's roots in the same way as nonlabeled phosphate. Because emissions from radioisotopes are easily detected by Geiger counters and other devices, the movement of phosphorus-32 through the plant can be detected. Such information is valuable in determining how plants grow, how they utilize fertilizers, and how their yields might be improved. In a similar way, other radioisotopes, or **tracers,** can be used to follow, or trace, the movement of pollutants through food chains or through water systems.

A commercially advantageous but still controversial application of nuclear radiation is in food preservation. Irradiation of food with gamma rays destroys insects and the microorganisms that cause spoilage. It also controls mold formation **(Fig. 5.12)** and retards sprouting in vegetables such as potatoes and onions. This procedure is routinely used in Europe, Canada, and Mexico. No residual radiation remains in food after irradiation, and no adverse effects on humans have been observed in countries where it is used. In the United States, however, because of concern that irradiation may produce as yet undiscovered chemicals capable of causing genetic damage, irradiation of crops is still restricted. In time, it is likely that irradiation will replace ethylene dibromide (EDB) and other potentially harmful chemicals that are now used to fumigate fruits and vegetables (Chapter 16).

In 1997, after several years of study, the Food and Drug Administration concluded that irradiation of food was a safe process. Recently, the Department of Agriculture established guidelines for the use of irradiation of meat to kill *Salmonella* and other bacteria that, according to the department, contaminate as much as 40% of all raw poultry. The guidelines apply to raw meat only and do not include processed and packaged products, such as hot dogs. To satisfy consumer groups, irradiated products must be labeled to indicate that they have been treated.

Irradiation offers an effective means of pest control because irradiated pests become sterile. If enough sterile individuals are released into the natural population, most matings will produce no offspring. Unlike chemical pesticides (Chapter 16), this method does not pollute the environment. In another agricultural application, new and improved strains of wheat, corn, and other crops have been produced by irradiating plants to produce intentional mutations.

5.10 Nuclear Fission

Until about 60 years ago, scientists believed that the only way an unstable nucleus such as uranium-238 could change into one that was more stable was by a series of successive steps. In each of these steps, a small nuclear particle—such as an alpha or beta particle, a proton, or a neutron—was emitted. As a result of such emissions, atomic numbers changed by no more than one or two units at a time (see Fig. 5.3).

In 1938, on the eve of World War II, these beliefs had to be completely revised. Two German chemists, Otto Hahn and Fritz Strassman, found that, if they bombarded uranium-238 with neutrons, they obtained not only the expected products—uranium-239 (atomic number 92) and neptunium-239 (atomic number 93)—but also small amounts of barium (atomic number 56), lanthanum (atomic number 57), and cerium (atomic number 58). Hahn and Strassman were astounded to find elements with such low atomic numbers among the products. They realized that the original uranium nucleus must have been split almost in half, in a process they named **nuclear fission.**

The implications of nuclear fission were immediately evident to Hahn's colleague, Lise Meitner, who, because she was Jewish, had fled to Sweden in 1938, when her homeland, Austria, was annexed by Hitler. Meitner and her nephew, Otto Frisch, used calculations based on Albert Einstein's famous equation, $E = mc^2$, to determine the tremendous amount of energy that would be released if an atomic nucleus was split. They realized that if this energy could be harnessed, it could be used to construct very powerful bombs.

The Energy Source in Nuclear Fission

As you learned earlier, despite repulsions between protons, a powerful force within the nucleus holds it together. This **binding energy** comes from the conversion of a very small amount of mass into energy that occurs when protons and neutrons are packed together to form nuclei. Einstein had discovered that energy and mass are two aspects of the same thing. In his equation, E represents energy, m represents mass, and c is the speed of light (300,000 km/s, or 186,000 mi/s). Because c^2 is a very large number, the equation means that even if the mass (m) converted is very small, the amount of energy produced ($m \times c^2$) will be enormous.

The binding energy in the nuclei of various elements is related to the mass of their nuclei as shown in **Figure 5.13**. Nuclei of iron atoms—and other nuclei of similar mass—are more stable and need less energy to hold them together than either lighter or heavier

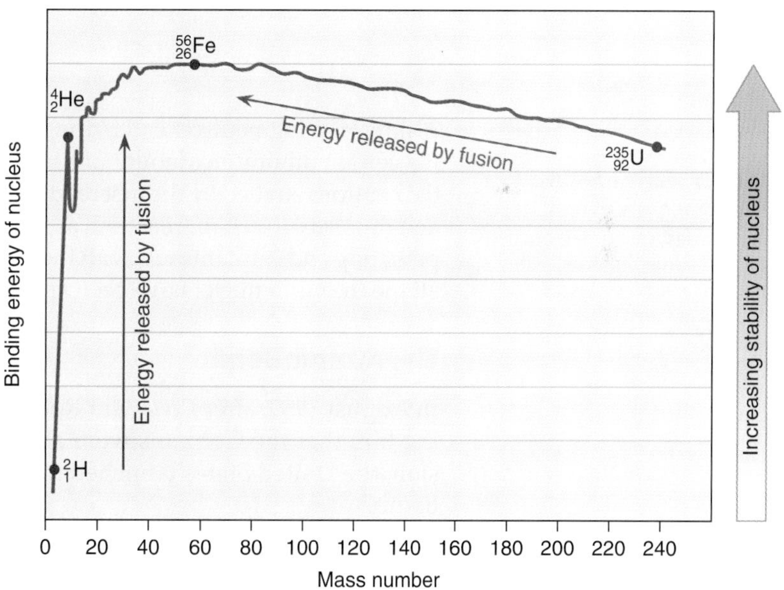

FIGURE 5.13

Intermediate-sized nuclei with mass numbers between about 30 and 65, such as $^{56}_{26}$Fe, need the least amount of energy to hold them together and are the most stable. During fission, nuclei of elements with high mass numbers, such as $^{235}_{92}$U, split apart to form smaller, more stable nuclei, and energy is released. Energy is also released during fusion when small nuclei such as $^{2}_{1}$H and $^{4}_{2}$He join to form larger, more stable nuclei.

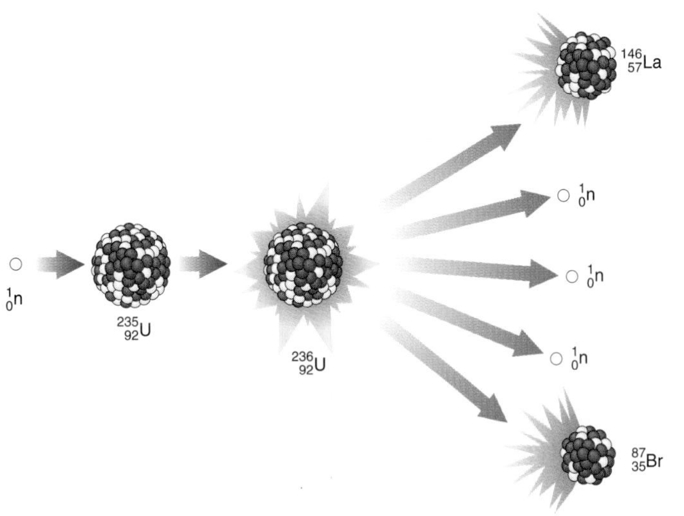

FIGURE 5.14

Nuclear fission occurs when $^{235}_{92}U$ atoms are bombarded with slow-moving neutrons. During fission, the uranium-235 atoms split apart to form two lighter atoms and from two to four neutrons. Because the new elements formed are more stable than uranium, a large amount of energy is released. Uranium atoms split in many different ways. Up to 35 elements have been identified among the fission products, including $^{146}_{57}La$ and $^{87}_{35}Br$.

nuclei do. When a heavy nucleus such as a uranium nucleus is split apart to form lighter, more stable nuclei, some of the binding energy is released. It has been calculated that fission of 1 g of uranium releases approximately 10 million times more energy than burning 1 g of coal.

Figure 5.13 also shows that energy is released when nuclei of very light elements such as hydrogen and helium join, or fuse, together. Fusion reactions, which can only occur at extremely high temperatures, occur continually in the sun, and provide the energy that sustains life on earth. Nuclear fusion also occurs in the explosion of a hydrogen bomb (see Chapter 13).

Fission Reactions

If a fissionable isotope such as uranium-235 or plutonium-239 is bombarded with neutrons, the nuclei can split apart in more than one way to produce a variety of lighter elements. Whatever way a nucleus splits, the fission reaction releases between two and four neutrons and a great deal of energy. One way a uranium-235 nucleus splits apart is shown in the following equation:

$$^{235}_{92}U + ^{1}_{0}n \longrightarrow ^{146}_{57}La + ^{87}_{35}Br + 3\,^{1}_{0}n + \text{energy}$$

A chain reaction is a self-sustaining reaction that continues unchecked once it has begun.

A slow-moving neutron enters the uranium-235 nucleus and causes fission (Fig. 5.14). If a certain minimum amount of fissionable nuclei, called the **critical mass,** is present, the neutrons emitted by the first reaction can cause the fission of more uranium-235 nuclei, and thus set off a **chain reaction,** as illustrated in Figure 5.15. Such a reaction is self-perpetuating and can continue, with the release of ever-increasing amounts of energy, until all the uranium nuclei have been split.

The Atomic Bomb

In August 1939, after President Roosevelt received a letter signed by Einstein warning him that the Germans were already working on a bomb based on nuclear fission, the United States launched the secret Manhattan Project to develop an atomic bomb.

To obtain a critical mass of uranium-235 (which makes up only 0.7% of natural uranium), the U.S. government built a top-secret facility for enriching uranium in Oak Ridge, Tennessee. At the same time, a facility was established in Hanford, Washington, to produce plutonium-239, a fissionable radioisotope that is not naturally occurring.

In 1942, the first sustained chain reaction, using uranium-235 as the starting material, was achieved in a laboratory under the football stadium at the University of Chicago. The buildup of neutrons was carefully controlled to prevent a dangerous explosion. By early 1945, sufficient plutonium-239 had been prepared for the construction of a bomb. In an atomic bomb, two separate masses of fissionable material, each incapable of sus-

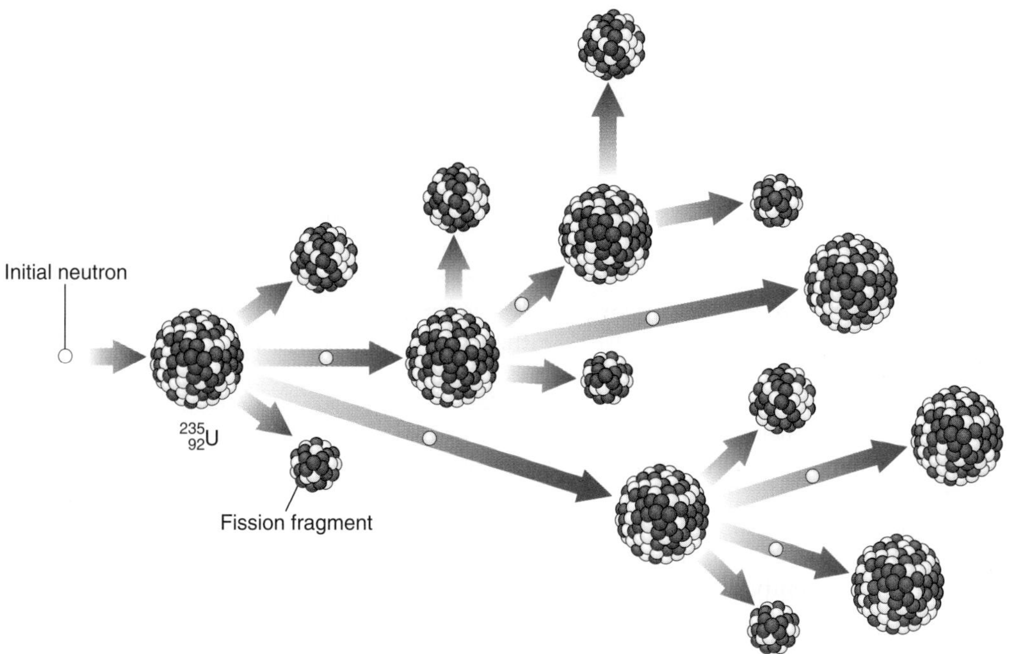

FIGURE 5.15

A nuclear chain reaction can begin when fission of a $^{235}_{92}U$ atom by a single neutron releases two to four neutrons from the uranium nucleus. Each of these neutrons can trigger fission of another uranium atom, and the release of more neutrons. As a result, an uncontrollable chain reaction occurs, releasing an enormous amount of energy.

The bomb dropped on Hiroshima was equivalent to about 20,000 tons of TNT.

taining a chain reaction separately, are brought together to form a critical mass at the moment of detonation.

In July 1945, the first atomic bomb was successfully tested in the desert near Alamogordo, New Mexico. Those who witnessed the explosion were awed by the magnitude of the blast; the brilliance of the light, the tremendous heat generated, and the huge mushroom cloud that formed were overwhelming **(Fig. 5.16)**. On August 6, 1945, President Harry Truman, compelled by his desire to avoid the millions of American and Allied casualties expected in an invasion of Japan, ordered a uranium bomb to be dropped on Hiroshima. Three days later, a plutonium bomb was dropped on Nagasaki. Approximately 200,000 Japanese were either killed or injured by the two bombs. The war ended with the surrender of Japan on August 14, 1945.

Peaceful Uses of Nuclear Fission

Nuclear fission is used to create bombs and other weapons of war, but it can also be harnessed for peaceful purposes. If fission is controlled so that the concentration of neutrons produced is sufficient to maintain the fission reactions but not great enough to allow an uncontrolled chain reaction, the process can be used as a source of energy for electrical power generation. Nuclear power plants generate much of the electricity used in Europe, but the United States is still heavily dependent on fossil fuels to meet its energy needs. Nuclear energy is discussed in more detail in Chapter 13.

FIGURE 5.16

The explosion of an atomic bomb (a bomb based on a nuclear fission reaction) produces a characteristic mushroom-shaped cloud.

EXPLORATIONS

Marie Curie—Wife, Mother, and Two-Time Nobel Prize Winner

Marie Curie was born Marya Sklodowska in Russian-dominated Warsaw, Poland, in 1867, the youngest of five children—one boy and four girls. She enjoyed a happy childhood until she was 9, when, tragically, her oldest sister died of typhus. Just 2 years later, her mother died of tuberculosis.

After graduating from school at 15 with top honors, Marya was anxious to continue her education, but because the Tsar's University of Warsaw was closed to women, her only choice was to study abroad. Her father, who had lost his position as a teacher of mathematics and physics for not being sufficiently pro-Russian, could not afford the expense, and Marya began to save toward her education by working as a tutor. At the same time, ardently patriotic, she joined the illegal "Floating University," well aware that discovery could mean imprisonment or deportation to Siberia. (This night school was dedicated to building a core of Polish intellectuals. Teachers gave free instruction to students, who, in turn, taught workers.)

At 18, to help pay for her sister Bronya's medical studies in Paris, Marya took a position as a governess, with the understanding that, when possible, Bronya would help her finance her further education. The bargain was kept, and in 1891, Marya registered for classes at the Sorbonne using the French form of her name, Marie. To be close to the university, she lived in a garret, existing mainly on buttered bread and tea, with an occasional treat of a piece of fruit or an egg. If she wanted heat, she had to carry coal for a small stove up six flights of stairs. Marie was completely absorbed in the joy of learning and, despite the hardships, was later to describe her student days as "a time of charm." In 1893, at the top of her class, she became the first woman to receive the *licence ès sciences physiques* from the Sorbonne. A year later, second in her class, she received the *licence ès sciences mathematiques.*

Marriage Combined with Two Careers

In the year she graduated, Marie met Pierre Curie, a shy, reserved man of 35, already recognized for his work on crystal symmetry, magnetism, and piezo electricity. Without any ambition for himself, Pierre was committed to a life devoted entirely to scientific research. Today, many physicists regard him as a much greater scientist than his far more famous wife.

There was an immediate rapport between Marie and Pierre, and they were married in July 1895. It was a difficult decision for Marie, because it meant not returning to her family and her beloved Poland, but she was persuaded to follow her heart. Always practical, she chose for her wedding a plain dark dress that she could wear afterwards in the laboratory. The couple spent their honeymoon bicycling through the French countryside.

Marie and Pierre Curie in the laboratory in 1904.

Neither marriage, the birth of her daughter, Irene, in 1897, nor the fact that no European woman had ever completed a doctorate in science deterred Marie from her pursuit of an advanced degree. For her research topic, she chose the rays spontaneously emitted by uranium compounds that had recently been discovered by Henri Becquerel. This fateful decision determined the course of her work for the rest of her life.

Studying Radioactivity

Becquerel's rays caused air to conduct electricity; using a sensitive electrometer that had been invented by Pierre and his brother, Jacques, Marie tested compounds of all the known elements for this property, which she named "radioactivity."

Working at her own expense in a small, damp, unheated, glassed-in room that Pierre's superior allowed her to use, she soon found that thorium, like uranium, was radioactive. She also found that, regardless of the nature of the radioactive sample, the intensity of its radiation was always proportional to the quantity of uranium or thorium it contained. This finding led her to the then startling conclusion that radiation must come from the atom itself, a fundamental discovery that was to be Marie's most important contribution to science.

Marie next discovered that, based on its uranium content, the ore pitchblende was much more radioactive than expected. She realized that it must contain a new and extremely radioactive element, present in an amount so small that it had previously escaped detection. The discovery was so important that Pierre decided to abandon his own work temporarily to help Marie isolate the new element and thus prove its existence. The Curies soon found that two new elements were present in the ore. One, Marie named polonium in honor of her homeland, but it was the second, far more radioactive element, which she named radium, that was to make her famous.

Her discovery that radiation had its origin in the atom was a milestone in nuclear physics.

Optimistically, the Curies thought radium might make up 1% of pitchblende; in fact, it accounted for only 0.0001%. Pitchblende was expensive, and the Curies needed tons of it. Fortunately, they found that they could use the cheap residue discarded after the valuable uranium had been extracted. Together, they worked with inadequate equipment, under the most primitive conditions, in an old shed with a leaky roof. To avoid inhaling noxious fumes, they often worked outside in the yard. Over and over again, Marie took 20-kilogram samples of pitchblende, the most she could handle at one time, and using huge cauldrons, separated the different elements following the standard analytical techniques of the day. With each step, the Curies obtained an increasingly pure and more radioactive radium fraction. Finally, in 1902, Marie had a 100-milligram sample of pure radium chloride, indisputable proof that she had isolated a new element with an atomic weight of 226.

Tragedy and Triumph

The work took a heavy toll on the Curies' health, and both Marie and Pierre suffered frequently from extreme fatigue and various other unexplained illnesses. They realized that their red, cracking, permanently damaged fingertips were the result of handling radium. But it never occurred to them—or to any one else, until many years later—that exposure to the radiation and the radon gas that radium emitted could be a health hazard. Thus, they took no precautions to protect themselves.

In 1903, with Henri Bequerel, the Curies received the Nobel Prize in physics for the discovery of radioactivity. In that same year, Marie received her doctoral degree, and in 1904, their second daughter, Eve, was born. Then, in 1906, tragedy struck. Pierre Curie, while crossing a street, was hit by a horse-drawn wagon and killed instantly. Marie was completely devastated by the loss of her beloved companion, but in time, she found solace in her work and her children. She was named to Pierre's chair at the Sorbonne and became the first woman in France ever to hold a position in higher education. She continued her work with radium and, in 1911, was awarded the Nobel Prize in chemistry for her isolation of pure radium metal.

Although Marie Curie made no further important discoveries, she made possible the work of others, including her daughter, Irene, who with her husband, Frederic Joliot, won the Nobel Prize in chemistry in 1935 for the synthesis of radioelements. Radium showed great promise as a cure for cancer and was hailed as a wonder drug. As radium's discoverer, Marie was honored around the world, particularly in the United States, where she raised huge sums of money for research into its applications.

In 1934, at the age of 67, Marie Curie died from leukemia, undoubtedly caused by years of exposure to radiation. Her discovery that radiation had its origin in the atom was a milestone in nuclear physics. That this discovery was made by a young woman in a field completely dominated by men makes it doubly remarkable. Other scientists elucidated the structure of the atom, but Marie Curie was the first to show that the atom is not immutable, and her seminal discovery ushered in the atomic age.

References: R. Pflaum, *Grand Obsession* (New York: Doubleday, 1989); E. Curie, *Madame Curie: A Biography by Eve Curie*, translated by V. Sheean (New York: Doubleday, 1937); Robert R. Reid, *Marie Curie* (New York: Saturday Review Press/E. P. Dutton, 1974).

Chapter Summary

1. Radiation emitted by naturally occurring radioisotopes is of three types: alpha (α) particles ($^{4}_{2}He$), beta (β) particles (identical to electrons), and gamma (γ) rays (short wavelength EMR).
2. The speed and penetrating power of the three types of radiation increase in the following order: alpha, beta, gamma.
3. In a nuclear transmutation, a radioisotope emits an alpha or a beta particle, and a new element is formed.
4. In a correctly written nuclear equation, mass numbers and atomic numbers balance on each side of the equation.
5. Artificial transmutations of one element into another can be achieved by bombarding stable atomic nuclei with alpha particles, protons, neutrons, and other subatomic particles.
6. The rate at which a radioisotope decays is expressed as its half-life: the time required for one-half of any given quantity of the isotope to decay. Half-lives range from billionths of a second to billions of years.
7. Radiation is harmful to living organisms because it can knock electrons from atoms, forming positive ions, which destroy tissue and can cause mutations and cancer.
8. Gamma rays are the most damaging form of radiation if the source is outside the body; alpha particles are the most damaging if the source is inside the body.
9. The average person is exposed to more radiation from natural sources than from artificial sources, such as medical procedures and occupational activities.
10. Scientist do not know for certain what dose of radiation is harmful.
11. The half-lives of naturally occurring radioisotopes can be used to determine the age of ancient objects. Carbon-14 is used to determine the age of artifacts containing carbon.
12. Artificially produced radioisotopes are used in medicine, for both diagnosis and therapy, and as tracers for following the pathways of elements in chemical reactions.
13. Gamma rays can be used to kill insects and microorganisms that cause food spoilage.
14. Einstein's equation, $E = mc^2$, accounts for the enormous amount of energy that is released in nuclear fission (the splitting of an atom almost in half).
15. A very small amount of the mass of the protons and neutrons in an atomic nucleus is in the form of the binding energy, which holds the nucleus together. This energy is released by either fission of atoms of elements with high atomic masses or fusion of atoms of very light elements.
16. A self-sustaining chain reaction occurs when, as a result of the bombardment of a fissionable nucleus with a neutron, several neutrons are produced; these, in turn, cause further fission reactions and the release of increasing amounts of energy.
17. The secret U.S. Manhattan Project was responsible for the two atomic bombs that were dropped on Japan in 1945 and ended World War II. One group of scientists produced the critical mass of fissionable uranium-235 needed to sustain a chain reaction; a second group produced fissionable plutonium-239.
18. Nuclear fission can be harnessed as a source of energy for electrical power generation.

Key Terms

alpha particle (p. 119)
artifical transmutation (p. 122)
beta particle (p. 119)
binding energy (p. 137)
chain reaction (p. 138)
cosmic rays (p. 130)
critical mass (p. 138)
curie (p. 127)
gamma ray (p. 119)
Geiger counter (p. 127)
half-life (p. 125)
ionizing radiation (p. 126)
nuclear fission (p. 137)
nuclear medicine (p. 133)
nuclear reaction (p. 117)
positron (p. 123)
rad (gray) (p. 127)
radioactive decay (p. 122)
radioactivity (p. 118)
radiocarbon dating (p. 132)
radioisotope (p. 118)
radon (p. 129)
rem (sievert) (p. 128)
terrestrial radiation (p. 130)
tracers (p. 136)
transmutation (p. 120)
X-rays (p. 118)

Questions and Problems

1. Write the symbols used to represent the following:
   a. proton b. electron c. neutron

2. Give the symbol for the following:
   a. beta particle b. alpha particle c. gamma ray

3. The nuclear particle with a 2+ charge is the ________.

4. How do beta particles differ from electrons?

5. The nuclear particle with an atomic number of 0 and an atomic mass of 1 is the ________.

6. Give the numbers of protons and neutrons in each of the following nuclei:
   a. chlorine-37 b. silver-115 c. oxygen-17
   d. cesium-136 e. molybdenum-99 f. carbon-13

7. How many neutrons are in each of the following nuclei?
   a. cadmium-113 b. neon-20
   c. plutonium-234 d. barium-137

8. How does the chemical reactivity of phosphorus-31, a radioactive isotope, compare with that of phosphorus-30?

9. Write balanced nuclear equations for:
   a. emission of a beta particle by a magnesium-28 nucleus
   b. emission of an alpha particle by a lawrencium-255 nucleus
   c. emission of a beta particle by a nickel-65 nucleus

10. Lead-210 is used to prepare eyes for corneal transplants. The product of the radioactive decay of lead-210 is bismuth-210. What particle is emitted as lead-210 decays?

11. Thorium-231 is the product when a certain radioisotope emits an alpha particle. Thorium-231 is radioactive and emits a beta particle. Write a nuclear equation for the reaction in which thorium-231 is produced. Write a second equation that shows the product of thorium-231 decay.

    $\_\_\_\_\_\_ \longrightarrow {}^{231}_{90}\text{Th} + {}^{4}_{2}\alpha$

    ${}^{231}_{90}\text{Th} \longrightarrow \_\_\_\_\_\_ + {}^{0}_{-1}\beta$

12. Fill in the blank in each of the following nuclear equations:
    a. ${}^{210}_{83}\text{Bi} \longrightarrow {}^{4}_{2}\alpha + \_\_\_\_\_\_$
    b. ${}^{15}_{8}\text{O} \longrightarrow {}^{15}_{7}\text{N} + \_\_\_\_\_\_$
    c. $\_\_\_\_\_\_ \longrightarrow {}^{4}_{2}\alpha + {}^{222}_{86}\text{Rn}$

13. Fill in the blank in each of the following:
    a. ${}^{9}_{4}\text{Be} + \_\_\_\_\_\_ \longrightarrow {}^{12}_{6}\text{C} + {}^{1}_{0}\text{n}$
    b. ${}^{27}_{13}\text{Al} + {}^{2}_{1}\text{H} \longrightarrow \_\_\_\_\_\_ + {}^{4}_{2}\text{He}$

14. What is the effect on the mass number and atomic number of the reacting isotope when the following transmutations occur?
    a. A beta particle is emitted.
    b. An alpha particle is emitted.
    c. A gamma ray is emitted.

15. Write balanced nuclear equations for the following:
    a. Nickel-65 emits a beta particle.
    b. Neodymium-150 emits an alpha particle.
    c. Magnesium-28 emits a beta particle.

16. A sample of sulfur-35 is radioactive and emits beta particles. Sulfur-32 is not radioactive. What

differences in chemical reactivity would you expect between sulfur-35 and sulfur-32? Explain your answer.

17. If a nucleus of an atom of the element actinium emits an alpha particle, a new element is created. What is that element?

18. When cobalt-60 emits a gamma ray, what is the other product of the reaction?

19. Define the following:

a. atomic number

b. mass number

c. nuclear transmutation

20. Potassium-42 is a beta-emitting radioisotope used to locate brain tumors.

a. Write a nuclear reaction for the decay of potassium-42.

b. If the half-life of potassium-42 is 12.4 h, what fraction of a sample of this radioisotope remains after 62 h?

c. Give two reasons why you think this particular radioisotope is chosen for this diagnostic test.

21. The half-life of barium-131 is 12.0 days. How many grams of barium-131 will remain after four half-lives if you begin with 100 g? How many days do four half-lives take?

22. The radioisotope americium-241 is used commercially in home smoke detectors. This isotope has a half-life of 432 years.

a. How long will it take for the radioactivity of this material to drop to less than 1% of its original level?

b. On the basis of your answer to part (a), should a 20-year-old smoke detector be disposed of by placing it in the municipal trash?

23. A 100-g sample of a radioactive material has a half-life of 6.0 h. How many grams of the radioactive material remains after 18 h?

24. Describe how a Geiger counter works and how radioactivity is detected.

25. The radioisotope that is most often used to date human artifacts is ________.

26. Could archeologists use radiocarbon dating to measure the age of metal tools used by people thousands of years ago?

27. If a radioactive isotope has a half-life of 4 years, it will be considered dangerous for a minimum of how many years?

28. The synthetic radioisotope technetium-99 (Tc), which is a beta emitter, is the most widely used isotope in nuclear medicine. The following data were obtained using a Geiger counter.

Disintegrations per minute	Time (h)
180	0.0
130	2.5
104	5.0
77	7.5
59	10.0
46	12.5
24	17.5

a. Make a graph of this information similar to Figure 5.4, and determine the half-life of technetium-99. (Remember that disintegrations per minute are directly proportional to the grams of radioisotope remaining.)

b. How long after injection into the patient would it take for 90% of the technetium-99 to decay?

29. Chemical reactions can often be used to change a toxic compound into another compound that is not as toxic. Why can't chemical treatment technology be applied to make nuclear waste harmless?

30. A so-called expert suggested that strontium-90 deposited in the Nevada desert during nuclear testing would undergo radioactive decay more rapidly than strontium-90 elsewhere because of the high desert temperatures. Do you agree with this assessment? Explain.

31. Prior to the Nuclear Test Ban Treaty of 1963, the average exposure to radioactive fallout from nuclear weapons testing amounted to 30 mrem/yr per person. In 1990, the average exposure was only 1 mrem/yr per person. Estimate the additional exposure for an average individual for the period 1990–2000, if open air nuclear bomb testing had not been banned. (Assume that the average exposure per year would continue to be 30 mrem per person.)

32. An oil painting alleged to be painted by Rembrandt (1606–1669) is subjected to radiocarbon dating. The carbon-14 content of the canvas used is 0.96 times that of a living tree. If the half-life of carbon-14 is 5720 years, could this painting be an original Rembrandt?

33. Would the radioactive decay of uranium-238 to lead-208 provide an accurate method for determining the age of a material thought to be about 500 years old?

34. Define the following:
   a. nuclear fission
   b. chain reaction
   c. critical mass

35. How is the equation $E = mc^2$ used to calculate the amount of energy that is released by an atomic bomb?

36. What naturally occurring isotope can be used to make an atomic bomb?

37. Name the nuclear reaction that produces the sun's energy.

38. What percentage of uranium-235 does naturally occurring uranium ore contain?

39. Describe the sources of everyday radiation to which the general public is exposed and indicate the percentage each source contributes to average yearly exposure.

40. What is the naturally occurring radioisotope that people are most likely to encounter?

41. **a.** Why is radioactivity used in treating cancer? How is treatment designed so that damage to healthy cells is minimized?
   b. Why is a critical mass of radioactive material necessary for an atomic bomb? Could a little bomb be made with less than the critical mass?
   c. Some food is irradiated prior to sale. Why doesn't the food become radioactive and poisonous?
   d. Where does radon come from? Where is it most likely to be found in the home? How can you lessen your exposure to it?
   e. Does the bulk of the radiation most of us are exposed to come from natural or artificial sources?
   f. What is the relative contribution of natural sources to the average person's total exposure to radiation?

42. Which of the following has the greatest penetrating ability—alpha particle, beta particle, or gamma ray?

43. What type of shield is necessary to stop each of the following?
   a. X-rays **b.** alpha particles
   c. beta particles **d.** gamma rays

44. Which tissues in your body are most easily damaged by radiation?

45. Why are cancer cells vulnerable to radiation therapy?

46. Why are alpha particles very dangerous when the source of the particles is inside your body and not very dangerous when it is outside your body?

47. Why are gamma rays but not alpha particles used to irradiate food?

48. Why is the rem the most useful unit to use when making an assessment of the impact of nuclear radiation on people?

49. One gram of uranium-234 produces as much energy as 14 barrels of crude oil. How many barrels of crude oil can be saved for every kilogram of uranium-235 used in a nuclear power plant?

50. Why does limiting the number of neutrons limit the rate of a nuclear reaction in an electrical power plant that uses uranium-235 as a fuel?

Answers to Practice Exercises

5.1 $^{93}_{40}\text{Zr} \longrightarrow {}^{0}_{-1}\beta + {}^{93}_{41}\text{Nb}$

5.2 $^{1}_{1}\text{H} + {}^{35}_{17}\text{Cl} \longrightarrow {}^{4}_{2}\alpha + {}^{32}_{16}\text{S}$

5.3 17,190 years

5.4 **a.** 0.0025 g; **b.** 0.00003 g

Chapter 6

Chemical Reactions

Chapter Objectives

In this chapter, you should gain an understanding of:

- How to write balanced chemical equations
- Calculations based on chemical equations
- The concept of a mole
- Entropy and the difference between exothermic and endothermic reactions
- Factors that influence the rate of a chemical reaction
- Reversible reactions and chemical equilibrium

In a forest fire, carbon-containing compounds in trees and undergrowth react vigorously with oxygen in the air. Carbon dioxide is produced, and the forest is reduced to ashes.

ONLINE FEATURES

www.jbpub.com/chemistry

- Chemistry and the Environment
- eLearning
- Chemistry in the News
- Special Topics
- Research & Reference Links
- Ask the Authors

SINCE ANCIENT TIMES, HUMANS HAVE BEEN initiating chemical reactions to produce new substances from natural materials—pottery from clay, medicines and dyes from plants, and beer and wine from fermenting grain. Many of these useful reactions were discovered accidentally. Today, chemists are able to predict the course of a chemical reaction, and they know how to manipulate conditions to increase both the speed of a reaction and the yield of the desired product.

Can we tell if a chemical reaction will occur when two substances are mixed together? Will the reaction occur spontaneously, or will it need an input of energy to get it started? How fast will the reaction go, and can it be made to go faster or slower? And how much of the product can be expected if we use a given amount of starting materials? This chapter answers these questions. As you will learn, all chemical reactions—whether occurring in a saucepan on the stove, in a chemical factory, in the body, or in the atmosphere, the ocean, or any other part of the environment—are governed by the same basic principles.

6.1 Chemical Equations

In any chemical reaction, one or more substances, called **reactants,** are changed into one or more different substances, called **products.** A chemical reaction can be represented in a shorthand form by a chemical equation.

Chemical symbols are equivalent to the letters of the alphabet, chemical formulas are equivalent to words, and chemical equations are equivalent to sentences.

Carbon (C), the main ingredient in coal and charcoal, burns in atmospheric oxygen to form carbon dioxide (CO_2). Any time we grill food over hot charcoal, the carbon in the charcoal is burning and producing carbon dioxide (**Fig. 6.1**). The chemical equation for this reaction is

$$\underset{\text{carbon}}{C} + \underset{\text{oxygen}}{O_2} \longrightarrow \underset{\text{carbon dioxide}}{CO_2}$$

The plus sign (+) is read as "and" and indicates that the reactants (C and O_2) are mixed together in some way. The arrow ($\longrightarrow$) means "react to yield."

A chemical equation tells us much more than just what substances are reacting and what substances are produced. It tells us how many atoms or molecules of each reactant take part in the reaction and how many atoms or molecules of each product are formed. The above equation tells us that one atom of carbon (C) reacts with one molecule of oxygen (O_2) to yield one molecule of carbon dioxide (CO_2).

A chemical equation is also a bookkeeping device: It keeps track of the numbers of atoms of the different elements taking part in a reaction. In accordance with the law of conservation of matter, which states that matter is neither created nor destroyed in a chemical reaction (Chapter 3), the number of atoms of each element present at the start of a chemical reaction always equals the number present at the end of the reaction. Therefore, in a correctly written, or *balanced,* chemical equation for a reaction, the total number of each kind of atom on the left side of the arrow equals the total number of each kind of atom on the right side of the arrow.

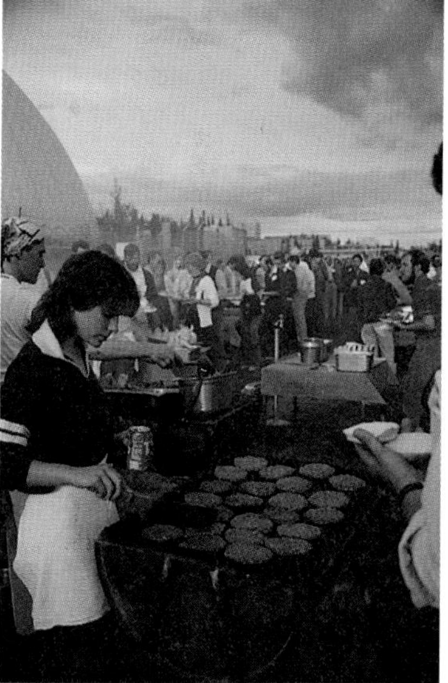

FIGURE 6.1

When food is grilled over hot charcoal, carbon in the charcoal combines with oxygen in the air to form carbon dioxide.

6.2 Writing Balanced Chemical Equations

The first step in writing a balanced chemical equation is to write the correct formulas for the reactants and products. The next step is to add the numbers of atoms

In a balanced chemical equation, the sum of the numbers of atoms of each element in the reactants (to the left of the arrow) is equal to the sum of the numbers of atoms of each element in the products (to the right of the arrow).

on both sides of the equation and see if they are equal. The equation for the burning of charcoal we considered earlier is properly balanced: The numbers of carbon and oxygen atoms on the left side of the equation are equal to the numbers on the right side.

carbon		oxygen		carbon dioxide
C	+	O_2	$\longrightarrow$	CO_2

	Reactants	Product
C	1	1
O	2	2

balanced

Not all chemical reactions can be so easily balanced. For example, it is not as simple to write the correct equation for the explosive reaction that occurs between hydrogen (H_2) and oxygen (O_2) to form water (H_2O) when an electric spark is passed through a mixture of the two gases. As the first step in writing the balanced equation for this reaction, we enter the correct formulas for the reactants and the product in the equation.

hydrogen		oxygen		water	
H_2	+	O_2	$\longrightarrow$	H_2O	(unbalanced)

	Reactants	Product
H	2	2
O	2	1

unbalanced

The number in front of a formula in a chemical equation is the *coefficient.*

We can see that this equation is not balanced. There are 2 atoms of oxygen on the left side of the equation and only 1 on the right side. We could balance the oxygen atoms by assuming that 2 molecules of water are formed and putting the coefficient 2 in front of the formula for the water molecule (remember that a coefficient in front of a formula multiplies everything in the formula).

$$H_2 + O_2 \longrightarrow 2\ H_2O \quad \text{(unbalanced)}$$

	Reactants	Product
H	2	$2 \times 2 = 4$
O	2	2

unbalanced

But this does not give us a balanced equation. In establishing an oxygen balance, we have created a hydrogen imbalance. This imbalance can be corrected by having 2 molecules of hydrogen react with 1 molecule of oxygen:

$$2\,H_2 + O_2 \longrightarrow 2\,H_2O \quad \text{(balanced)}$$

	Reactants	Products
H	2 × 2 = 4	2 × 2 = 4
O	2	2

balanced

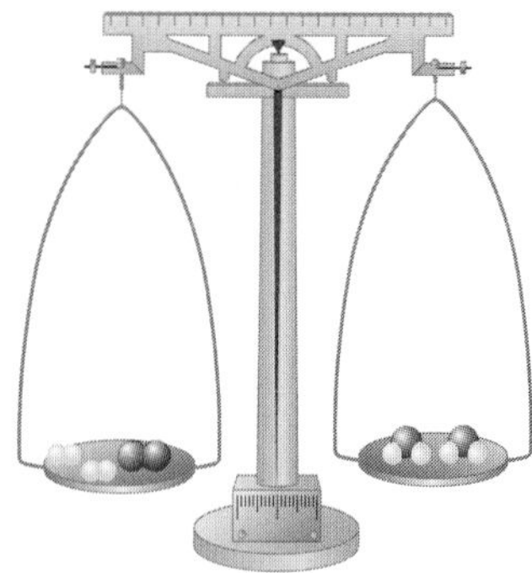

Always remember that the *subscripts* in the formulas for the reactants or products cannot be changed to balance an equation because such a change would alter the *identity* of one or more of the substances taking part in the reaction. For example, the reaction between hydrogen and oxygen to form water cannot be written as follows:

$$\underset{\text{hydrogen}}{H_2} + \underset{\text{oxygen}}{O_2} \longrightarrow \underset{\text{hydrogen peroxide}}{H_2O_2} \quad \text{(incorrect)}$$

This equation is balanced, but it is incorrect for the reaction we are working with; the formula of the product, H_2O_2, represents hydrogen peroxide rather than water, H_2O.

6.3 Some Environmentally Important Reactions

For practice in balancing equations, we can consider several chemical reactions that are important for the environment.

Oxides of Nitrogen and Smog

Automobiles are a major source of air pollution (Chapter 11). At ordinary temperatures, nitrogen (N_2) and oxygen (O_2) in the atmosphere do not combine, but at the high temperatures that exist inside running automobile engines, they react to produce the colorless gas nitric oxide (NO). As a first step, we can write the equation for this reaction as follows:

Nitrogen and oxygen make up approximately 78% and 21% by volume of the air we breathe.

$$\underset{\text{nitrogen}}{N_2} + \underset{\text{oxygen}}{O_2} \longrightarrow \underset{\text{nitric oxide}}{NO} \quad \text{(unbalanced)}$$

	Reactants	Product
N	2	1
O	2	1

unbalanced

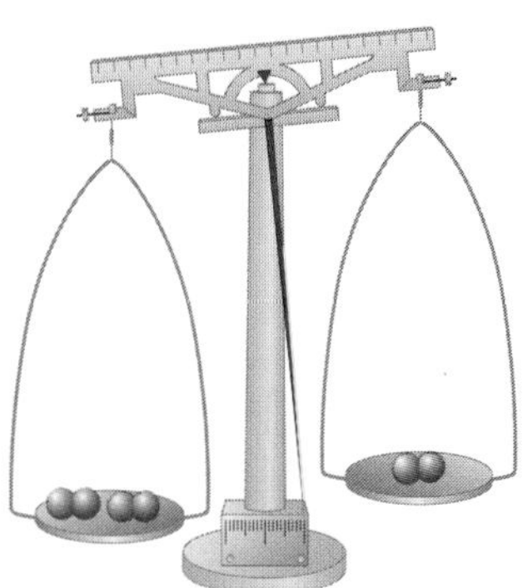

The equation can be balanced if we have 2 molecules of nitric oxide.

$$N_2 + O_2 \longrightarrow 2\ NO \quad \text{(balanced)}$$

	Reactants	Product
N	2	2
O	2	2

balanced

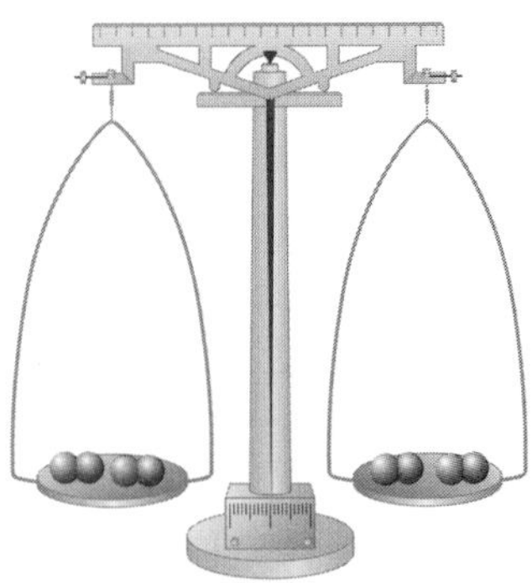

Once in the atmosphere, nitric oxide combines rapidly with atmospheric oxygen to form nitrogen dioxide (NO_2), the red-brown toxic gas that is responsible for the color of the yellow-brown smog that often settles over Los Angeles and Denver.

nitric oxide + oxygen → nitrogen dioxide

$$NO + O_2 \longrightarrow NO_2 \quad \text{(unbalanced)}$$

	Reactants	Product
N	1	1
O	1 + 2 = 3	2

unbalanced

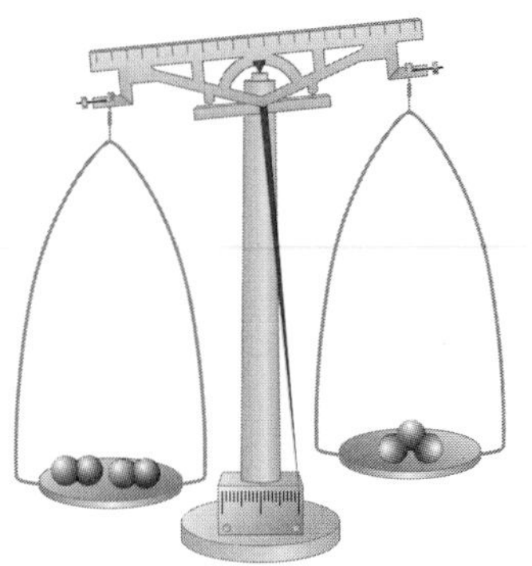

The equation can be balanced if we have 2 molecules of nitric oxide and 2 molecules of nitrogen dioxide.

$$2\ NO + O_2 \longrightarrow 2\ NO_2 \quad \text{(balanced)}$$

	Reactants	Product
N	2	2
O	2 + 2 = 4	2 × 2 = 4

balanced

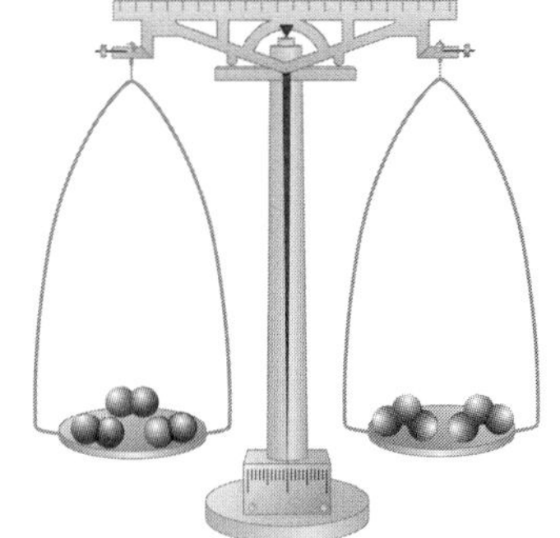

Much of the coal burned in the United States contains sulfur in the form of the mineral pyrite (FeS_2).

Oxides of Sulfur and Acid Rain

One of the main causes of acid rain is the burning of coal that has a high sulfur content (Chapter 11). When such coal is burned, the sulfur (S) is converted to the pungent, choking gas sulfur dioxide (SO_2).

sulfur		oxygen		sulfur dioxide	
S	$+$	O_2	$\longrightarrow$	SO_2	(balanced)

	Reactants	**Product**
S	1	1
O	2	2

balanced

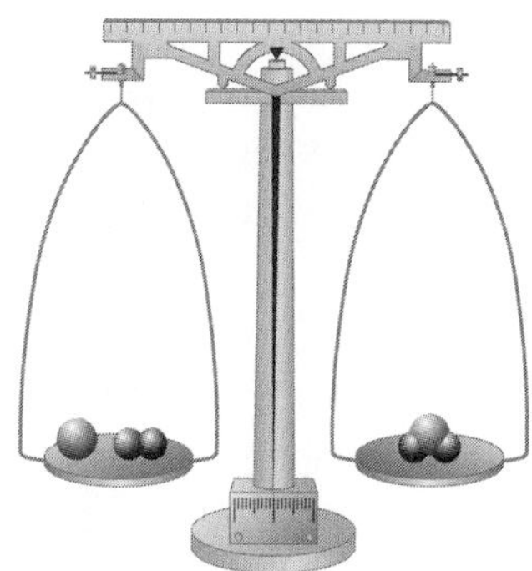

If released into the atmosphere, sulfur dioxide combines with atmospheric oxygen to form sulfur trioxide (SO_3).

sulfur dioxide		oxygen		sulfur trioxide	
SO_2	$+$	O_2	$\longrightarrow$	SO_3	(unbalanced)

	Reactants	**Product**
S	1	1
O	2 + 2 = 4	3

unbalanced

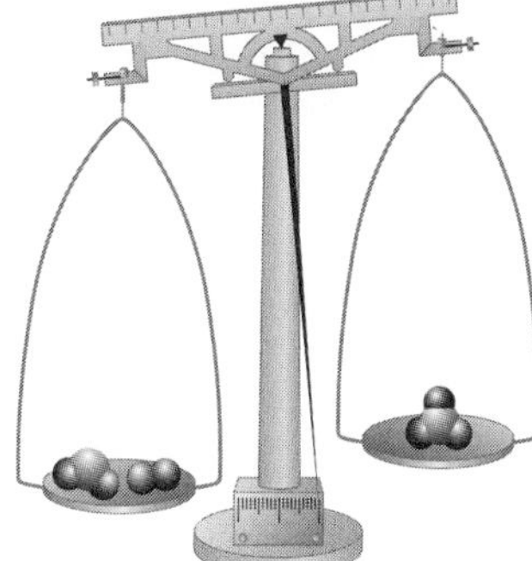

This equation can be balanced by having 2 molecules of sulfur dioxide and 2 molecules of sulfur trioxide.

$$2\ SO_2 + O_2 \longrightarrow 2\ SO_3 \quad \text{(balanced)}$$

	Reactants	**Product**
S	2	2
O	(2 × 2) + 2 = 6	2 × 3 = 6

balanced

Sulfur trioxide combines very rapidly with water vapor in the atmosphere to form sulfuric acid (H_2SO_4), which reaches the earth in raindrops.

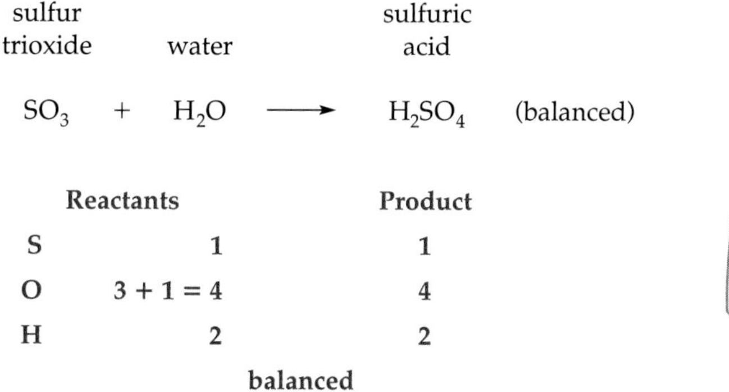

sulfur trioxide		water		sulfuric acid	
SO_3	$+$	H_2O	$\longrightarrow$	H_2SO_4	(balanced)

	Reactants	**Product**
S	1	1
O	3 + 1 = 4	4
H	2	2

balanced

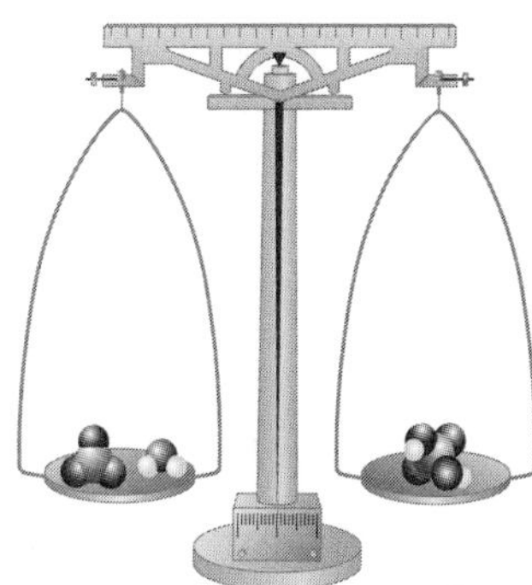

FIGURE 6.2

Green plants are fundamental to all life on earth. In the process of photosynthesis, they use light energy from the sun to convert carbon dioxide and water to the sugar glucose, releasing oxygen as a by-product.

Photosynthesis

Practically all living creatures are directly or indirectly dependent on photosynthesis for their food.

In the process of photosynthesis, green plants absorb solar energy and use it to convert carbon dioxide and water into glucose (a simple sugar) and oxygen (Fig. 6.2). As we saw in Chapter 2, if green plants did not store energy from the sun and replenish the atmosphere with oxygen, life on earth as we know it could not exist. Photosynthesis is a complex process that occurs in many steps. The overall reaction is given by the following equation:

water	carbon dioxide		glucose	oxygen	
H_2O +	CO_2	$\xrightarrow{\text{solar energy}}$	$C_6H_{12}O_6$ +	O_2	(unbalanced)

	Reactants	Products
H	2	12
O	1 + 2 = 3	6 + 2 = 8
C	1	6

unbalanced

To match the 6 carbon atoms and 12 hydrogen atoms in the sugar, 6 molecules of carbon dioxide and 6 molecules of water are required. To balance the oxygen atoms, 6 molecules of oxygen are needed.

$$6\ H_2O + 6\ CO_2 \longrightarrow C_6H_{12}O_6 + 6\ O_2 \quad \text{(balanced)}$$

	Reactants	Products
H	6 × 2 = 12	12
O	6 + (6 × 2) = 18	6 + (6 × 2) = 18
C	6	6

balanced

6.4 Chemical Arithmetic

Chemical equations, in addition to showing the numbers of atoms and molecules that are involved in a reaction, tell us the *relative masses* (weights) of the reactants and the products. Such information is essential not only to the laboratory chemist who wants to make a few grams of a particular compound, but also to the plant manager who is planning for large-scale production of an industrial chemical.

Mass Relationships in Chemical Equations

As an example, let's look at the mass (weight) relationships in the equation for the reaction of carbon with oxygen.

$$C + O_2 \longrightarrow CO_2$$

The equation shows that one atom of carbon reacts with one molecule of oxygen to form one molecule of carbon dioxide.

Recall from Chapter 3 that the atomic mass unit (amu) of the most common isotope of carbon is exactly 12 and the masses of the atoms of all other elements are compared to that value. (The atomic masses, or atomic weights, of the elements are listed inside the back cover of this book.)

Since one atom of carbon weighs 12 amu and one atom of oxygen weighs 16 amu, the equation also shows that, in agreement with the law of conservation of mass:

> 12 amu of carbon combine with 32 (2×16) amu of oxygen to form 44 ($12 + 32$) amu of carbon dioxide.

That is, the sum of the masses of the reactants equals the mass of the product.

$$\begin{array}{ccccc} C & + & O_2 & \longrightarrow & CO_2 \\ 12 & & 32\ (2 \times 16) & & 44\ (12 + 32) \end{array}$$

44 is the formula mass of carbon dioxide.

Determining Formula Masses

The formula mass of a compound such as carbon dioxide (CO_2) or water (H_2O) is often called its *molar mass.* For an ionic compound such as sodium chloride (NaCl), which does not exist as a discrete molecule, the term *formula mass* is used.

The **formula mass** of a substance is the sum of the atomic masses of all the atoms in the substance's chemical formula.

The first step in analyzing the mass relationships in any chemical equation is to determine the formula masses of the reactants and the products. As we have seen for the reaction of carbon with oxygen, this value is easily obtained for simple molecules like O_2 and CO_2, but it can be more difficult for larger and more complex molecules. The following examples describe the steps to be followed.

EXAMPLE 6.1 Determining formula masses

Calculate the formula mass of silver nitrate ($AgNO_3$), the compound from which most other silver compounds are produced.

Solution A formula mass is calculated by adding together the atomic masses of the constituent atoms. Refer to the table inside the back cover of this book to obtain the atomic mass of each type of atom.

The formula $AgNO_3$ contains

1 atom of Ag		= 108 amu
1 atom of N		= 14 amu
3 atoms of O	3×16	= 48 amu
Formula mass of $AgNO_3$		= 170 amu

Practice Exercise 6.1

Calculate the formula mass of sulfur dioxide (SO_2), the main cause of acid rain.

EXAMPLE 6.2

Calculate the formula mass of ammonium phosphate, $(NH_4)_3PO_4$, which is an important fertilizer.

Solution The formula $(NH_4)_3PO_4$ contains

3 atoms of N	$3 \times 14 =$	42 amu
$3 \times 4 = 12$ atoms of H	$12 \times 1 =$	12 amu
1 atom of P	=	31 amu
4 atoms of O	$4 \times 16 =$	64 amu
Formula mass of $(NH_4)_3PO_4$		= 149 amu

Practice Exercise 6.2

Calculate the formula mass of each of the following: **a.** $KMnO_4$; **b.** $Ca(OH)_2$

Mass relationships for an equation can be expressed in mass units other than atomic mass units; they can be expressed in grams, kilograms, pounds, tons, and so on. For example, we can also express the equation $C + O_2 = CO_2$ as follows:

12 lb of carbon combine with 32 lb of oxygen to form 44 lb of carbon dioxide.
12 g of carbon combine with 32 g of oxygen to form 44 g of carbon dioxide.

Calculations Based on Chemical Equations

The law of conservation of mass is the basis for solving mathematical problems based on chemical equations.

If the balanced chemical equation for a reaction is known, it is possible to calculate the weight of product that can be obtained from a given amount of reactant or the amount of reactant required to produce a desired amount of product. As Examples 6.3 and 6.4 will show, we can, for example, find answers to the following questions:

1. How many kilograms of aluminum can be produced from 10 kg of the mineral bauxite (Al_2O_3)?
2. How many kilograms of nitrogen are required to react with 5 kg of hydrogen to form ammonia, the starting material for the manufacture of synthetic fertilizers?

Approximately ten times as much energy is required to produce a ton of aluminum as is needed to produce a ton of steel.

ALUMINUM FROM BAUXITE As you learned in Chapter 2, aluminum is one of the most essential metals in modern society. Aluminum is an abundant element that constitutes approximately 8% of the earth's crust. However, the metal is tied up in ores, and until a little over a century ago, there was no feasible, cost-effective way to extract it. Then, in 1886, two young chemists, American Charles M. Hall **(Fig. 6.3)** and Frenchman Paul Heroult simultaneously but independently discovered a way to extract aluminum from bauxite, a mineral composed mainly of Al_2O_3. They discovered that when an electric current was passed through a mixture of purified Al_2O_3 and molten cyrolite (a high-melting aluminum mineral, Na_3AlF_6), molten aluminum metal separated out and settled at the bottom of the reaction vessel **(Fig. 6.4)**.

FIGURE 6.3

While a student, Charles M. Hall (1863–1914) began looking for a cheap way to produce aluminum, and at age 22, he invented the electrolytic process that is still used today. He founded the Aluminum Corporation of America (ALCOA) and became a very wealthy man. Generally overlooked is the contribution his elder sister Julia made to the discovery. She, like Charles, studied chemistry at Oberlin College, and she worked with her brother in the laboratory in the family's woodshed. Julia's records were crucial in establishing that Charles had discovered the electrolytic process about a month before Paul Heroult.

EXAMPLE 6.3 Calculating amounts of products or reactants

How many kilograms of aluminum (Al), can be obtained from 10 kg of bauxite (Al_2O_3)?

Solution

1. The first step in solving any problem based on a chemical reaction is to write the balanced equation for the reaction.

   The balanced equation for the formation of aluminum from bauxite is

$$2\ Al_2O_3 \longrightarrow 4\ Al + 3\ O_2$$

2. Next, determine the formula mass of Al_2O_3, as shown in Example 6.1. (The formula mass of O_2 is not needed because it is not mentioned in the question.)

Atomic mass of Al		= 27 amu
Formula mass of Al_2O_3:		
2 atoms of Al	$2 \times 27 =$	54 amu
3 atoms of O	$3 \times 16 =$	48 amu
Formula mass of Al_2O_3		= 102 amu

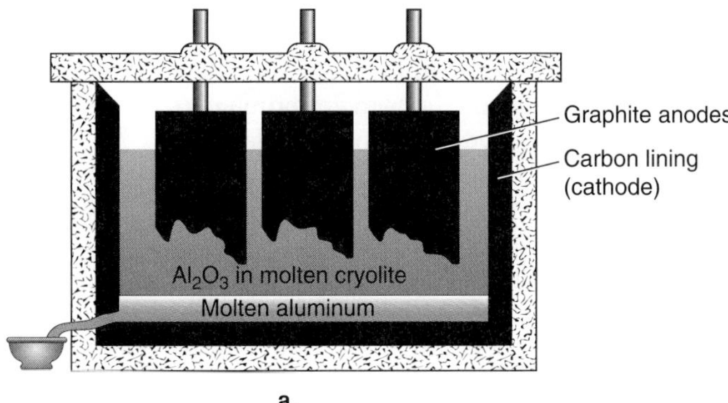

a.

b.

FIGURE 6.4

(a) An electrolysis cell for the production of aluminum from bauxite. Molten aluminum is denser than the molten mixture of Na_3AlF_6 and Al_2O_3 and collects at the bottom of the cell. (b) Thin sheets of aluminum metal are used to make cans for soft drinks and other products.

3. Substitute these values in the balanced chemical equation:

$$
\begin{array}{ccccc}
2\ Al_2O_3 & \longrightarrow & 4\ Al & + & 3\ O_2 \\
2 \times 102 \text{ amu} & & 4 \times 27 \text{ amu} & & \\
204 \text{ amu} & & 108 \text{ amu} & &
\end{array}
$$

4. State the mass (weight) relationship between Al_2O_3 and Al in terms of the given mass unit (kg).

$$204 \text{ kg } Al_2O_3 \quad \text{yield} \quad 108 \text{ kg Al}$$

5. Solve the problem using dimensional analysis, as explained in Appendixes C and E.
   a. Given quantity: 10 kg Al_2O_3
   b. Required quantity: ? kg Al
   c. The conversion factor for solving the problem is the mass relationship between Al_2O_3 and Al as determined from the balanced equation:

$$\text{conversion factor} = \frac{108 \text{ kg Al}}{204 \text{ kg } Al_2O_3}$$

   d. Solution:

$$10 \cancel{\text{kg } Al_2O_3} \times \frac{108 \text{ kg Al}}{204 \cancel{\text{kg } Al_2O_3}} = 5.92 \text{ kg Al}$$

   That is, 5.92 kg of aluminum can be obtained from 10 kg of bauxite.

Practice Exercise 6.3

How many kilograms of iron (Fe) can be obtained from 5 kg of the mineral hematite (Fe_2O_3)?

FIGURE 6.5

German chemist Fritz Haber (1868–1934). At the beginning of World War I (1914–1918), Germany relied on nitrates imported from South America to make munitions. After a British naval blockade cut off this supply, Germany was able to maintain weapons production by industrializing the Haber process and using ammonia to make the needed nitrogen compounds. It is ironic that Haber, who contributed so much to Germany's war effort, was dismissed from his position as director of the Kaiser Wilhelm Institute for Physical Chemistry in 1933 because he was a Jew.

Ammonia from Nitrogen and Hydrogen: The Haber Process In 1908, the German chemist Fritz Haber (Fig. 6.5), established that under certain conditions, including high temperature and high pressure, atmospheric nitrogen (N_2) combines with hydrogen gas (H_2) to produce ammonia (NH_3). Today, in the United States alone, this process is used to produce over 10 million tons of ammonia for use as fertilizer (Fig. 6.6). Before the development of the Haber process, farmers relied almost entirely on animal manure, crop rotation, and dwindling natural deposits of sodium nitrate (Chilean saltpeter) to replenish the nitrogen in the soil. Synthetic fertilizers made from ammonia play a crucial role in providing the food needed by the world's ever-growing population.

Example 6.4 Calculating amounts of products or reactants

Five kilograms of hydrogen (H_2) react with nitrogen (N_2) to form ammonia (NH_3).

a. How many kilograms of nitrogen are required?

b. How many kilograms of ammonia are formed?

Solution

1. Write the balanced chemical equation for the reaction:

$$N_2 \quad + \quad 3\ H_2 \quad \longrightarrow \quad 2\ NH_3$$

2. Determine the formula masses for N_2, H_2, and NH_3:

Formula mass of N_2	$2 \times 14 = 28$ amu
Formula mass of H_2	$2 \times 1 = 2$ amu
Formula mass of NH_3	$14 + (3 \times 1) = 17$ amu

3. Substitute these values in the balanced equation:

$$N_2 \quad + \quad 3\ H_2 \longrightarrow 2\ NH_3$$

N_2	$3\ H_2$	$2\ NH_3$
28 amu	3×2 amu	2×17 amu
	6 amu	34 amu

4. State the mass relationships between reactants and product in terms of the given mass units (kg).

28 kg of N_2 react with 6 kg of H_2 to give 34 kg of NH_3

5. a. Use dimensional analysis to determine the *number of kilograms of N_2* that will react with 5 kg of H_2.

Given quantity: 5 kg H_2
Required quantity: ? kg N_2

Conversion factor (from step 4): $\dfrac{28 \text{ kg } N_2}{6 \text{ kg } H_2}$

Solution: $5 \cancel{\text{kg } H_2} \times \dfrac{28 \text{ kg } N_2}{6 \cancel{\text{kg } H_2}} = 23.3 \text{ kg } N_2$

Thus, 23.3 kg of N_2 are required for the formation of NH_3 using 5 kg of H_2.

b. Similarly, determine the *number of kilograms of NH_3* that can be formed.

Given quantity: 5 kg H_2
Required quantity: ? kg NH_3

Conversion factor (from step 4): $\dfrac{34 \text{ kg } NH_3}{6 \text{ kg } H_2}$

Solution: $5 \cancel{\text{kg } H_2} \times \dfrac{34 \text{ kg } NH_3}{6 \cancel{\text{kg } H_2}} = 28.3 \text{ kg } N_2$

Thus, 28.3 kg of NH_3 are produced using 5 kg of H_2.

FIGURE 6.6

A chemical plant that manufactures ammonia from nitrogen gas and hydrogen gas using the Haber process. Ammonia is one of the top ten chemicals produced in the United States.

Practice Exercise 6.4

Oxygen masks used in emergencies contain potassium superoxide (KO_2), which reacts with exhaled carbon dioxide and water to produce oxygen (O_2).

$$4\ KO_2 + 2\ H_2O + 4\ CO_2 \longrightarrow 4\ KHCO_3 + 3\ O_2$$

If a person wearing such a mask exhales 0.70 g of CO_2 per minute, how many grams of KO_2 must the mask contain to remove all exhaled CO_2 for 10 minutes?

6.5 The Mole

Let's return once again to the equation for the burning of charcoal or coal:

$$C + O_2 \longrightarrow CO_2$$

As we saw earlier, the equation gives the following information:

$$\begin{array}{ccccc} 1 \text{ atom C} & + & 1 \text{ molecule } O_2 & \longrightarrow & 1 \text{ molecule } CO_2 \\ 12 \text{ g C} & + & 32 \text{ g } O_2 & \longrightarrow & 44 \text{ g } CO_2 \end{array}$$

In addition to the above information, chemists sometimes want to know the quantities of substances that take part in a reaction in terms of *numbers* of atoms and *numbers* of molecules. For example, how many atoms of carbon are there in 12 g of carbon? Because atoms are incredibly small, it is obviously impossible to count out numbers of atoms or molecules. Even the very smallest amount of carbon visible to the naked eye is made up of billions of carbon atoms. Instead, chemists use the **mole** as their counting unit to keep track of numbers of atoms or molecules.

In our daily lives, we use a variety of counting units. At the grocery store, eggs are purchased by the dozen (12 eggs) or half-dozen, not as individual eggs. In offices and schools, pencils are often purchased by the gross (144 pencils) and paper by the ream (500 sheets) **(Fig. 6.7)**. In the same way that 1 dozen is understood to equal 12, 1 gross to be 144, and 1 ream to equal 500, it is understood by chemists that

A mole is a specific number of structural units (atoms, molecules, or formula units).

$$\textbf{1 mole} = \mathbf{6.02 \times 10^{23}} \textbf{ structural units}$$

The structural units may be atoms, molecules, or formula units.

A mole of one compound or element contains the same number of structural units (Avogadro's number) as a mole of another compound or element.

$$\begin{array}{ll} \text{1 mole of C atoms} & = 6.02 \times 10^{23} \text{ C atoms} \\ \text{1 mole of } O_2 \text{ molecules} & = 6.02 \times 10^{23}\ O_2 \text{ molecules} \\ \text{2 moles of } CO_2 \text{ molecules} & = 2(6.02 \times 10^{23})\ CO_2 \text{ molecules} \\ \text{1 mole of NaCl formula units} & = 6.02 \times 10^{23} \text{ NaCl formula units} \end{array}$$

The number 6.02×10^{23} is called **Avogadro's number,** in honor of the Italian scientist Amadeo Avogadro (1776–1856), whose pioneering work with gases led to the discovery of this basic number. The theoretical basis for Avogadro's number is beyond the scope of this book and will not be discussed.

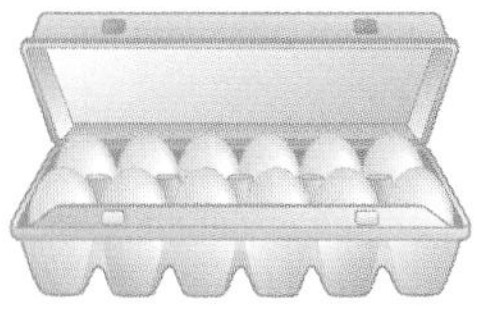

12 eggs = 1 dozen

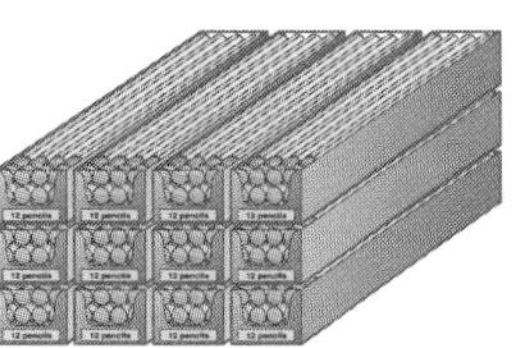

144 pencils = 1 gross

500 sheets of paper = 1 ream

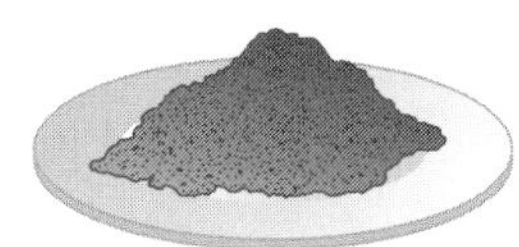

6.02×10^{23} carbon atoms = 1 mole

FIGURE 6.7

Convenient counting units for various objects. Chemists use the mole as the counting unit for atoms, molecules, ions, and formula units. In the same way that a dozen eggs means 12 eggs, a mole of carbon atoms means 6.02×10^{23} carbon atoms.

FIGURE 6.8

The mass in grams of a mole of a pure substance is calculated from its formula mass. Shown are mole amounts of several common substances: 342.30 g of sucrose (back), 207.21 g of lead (lead shot, middle left), 294.21 g of potassium dichromate (center), 200.61 g of mercury (middle right), 18.016 g of water (middle far right), 63.54 g of copper (pennies, front left), 58.45 g of sodium chloride (front center), and 32.066 g of sulfur (front right).

Because atoms and molecules are incredibly small, Avogadro's number is almost incomprehensibly large:

$$6.02 \times 10^{23} = 602{,}000{,}000{,}000{,}000{,}000{,}000{,}000$$

If the entire population of the United States (some 250 million people) spent \$1 bills at the rate of one per second, 12 hours a day, every day of the year, it would take 170 million years for Avogadro's number of \$1 bills to be used up.

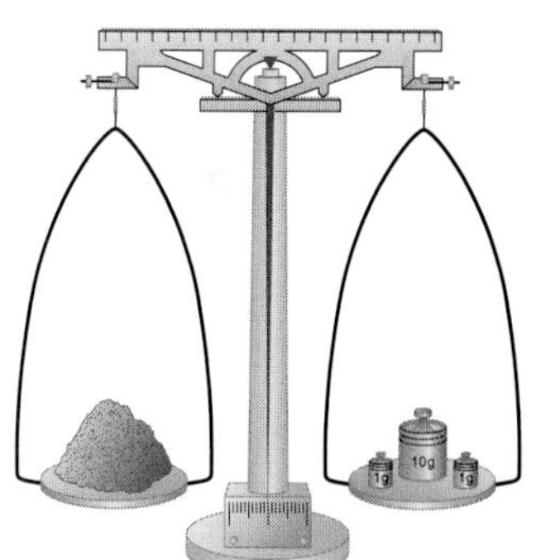

FIGURE 6.9

The mass of a mole (the molar mass) of a chemical is the atomic mass or formula mass of the chemical expressed in grams (g). Thus, the mass (or weight) of a mole (6.02×10^{23}) of carbon atoms is 12.0 g.

Molar Mass

Even though a mole is such a large number of atoms or molecules, a mole of a substance is a convenient amount to work with in the laboratory. The mass of a mole, like the mass of a dozen, depends on the identity of the structural units being weighed. In the same way that a dozen grapes do not weigh as much as a dozen apples, a mole of carbon atoms do not weigh as much as a mole of gold atoms. The mass of a mole (or molar mass) of a substance varies depending on the atomic masses of the elements that make it up. A mole of a few common substances is shown in **Figure 6.8**.

The **molar mass** of a substance is defined as *the atomic or formula mass expressed in grams*. Thus, for carbon, the molar mass is 12 g (**Fig. 6.9**). Molar masses of some of the substances taking part in the reactions we have just studied are listed in **Table 6.1**.

Table 6.1 Some Formula Masses and Molar Masses

Chemical	Formula Mass	Molar Mass
C	12 amu	12 g
O_2	32 amu	32 g
NH_3	17 amu	17 g
Al_2O_3	102 amu	102 g

EXAMPLE 6.5 Using molar mass in calculations

How much would 1 mole of calcium chloride ($CaCl_2$) weigh in grams?

Solution Calculate the formula mass of calcium chloride:

$$\begin{aligned} \text{1 atom of Ca} &= 1 \times 40.1 \text{ amu} = 40.1 \text{ amu} \\ \text{2 atoms of Cl} &= 2 \times 35.5 \text{ amu} = \underline{71.0 \text{ amu}} \\ \text{Formula mass of } CaCl_2 &= 111.1 \text{ amu} \end{aligned}$$

The molar mass of $CaCl_2$ is 111.1 g. So 1 mole of $CaCl_2$ weighs 111.1 g.

Practice Exercise 6.5

How much would 1 mole of glucose ($C_6H_{12}O_6$) weigh in grams?

EXAMPLE 6.6

How many moles of calcium chloride ($CaCl_2$) are there in 22.2 g of pure $CaCl_2$?

Solution From the molar mass of $CaCl_2$ calculated in Example 6.5, we know that 1 mole of calcium chloride weighs 111.1 g.

$$22.2 \text{ g } \cancel{CaCl_2} \times \frac{1 \text{ mole } CaCl_2}{111.1 \text{ g } \cancel{CaCl_2}} = 0.20 \text{ mole } CaCl_2$$

Practice Exercise 6.6

Calculate the number of moles in each of the following:

a. 47.45 g CH_3Br **b.** 16.17 g NaCN

Calculations Using the Concepts of Mole and Molar Mass

A balanced chemical equation shows both the mole relationships and the mass relationships between the substances involved in a particular reaction. For example, from the equation for the reaction of hydrogen with nitrogen to produce ammonia, we know the following:

Balanced equation:	N_2 +	$3\ H_2$	$\longrightarrow$	$2\ NH_3$
Number of moles:	1 mole	3 moles		2 moles
Molar mass:	28 g	2 g		17 g
Molar mass × number of moles:	28 g	6 g		34 g

We can express this in words as follows:

1 mole of nitrogen reacts with 3 moles of hydrogen to produce 2 moles of ammonia
28 g of nitrogen react with 6 g of hydrogen to produce 34 g of ammonia

Thus, from the correctly balanced equation for a reaction, we can calculate the number of moles, or grams, of a particular reactant needed to produce a certain number of moles, or grams, of a product.

EXAMPLE 6.7 Expressing molar and mass relationships

Ozone (O_3) is formed in the stratosphere from oxygen (O_2):

$$3\ O_2 \xrightarrow{\text{sunlight}} 2\ O_3$$

Express this equation in words in terms of moles and grams.

Solution

1. Three moles of O_2 react to form 2 moles of O_3.
2. The formula mass of O_2 is 2×16.0 amu $= 32.0$ amu

   Therefore, 1 mole of O_2 weighs 32.0 g.

   $$3\ \cancel{\text{moles } O_2} \times \frac{32.0\text{ g } O_2}{1\ \cancel{\text{mole } O_2}} = 96.0\text{ g}$$

   The formula mass of O_3 is 3×16.0 amu $= 48.0$ amu.

   $$2\ \cancel{\text{moles } O_3} \times \frac{48.0\text{ g } O_3}{1\ \cancel{\text{mole } O_3}} = 96.0\text{ g } O_3$$

Thus, 96.0 g of O_2 react in sunlight to form 96 g of O_3.

Practice Exercise 6.7

Iron reacts with oxygen in air to produce iron oxide, Fe_2O_3, more commonly known as rust:

$$4\ Fe + 3\ O_2 \longrightarrow 2\ Fe_2O_3$$

Express the molar and mass relationships in words.

EXAMPLE 6.8

How many moles of carbon dioxide (CO_2) will be produced in the complete combustion of 10 moles of octane (C_8H_{18}), a component of gasoline?

$$2\ C_8H_{18} + 25\ O_2 \longrightarrow 16\ CO_2 + 18\ H_2O$$

Solution

1. Write in the number of moles of C_8H_{18} and CO_2 that take part in the reaction (the number of moles of O_2 and of H_2O need not be included since these substances are not mentioned in the question).

   $$\underset{2\text{ moles}}{2\ C_8H_{18}} + 25\ O_2 \longrightarrow \underset{16\text{ moles}}{16\ CO_2} + 18\ H_2O$$

2. Use this information to calculate the number of moles of CO_2 produced from 1 mole of C_8H_{18}. You can probably solve this problem by just looking at the equation, but for more difficult problems you may want to use dimensional analysis, as follows:
   a. Given quantity: 10 moles of C_8H_{18}
   b. Required quantity: ? moles of CO_2
   c. The conversion factor is the relationship between moles of octane and moles of CO_2 as shown in the balanced equation:

$$\text{conversion factor} = \frac{16 \text{ moles of } CO_2}{2 \text{ moles of octane}}$$

   d. Solution:

$$10 \cancel{\text{ moles of octane}} \times \frac{16 \text{ moles of } CO_2}{2 \cancel{\text{ moles of octane}}} = 80 \text{ moles of } CO_2$$

That is, 80 moles of CO_2 will be produced in the complete combustion of 10 moles of C_8H_{18}.

Practice Exercise 6.8

One way to remove the air pollutant nitric oxide (NO) from the gases discharged from smokestacks is to react it with ammonia (NH_3).

$$NH_3 \quad + \quad NO \quad \longrightarrow \quad N_2 \quad + \quad H_2O \qquad \text{(unbalanced)}$$

How many moles of ammonia need to be added for every 6 moles of nitric oxide in the smokestack gases?

Example 6.9

In the reaction shown in Example 6.8, how many moles of water are produced by the complete combustion of 342 g of octane?

Solution In this problem, you are asked to determine the number of *moles* of water that are produced from 342 *grams* of octane. Therefore, to solve the problem, you need a relationship between moles of water and grams of octane.

1. From the balanced equation, we know the following:

$$\underset{2 \text{ moles}}{2\ C_8H_{18}} \quad + \quad 25\ O_2 \quad \longrightarrow \quad 16\ CO_2 \quad + \quad \underset{18 \text{ moles}}{18\ H_2O}$$

2. We also know that 1 mole of a substance is equal to its formula mass in grams. To obtain the relationship between moles of H_2O and grams of C_8H_{18} in the above reaction, convert moles of C_8H_{18} to grams of C_8H_{18}. Calculate the formula mass of C_8H_{18}.

8 atoms of C	$8 \times 12 =$	96
18 atoms of H	$18 \times 1 =$	18
	formula mass of C_8H_{18} =	114

   Therefore,

   1 mole of C_8H_{18} weighs 114 g
   2 moles of C_8H_{18} weigh 228 g

Substitute this value in the equation.

$$\underset{\substack{\text{2 moles} \\ \text{228 g}}}{2\ C_8H_{18}} + 25\ O_2 \longrightarrow 16\ CO_2 + \underset{\text{18 moles}}{18\ H_2O}$$

3. Solve the problem using dimensional analysis.
   a. Given quantity: 342 g of C_8H_{18}
   b. Required quantity: ? moles of H_2O
   c. The conversion factor is the relationship between moles of water and grams of octane:

$$\text{conversion factor} = \frac{18 \text{ moles of } H_2O}{228 \text{ g of octane}}$$

   d. Solution:

$$342 \cancel{\text{g octane}} \times \frac{18 \text{ moles of } H_2O}{228 \cancel{\text{g of octane}}} = 27 \text{ moles of } H_2O$$

In words: 27 moles of H_2O are produced by the complete combustion of 342 g of C_8H_{18}.

Practice Exercise 6.9

Copper metal can be produced from ores rich in copper sulfide (CuS) by heating the ore with carbon in air. How many grams of copper metal can be made by heating 3.0 moles of copper sulfide?

$$CuS + C + O_2 \longrightarrow Cu + SO_2 + CO_2 \quad \text{(unbalanced)}$$

6.6 Chemical Reactions: What Makes Them Happen?

Spontaneous Reactions

A spring-wound toy car keeps running until the spring has unwound. Water flows downhill until it reaches level ground. These are **spontaneous processes;** once started, they proceed without the intervention of any outside agency. On the other hand, we cannot make the toy car start to run again or get the water back to the top of the hill without some outside intervention and expenditure of energy. Rewinding the spring and carrying the water to the top of the hill are both **nonspontaneous processes.**

A spring as it unwinds and water as it flows downhill lose potential energy (Chapter 2) and achieve greater stability. In these and other spontaneous mechanical processes, the driving force is the *tendency of the system to move to a lower energy state.* The same principle applies to most—but not all—chemical reactions. As the reaction proceeds, energy in the form of heat is released, and the energy of the products is less than the energy of the reactants.

In the burning of coal, the products (carbon dioxide and ashes) have less energy than the reactants (coal and oxygen) from which they are formed.

In a chemical substance, potential energy is stored in the chemical bonds that hold the constituent atoms together. In any chemical reaction, bonds between atoms in the reactants break, and the atoms then recombine in a different way to form the product(s). We can use the following equation to represent a spontaneous chemical reaction:

$$A{-}B + C{-}D \longrightarrow A{-}C + B{-}D$$

If the energy stored in the bonds of the reactants in a particular chemical reaction is *greater* than the energy stored in the bonds of the products, the difference in energy is released as heat (exothermic reaction). Conversely, if the energy stored in the bonds of the reactants is *less* than the energy stored in the bonds of the products, the difference in energy, in the form of heat, is absorbed from the surroundings (endothermic reaction).

If the principle that explains spontaneous mechanical processes also applies to spontaneous chemical reactions, we expect the potential energy in the bonds in the products (A—C and B—D) to be lower than the potential energy in the bonds in the reactants (A—B and C—D). And, in accordance with the first law of thermodynamics, which states that energy can neither be created nor destroyed (Chapter 2), energy should be released in the reaction. In the great majority of spontaneous chemical changes, energy is indeed released.

In the burning of coal, for example, large amounts of heat energy are produced.

$$C + O_2 \longrightarrow CO_2 + \text{heat energy}$$

Another less obvious example is the rusting of iron. In this process, heat is produced, but the reaction proceeds so slowly that the heat can be detected only with very sensitive instruments. Chemical reactions in which heat is produced are called **exothermic reactions.**

Although the majority of spontaneous chemical reactions are exothermic, we do not have to look far to find spontaneous reactions in which energy is *absorbed.* For example, when steam is passed over red hot coals, carbon monoxide and hydrogen gas are produced, and heat is taken up from the surroundings.

$$2\ C + 2\ H_2O + \text{heat energy} \longrightarrow 2\ CO + 2\ H_2$$

Reactions in which heat is absorbed are called **endothermic reactions.** A dramatic example of a spontaneous endothermic reaction is readily demonstrated in the laboratory. When barium hydroxide and ammonium chloride are mixed together, a precipitous drop in temperature is observed as heat is absorbed from the surroundings (**Fig. 6.10**). Barium chloride and ammonia are formed according to the following equation:

$$Ba(OH)_2 \cdot 8\,H_2O + 2\ NH_4Cl + \text{energy} \longrightarrow BaCl_2 + 2\ NH_3 + 10\ H_2O$$

Obviously, the tendency to achieve a lower energy state is not the sole driving force for a spontaneous chemical reaction.

The melting of ice at room temperature provides a clue to a second driving force. As an ice cube melts, it absorbs heat from its surroundings, and, as in all endothermic changes, the energy of the system increases. At the same time, the well-ordered crystal state of the ice (see Chapter 10) changes to the less orderly liquid state of water (**Fig. 6.11**). This tendency to achieve a more disordered, or random, state is the second driving force for spontaneous chemical reactions.

The term **entropy** is used to describe the degree of disorder, or randomness, of a system. The greater the disorder, the higher the entropy. Liquid water has higher entropy than ice. A shuffled deck of cards has more entropy than one arranged in suits. In our

FIGURE 6.10

When barium hydroxide [$Ba(OH)_2$] is mixed with ammonium chloride [NH_4Cl], an endothermic reaction occurs. Heat is absorbed from the surroundings, and the temperature of the mixture falls rapidly from room temperature (about 25°C or 77°F) to well below the freezing point of water. (If you look closely at the thermometer in the photo, you can see that the temperature of the mixture is −3°C. Condensation on the beaker has frozen, making the piece of wood stick to the beaker.)

everyday lives, we notice a general tendency toward disorder. An untended garden loses orderliness as weeds take over, and a house, if not maintained and painted regularly, falls into disrepair **(Fig. 6.12)**. From your own experience, you can appreciate that a room becomes increasingly disordered if clothes, books, and other possessions are not tidied away.

You can therefore see that natural processes are driven in two ways: toward *lower energy* and toward *higher entropy*. If these two tendencies oppose each other, the dominant tendency determines the course of the reaction. If an ice cube is kept at a temperature below 0°C, it does not melt (see Fig. 6.11). In this situation, the tendency to move toward a more disordered state (liquid) is weaker than the tendency to remain in a lower energy state (ice). In the reaction of steam with coal, the reverse is true.

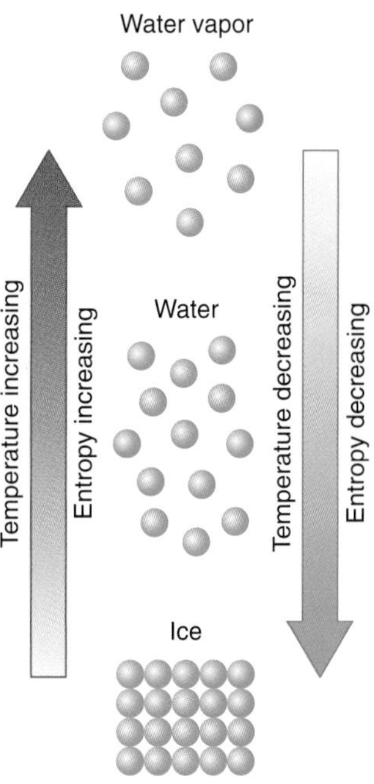

FIGURE 6.11

An increase in disorder occurs as a substance such as ice changes from a solid to a liquid to a gas. The well-ordered ice structure becomes increasingly disordered, and the entropy of the system increases. When the reverse process occurs and water vapor condenses to a liquid and then freezes, the system becomes well-ordered again, and entropy decreases.

6.7 Rates of Chemical Reactions

Some chemical reactions are practically instantaneous; others take months or even years to complete. In the tragic 1986 accident in which the space shuttle *Challenger's* fuel tanks ruptured, a spark caused the escaping hydrogen to react instantly and explosively with oxygen in the air:

$$2\ H_2 \quad + \quad O_2 \quad \longrightarrow \quad 2\ H_2O \quad + \quad \text{energy}$$

In contrast, it may take years for a discarded iron nail to rust away completely.

The **rate of a chemical reaction** is essentially the speed at which reactants are converted to products. It is measured by the amount of reactant used up or the amount of product formed in a specified period of time. How fast or how slowly a chemical reaction proceeds depends on a number of factors.

FIGURE 6.12

There is a natural tendency toward increasing disorder, or entropy. If a well-cared-for house is not maintained, it deteriorates and becomes increasingly shabby.

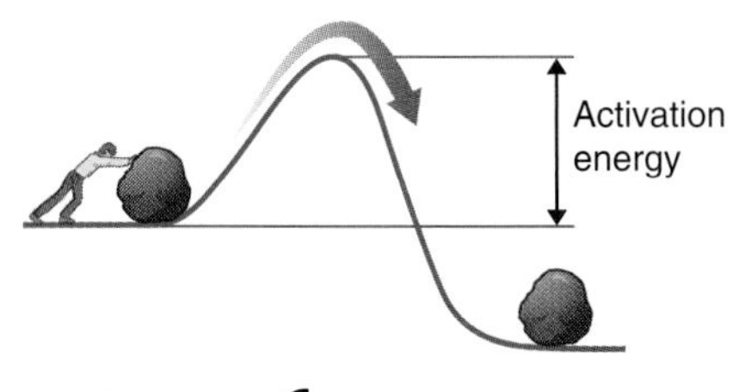

FIGURE 6.13

Activation energy is comparable to the energy required to push a boulder over a rise. Once over the top, the boulder will roll down the far slope of its own accord.

Getting a Reaction Started

Before any reaction between two or more substances can occur, the reactant particles (atoms, molecules, or ions) must make contact with each other by colliding. The more frequently these collisions occur, the more rapidly the product is formed. However, not every collision produces product. To be successful, a collision must have sufficient energy to break the bonds that hold atoms together in the reactants. Another factor is the orientation of the reactants when they make contact. Head-on collisions are more likely to be effective in breaking bonds than are glancing blows.

The minimum amount of energy that reactant particles must possess in order to react is called the **activation energy.** This can be likened to the energy required to lift or push a boulder over a barrier or hill before it will roll freely down the other side **(Fig. 6.13)**. Lighting a kitchen match provides another example. The match will not burst into flame until it is drawn quickly over the rough striking surface. The heat generated by friction is necessary to supply the activation energy. Once lit, the match keeps on burning by obtaining energy from the burning wood of the match stick. Similarly, if hydrogen and oxygen are mixed together at room temperature, they do not react. But if a spark is introduced into the mixture, an explosive reaction occurs immediately.

6.8 Factors That Influence the Rate of a Reaction

There are many ways to increase or decrease the rate of a chemical reaction. The chemical industry spends a great deal of money researching ways to make products more quickly. The food industry, on the other hand, is interested in developing ways to slow reactions that lead to spoilage.

The Effect of Temperature

For many reactions, a 10°C increase in temperature will double the rate.

EXPLORATIONS presents a fascinating case in which chemical reactions were drastically slowed by low temperatures (see pp. 168–169).

The rate of a chemical reaction can be increased considerably by raising the temperature. A higher temperature gives the reactant particles more energy. They collide more frequently, and more of the collisions have sufficient energy to provide the necessary activation energy. From your own experience, you know that foods cook more rapidly (if often less satisfactorily) in a hot oven than they do in a cooler one.

Conversely, lowering the temperature decreases the rate of a reaction. Milk turns sour much more slowly in the refrigerator than it does at room temperature, and many foods keep fresh for months in the freezer. Some animals can withstand long periods of hibernation in winter because their metabolic processes operate so slowly in the cold temperature that they can survive on stored fat.

The Effect of Concentration

When a given amount of a solid is subdivided into small pieces, the total surface area increases. The chance of ignition is increased because more atoms or molecules of the substance are in contact with oxygen molecules in the atmosphere.

In the same way that increasing the number of cars on the highway increases the chance of collisions, an increase in the concentration of reactant particles increases the frequency of collisions, and thus the rate of a chemical reaction. For example, a lighted splint of wood will burn much more vigorously in pure oxygen than in air, which has only a 21% concentration of oxygen.

Related to the concentration of reacting substances is their state of subdivision. When solid reactants are present in very small pieces, reaction rates can increase dramatically. A piece of coal does not ignite very easily, but coal dust ignites explosively, a reaction that has led to many mining accidents. Similarly, grain dust explosions caused by an accidental spark are a constant threat at grain elevators **(Fig. 6.14)**.

FIGURE 6.14

Explosive ignition of grain dust by an accidental spark destroyed this grain elevator.

The Effect of a Catalyst

The rate of a chemical reaction can often be increased by adding a substance called a *catalyst* to the reaction mixture. For example, catalysts are the key to the success of emission control systems for cars (see Chapter 11), in which carbon monoxide (CO) and nitric oxide (NO)—toxic gases present in engine exhaust—are converted to carbon dioxide (CO_2) and nitrogen gas (N_2) by passing them over the catalysts palladium and rhodium.

$$2\ CO \ + \ 2\ NO \xrightarrow{Pd/Rh} 2\ CO_2 \ + \ N_2$$

A **catalyst** is defined as *a substance that increases the rate of a chemical reaction without itself being consumed in the process.* It increases the rate by lowering the activation energy for the reaction **(Fig. 6.15)**, thus increasing the number of collisions that are effective. The working of a catalyst is somewhat analogous to driving through a tunnel in a mountain instead of taking the road over the mountain. The tunnel, like a catalyst, provides a route that saves time and requires less energy.

A catalyst speeds up a chemical reaction without itself undergoing any permanent change.

Catalysts are used extensively in industry. They are important in petroleum refining and in the manufacture of ammonia, sulfuric acid, and many other important chemicals. Special kinds of catalysts called *enzymes* are at work in all biological systems (Chapter 14). Enzymes, which are complex proteins, make it possible for reactions in living cells to proceed rapidly at body temperature. Outside the body, in the absence of enzymes, these same reactions occur very slowly.

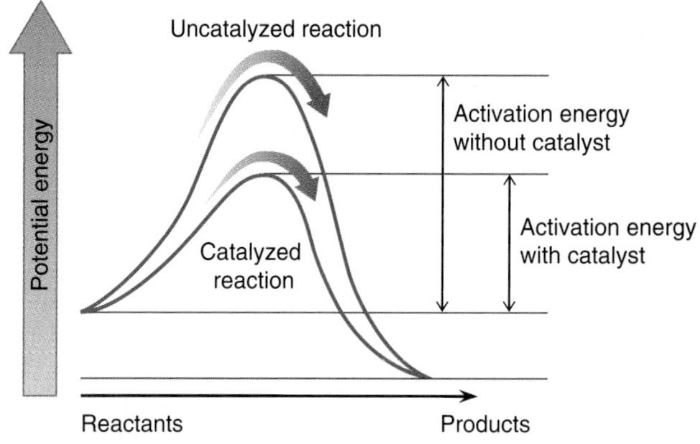

FIGURE 6.15

A catalyst lowers the activation energy for a reaction. As a result, more reactant particles have sufficient energy to overcome the energy barrier.

6.9 Reversible Reactions and Chemical Equilibrium

Reactions That Go in Either Direction

Many chemical reactions are reversible. Depending on conditions, they can be made to proceed in either direction. For example, the reaction between hydrogen and iodine to form hydrogen iodide

$$H_2 \ + \ I_2 \longrightarrow 2\ HI$$

EXPLORATIONS

The Man in the Ice—A 5300-Year-Old Ancestor

On September 19, 1991, a German couple hiking on an Alpine glacier near a ridge marking the Italian-Austrian border were startled to see the head and shoulders of a human body protruding from the ice in front of them. The Austrian forensic team sent to retrieve the corpse discovered an ax nearby and suspected foul play. As they carelessly hacked the man from the ice, they damaged one hip, tore his clothing, and castrated him. It was not until Innsbruck University archeologist Konrad Spindler saw the body and the ax head that anyone knew that an unprecedented discovery had been made. Spindler immediately realized that the brown, shriveled, but remarkably preserved body was that of a man who had lived at least 4000 years ago. Radiocarbon dating proved that the corpse was approximately 5300 years old.

Over the centuries, many people have been lost in the Alps, but most of the bodies have reappeared within 100 years, after glacial flow has transported them to lower altitudes and the ice around them has melted. Usually, these bodies have been mutilated by the moving ice and have a white waxy appearance as a result of chemical reactions that changed their body fat.

Special circumstances saved Ice Man's body from this fate. Scientists believe that a sudden autumn storm must have occurred at the time of his death. They conclude that it was fall because a ripe sloe, a fruit that ripens then, was found with the body. Cold, dry winds, common in the Alps at that time of year, quickly desiccated the body before snow and ice buried it. The very low temperature under the ice slowed down the chemical reactions that would normally decompose a body at higher temperatures. Also, because the body was dehydrated and not exposed to oxygen in the air, chemical reactions that would have occurred if it had remained at the surface did not take place. As a result, the man and his belongings—things made of wood, leather, and other animal or plant materials—were almost perfectly preserved.

For over 50 centuries, glaciers must have moved above the body without disturbing it. Then, in 1991, freak storms in North Africa blew desert sand onto the Alps. Because a dark layer of sand absorbs more solar energy from the sun than pristine white snow, more than the usual amount of snow melted during the summer months. The glacier retreated, and the body was exposed. Had the body not been discovered at that time, it would have been reburied under 2 feet of snow by the end of September.

Archeologists were particularly excited about the find because, unlike most prehistoric remains, this corpse was found not in a grave but at a campsite.

Drawing of Ice Man, based on analysis of his recovered body and belongings

Ice Man was about 30 years old and 5 feet 2 inches tall, and his features resembled those of the average European alive today. His skin was tattooed with numerous groups of small parallel lines—proof that this practice had begun some 2500 years earlier than was previously believed. His brownish-black hair, which had fallen from his head, was only 3 inches long, evidence that it had been cut. From the fragments of garments that were retrieved, it appears that he wore a tunic made from pieces of animal skins. The pieces were neatly whipstitched together with sinew, except in a few places where they had been crudely repaired with twisted dried grass. Ice Man wore leather boots stuffed with dried grass for warmth. He had a fur cap and a long woven grass cape, and he probably wore some kind of leggings. Surprisingly, no wool or linen was found, although both were used for clothing at that time. Because grains of domesticated wheat were found in Ice Man's clothing, experts believe that he came from a farming community.

Ice Man was well-equipped. He had a small, still razor-sharp flint dagger with an ash wood handle that was probably used as a cutting tool rather than a weapon and a leather pouch, most likely worn attached to a belt, that contained several useful items, including two flint scrapers, a gouging tool, and some pieces of iron pyrite (FeS_2), a valuable mineral that could be used to strike sparks to kindle a fire. He also had a U-shaped piece of hazel wood and two thin flat pieces of larch that must have formed the frame of a backpack.

Ice Man and his extensive possessions provide the most complete picture of the daily life of ancient humans that has ever been revealed.

Several artifacts had never been seen before. These included two mushrooms of a type known to have antibiotic properties, strung on a leather cord; a pencil-shaped piece of wood with a hard, sharp piece of material inserted in the tip, which may have been used to sharpen flint blades; and two birchbark canisters. One canister contained lumps of charcoal, indicating that it was probably used to carry glowing charcoal embers from one campsite to the next as a means of starting a fire. Still unexplained is a small marble disc with a tasseled leather thong threaded through a hole in its center that may have been a talisman.

The man had a bow and a unique deerskin quiver containing 14 arrows with $2^1/_2$-foot shafts made of viburnum and dogwood. Strangely, the bow and 12 of the arrows were unfinished and could not have been used. The 6-foot-long bow, made of yew, was only roughly whittled, and no grooves had been cut for the bowstring. The two completed arrows had flint heads and feathers arranged at angles that indicate a good understanding of ballistic principles.

The ax was particularly intriguing to archeologists. Initially, it was assumed to be bronze, but analysis showed that it was pure copper—evidence that Ice Man lived during the relatively short transition period when Europeans were emerging from the Stone Age and beginning to work with metals but had not yet entered the Bronze Age. Bronze, an alloy of copper with tin, is harder and easier to work with than copper; once discovered, it soon became the metal of choice for weapons. Numerous ancient bronze axes have been found in Europe, but copper ones are very rare.

Because initial examination of Ice Man's body had revealed no signs of disease or injury, scientists had concluded that he was killed in a fall or died of hypothermia when the snowstorm came up suddenly while he slept. However, a later and more extensive study showed that Ice Man suffered a more violent death. Computerized tomography (a multidimensional X-ray technique) revealed an arrowhead embedded under his left shoulder. Although the arrowhead damaged no vital organ, scientists now believe that blood loss was the likely cause of death.

Ice Man's DNA has been compared to the DNA of contemporary populations around the world and has been found to be most closely related to that of Northern Europeans alive today. Other studies have indicated that Ice Man suffered from intestinal parasites and a lung fungus, and examination of the growth pattern of his nails has revealed that he must have experienced frequent bouts of infection. It has been suggested that he was carrying the antibiotic mushrooms in case he needed to use them to treat a recurrence of infection.

Ice Man's body has provided scientists with a unique opportunity to open a window on the distant past. A great deal of information has been obtained from this incredible relic, but the exact circumstances of the man's life and death will always be a mystery.

References: B. Rensberger, "'Iceman' Yields Details of Stone Age Transition," *Washington Post* (October 15, 1992), p. A1; L. Jaroff, "Iceman," *Time*, (October 26, 1992), p. 62; O. Handt et al., "Molecular Genetic Analyses of the Tyrolean Ice Man," *Science* (June 1994), p. 1775; K. B. Richburg, "Case Closed on the Iceman Mystery," *Washington Post* (July 26, 2001), p. A19.

can be made to go in the opposite direction by applying heat:

$$2\ HI \xrightarrow{\text{heat}} H_2 + I_2$$

Equations describing reversible reactions are written with a double arrow ($\rightleftharpoons$):

$$H_2 + I_2 \rightleftharpoons 2\ HI$$

Many of the chemical reactions that occur in the body are reversible. Oxygen in the air we breathe combines with a protein called *hemoglobin* (Hb) in the lungs to form oxyhemoglobin (HbO_2). Oxyhemoglobin is carried in the bloodstream to different parts of the body, where the oxygen is released for use in various metabolic processes.

$$O_2 + Hb \rightleftharpoons HbO_2$$

Reactions That Go Partway

If the products of a chemical reaction are much more stable than the reactants, the reaction, once started, will continue until the reactants are used up.

In writing balanced chemical equations in this chapter, we have assumed that, once started, the reactions go to completion: That is, reactants are completely converted to products. Completion occurs in many reactions. For example, in the presence of sufficient air, coal continues to burn until all the carbon has been converted to carbon dioxide.

$$C + O_2 \longrightarrow CO_2$$

There is no tendency for the carbon dioxide to break down into oxygen and carbon. However, the reaction between nitrogen and hydrogen to produce ammonia does not go to completion. At some point, the reverse reaction starts to occur, and ammonia molecules decompose to give nitrogen and hydrogen.

$$N_2 + 3\ H_2 \rightleftharpoons 2\ NH_3$$

The point at which dynamic equilibrium is reached is a characteristic property of a particular reaction under a given set of conditions.

At the point when the rate at which product is being formed equals the rate at which product is being decomposed, the reaction is said to have reached a state of **dynamic equilibrium.** The word *dynamic* is used because even though no further *net* change occurs in the concentrations of reactants or products once equilibrium has been established, individual molecules are still being changed. Reactant molecules continue to combine, and product molecules continue to decompose.

An equilibrium reaction can be forced to go in one direction more than in the other by adjusting the conditions (temperature, pressure, or concentration). In the commercial production of ammonia by the Haber process, the reaction is carried out at about 550°C and 250 atm of pressure, in the presence of a catalyst, and the ammonia is removed as it is formed. Under these conditions, the reaction goes almost entirely toward the right; ammonia production is at a maximum, and very little ammonia decomposes back to nitrogen and hydrogen.

A great many of the reactions that occur in the natural environment are equilibrium reactions. For example, the concentration of calcium ions in natural waters is largely controlled by an equilibrium reaction between calcium carbonate minerals, atmospheric carbon dioxide, and water.

$$CaCO_3 + CO_2 + H_2O \rightleftharpoons 2\ HCO_3^- + Ca^{2+}$$

Acid rain can seriously upset this and other natural equilibria (Chapter 11). Metabolic processes in living organisms are also controlled primarily by equilibrium reactions (Chapter 14).

Chapter Summary

1. A chemical reaction can be represented in shorthand form by a chemical equation.
2. In a balanced chemical equation, the numbers of atoms of any elements on the left side of the equation equal the numbers of atoms of those elements on the right side.
3. A chemical equation shows the relative masses of the atoms and molecules that take part in the chemical reaction.
4. The formula mass of a substance is the sum of the atomic masses of all the atoms in the substance's chemical formula.
5. If the balanced equation for a chemical reaction is known, it is possible to calculate both the weight of product that can be obtained from a given amount of reactant and the weight of reactant required to produce a desired amount of product.
6. A mole is a number called Avogadro's number. It is equal to 6.02×10^{23} structural units, which may be atoms, molecules, or formula units. For example, 1 mole of oxygen molecules equals 6.02×10^{23} oxygen molecules.
7. The molar mass (mass of a mole) of a substance is defined as the atomic or formula mass of the substance expressed in grams (g). For example, the molar mass of carbon (C) is 12.00 g; that of sodium chloride (NaCl) is 58.44 g.
8. A balanced equation for a chemical reaction shows both the mole relationships and the mass relationships between the substances involved in the reaction.
9. Spontaneous reactions proceed without any outside intervention; nonspontaneous reactions cannot proceed without some outside intervention.
10. In the majority of spontaneous chemical reactions, energy is released in the form of heat.
11. Reactions in which heat is produced are exothermic reactions; those in which heat is absorbed are endothermic reactions.
12. There are two driving forces in a spontaneous chemical reaction: a tendency to move to a lower energy state, and a tendency to achieve a more random or disordered state.
13. The degree of disorder of a system is the entropy of the system. The greater the disorder, the higher the entropy.
14. In order to react, reactants must possess a minimum amount of energy called the activation energy.
15. The rate of a chemical reaction (the rate at which reactants are converted to products) can be increased in three ways: by raising temperature, by increasing the concentration of the reactants, and by adding a catalyst.
16. A catalyst is a substance that increases the rate of a chemical reaction without itself being consumed in the process.
17. Not all chemical reactions go to completion. Some reach a point of dynamic equilibrium, when the rate at which product is being formed equals the rate at which product is being decomposed to form the reactants.

Key Terms

activation energy (p. 166)
Avogadro's number (p. 158)
catalyst (p. 167)
dynamic equilibrium (p. 170)
endothermic reaction (p. 164)
entropy (p. 164)
exothermic reaction (p. 164)
formula mass (p. 153)
molar mass (p. 159)
mole (p. 158)
nonspontaneous process (p. 163)
products (p. 147)
rate of a chemical reaction (p. 165)
reactants (p. 147)
spontaneous process (p. 163)

Questions and Problems

1. Why is it important to make sure that a chemical equation is properly balanced?

2. Balance the following chemical equations:
   a. $Al + O_2 \longrightarrow Al_2O_3$
   b. $Fe + O_2 + H_2O \longrightarrow Fe(OH)_3$
   c. $CH_3OH + O_2 \longrightarrow CO_2 + H_2O$
   d. $CH_4 + O_2 \longrightarrow CO_2 + H_2O$

3. Balance these chemical equations:
   a. $KClO_3 \longrightarrow KClO_4 + KCl$
   b. $PbO \longrightarrow PbO + O$
   c. $NO_2 + H_2O \longrightarrow HNO_3 + NO$

4. Balance the following equations:
   a. $Na + Cl_2 \longrightarrow NaCl$
   b. $H_2 + I_2 \longrightarrow HI$
   c. $Si + Br_2 \longrightarrow SiBr_4$

5. Balance these equations:
   a. $N_2 + H_2 \longrightarrow NH_3$
   b. $C_2H_4 + O_2 \longrightarrow CO_2 + H_2O$
   c. $KClO_3 \longrightarrow KCl + O_2$

6. Write a balanced equation for each of the following:
   a. the reaction of magnesium (Mg) with solid CO_2 (dry ice) to form MgO and carbon (C)
   b. the reaction of bromine (Br_2) with potassium (K)
   c. the combustion of silane gas (SiH_4) to produce water vapor and solid silicon dioxide (SiO_2)

7. Write balanced equations for the reaction of bromine (Br) with each of the following metals to form a solid:
   a. potassium (K)
   b. calcium (Ca)
   c. aluminum (Al)

8. How many kilograms of copper are needed to make 100.0 kg of copper sulfide?

   $$Cu + S \longrightarrow CuS$$

9. How many grams of Cl_2 are needed to make 73.0 g of HCl?

   $$H_2 + Cl_2 \longrightarrow 2\,HCl$$

10. One way to remove nitric oxide (NO) from smokestack emissions is to react it with ammonia:

    $$4\,NH_3 + 6\,NO \longrightarrow 5\,N_2 + 6\,H_2O$$

    Use this equation to fill in the blanks.
    a. 10 moles of NO react with _____ moles of NH_3.
    b. 5 moles of NO produce _____ moles of N_2.
    c. The production of 6 moles of N_2 requires _____ moles of NO.
    d. 10 moles of NO produce _____ moles of H_2O.
    e. _____ grams of H_2O are formed from 3 moles of NH_3.

11. The combustion of hexane (C_6H_{14}) in air yields carbon dioxide (CO_2) and water (H_2O).

    $$2\,C_6H_{14} + 19\,O_2 \longrightarrow 12\,CO_2 + 14\,H_2O$$

    a. How many moles of oxygen (O_2) react with 4 moles of hexane?
    b. How many moles of water are produced by burning 4 moles of hexane?
    c. How many moles of carbon dioxide, the greenhouse gas, are produced for every mole of hexane burned?

12. What is the molar mass of each of the following in grams?
    a. NH_3
    b. $MgSO_4$
    c. C_2H_5OH
    d. Fe_2O_3
    e. H_2SO_4

13. What is the molar mass of each of the following in grams?

a. xenon hexafluoride

b. silicon tetrachloride

c. nitrogen trifluoride

14. What is the formula of a hydrocarbon that has a formula mass of 26? Hydrocarbons contain only carbon and hydrogen.

15. Complete the following table for ethyl alcohol (CH_3CH_2OH).

Number of grams	Number of moles
46.0	___
___	0.25
115.0	___
___	0.50

16. How much does 1.0 mole of each of the following substances weigh in grams?

a. NaCl **b.** $C_{12}H_{22}O_{11}$
c. $Pb(OH)_2$ **d.** $Mg_3(PO_4)_2$

17. What is the formula of a compound that consists of 50% sulfur and 50% oxygen by mass?

18. A compound contains only carbon and hydrogen, and 40 grams of the compound contains 30 grams of carbon. What is the formula of the compound?

19. Calculate the number of moles of each type of atom present in each of the following:

a. 4 moles of N_2O **b.** 2 moles of $Al(OH)_3$
c. 5 moles of $Na_2S_2O_3$

20. Calculate the number of moles of oxygen (O_2) in each of the following:

a. 3 moles of $C_6H_8O_6$ (Vitamin C)

b. 5 moles of H_2SO_4

c. 2 moles of SO_3

21. The combustion of butane gas (C_4H_{10}) in air yields carbon dioxide and water.

a. Write a balanced equation for the reaction.

b. How many moles of C_4H_{10} are required to form 10 moles of CO_2?

c. How many moles of H_2O are formed from each mole of C_4H_{10}?

d. How many grams of H_2O are formed from 4 moles of C_4H_{10}?

22. Arsenic (As) reacts with chlorine (Cl_2) to form a chloride. If 3.17 g of arsenic reacts with 7.51 g of chlorine, what is the formula of the chloride?

23. How many moles of nitrogen gas are needed to produce 12 moles of ammonia?

$$N_2 + H_2 \longrightarrow 2\,NH_3$$

24. How many moles of methane (CH_4) are needed to produce 4 moles of water?

$$CH_4 + 2\,O_2 \longrightarrow CO_2 + 2\,H_2O$$

25. Iron ore consists mainly of iron oxide (Fe_2O_3). When iron oxide is heated with an excess of coke (carbon), iron metal and carbon monoxide are produced.

a. Write a balanced equation for the reaction.

b. How many moles of iron oxide are required to form 10 moles of iron?

c. How many grams of carbon monoxide are formed from 12.0 g of coke?

26. Crude oil burned in an electrical generating plant contains about 1.2% sulfur by mass. When the oil burns, the sulfur reacts with oxygen and forms sulfur dioxide, a major air pollutant.

$$S + O_2 \longrightarrow SO_2 \quad \text{(unbalanced)}$$

How many grams of sulfur dioxide are formed when 1 metric ton (1000 kg) of coal is burned?

27. The catalytic converter, required on American automobiles, converts poisonous carbon monoxide (CO) to carbon dioxide (CO_2).

$$2\,CO + O_2 \longrightarrow 2\,CO_2$$

a. What mass of CO_2, in grams, is produced when 100 g of CO react?

b. What mass of O_2 is needed to react with 50 g of CO?

28. Which of the following processes occur spontaneously?

a. Iron reacting with oxygen to form Fe_2O_3 (rust)

b. Salt dissolving in water

c. Water vaporizing at 0 °C

d. You making an outline of your class notes

29. Explain each of the following:

a. Not all exothermic reactions are spontaneous.

b. Not all endothermic reactions are spontaneous.

c. If an equilibrium reaction is spontaneous in one direction, it is not spontaneous in the opposite reaction.

30. Define each of the following in your own words.
    a. rate of reaction
    b. activation energy
    c. entropy
    d. exothermic reaction

31. Rank the three physical states of matter—solid, liquid, and gas—in order of increasing entropy. Explain your answer.

32. For each of the following processes, indicate whether entropy increases or decreases.
    a. the freezing of water
    b. weeding a garden
    c. butter melting
    d. sugar dissolving in coffee

33. In the 1950s, the insecticide DDT was sprayed over wide areas. After application, DDT moves into the soil and from there into plants and water. In lakes, it concentrates in the fatty tissue of fish. Describe the entropy change associated with each of the processes mentioned. That is, does entropy increase or decrease in each step? In order to remove DDT from the environment, what entropy change is required? How can such an entropy change be brought about?

34. Define *activation energy.*

35. When the gas is turned on in a gas stove, the reaction of methane and oxygen does not begin until a flame or a spark starts the reaction. What is the flame or spark providing to the reaction?

36. List three factors that influence the rate of a chemical reaction.

37. What is a catalyst? Explain how a catalyst changes the rate of a chemical reaction. What effect does it have on the activation energy of a reaction?

38. Milk will sour in a day or two if left at room temperature. When it is refrigerated, however, it will keep for 2 weeks. Explain.

39. Combustible materials burn more rapidly in oxygen than in air. Explain.

40. Explain in your own words.
    a. A flame lights a candle, but the candle continues to burn after the flame is removed.
    b. A decrease in temperature slows the rate of a reaction.
    c. Increasing the concentration of a reactant increases the rate of a reaction.

41. Cold-blooded animals decrease their body temperature in cold weather to match that of their environment. In light of what you have learned in this chapter, explain why they do this and how it affects their need for energy (food).

42. Why does increasing the concentration of the reactants in a chemical reaction increase the reaction rate?

43. Endothermic reactions require ______ bonds to be broken and ______ bonds to be formed.

44. Give two reasons why increasing the temperature will increase the rate of a chemical reaction.

45. If a substance can exist as a solid, a liquid, or a gas, in which state will it be most chemically reactive?

46. Give two reasons why cooling the reaction environment slows down the rate of a chemical reaction.

47. Why does a reversible reaction of gas molecules have to occur in a sealed container?

48. Explain the following:
    a. Bulk flour is hard to burn, yet grain elevators in which flour is stored have been known to explode.
    b. Solids dissolve in water more rapidly if they are first ground into small particles.

49. How does a catalyst affect the activation energy so that a reaction will take place at a faster rate?

50. Describe how the rate of reaction relates to whether a reaction goes to completion.

51. Fire fighters fighting a natural gas (CH_4) fire try to reduce the rate of the following reaction:

$$CH_4 + 2\,O_2 \longrightarrow CO_2 + 2\,H_2O + \text{heat}$$

    Explain how each of the following slow the reaction:
    a. limiting the fuel supply (CH_4)
    b. "smothering" the fire
    c. applying water
    d. using a carbon dioxide fire extinguisher

Answers to Practice Exercises

6.1 64 amu

6.2 **a.** 158 amu; **b.** 74 amu

6.3 3.5 kg of Fe

6.4 11.3 g of KO_2

6.5 180 g

6.6 **a.** 0.5 mole; **b.** 0.33 mole

6.7 4 moles of Fe react with 3 moles of O_2 to give 2 moles of Fe_2O_3.

223.2 g of Fe react with 96.0 g of O_2 to give 319.2 g of Fe_2O_3.

6.8 4 moles of NH_3

6.9 190.5 g Cu

Chapter 7

Reactions in Solution

Acids and Bases and Oxidation-Reduction Reactions

Chapter Objectives

In this chapter, you should gain an understanding of:

- Aqueous solutions and the units used to measure their concentrations
- The chemical properties that distinguish acids and bases
- The definitions of acids and bases in terms of protons
- The difference between a weak acid and a strong acid
- Acid-base neutralization reactions
- The pH scale and measuring acidity
- Acid rain and its effect on the environment
- The definitions of oxidation and reduction
- Oxidation-reduction reactions

Rainwater, which is naturally slightly acidic, dissolved underground deposits of limestone very slowly over thousands of years to form these spectacular caverns.

ONLINE FEATURES

www.jbpub.com/chemistry

- Chemistry and the Environment
- eLearning
- Chemistry in the News
- Special Topics
- Research & Reference Links
- Ask the Authors

NEARLY ALL THE CHEMICAL REACTIONS that occur in the natural environment occur in aqueous solution (in water), and nearly all these reactions are either acid-base reactions (reactions between acids and bases) or oxidation-reduction reactions. In the living world, photosynthesis and metabolic processes such as digestion of food and elimination of wastes are brought about by either acid-base or oxidation-reduction reactions. In the nonliving world, these same types of reactions are responsible for the weathering of rocks and soil, the rusting of iron, and the decomposition of dead plants and animals.

Almost everyone is familiar with some acids and bases; we use them every day in our homes, and we have become increasingly aware of the problem of acid rain.

In this chapter, we see that certain features common to all acids and other features common to all bases explain the distinctive properties that characterize these two classes of compounds. We discover that acids and bases are interrelated compounds and that acid-base reactions involve the transfer of protons. We discuss the causes of acid rain and the impact that increased acidity has on the environment. We also consider some important oxidation-reduction reactions and see how they can be understood in terms of electron transfer. But first we look at some general properties of aqueous solutions.

More than one solute can be present in a solution. For example, both instant coffee and sugar can be dissolved in a cup of hot water.

In addition to temperature, pressure and the presence of other solutes in solution affect the solubility of a solute.

7.1 Aqueous Solutions

A **solution** is a homogeneous mixture of two or more substances. The substance present in larger quantity is termed the **solvent;** the substance dissolved in the solvent is called the **solute.** Solutions in which the solvent is water are called **aqueous solutions.**

Water is an exceptionally good solvent that can dissolve a wide variety of ionic and polar substances. In seawater, sodium chloride is the main solute, but many other solutes are present in smaller quantities. In our bodies and in plants, nutrients are transported in aqueous solution to tissues where they are needed, and waste products of metabolism are carried away in aqueous solution. Oxygen gas in solution in water is essential for the survival of fish and other aquatic organisms.

The maximum quantity of a solute that will dissolve in a given quantity of solvent is its **solubility.** For most common salts in water, solubility increases with increasing temperature (**Fig. 7.1**). For many gases, however, the opposite is true. Both oxygen and carbon dioxide are less soluble in hot water than in cold water, which has important implications for aquatic life (Chapter 10).

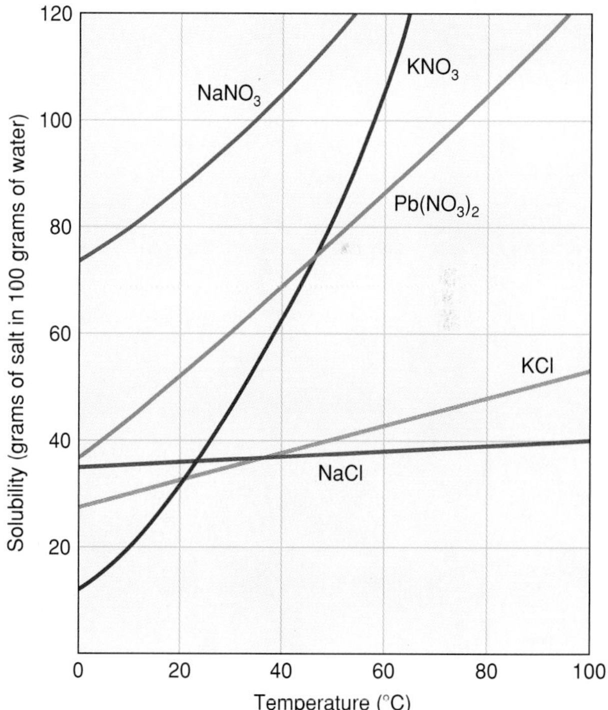

FIGURE 7.1

This graph shows the effect of temperature on the solubilities of several common salts in water: sodium nitrate ($NaNO_3$), potassium nitrate (KNO_3), lead nitrate [$Pb(NO_3)_2$], potassium chloride (KCl), and sodium chloride (NaCl).

7.2 The Formation of Aqueous Solutions

When any solution is formed, solute particles (molecules or ions) become uniformly dispersed among solvent particles, as shown diagrammatically in **Figure 7.2** for a solution of sugar in water. For this uniform dispersal to occur, attractions between solute particles (sugar molecules) and between solvent particles (water molecules) must first be overcome. A rough rule of thumb for determining whether a given solute will dissolve in a particular solvent is "like dissolves like." In this context, "like" means "similar in polarity." Polar solutes such as sodium chloride (common table salt) and sugar tend to dissolve readily in polar solvents such as water but not in nonpolar solvents such as oils. Conversely, nonpolar solutes such as fats dissolve readily in nonpolar solvents but not in polar solvents.

Dry cleaners apply the "like dissolves like" principle when they use nonpolar solvents such as perchloroethylene rather than water to remove greasy (nonpolar) dirt from clothes. Grease will not dissolve in water but will dissolve in a nonpolar solvent.

To understand how and why this principle applies, let's consider the dissolution of sodium chloride in water. In a sodium chloride (NaCl) crystal, sodium ions (Na^+) and chloride ions (Cl^-) are held together by electrostatic forces (Chapter 4). These forces must be overcome before dissolution can occur. When a sodium chloride crystal is put into water, the positive ends of the polar water molecules are attracted toward negatively charged chloride ions at the surface of the crystal **(Fig. 7.3)**. These chloride ions, because they are not completely surrounded by positively charged sodium ions, are less strongly held in the crystal than are chloride ions located in the interior. The pull of the water molecules is sufficient to detach chloride ions from the crystal, and these ions, surrounded

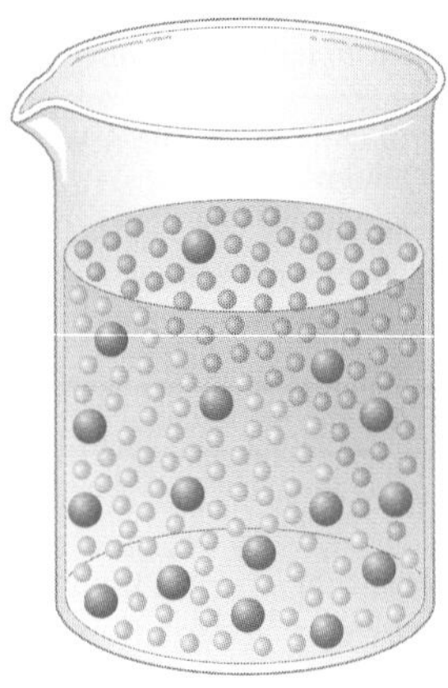

FIGURE 7.2

Diagrammatic representation of sugar dissolved in water (not to scale): The solute particles (sugar molecules) are uniformly distributed among the solvent particles (water molecules).

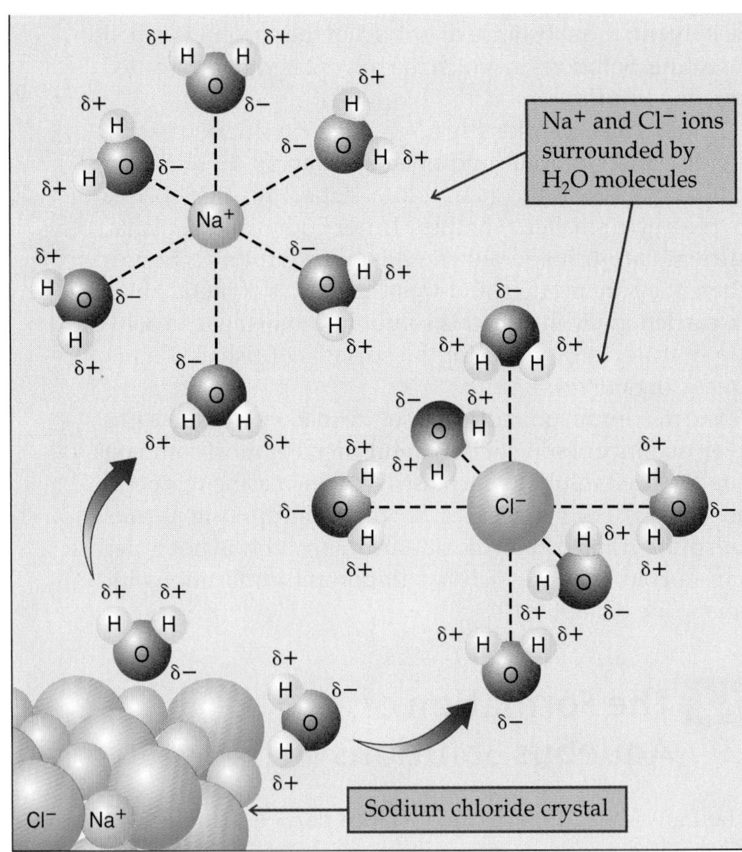

FIGURE 7.3

The formation of an aqueous solution of sodium chloride (NaCl): The positive ends of the polar water molecules (δ+) are attracted to negatively charged chloride ions (Cl^-); their negative ends (δ−) are attracted to positively charged sodium ions (Na^+). The attractive forces dislodge the Na^+ and Cl^- ions from the solid crystal, and the ions become surrounded by water molecules.

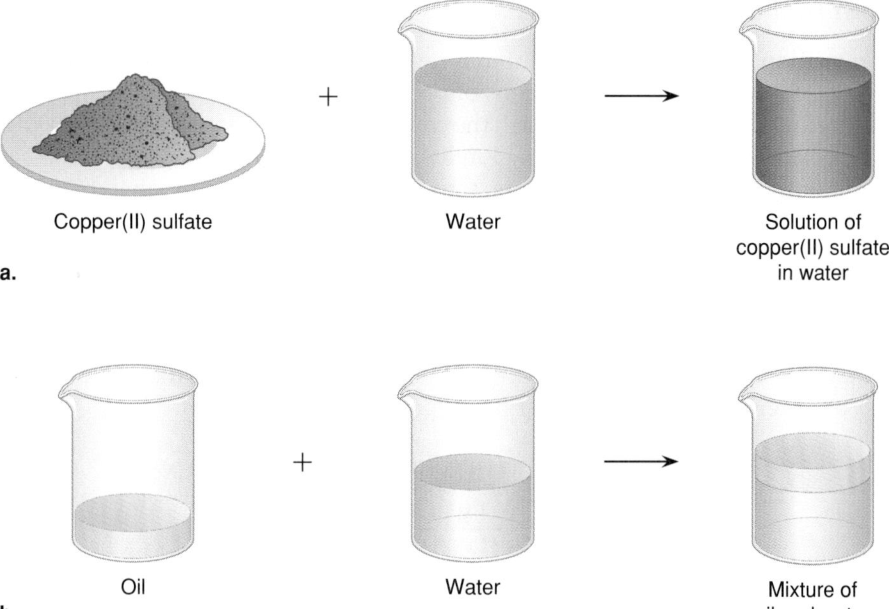

FIGURE 7.4

(a) Copper(II) sulfate, an ionic substance, dissolves in water. (b) Oil, a nonpolar substance, does not dissolve in water. Because it is less dense than water, oil floats on the surface.

by water molecules, move away from the crystal. In a similar manner, the negative ends of polar water molecules are attracted to positively charged sodium ions at the crystal surface, and these sodium ions, surrounded by water molecules, leave the crystal. The removal of surface chloride and sodium ions leaves other such ions in the crystal exposed; these, in turn, become hydrated and dispersed. The process continues until the entire crystal has dissolved.

When ions in solution are surrounded by water molecules, they are said to be *hydrated.*

The polar nature of water thus makes it a good solvent for other polar substances but not for nonpolar substances such as oil (**Fig. 7.4**). Oil does not dissolve in water because oil molecules have no charge with which to attract polar water molecules. Attractive forces between oil and water molecules are therefore very weak, definitely not strong enough to break the hydrogen bonds between adjacent water molecules.

7.3 Concentration Units

Chemists often need to know the exact **concentration** of a solute in a given volume of a solvent. There are various ways of expressing concentration; we'll consider molarity and parts per million and parts per billion.

Molarity and Molar Solutions

For aqueous solutions, the unit of concentration most often used by chemists is **molarity**—the number of moles (Chapter 6) of solute per liter of solution.

$$\text{molarity} = \text{M} = \frac{\text{moles of solute}}{\text{liter of solution}}$$

A 1 molar (1 M) solution contains the molecular (or formula) mass of the solute dissolved in 1 liter (1 L) of solution.

Parts per million (ppm) means the number of milligrams of solute dissolved in 1 liter of solution.

Parts per Million

Parts per million (ppm) is a convenient unit for describing very dilute solutions. It is frequently used for stating concentrations of pollutants in water.

Let's consider drinking water in which the concentration of dissolved lead is 1 ppm. This means there is 1 part of lead in every 1 million parts of water. Parts can be expressed in any unit of mass (for example, ounces, tons, micrograms), but the same unit must be used for both solute and solvent. We will use grams.

$$1 \text{ ppm} = \frac{1 \text{ g of solute}}{1 \text{ million g of water}}$$

Because it is more convenient to measure liquids by volume than by mass, we change the mass of water to a volume of water:

1 g of water has a volume of 1 mL

Therefore, $$1 \text{ ppm} = \frac{1 \text{ g of solute}}{1 \text{ million } (1{,}000{,}000) \text{ mL of water}}$$

We change milliliters to liters:

$$1 \text{ ppm} = \frac{1000 \cancel{\text{mL}}}{1 \text{ L}} \times \frac{1 \text{ g}}{1{,}000{,}000 \cancel{\text{mL}}} = \frac{1 \text{ g}}{1000 \text{ L}}$$

We change grams to milligrams:

$$1 \text{ ppm} = \frac{1000 \text{ mg}}{1 \cancel{\text{g}}} \times \frac{1 \cancel{\text{g}}}{1000 \text{ L}}$$

Therefore, $$1 \text{ ppm} = 1 \text{ mg per L}$$

Each liter of the drinking water contains 1 mg of lead.

It is important to remember that a concentration of 1 ppm is the same as a concentration of 1 mg/L. A solution containing 1 ppm of solute is a very dilute solution. For example, if 1 teaspoon of salt were dropped into a swimming pool, the concentration of salt in the water would be approximately 1 ppm.

Parts per Billion

For certain solutions, particularly water samples containing minute traces of contaminants, it is often more convenient to express concentration in **parts per billion (ppb)** rather than parts per million.

$$1 \text{ ppb} = \frac{1 \text{ ppm}}{1000} = \frac{1 \text{ mg}}{1 \text{ L}} \times \frac{1}{1000}$$

We change milligrams to micrograms:

$$1 \text{ ppb} = \frac{1000 \text{ µg}}{1 \cancel{\text{mg}}} \times \frac{1 \cancel{\text{mg}}}{1 \text{ L}} \times \frac{1}{1000}$$

Therefore, $$1 \text{ ppb} = 1 \text{ µg per L}$$

EXAMPLE 7.1 Using concentration units

The U.S. Environmental Protection Agency (EPA) set a limit for the concentration of lead in drinking water at 15 ppb. A laboratory finds the concentration of lead in a sample taken from a water fountain to be 18 μg/100 mL. Is this above or below the EPA limit? By how much?

Solution

1. Given the concentration of lead in the sample:

$$\text{concentration} = \frac{18\ \mu\text{g}}{100\ \text{mL}}$$

2. To convert the concentration to micrograms per liter (ppb), multiply top and bottom by 10 to make the denominator 1000 mL (1 L)

$$\text{concentration} = \frac{18\ \mu\text{g}}{100\ \text{mL}} \times \frac{10}{10} = \frac{180\ \mu\text{g}}{1000\ \text{mL}} = \frac{180\ \mu\text{g}}{1\ \text{L}}$$

Remember, 1 ppb = 1 μg/L. Therefore,

$$\frac{180\ \mu\text{g}}{1\ \text{L}} = 180\ \text{ppb}$$

3. This concentration is above the EPA limit of 15 ppb. How much above?

$$180 - 15 = 165\ \text{ppb}$$

The concentration of 180 ppb is 165 ppb above the EPA limit.

Practice Exercise 7.1

The concentration of arsenic in the water of a stream is found to be 0.02 mg per 100 mL. Express the concentration in ppm and ppb. (The EPA limit for arsenic in drinking water is 20 ppb.)

7.4 Acids and Bases: Properties and Definitions

Acids and bases are familiar substances that can be found around the home (**Fig. 7.5**). Vinegar and citrus fruits (e.g., grapefruit, oranges, and lemons) taste sour because they

a.

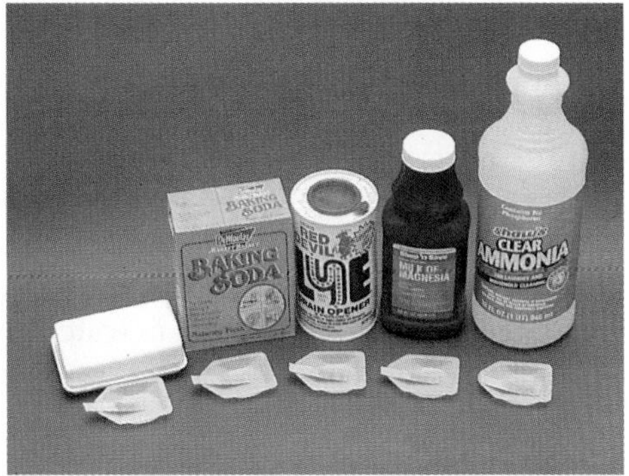

b.

FIGURE 7.5 (a) Acids and (b) bases are present in many consumer products. Acidic substances turn litmus paper red, and basic substances turn litmus paper blue.

contain acids. Hydrochloric acid is the main constituent of the gastric fluid in your stomach. We clean house with products that contain bases such as ammonia and sodium hydroxide. The bases sodium bicarbonate and calcium carbonate are ingredients in the antacids taken for relief of an upset stomach.

Properties of Acids and Bases

Long before the currently accepted definitions of acids and bases were established, solutions of acids and bases were identified by their characteristic properties.

Properties of Acids

The word *acid* is derived from *acidus*, the Latin word for "sour."

1. Solutions of acids in water have a sour taste. You can safely recognize the sour taste of acids in certain everyday foods and beverages. For example, the sour taste of vinegar is caused by acetic acid. Lemons and other citrus fruits taste tart because they contain citric acid. Lactic acid gives unsweetened yogurt its characteristic tang.

The symbol (g), (l), (s), or (aq) is sometimes added after the formula of an element or compound in a chemical equation to indicate whether the substance is in a gaseous, liquid, or solid state or is in aqueous solution.

2. When an aqueous solution of an acid is added to certain metals, such as tin, magnesium, or zinc, the acid dissolves the metal, and hydrogen gas is produced. The equation for the reaction of magnesium with hydrochloric acid is as follows:

$$2\ HCl(l) \quad + \quad Mg(s) \quad \longrightarrow \quad MgCl_2(aq) \quad + \quad H_2(g)$$

3. Acids in aqueous solution change the color of litmus from blue to red. Using litmus paper is the simplest and safest way to test for an acid (Fig. 7.5a).
4. Acids react with bases to form water and ionic compounds called *salts*. (We will consider these reactions in more detail later, in Section 7.6.)

Litmus is extracted from certain lichens containing *erythrolitmin*, a substance that turns red in acid and blue in base.

Properties of Bases

1. Solutions of bases in water have a bitter taste. Since bases, unlike acids, are not present in common foods or beverages, this property is not easily tested. However, you can recognize the characteristic taste of a base in certain antacids. For example, the bitter taste of Alka Seltzer is due to the base sodium bicarbonate.
2. Bases feel slippery on the skin. Soaps, which contain bases, exhibit this property.
3. Bases in aqueous solution change litmus from red to blue (Fig. 7.5b). Again, using litmus paper is the simplest and safest way to test for a base.
4. Bases react with acids to form water and salts.

Early Definitions of Acids and Bases

Why do acids and bases have the properties just described? By the end of the 19th century, it was generally agreed that the presence of hydrogen ions (H^+) in aqueous acid solutions and of hydroxide ions (OH^-) in aqueous basic solutions accounted for these characteristic properties. That is, an *acid* was defined as any substance that produced hydrogen ions when dissolved in water, and a *base* was a substance that yielded hydroxide ions when dissolved in water.

For example, hydrochloric acid (HCl) in water provides hydrogen ions:

$$HCl(g) \xrightarrow{\text{water}} H^+(aq) \quad + \quad Cl^-(aq)$$

A hydrogen ion (H^+) is a hydrogen atom from which the negatively charged electron has been removed. Recall from Chapter 3 that a hydrogen atom—the simplest of all atoms—consists of one electron with a 1− charge and one proton with a 1+ charge. When the electron is removed from a hydrogen atom, a positively charged nucleus consisting of a single proton is left. For this reason, the hydrogen ion (H^+) is often called a **proton.**

The base sodium hydroxide (NaOH) in water provides hydroxide ions (OH^-):

$$NaOH(s) \xrightarrow{\text{water}} OH^-(aq) + Na^+(aq)$$

The Hydronium Ion

Chemists now know that it is the **hydronium ion** (H_3O^+), rather than the hydrogen ion (H^+), that is responsible for the observed properties of acids in water. The hydroxide ion (OH^-) is responsible for the observed properties of bases.

Free H^+ ions (protons) do not exist in water because they immediately combine with water molecules to form hydronium ions. The slightly negatively charged oxygen atom in a water molecule attracts a positively charged proton. The proton attaches itself to one of the unshared pairs of electrons surrounding the oxygen atom.

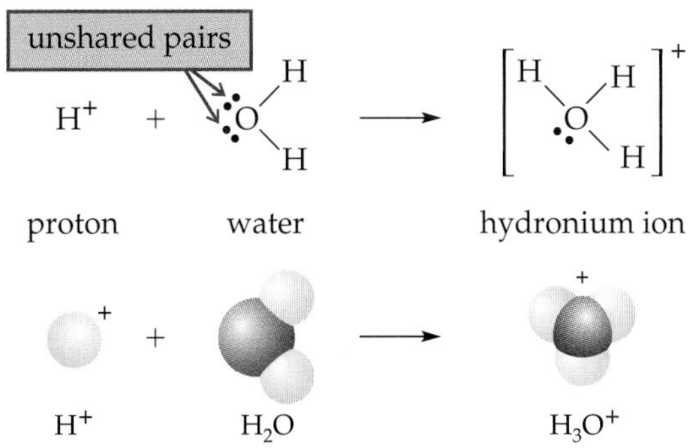

When the hydronium ion is formed, the oxygen atom in the water molecule retains its stable octet of electrons (Chapter 4), and all three hydrogen atoms achieve the stable two-electron helium electron configuration. Thus, the hydronium ion is very stable. (In this text, for the sake of simplicity, we will often use H^+ instead of the more accurate H_3O^+. For aqueous solutions, whenever H^+ is written, H_3O^+ should be understood.)

The Brønsted-Lowry Definitions of Acids and Bases

The early definitions of acids and bases, although still frequently used, are limited. They apply only to acids and bases in solution in water. In 1923, expanded definitions of acids and bases were proposed almost simultaneously by the Danish scientist Nicolas Brønsted and the British scientist Thomas Lowry. The Brønsted-Lowry definitions are as follows:

An **acid** is any substance that can *donate a proton* to some other substance.
A **base** is any substance that can *accept a proton* from some other substance.

The Brønsted-Lowry definitions apply to all the acids and bases covered by the earlier definitions. For example, in reacting with water to form a hydronium ion, hydrochloric acid donates its proton to a water molecule:

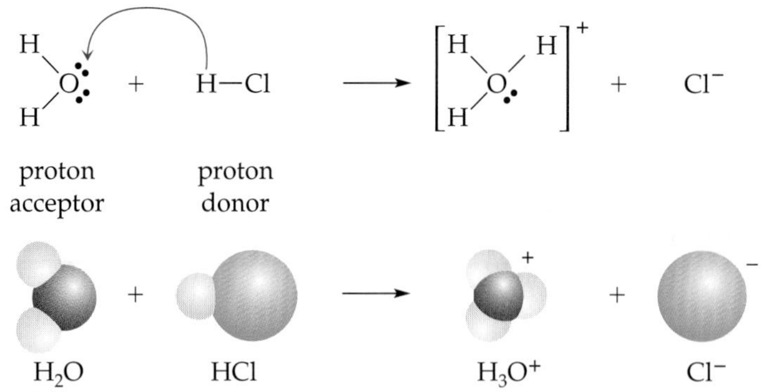

$$H_2O + H{-}Cl \longrightarrow H_3O^+ + Cl^-$$

proton acceptor (H₂O) — proton donor (HCl)

H_2O HCl H_3O^+ Cl^-

At the same time, the water molecule accepts a proton. In this reaction, therefore, water is acting as a base, according to the Brønsted-Lowry definition.

The base sodium hydroxide (NaOH) also fits the Brønsted-Lowry definition. As shown in the following equation, the hydroxide ion (OH^-) can accept a proton to form water:

$$OH^- + H^+ \longrightarrow H_2O$$

proton acceptor — proton donor

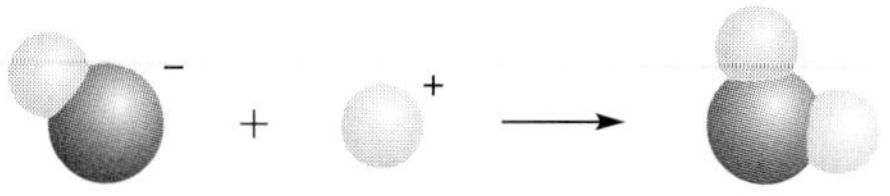

An aqueous solution of ammonia (NH_3) has all the characteristic properties of a typical base. This, at first, seems surprising because ammonia does not have an OH group in its formula. However, the following equation makes it clear why ammonia is a base:

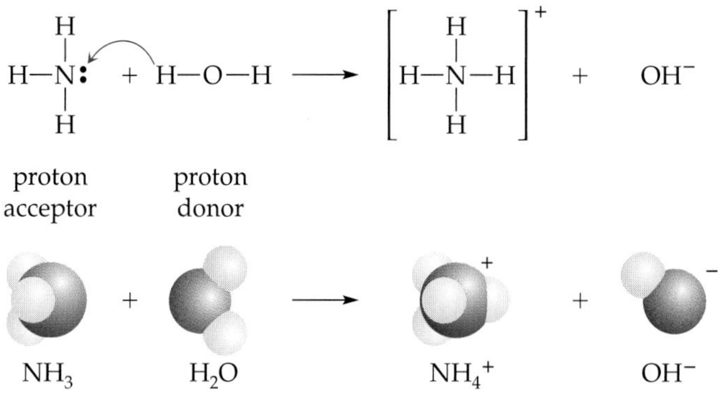

$$NH_3 + H{-}O{-}H \longrightarrow NH_4^+ + OH^-$$

proton acceptor — proton donor

NH_3 H_2O NH_4^+ OH^-

When ammonia dissolves in water, it accepts a proton from water; a positively charged ammonium ion (NH_4^+) and a negatively charged hydroxide ion (OH^-) are formed. Ammonia therefore fits both the older definition of a base and the Brønsted-Lowry definition.

In addition to NaOH, KOH, and NH_3, other common bases are the ionic compounds sodium carbonate (Na_2CO_3), sodium bicarbonate ($NaHCO_3$), sodium oxide (Na_2O), and other metal oxides, all of which form hydroxide ions in aqueous solution. In each case, the negative ion in solution accepts a proton from a water molecule, as shown below for bicarbonate (HCO_3^-) and oxide (O^{2-}) ions.

$$HCO_3^- + H{-}O{-}H \longrightarrow H_2CO_3 + OH^-$$

$$O^{2-} + H{-}O{-}H \longrightarrow 2\ OH^-$$

EXAMPLE 7.2 Recognizing acid behavior

Show why nitric acid (HNO_3) is an acid.

Solution When nitric acid is placed in water, the following reaction occurs:

$$HNO_3 + H_2O \longrightarrow H_3O^+ + NO_3^-$$

The nitric acid acts as a Brønsted-Lowry acid. It donates a proton to water, forming the hydronium ion, H_3O^+.

Practice Exercise 7.2

Show why hydrobromic acid (HBr) is an acid.

7.5 The Strengths of Acids and Bases

Strength of Acids

Hydrochloric acid, nitric acid, and sulfuric acid are classified as **strong acids,** whereas acetic acid, carbonic acid, and citric acids are classified as **weak acids** (Table 7.1). All these acids contain one or more hydrogen atoms that form hydronium ions when the acids are dissolved in water, but the extent to which the acids do this differs in the two groups. That is, the degree of **dissociation** determines whether an acid is classified as strong or weak.

A strong acid is defined as one that completely—or almost completely—dissociates into ions in water. For example, hydrochloric acid in water exists almost entirely as H_3O^+ and Cl^- ions:

$$HCl + H_2O \longrightarrow H_3O^+ + Cl^-$$

The boldfaced arrow indicates that the reaction proceeds almost completely to the right; that is, HCl dissociates almost completely into ions. Nitric acid and sulfuric acid also dissociate almost completely.

Table 7.1 Some Common Acids

Name of Acid*	Formula†	Classification
Sulfuric	H_2SO_4	Strong
Nitric	HNO_3	Strong
Hydrochloric	HCl	Strong
Phosphoric	H_3PO_4	Weak
Tartaric	CH(OH)COOH–CH(OH)COOH	Weak
Lactic	$CH_3CH(OH)COOH$	Weak
Citric	CH_2COOH–C(OH)COOH–CH_2COOH	Weak
Acetic	CH_3COOH	Weak
Carbonic	H_2CO_3	Weak
Hydrocyanic	HCN	Weak

* In decreasing order of acid strength.
† The hydrogen atoms that form hydronium ions in solution are shaded.

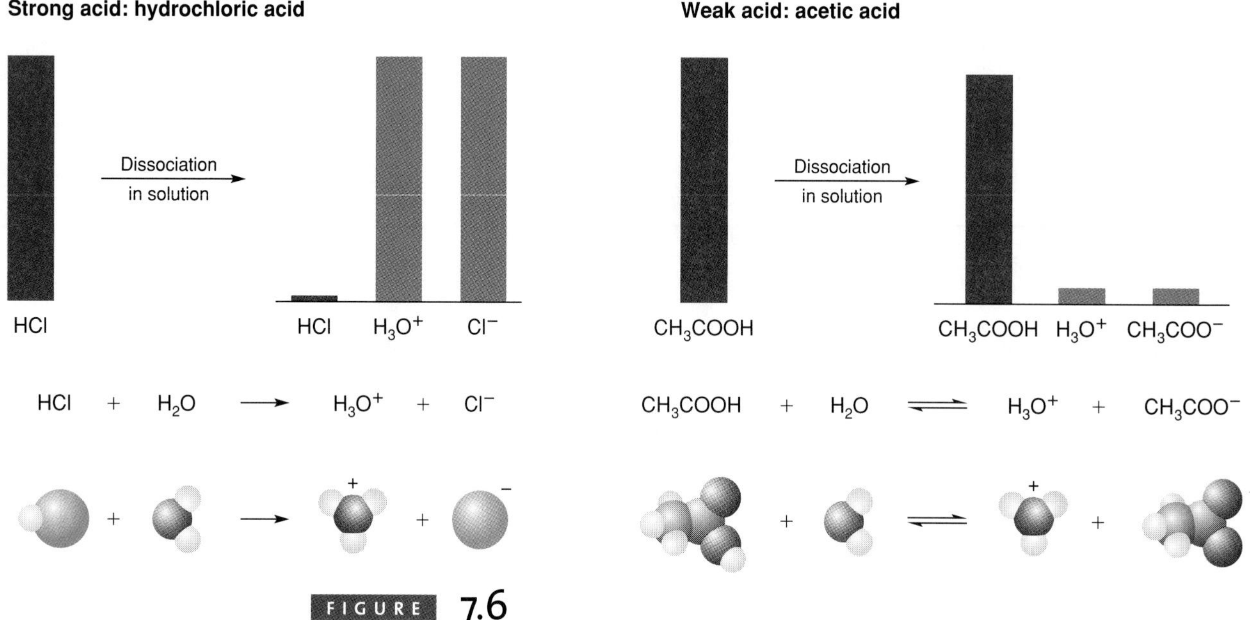

FIGURE 7.6

In aqueous solution, a strong acid such as hydrochloric acid (HCl) is almost completely dissociated into ions; a weak acid such as acetic acid (CH_3COOH) dissociates very little.

Weak acids, on the other hand, dissociate very little in aqueous solution—often less than 5%. They exist in solution mainly as undissociated molecules. We can write the equation for the dissociation of acetic acid (CH_3COOH), a typical weak acid, as follows:

$$CH_3COOH + H_2O \rightleftharpoons H_3O^+ + CH_3COO^-$$

The boldfaced arrow indicates that the reaction lies toward the left and that relatively few H_3O^+ and CH_3COO^- ions are present in solution. The use of two arrows—one pointing to the left and the other to the right—also indicates that an equilibrium exists in the solution: The rate of formation of ions is equal to the rate at which they recombine to form undissociated molecules. The difference between the dissociation of a strong acid and that of a weak acid is illustrated for hydrochloric acid and acetic acid in **Figure 7.6**.

Strength of Bases

Bases, like acids, can be classified as strong or weak. Common strong bases are sodium hydroxide (NaOH) and potassium hydroxide (KOH).

A **strong base,** like a strong acid, dissociates completely—or almost completely—into ions in water.

$$NaOH \longrightarrow Na^+ + OH^-$$

Calcium hydroxide [$Ca(OH)_2$] and magnesium hydroxide [$Mg(OH)_2$] are other examples of strong bases. However, they are only slightly soluble in water. The relatively little that goes into solution dissociates almost completely, but the total number of ions present is small.

A **weak base,** like a weak acid, dissociates very little in water. A common example of a weak base is ammonia. Although ammonia dissolves readily in water, it dissociates only slightly to give ammonium (NH_4^+) ions and hydroxide (OH^-) ions:

$$NH_3 + H_2O \rightleftharpoons NH_4^+ + OH^-$$

As occurs in an aqueous solution of a weak acid, an equilibrium is established when a weak base dissolves in water.

7.6 Neutralization Reactions

If a solution containing hydronium ions (an acid) is mixed with a solution containing an *equal* number of hydroxide ions (a base), the resulting solution has none of the properties of an acid and none of the properties of a base: It tastes neither sour nor bitter, and it has no effect on litmus paper. The acid and base have *neutralized* each other, and water and a **salt** have been formed in the **neutralization reaction.**

For example, a solution of hydrochloric acid (HCl) reacts with a solution of sodium hydroxide (NaOH) to produce water and sodium chloride (NaCl):

$$\underset{\text{acid}}{HCl} + \underset{\text{base}}{NaOH} \longrightarrow \underset{\text{water}}{H_2O} + \underset{\text{salt}}{NaCl}$$

Sodium ions (Na^+) are associated with a salty taste.

Or, more accurately, remembering that in aqueous solution, hydrogen ions react with water molecules to form hydronium ions:

$$H_3O^+ + Cl^- + Na^+ + OH^- \longrightarrow 2\ H_2O + Na^+ + Cl^-$$

Other examples of neutralization reactions are shown below.

$$\underset{\text{nitric acid}}{HNO_3} + \underset{\text{potassium hydroxide}}{KOH} \longrightarrow \underset{\text{water}}{H_2O} + \underset{\text{potassium nitrate}}{KNO_3}$$

$$\underset{\text{acetic acid}}{CH_3COOH} + \underset{\text{sodium hydroxide}}{NaOH} \longrightarrow \underset{\text{water}}{H_2O} + \underset{\text{sodium acetate}}{CH_3COONa}$$

In each reaction, a salt is formed, and the hydronium ion from the acid reacts with the hydroxide ion from the base to form water as follows:

$$\underset{\text{proton donor}}{H_3O^+} + \underset{\text{proton acceptor}}{OH^-} \longrightarrow \underset{\text{water}}{2\ H_2O}$$

EXAMPLE 7.3 Writing equations for neutralization reactions

Acid rain containing nitric acid (HNO_3) has fallen in lakes in the Adirondack mountains in New York. The nitric acid can be neutralized by adding calcium hydroxide [$Ca(OH)_2$] to the lake water. Write the balanced equation for the neutralization of nitric acid with calcium hydroxide.

Solution

$$2\ HNO_3 + Ca(OH)_2 \longrightarrow 2\ H_2O + Ca(NO_3)_2$$

Practice Exercise 7.3

Write the balanced equation for the neutralization reaction between potassium hydroxide (NaOH) and hydrobromic acid (HBr).

7.7 The Dissociation of Water

Water is generally thought of as H_2O, a covalent undissociated molecule. In fact, a very small percentage of water molecules dissociate to give hydronium and hydroxide ions:

$$\text{H—O—H} \quad + \quad \text{H—O—H} \quad \longrightarrow \quad H_3O^+ \quad + \quad OH^-$$

Only about one water molecule in every 500 million dissociates. In this dissociation, a proton is transferred from one water molecule to another; thus, water is acting as both an acid and a base. Although the dissociation of water occurs to such a slight degree, it is important for understanding pH, the scale that is used to determine how acidic or basic a solution is.

7.8 The pH Scale

Most people are familiar with the abbreviation **pH,** and, although they may not know exactly what it means, they know that it is used for measuring acidity. Chemists in laboratories are not the only ones who measure pH values of solutions. The pH of rain and water in lakes is monitored regularly by environmental scientists to check for increased acidity, and gardeners measure the pH of soils to determine if lime (CaO) should be added to neutralize excess acidity. The CaO reacts with water in the soil to form the base $Ca(OH)_2$.

To see more precisely what is meant by pH, we need to consider the concentrations of H_3O^+ and OH^- ions in aqueous solutions, and in water itself, in terms of molarity. Hydrochloric acid (HCl), a strong acid, dissociates almost completely into ions in water. Thus, a 1 molar (1 M) solution of hydrochloric acid contains 1 mole of H_3O^+ ions (and 1 mole of Cl^- ions) per liter of solution.

$$\underset{1\ \text{mole/L}}{HCl} \quad + \quad H_2O \quad \longrightarrow \quad \underset{1\ \text{mole/L}}{H_3O^+} \quad + \quad \underset{1\ \text{mole/L}}{Cl^-}$$

Water, however, dissociates very little, and 1 liter of pure water contains just 1.0×10^{-7} (or 0.0000001) mole of H_3O^+ and 1.0×10^{-7} (or 0.0000001) mole of OH^- ions.

A solution that contains more than 1.0×10^{-7} mole of H_3O^+ ions per liter is described as *acidic.* A solution that contains fewer than 1.0×10^{-7} mole of H_3O^+ ions per liter is described as *basic.*

It was the Danish biochemist S. P. L. Sorenson who proposed that the number in the exponent be used to express the hydronium ion concentration.

The abbreviation *pH* is derived from the French *pouvoir hydrogène,* which means "hydrogen power."

The pH scale was introduced to provide a convenient and concise way of expressing small concentrations of H_3O^+ ions in solution. pH can be defined as

$$[H_3O^+] = 10^{-pH}$$

(The brackets [ ] are a shorthand way of indicating the concentration of H_3O^+ in moles per liter.) Thus, on the pH scale, a concentration of 1.0×10^{-7} mole per liter of H_3O^+ ions becomes a pH of 7. Similarly, a concentration of 1.0×10^{-4} mole per liter becomes a pH of 4. It therefore follows that:

A solution with a *pH below 7 is acidic.*
A solution with a *pH of 7 is neutral.*
A solution with a *pH above 7 is basic.*

The relationship between pH and the concentrations of H_3O^+ and OH^- ions is shown in **Figure 7.7**. In pure water, the numbers of H_3O^+ and OH^- ions are equal, and the pH is 7. The lower the pH of a solution, the more acidic it is and the higher its concentration of H_3O^+ ions. The higher the pH of a solution, the more basic it is and the higher its concentration of OH^- ions. The exponent on 10 in the equation defining pH changes from -1 in a strong acidic solution to -7 in a neutral solution and to -14 in a strong basic solution. Figure 7.7 also shows the pH values of some common solutions.

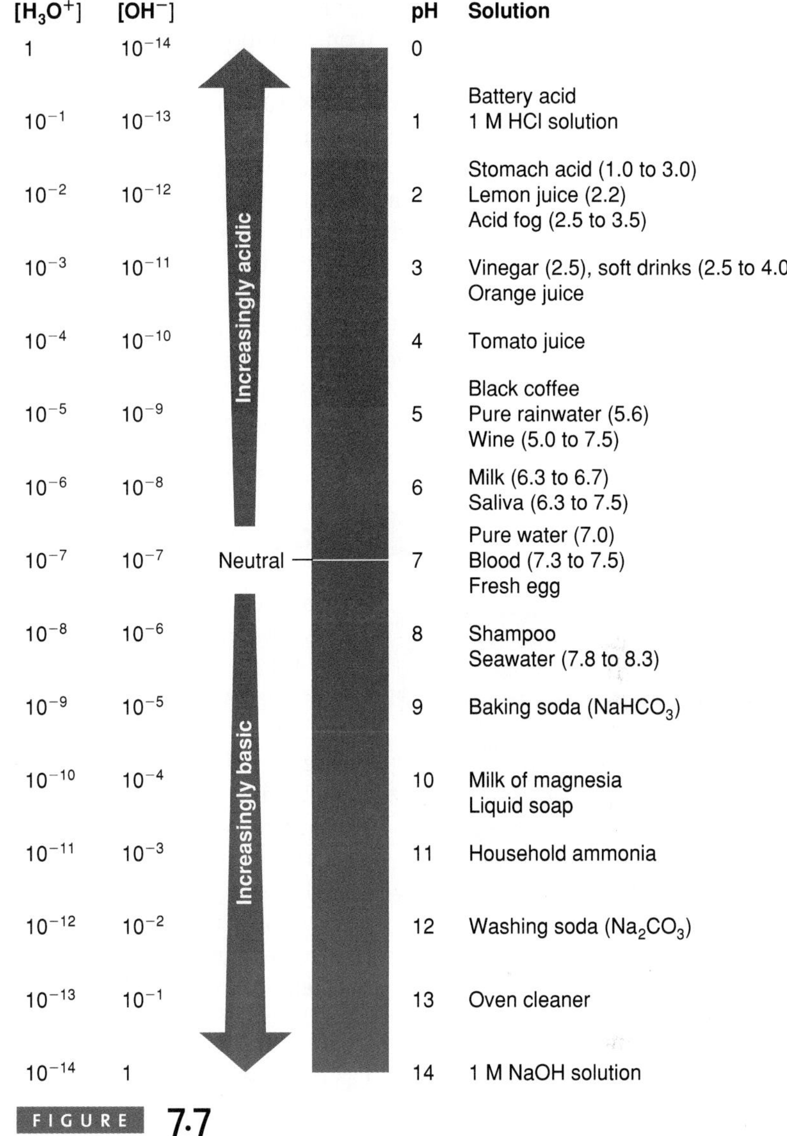

[H₃O⁺] [OH⁻] pH Solution
1 10⁻¹⁴ 0
10⁻¹ 10⁻¹³ 1 Battery acid
1 M HCl solution
10⁻² 10⁻¹² 2 Stomach acid (1.0 to 3.0)
Lemon juice (2.2)
Acid fog (2.5 to 3.5)
10⁻³ 10⁻¹¹ 3 Vinegar (2.5), soft drinks (2.5 to 4.0)
Orange juice
10⁻⁴ 10⁻¹⁰ 4 Tomato juice
10⁻⁵ 10⁻⁹ 5 Black coffee
Pure rainwater (5.6)
Wine (5.0 to 7.5)
10⁻⁶ 10⁻⁸ 6 Milk (6.3 to 6.7)
Saliva (6.3 to 7.5)
10⁻⁷ 10⁻⁷ Neutral 7 Pure water (7.0)
Blood (7.3 to 7.5)
Fresh egg
10⁻⁸ 10⁻⁶ 8 Shampoo
Seawater (7.8 to 8.3)
10⁻⁹ 10⁻⁵ 9 Baking soda (NaHCO₃)
10⁻¹⁰ 10⁻⁴ 10 Milk of magnesia
Liquid soap
10⁻¹¹ 10⁻³ 11 Household ammonia
10⁻¹² 10⁻² 12 Washing soda (Na₂CO₃)
10⁻¹³ 10⁻¹ 13 Oven cleaner
10⁻¹⁴ 1 14 1 M NaOH solution
Increasingly acidic
Increasingly basic

FIGURE 7.7

The pH scale. The diagram shows the relationship between pH and the concentrations of H_3O^+ and OH^- ions in solution in water at 25°C, as well as the pH values for some common aqueous solutions.

EXAMPLE 7.4 Converting between pH and hydronium ion concentration

What is the pH of a solution in which the hydronium ion concentration is

a. 1.0×10^{-6} M; **b.** 0.00001 M?

Solution

a. The exponent on 10 is -6. Therefore, the pH is 6.

b. Expressing the concentration in scientific notation gives 1.0×10^{-5} M. The exponent on 10 is -5. Therefore, the pH is 5.

Practice Exercise 7.4

What is the pH of a solution in which the hydronium ion concentration is **a.** 1.0×10^{-4} M; **b.** 0.001 M?

EXAMPLE 7.5

What is the hydronium ion concentration of a solution that has a pH of 2?

Solution If the pH is 2, the exponent on 10 in the expression for the concentration must be -2. Therefore, the hydronium ion concentration is 1.0×10^{-2} M, or 0.01 M.

Practice Exercise 7.5

What is the hydronium ion concentration of a solution that has a pH of 1?

It is important to note that a change in pH of 1 unit corresponds to a *10-fold* change in H_3O^+ concentration. For example, a pH of 3 corresponds to a H_3O^+ concentration of 1.0×10^{-3} (or 0.001) mole per liter, and a pH of 2 corresponds to a H_3O^+ concentration of 1.0×10^{-2} (or 0.01) mole per liter. Since 0.01 is 10 times as great as 0.001, a change in pH from 3 to 2 represents a 10-fold increase in the number of H_3O^+ ions in 1 L of solution. The concentration of hydrogen ions in battery acid (pH 1) is 400 times the concentration of hydrogen ions in tomato juice (pH 4).

In the examples above, we considered only solutions with whole-number pH values. Most solutions, as shown in Figure 7.7, have pH values that are not whole numbers. The pH of lemon juice, for example, is 2.3, and that of milk lies between 6.3 and 6.6. Solving problems with pH values that are not whole numbers is more complex and involves the use of logarithms. For our purposes, it is enough for you to know that a pH of 7.4, for example, lies between pH 7.0 and pH 8.0 and thus represents a H_3O^+ concentration between 1.0×10^{-7} M and 1.0×10^{-8} M.

7.9 Antacids

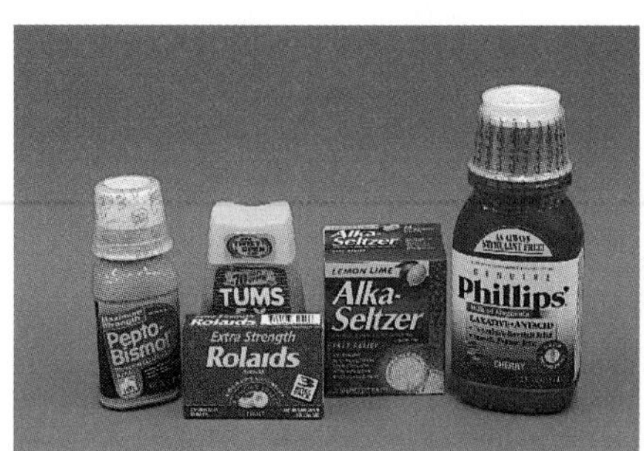

FIGURE 7.8

Some common antacids.

Compared with most body fluids such as blood, saliva, spinal fluid, and bile, which have a pH near 7, stomach fluid is very acidic; its pH is between 1 and 3. This acidity is due to hydrochloric acid (HCl), which is released constantly in small amounts by the cells in the inner lining of the stomach. When food enters the stomach—or even when it is smelled—acid production increases. The acid assists in the digestion of food and also prevents the growth of bacteria. Overeating, eating spicy foods, and feeling strong emotions can stimulate overproduction of acid, which causes pain or discomfort, variously referred to as acid indigestion, heartburn, or upset stomach. This condition can be relieved by consuming an **antacid** (Fig. 7.8), an over-the-counter drug containing one or more bases that neutralize the excess acid. (Newer drugs such as Tagamet, Pepcid AC, and Zantac do not neutralize stomach acid. Instead, they inhibit its production.)

The bases present in commercial antacids include sodium bicarbonate, calcium carbonate, and magnesium and aluminum hydroxides (Table 7.2). They react with acid as follows:

$$H^+ + NaHCO_3 \longrightarrow Na^+ + H_2O + CO_2$$

$$2\ H^+ + CaCO_3 \longrightarrow Ca^{2+} + H_2O + CO_2$$

$$2\ H^+ + Mg(OH)_2 \longrightarrow Mg^{2+} + 2\ H_2O$$

$$3\ H^+ + Al(OH)_3 \longrightarrow Al^{3+} + 3\ H_2O$$

Sodium bicarbonate and calcium carbonate produce carbon dioxide gas as they neutralize the acid.

Antacids are designed to decrease stomach acidity rapidly to give "instant relief" but not to neutralize all the acid present. If the stomach fluid reached pH 7, digestion would stop, and the stomach lining would begin to release more acid, a condition called *acid rebound.*

Extended use of antacids can cause side effects. Calcium carbonate and aluminum hydroxide can cause constipation; magnesium hydroxide acts as a laxative.

Table 7.2 Commercial Antacids

Brand Name	Active Ingredient(s)	Formula
Alka-Seltzer	Sodium bicarbonate	$NaHCO_3$
DiGel	Aluminum hydroxide	$Al(OH)_3$
Phillip's Milk of Magnesia	Magnesium hydroxide	$Mg(OH)_2$
Maalox	Magnesium hydroxide and aluminum hydroxide	$Mg(OH)_2$ and $Al(OH)_3$
Mylanta	Magnesium hydroxide and aluminum hydroxide	$Mg(OH)_2$ and $Al(OH)_3$
Rolaids	Dihydroxy aluminum sodium carbonate	$Al(OH)_2NaCO_3$
Tums	Calcium carbonate	$CaCO_3$

7.10 Acid-Base Buffers

In many biological systems, it is important that pH be maintained within a narrow range. You will notice in Figure 7.7 that blood is slightly basic, with a pH between 7.3 and 7.5. Even a small deviation below or above these values can be life-threatening.

The pH of blood is maintained within its critical range by **buffers,** which are substances that resist changes in pH. The primary buffer system in blood is the bicarbonate system. Blood contains dissolved carbon dioxide and bicarbonate ions (HCO_3^-). When acid enters the blood as a result of metabolic processes, it reacts with bicarbonate ions as shown below: Hydrogen ions are removed from solution, and a drop in pH is prevented.

$$H^+ + HCO_3^- \rightleftharpoons H_2CO_3 \rightleftharpoons CO_2 + H_2O$$

If base is added, the following reaction occurs: Hydroxide ions are removed, and a rise in pH is prevented.

$$OH^- + H_2CO_3 \rightleftharpoons HCO_3^- + H_2O$$

This bicarbonate buffer system is also important in regulating the pH of seawater.

7.11 Naturally Occurring Acids

With few exceptions, naturally occurring acids are weak acids. The most common acid in the environment is carbonic acid, which forms when carbon dioxide gas dissolves in water. Pure water is neutral (pH 7.0), but rainwater is naturally slightly acidic (pH 5.6) because, as it falls, it dissolves carbon dioxide, a natural component of the atmosphere. A dilute solution of carbonic acid is formed. Although carbonic acid dissociates very little, it does so sufficiently to provide enough hydrogen ions to lower the pH of rainwater.

Citrus fruits, grapes, tomatoes, and rhubarb all contain weak acids.

$$CO_2 + H_2O \rightleftharpoons H_2CO_3 \rightleftharpoons H^+ + HCO_3^-$$

Other weak acids are formed naturally in soil when plant materials decay. These weak organic acids dissociate slightly to give hydrogen ions (R represents a large organic group; see Chapter 14).

$$\underset{\text{organic acid}}{\text{R—COOH}} \rightleftharpoons H^+ + \text{R—COO}^-$$

Complete decomposition of the remains of plant and animal materials results in the formation of carbon dioxide, which adds additional hydrogen ions to soil as it dissolves in groundwater. Soils rich in organic matter are therefore naturally acidic.

A strong acid that is produced naturally in the environment is sulfuric acid. In many volcanic eruptions—including the eruption of Mount Pinatubo in the Philippines in 1991—large quantities of sulfur dioxide (SO_2) are emitted and carried great distances by the wind. In the atmosphere, the SO_2 is converted to sulfur trioxide (SO_3), which dissolves in rainwater to yield sulfuric acid (H_2SO_4).

7.12 Uses of Acids and Bases

Acids and bases have many important industrial uses (**Table 7.3**). They also rank high among the chemicals that are produced in greatest quantities in the United States (**Table 7.4**).

Sulfuric acid, nitric acid, and phosphoric acid are all used in the production of fertilizers. Sulfuric acid, which is produced in the United States in greater quantity than any other chemical, is also used in the petroleum industry, in metal and ore processing, and in automobile batteries. It is important in the manufacture of dyes, drugs, plastics, and many other products. Nitric acid is used in the manufacture of explosives and

Table 7.3 Common Acids and Bases and Their Uses

	Formula	Uses
Acids		
Sulfuric acid	H_2SO_4	Manufacture of fertilizers, plastics, dyes, paper; petroleum refining; steel production; liquid in automobile batteries
Nitric acid	HNO_3	Manufacture of fertilizers, explosives, plastics, dyes; steel production
Hydrochloric acid	HCl	In building industry for removal of mortar from brick and rust scale from metals
Phosphoric acid	H_3PO_4	Manufacture of fertilizers, detergents, baking powder, fire-resistant fabrics, carbonated drinks
Bases		
Sodium hydroxide	NaOH	Petroleum industry; manufacture of aluminum, synthetic fibers, paper, dyes, detergents, soaps; drain cleaners
Potassium hydroxide	KOH	Production of fertilizers, soaps, detergents
Calcium hydroxide	$Ca(OH)_2$	Production of cement, mortar, plaster, paper, bleaching powder; water softening; reducing soil acidity
Ammonia	NH_3	Production of fertilizer, explosives, plastics, synthetic fibers, paper, rubber, detergents; household cleaners

in steel refining; like sulfuric acid, it plays a role in the manufacture of dyes and plastics. Hydrochloric acid (also known as *muriatic acid*) is important in the building industry. It is used to remove excess mortar from bricks and scale and rust from metals. It is also used to clean automobile radiators. Phosphoric acid and its salts are used in the manufacture of many commercial products, including carbonated beverages, detergents, baking powder, and fire resistant textiles.

The base produced in greatest quantity in the United States is lime (calcium oxide, CaO). It is made from calcium carbonate ($CaCO_3$), or limestone, and is used in manufacturing cement, mortar, and plaster. It is also used for reducing soil acidity and softening hard water. Ammonia, which is also produced in very large quantities is essential for the production of fertilizers and is also used in the manufacture of plastics, explosives, paper, and many other products. Another base produced in great quantity is sodium hydroxide, which is used in the manufacture of synthetic fibers, dyes, detergents, and household cleaners. Potassium hydroxide is used in the production of fertilizers and soaps.

Table 7.4 Acids and Bases Produced in the United States (1995)

	Rank Among All Chemicals Produced	Billions of Pounds
Acids		
Sulfuric acid	1	95.5
Phosphoric acid	7	26.2
Nitric acid	14	17.3
Hydrochloric acid	27	7.3
Bases		
Lime	5	41.3
Ammonia	6	35.6
Sodium hydroxide	8	26.2

7.13 Acid Rain

Acid rain is a serious environmental problem. As we have seen, pure rainwater is naturally slightly acidic, with a pH of 5.6. Rainwater with a pH less than 5.6 is characterized as **acid rain.** In recent years, the average pH of rainfall in many parts of northeastern North America has fallen below 4.6 (Fig. 7.9), and rain with a pH as low as 2.9 has been recorded.

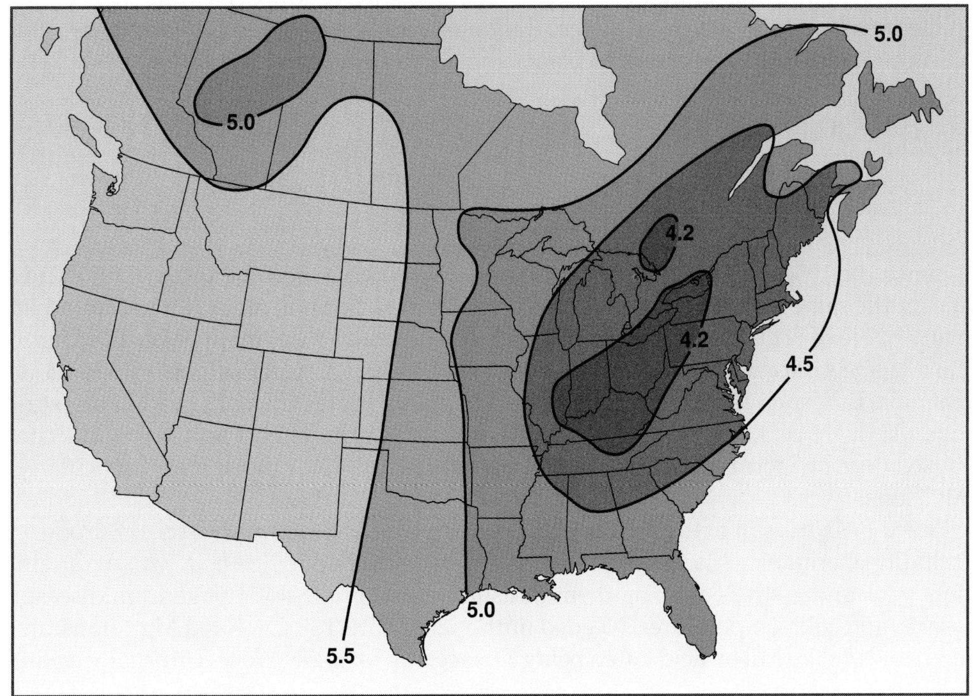

FIGURE 7.9 Average pH of precipitation in the United States and Canada. In many parts of North America, especially in the northeast, precipitation is abnormally acidic (pH below 5.6).

FIGURE 7.10

These cave formations are the result of the slow dissolution, over thousands of years, of underground deposits of limestone by naturally acidic rain.

The Effects of Acid Rain

The weathering of rock is in large part caused by the acid naturally found in pure rainwater and soils. If rain is unusually acidic (pH less than 5.6), the normal weathering process is accelerated, and other chemical reactions also occur. The effects on the environment of both pure rain and acid precipitation depend on the type of soil and bedrock that the precipitation encounters.

In the normal process of weathering, when naturally acidic rain falls on soils derived from limestone ($CaCO_3$), some limestone dissolves, and the acid is neutralized:

$$CaCO_3 + H^+ \longrightarrow Ca^{2+} + HCO_3^-$$

Although this reaction occurs very slowly, over thousands of years, great quantities of rock are dissolved. The extensive Carlsbad Caverns in New Mexico and many other caves were formed in this way **(Fig. 7.10)**.

When acid rain falls in regions where limestone is common, such as the Midwest and the Great Lakes area, the acid is neutralized according to the above reaction, and adverse effects are minimized. However, in many parts of the United States and Canada, the underlying rock is primarily granite and basalt, igneous rock composed of silicate minerals such as feldspar, which provide little buffering action. In the normal process of weathering, feldspar is partially dissolved by naturally acidic rain and soils and converted to kaolinite (a clay) and silica, with potassium ions going into solution:

$$\underset{\text{feldspar}}{2\,KAlSi_3O_8} + \underset{\text{rainwater}}{2\,H^+ + H_2O} \longrightarrow \underset{\text{kaolinite (clay)}}{Al_2Si_2O_5(OH)_4} + \underset{\text{silica}}{4\,SiO_2} + \underset{\text{potassium ions}}{2\,K^+}$$

Other silicate minerals weather in the same general way, yielding clays and metal ions. In the presence of an increased concentration of acid, the clay partially dissolves, and aluminum ions, which are toxic to plants and aquatic life, go into solution:

$$\underset{\text{clay}}{Al_2Si_2O_5(OH)_4} + \underset{\text{acid}}{6\,H^+} \longrightarrow \underset{\text{aluminum ions}}{2\,Al^{3+}} + \underset{\text{silica}}{2\,SiO_2} + \underset{\text{water}}{5\,H_2O}$$

Since the 1950s, lakes in many parts of the world have become increasingly acid, and, at the same time, fish populations have declined. Salmon are no longer found in many of Nova Scotia's rivers, and fish have disappeared from many lakes in eastern Canada, the Adirondacks, and Scandinavia. When the pH of water falls below 5.5, many desirable fish, such as trout and bass, die. At a pH of 5.0, few fish of any kind can survive. At a pH of 4.5, lakes become virtually sterile. The effects on aquatic life are caused both by increased acidity and by the presence of toxic metal ions, particularly Al^{3+} ions.

Since 1980, trees in many forests in the eastern United States and parts of Europe—including Germany's Black Forest—have suffered severe damage **(Fig. 7.11)**. Acid rain puts trees under stress, making them unusually susceptible to damage from disease, insects, and cold temperatures. Needed nutrients, such as K^+, Ca^{2+} and Mg^{2+} ions, are leached from the soil by acid, often being replaced by toxic Al^{3+} ions. Unusually acidic

FIGURE 7.11

In many parts of the eastern United States, the growth of forests at high elevations has declined and trees are dying. Acid deposition is blamed for much of this damage.

soil damages fine root hairs and destroys beneficial microorganisms. Acid rain can damage surface structures on leaves and pine needles, causing them to wither and drop.

Building materials, particularly limestone and marble, which are composed of $CaCO_3$, are readily eroded by acid. As shown in **Figure 7.12**, many ancient statues in Europe, particularly those on historic cathedrals, have been severely and rapidly eroded over the last 50 years. Acid rain is considered to be responsible for the accelerated weathering of these structures.

The Causes of Acid Rain

The primary causes of acid rain that are due to human activity are emissions of sulfur dioxide (SO_2) and nitrogen dioxide (NO_2) from coal- and oil-burning power plants and emissions of oxides of nitrogen from automobiles. In the atmosphere, these emissions are chemically converted into sulfuric acid (H_2SO_4) and nitric acid (HNO_3), which accumulate in droplets of water in clouds and fall to earth in rain **(Fig. 7.13)**. Because these acids can be present in fog, sleet, snow, and fine particulates as well as in rain, this type of pollution is more accurately termed **acid deposition.** (The sources of acid rain and the steps being taken to control them will be discussed in more detail in Chapter 11.)

The damage to lakes and forests in the northeastern United States and Canada has usually been blamed on emissions from industrial centers in the Midwest, which are carried from their source by the prevailing winds. However, recent research sug-

FIGURE 7.12

Over time, many ancient limestone and marble monuments have been severely eroded by acid deposition.

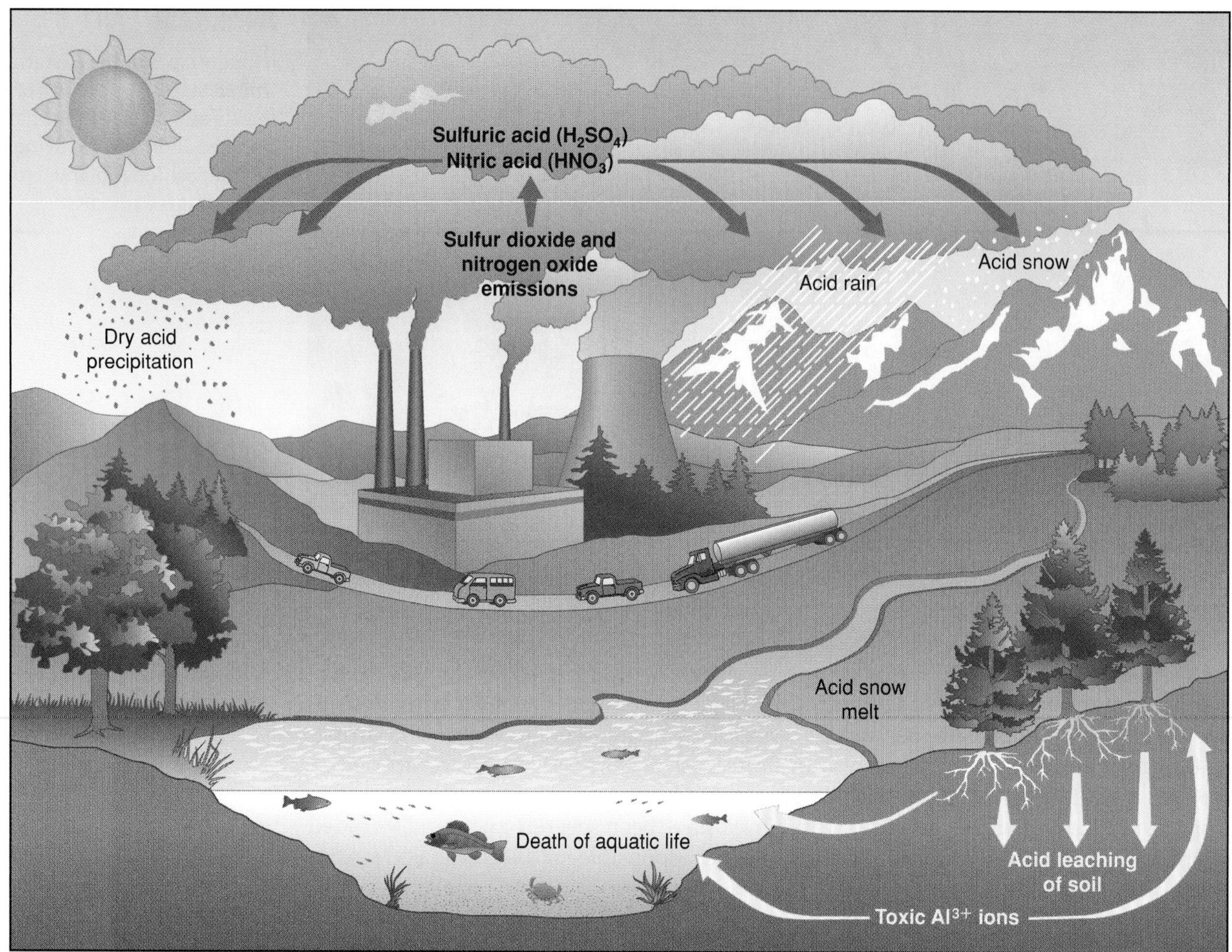

FIGURE 7.13

The main sources of acid deposition are emissions from oil- and coal-burning power plants and automobiles. Sulfur dioxide and nitrogen oxides from these sources combine with water vapor in the atmosphere to form sulfuric acid (H_2SO_4) and nitric acid (HNO_3). The acids collect in clouds and fall to earth in rain and snow, causing many adverse effects.

Industrial centers in Europe are believed to be responsible for increased acid deposition and environmental damage in Scandinavia.

gests that the natural process of soil formation may be an important factor in causing acidification of lakes and soil. Changes in land use, such as have occurred in the northeastern part of the United States and in Scandinavia, are known to acidify soils. Lowering acid-causing industrial emissions may not, therefore, be as effective as hoped in saving lakes and forests.

7.14 Acid Mine Drainage

Acid released at certain types of mines can cause local increases in the acidity of soils, streams, and rivers. **Acid mine drainage** is associated primarily with the mining of coal deposits rich in the mineral pyrite (FeS_2). When underground sources of pyrite-rich coal are mined, the coal becomes exposed to water and the atmosphere. The pyrite reacts with oxygen and water, and acid is formed according to the following equations:

$$2\ FeS_2 + 7\ O_2 + 2\ H_2O \longrightarrow 2\ Fe^{2+} + 4\ SO_4^{2-} + 4\ H^+$$

$$4\ Fe^{2+} + O_2 + 10\ H_2O \longrightarrow 4\ Fe(OH)_3 + 8\ H^+$$

Rainwater seeping through sulfur-rich wastes left at mine sites also produces acid. Although not widespread, the effects of acid runoff from mines into nearby streams can be severe and, in some instances, have caused sizeable fish kills.

Regulations now limit acid discharge from mine sites, and at most active mines, acid is neutralized with lime before it is released into natural water systems. Many abandoned mines have been sealed in an effort to prevent acid formation by excluding water and oxygen. These measures have not always been successful, and acid mine drainage remains a problem in some areas.

7.15 Oxidation and Reduction

Among the most important reactions that occur in the environment, and in industrial processes, are **oxidation-reduction reactions.** In our bodies, cells obtain energy by the oxidation of food. The oxidation of fossil fuels provides 90% of the energy used in the world today. Metals are obtained from their ores by reduction reactions. Photosynthesis by green plants involves reduction of carbon dioxide obtained from the atmosphere. Oxidation and reduction are complementary processes; they always occur together. In any oxidation-reduction reaction, one substance is oxidized, while, at the same time, another is reduced.

Reactions were described as oxidation or reduction reactions long before it was possible to explain what was occurring at the molecular level. Like the definitions of acids and bases, the definitions of oxidation and reduction broadened once the structure of the atom was understood. Oxidation and reduction can be defined narrowly in terms of gaining or losing oxygen and hydrogen atoms or, more generally, in terms of gaining and losing electrons.

All oxidation-reduction reactions that occur in living organisms and many that occur in the physical environment take place in aqueous solution. We will consider some important oxidation-reduction reactions that occur in the absence of water before we discuss oxidation-reduction reactions in aqueous solution.

7.16 Gaining and Losing Oxygen Atoms

Early chemists, who studied many reactions involving oxygen, defined *oxidation* as the combination of oxygen with some other element or compound. The element or compound that gained oxygen was said to be *oxidized.*

Gaining Oxygen Atoms

Oxygen is a very reactive element, which combines with every other element except helium, argon, and neon to form compounds called *oxides.* For example, when iron rusts, it combines with oxygen in the air to form iron(III) oxide:

$$4\ Fe + 3\ O_2 \longrightarrow 2\ Fe_2O_3$$

Sulfur combines with oxygen to form sulfur dioxide:

$$S + O_2 \longrightarrow SO_2$$

The reactions in which iron and sulfur combine with oxygen are oxidation reactions; iron and sulfur have been oxidized.

One oxidation reaction that we have discussed many times is combustion (or burning). When wood, coal, or other carbon-containing fuels are burned, the carbon atoms combine with oxygen in the air to form carbon dioxide:

$$C + O_2 \longrightarrow CO_2$$

If the supply of oxygen is limited, carbon monoxide is formed:

$$2\ C + O_2 \longrightarrow 2\ CO$$

In both reactions, carbon is oxidized. In most oxidation reactions, heat is produced; in combustion reactions, the heat is given off rapidly. This emission of heat is evident when wood is burned in the fireplace or when a forest fire is started.

Many oxidation reactions that occur naturally in the environment proceed very slowly. The breakdown of organic material in dead plants and animals by aerobic bacteria and other decomposers (Chapter 2) is an example. Carbon in the organic matter is gradually oxidized, and the end product, as in combustion, is carbon dioxide.

EXAMPLE 7.6 Writing equations for oxidation reactions

When chromium oxide (Cr_2O_3) reacts with aluminum metal (Al), the aluminum is oxidized to aluminum oxide (Al_2O_3). Write the balanced equation for the oxidation reaction.

Solution

$$Cr_2O_3 + 2\ Al \longrightarrow 2\ Cr + Al_2O_3$$

Practice Exercise 7.6

Boron hydride (BH_3) is pyrophoric, meaning that it reacts spontaneously with oxygen in the air, producing a flame. In this oxidation reaction, boric acid (H_3BO_3) is formed. Write a balanced equation for the reaction.

Losing Oxygen Atoms

Originally, the term *reduction* referred to removal of oxygen from a compound. Many of the early reactions that were studied were reduction reactions in which metals were produced from their oxides. One of Lavoisier's experiments (Chapter 3), which led to the formulation of the law of conservation of matter and the identification of oxygen, was a reduction reaction. Lavoisier heated mercury(II) oxide and obtained mercury metal and oxygen gas:

$$2\ HgO \longrightarrow 2\ Hg + O_2$$

Other reduction reactions in which a metal oxide loses oxygen to yield the metal are shown in the following two equations:

$$CuO + H_2 \longrightarrow Cu + H_2O$$

$$2\ Fe_2O_3 + 3\ C \longrightarrow 4\ Fe + 3\ CO_2$$

In these reduction reactions, the mass of the starting material, the metal oxide, is *reduced*. This explains the origin of the term *reduction* for reactions in which oxygen atoms are lost.

Substances that are in a reduced form, such as coal, have a high energy content. The energy is released when the substance is oxidized. Conversely, oxidized substances, such as carbon dioxide, are low in energy.

EXAMPLE 7.7 Identifying oxidation-reduction reactions

Chlorine dioxide (ClO_2), a very powerful disinfecting agent, is prepared from sodium chlorate ($NaClO_3$) and sulfuric acid. Is the $NaClO_3$ oxidized or reduced?

Solution

$$NaClO_3 \longrightarrow ClO_2 \quad \text{(not a complete equation)}$$

1. Even though all the products of the reaction are not given, it can be seen that the $NaClO_3$ has lost oxygen.
2. Since oxygen has been lost, the $NaClO_3$ is reduced.

Practice Exercise 7.7

Permanganate ion (MnO_4^-) can be oxidized in some reactions and reduced in others. In the following reaction, is permanganate ion oxidized or reduced?

$$MnO_4^- \longrightarrow MnO \quad \text{(not a complete equation)}$$

7.17 Gaining and Losing Hydrogen Atoms

In the past, oxidation and reduction were also defined in terms of losing and gaining *hydrogen* atoms. A substance was said to be *oxidized* if it lost hydrogen atoms and *reduced* if it gained hydrogen atoms.

Gaining Hydrogen Atoms

An important reduction reaction is the combination of carbon monoxide with hydrogen, in the presence of a catalyst, to form methanol:

$$CO + 2\ H_2 \xrightarrow{\text{catalyst}} CH_3OH$$

Carbon monoxide gains four hydrogen atoms and is reduced. This reaction is important in the production of methanol as a fuel for automobiles (Chapter 13).

Losing Hydrogen Atoms

A reaction in which a substance loses one or more hydrogen atoms is termed an oxidation reaction. Many such reactions occur in metabolic processes in the body (discussed in Chapter 14). An example is the enzyme-catalyzed conversion of ethanol (C_2H_5OH) to acetaldehyde (CH_3CHO) in the liver:

$$\underset{\text{6 H atoms}}{C_2H_5OH} \xrightarrow[\text{enzyme}]{-2\,H} \underset{\text{4 H atoms}}{CH_3CHO}$$

In this reaction, ethanol has lost two hydrogens and has thus been oxidized.

EXAMPLE 7.8 Identifying oxidation-reduction reactions

In which of the following is the reactant undergoing oxidation?

a. $CH_3OH \longrightarrow CH_2O$ (not a complete equation)

b. $C_4H_{10} + H_2 \longrightarrow C_4H_{12}$

Solution

a. In this reaction, methyl alcohol (CH_3OH) loses two H atoms, forming CH_2O; therefore, it is oxidized.

b. In this reaction butene (C_4H_{10}) gains two H atoms, forming C_4H_{12}; therefore, it is reduced.

Practice Exercise 7.8

When the very reactive chemical reagent lithium aluminum hydride ($LiAlH_4$) reacts with an organic compound, hydrogen atoms are usually added to the organic compound. When it reacts with acetaldehyde (CH_3CO), the following occurs:

$$LiAlH_4 + CH_3CHO \longrightarrow CH_3CH_2OH$$

Is acetaldehyde oxidized or reduced?

7.18 Gaining and Losing Electrons

After the discovery of the electron, oxidation and reduction were defined in much broader terms. According to the newer, more general definition, **oxidation** *occurs if a substance involved in a chemical reaction loses one or more electrons.* **Reduction** *occurs if a substance gains one or more electrons.* The earlier definitions limited oxidation and reduction to reactions in which oxygen or hydrogen were involved. The newer definitions broaden the concepts to include many reactions involving compounds that contain neither oxygen nor hydrogen.

Oxidation-Reduction (Redox) Reactions

Oxidation cannot occur without reduction. For simplicity and to emphasize their dual nature, oxidation-reduction reactions are often called *redox reactions.*

We will first consider an oxidation-reduction reaction in aqueous solution that demonstrates experimentally, and in an obvious way, that electrons are being transferred. If a strip of zinc (Zn) is immersed in a solution of copper(II) sulfate ($CuSO_4$), a spontaneous redox reaction occurs (**Fig. 7.14**). Copper (II) sulfate in solution dissociates into Cu^{2+} ions, which impart a blue color to the solution, and SO_4^{2-} ions, which are colorless. The zinc strip in the solution gradually becomes covered with a deposit, which can be shown to be elemental copper (Cu). At the same time, the blue color of the solution fades, indicating a decrease in the number of Cu^{2+} ions in solution (see Fig. 7.14). The Cu^{2+} ions have been reduced. They have gained electrons to become uncharged Cu atoms:

$$Cu^{2+}(aq) + 2\,e^- \longrightarrow Cu(s)$$

$$Zn + Cu^{2+} \longrightarrow Zn^{2+} + Cu$$

FIGURE 7.14

When a strip of zinc is placed in a solution of copper(II) sulfate, a spontaneous oxidation-reduction (redox) reaction occurs. (a) In solution, copper(II) sulfate dissociates into Cu^{2+} and SO_4^{2-} ions. The Cu^{2+} ions give the solution a blue color. (b) Electrons are transferred from the zinc (Zn) to the Cu^{2+} ions, and Zn^{2+} ions and copper metal (Cu) are formed. (c) The zinc strip gradually dissolves as more and more Zn^{2+} ions go into solution. The blue color of the solution fades as Cu^{2+} ions are removed from solution and deposited as Cu on the zinc strip. (Note that the metal ions are smaller than their corresponding atoms.)

The electrons have been supplied by the zinc, which gradually dissolves in the solution (Fig. 7.14). In giving up electrons, the elemental zinc (Zn) is oxidized and converted to colorless Zn^{2+} ions:

$$Zn(s) \longrightarrow Zn^{2+}(aq) + 2\,e^-$$

Combining the two reactions, we obtain

$$Cu^{2+}(aq) + Zn(s) + \cancel{2e^-} \longrightarrow Cu(s) + Zn^{2+}(aq) + \cancel{2e^-}$$

The electrons cancel out, and the overall redox reaction becomes

$$Cu^{2+}(aq) + Zn(s) \longrightarrow Cu(s) + Zn^{2+}(aq)$$

The SO_4^{2-} ions take no part in the reaction and are not included in the equation.

The reaction illustrates the complementary nature of oxidation and reduction. Neither can occur without the other. As one substance (in this case, Cu^{2+}) is reduced, another (Zn) is oxidized.

The substance that causes oxidation by accepting electrons is termed an *oxidizing agent;* in the process of oxidation, the oxidizing agent is itself reduced. Similarly, the substance that causes reduction by providing electrons is termed a *reducing agent;* in the process of reduction, the reducing agent is oxidized.

EXAMPLE 7.9 Identifying oxidation-reduction reactions

In the following reactions, are the metals oxidized or reduced?

a. $Zn + 2\ HCl \longrightarrow ZnCl_2 + H_2$

b. $Fe^{3+} \longrightarrow Fe^{2+}$ (not a complete equation)

Solution

a. The zinc metal (Zn) is converted to its ion, Zn^{2+}. This ion must have a 2+ charge because we know that each chloride ion carries a 1− charge, and electrical neutrality must be maintained. To change Zn to Zn^{2+}, the metal atom must *lose* 2 electrons. Therefore, zinc metal is oxidized.

b. To change from a 3+ ion to a 2+ ion, iron must *gain* an electron. Therefore, iron is reduced.

Practice Exercise 7.9

Are the following reactions oxidation or reduction reactions?

a. $CuSO_4 \longrightarrow Cu$ (not a complete equation)

b. $2\ Ag + 2\ HNO_3 \longrightarrow 2\ AgNO_3 + H_2$

7.19 Oxidation and Reduction: Three Definitions

Let's summarize the three definitions of oxidation and reduction:

	Oxidation	Reduction
In terms of oxygen:	gain of oxygen	loss of oxygen
In terms of hydrogen:	loss of hydrogen	gain of hydrogen
In terms of electrons:	loss of electrons	gain of electrons

A simple mnemonic to help you remember the definitions of oxidation and reduction is "LEO (the lion) says GER": *L*ose *E*lectrons, *O*xidation; *G*ain *E*lectrons, *R*eduction.

Although it may not be immediately obvious, electrons were transferred in all the reactions used to illustrate oxidation and reduction in terms of the gain and loss of oxygen and hydrogen.

7.20 Batteries: Energy from Oxidation-Reduction Reactions

EXPLORATIONS offers a look at one kind of vital battery-powered device—the pacemaker (see pp. 204–205).

Redox reactions can be used to produce electrical energy. For example, in the spontaneous transfer of electrons between copper ions and zinc atoms, which we discussed earlier, chemical energy is converted to electrical energy that can be used to power a battery:

$$Zn(s) + Cu^{2+}(aq) \longrightarrow Zn^{2+}(aq) + Cu(s)$$

If a strip of zinc is placed in a solution containing Cu^{2+} ions, as shown in Figure 7.14, energy is released as heat, which warms the solution. If, however, the apparatus is modified as shown in **Figure 7.15**, electrical current is produced and can be used to light an electric lightbulb. In one compartment of the apparatus, a strip of zinc is immersed in a solution of zinc sulfate (which dissociates to give Zn^{2+} and SO_4^{2-} ions). In the other compartment, a strip of copper is immersed in a solution of copper(II) sulfate (which dissociates to give Cu^{2+} and SO_4^{2-} ions). The two compartments are separated by a porous disc that prevents the two solutions from mixing but allows the passage of ions. The metal strips, which serve as electrodes, are connected through a voltmeter.

As the reaction proceeds, electrons pass through the wire from the zinc electrode to the copper electrode. They pass down through the copper electrode (cathode), and copper is deposited on the electrode as electrons combine with Cu^{2+} ions in solution. Zinc atoms in the zinc electrode (anode) lose electrons, and Zn^{2+} ions are formed and pass into solution. Sulfate ions take no part in the electron transfer, but they migrate through the porous partition to maintain a balance of positive and negative ions in solution. The electric current passing through the wire is sufficient to light a lightbulb.

An apparatus like the one just described, which produces an electric current, is called an **electrochemical cell.** A **battery** is made up of a series of electrochemical cells. Commercial batteries such as flashlight batteries, button batteries, and automobile storage batteries **(Fig. 7.16)** all operate on a similar principle.

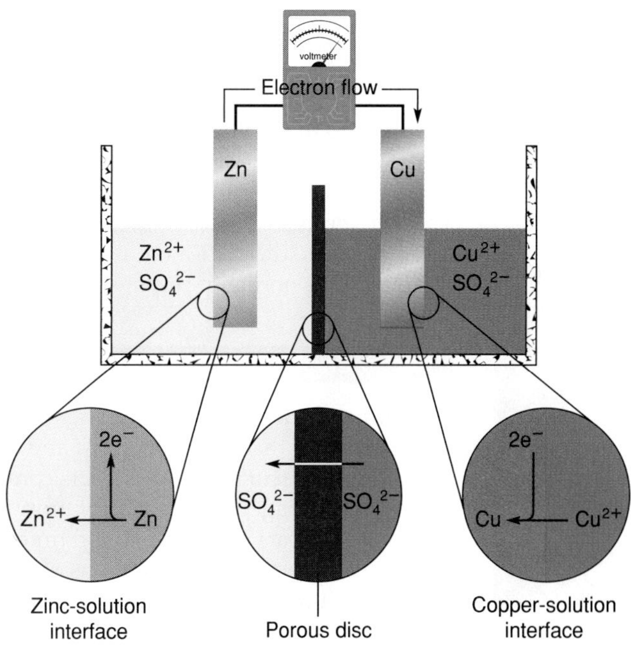

FIGURE 7.15

An electrochemical cell uses a redox reaction to produce an electric current. The two solutions are separated by a porous material that minimizes their mixing but allows ions to pass through to maintain a charge balance. During the reaction, zinc atoms in the zinc strip (the anode) give up electrons and go into solution as Zn^{2+} ions; Cu^{2+} ions in the copper(II) sulfate solution combine with electrons that have passed through the wire and are deposited as copper atoms on the copper strip (the cathode). The flow of electrons forms an electric current. Sulfate ions pass through the porous material to maintain balance between the charges in the solution.

Single-Use Batteries

DRY CELL BATTERIES A cross-section of a standard flashlight battery is shown in **Figure 7.17**. The outer case, which is made of zinc, serves as the negative electrode (or anode). A carbon rod surrounded by a paste containing powdered manganese dioxide (MnO_2) serves as the positive electrode (or cathode). Surrounding the MnO_2 paste is a moist mixture of ammonium chloride (NH_4Cl) and zinc chloride ($ZnCl_2$). The battery is called a *dry cell* because no free liquid is present.

As current flows through the cell and through the flashlight bulb, the zinc anode loses electrons and is oxidized:

$$Zn \longrightarrow Zn^{2+} + 2\,e^-$$

The reaction at the carbon cathode is complex and involves the reduction of manganese dioxide. The overall reaction can be written as follows:

$$Zn + 2\,MnO_2 + 2\,NH_4Cl \longrightarrow ZnCl_2 + Mn_2O_3 + 2\,NH_3 + H_2O + \text{electrical energy}$$

Dry cells do not last very long if stored because the acidic NH_4Cl gradually corrodes the zinc container.

In any electrical circuit, the anode is the electrode that releases electrons and the one at which oxidation occurs. The cathode is the electrode that receives electrons and the one at which reduction occurs. To remember this distinction, note that the words *anode* and *oxidation* begin with vowels while *cathode* and *reduction* begin with consonants.

Batteries are portable sources of electrical energy.

EXPLORATIONS

Cardiac Pacemakers—Battery-Powered Devices That Save Lives

Today, thousands of people with failing hearts, who in the past would have died, are leading normal lives thanks to tiny electronic cardiac pacemakers implanted in their bodies. These ingenious devices send out electrical impulses that stimulate a failing heart and reestablish its normal pumping action.

Functioning of the Heart

The human heart is an amazing organ. Slightly bigger than a clenched fist and weighing less than a pound, it beats continuously—on average, 72 times every minute. Each year, it beats about 38 million times and, in doing so, pumps 700,000 gallons of blood around the body.

The heart is a hollow muscular organ with four chambers: the right and left atria and the right and left ventricles. Blood collected into large veins from all parts of the body except the lungs, enters the heart through the right atrium, and from there flows into the right ventricle. From the right ventricle, it is pumped to the lungs to be reoxygenated. Oxygenated blood is then returned to the heart through the left atrium and from there enters the left ventricle. From the left ventricle it is pumped to all parts of the body (except the lungs) via the arteries.

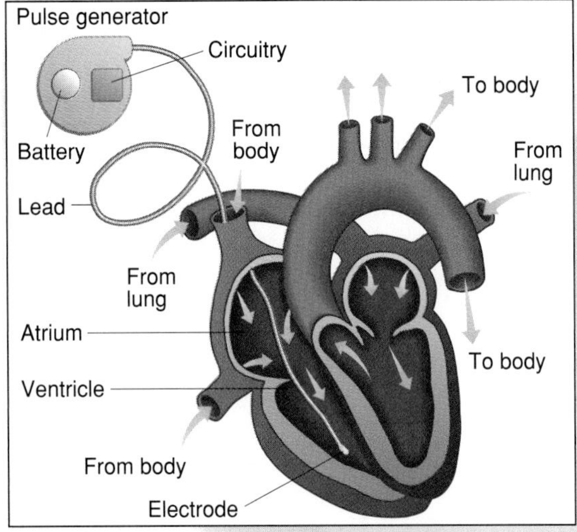

Circulation of blood through the heart and placement of a pacemaker.

The heart has a natural pacemaker that normally initiates and controls its rhythmic pumping action. At regular intervals, this pacemaker, which consists of a group of specialized cells located in the upper part of the right atrium, spontaneously generates a nerve impulse to initiate a heartbeat. The impulse spreads through the heart muscle, causing the atria and ventricles to contract in proper sequence. The coordinated rhythmic contractions keep the blood circulating through the body.

In certain pathological disorders, the heart's own pacemaker is impaired, and the heart ceases to beat with a normal rhythm. For example, in some conditions the heart beats abnormally slowly. Cardiac output is diminished, and congestive heart failure or other serious problems may result. In other disorders, the heart beats more rapidly than normal or fluctuates between a slow and a fast rate. Most of these potentially fatal disorders can now be treated successfully with an artificial pacemaker.

Early Pacemakers

Electrical impulses have been used for many years to stimulate and control a patient's heartbeat. In early procedures, electrodes placed on a patient's chest were connected to a bulky external pacemaker plugged into an external circuit. Later, battery-powered units that could be attached to patients' clothing were introduced. The first completely implantable pacemaker was placed in a human by a Swedish surgeon in 1958. A year later, an American team headed by Dr. William Chardack implanted a pacemaker designed by the engineer and inventor Wilson Greatbatch.

Wilson Greatbatch had been so convinced of the life-saving potential of an implantable device that he had given up his job so that he could work in a barn behind his house to design one. Within 2 years, he had a working model that he demonstrated to a number of cardiac specialists. Only Dr. Chardack immediately appreciated its possibilities. The Chardack-Greatbatch pacemaker soon became the most widely used type in the world.

An implantable pacemaker has three main parts: the pulse generator containing the battery and the electronic circuit; the pacing leads, which carry the electrical impulses; and the metal electrodes at the end of the leads, which make contact with the heart muscle. Pacemakers are installed in the body in one of two ways:

Either the electrodes are attached to the outside surface of the heart, and the pulse generator is implanted beneath the soft tissue of the abdominal wall; or the electrodes are passed through a vein in the neck directly into the heart, and the pulse generator is implanted under the skin just below the collarbone. Depending on the nature of the heart impairment, the electrical stimuli are applied to the ventricle (the most common procedure), the atrium, or the two chambers in sequence. The pacing systems are programmable: Parameters such as timing, duration, and voltage of the electrical stimuli can be adjusted to meet a patient's needs from outside the body.

The early implantable pacemakers had many problems. They were quite bulky, often causing discomfort to the patient, and, almost invariably, they failed within 1–2 years. The greatest problem was the battery. The mercury-zinc batteries used at that time had a life expectancy of about 5 years, but when placed in pacemakers, they failed much sooner, primarily because the pulse generator could not be hermetically sealed. Body fluids seeped in and caused the device to short circuit. Various alternative power sources were studied, including biological and nuclear sources, but the real breakthrough came with the introduction of lithium batteries in the 1970s.

Developments in microelectronics that revolutionized the computer industry also revolutionized the pacemaker industry.

Current Technology

Lithium makes a good anode for a battery because it loses electrons easily. It is the lightest of all the metals and, compared weight for weight with other metals, produces a large quantity of energy. Thus, lithium-iodine batteries—the type most commonly used in pacemakers—are very small and lightweight. Today's pacemakers weigh approximately 1 ounce (28 g). The redox reaction in a lithium-iodine battery is as follows:

$$2\,Li \longrightarrow 2\,Li^+ + 2\,e^-$$
(anode reaction/oxidation)

$$2\,Li^+ + 2\,e^- + I_2 \longrightarrow 2\,LiI$$
(cathode reaction/reduction)

$$2\,Li + I_2 \longrightarrow 2\,LiI$$
(overall reaction)

Pulse generators can now be sealed efficiently, and implanted pacemakers generally function reliably for 6–10 years, and sometimes longer. Often the cause of failure is not the battery but dislodgement of the leads.

Tremendous advances have been made in the electronic circuitry. Developments in microelectronics that revolutionized the computer industry also revolutionized the pacemaker industry. With each advance, pacemaker circuits became smaller, more sophisticated, and more reliable. The first pulse generators delivered electrical stimuli at a fixed rate that was unrelated to the patient's own heart rhythm. The complex circuits in modern units are able to sense a patient's own heartbeat and adjust the stimuli accordingly; if the circuit senses a natural beat, the generator does not fire, but if there is a longer than normal pause between beats, it fires. Some pacemakers also sense and respond to other body signals. For example, some can sense increases in body temperature and respiratory rate that occur during exercise, and then adjust the heart rate to meet the body's changing needs.

Pacemaker patients rely primarily on electrocardiograms to check that their hearts and pacemakers are functioning properly. Using a telephone transmitter in the comfort of their homes, patients can relay tracings to their doctors. Some sophisticated pacemakers are monitored by telemetry; that is, the unit not only receives instructions from but also relays information to an outside receiver. A built-in microprocessor in the unit records and stores data about cardiac function and the condition of the different parts of the pacemaker; it then relays this information to the doctor's office.

Despite all the advances, a pacemaker is still limited by its battery, which inevitably runs down. In the future, it is expected that this problem will be solved by taking advantage of a naturally occurring oxidation-reduction reaction in the body. If such an inexhaustible source of energy can be tapped, a pacemaker will last a patient for as long as it is needed.

References: D. Sonnenburg, M. Birnbaum, and E. A. Naclerio, *Understanding Pacemakers* (New York: Michael Kesend Publishing, 1982); H. W. Moses, J. A. Schneider, B. D. Miller, and G. J. Taylor, *A Practical Guide to Cardiac Pacing* (New York: Little Brown, 3rd ed., 1991).

FIGURE 7.16

Batteries come in many sizes and depend on a variety of redox reactions to produce electric current.

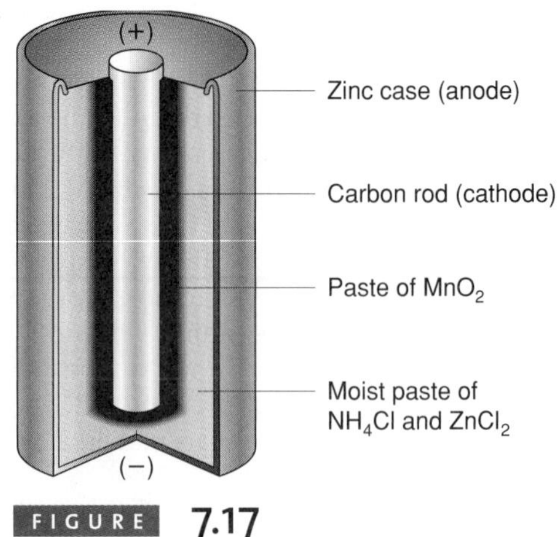

FIGURE 7.17

Cross section of a dry cell flashlight battery.

The dry cell battery was invented by Georges Leclanche in 1866.

Alkaline Batteries Alkaline batteries are also dry cells. The main difference is that the NH_4Cl is replaced with the alkaline substance potassium hydroxide (KOH), which gives the battery its name. Alkaline batteries are more expensive than regular dry cells, but they last longer because corrosion occurs much more slowly under alkaline conditions. All sizes of alkaline batteries, from D to small AAA, produce about 1.5 volts of electrical current.

The standard carbon-zinc dry cell battery used in flashlights and portable radios is relatively inexpensive. Alkaline, lithium, and mercury batteries are more expensive but last longer.

Mercury Batteries Many of the miniature button batteries used in watches, hearing aids, cameras, and electronic calculators are mercury cells **Figure 7.18**, which depend on a redox reaction in which zinc is oxidized and mercury is reduced:

$$HgO + Zn + H_2O \longrightarrow Hg + Zn(OH)_2 + \text{electrical energy}$$

Mercury batteries produce a very constant current of 1.34 volts.

Mercury is toxic (Chapters 10 and 18), and if mercury batteries are discarded in household trash which is then incinerated, deadly mercury vapor is released. To avoid this danger, these batteries should be recycled to recover the mercury.

Lithium Batteries Lithium batteries are very lightweight and powerful, producing 3.0 volts. Lithium-iodine batteries, which are long-lasting and very reliable, are used in pacemakers. Batteries that employ a lithium anode and a metal oxide or metal sulfide cathode are used in cameras and watches.

With the exception of some newly designed alkaline batteries, the commercial batteries just described are not rechargeable because the oxidation-reduction reactions on which they depend are not easily reversed.

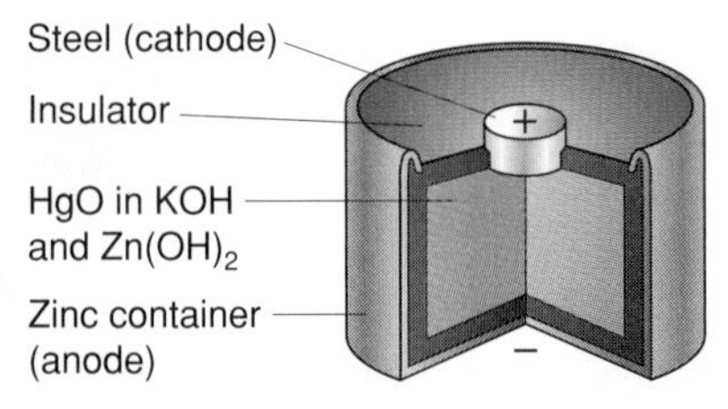

FIGURE 7.18

Cross section of a mercury battery.

Rechargeable Batteries

Nickel-Cadmium (Nicad) Batteries Nicad batteries are not quite as powerful as alkaline batteries and do not last as long if stored, but they are popular because they can be readily recharged and used over and over again. During recharging, the discharging oxidation-reduction reaction is reversed.

$$NiO_2 + Cd + 2\,H_2O \underset{\text{charging}}{\overset{\text{discharging}}{\rightleftharpoons}} Ni(OH)_2 + Cd(OH)_2$$

Nicad batteries are used in portable radios, battery-powered children's toys, tools, and calculators. Because cadmium is toxic (see Chapter 10), nicad batteries should not be discarded with household trash but should be recycled.

LEAD-ACID BATTERIES The familiar 12-volt car battery consists of six electrochemical cells joined in series (**Fig. 7.19**). The electrodes are made of lead and are immersed in dilute sulfuric acid. The negative electrodes (anodes) are impregnated with spongy lead metal; the positive electrodes (cathodes), with lead oxide (PbO_2). As the battery discharges, metallic lead is oxidized to lead sulfate ($PbSO_4$) at the anodes, and lead oxide is reduced at the cathode. The net reaction is

$$Pb + PbO_2 + 2\,HSO_4^- + 2\,H^+ \underset{\text{recharging}}{\overset{\text{discharging}}{\rightleftharpoons}} 2\,PbSO_4 + 2\,H_2O + \text{electrical energy}$$

A lead-acid battery can be recharged if electrons are forced to flow in the opposite direction by connecting the battery to an external source of electricity. An automobile battery is normally recharged during driving.

Lead storage batteries are reliable and relatively inexpensive. Electric cars such as General Motors EV1 (see Explorations in Chapter 9) are powered by arrays of lead-acid batteries. Disadvantages are their weight and the problem of disposal. If carelessly discarded, both the lead and the sulfuric acid can pollute the environment. In the United States, most automobile batteries are now being recycled.

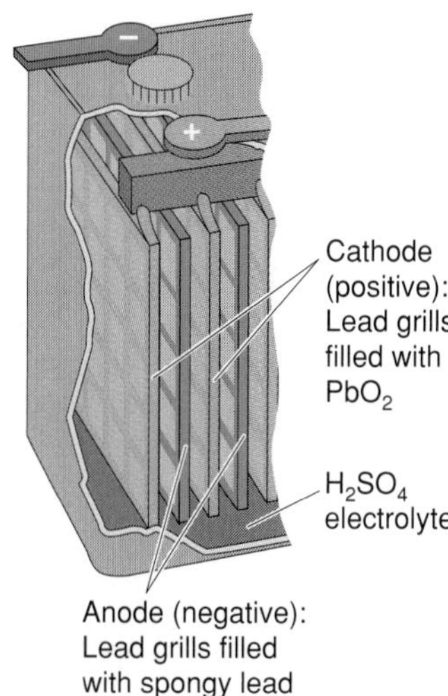

FIGURE 7.19

Cross section of a lead storage battery.

Chapter Summary

1. A given solute will dissolve in a particular solvent if its polarity is similar to that of the solvent; that is, "like dissolves like."

2. A 1 molar (1 M) solution contains one molecular (or formula) mass of solute dissolved in 1 liter (1 L) of solution:

$$\text{molarity} = \text{M} = \frac{\text{moles of solute}}{\text{liters of solution}}$$

3. Concentration units used for very dilute solutions are parts per million and parts per billion: 1 ppm = 1 mg/L and 1 ppb = 1 μg/L.

4. An acid is a substance that donates a proton. Solutions of acids in water taste sour, dissolve certain metals with the release of hydrogen gas, change the color of litmus from blue to red, and react with bases to form salts and water.

5. A base is a substance that accepts a proton. Solutions of bases in water taste bitter, feel slippery to the skin, change the color of red litmus to blue, and react with acids to form salts and water.

6. In aqueous solution, acids dissociate to form hydrogen ions (H^+ ions, protons), which immediately combine with water to form hydronium ions (H_3O^+). Hydronium ions are responsible for the properties of acids.

7. In aqueous solution, bases dissociate to form hydroxide ions (OH^-), which are responsible for the properties of bases.

8. In aqueous solution, strong acids (such as HCl, H_2SO_4, and HNO_3) and strong bases (such as NaOH and KOH) dissociate completely into ions. Weak acids (such as CH_3COOH and H_2CO_3) and weak bases (such as NH_3) do not dissociate into ions to any great extent.

9. A neutralization reaction occurs when H_3O^+ ions in an acidic solution react with an equal number of OH^- ions in a basic solution to form a salt and water.

10. Water dissociates very little. One liter of water contains 1×10^{-7} mole of H_3O^+ ions and 1×10^{-7} mole of OH^- ions.

11. The pH scale is used to express the acidity of a solution, that is, the concentration of H_3O^+ ions in the solution:

$$[H_3O^+] = 10^{-pH}$$

12. A solution with a pH of 7 is neutral. One with a pH below 7 is acidic, and one with a pH above 7 is basic.

13. A change in pH of 1 unit corresponds to a 10-fold change in H_3O^+ concentration.

14. Pure rainwater is slightly acidic (pH 5.6) because it dissolves carbon dioxide as it falls.

15. Acid deposition is caused by emissions of sulfur dioxide and nitrogen dioxide from coal- and oil-burning power plants and of oxides of nitrogen from automobiles. Acid rain damages trees, kills fish, and attacks structures made of limestone.

16. When a substance gains an oxygen atom or loses a hydrogen atom, it is oxidized. When a substance gains a hydrogen atom or loses an oxygen atom, it is reduced.

17. When a substance loses an electron, it is oxidized; when it gains an electron, it is reduced (LEO says GER).

18. Oxidation and reduction are complementary processes. In a redox reaction, one substance is oxidized while another is reduced.

19. In an electrochemical cell, an electric current is generated. Electrons pass through a wire from the anode, where oxidation occurs, to the cathode, where reduction occurs.

20. Commercial batteries are made up of electrochemical cells.

Key Terms

acid (p. 183)
acid deposition (p. 195)
acid mine drainage (p. 196)
acid rain (p. 193)
antacid (p. 190)
aqueous solution (p. 177)
base (p. 183)
battery (p. 203)
buffer (p. 191)
concentration (p. 178)
dissociation (p. 185)
electrochemical cell (p. 203)
hydronium ion (p. 183)
molarity (p. 179)
neutralization reaction (p. 187)
oxidation (p. 200)
oxidation-reduction (redox) reactions (p. 197)
parts per billion (ppb) (p. 180)
parts per million (ppm) (p. 180)
pH (p. 188)
proton (p. 183)
reduction (p. 200)
salt (p. 187)
solubility (p. 177)
solute (p. 177)
solution (p. 177)
solvent (p. 177)
strong acid (p. 185)
strong base (p. 186)
weak acid (p. 185)
weak base (p. 186)

Questions and Problems

1. Define the following:
   a. solution b. solvent c. solute d. solubility
2. What ions are present in aqueous solutions of the following salts?
   a. KBr b. $MgCO_3$ c. Na_2SO_4 d. NH_4Cl
   e. $Mg(NO_3)_2$
3. What ions are present in aqueous solutions of the following salts?
   a. $PbCl_2$ b. NaF c. $Ba(OH)_2$ d. AgI e. $SrCl_2$
4. What is the molar concentration (molarity) of a 1.0 L solution that contains 156 g of nitric acid (HNO_3)? Of a 1.0 L solution that contains 7.8 g?
5. What is the molarity of a 20.0 L solution that contains 40 g of sodium hydroxide (NaOH)?
6. Waste discharge water from a paper mill is sampled for dioxin, a toxic substance that is formed in the bleaching of paper pulp. Analysis of the waste water shows a dioxin concentration of 0.010 μg/1.0 mL. Express this concentration in parts per billion. The EPA standard for dioxin is 1 ppb. Is this waste water in violation of the EPA standard?
7. Which is the higher concentration in each pair?
   a. 100 ppb or 0.05 ppm
   b. 500 ppb or 250 ppm
   c. 10 ppb or 1 ppm
8. If you dropped 25 g of salt (NaCl) into a 120,000 L swimming pool, what would be the concentration of NaCl in the water in ppm and ppb?
9. If a 100 g sample of water contains 1.5 mg of arsenic, what is the concentration of arsenic in the sample in parts per million?
10. Litmus is turned ____________ by acids and ____________ by bases.
11. a. Name a base present in household products that is used to remove greasy dirt.
    b. Name an acid that is commonly used in food preparation.
12. If a substance has a bitter taste and feels slippery, is it more likely to be an acid or a base?
13. Name and give the formula for the ion that is responsible for:
    a. the properties of an acid in water.
    b. the properties of a base in water.
14. Identify the Brønsted-Lowry acid and base in the following reactions:
    a. $HF + H_2O \longrightarrow H_3O^+ + F^-$
    b. $S^{2-} + H_2O \longrightarrow HS^- + OH^-$
    c. $H_2CO_3 + H_2O \longrightarrow H_3O^+ + HCO_3^-$
    d. $CH_3NH_2 + H_2O \longrightarrow CH_3NH_3^+ + OH^-$

15. Complete the following reaction:

$$HCl + NaOH \longrightarrow$$

a. Name the products.

b. This reaction is a ________ reaction.

16. Which of the following are weak acids?

a. acetic acid

b. hydrochloric acid

c. nitric acid

17. Which of the following are strong acids?

a. hydrocyanic acid

b. carbonic acid

c. sulfuric acid

18. Write an equation to show how water dissociates into ions.

19. What is the pH of a neutral solution?

20. Indicate whether solutions having the following pH values are acidic, basic, or neutral.

a. 3.5 b. 7.3 c. 6.5

21. Which of the solutions having the following pH values are basic?

a. 9.4 b. 7.0 c. 12.5

22. What is the pH of a solution that has a hydronium ion concentration, $[H_3O^+]$, of:

a. 1.0×10^{-4} M

b. 1.0×10^{-8} M

c. 1.0×10^{-1} M

23. What is the pH of a solution that has a hydronium ion concentration of:

a. 0.001 M

b. 0.1 M

c. 0.000000000001 M

24. What is the hydronium ion concentration of a solution that has a pH of:

a. 10 b. 7 c. 4 d. 2

25. What pH range is most likely to describe a salad dressing that contains vinegar?

26. Solution A has a pH of 2. Solution B has a pH of 5.

a. Which solution is more acidic?

b. How many times more acidic is that solution than the other?

27. Describe four general properties of aqueous solutions of:

a. acids b. bases

28. Name some acids and bases found in the average American home, and explain briefly how they are used.

29. What is the hydronium ion? Explain how it is formed in an aqueous solution of acid. Is it a stable ion? Explain your answer.

30. Explain how ammonia acts as a base in water. Is it a strong or a weak base? Give equations.

31. Using HA to represent an acid, write equations that show the difference between a strong acid and a weak acid. Give the names and formulas of two strong acids and two weak acids.

32. Explain how acids and bases neutralize each other. What products are formed? Give two equations to illustrate your answer.

33. Explain, using an equation, how water acts as both an acid and a base.

34. Why was the pH scale introduced? Explain why the pH of a neutral solution is 7.

35. If the pH of a solution increases from 2 to 5, the concentration of hydronium ions will be ________ times (greater/less) than the original value.

36. What is a buffer? How does a buffer work?

37. List three uses of sulfuric acid.

38. When an acid neutralizes a base, the products of the reaction are water and ________.

39. In the following reactions, is the bicarbonate ion (HCO_3^-) acting as an acid or a base?

a. $HCO_3^- + H^+ \longrightarrow H_2CO_3$

b. $HCO_3^- + OH^- \longrightarrow H_2O + CO_3^{2-}$

40. Name and give the formula of the most common acid in the natural environment.

41. Write equations showing:

a. how water dissolves atmospheric carbon dioxide to form an acidic solution.

b. how the acid formed dissociates into ions.

42. What is the pH of physiological liquids such as blood?

43. Write equations to show how limestone is weathered by rainfall.

44. What is the pH of pure rain? Why is it acidic? Give equations to explain your answers.

45. Describe two natural sources of sulfuric acid.

46. Write equations for the reactions responsible for the natural acid weathering of:

a. limestone

b. feldspar

47. **a.** Define *acid rain*.

b. What are the effects of acid rain on trees and other plant life?

c. At what pH will a lake become sterile?

48. Why does acid rain create more serious problems in lakes in eastern North America than in the Great Lakes? Include chemical equations to illustrate your answer.

49. What is the main cause of acid mine drainage? Give an equation to show how this acid is formed.

50. What steps have been taken to control acid mine drainage? Have these methods been successful?

51. Classify each of the following as either an oxidation or reduction:

a. $FeO \longrightarrow Fe_2O_3$

b. $ClO^- \longrightarrow Cl_2$

c. $OH^- \longrightarrow O_2$

d. $MnO_4^- \longrightarrow MnO_2$

52. Which of the following is an oxidation?

a. $H_2S \longrightarrow S$

b. $Cl_2 \longrightarrow ClO_4^-$

c. $Fe(OH)_3 \longrightarrow Fe$

53. Which of the following is a reduction?

a. $NO_2^- \longrightarrow NO_3^-$

b. $Cl_2 \longrightarrow ClO_3^-$

c. $Cr_2O_3 \longrightarrow Cr$

54. Do the following statements describe an oxidation, a reduction, or neither?

a. Hydrogen ion is lost.

b. Electrons are gained.

c. Hydrogen atom is lost.

d. Electrons are lost.

55. Do the following statements describe an oxidation, a reduction, or neither?

a. Hydrogen atom is gained.

b. Oxygen atom is gained.

c. Hydroxide ion is gained.

d. Hydroxide ion is lost.

56. Zinc metal is the source of electrons in a dry cell battery. Write the equation for the reaction zinc metal undergoes.

57. Lead metal forms one electrode in a car battery. What is the other electrode made of? Which of the two is oxidized?

58. The button batteries that are used in watches and calculators produce electricity by means of a chemical reaction that involves two metals. Name the two metals.

59. Which substance is oxidized in the following reactions?

a. $Mg + 2\,AgNO_3 \longrightarrow 2\,Ag + Mg(NO)_2$

b. $2\,Na + 2\,H_2O \longrightarrow 2\,NaOH + H_2$

60. Which substance is reduced in the following reactions?`

a. $FeCl_2 + SnCl_4 \longrightarrow 2\,FeCl_3 + SnCl_2$

b. $2\,H_2 + O_2 \longrightarrow 2\,H_2O$

61. Which of the following are oxidation reactions?

a. $Cl_2 \longrightarrow 2\,Cl^-$

b. $Cu \longrightarrow Cu^{2+}$

c. $C_2H_4 \longrightarrow C_2H_6$

d. $2\,H^+ \longrightarrow H_2$

62. In the following reactions, which metal is oxidized and which is reduced?

a. $4\,H^+ + Sn^{2+} + O_2 \longrightarrow Sn + 2\,H_2O$

b. $4\,Zn + NO_3^- + 10\,H^+ \longrightarrow 4\,Zn^{2+} + NH_4^+ + 3\,H_2O$

c. $6\,Fe^{2+} + Cr_2O_7^{2-} + 14\,H^+ \longrightarrow 6\,Fe^{3+} + 2\,Cr^{3+} + 7\,H_2O$

d. $MnO_4^- + Cl^- \longrightarrow Mn^{2+} + ClO^-$

Answers to Practice Exercises

7.1 0.2 ppm = 200 ppb

7.2 $HBr + H_2O \longrightarrow H_3O^+ + Br^-$

7.3 $NaOH + HBr \longrightarrow NaBr + H_2O$

7.4 **a.** pH = 4 **b.** pH = 3

7.5 $[H_3O^+] = 0.1\ M$

7.6 $2\ BH_3 + 3\ O_2 \longrightarrow 2\ H_3BO_3$

7.7 Since the MnO_4^- has lost oxygen, it is reduced.

7.8 Acetaldehyde is reduced.

7.9 **a.** reduction
b. oxidation

Chapter 8

Carbon Compounds

An Introduction to Organic Chemistry

Chapter Objectives

In this chapter, you should gain an understanding of:

- Why carbon forms many more compounds than any other element
- The different classes of hydrocarbons
- Structural isomerism and *cis-trans* isomerism of carbon compounds
- The structure of benzene
- Nomenclature and properties of different classes of organic compounds based on functional groups

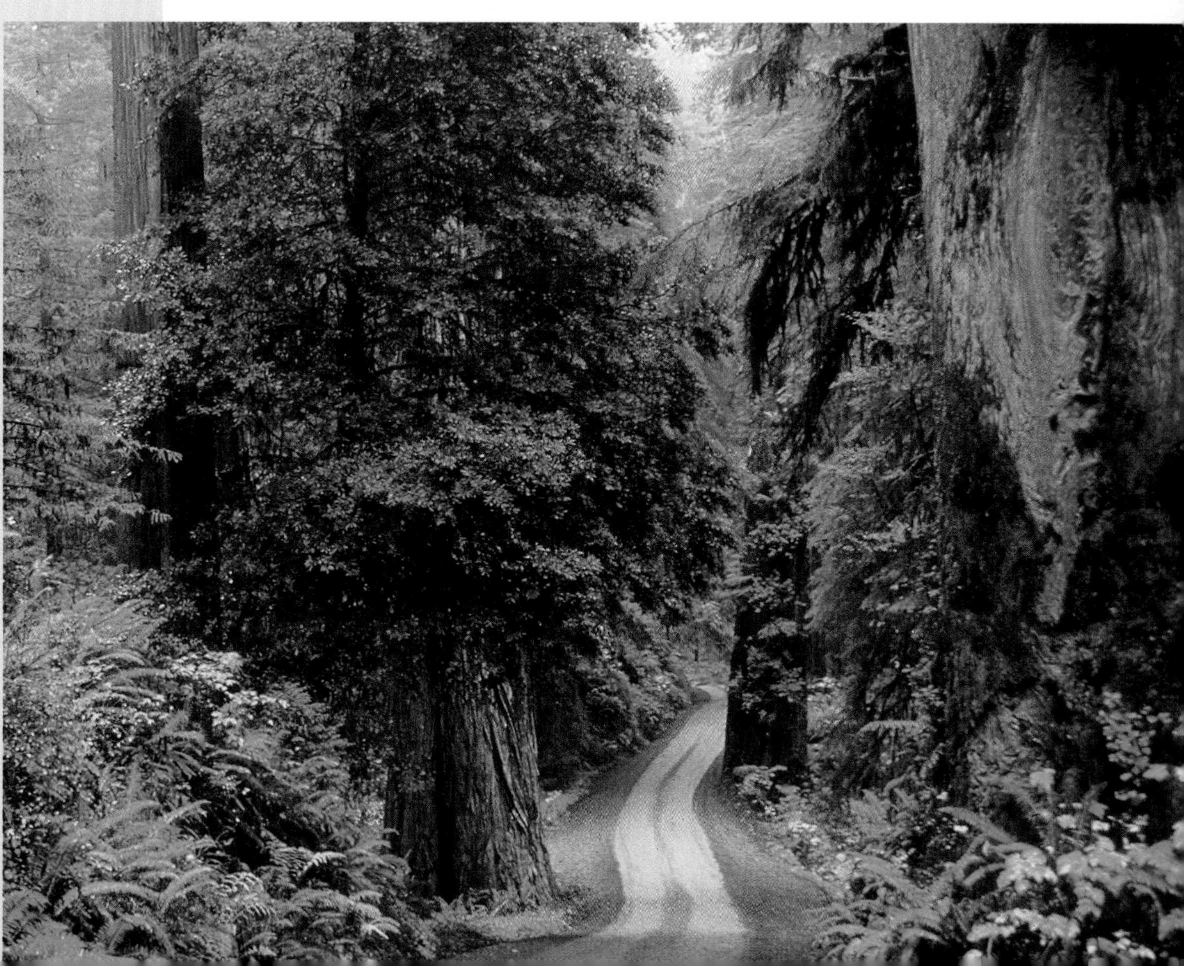

The huge redwood trees in this California forest, like every living thing on earth, are composed primarily of carbon-containing compounds.

ONLINE FEATURES

www.jbpub.com/chemistry

- Chemistry and the Environment
- eLearning
- Chemistry in the News
- Special Topics
- Research & Reference Links
- Ask the Authors

COMPOUNDS OF CARBON ARE CENTRAL to life on earth. Deoxyribonucleic acid (DNA), the huge molecule that carries the genetic code for a given species, is a carbon compound. Skin, muscle, hair, blood, leaves, roots, and all other living tissues are made from carbon compounds. Not only are our bodies and everything in the living world around us made from carbon compounds, but we depend on these compounds to support us in our daily lives and to enhance our standard of living. Our clothing—whether it contains natural fibers, such as wool and cotton, or synthetic fibers, such as nylon and polyester—is made from carbon compounds. Numerous other consumer goods—foods, medicines, paper, plastics, pesticides, and tires—are also made of carbon compounds. The combustion of carbon compounds in natural gas, petroleum, and coal generates the energy that fuels modern society.

EXPLORATIONS looks at a natural dye that can be made from a carbon compound and was once reserved for the use of high-ranking Romans (see pp. 242–243).

With the exception of ionic compounds such as carbonates, cyanides, and oxides of carbon, which are considered to be inorganic, compounds containing carbon are described as organic compounds (Chapter 2). Over 10 million chemical compounds are known, and thousands more are being synthesized each year. Of these, 9 million are organic compounds, and the remainder are inorganic. Although there are 92 naturally occurring elements (Chapter 4), only carbon can form so many compounds. Organic chemistry is devoted to the study of carbon compounds; inorganic chemistry covers all the other elements and their compounds. This chapter covers some of the basics of organic chemistry and introduces you to some interesting carbon compounds.

8.1 The Carbon Atom and Chemical Bonding

Valence electrons are the electrons in the highest energy level in an atom.

The octet rule (or rule of eight) states that in forming compounds, atoms gain, lose, or share electrons until they are surrounded by 8 electrons.

Why does carbon form so many compounds? The answer lies in its atomic structure. Carbon, atomic number 6, has four valence electrons. When it forms compounds by sharing these valence electrons (Chapter 4) with other carbon atoms, or with atoms of other elements, it obeys the octet rule. It forms carbon-carbon single bonds by sharing pairs of electrons:

$$\cdot\underset{\cdot}{\dot{C}}\cdot \;+\; \cdot\underset{\cdot}{\dot{C}}\cdot \;+\; \cdot\underset{\cdot}{\dot{C}}\cdot \;\longrightarrow\; \cdot\underset{\cdot}{\dot{C}}:\underset{\cdot}{\dot{C}}:\underset{\cdot}{\dot{C}}\cdot$$

When a carbon atom combines with four hydrogen atoms to form methane, it shares its four valence electrons with the hydrogen atoms, thus forming four stable covalent bonds:

Remember that even though the electrons in the carbon atom and in the hydrogen atoms are represented differently here, all electrons are identical, regardless of source.

$$\cdot\underset{\cdot}{\dot{C}}\cdot \;+\; 4\,H\cdot \;\longrightarrow\; \begin{array}{c}H\\ H:\overset{..}{\underset{..}{C}}:H\\ H\end{array} \;\text{or}\; \begin{array}{c}H\\ |\\ H-C-H\\ |\\ H\end{array} \;\text{or}\; CH_4$$

methane

Carbon forms more compounds than any other element primarily because carbon atoms link up with each other in so many different ways. A molecule of a carbon compound may contain one carbon-carbon bond or thousands of such bonds. The carbon atoms can link together in straight chains, branched chains, or rings (**Fig. 8.1**). In addition to a single bond, two carbon atoms can form a double or triple bond. In each of these different bonding patterns, the carbon atoms form four covalent bonds.

Carbon is the only element whose atoms are able to form long chains. Some elements, including oxygen (O), nitrogen (N), and chlorine (Cl), form stable two-atom molecules; sulfur (S), silicon (Si), or phosphorus (P) atoms can form unstable chains containing from four to eight atoms, but no element can form chains as long as carbon. Even silicon and germanium (Ge), which are in the same group of the periodic table as carbon, do not form long chains of their atoms.

straight chain

branched chain

rings

double bond

triple bond

FIGURE 8.1

Carbon atoms can join to form straight chains, branched chains, or rings. In addition to carbon-carbon single bonds, double and triple bonds are also formed.

Different Forms of Carbon

Elemental carbon exists in two very different crystalline forms: diamond and graphite. In diamond, each carbon atom is joined by strong covalent bonds to four other carbon atoms. Each of these carbon atoms is also joined to four more carbon atoms, and so on, throughout a huge three-dimensional interlocking network of carbon atoms **(Fig. 8.2a)**. This carbon-carbon bonding pattern accounts for the stability and extreme hardness of diamond. For a diamond to undergo a chemical change, many strong bonds within the crystalline structure must be broken.

Graphite is completely different from diamond. It is a soft, black, slippery material made up of hexagonal arrays of carbon atoms arranged in sheets **(Fig. 8.2b)**. Each carbon atom in the array is joined to three other carbon atoms by forming two single bonds and one double bond. Graphite is slippery because the intermolecular attractive forces holding the sheets together are relatively weak, and the sheets easily slide past each other. Graphite is used as a dry lubricant and, combined with a binder, forms the "lead" in pencils.

Two or more forms of an element that differ in their bonding or molecular structure are called *allotropes*. Diamond and graphite are allotropes of carbon.

Compounds of Carbon with Other Elements

In addition to forming carbon-carbon bonds, carbon atoms form strong covalent bonds with other elements, particularly with the nonmetal elements hydrogen (H), oxygen (O),

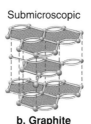

Submicroscopic

a. Diamond

Submicroscopic

b. Graphite

FIGURE 8.2

Two very different forms of carbon: (a) In diamond, the carbon atoms form a strong, three-dimensional network. (b) In graphite, the carbon atoms form sheets held together by weak attractive forces.

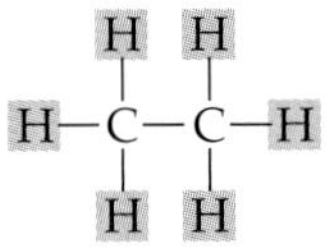

H H
H—C—C—H
H H

ethane

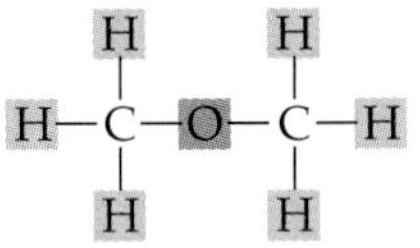

H H
H—C—O—C—H
H H

dimethyl ether

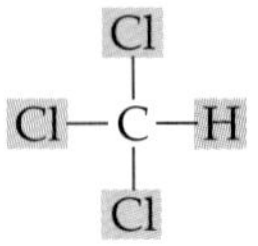

Cl
Cl—C—H
Cl

chloroform

H
H—C—N—H
H H

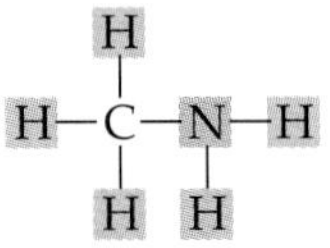

methylamine

FIGURE 8.3

Some compounds in which carbon atoms are covalently bonded to other kinds of atoms.

nitrogen (N), fluorine (F), and chlorine (Cl). Some simple compounds in which carbon is combined with other elements are shown in **Figure 8.3**. Carbon also bonds with phosphorus (P), sulfur (S), silicon (Si), boron (B), and the metals sodium (Na), potassium (K), calcium (Ca), and magnesium (Mg).

The enormous number of carbon compounds that exist can be divided into a relatively small number of classes according to the functional groups they contain. A **functional group** is a particular arrangement of atoms that is present in each molecule of a class of compounds and that largely determines the chemical behavior of the class. Examples of functional groups are —Cl, —OH, —COOH, and $-NH_2$. Before examining the functional groups, we will consider the hydrocarbons, the parent compounds from which all other carbon compounds are derived.

8.2 Hydrocarbons

Hydrocarbons are composed of the two elements carbon and hydrogen. There are fourclasses of hydrocarbons: **alkanes,** which contain only carbon-carbon single bonds (—C—C—); **alkenes,** which contain one or more carbon-carbon double bonds (—C=C—); **alkynes,** which contain one or more triple bonds (—C≡C—); and **aromatic hydrocarbons,** which contain one or more benzene rings. Complex mixtures of many different hydrocarbons, which are present in natural gas and oil (petroleum), are the source of many of the organic compounds used by industry.

Alkanes

The simplest alkane is *methane* (CH_4), the major component of natural gas. The structure of the methane molecule can be represented in a number of different ways, as illustrated in **Figure 8.4**. The expanded structural formula includes the four covalent bonds;

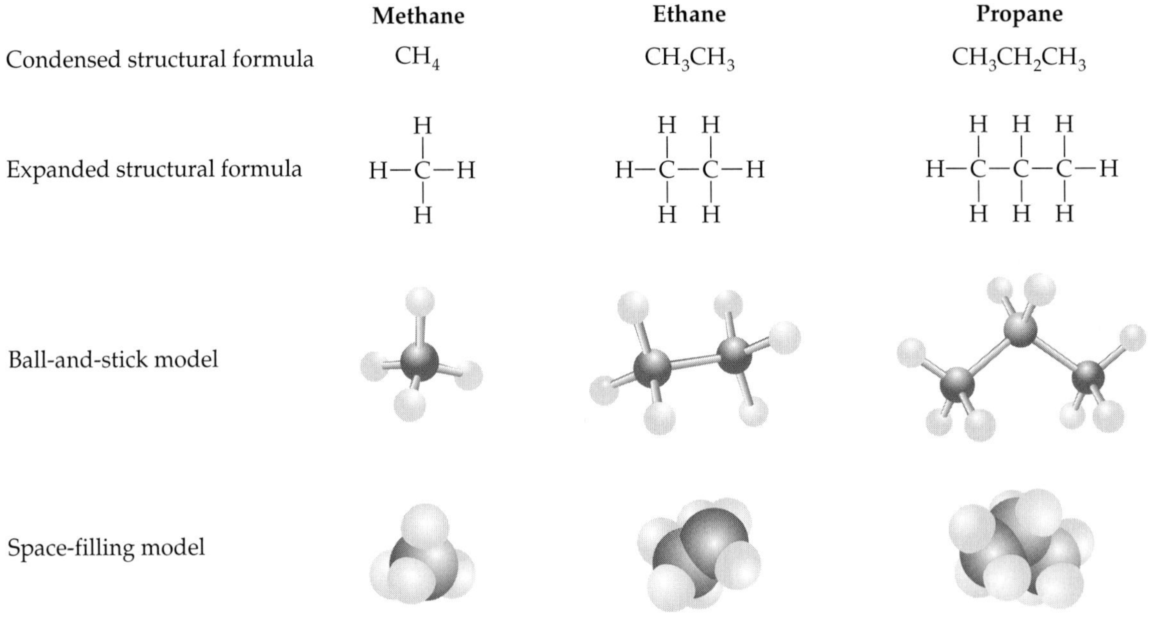

	Methane	Ethane	Propane
Condensed structural formula	CH_4	CH_3CH_3	$CH_3CH_2CH_3$
Expanded structural formula	H H—C—H H	H H H—C—C—H H H	H H H H—C—C—C—H H H H
Ball-and-stick model			
Space-filling model			

FIGURE 8.4

Organic molecules can be represented in several ways, as shown here for the first three members of the alkane series.

the ball-and-stick and space-filling models show the spatial arrangement of the atoms. As the ball-and-stick model indicates, the methane molecule is in the shape of a tetrahedron. The space-filling model gives the most accurate representation of the actual shape of the molecule. For simplicity, the condensed structural formula, which does not show the bonds or the bond angles, is usually used to represent the molecule.

The two alkanes that follow methane are *ethane* (C_2H_6) and *propane* (C_3H_8) (Fig. 8.4). Propane is a major component of bottled gas. The next member of the series is *butane* (C_4H_{10}), and for this alkane, two structures are possible **(Fig. 8.5)**. The four carbon atoms can be joined in a straight line (*n*-butane), or the fourth carbon can be added to the middle carbon atom in the —C—C—C— chain, forming a branch (isobutane).

Alkanes are known as **saturated hydrocarbons,** because they contain the largest possible number of hydrogen atoms for the given number of carbon atoms. The first 10 straight-chain saturated alkanes are shown in **Table 8.1**. The *n*- stands for "normal," which signifies a straight chain. Notice that each alkane differs from the one preceding it by the addition of a $-CH_2$ group. A series of compounds in which each member differs from the preceding member by a constant increment is called a **homologous series.** The general formula for the alkane series is C_nH_{2n+2}, where n is the number of carbon atoms in a member of the series. In a homologous series, the properties of the members change systematically with increasing molecular weight. As Table 8.1 shows, in the

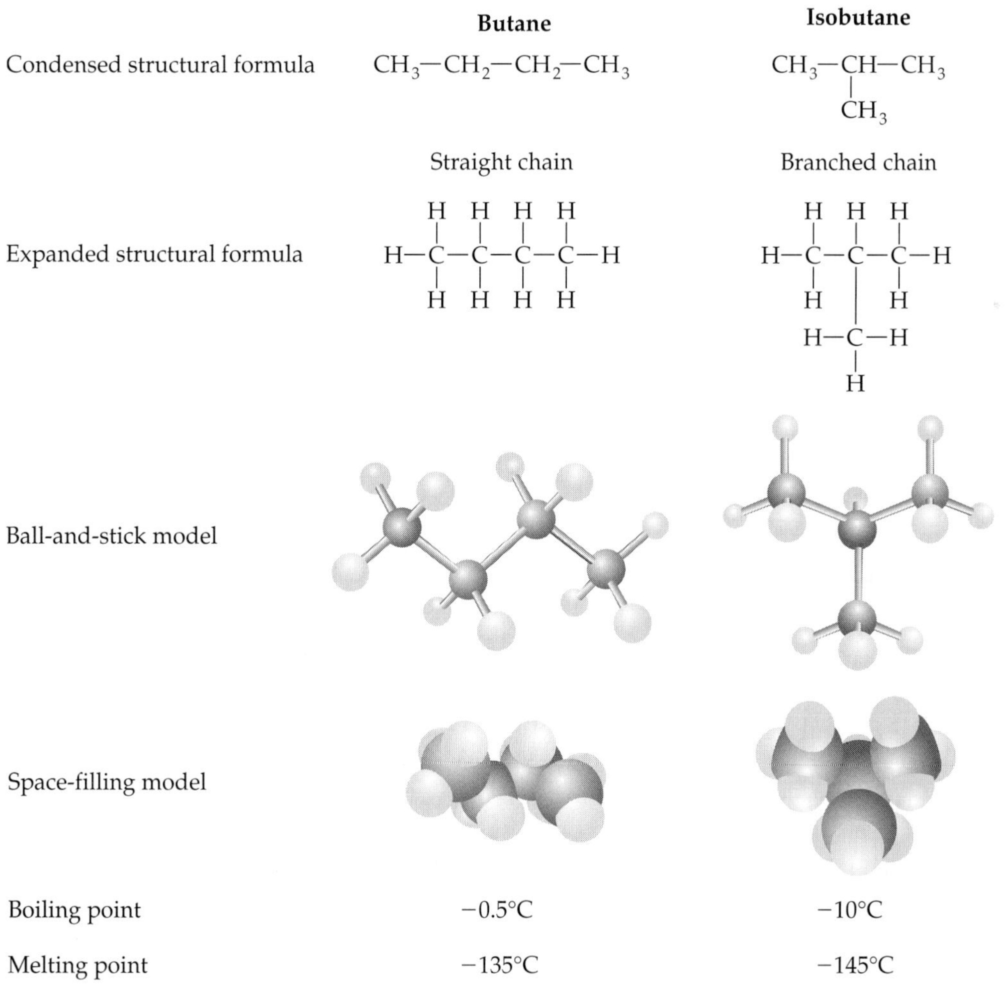

	Butane	Isobutane
Condensed structural formula	$CH_3—CH_2—CH_2—CH_3$	$CH_3—CH(CH_3)—CH_3$
	Straight chain	Branched chain
Expanded structural formula	H—C(H)(H)—C(H)(H)—C(H)(H)—C(H)(H)—H	H—C(H)(H)—C(H)(C(H)(H)H)—C(H)(H)—H
Ball-and-stick model		
Space-filling model		
Boiling point	−0.5°C	−10°C
Melting point	−135°C	−145°C

FIGURE 8.5

The two structural isomers of butane (C_4H_{10}).

Table 8.1 The First Ten Straight-Chain Alkanes

Name	Formula	Boiling Point (°C)	Structural Formula
Methane	CH_4	−162	H H—C—H H
Ethane	C_2H_6	−88.5	H H H—C—C—H H H
Propane	C_3H_8	−42	H H H H—C—C—C—H H H H
n-Butane	C_4H_{10}	0	H H H H H—C—C—C—C—H H H H H
n-Pentane	C_5H_{12}	36	H H H H H H—C—C—C—C—C—H H H H H H
n-Hexane	C_6H_{14}	69	H H H H H H H—C—C—C—C—C—C—H H H H H H H
n-Heptane	C_7H_{16}	98	H H H H H H H H—C—C—C—C—C—C—C—H H H H H H H H
n-Octane	C_8H_{18}	126	H H H H H H H H H—C—C—C—C—C—C—C—C—H H H H H H H H H
n-Nonane	C_9H_{20}	151	H H H H H H H H H H—C—C—C—C—C—C—C—C—C—H H H H H H H H H H
n-Decane	$C_{10}H_{22}$	174	H H H H H H H H H H H—C—C—C—C—C—C—C—C—C—C—H H H H H H H H H H H

straight-chain alkane series, the boiling points rise quite regularly as the number of carbon atoms increases. The first four straight-chain alkanes are gases, and the next six are liquids.

EXAMPLE 8.1 Writing structural formulas for alkanes

Write the structural formula for the straight-chain alkane C_5H_{12}.

Solution Write five carbon atoms linked together to form a chain,

```
C—C—C—C—C
```

Attach hydrogen atoms to the carbon atoms so that each carbon atom forms four covalent bonds.

```
    H   H   H   H   H
    |   |   |   |   |
H—C—C—C—C—C—H
    |   |   |   |   |
    H   H   H   H   H
```

Practice Exercise 8.1

Write the structural formula for the straight-chain hydrocarbon represented by C_nH_{2n+2}, where $n = 7$.

We have described unbranched chains as straight chains. In fact, because of tetrahedral bonding, the carbon atoms are not in a straight line as shown in Table 8.1 but are staggered as shown below and in the ball-and-stick models for propane and *n*-butane in Figures 8.4 and 8.5.

```
     \ /   \ /   \ /   |
      C     C     C    C
\   /   \ /   \ /   \ /  \
 C       C     C     C
/ |     / \   / \   / \
```

Structural Isomerism

Compounds such as butane and isobutane, which have the same molecular formula (C_4H_{10}) but different structures, are called **structural isomers.** Because their structures are different (Fig. 8.5), their properties—for example, boiling point and melting point—are also different.

All alkanes containing four or more carbon atoms form structural isomers. The number of possible isomers increases rapidly as the number of carbon atoms in the molecule increases. For example, butane, C_4H_{10}, forms two isomers; decane, $C_{10}H_{22}$, theoretically forms 75; and $C_{30}H_{62}$ forms over 400 million **(Table8.2)**. Most of these isomers do not exist naturally and have not been synthesized, but the large number of possibilities helps explain the natural abundance of carbon compounds. For isomers with large numbers of carbon atoms, the crowding of atoms makes some of the possible structures too unstable to exist.

Table 8.2 Number of Possible Isomers for Selected Alkanes

Molecular Formula	Number of Possible Isomers
C_4H_{10}	2
C_5H_{12}	3
C_6H_{14}	5
C_7H_{16}	9
C_8H_{18}	18
C_9H_{20}	35
$C_{10}H_{22}$	75
$C_{15}H_{32}$	4,347
$C_{20}H_{42}$	336,319
$C_{30}H_{62}$	4,111,846,763

EXAMPLE 8.2 Identifying structural isomers of alkanes

Give the structural and condensed formulas of the isomers of pentane.

Solution Pentane has five carbon atoms. The carbon skeletons of the three possible isomers are

```
                                      C
                                      |
C—C—C—C—C     C—C—C—C        C—C—C
                  |                   |
                  C                   C
```

The structural formulas are written by adding hydrogen atoms so that each carbon atom forms four bonds. Each isomer has the same condensed formula, C_5H_{12}.

Practice Exercise 8.2

Give the structural formulas of the possible isomers of hexane (C_6H_{14}).

Nomenclature of Alkanes

Pent-, hex-, hept-, oct-, non-, and *dec-* are derived from the Greek or Latin words for five, six, seven, eight, nine, and ten.

Organic compounds are named according to rules established by the International Union of Pure and Applied Chemistry (IUPAC). As shown in Table 8.1, except for the first four members of the family, the first part of the name of an alkane is derived from the Greek name for the number of carbon atoms in the molecule. The suffix *-ane* means that the compound is an alkane.

Before you can follow the IUPAC rules for naming branched-chain alkanes, you must know the names of the groups that are formed when one hydrogen atom is removed from an alkane. For example, when a hydrogen atom (H) is removed from methane (CH_4), a *methyl group* ($-CH_3$) is formed. Similarly, removal of a hydrogen atom from ethane (C_2H_6) gives an *ethyl group* ($-C_2H_5$). Because these groups are derived from alkanes, they are called **alkyl groups.** Examples of some common alkyl groups are given in **Table 8.3.**

Branched-chain alkanes are named by applying the following rules:

1. *Determine the name of the parent compound by finding the longest continuous chain of carbon atoms.* Consider the following example:

```
a. H3C—CH2—CH2—CH—CH3
                |
                CH3
```

The longest chain contains five carbon atoms; therefore, the name of the parent compound is *pentane.*

Sometimes, because of the way in which a structural formula is written, it is not easy to recognize the longest chain. For example, the alkane shown below is not a hexane (six carbon atoms in the longest chain) as at first might be supposed, but a *heptane* (seven carbon atoms in the longest chain).

```
b. H3C—CH2—CH2—CH2—CH—CH3
                   |
                   CH2
                   |
                   CH3
```

Table 8.3 Some Common Alkyl Groups

Name	Group
Methyl	$-CH_3$
Ethyl	$-CH_2-CH_3$
n-Propyl	$-CH_2-CH_2-CH_3$
n-Butyl	$-CH_2-CH_2-CH_2-CH_3$
Isopropyl	$-\overset{CH_3}{\overset{\vert}{C}}-H$ with CH_3 below (i.e., $-CH(CH_3)_2$ drawn as CH_3 above, $-C-H$, CH_3 below)

2. *Number the longest chain beginning with the end closest to the branch. Use these numbers to designate the location of the alkyl group (or substituent) at the branch.*

Applying this rule to the above examples, the carbon atoms are numbered and named as follows:

In the examples in this section, the left-hand terminal methyl group has been written H_3C—; in condensed formulas, it is normally written CH_3.

a. 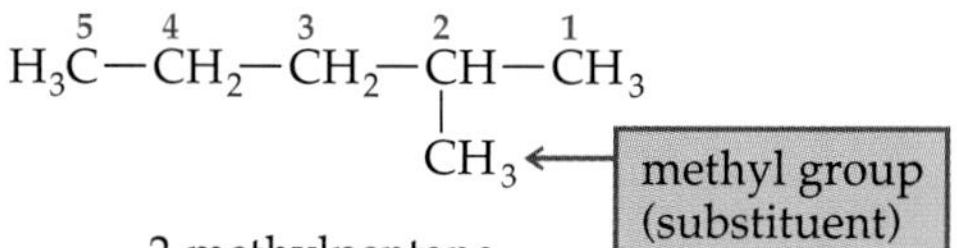 $\overset{5}{H_3C}-\overset{4}{CH_2}-\overset{3}{CH_2}-\overset{2}{CH}(CH_3)-\overset{1}{CH_3}$ ← methyl group (substituent)

2-methylpentane

b. 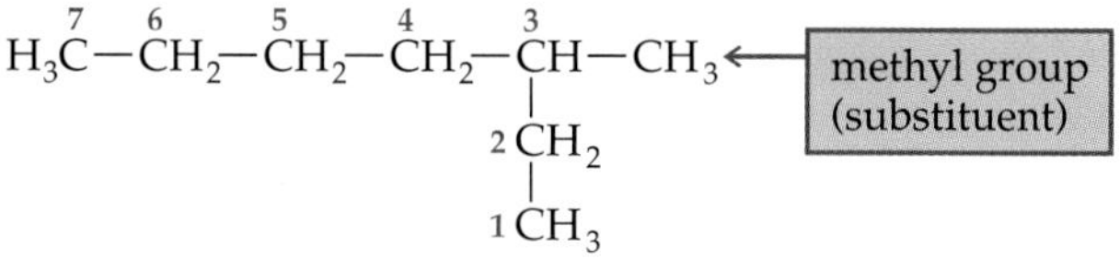 $\overset{7}{H_3C}-\overset{6}{CH_2}-\overset{5}{CH_2}-\overset{4}{CH_2}-\overset{3}{CH}(-\overset{2}{CH_2}-\overset{1}{CH_3})-CH_3$ ← methyl group (substituent)

3-methylheptane

In the names of carbon compounds, the numbers are separated from words by a hyphen, and the parent name is placed last. In compound **a**, 2-methylpentane, a methyl group is attached to carbon number 2 of pentane; in compound **b**, 3-methylheptane, a methyl group is attached to carbon number 3 of heptane.

EXAMPLE 8.3 Naming alkanes

Name the following compound:

$$H_3C-CH(CH_3)-CH_2-CH_2-CH_3$$

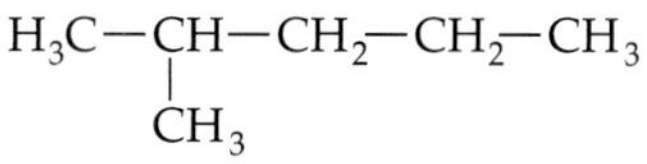

Solution

1. Determine the name of the parent compound by finding the longest continuous chain of carbon atoms.

   The longest chain is five carbons long. Therefore, the parent compound is a pentane.

2. Number the longest chain beginning with the end closest to the branch. Use these numbers to designate the location of the alkyl group (or substituent) at the branch.

   The methyl group is attached to carbon number 2.

3. The compound is 2-methylpentane.

Practice Exercise 8.3

Name the following compound:

$$H_3C-CH_2-CH_2-CH_2-CH(CH_3)-CH_3$$

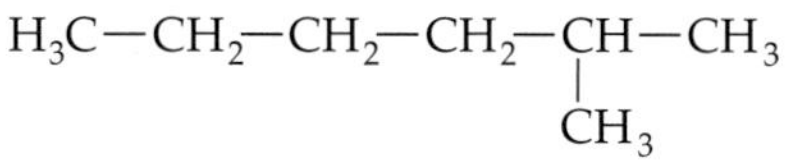

3. *When two or more substituents are present on the same carbon atom, use the number of that carbon atom twice. List the substituents alphabetically.*

   Thus, the compound shown below is 3-ethyl-3-methylhexane.

```
                  CH3
     1     2     3|    4      5      6
   H3C—CH2—C—CH2—CH2—CH3
                   |
                  CH2
                   |      ←  ethyl group
                  CH3
```

3-ethyl-3-methylhexane

4. *When two or more substituents are identical, indicate this by the use of the prefixes di-, tri-, tetra-, and so on. Commas are used to separate numbers from each other.*

   Thus, the following compounds are 2,2,4-trimethylhexane and 3-ethyl-2,4-dimethyloctane:

```
       CH3        CH3
  1  2|    3     4|     5      6
 H3C—C—CH2—CH—CH2—CH3
       |
      CH3
```

2,2,4-trimethylhexane

```
                                                        CH3
  8      7      6      5      4     3     2|     1
 H3C—CH2—CH2—CH2—CH—CH—CH—CH3
                                 |     |
                                CH3  CH2
                                       |
                                      CH3
```

3-ethyl-2,4-dimethyloctane

EXAMPLE 8.4 Naming alkanes

Name the following compound:

```
 H3C—CH—CH2—CH—CH2—CH—CH3
       |          |          |
      CH3        CH3        CH3
```

Solution

1. Determine the name of the parent compound by finding the longest continuous chain of carbon atoms.

   The longest chain is seven carbon atoms long. Therefore, the parent compound is a heptane.

2. Number the longest chain beginning with the end closest to a branch. Use these numbers to designate the locations of the alkyl groups (or substituents) at the branches.

   There are methyl groups on carbons 2, 4, and 6.
   Trimethyl- indicates three methyl groups.

3. The name of the compound is 2,4,6-trimethylheptane.

Practice Exercise 8.4

Name the following compound:

```
       CH3               CH3
        |                  |
 H3C—C—CH2—CH2—C—CH3
        |                  |
       CH3               CH3
```

Reactions of Alkanes

Alkanes are relatively unreactive. Like all hydrocarbons, alkanes undergo combustion, but for this to occur, a flame or spark is required to initiate bond cleavage. When burned in air, alkanes form carbon dioxide and water, as shown below for methane and butane.

> Combustion is a chemical reaction in which heat is produced. It usually refers to the combination of a fuel with atmospheric oxygen.

$$CH_4 + 2\ O_2 \longrightarrow CO_2 + 2\ H_2O + \text{heat}$$

$$2\ C_4H_{10} + 13\ O_2 \longrightarrow 8\ CO_2 + 10\ H_2O + \text{heat}$$

Once initiated, these combustion reactions are exothermic (Chapter 6) and produce a considerable amount of heat. They are the source of the energy produced when natural gas, gasoline, or fuel oil is burned (Chapter 12).

Alkenes

Alkenes are hydrocarbons that have one or more carbon-carbon double bonds ($-C{=}C-$). The two simplest are *ethene* (C_2H_4), commonly called *ethylene,* and *propene* (C_3H_6), commonly called *propylene.* In forming the double bond, the two carbon atoms share two pairs of electrons to acquire the stable electron configuration of an octet.

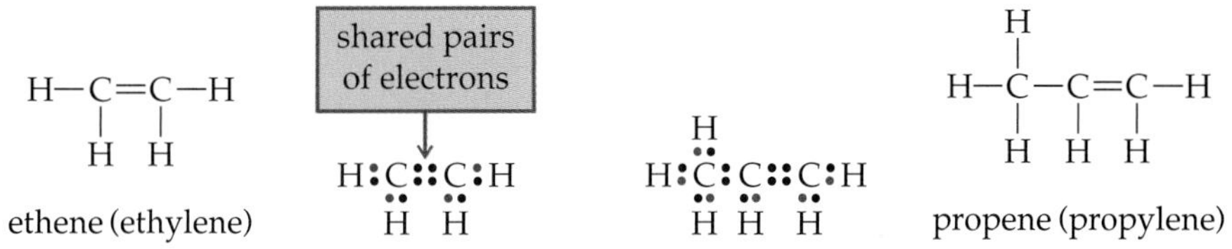

```
H—C=C—H        [shared pairs          H          H
  |  |          of electrons]         ..         |
  H  H              ↓           H:C:C::C:H    H—C—C=C—H
               H:C::C:H           ..  ..  ..     |  |  |
                 ..  ..           H   H   H      H  H  H
                 H   H
```

ethene (ethylene) — propene (propylene)

Different ways of representing the structures of alkenes are shown in **Figure 8.6.**

	Ethene (ethylene)	Propene (propylene)
Condensed structural formula	C_2H_4 $CH_2{=}CH_2$	C_3H_6 $CH_3CH{=}CH_2$
Expanded structural formula	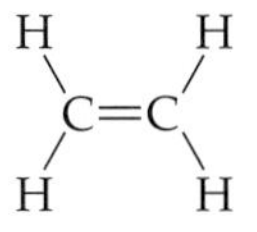 (H)(H)C=C(H)(H)	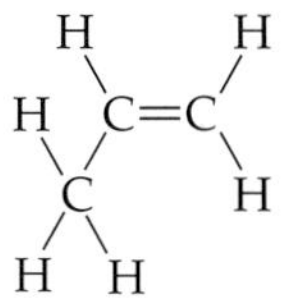 (H)(H)(H)C–C(H)=C(H)(H)
Ball-and-stick model	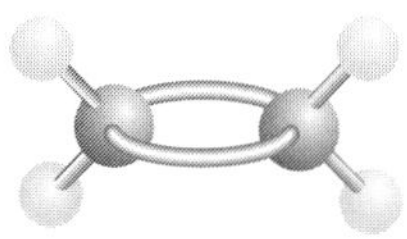	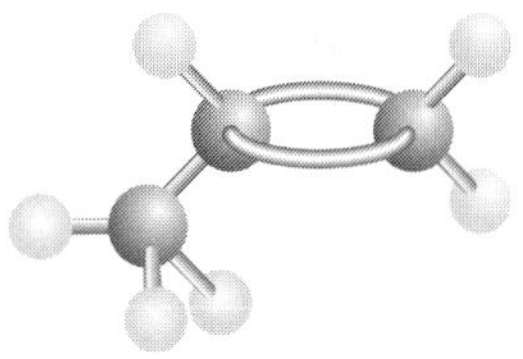
Space-filling model	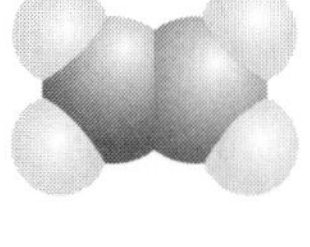	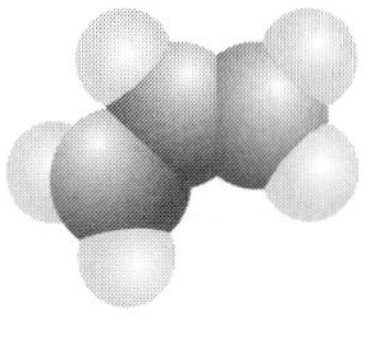

FIGURE 8.6

The first two members of the alkene series.

Alkenes are named using the suffix *-ene*. Like alkanes, they form a homologous series. The general formula is C_nH_{2n}.

Alkenes are called **unsaturated hydrocarbons** because their molecules do not contain the maximum possible number of hydrogen atoms. In the presence of a catalyst such as platinum (Pt), an alkene reacts with hydrogen to form the saturated parent alkane.

$$\underset{\text{ethene}}{H_2C{=}CH_2} + H_2 \xrightarrow{Pt} \underset{\text{ethane}}{H_3C{-}CH_3}$$

The third member of the alkene series is *butene* (C_4H_8). There are two possible positions for the double bond, and two isomers exist:

$$\underset{\text{1-butene}}{H_3\overset{1}{C}{=}\overset{2}{C}H{-}\overset{3}{C}H_2{-}\overset{4}{C}H_3} \qquad \underset{\text{2-butene}}{H_3\overset{1}{C}{-}\overset{2}{C}H{=}\overset{3}{C}H{-}\overset{4}{C}H_3}$$

In naming alkenes, the first rule used for naming alkanes is modified: *The longest hydrocarbon chain is numbered from the end that will give the carbon-carbon double bond (—C=C—) the lowest number. The ending -ene is used for the parent name instead of -ane. Substituents are then numbered and named as for alkanes.* Thus, the compounds shown below are 7-chloro-3-heptene and 4-ethyl-5-methyl-2-hexene.

$$\underset{\text{7-chloro-3-heptene}}{H_3\overset{1}{C}{-}\overset{2}{C}H_2{-}\overset{3}{C}H{=}\overset{4}{C}H{-}\overset{5}{C}H_2{-}\overset{6}{C}H_2{-}\overset{7}{C}H_2{-}Cl}$$

$$\underset{\text{4-ethyl-5-methyl-2-hexene}}{H_3\overset{1}{C}{-}\overset{2}{C}H{=}\overset{3}{C}H{-}\overset{4}{C}H(CH_2CH_3){-}\overset{5}{C}H(CH_3){-}\overset{6}{C}H_3}$$

Notice that in all the formulas shown, every carbon atom has four bonds.

In the United States, ethene is produced in greater quantity than any other organic chemical. Approximately 8 million tons are produced annually.

Ethene is the most important raw material in the organic chemical industry. It is produced in enormous quantity by the "cracking" of petroleum (Chapter 12) and is used primarily to make the plastic polyethylene (Chapter 9). It is also the starting material for the manufacture of ethylene glycol (the major component of antifreeze in automobile radiators), other plastics, and many chemicals. Propene is important in the production of polypropylene and other plastics.

Ethene has a natural source: It is produced by fruits and causes the change in skin color that occurs with ripening, for example, when tomatoes change from green to red. To avoid damage in transit, many producers pick fruit when it is green, ship it to its destination, and then treat it with ethene gas to produce the proper ripe color. The flavor, however, is generally considered to be inferior to naturally ripened fruit.

EXAMPLE 8.5 Naming alkenes

Name the following compound:

$$H_2C{=}CH{-}CH_2{-}CH(CH_3){-}CH_3$$

Solution

1. The longest chain has five carbon atoms. The parent compound is a pentene.
2. The double bond is between carbon 1 and carbon 2.

3. The methyl group is on carbon 4.
4. The name of the compound is 4-methyl-1-pentene.

Practice Exercise 8.5

Name the following compound:

$$H_2C{=}CH{-}CH_2{-}CH_2{-}CH_2{-}CH_2{-}CH_2{-}CH_3$$

Cis-Trans Isomerism

A carbon-carbon single bond allows the groups on either side of it to rotate freely with respect to each other, at ordinary temperatures. For example, the two methyl groups in ethane can rotate freely about the single C—C bond. Groups joined by a C=C double bond, however, are restricted and cannot rotate. In an ethene molecule, the two carbons and their attached hydrogens lie in the same plane and remain fixed in that arrangement.

If one of the hydrogen atoms attached to each of the two carbon atoms in ethene is replaced by a chlorine atom, two distinct compounds are formed:

```
Cl      Cl            Cl      H
  \    /                \    /
   C=C                   C=C
  /    \                /    \
H       H             H       Cl
```

cis-1,2-dichloroethene   *trans*-1,2-dichloroethene

These two compounds have basically the same name, 1,2-dichloroethene, and the same condensed formula, $C_2H_2Cl_2$, but they are not identical (the two molecules cannot be superimposed upon one another so that all atoms and bonds coincide), and their physical properties are different.

	cis-1,2-dichloroethene	*trans*-1,2-dichloroethene
Boiling point	60.3°C	47.5°C
Melting point	−80.5°C	−50°C

The prefixes *cis*- and *trans*- are derived from the Latin for "on this side" and "across."

The two compounds are not structural isomers because the atoms attached to each carbon atom in the two compounds are the same. That is, in both isomers, each carbon atom is joined to another carbon atom, a chlorine atom, and a hydrogen atom. Only the spatial arrangement is different. This type of isomerism is called ***cis-trans* isomerism.** The isomers are distinguished by the prefixes *cis*- and *trans*-.

If two identical groups are attached to the same carbon atom involved in the double bond, *cis-trans* isomerism is not possible. For example, 1,1-dichloroethene is the *structural* isomer of the *cis* and *trans* isomers shown above but exists in just one form (it cannot form *cis* and *trans* isomers).

```
H       Cl
  \    /
   C=C
  /    \
H       Cl
```

1,1-dichloroethene

	Ethyne (acetylene)
Condensed structural formula	C_2H_2 or HC≡CH
Expanded structural formula	H—C≡C—H
Ball-and-stick model	
Space-filling model	

FIGURE 8.7

Different ways of representing the alkyne ethyne (acetylene).

Alkynes

Alkynes are hydrocarbons that have one or more carbon-carbon triple bonds (—C≡C—). The simplest alkyne is *ethyne* (C_2H_2), better known by its common name, *acetylene*. The next member of the series is *propyne* (C_3H_4). In forming a triple bond, two carbon atoms share three pairs of electrons so that each attains an octet:

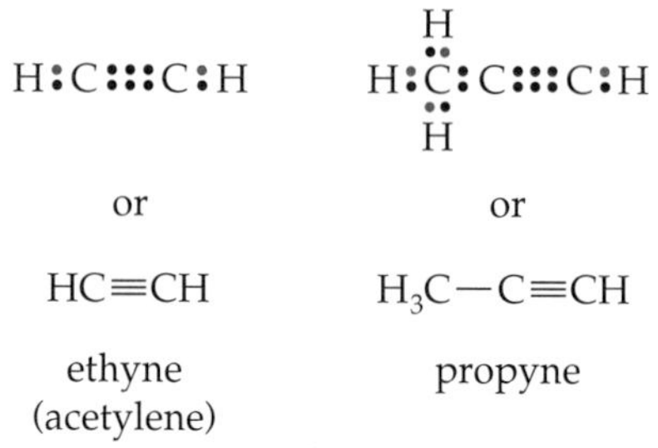

ethyne (acetylene)	propyne
H:C:::C:H	H:C(H)(H):C:::C:H
or	or
HC≡CH	H_3C—C≡CH

Different ways of representing ethyne are shown in **Figure 8.7**.

Acetylene has many industrial uses. When burned in oxyacetylene torches, it produces a very hot flame (about 3000°C, or 5400°F) that is used for cutting and welding metals **(Fig. 8.8)**. It is also an important starting material for the production of plastics and synthetic rubber.

Hydrocarbon Rings

Carbon atoms can join together to form rings as well as straight chains and branched chains. Alkanes in the form of rings are called **cycloalkanes.** The structures of the four smallest cycloalkanes are shown in **Figure 8.9**. In the simplified depictions of the struc-

FIGURE 8.8

The high temperature of an oxyacetylene torch can be used to cut steel.

tural formulas, each corner represents a carbon atom attached to two hydrogen atoms, and the lines represent carbon-carbon single bonds.

Stability increases from cyclopropane to cyclohexane as the angles between the bonds connecting adjacent carbon atoms increase. Cyclohexane is important as the parent compound of glucose and many other carbohydrates (sugar and starches) that are vital for animal metabolism (Chapter 14).

Hydrocarbons can also form rings in which some of the carbon-carbon bonds are double bonds, as illustrated by the following six-carbon unsaturated rings.

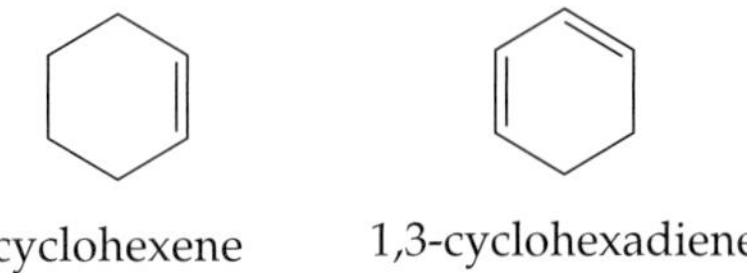

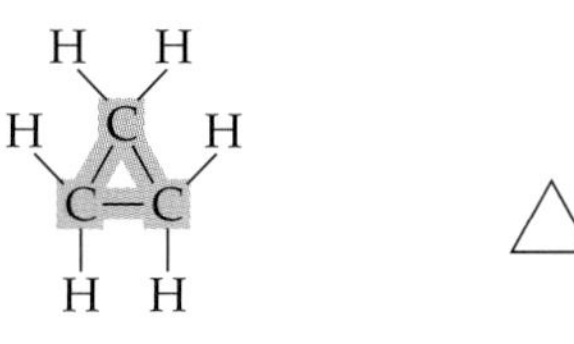

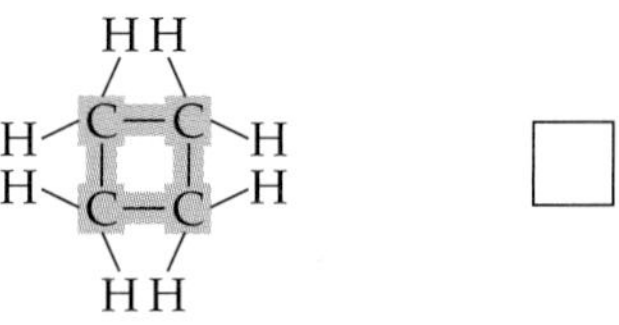

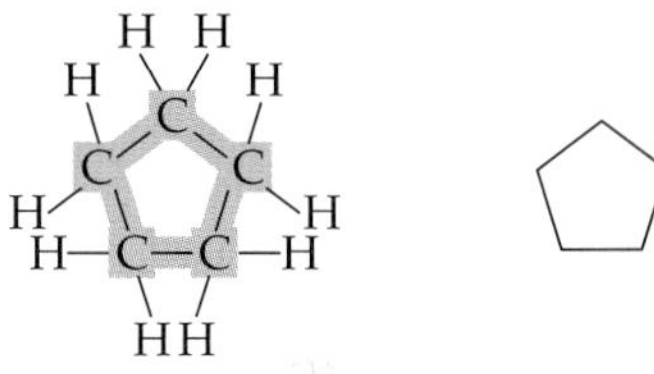

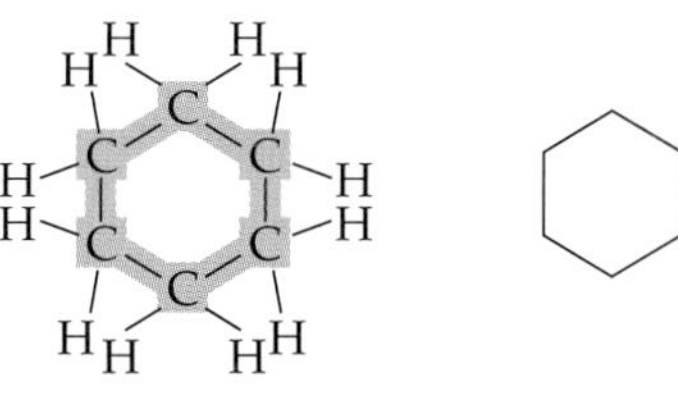

FIGURE 8.9

Expanded and simplified structural formulas of cycloalkanes.

8.3 Aromatic Hydrocarbons

By far the most important unsaturated hydrocarbon ring compound is **benzene** (C_6H_6). Compounds containing a benzene ring are called *aromatic compounds* because many of the first ones isolated had distinctive pleasant scents (or aromas).

The Structure of Benzene

Benzene was discovered by Michael Faraday in 1826, but its structure remained a mystery for 40 years. The molecular formula, C_6H_6, indicated a highly unsaturated hydrocarbon, but benzene did not behave like the known alkenes or alkynes. Then, in 1865, the German chemist August Kekulé proposed a ring of six carbon atoms joined by alternating single and double bonds, with one hydrogen atom bonded to each carbon:

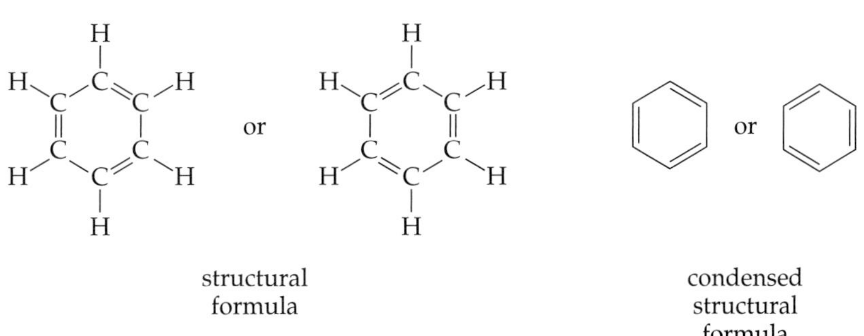

Although these structures explain much of the chemistry of benzene, they cannot accurately explain all the established experimental facts. We now know that the bonds joining the carbon atoms in the ring are neither single bonds nor double bonds. They are shorter than normal single bonds, but longer than normal double bonds, and all are identical. The six "extra" electrons that would be involved in forming three double bonds are shared equally among all the carbon atoms in the ring, as depicted in **Figure 8.10a** or more simply in **Figure 8.10b**. These shared electrons are said to be *delocalized*. They can be visualized as moving above and below the flat plane of the carbon ring, as shown in **Figure 8.10c**. This arrangement makes the benzene ring very stable. The formula for benzene is usually written without the carbon and hydrogen atoms and with a circle in the center of the ring to represent the delocalized electrons **(Fig. 8.10d)**.

Kekulé is said to have solved the riddle of the structure of benzene after he had a dream in which he saw rows of atoms twisting and turning in snakelike motion. Suddenly, he saw one snake whirling with its tail in its mouth, giving him the idea that the carbon atoms in benzene must be joined in a ring.

• electrons shared equally by all six carbon atoms

a. b. c. d.

FIGURE 8.10

The benzene molecule: (a) The six carbon-carbon bonds in benzene are identical. The six electrons associated with the three double bonds are shared equally among the six carbon atoms of the ring. (b) The circle in the center of the ring represents the six shared electrons. (c) The six delocalized electrons can be represented as occupying a space above and below the ring of carbon atoms. (d) Benzene is usually written in this simplified form.

Alkyl Derivatives of Benzene

Replacement of any of the hydrogen atoms of a benzene ring with an alkyl group gives rise to numerous other aromatic hydrocarbons. The simplest member of the family, methylbenzene (common name, *toluene*), is formed when one hydrogen atom is replaced with a methyl group ($-CH_3$). If two hydrogens are replaced by methyl groups, three different dimethylbenzenes (commonly called *xylenes*) can be formed **(Fig. 8.11)**.

If there are two substituents on a benzene ring, their relative positions are indicated by numbering the ring carbons. If two substituents are present, as in the xylenes, the prefix *ortho* (*o*-), *meta* (*m*-), or *para* (*p*-) is often used instead of the numbers 1,2 or 1,3 or 1,4, respectively. For more than two substituents, numbers must be used. The numbers are assigned to give the lowest possible numbers to the substituents.

Benzene, toluene, and the xylenes are used extensively as raw materials for the manufacture of plastics, pesticides, drugs, and hundreds of other organic chemicals. The use of benzene as a laboratory solvent was discontinued in the 1970s after it was discovered that inhalation of its fumes lowered the white blood cell count in humans and caused leukemia in laboratory rats.

Polycyclic Aromatic Compounds

Benzene rings can fuse, or join together, to form **polycyclic aromatic compounds (Fig. 8.12)**. The simplest one is naphthalene, which is used as a moth repellent. Additional

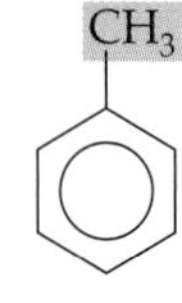

methylbenzene
(toluene)

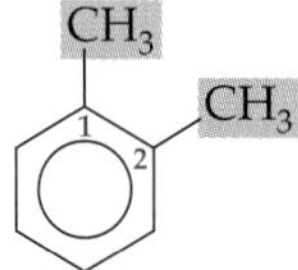

1, 2-dimethylbenzene
(*ortho*-xylene)

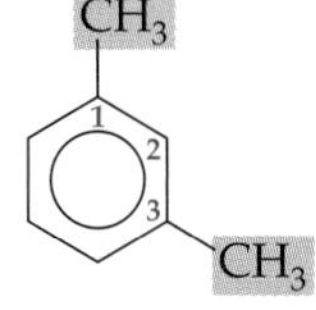

1, 3-dimethylbenzene
(*meta*-xylene)

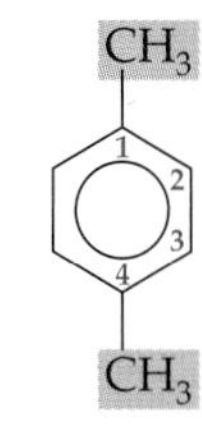

1, 4-dimethylbenzene
(*para*-xylene)

FIGURE 8.11

Replacement of one hydrogen atom of benzene with a methyl group yields toluene. Replacement of two hydrogen atoms with methyl groups gives one of the three xylenes.

benzene rings can be added to form a wide variety of fused-ring aromatic compounds, including benzo(a)pyrene, a known carcinogen that is present in tobacco smoke.

naphthalene
($C_{10}H_8$)

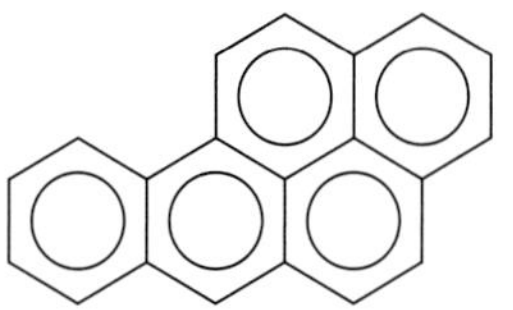

benzo(a)pyrene
($C_{20}H_{12}$)

FIGURE 8.12

Two polycyclic aromatic compounds.

8.4 Functional Groups

The majority of organic compounds are derivatives of hydrocarbons, obtained by replacing one or more hydrogen atoms in a parent hydrocarbon molecule with a different atom or group of atoms. These atoms or groups of atoms, called *functional groups*, form the basis for classifying organic compounds. Most chemical reactions involving an organic molecule take place at the site of a functional group. The functional group, as the name implies, determines how the molecule functions as a whole. Molecules containing the same functional groups exhibit similar chemical behavior. The major classes of organic compounds that result from replacing hydrogen atoms in hydrocarbons with functional groups are shown in **Table 8.4**, with examples from each class.

Table 8.4 Classes of Organic Compounds Based on Functional Groups

General Formula of Class*	Name of Class	Example	Name of Compound
R—X	Halide	CH_3—Cl	Chloromethane (methyl chloride)
R—OH	Alcohol	C_2H_5—OH	Ethanol (ethyl alcohol)
R—O—R′	Ether	C_2H_5—O—C_2H_5	Diethyl ether (ethyl ether)
R—C(=O)—H	Aldehyde	CH_3—C(=O)—H	Ethanal (acetaldehyde)
R—C(=O)—R′	Ketone	CH_3—C(=O)—CH_3	Propanone (acetone)
R—C(=O)—OH	Carboxylic acid	CH_3—C(=O)—OH	Ethanoic acid (acetic acid)
R—O—C(=O)—R′	Ester	C_2H_5—O—C(=O)—CH_3	Ethyl ethanoate (ethyl acetate)
R—N(H)—H	Primary amine	C_2H_5—N(H)—H	Ethylamine
R—C(=O)—N(H)—H	Simple amide	CH_3—C(=O)—N(H)—H	Acetamide

*R represents an H atom or a carbon-containing group, often an alkyl group such as —CH_3 or —C_2H_5. R′ may be the same as R or different.

trichloromethane
(chloroform)

tetrachloromethane
(carbon tetrachloride)

dichlorodifluoromethane
(Freon-12)

chloroethane
(vinyl chloride)

dichlorodiphenyl trichloroethane
(DDT)

FIGURE 8.13

Examples of organic halides.

Organic Halides

Organic halides are hydrocarbon derivatives in which one or more hydrogen atoms in the parent compound have been replaced by halogen atoms: fluorine, chlorine, bromine, or iodine. An atom of any of the halogens, which are in Group VIIA of the periodic table, forms a stable covalent bond by sharing one of its seven valence electrons, as shown below for chloroethane:

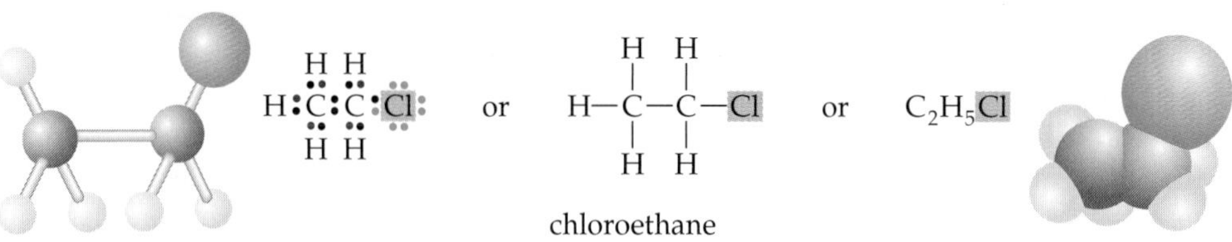

According to the IUPAC system, organic halides are named by writing the name of the parent hydrocarbon preceded by the appropriate designation for the halogen: fluoro-, chloro-, bromo-, or iodo-. In many cases, common names are used. For example, trichloromethane is usually called *chloroform* **(Fig. 8.13)**.

Organic halides are very versatile compounds and have been used in numerous ways. Examples are shown in Figure 8.13. In recent years, some of these halides have been shown to be health hazards, and their use has been discontinued. For years, chloroform ($CHCl_3$) and carbon tetrachloride (CCl_4), both excellent solvents for grease and oils, were used in the dry-cleaning industry and in chemical laboratories. Since they were shown to be toxic and carcinogenic, they have been replaced by safer solvents, such as perchlorethylene. We will see in Chapter 11 that chlorofluorocarbons (CFCs)—substituted alkanes such as Freon-12, which contain both fluorine and chlorine—were widely used as coolants and aerosol propellants before it was discovered that their release into the atmosphere was a factor in the depletion of the ozone layer. Other organic halides that have caused environmental problems are DDT and the polychlorinated biphenyls (PCBs) (Chapter 10). The two plastics polyvinyl chloride (PVC) and Teflon (the non-stick coating on cookware) (Chapter 9) are made from the chlorine and fluorine derivatives of ethylene, respectively.

EXAMPLE 8.6 Naming organic halides

Give the IUPAC name of the following organic halide.

$$H_3C-CH_2-CH_2-Cl$$

Solution

1. The parent hydrocarbon has three carbon atoms. The compound is a propane.
2. The chlorine (Cl) atom is on a terminal carbon atom.
3. Carbon atoms are numbered so that the carbon attached to the chlorine substituent is given the lowest possible number.
4. The compound is 1-chloropropane.

Practice Exercise 8.6

Write the structural formula of bromomethane (or methyl bromide, CH_3Br).

ethanol
(ethyl alcohol)

1-propanol
(*n*-propyl alcohol)

2-propanol
(isopropyl alcohol)

1, 2-ethanediol
(ethylene glycol)

1, 2, 3-propanetriol
(glycerol, glycerin)

phenol

2-methyphenol
(*o*-cresol)

a.

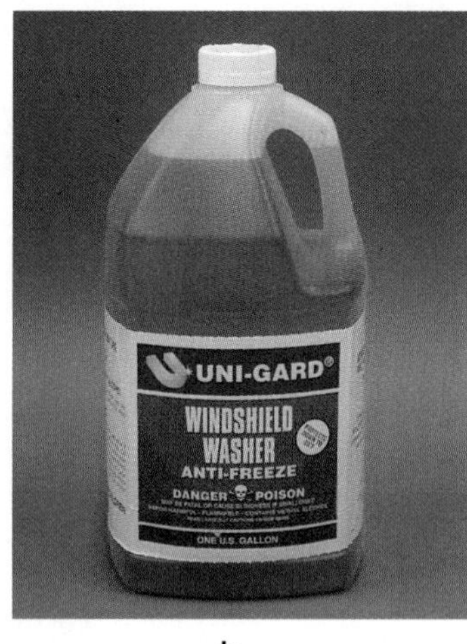

b.

FIGURE 8.14

(a) Examples of alcohols. (b) Windshield wiper fluid contains the alcohol methanol, which acts as an antifreeze.

Alcohols

Alcohols are hydrocarbon derivatives in which one or more hydrogen atoms in the parent hydrocarbon have been replaced by hydroxyl groups (—OH) (**Fig. 8.14**). Oxygen has six valence electrons, and in forming an alcohol, it acquires a stable octet of electrons by forming one covalent bond with a carbon atom and one with a hydrogen atom:

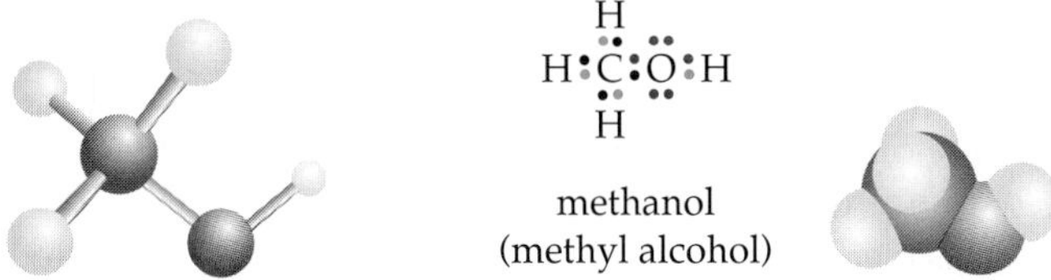

methanol
(methyl alcohol)

The IUPAC name of an alcohol is obtained from the parent alkane name by replacing the final *-e* with the suffix *-ol*. For example, methane becomes methanol, ethane becomes ethanol, and propane becomes propanol.

Alcohols, like water, contain a polar —OH group, and as a result, hydrogen bonding (Chapter 4) can occur between hydrogen and oxygen atoms in adjacent molecules, as shown for methanol in **Figure 8.15**. When alcohols are converted from the liquid state to the gaseous state, energy is needed to break the hydrogen bonds. As a result, the boiling points of alcohols are higher than the boiling points of the alkanes from which they are derived (**Table 8.5**).

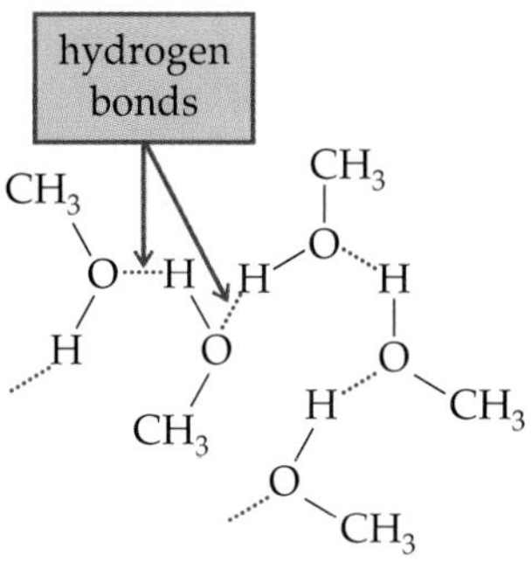

FIGURE 8.15

Hydrogen bonding between methanol molecules.

Table 8.5 Comparison of Boiling Points of Alcohols and Their Corresponding Alkanes

Alcohol	Boiling Point (°C)	Alkane	Boiling Point (°C)
Methanol	65.0	Methane	−162
Ethanol	78.5	Ethane	−88.5
1-Propanol	97.4	Propane	−42
2-Propanol	82.4		

Alcohols with low molecular weights such as methanol and ethanol are very soluble in water because hydrogen bonding occurs between water and alcohol molecules. Alcohols with higher molecular weights are less soluble in water because the increasing size of the nonpolar part of the molecule disrupts the hydrogen bonding network.

Methanol is also known as *wood alcohol* because for many years it was produced by heating hardwoods such as maple, birch, and hickory to high temperatures in the absence of air. (The current methods of production from coal and coal gas will be examined in Chapter 12.) Methanol has many industrial uses. It is used in the manufacture of plastics and other polymers, as an antifreeze in windshield washer fluid, and as a gasoline additive. Its use as a fuel to replace gasoline will be discussed in Chapter 13. Methanol is extremely toxic. Drinking quite small quantities can result in temporary blindness, permanent blindness, or even death.

Ethanol, or ethyl alcohol (Fig. 8.14), is the active ingredient in alcoholic beverages. It is produced by the fermentation of carbohydrates (sugars and starches) in fruits and other plant sources, using a process that was discovered thousands of years ago. During fermentation, microorganisms such as yeasts convert sugar to alcohol and carbon dioxide:

$$C_6H_{12}O_6 \longrightarrow 2\ C_2H_5OH + 2\ CO_2$$

Ethanol is sometimes called *grain alcohol* because it can be produced from grains such as barley, wheat, corn, and rice. Beer is usually made from barley with the addition of hops (Chapter 14). Wine is made from grapes and other fruits.

The word *proof* originated in England in the 17th century. To prove to their customers that their product was not diluted, whiskey dealers would pour a small amount of it onto gunpowder. If the mixture could be ignited, that was proof that water had not been added to the whiskey.

The percentage of alcohol in the fermentation product is determined by the enzymes present. At alcohol concentrations above about 15%, the enzymes begin to be destroyed, and the reaction can proceed no further. The alcohol content of a fermentation product can be increased by distillation. For example, whiskey is made by distilling fermented grains. The *proof* of an alcoholic beverage is equal to twice the percentage (by volume) of the ethanol. Thus, 90-proof rum is 45% ethanol.

Ethanol is not as toxic as methanol, but one pint of pure ethanol, if ingested rapidly, is likely to prove fatal for most people. Ethanol is quickly absorbed into the bloodstream; excessive use causes damage to the liver and the neurological system and leads to physiological addiction.

In addition to being a constituent of alcoholic beverages, ethanol has many industrial uses. It is used in the preparation of many organic chemicals and is an important industrial solvent. It is also used by the pharmaceutical industry as an ingredient in cough syrups and other medications and as an antiseptic for skin disinfection.

Ethanol for industrial use is produced by the high-pressure hydration of ethylene, not by fermentation:

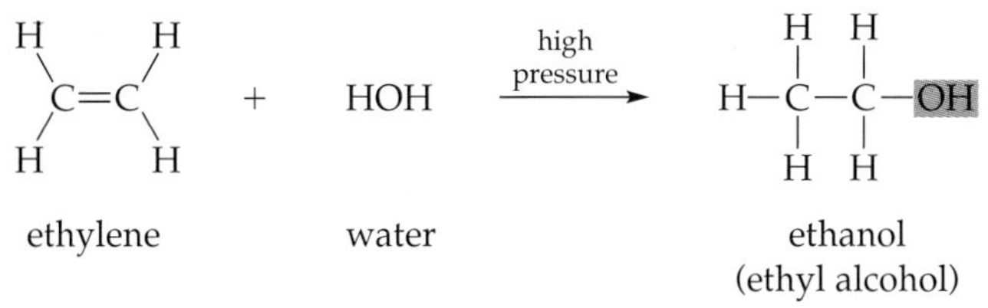

$$\underset{\text{ethylene}}{\mathrm{H_2C{=}CH_2}} + \underset{\text{water}}{\mathrm{HOH}} \xrightarrow[\text{pressure}]{\text{high}} \underset{\text{ethanol (ethyl alcohol)}}{\mathrm{H{-}CH_2{-}CH_2{-}OH}}$$

Alcoholic beverages are heavily taxed, but alcohol intended for industrial use is not. To avoid the tax, manufacturers must *denature* the alcohol; that is, render it unfit for human consumption by the addition of a small amount of a toxic substance. Methanol or benzene is often added because these substances do not alter the solvent properties of the alcohol.

Propanol, the next member of the series following ethanol, can exist in two forms: 1-propanol (*n*-propyl alcohol) and 2-propanol (isopropyl alcohol) (Fig. 8.14). Rubbing alcohol, a 70% solution of 2-propanol in water, is an effective disinfectant.

Important alcohols with more than one hydroxyl group in each molecule include ethylene glycol (1,2-ethanediol) and glycerol (glycerin, or 1,2,3-propanetriol) (Fig. 8.14). Ethylene glycol, with its two polar —OH groups, is very soluble in water and can be mixed with it in any desired proportion. Because of hydrogen bonding, ethylene glycol has a relatively high boiling point. These properties make it an ideal choice for use as the main ingredient in permanent antifreeze. Glycerol is a nontoxic, clear, syrupy liquid. With its three —OH groups, it has an affinity for moisture. For this reason, it is often added to personal hygiene products such as hand lotions, shaving creams, and soaps to help keep skin moist and soft. Glycerol is also used as a lubricant and in the manufacture of explosives and a variety of other chemicals.

The simplest aromatic alcohol is phenol (Fig. 8.14), which was the first widely used disinfectant. It was introduced in 1867 by the British surgeon Joseph Lister. However, because it could burn skin if not in very dilute solution, it was soon replaced by safer related compounds. A methyl derivative of phenol, *o*-cresol, is the main ingredient in the wood preservative creosote.

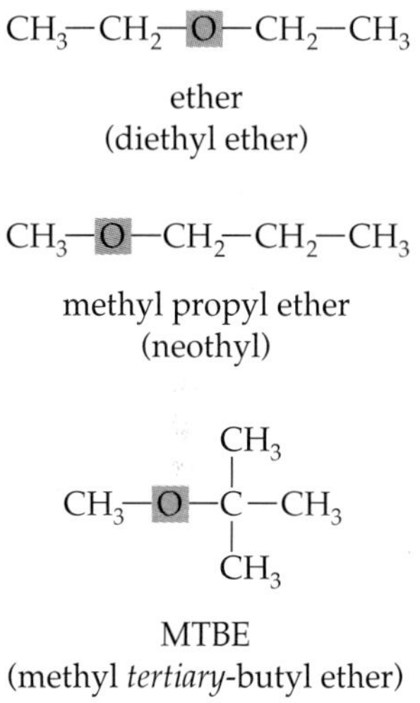

$CH_3{-}CH_2{-}O{-}CH_2{-}CH_3$

ether
(diethyl ether)

$CH_3{-}O{-}CH_2{-}CH_2{-}CH_3$

methyl propyl ether
(neothyl)

$CH_3{-}O{-}C(CH_3)_2{-}CH_3$

MTBE
(methyl *tertiary*-butyl ether)

FIGURE 8.16

Examples of ethers.

Ethers

Compounds that have two hydrocarbon groups attached to an oxygen atom are called **ethers** (Table 8.4). They have the general formula R—O—R′, where R and R′ may or may not be the same **(Fig. 8.16)**. The best-known ether is diethyl ether, which is commonly called ether.

The word *ether* is derived from the Greek word *aither* meaning "to ignite."

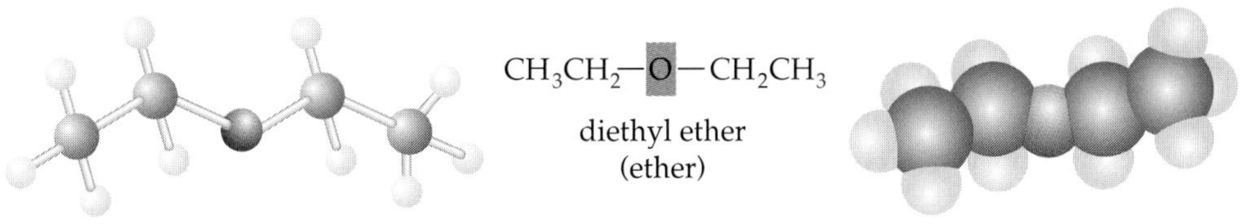

$CH_3CH_2{-}O{-}CH_2CH_3$

diethyl ether
(ether)

Ether acts as a general anesthetic; its introduction for this purpose in the mid-1840s revolutionized medicine **(Fig. 8.17)**. Surgical procedures that previously could not be performed because of the pain and shock involved became possible. General anesthetics cause unconsciousness and insensitivity to pain by temporarily depressing the central nervous system. Ether is not an ideal anesthetic; it is highly flammable, and inhalation produces some unpleasant side effects, such as nausea and irritation of the respiratory tract. It has been replaced by more efficient anesthetics, including a related ether, neothyl (Fig. 8.16).

FIGURE 8.17

In 1846, the Boston dentist William Morton gave a public demonstration of the use of ether as an anesthetic.

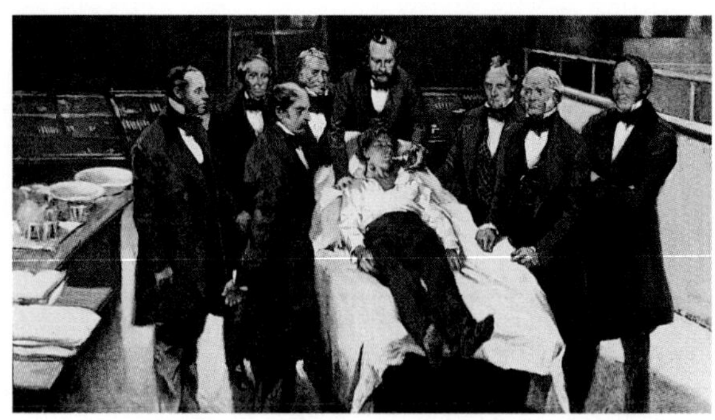

A commercially important ether is methyl tertiary-butyl ether (MTBE) (Fig. 8.16), which, since the phaseout of the antiknock agent tetraethyl lead, has become the agent most frequently added to gasoline to increase octane rating (see Chapter 12).

EXAMPLE 8.7 Identifying alcohols and ethers

Indicate the functional group in each of the following. Identify each compound as an ether, alcohol, or phenol.

a. (benzene ring with OH and adjacent Cl)

b. $CH_3—O—CH_3$

c. $CH_3CH(OH)CH_3$

Solution

a. The functional group is the —OH group attached to the benzene ring. This compound is a phenol. The benzene ring also has a chlorine substituent. The name of the compound is 2-chlorophenol (or *o*-chlorophenol).

b. There is an oxygen atom between two carbon atoms. This compound is an ether. It is dimethyl ether.

c. The —OH group is attached to a carbon atom. This compound is an alcohol. Since the —OH group is attached to carbon 2, the name is 2-propanol.

Practice Exercise 8.7

Draw the structure of 1-propanol.

Aldehydes and Ketones

Aldehydes and ketones are related families of compounds (Table 8.4). Both contain a **carbonyl group,** in which a carbon atom is joined by a double bond to an oxygen atom. In **aldehydes,** the carbonyl carbon is attached to at least one hydrogen atom; in **ketones,** it is attached to two carbon atoms:

$—C(=O)—$	$R—C(=O)—H$ or RCHO	$R—C(=O)—R$ or RCOR
carbonyl group	aldehyde	ketone

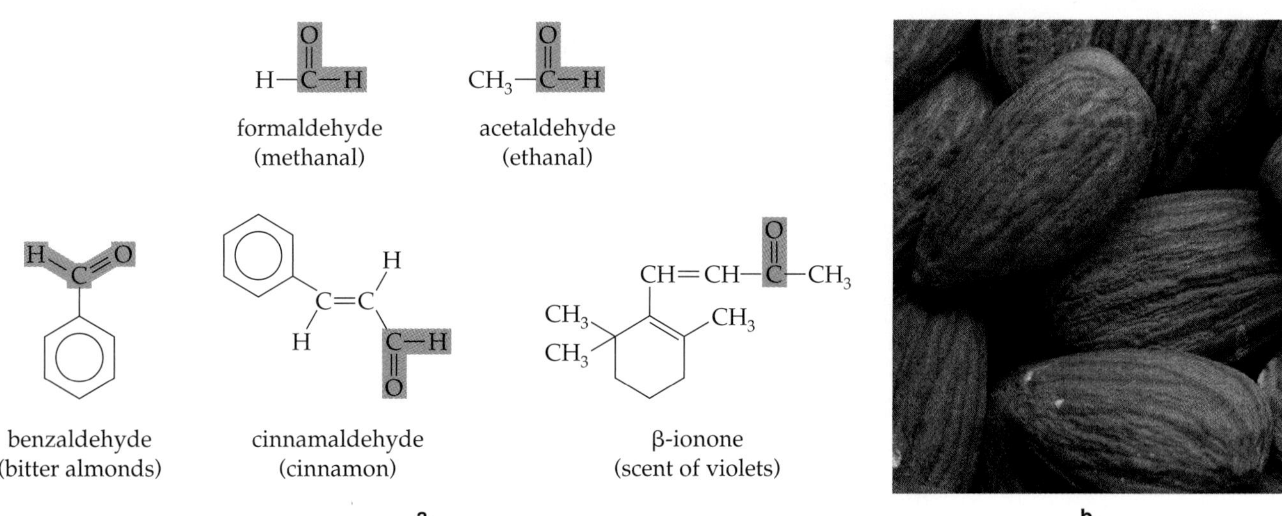

FIGURE 8.18

(a) Examples of aldehydes and ketones. (b) Almonds contain benzaldehyde.

In the IUPAC naming system, the suffixes *-al* and *-one* are used to identify aldehydes and ketones, but many are better known by their common names.

The simplest aldehyde is formaldehyde **(Fig. 8.18)**. As a 40% solution in water called *formalin*, formaldehyde is used as a preservative for biological specimens and as an enbalming fluid. It is also a starting material for the manufacture of certain polymers. Formaldehyde has a disagreeable odor and is a suspected carcinogen.

Aldehydes and ketones are made by oxidizing the appropriate alcohols, as shown below for the preparation of propanal and propanone (acetone).

$$CH_3-CH_2-CH_2-OH \xrightarrow{\text{oxidation}} CH_3-CH_2-\overset{\overset{\displaystyle O}{\|}}{C}-H$$

or $CH_3CH_2CH_2OH$ — propanol (propyl alcohol) → or CH_3CH_2CHO — propanal

$$CH_3-\underset{}{\overset{\overset{\displaystyle OH}{|}}{C}H}-CH_3 \xrightarrow{\text{oxidation}} CH_3-\overset{\overset{\displaystyle O}{\|}}{C}-CH_3$$

or $CH_3CHOHCH_3$ — isopropanol (isopropyl alcohol) → or CH_3COCH_3 — propanone (acetone)

Acetone, the simplest ketone, is an excellent solvent. It is miscible (soluble in all proportions) with both water and nonpolar liquids and is used to remove varnish, paint, and fingernail polish. Many aromatic aldehydes and ketones have pleasant aromas (Fig. 8.18), and some are used as food flavorings. The scents of many flowers—violets, for example—are due to the release of small amounts of ketones.

EXAMPLE 8.8 Identifying aldehydes and ketones

Identify each of the following as an aldehyde or ketone.

a. (cyclohexanone ring structure) or H_2C–CH_2–CH_2–CH_2–CH_2–C(=O) (ring)   b. $CH_3{-}CH_2{-}\overset{\overset{\large O}{\|}}{C}{-}CH_3$ c. $H{-}\overset{\overset{\large O}{\|}}{C}{-}C_6H_5$

Solution

a. The carbon of the C=O group is attached to two other carbon atoms. This compound is a ketone, cyclohexanone.

b. The carbon of the C=O group is attached to two other carbon atoms. This compound is a ketone, 2-butanone.

c. A hydrogen atom is attached to the carbon of the C=O group. This compound is an aldehyde, benzaldehyde.

Practice Exercise 8.8

Identify each of the following as a ketone or an aldehyde:

a. (cyclopentanone ring structure) or C(=O) with CH_2, CH_2, $CH_2{-}CH_2$ (ring)   b. $CH_3{-}CH_2{-}CH_2{-}\overset{\overset{\large O}{\|}}{C}{-}H$

c. $CH_3{-}\underset{\underset{\large CH_3}{|}}{CH}{-}CH_2{-}CH_2{-}\overset{\overset{\large O}{\|}}{C}{-}CH_3$

Carboxylic Acids

Organic acids are called **carboxylic acids** because they contain the **carboxyl group,** a group consisting of a carbon atom that is double-bonded to an oxygen atom and single-bonded to a hydroxyl group:

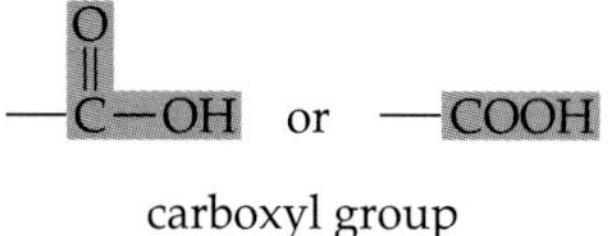

carboxyl group

Examples of carboxylic acids are shown in Figure 8.19.

In naming carboxylic acids, the final *-e* in the parent alkane is replaced with the suffix *-oic,* and the word *acid* is added. The two simplest carboxylic acids are methanoic acid and ethanoic acid, better known by their common names, formic acid and acetic acid.

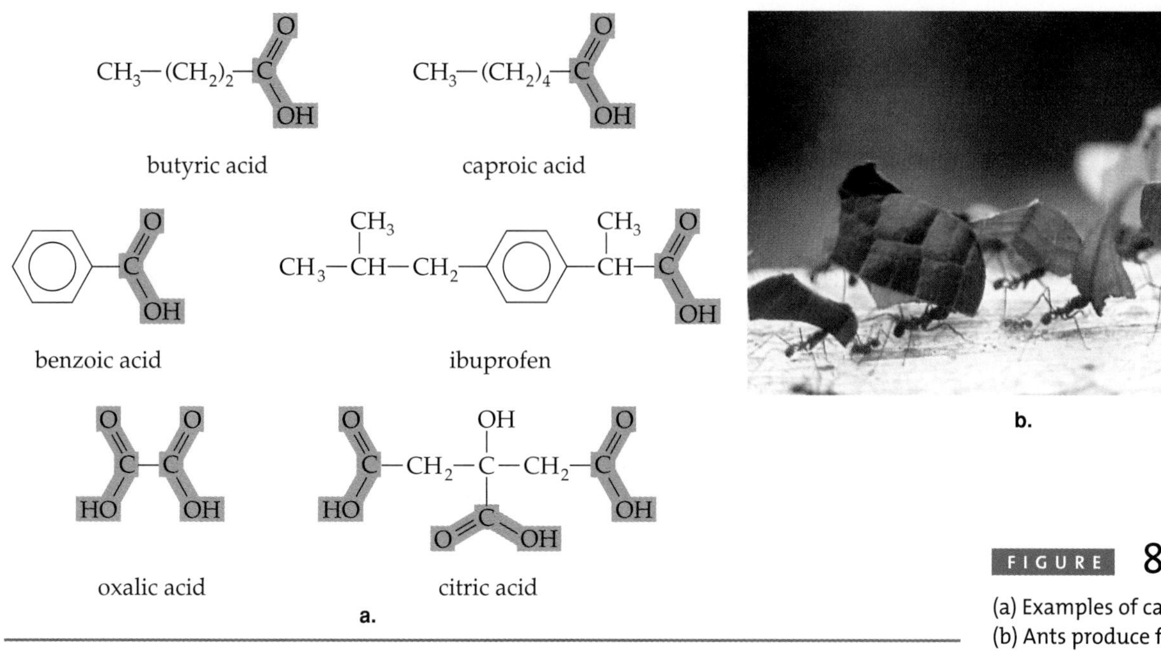

FIGURE 8.19

(a) Examples of carboxylic acids.
(b) Ants produce formic acid.

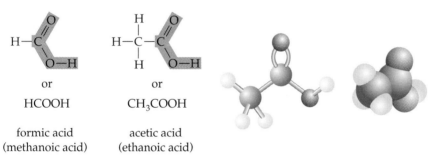

formic acid	acetic acid
$HCOOH$	CH_3COOH
formic acid (methanoic acid)	acetic acid (ethanoic acid)

Compared with inorganic acids such as hydrochloric acid (HCl) and nitric acid (HNO_3), carboxylic acids are weak acids (Chapter 7) that dissociate very little to form ions. Most are less than 2% ionized. The acidic hydrogen is the one in the $-OH$ of the carboxyl group.

$$\mathrm{R{-}\overset{\overset{\displaystyle O}{\|}}{C}{-}OH} \rightleftharpoons \mathrm{R{-}\overset{\overset{\displaystyle O}{\|}}{C}{-}O^-} + \mathrm{H^+}$$

$$\text{or}\qquad\qquad\qquad \text{or}$$

$$\mathrm{RCOOH} \qquad\qquad \mathrm{RCOO^-}$$

The arrows $\rightleftharpoons$ indicate that the equilibrium favors the undissociated form. Like inorganic acids, carboxylic acids are neutralized by bases to form salts:

$$\underset{\text{carboxylic acid}}{\mathrm{R{-}\overset{\overset{\displaystyle O}{\|}}{C}{-}OH}} + \mathrm{Na^+} + \mathrm{OH^-} \longrightarrow \underset{\text{salt}}{\mathrm{R{-}\overset{\overset{\displaystyle O}{\|}}{C}{-}O^-\ Na^+}} + \underset{\text{water}}{\mathrm{H_2O}}$$

$$\text{or}\qquad\qquad\qquad\qquad \text{or}$$

$$\mathrm{RCOOH} \qquad\qquad\qquad \mathrm{RCOO^-\ Na^+}$$

Sodium benzoate, the sodium salt of benzoic acid, is used as a food preservative.

The Latin word for ant is *formica.*

Carboxylic acids have many natural sources. Formic acid is present in ants, various other insects, and nettles. It is the irritant injected under the skin by red ants, bees, and other insects when they sting.

Vinegar, which is a 5% aqueous solution of acetic acid, is the most commonly encountered organic acid. It can be prepared by the aerobic (in the presence of air) fermentation of ethyl alcohol. This is an oxidation process in which acetaldehyde is formed as an intermediate:

$$\underset{\text{ethyl alcohol}}{CH_2CH_3OH} \xrightarrow{\text{oxidation}} \underset{\text{acetaldehyde}}{CH_3CHO} \xrightarrow{\text{oxidation}} \underset{\text{acetic acid}}{CH_3COOH}$$

Acetic acid is used as a starting material in the manufacture of textiles and plastics.

The Latin words for butter and goat are *butyrum* and *caprer.*

Some of the simple carboxylic acids have very unpleasant odors. Butyric acid is responsible for the smell of rancid butter and, in part, for body odor. The powerful odors of certain cheeses and of goats are due to longer-chain carboxylic acids, such as caproic acid. Carboxylic acids with 12 or more carbon atoms are called *fatty acids* (these will be discussed in Chapters 14 and 15).

Many naturally occurring carboxylic acids contain more than one carboxyl group, and some also contain additional hydroxyl groups. Citric acid, which is found in all citrus fruits and gives them their tart flavor, contains three carboxyl groups. Oxalic acid, with two carboxyl groups, is toxic in large doses and is present in rhubarb, tomatoes, and other vegetables. It is a useful oxidizing agent for removing rust and ink stains.

Esters

Esters (Table 8.4) are derivatives of carboxylic acids in which the hydrogen of the carboxyl group (—COOH) is replaced by an alkyl group (—R′).

$$R{-}\overset{\overset{\displaystyle O}{\|}}{C}{-}OR' \quad \text{or} \quad RCOOR'$$

ester

The reaction between an alcohol and a carboxylic acid to yield an ester and water is called an *esterification reaction.*

For example, the ester ethyl acetate is formed by replacing the hydrogen in the —COOH group of acetic acid with an ethyl group ($-CH_2CH_3$).

$$CH_3{-}\overset{\overset{\displaystyle O}{\|}}{C}{-}OH$$

or

CH_3COOH

acetic acid

$$CH_3{-}\overset{\overset{\displaystyle O}{\|}}{C}{-}O{-}CH_2CH_3$$

or

$CH_3COOC_2H_5$

ethyl acetate

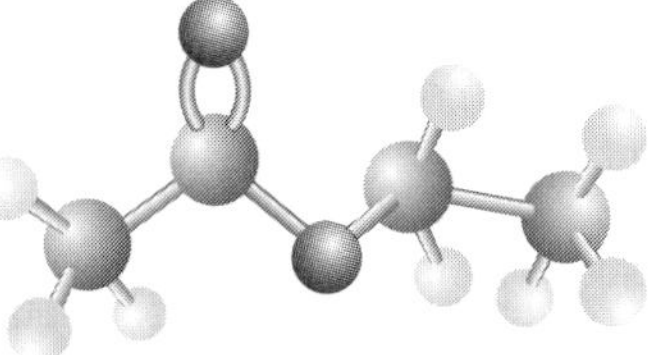

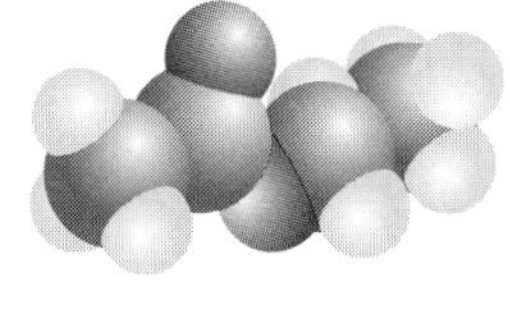

The names of esters are based on the acids from which they are derived.

Esters are widely distributed in nature and are responsible for the tastes of many fruits and the scents of many flowers. Many esters have been synthesized for use as flavorings in foods and beverages. It is interesting to note that just a small change in an R group can significantly alter the flavor sensation produced by an ester. The esters responsible for apple and pineapple flavors, for example, differ from each other by a single $-CH_2-$ group **(Fig. 8.20)**.

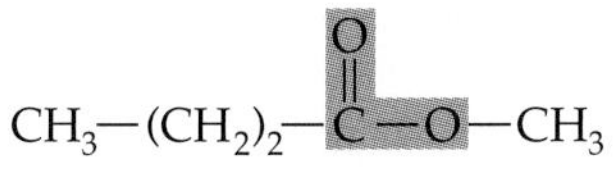

$$CH_3-(CH_2)_2-\overset{\overset{\displaystyle O}{\|}}{C}-O-CH_3$$

methyl butanoate
(apple)

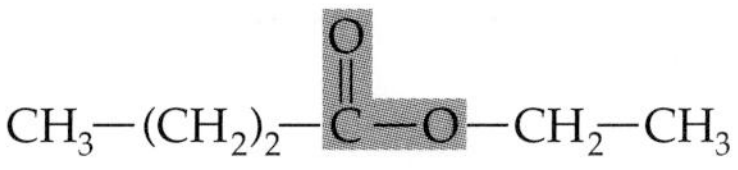

$$CH_3-(CH_2)_2-\overset{\overset{\displaystyle O}{\|}}{C}-O-CH_2-CH_3$$

ethyl butanoate
(pineapple)

FIGURE 8.20

The two esters responsible for apple and pineapple flavors differ by a single $-CH_2-$ group.

Many fragrances used in perfumes are esters. For example, benzyl acetate is the main constituent of oil of jasmine. Ethyl acetate is a useful solvent for removing lacquers, paints, and nail polish.

EXAMPLE 8.9 Identifying carboxylic acids and esters

Identify each of the following as a carboxylic acid or an ester.

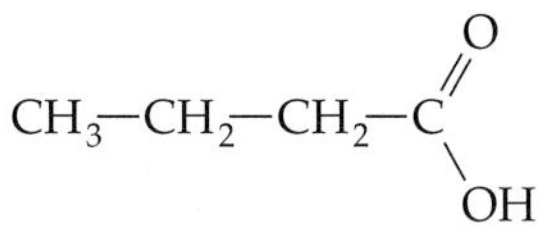

a. 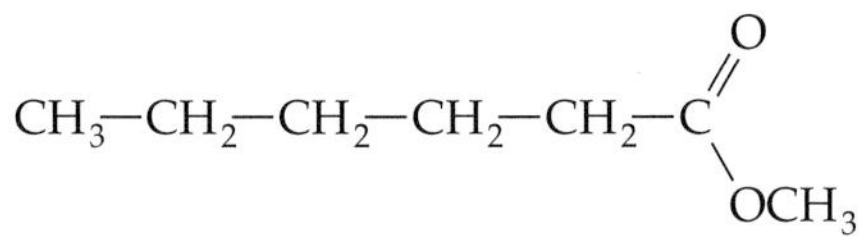 $CH_3-CH_2-CH_2-CH_2-CH_2-C(=O)-OCH_3$ **b.** $CH_3-CH_2-CH_2-C(=O)-OH$ **c.** $C_6H_5-C(=O)-OH$

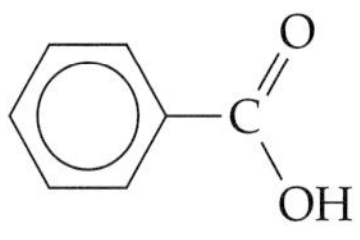

Solution

a. A methyl group is attached to the $-\overset{\overset{\displaystyle O}{\|}}{C}-O$ group. This compound is a methyl ester, methyl hexanoate.

b. A hydrogen atom is attached to the $-\overset{\overset{\displaystyle O}{\|}}{C}-O$ group. This compound is a carboxylic acid, butyric acid.

c. This compound is also a carboxylic acid, benzoic acid.

Practice Exercise 8.9

Write the condensed formula of each of the following:

a. acetic acid **b.** methyl acetate **c.** pentanoic acid

Amines and Amides

Many important organic compounds—both naturally occurring and synthetic—contain nitrogen. Two important families of nitrogen-containing compounds are amines and amides.

Amines can be thought of as being derived from ammonia (NH_3) by replacement of one, two, or all three of the hydrogen atoms with alkyl groups. A *primary* amine has two hydrogen atoms and one alkyl group ($-R$) attached to the nitrogen atom; a *secondary* amine has one hydrogen atom and two alkyl groups attached to the nitrogen atom; and a *tertiary* amine has three alkyl groups and no hydrogen atoms attached to the nitrogen atom. The general structural formulas for these three types of amines are shown below:

H | H—N—H	H | R—N—H	H | R—N—R′	R″ | R—N—R′
ammonia	primary amine	secondary amine	tertiary amine

The two simplest amines are methylamine and ethylamine.

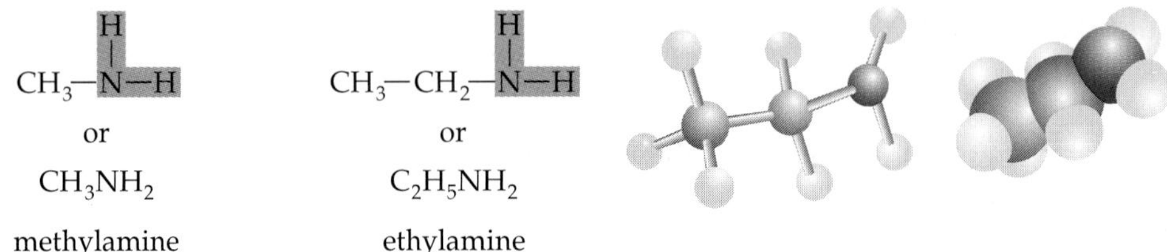

Most decaying organic matter produces amines, many of which have disagreeable fishy odors (Chapter 2). The smells associated with sewage-treatment plants and meat-packing plants are due primarily to amines. Two particularly unpleasant smelling amines are cadaverine and putrescine, which are products of the bacterial decomposition of proteins and are partly responsible for the smell of decaying flesh **(Fig. 8.21)**.

Important amines that include a benzene ring in their structure are aniline, amphetamine, and epinephrine (commonly called *adrenaline*) (Fig. 8.21). Aniline is used as a starting material for the manufacture of dyes and certain drugs. Amphetamines are stimulants that have been widely used by people who need to stay alert. They are also used to treat mild depression and as appetite suppressants. Epinephrine and the related compound norepinephrine are secreted by the adrenal gland and play an important role in controlling blood pressure and cardiac output.

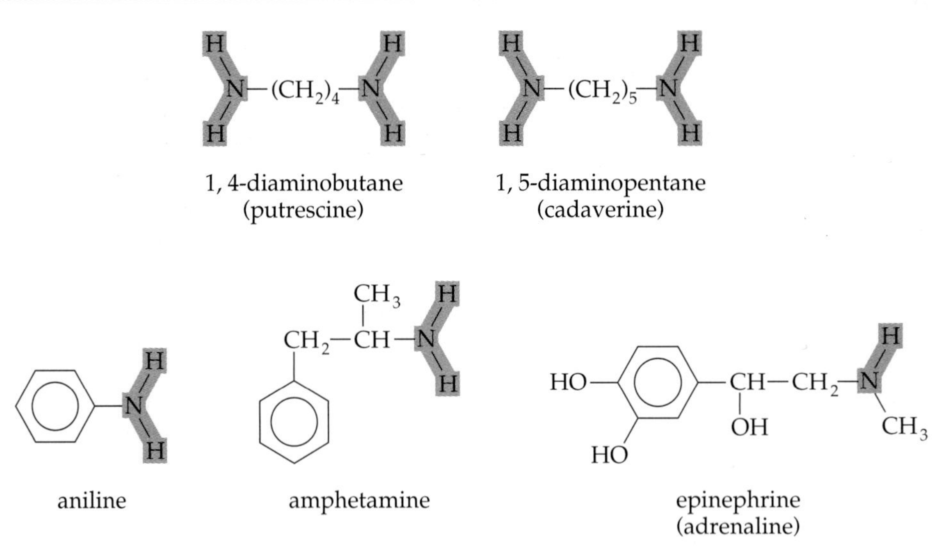

FIGURE 8.21

Examples of amines.

EXAMPLE 8.10 Drawing and classifying amines

Draw the structure of each of the following and indicate whether it is a primary, secondary, or tertiary amine:

a. propylamine ($C_3H_7NH_2$)

b. methylbutylamine ($CH_3NHC_4H_9$)

c. trimethylamine [$(CH_3)_3N$]

Solution

a. This amine has a propyl group attached to the nitrogen atom. Nitrogen forms three bonds. The other two bonds must be attached to hydrogen atoms. Since there is only one alkyl group, the compound is a primary amine.

$$CH_3CH_2CH_2-\underset{\displaystyle H}{\underset{|}{N}}-H$$

b. This amine has a methyl group and a butyl group attached to the nitrogen atom. It is a secondary amine.

$$CH_3-\underset{\displaystyle H}{\underset{|}{N}}-CH_2CH_2CH_2CH_3$$

c. This amine has three methyl groups attached to the nitrogen atom. It is a tertiary amine.

$$CH_3-\underset{\displaystyle CH_3}{\underset{|}{N}}-CH_3$$

Practice Exercise 8.10

Is each of the following a primary, secondary, or tertiary amine?

a. trioctylamine **b.** dibutylamine **c.** isopropylamine

Amides are derivatives of carboxylic acids in which the hydroxyl group (—OH) of the carboxyl group is replaced with an amino group ($-NH_2$) or a substituted amino group (—NHR′ or —NR′R″). The nitrogen atom in an amide is attached to a —C=O group. The general formulas for the three types of amides are shown below:

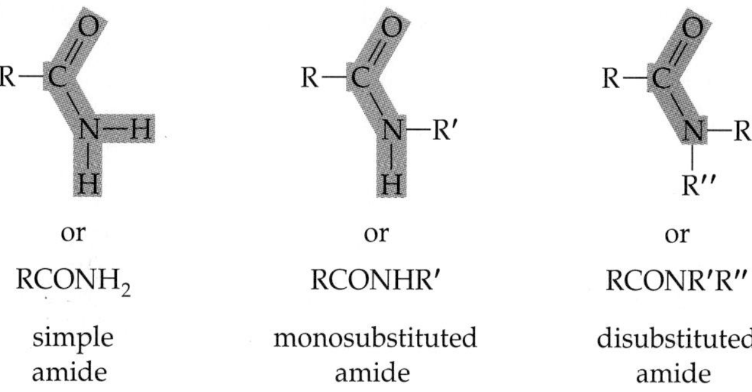

or	or	or
$RCONH_2$	RCONHR′	RCONR′R″
simple amide	monosubstituted amide	disubstituted amide

EXPLORATIONS

Tyrian Purple—A Color Fit for a King

In imperial Rome, wearing a toga dyed with Tyrian purple could cost you your life—unless you were the emperor. The royal purple dye, obtained at enormous expense from shellfish, was the most renowned dye in the ancient world.

Tyrian purple was used to dye Roman togas.

In republican Rome, purple robes were a sign of rank and prestige. Only military commanders who had been triumphant in battle and the two magistrates (or censors) who supervised public morals and kept the census were permitted to wear garments dyed with Tyrian purple. Consuls and praetors were allowed to wear clothing trimmed with purple, but the dye was prohibited to lesser mortals. Under the emperors, the wearing of purple was further restricted. A decree issued by Nero in the 1st century A.D. gave to the emperor the exclusive right to wear the royal color. Purple dyeing became a state monopoly, and the method of production as well as the workers were tightly controlled. By imperial edict, it became a crime punishable by death to manufacture Tyrian purple anywhere but in the imperial dyeworks. The association of purple with emperors and kings gave rise to the phrase "born to the purple" to describe infants of royal birth.

Until the middle of the 19th century, when the explosive growth in the synthetic dye industry began, the only dyes available were of natural origin. The yellow, red, and brown earth pigments were generally quite stable, but the colors derived from plants and animals were more elusive. Most were not true dyes, and the colors washed out or faded in sunlight. What made Tyrian purple so desirable was that, in addition to being a brilliant color, it was a permanent dye. According to Ronulph of Chester, writing in the 14th century, Tyrian purple was "wonder fair and stable, staineth never with cold or with heat, nor with wet, nor with dry, but ever the older the color is fairer."

The Stuff of Legend

Legend attributes the discovery of Tyrian purple to the Phoenician god Melkarth, known to the Romans as Hercules. According to the ancient story, Hercules was walking along the seashore one day with his mistress, the nymph Tyros, and his sheep dog. The dog, nosing about among the rocks, found a shellfish and crushed it between his teeth. Juice from the shellfish stained the dog's jaw a bright purplish red. Tyros admired the color, and, to please his mistress, Hercules dyed a gown for her using it. The legend lends support to the belief that the dye originated with the Phoenicians. We know that these enterprising seafarers understood the art of purple dyeing and set up many dye factories in the colonies they established along the Mediterranean coast.

According to some authorities, a dye made from shellfish was used in Crete as early as 1600 B.C., and shellfish were the source of the purple mentioned in the Bible. Ezekiel (27:16), describing trade with Syria in the 8th century B.C., wrote: "They exchanged for your wares emeralds, purple, embroidered work, fine linen, coral, and agate." Evidence that the Greeks and Romans carried on the art of purple dyeing is found in the huge piles of mollusk shells that have been found close to sites of ancient dyeworks in Athens and Pompeii.

Secretions of Mollusks

The dye used by the ancients was extracted from two species of mollusks, *Murex brandaris* and *Thais haemastoma,* which were common along the shores of the Mediterranean. A small gland close to the mollusk's head secretes a whitish fluid containing the precursors of the dye. When the fluid is exposed to the oxygen of the atmosphere and bright sunlight, a series of reactions occurs, and the color of the fluid changes gradually from white to yellow and then to green, blue, and finally to a deep purplish red.

The origin of the name *Tyrian purple*, is uncertain, but it probably derived from the city of Tyre, where, according to the Roman scholar Pliny the Elder, the best dye was made. The word *purple* is derived from the Latin word *purpura*, the name by which the mollusk was known in classical Rome. Perception of the color purple has changed over the centuries: Today, we would describe the famous color as crimson rather than purple. In 14th-century England, it was described simply as red.

The Dyeing Process

Pliny, in his *Historia Naturalis* written in the midde of the 1st century A.D., gives a detailed account of the dyeing process. According to him, the best time to collect the mollusks was in the spring after the dog star, Sirius, had risen. Fishermen captured the creatures in baited wicker baskets, and when enough had been collected, the glands were carefully removed. After they were crushed, the glands were left to stand over salt for several days before being boiled with water in a large vat. Boiling continued for several weeks, during which time debris rising to the surface was removed, and, at intervals, the dyeing properties of the liquid were tested. When the liquid was of the right strength to produce the desired deep color, fabric was added. In the vat, the fabric acquired a yellow color, but when it was removed and spread out to dry in the sun, like magic, the brilliant royal color appeared. A second dipping in the vat produced an even deeper, richer color.

Because thousands of mollusks were needed to obtain enough dye to color a single garment, Tyrian purple was extremely costly. It has been estimated that in 301 A.D., a pound of dyed wool cost about three times as much as a baker would earn in a year, and a pound of dyed silk was several times more expensive. At certain times, Tyrian purple was worth as much as 10–20 times its weight in gold.

At certain times, Tyrian purple was worth as much as 10–20 times its weight in gold.

The Dye's Disappearance

Mysteriously, by the end of the 15th century, the art of making Tyrian purple appears to have been forgotten and the dye is rarely mentioned again. One probable reason was the fall of Constantinople (Istanbul) to the Turks in 1453, which ended the rule of the Byzantine emperors. No doubt, in the resulting turmoil, the imperial dyeworks ceased to function. From time to time during the ensuing centuries, the dyeing properties of the mollusks were accidentally rediscovered. There are occasional reports of local fishermen dyeing their shirts red, but the royal dye never again attained the prestige it had known in classical times.

Identification of Structure

In 1908, P. Friedlander made a scientific study of the dye. He extracted 1.4 grams of dye from 12,000 mollusks of the species *Murex brandaris* and determined that the purple dye so prized in antiquity was the organic compound 6,6′-dibromoindigo:

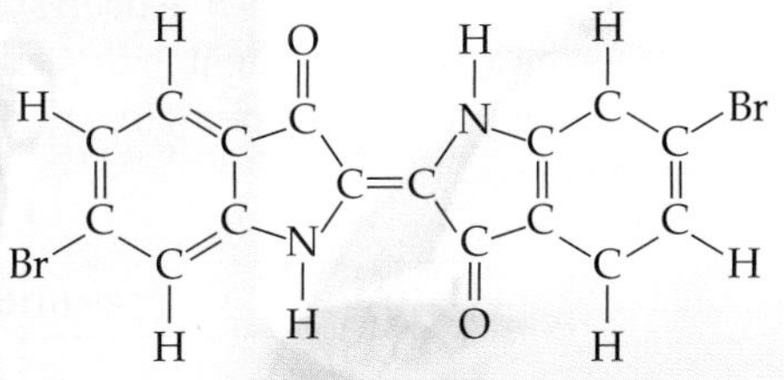

It is not surprising that the pigment contains bromine since there is a considerable amount of bromide in seawater (67 ppm). The crimson color is explained by the large number of alternating double bonds in the molecule, but that's another story.

References: M. R. Fox, *Vat Dyestuffs and Vat Dyeing* (New York: John Wiley, 1947); L. S. Pratt, *The Chemistry and Physics of Organic Pigments* (New York: John Wiley, 1947); S. Robinson, *A History of Dyed Textiles* (Cambridge, MA: MIT Press, 1969); P. E. McGovern and R. H. Michel, "Royal Purple Dye: The Chemical Reconstruction of the Ancient Mediterranean Industry," American Chemical Society, *Accounts of Chemical Research* (1990), 23, 152.

FIGURE 8.22

Examples of amides.

$$-\mathrm{CH(NH_2)}-\mathrm{C(=O)}-\mathrm{N(H)}-\mathrm{CH(R')}-\mathrm{C(=O)}-\mathrm{N(H)}-\mathrm{CH(R'')}-$$

protein segment

$$\mathrm{H_2N-C(=O)-NH_2}$$

urea

$$\mathrm{CH_3-C(=O)-NH-C_6H_4-OH}$$

acetaminophen

The simplest amide is acetamide. Amides are named by replacing the ending *-ic* (or *-oic*) and the word *acid* in the name of the corresponding acid with *-amide.*

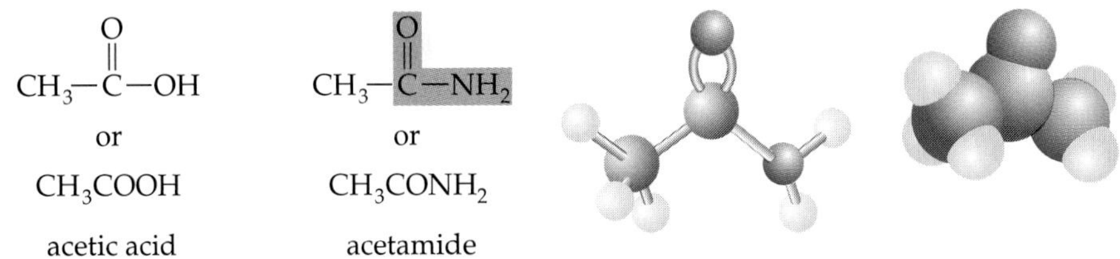

The amide functional group is present in many important biological compounds, including proteins and urea, the end product of human metabolism of proteins **(Fig. 8.22)**. These compounds will be studied in more detail in Chapters 14 and 15.

Many amides have been synthesized for use as drugs. For example, acetaminophen, the active ingredient in the pain reliever Tylenol, is an amide (Fig. 8.22). Amides are also important intermediates in the manufacture of nylon and other polymers.

EXAMPLE 8.11 Drawing structures of organic compounds

Draw the structure for each of the following compounds, which may belong to any of the classes of organic compounds we have studied:

a. methylethylamine **b.** 3-pentanone **c.** butyric acid **d.** methylacetamide

Solution

a. This compound has a methyl group and an ethyl group attached to the nitrogen atom. Nitrogen forms three bonds; therefore, the third bond is attached to a hydrogen atom.

$$\mathrm{CH_3-N(H)-CH_2-CH_3}$$

b. This compound contains five carbon atoms (penta-), and the oxygen of the ketone is attached to carbon 3.

$$\mathrm{CH_3-CH_2-C(=O)-CH_2-CH_3}$$

c. This compound contains four carbon atoms (butyric) and is a carboxylic acid.

$$CH_3-CH_2-CH_2-\overset{\overset{\displaystyle O}{\|}}{C}-OH$$

d. This compound is formed from acetamide by the addition of a methyl group.

$$CH_3-\overset{\overset{\displaystyle O}{\|}}{C}-\underset{\underset{\displaystyle H}{|}}{N}-CH_3$$

Practice Exercise 8.11

Identify all the functional groups in epinephrine, a compound used to control allergic reactions.

$$HO-CH-CH_2-NH-CH_3$$

(benzene ring attached to CH, bearing two OH groups)

OH

OH

This chapter has provided a brief introduction to organic chemistry by discussing hydrocarbons and the major functional group classes. There are many other organic compounds that have not been mentioned. In Chapters 14 and 15, we will examine in more detail the naturally occurring organic compounds that constitute living tissues and also make up the food on which we depend for survival. Fossil fuels, the starting materials for the synthesis of a great many of the organic compounds produced by industry, will be studied in Chapter 12.

Chapter Summary

1. Carbon forms more compounds than any other element. It does so by sharing its four valence electrons.
2. Carbon atoms can link together in straight chains, branched chains, or rings.
3. Hydrocarbons can be divided into four classes: Alkanes contain carbon-carbon single bonds (suffix -ane); alkenes contain one or more carbon-carbon double bonds (suffix -ene); alkynes have one or more carbon-carbon triple bonds (suffix -yne), and aromatic hydrocarbons contain one or more benzene rings. Hydrocarbons are the main constituents in petroleum and natural gas.
4. Alkanes form a homologous series with the general formula C_nH_{2n+2}, where n equals the number of carbon atoms in a member of the series.
5. Alkanes containing more than four carbon atoms form structural isomers, compounds that have the same formula but different structures.
6. Alkenes form a homologous series with the general formula C_nH_{2n}, where n equals the number of carbon atoms in a member of the series.

7. A pair of *cis-trans* isomers has the same condensed formula. The isomers differ in the way atoms or groups of atoms are attached to different sides of the double bond (see page 225).

8. Benzene, C_6H_6, is a flat six-membered ring of carbon atoms with a hydrogen atom attached to each carbon. The carbon-carbon bonds, which are all identical, are intermediate between single and double bonds (see Fig. 8.10).

9. Most organic compounds are derivatives of hydrocarbons in which one or more hydrogen atoms in the parent hydrocarbon have been replaced with a different atom or a group of atoms. The substituents, which are called functional groups, determine the properties of the compound.

10. Organic halides (R—X) are hydrocarbon derivatives containing one or more halogen atoms (F, Cl, Br, I). Examples are chloroform, DDT, PCBs, and CFCs, all of which have caused health and/or environmental problems.

11. Alcohols (R—OH) contain a hydroxyl group (—OH). Examples are methanol, ethanol, ethylene glycol (antifreeze), and the aromatic alcohol phenol.

12. Ethers (R—O—R′) have two hydrocarbon groups attached to an oxygen atom. Examples are ethyl ether (used as an anesthetic) and MTBE (used as an octane booster).

13. Both aldehydes ($R-\overset{\overset{\displaystyle O}{\|}}{C}-H$) and ketones ($R-\overset{\overset{\displaystyle O}{\|}}{C}-R'$) contain a carbonyl group (—C=O). Examples include formaldehyde (a preservative) and acetone (an excellent solvent).

14. Carboxylic acids (R—COOH) contain a carboxyl group ($R-\overset{\overset{\displaystyle O}{\|}}{C}-OH$). Examples include formic acid (present in ants and bee stings), acetic acid (present in vinegar), and citric acid (present in fruits).

15. Esters ($R-\overset{\overset{\displaystyle O}{\|}}{C}-OR'$) are derivatives of carboxylic acids in which the hydrogen of the carboxyl group has been replaced with an alkyl group (R′). Esters are used as flavorings and fragrances.

16. Amines ($R-NH_2$) are hydrocarbon derivatives of ammonia (NH_3) in which one, two, or all three of ammonia's hydrogen atoms have been replaced with an alkyl group. Decaying organic matter produces unpleasant smelling amines. Aromatic amines include amphetamine and epinephrine (adrenaline).

17. Amides ($R-\overset{\overset{\displaystyle O}{\|}}{C}-NH_2$) are derivatives of carboxylic acids in which the hydroxyl group (—OH) has been replaced with an amino group ($-NH_2$) or a substituted amino group (—NHR′ or —NR′R″). Examples are acetaminophen (in Tylenol) and amides used to manufacture nylon and other polymers.

Key Terms

alcohol (p. 231)
aldehyde (p. 234)
alkane (p. 216)
alkene (p. 216)
alkyl group (p. 220)
alkyne (p. 216)
amide (p. 241)
amine (p. 239)
aromatic hydrocarbon (p. 216)
benzene (p. 227)
carbonyl group (p. 234)
carboxyl group (p. 236)
carboxylic acid (p. 236)
cis-trans isomerism (p. 225)
cycloalkane (p. 226
ester (p. 238)
ether (p. 233)
functional group (p. 216)
homologous series (p. 217)
hydrocarbon (p. 216)
ketone (p. 234)
organic halide (p. 230)
polycyclic aromatic compound (p. 228)
saturated hydrocarbon (p. 217)
structural isomers (p. 219)
unsaturated hydrocarbon (p. 224)

Questions and Problems

1. Write the names of the following alkanes:
   a. $CH_3-CH_2-CH_2-CH_2-CH_3$
   b. $CH_3-CH_2-CH_2-CH_2-CH_2-CH_2-CH_2-CH_3$
   c. $CH_3-CH_2-CH_3$ **d.** $CH_3-CH_2-CH_2-CH_3$

2. Write the names of the following branched alkanes:
   a. $CH_3-\underset{\displaystyle CH_3}{\underset{|}{CH}}-\underset{\displaystyle CH_3}{\underset{|}{CH}}-CH_3$
   b. $CH_3-CH_2-\underset{\displaystyle CH_3}{\underset{|}{\underset{\displaystyle CH_2}{\underset{|}{CH}}}}-CH_2-CH_3$
   c. $CH_3-\underset{\displaystyle CH_3}{\underset{|}{CH}}-CH_2-CH_3$
   d. $CH_3-CH_2-CH_2-\underset{\displaystyle CH_3}{\underset{|}{CH}}-CH_2-CH_2-CH_3$

3. Why does carbon form so many compounds with oxygen, nitrogen, and hydrogen?

4. Write the structural formulas and names for the members of the homologous series of saturated hydrocarbons starting with two carbon atoms and ending with six carbon atoms.

5. Which is the smallest alkane to have an isomer?

6. Draw and name all the isomers of pentane.

7. Draw and name the isomers of hexane that are butanes.

8. Write the structural formulas for the following compounds:
   a. *cis*-2-butene **b.** 1-hexene
   c. propene **d.** 1-pentene

9. Write the structural formulas for the following compounds:
   a. cyclohexene **b.** 3-methyl-1-butene
   c. 3,6-dimethyl-2-octene **d.** 2-methyl-2-butene

10. Draw five different structural isomers of a hydrocarbon with the formula C_6H_{14}.

11. What is the common name for ethyne?

12. Which of the following alkenes show *cis-trans* isomerism? Draw the isomeric structures.
    a. 1-butene **b.** 2-methyl-2-butene
    c. 2-butene **d.** 1-chloropropene

13. Which of the following alkenes show *cis-trans* isomerism? Draw the isomeric structures.
    a. 3-methyl-4-ethyl-3-hexene **b.** 2-pentene
    c. 2,3-dichloro-2-butene **d.** 1-pentene

14. Write a structural formula for each of the following compounds:
    a. 2,3-dichloropentane
    b. 1,4-dichlorocyclohexane
    c. 3-ethylpentane
    d. 1-bromobutane

15. Write a structural formula for each of the following compounds:
    a. 2,3-dimethyl-2-butene
    b. *cis*-2-butene
    c. 1,1-dimethylcyclopentane
    d. 2,3-dichlorobutane

16. Identify the functional group in each of the following compounds:
    a. acetone, CH_3COCH_3
    b. butyric acid, $CH_3CH_2CH_2COOH$
    c. methyl acetate, CH_3COOCH_3

17. Identify all of the functional groups in each of the following compounds:

a. glutaraldehyde, $OHCCH_2CH_2CH_2CH_2CHO$

b. glycine, H_2NCH_2COOH

c. oxalic acid, HOOCCOOH

18. Identify all of the functional groups in each of the following compounds:

a. A sex attractant of the female tiger moth

$$CH_3{-}\underset{\displaystyle CH_3}{\underset{|}{C}}H(CH_2)_{12}{-}CH_3$$

b. Component of peppermint oil

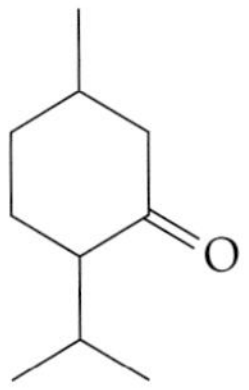

c. Vitamin A

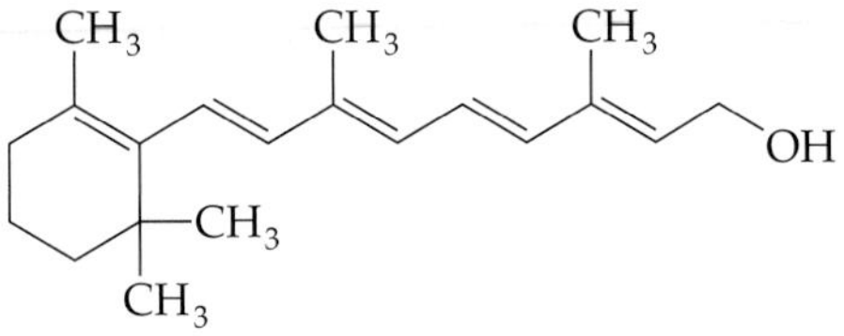

d. Testosterone

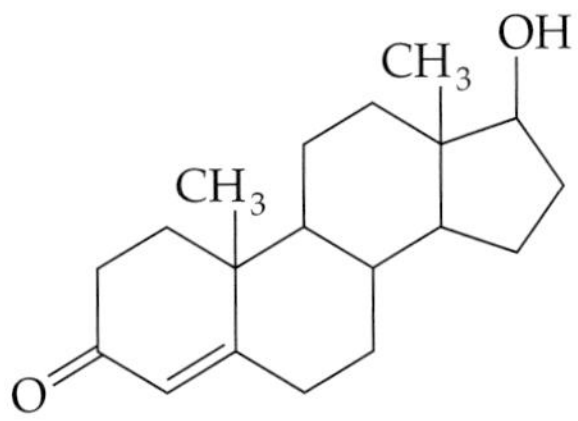

e. Nepetalactone, a constituent of catnip

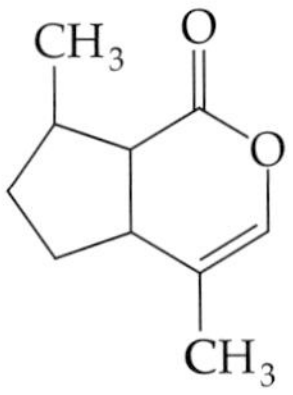

f. An anesthetic

$$CH_2{=}CH{-}O{-}CH{=}CH_2$$

19. Cyclohexane and 1-hexene have the same molecular formula, C_6H_{10}. Write the structural formulas.

20. Name the following:

a. $(CH_3)_3CCH{=}CHCH_2CH_3$

b. $CH_3CH_2CH{=}CH(CH_2)_4CH_3$

c. $CH_3CH_2CH{=}CHCH_2CH_3$

21. Write a structural formula for each of the following:

a. methylacetylene

b. methylpropylacetylene

c. methylbutylacetylene

22. Write the general formula for:

a. a carboxylic acid **b.** a ketone

c. an alcohol **d.** an ester

23. Write the general formula for:

a. an amine **b.** an amide

c. an ether **d.** an aldehyde

24. What are the chemical names for the following materials?

a. rubbing alcohol **b.** wood alcohol

c. grain alcohol **d.** antifreeze

25. Write the structural formula for the aromatic alcohol phenol.

26. Methanol is oxidized in two steps. What is the product of the first oxidation step?

27. What two chemicals would you use to make the ester methyl acetate?

28. What kind of compound would you react with an amine to make an amide?

29. What kind of compound would you react with an alcohol to make an ester?

30. What kind of compound is responsible for the fruity smell of a banana?

31. What compound makes certain insect bites sting?

32. How is ethylene glycol different from ethanol?

33. For what is isopropyl alcohol used?

34. Write the structural formula for the aromatic carboxylic acid benzoic acid.

35. What is the percentage of alcohol by volume in 96-proof whiskey?

36. Ethyl alcohol is produced by fermentation. Describe the two materials that react to form ethyl alcohol.

37. Why is ethyl alcohol used in cough syrup preparations rather than methyl alcohol?

38. What is the product of the oxidation of ethyl alcohol?

39. Write a chemical equation that shows why acetic acid is a Brønsted-Lowry acid as defined in Chapter 7.

40. What is the difference between an aldehyde and a ketone?

41. Acetaminophen is often taken in place of aspirin for a headache. Aspirin is a carboxylic acid and can cause stomach trouble. After reviewing the structure of acetaminophen in Figure 8.22, describe which functional groups it contains.

42. Draw the structure of:

a. formaldehyde **b.** ethyl acetate
c. ethyl alcohol **d.** acetone

43. Draw the structure of:

a. acetic acid **b.** methylamine
c. methyl alcohol **d.** phenol

44. Compare the odor of butyric acid with that of the ester methyl butyrate.

45. Identify each of the following as a primary, secondary, or tertiary amine.

a. ethylamine **b.** methylamine
c. dimethylamine **d.** diethylamine
e. trimethylamine **f.** triethylamine

46. Name the following functional groups:

a. $-NH_2$ **b.** $-CHO$
c. $-COOH$ **d.** $-COOCH_3$

47. Name the following functional groups:

a. $-CONH_2$ **b.** $-OH$
c. $-CO-$ **d.** $-O-$

48. Write the structural formula for each of the following:

a. four-carbon alcohol with a double bond at carbon 2

b. ketone containing three carbon atoms

c. 12-carbon carboxylic acid with a double bond at carbon 6

49. Write the structural formula for each of the following:

a. methyl ester of a four-carbon, straight-chain, carboxylic acid

b. tertiary amine with three ethyl groups

c. two-carbon aldehyde

d. methyl amide of propionic acid

50. Write structural formulas for the following aromatic compounds:

a. toluene

b. 1,3-dimethylbenzene (Are the methyl groups in the *ortho-*, *meta-*, or *para-* positions?)

c. *p*-dimethylbenzene

51. Draw structural formulas for the following aromatic compounds

a. *m*-dichlorobenzene

b. 1,2-dimethylbenzene

c. 3-bromotoluene

52. How many structural formulas can be written for tetrachlorobenzene?

53. How does benzene differ from cyclohexane?

54. When one of benzene's hydrogens is replaced by a methyl group, what is the name of the resulting molecule?

55. Draw and name the molecule that is formed by fusing two benzene rings together.

56. When ethyl groups are placed at the opposite ends of a benzene ring what position are they in?

Answers to Practice Exercises

8.1

```
    H   H   H   H   H   H   H
    |   |   |   |   |   |   |
H — C — C — C — C — C — C — C — H
    |   |   |   |   |   |   |
    H   H   H   H   H   H   H
```

8.2 The carbon skeletons are shown below. Add the hydrogen atoms and check that each isomer has 14 hydrogen atoms.

```
C—C—C—C—C—C      C—C—C—C—C
                       |
                       C

                         C
                         |
C—C—C—C—C       C—C—C—C
    |                    |
    C                    C

C—C—C—C
  |  |
  C  C
```

8.3 2-methylhexane

8.4 2,2,5,5-tetramethylhexane

8.5 1-octene

8.6

```
    H
    |
H — C — Br
    |
    H
```

8.7 HO—CH_2—CH_2—CH_3

8.8 **a.** ketone **b.** aldehyde **c.** ketone

8.9 **a.** CH_3COOH **b.** CH_3COOCH_3
c. $CH_3CH_2CH_2CH_2COOH$

8.10 **a.** tertiary **b.** secondary **c.** primary

8.11 Three alcohol groups (—OH) and one secondary amine

```
     H
    /
( —N      )
    \
     CH3
```

Chapter 9

Synthetic Polymers

Materials for the Modern Age

Chapter Objectives

In this chapter, you should gain an understanding of:

- Similarities and differences between natural and synthetic polymers
- The synthesis of macromolecules by addition, condensation, and rearrangement polymerization
- The structures and properties of common synthetic polymers
- Differences between silicon-based and carbon-based polymers
- The problems associated with the disposal of plastics

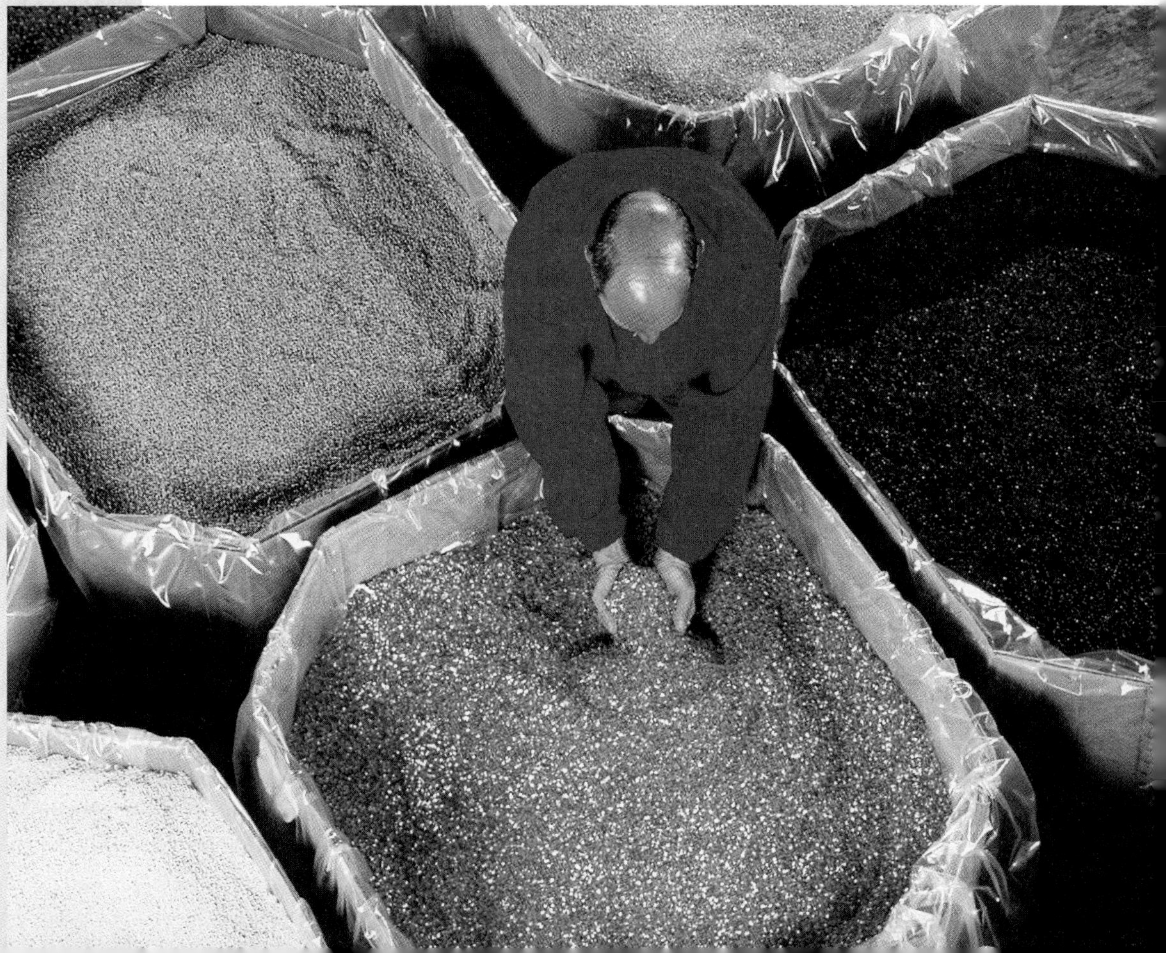

Synthetic polymers are found everywhere in modern society. These giant molecules are made into fabrics, tile, pipes, kitchenware, tires, toys, and numerous other products.

ONLINE FEATURES

www.jbpub.com/chemistry

- Chemistry and the Environment
- eLearning
- Chemistry in the News
- Special Topics
- Research & Reference Links
- Ask the Authors

DURING THE LAST 80 YEARS in the United States, the production of synthetic polymers—known to the general public as plastics—has developed into a multibillion-dollar industry that has revolutionized the way we live. Polyester, nylon, Dacron, Styrofoam, Teflon, Formica, and Saran are all very familiar. We use these, and many other synthetic polymers, in almost every aspect of our daily lives. In the home, synthetic polymers are used in roofing, wall paneling, carpeting, furniture, paint, appliances, dishes, and food packaging. The steering wheel, instrument panel, upholstery, tires, and other parts of our cars are made of polymers; many of the clothes we wear are made from synthetic polymer fibers. In the United States, annual production of synthetic polymers is more than 90 kilograms (200 pounds) per person, an amount that has exceeded the average annual production of steel since 1976. We are indeed living in the "age of polymers."

The great surge in the production of polymers occurred in response to several factors, including the need to find alternatives for dwindling supplies of natural raw materials, such as wood, rubber, and some metal ores, and the increasing ability of chemists to make polymers that, for many purposes, were superior to materials normally used. Certain polymers, for example, are stronger than steel, and many have specialized properties that natural materials do not possess. The cheap and plentiful supply of petroleum, which is the main source of raw materials for polymer manufacturing, made it less costly to make products by molding plastics than by shaping and joining wood or metals.

In this chapter, we examine the basic chemical reactions that lead to the formation of different types of polymers. We consider the structural differences that account for the diverse properties of polymers and examine some of the ways in which common polymers are used. We also discuss the problem of disposing of the mounting volume of discarded plastic products.

9.1 Polymers: Characteristics and Types

The word *polymer* is derived from the Greek words *poly*, meaning "many," and *meros*, meaning "parts."

Polymers are giant molecules with formula masses that range from several thousand to over a million atomic mass units (amu). These macromolecules are composed of thousands of smaller repeating units called **monomers** that are linked together to form long chains and, in some cases, complex three-dimensional networks. Varying the building blocks (the monomers) allows chemists to synthesize an incredible number of polymers with a wide range of properties.

Homopolymers and Copolymers

A polymer made from one kind of monomer is called a **homopolymer** (Fig. 9.1a). A polymer made from two or more different monomers is called a **copolymer.** The monomers in a copolymer may be linked in regular or irregular patterns. For example, depending on the reaction conditions for polymerization, two different monomers can alternate regularly or randomly. In still other arrangements, blocks of one monomer can be introduced at intervals along a chain of another monomer, or they can be grafted on as side chains (Fig. 9.1b).

The word *plastic* is derived from the Greek word *plastikos*, meaning "capable of being molded or shaped."

Plastics

Polymers are frequently—but often incorrectly—called *plastics.* While all plastics are polymers, not all polymers are plastics. A **plastic,** in polymer terminology, is a material that, at some point in its manufacture, is in a sufficiently fluid state to allow it to be molded into various shapes.

a. Homopolymer

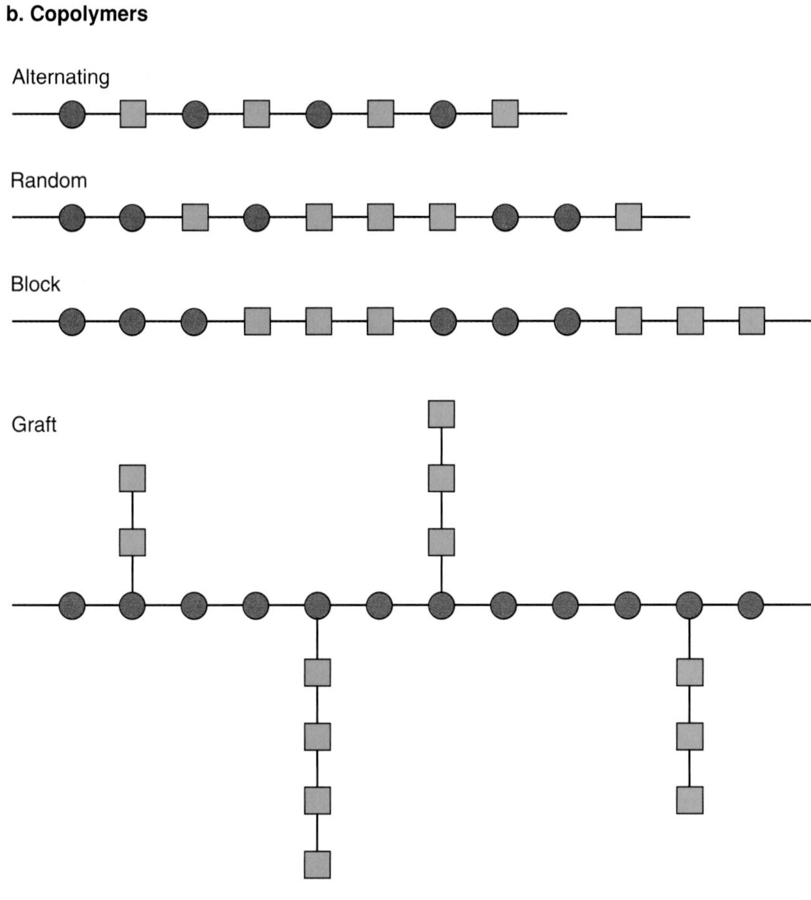

FIGURE 9.1

The basic structures of (a) homopolymers and (b) copolymers.

Plastics can be classified as either thermoplastic or thermosetting. **Thermoplastics** can be repeatedly softened by heating and molded into shapes that harden on cooling. **Thermosetting plastics** also soften when first heated and can be molded into shapes that they retain on cooling. When reheated, however, they do not become soft again but instead degrade or decompose.

Natural Polymers

Many polymers, both inorganic and organic, are found in the natural world. Silicates (Chapter 2), which make up so much of the earth's rocky surface, are polymers constructed from silicon dioxide (SiO_2) monomers. Diamond and graphite (Chapter 8) are polymers in which the monomer is a carbon atom. Other naturally occurring polymers include starches, cellulose, proteins, the nucleic acids DNA and RNA (Chapter 14), and natural rubber.

Until the advent of synthetic polymers, building materials, furnishings, and clothing were mostly made of natural materials. These materials were primarily wood, cotton, and linen (cellulose polymers), wool, silk, and leather (protein polymers), and natural rubber.

Early Synthetic Polymers

By the latter half of the 19th century, huge numbers of elephants were being slaughtered to meet the enormous demand for ivory. In 1863, a New England billiard-ball manufacturer offered a $10,000 reward to the person who could develop an ivory substitute. Hyatt was pursuing this prize when he developed celluloid, but ironically he never won it.

The early synthetic polymers were designed as substitutes for natural materials and were manufactured using natural starting materials. The first synthetic plastic was developed in 1869 by the American inventor John Wesley Hyatt. His product, which he called *celluloid*, was made from cellulose nitrate obtained by treating natural cellulose with nitric acid. Celluloid was used to make billiard balls, piano keys, combs, stiff collars, toys, and motion picture film. It was regarded primarily as a substitute for ivory and tortoise shell. Despite its flammability, celluloid remained popular until the end of the 19th century, when it began to be replaced gradually with safer synthetic materials.

The first artificial fibers that could be woven into fabric were also made from cellulose nitrate. These shiny fibers were developed as a replacement for silk and given the name *rayon*. In modified form, rayon is still produced today.

By the early part of the 20th century, the chemical industry had discovered that synthetic polymers could be made more efficiently by using small molecules as starting materials than by altering natural polymers. The first completely synthetic plastic, Bakelite, was made in 1909. It will be discussed later, in the section on condensation polymers.

General Types of Polymers

Based on how they are formed, there are three general types of polymers: (1) addition polymers, (2) condensation polymers, and (3) rearrangement polymers. **Addition polymers** are formed when monomers link together without the loss of any of their atoms. Important addition polymers that we will consider are polyethylene, poly(vinyl chloride) (PVC), polystyrene (Styrofoam), polyacrylonitrile (Dacron, Acrilan), polytetrafluoroethylene (Teflon), and synthetic rubber. **Condensation polymers** are formed when monomer units, in linking together, eliminate a small molecule, usually water. Examples of condensation polymers are polyesters, polyamides (nylon), and formaldehyde resins (Bakelite, Formica). In forming **rearrangement polymers,** monomers link together without losing any atoms; in contrast to the formation of addition polymers, however, some of the monomers' atoms are rearranged in the process of linking. The polyurethane known as *foam rubber* is the most common example of a rearrangement polymer. We will consider each of the three general types of polymers in turn.

9.2 Addition Polymers

Polyethylene

The simplest addition polymer is **polyethylene,** which is produced in greater quantity than any other polymer. In 1996, more than 11 billion kilograms (24 billion pounds) were produced in the United States, or approximately 75 pounds per person. Polyethylene is a very stable material that is resistant to chemical attack, and it is cheap to produce.

Polyethylene is made from ethylene (ethene) (Chapter 8), which is obtained from the cracking of petroleum (Chapter 12). At high temperature and pressure, and in the presence of a catalyst, ethylene molecules link together to form long chains of single-bonded carbon. No atoms in the monomers are lost as polymerization occurs:

$$\mathrm{H_2C{=}CH_2} \xrightarrow[\text{catalyst, high temperature, high pressure}]{\text{polymerization}} -\mathrm{CH_2{-}CH_2{-}CH_2{-}CH_2{-}CH_2{-}CH_2{-}CH_2{-}CH_2}- \quad \text{or} \quad \left(\mathrm{CH_2{-}CH_2}\right)_n$$

ethylene monomer — polyethylene — repeating unit

The double bond in ethylene contains two shared pairs of electrons. As polymerization occurs, one pair continues to be shared by the two carbon atoms in the monomer, forming a single bond between them; the other electron pair is used to bond with another monomer unit.

It is not possible to write an exact formula for polyethylene (or any other polymer) because the number of carbon atoms in the chain can vary from several hundred to as many as 50,000. The notation $-(CH_2-CH_2)_n-$ indicates that the unit inside the parentheses is repeated n times, where n is a very large number.

Polyethylene can take various forms, depending on the conditions used for processing. Varying the catalyst and altering the pressure or temperature allow polyethylene to take the form of long filaments, thin flexible sheets, or solid shapes. The very different properties of these various forms of polyethylene reflect structural differences in the polymer chains. At relatively low temperature and pressure and in the presence of a chromium oxide catalyst, long, unbranched chains (n = approximately 10,000) predominate. These zig-zag chains, shown in **Figure 9.2a**, can pack closely together to give an orderly, uniform arrangement. The product is **high-density polyethylene (HDPE),** which is hard, tough, and rigid.

During World War II, polyethylene was used to manufacture electrical wiring in radar sets. The invention of radar enabled British pilots to detect enemy aircraft before they became visible and was a major factor in the Allied defeat of Germany.

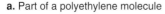

HDPE accounts for one-third of all polyethylene produced in the United States. It is used to make toys, barrels and plungers for syringes, pipe, bottles, and containers for many liquid household products such as bleach and detergent. Nearly all 1-gallon milk and juice containers in the United States are made of HDPE **(Fig. 9.2b)**. HDPE is quite strong and resistant to breakage, but it can be crushed fairly easily.

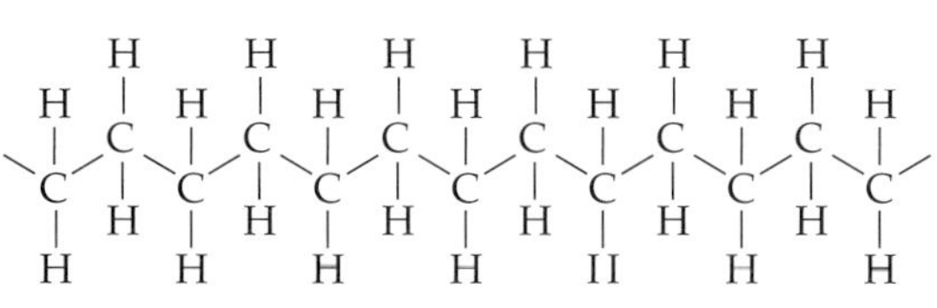
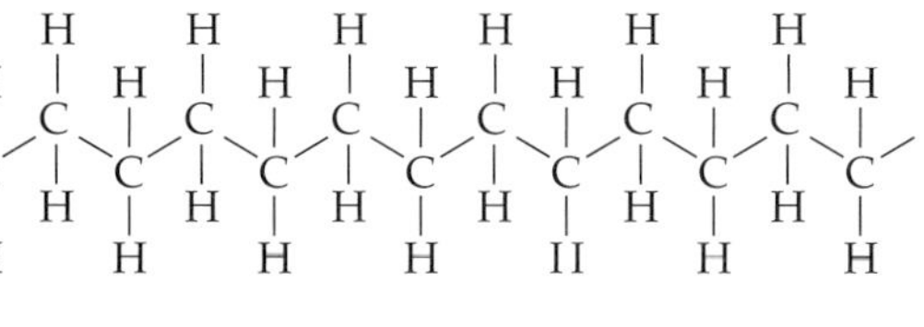

a. Part of a polyethylene molecule

b.

FIGURE 9.2

(a) High-density polyethylene (HDPE) consists of linear (unbranched) chains. (b) Containers for orange juice and milk are made from HDPE.

At high temperature and pressure with a different catalyst, shorter chains (n = approximately 500) with irregular branching are formed **(Fig. 9.3a)**. Because of the branching, these chains cannot pack as closely together as those of HDPE, and the product is **low-density polyethylene (LDPE),** a material that is waxy, transparent (or translucent), and very flexible. It can be formed into clear thin sheets that are used as food wrap. It is also made into trash bags, grocery bags **(Fig. 9.3b)**, shopping bags of various kinds, squeeze bottles, and flexible tubing. When this material is stretched, the short polymer chains move easily past each other; as they are pulled apart, the sheets tear.

Reaction conditions can be adjusted to give a type of LDPE in which the carbon chains are frequently cross-linked **(Fig. 9.3c)**. This weblike structure makes the material extremely strong and very difficult to tear. It is used to make threaded bottle caps and plastic ice at skating rinks.

In the same way that $-CH_3$ represents a methyl group, $-CH{=}CH_2$ represents a vinyl group.

Substituted Polyethylenes

So far, we've seen that changing structural features such as chain length and degree of branching and cross-linking produces polyethylene with different properties. Polymers with an even wider range of properties can be made if one or more of the hydrogen atoms in the ethylene monomer are replaced by other atoms or groups of atoms. As is the case for the ethylene-based polymers, the carbon-carbon double bond ($-C{=}C-$) is the key to the formation of substituted polyethylenes.

The carcinogenicity of vinyl chloride was first suspected in the 1960s when a high incidence of cancer was reported among workers exposed to it. The risk increased with the length of time of exposure.

The three substituted polyethylenes produced in greatest quantity are (1) poly(vinyl chloride) (PVC), (2) polypropylene, and (3) polystyrene. Others produced in smaller quantities, are best known by their trade names: Saran, Teflon, Dacron, and Acrilan.

Poly(vinyl chloride), or **PVC,** is produced from vinyl chloride (ethylene with one chlorine substituted for a hydrogen).

a. Short, branched chains

b.

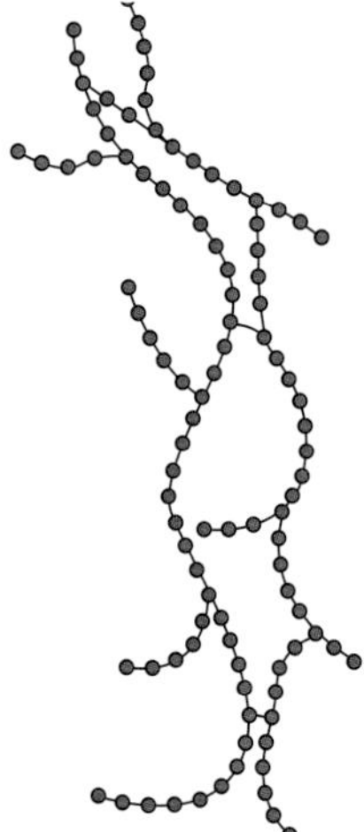

c. Branched cross-linked chains

FIGURE 9.3

(a) Low-density polyethylene (LDPE) with short, branched chains is very flexible. (b) Grocery bags and bread wrappers are made of this type of LDPE. (c) LDPE made with cross-linked branched chains is extremely hard and strong.

$$\mathrm{CH_2{=}CHCl} \xrightarrow{\text{polymerization}} \mathrm{-CH_2-CHCl-CH_2-CHCl-CH_2-CHCl-} \quad \text{or} \quad \left(\mathrm{-CH_2-CHCl-}\right)_n$$

vinyl chloride monomer; poly(vinyl chloride) (PVC); repeating unit

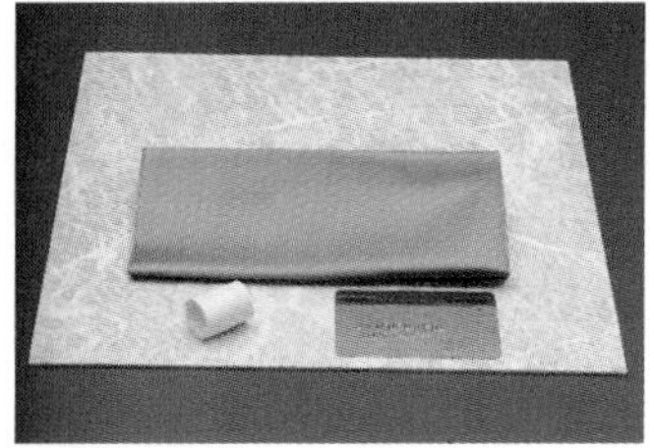

FIGURE 9.4

Items made from poly(vinyl chloride) (PVC) include floor tiles and credit cards.

Poly(vinyl chloride), the second most widely used polymer, is lightweight, hard, and resistant to chemicals. It is used to make plumbing pipes, floor tiles, credit cards, garden hoses, trash bags, and imitation leather **(Fig. 9.4)**.

If vinyl chloride is polymerized along with an excess of dichloro-substituted ethylene, known as vinylidene chloride (1,1-dichloroethene), a copolymer, which is marketed as Saran and other kinds of plastic wrap, is produced. The two monomers alternate randomly along the polymer chain. This polymer is used primarily to wrap food **(Fig. 9.5)**.

$$m\,\mathrm{CH_2{=}CCl_2} + n\,\mathrm{CH_2{=}CHCl} \xrightarrow{\text{polymerization}} \left(\mathrm{-CH_2-CCl_2-}\right)_m\left(\mathrm{-CH_2-CHCl-}\right)_n$$

vinylidene chloride (1,1-dichloroethene); vinyl chloride; Saran

FIGURE 9.5

The plastic film Saran, used extensively for wrapping food, is a copolymer of vinylidene chloride and vinyl chloride.

Polypropylene, made from methyl-substituted ethylene, is an exceptionally strong material.

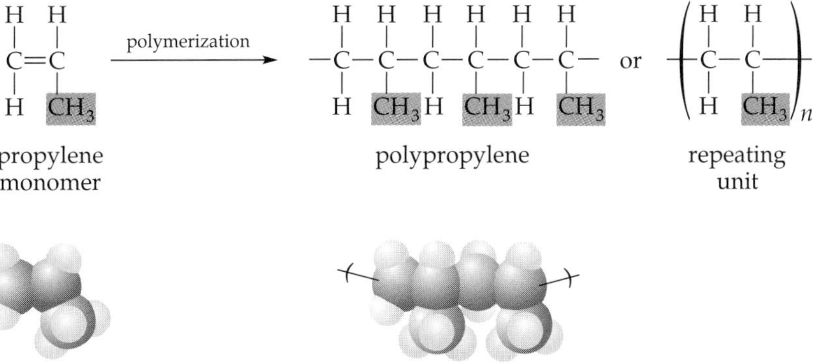

It is used for making auto battery casings and many household items **(Fig. 9.6)**. Because of its strength, fibers made from it are also used in outdoor carpeting, fishing nets, and rope.

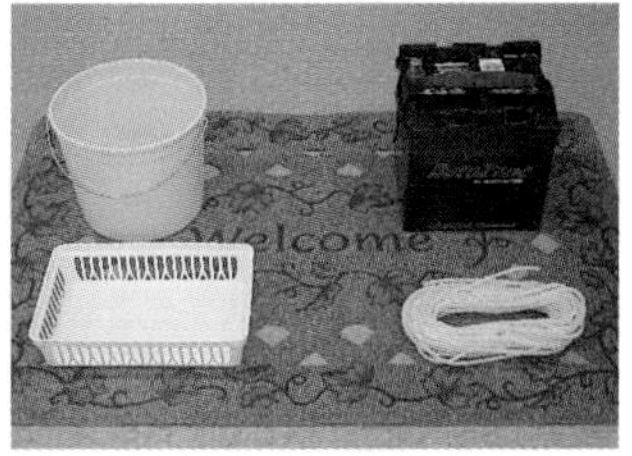

FIGURE 9.6

The exceptionally strong polymer polypropylene is used to make auto battery casings and other items.

Polystyrene, which was first prepared around 1920 by the German chemist Hermann Staudinger, is made by the polymerization of styrene. Styrene is a derivative of ethylene in which one hydrogen atom is replaced by a benzene ring.

$$\underset{\text{styrene monomer}}{CH_2{=}CH{-}C_6H_5} \xrightarrow{\text{polymerization}} \underset{\text{polystyrene}}{-CH_2{-}CH(C_6H_5){-}CH_2{-}CH(C_6H_5){-}CH_2{-}CH(C_6H_5){-}} \text{ or } \underset{\text{repeating unit}}{\left(\!-CH_2{-}CH(C_6H_5){-}\!\right)_n}$$

The benzene rings in polystyrene make it more rigid than polyethylene. As a result, polystyrene tends to be brittle and easily shattered. Clear, hard polystyrene is used to make plastic glasses, containers, wall tile, and appliance parts **(Fig. 9.7a)**.

Polystyrene can also be made into lightweight solid foams by adding a chemical agent that produces gas bubbles in molten styrene. These foams make good insulating and packing materials and are used in picnic coolers, egg cartons, and cups for keeping drinks hot and cold. Foam sold under the trade name Styrofoam is used as a cushioning material for shipping containers **(Fig. 9.7b)** and for insulating buildings.

Polytetrafluoroethylene (marketed as Teflon) is formed from tetrafluoroethylene, a derivative of ethylene in which all four hydrogen atoms are replaced with fluorine atoms:

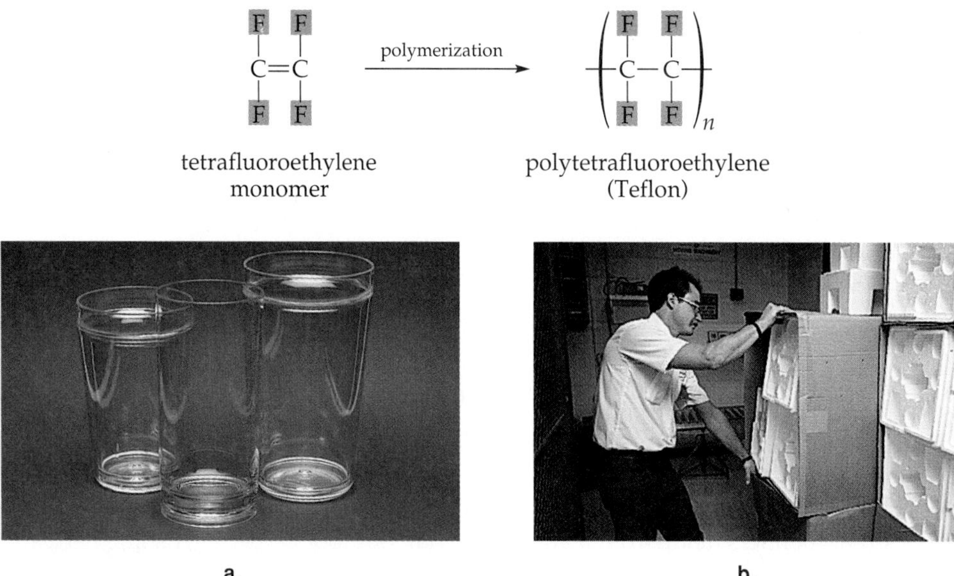

FIGURE 9.7

(a) Clear polystyrene glasses. (b) The white packing material used to protect computers and other items is made of polystyrene foam.

The presence of the fluorine atoms makes this polymer much more resistant to heat and chemical attack than polyethylene is. These properties make Teflon very suitable for use in gaskets, bearings, and machine parts that must operate in corrosive environments. Its best-known use, however, is as a nonstick coating for cooking utensils.

Polyacrylonitrile is a polymer of acrylonitrile, a molecule in which a nitrile group ($-C{\equiv}N$) is substituted for one of the hydrogen atoms in ethylene.

$$\underset{\text{acrylonitrile monomer}}{H_2C{=}CH{-}C{\equiv}N} \xrightarrow{\text{polymerization}} \underset{\text{polyacrylonitrile}}{\left(CH_2{-}CH(CN) \right)_n}$$

Polyacrylonitrile is processed into tough fibers for carpeting and fabrics. Because the nitrile group is polar, the polymer chains are attracted to one another; this attraction makes the fibers stronger than those formed from polymers made of nonpolar ethylene. The acrylic fibers sold under the trade names Orlon, Acrilan, and Creslan are copolymers of acrylonitrile with varying percentages (25–85%) of vinyl chloride.

Some important addition polymers are listed in **Table 9.1**, which shows the monomers from which they are formed and some of their uses.

Table 9.1 Some Addition Polymers

Monomer	Polymer	Typical Uses
$CH_2{=}CH_2$ ethylene (ethene)	$\left(CH_2{-}CH_2 \right)_n$ polyethylene	Containers, pipes, bags, toys, wire insulation, bottle caps
$CH_2{=}CHCH_3$ propylene (propene)	$\left(CH_2{-}CH(CH_3) \right)_n$ polypropylene	Fibers for carpets, artificial turf, rope, fishing nets, automobile trim
$CH_2{=}CHCl$ vinyl chloride (chloroethene)	$\left(CH_2{-}CH(Cl) \right)_n$ poly(vinyl chloride) (PVC)	Garden hoses, floor tiles, plumbing, artificial leather, food wrap, credit cards
$CH_2{=}CHCN$ acrylonitrile	$\left(CH_2{-}CH(CN) \right)_n$ polyacrylonitrile (Orlon, Acrilan)	Fibers for cloth, carpets, upholstery
$CH_2{=}CH{-}C_6H_5$ styrene	$\left(CH_2{-}CH(C_6H_5) \right)_n$ polystyrene	Styrofoam, hot-drink cups, insulation, packaging
$CF_2{=}CF_2$ tetrafluoroethylene	$\left(CF_2{-}CF_2 \right)_n$ Teflon	Nonstick coating for kitchen utensils

EXAMPLE 9.1 Drawing polymer structures

Draw the structure of the addition polymer formed from the monomer CH_2CH-R (R can be an atom or group, for example, Cl or CH_3). Show three segments.

Solution The structural formula of the monomer is

```
H   H
|   |
C = C
|   |
H   R
```

(Each carbon atom forms four bonds.) No atoms are lost in the formation of an addition polymer. Two of the electrons in each carbon-carbon double bond of the monomer are used to bond to adjacent monomers to give a polymer with only single bonds. The polymer is

```
  H   H   H   H   H   H              /H   H\
  |   |   |   |   |   |              ||   ||
 -C - C - C - C - C - C-     or     -|C - C|-
  |   |   |   |   |   |              ||   ||
  H   R   H   R   H   R              \H   R/n
```

Practice Exercise 9.1

Draw the structure showing four segments of the addition polymer that can be constructed from the monomer CH_2CCl_2.

9.3 Elastomers

Elastomers are synthetic polymers with properties similar to those of natural rubber. As their name implies, they are elastic; like rubber, they have the ability to return quickly to their original form after being stretched.

Natural Rubber

An emulsion is a suspension of submicroscopic particles evenly dispersed in a liquid.

Natural rubber is an addition polymer that is obtained as a milky white fluid called *latex* from the tropical rubber tree *Hevea brasiliensis*. The latex, which oozes out when the tree is cut, is an emulsion of rubber particles in water. The rubber can be precipitated from the latex and then processed.

Natural rubber is a polymer of the monomer *isoprene* (2-methyl-1,3-butadiene).

```
CH2     CH2
  \\    //
    C - C
   /     \
  H       CH3
```

isoprene

The people of Central and South America were using rubber balls for games long before Christopher Columbus observed the practice in 1496. According to one astonished witness, the balls rebounded so much that they appeared to be alive.

Isoprene has two double bonds. When it polymerizes, the repeating units have one double bond. The monomers can link together in a *cis-* or *trans-* arrangement (Chapter 8) with the linked $-CH_2-$ groups on the same or on opposite sides of the double bond.

$$-CH_2 \quad CH_2-CH_2 \quad CH_2-CH_2 \quad CH_2-$$
$$C=C \qquad C=C \qquad C=C$$
$$CH_3 \quad H \quad CH_3 \quad H \quad CH_3 \quad H$$

poly-*cis*-isoprene

$$-CH_2 \quad H \quad CH_3 \quad CH_2-CH_2 \quad H$$
$$C=C \qquad C=C \qquad C=C$$
$$CH_3 \quad CH_2-CH_2 \quad H \quad CH_3 \quad CH_2-$$

poly-*trans*-isoprene

In natural rubber, the linkages are all *cis*- in their orientation.

Natural rubber is a soft, water-repellent, elastic material that becomes sticky when warmed and stiff when cooled. If untreated, it is of limited use, and until the middle of the 19th century, it was used primarily in pencil erasers.

In 1839, Charles Goodyear (1800–1860), who had been working for many years to find ways to improve the qualities of rubber, accidentally spilled a mixture of rubber and sulfur on a hot surface. To his surprise, he found that the rubber had become tougher and more resistant to heat and cold, and its elasticity had improved. The process Goodyear discovered was later named **vulcanization** after Vulcan, the Roman god of fire. Vulcanization made natural rubber suitable for numerous applications, including automobile tires, which made it a major factor in the development of the automobile industry.

Rubber owes its elasticity to the flexibility of its randomly intertwined long chains. When rubber is stretched, the chains straighten; when the tension is released, the chains recoil. If the rubber is stretched enough, the adjacent chains slip past each other, and the rubber loses its elasticity. During vulcanization, short chains of sulfur atoms of varying lengths form links between the polymer chains **(Fig. 9.8a)**. When stretched, the chains can still untwine and straighten, but when the tension is released, the sulfur cross-links ensure that the chains spring back to their original positions **(Fig. 9.8b)**. The sulfur linkages hold the chains together and prevent them from slipping past each other when stretched.

Poly-*trans*-isoprene is present in *gutta percha*, a natural material found in Malaysia and South America. Not as elastic as rubber, it is used in golfball covers and as insulation for underwater electrical equipment and cables.

Rubber is said to have been given its name by Joseph Priestley (Chapter 3), who found that it could be used to rub out (erase) lead pencil marks.

EXPLORATIONS looks at a type of car that may be key to the future of the automobile industry (see pp. 274–275).

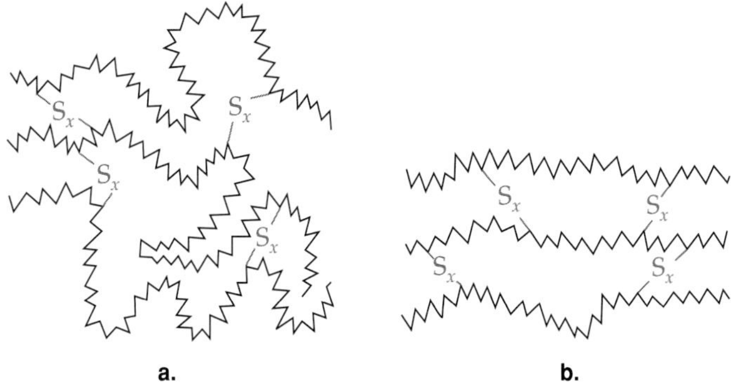

FIGURE 9.8

(a) When rubber is vulcanized, short chains consisting of varying numbers of sulfur atoms (indicated by the subscript *x*) form cross-links between adjacent hydrocarbon chains. (b) When stretched, the hydrocarbon chains straighten out but cannot slip past each other because of the sulfur cross-links. When the tension is released, the chains become coiled again, and the rubber resumes its original shape.

Synthetic Rubber

The first commercially successful synthetic elastomer to be produced in the United States was **neoprene.** Neoprene is an addition polymer of 2-chlorobutadiene, a derivative of isoprene in which a chlorine atom is substituted for the methyl group.

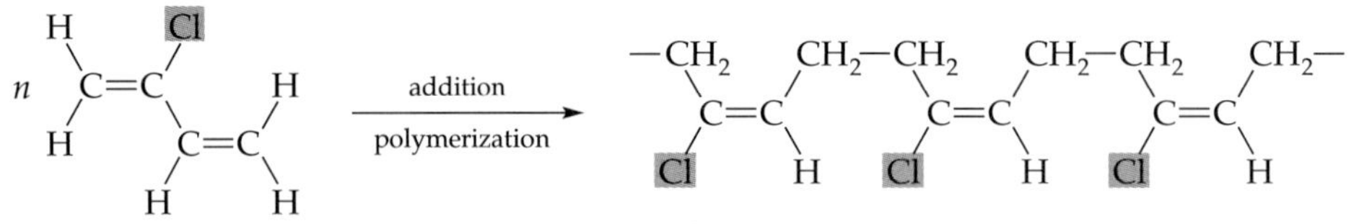

addition polymerization

neoprene

2-chlorobutadiene

Neoprene is resistant to abrasion and to oils, gasoline, and other solvents. It is a useful material for gasoline hoses, garden hoses, gaskets, shoe soles, adhesives, and conveyor belts, but it is not suitable for automobile tires.

The most important synthetic rubber is styrene-butadiene rubber (SBR), which was developed during World War II when supplies of imported natural rubber were cut off. SBR is a block copolymer made by mixing styrene and 1,3-butadiene in a 1 to 3 ratio.

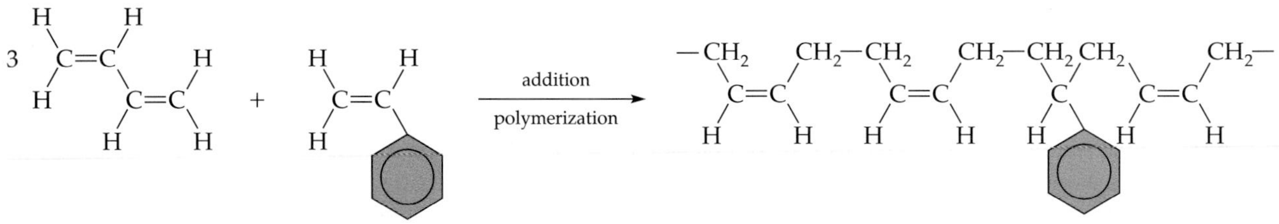

addition polymerization

1,3-butadiene

styrene

styrene-butadiene rubber (SBR)

SBR is more resistant to abrasion and oxidation than natural rubber is. Like natural rubber, it can be vulcanized to improve its properties. SBR, which accounts for more than 40% of synthetic rubber production, is used primarily for manufacturing tires.

Worldwide, tires and tire products account for more than half the combined production of natural and synthetic rubber. A typical tire is only about 60% rubber. The other main ingredient is carbon black, which is added to increase strength and resistance to abrasion. Steel belts, nylon cord, and fiberglass give tires added strength.

Synthetic poly-*cis*-isoprene, structurally identical to natural rubber, can be produced on an industrial scale. However, because of the cost of the raw materials, which are obtained from petroleum, this synthetic material is more expensive to produce than is natural rubber.

9.4 Polymers in Paints

An important result of the tremendous growth of the plastics industry after World War II was the development of water-based paints, which revolutionized the paint industry. Water-based and oil-based paints are composed of three major kinds of ingredients: (1) a pigment, (2) a solvent, and (3) a binder.

The *pigment* provides the desired color. The primary pigment in a paint is usually titanium oxide (TiO_2) or zinc oxide (ZnO), compounds that are white and opaque and thus serve to cover any existing color on the surface to be painted. White lead [$2\,PbCO_3 \cdot Pb(OH)_2$] had been used extensively as a primary pigment until 1977, when it was banned for interior use because of its toxicity (Chapters 10 and 18). Pigments such as carbon black, iron oxide (Fe_2O_3, brown and red), cadmium sulfide (CdS, orange), chromium oxide (Cr_2O_3, green), and various organic compounds are mixed with the primary pigment to obtain the desired color.

Carbon black is simply very small particles of carbon; it is prepared by burning various materials, including fats, oils, wood, and vegetables, under suitable conditions.

The *solvent* in a paint is the liquid in which the pigment, binder, and other ingredients are suspended, usually in the form of an emulsion. Once the paint has been spread, the solvent evaporates, and the paint dries. In oil-based paints, the solvent is usually a mixture of hydrocarbons obtained from petroleum. A compatible organic solvent must be used for thinning the paint and cleaning up after painting. This is usually turpentine, which, like most other suitable solvents, gives off hazardous fumes. In water-based paints, the solvent is water, which can be used both for thinning and for cleaning up.

The *binder* is a material that polymerizes and hardens as the paint dries, forming a continuous film that holds the paint to the painted surface. The pigment becomes trapped within the polymer network. In oil-based paints, the binder is usually linseed oil, but coconut, soybean, and castor oils are also used. As the solvent evaporates and the paint hardens, the oil reacts with atmospheric oxygen to form a cross-linked polymeric material.

The binder in water-based paints is made by mixing a partially polymerized material with some of its unreacted monomer. The slightly rubbery nature of the partially polymerized material explains why water-based paints are called *latex paints.* As the water evaporates, further polymerization occurs, and the paint hardens.

Several synthetic polymers are used as binders in latex and acrylic paints. Paints with poly(vinyl acetate) binders account for about 50% of the paint used for interior work. Those with poly(styrene-butadiene) binders are less expensive but do not adhere quite as well and have a tendency to yellow. Acrylic paints, made with acrylonitrile binders, are considerably more expensive than other types of paints, but they are excellent for exterior work. They adhere well and are washable, very durable, and resistant to damage by sunlight. In most respects, they are superior to oil-based paints, which they have largely replaced.

9.5 Condensation Polymers

A *condensation reaction* is one in which two molecules combine to form a larger molecule with the elimination of a small molecule, such as water. An example is the condensation reaction of a carboxylic acid and an alcohol to form an ester and water.

$$CH_3-\overset{\overset{\displaystyle O}{\|}}{C}-O-H + HO-CH_2-CH_3 \longrightarrow CH_3-\overset{\overset{\displaystyle O}{\|}}{C}-O-CH_2-CH_3 + H_2O$$

ester linkage

acetic acid (carboxylic acid) | ethanol (alcohol) | ethyl acetate (ester) | water

This type of reaction depends on the presence of two different functional groups on the reacting molecules—a carboxylic acid group (—COOH) and a hydroxyl group (—OH), in the above example. If the reacting molecules are bifunctional (if they have chemically reactive groups on *both* ends), they can polymerize and become linked end-to-end to form long chains. Both polyesters and polyamides are formed in this way.

Polyesters

The term *polyester* is commonly used to describe fabric made from PET fibers.

Polyesters are condensation copolymers of a diol (an alcohol with two —OH groups) and dicarboxylic acid (a carboxylic acid with two —COOH groups). The most important polyester is *poly(ethylene terephthalate) (PET)*, which is an alternating copolymer of ethylene glycol (Chapter 8) and terephthalic acid. The formation of the polymer occurs as follows:

$$n\ \text{H—O—CH}_2\text{—CH}_2\text{—O—H} + n\ \text{HO—}\overset{\text{O}}{\overset{\|}{\text{C}}}\text{—C}_6\text{H}_4\text{—}\overset{\text{O}}{\overset{\|}{\text{C}}}\text{—OH}$$

ethylene glycol — terephthalic acid

↓ condensation polymerization

$$\text{H—O—CH}_2\text{—CH}_2\text{—O}\left(\overset{\text{O}}{\overset{\|}{\text{C}}}\text{—C}_6\text{H}_4\text{—}\overset{\text{O}}{\overset{\|}{\text{C}}}\text{—O—CH}_2\text{—CH}_2\text{—O}\right)_n\overset{\text{O}}{\overset{\|}{\text{C}}}\text{—C}_6\text{H}_4\text{—}\overset{\text{O}}{\overset{\|}{\text{C}}}\text{—OH} + n\ \text{H}_2\text{O}$$

ester linkage

poly(ethylene terephthalate) (PET)

FIGURE 9.9
Soft drink bottles and other items made of poly(ethylene terephthalate) (PET).

As each linkage is formed, a water molecule is eliminated.

PET can be processed into fibers, rigid solids, and film. The textile fibers marketed as Dacron account for 50% of all synthetic fibers and are used primarily in clothing. Blended with cotton, Dacron forms a fabric that requires no ironing after washing. Dacron tubing can be used to replace diseased blood vessels in open-heart surgery.

Cross-linked PET is used in making soft drink bottles. The cross-links make the polymer strong enough to withstand the pressure exerted by the carbon dioxide gas in the soft drink **(Fig. 9.9)**. PET can also be made into a very strong, very thin film sold as Mylar, which is used for packaging frozen food and as a skin substitute in the treatment of burn victims. Magnetized Mylar is used for recording tape.

> Nylon 66 was given the number 66 because each of the two monomer molecules from which it is made has six carbon atoms.

Polyamides

Polyamides, which are better known as *nylons,* are condensation copolymers of a diamine and a dicarboxylic acid. One of the most important of these copolymers is nylon 66, which was the first completely synthetic fiber to be produced. It was discovered in 1931 by Wallace H. Carothers, a chemist working for DuPont.

$$\text{HO—}\overset{\text{O}}{\overset{\|}{\text{C}}}\text{—(CH}_2)_4\text{—}\overset{\text{O}}{\overset{\|}{\text{C}}}\text{—OH} + \text{H—}\overset{\text{H}}{\overset{|}{\text{N}}}\text{—(CH}_2)_6\text{—}\overset{\text{H}}{\overset{|}{\text{N}}}\text{—H} + \text{HO—}\overset{\text{O}}{\overset{\|}{\text{C}}}\text{—(CH}_2)_4\text{—}\overset{\text{O}}{\overset{\|}{\text{C}}}\text{—OH} + \text{H—}\overset{\text{H}}{\overset{|}{\text{N}}}\text{—(CH}_2)_6\text{—}\overset{\text{H}}{\overset{|}{\text{N}}}\text{—H}$$

adipic acid — 1,6-diaminohexane — adipic acid — 1,6-diaminohexane

↓

$$\text{—}\overset{\text{O}}{\overset{\|}{\text{C}}}\text{—(CH}_2)_4\text{—}\overset{\text{O}}{\overset{\|}{\text{C}}}\text{—}\overset{\text{H}}{\overset{|}{\text{N}}}\text{—(CH}_2)_6\text{—}\overset{\text{H}}{\overset{|}{\text{N}}}\text{—}\overset{\text{O}}{\overset{\|}{\text{C}}}\text{—(CH}_2)_4\text{—}\overset{\text{O}}{\overset{\|}{\text{C}}}\text{—}\overset{\text{H}}{\overset{|}{\text{N}}}\text{—(CH}_2)_6\text{—}\overset{\text{H}}{\overset{|}{\text{N}}}\text{—} + 3\ \text{H}_2\text{O}$$

amide linkage

nylon 66

FIGURE 9.10

When a solution of adipoyl chloride (a derivative of adipic acid) in hexane is poured gently onto a solution of 1,6-diaminohexane in water, a white film of nylon forms at the interface between the two liquid layers. The film can be pulled up as a string and wound around a stirring rod.

Nylon 66 is prepared from the diamine 1,6-diaminohexane and the dicarboxylic acid adipic acid. As the two monomers react, a molecule of water is eliminated, and an amide linkage is formed. With some modifications, the polymerization reaction can be carried out in the laboratory, as shown in **Figure 9.10**.

The second most important polyamide is nylon 6, a polymer made from a single repeating monomer—aminocaproic acid (which has six carbon atoms). The amino end of one molecule reacts with the carboxylic acid end of the next monomer:

$$\mathrm{H{-}\overset{\overset{H}{|}}{N}{-}(CH_2)_5{-}\overset{\overset{O}{\|}}{C}{-}OH + H{-}\overset{\overset{H}{|}}{N}{-}(CH_2)_5{-}\overset{\overset{O}{\|}}{C}{-}OH + H{-}\overset{\overset{H}{|}}{N}{-}(CH_2)_5{-}\overset{\overset{O}{\|}}{C}{-}OH}$$

aminocaproic acid monomers

$$\downarrow$$

$$\mathrm{{-}\overset{\overset{H}{|}}{N}{-}(CH_2)_5{-}\overset{\overset{O}{\|}}{C}{-}\overset{\overset{H}{|}}{N}{-}(CH_2)_5{-}\overset{\overset{O}{\|}}{C}{-}\overset{\overset{H}{|}}{N}{-}(CH_2)_5{-}\overset{\overset{O}{\|}}{C}{-}} + 2\,\mathrm{H_2O}$$

amide linkage

nylon 6

Nylons can be made into fibers that are stronger, more durable, more chemically inert, and cheaper to produce than wool, silk, or cotton. Nylons are used for carpeting, clothing (particularly hosiery), tire cords, rope, surgical sutures, and parachutes. Nylons can also be molded to make electrical parts, valves, and various machine parts.

When first introduced to the public in 1939, nylon stockings were an immediate success. However, production of hosiery had to be discontinued during World War II, when all nylon was reserved for military use, particularly for making parachutes. Once the war ended, the industry had difficulty keeping up with the enormous demand for nylon. Nylon stockings were sheer and more durable and less expensive than silk stockings; after the restrictions of the war years, women were eager to buy them.

The natural protein fibers silk and wool, like the synthetic nylons, are polyamides and are chemically very similar to nylon 6 **(Fig. 9.11)**. As we shall see in Chapter 14, proteins are polymers of amino acids linked by amide bonds, which in protein chemistry are called *peptide bonds*. Thus, all proteins are polyamides.

The extensively cross-linked polyamide Kevlar is 20 times stronger than steel, and Kevlar ropes have replaced steel ropes in many applications.

Unlike natural fibers, nylon, Dacron, Orlon, and other synthetic fibers are resistant to moths, mildew, and wrinkling.

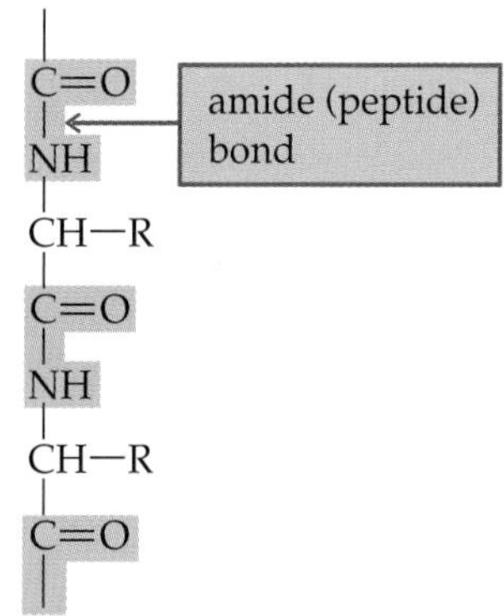

FIGURE 9.11

The protein keratin in wool is a natural polymer. Like nylon, it is formed from monomers joined via amide bonds.

EXAMPLE 9.2 Identifying Monomer Units in Copolymers

A polyester is a condensation polymer formed from a diol and a dicarboxylic acid. The structure of a typical polyester is

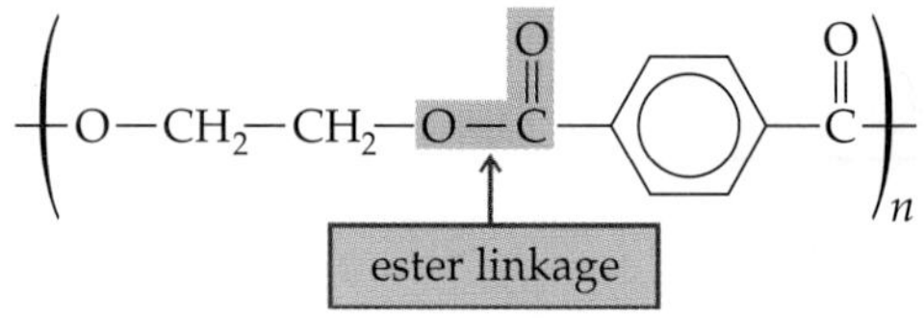

$$\left[O-CH_2-CH_2-O-\overset{\overset{\Large O}{\|}}{C}-C_6H_4-\overset{\overset{\Large O}{\|}}{C} \right]_n$$

ester linkage

What two monomers are used to produce this polyester?

Solution A diol has —OH groups at both ends of the molecule; a dicarboxylic acid has —COOH groups at both ends of the molecule. A water molecule is eliminated in the formation of the condensation polymer. To obtain the monomers, add the parts of the water molecule, —H and —OH, at the ester linkage where the monomers are joined. The monomers are

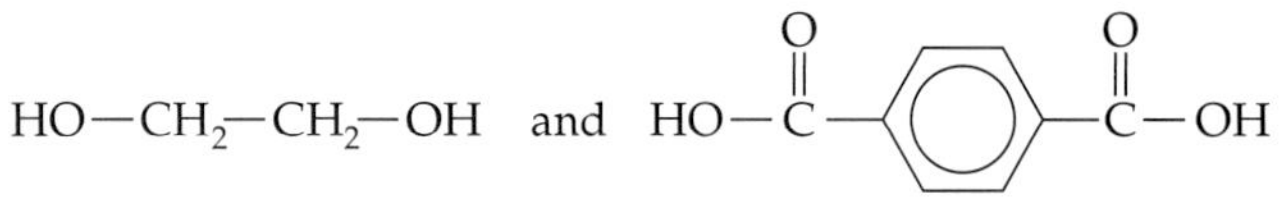

$$HO-CH_2-CH_2-OH \quad \text{and} \quad HO-\overset{\overset{\Large O}{\|}}{C}-C_6H_4-\overset{\overset{\Large O}{\|}}{C}-OH$$

Practice Exercise 9.2

What monomers are used to make the following polyamide, a condensation polymer?

$$\left[\overset{\overset{\Large O}{\|}}{C}-(CH_2)_x-\overset{\overset{\Large O}{\|}}{C}-\overset{\overset{\Large H}{|}}{N}-(CH_2)_y-\overset{\overset{\Large H}{|}}{N} \right]_n$$

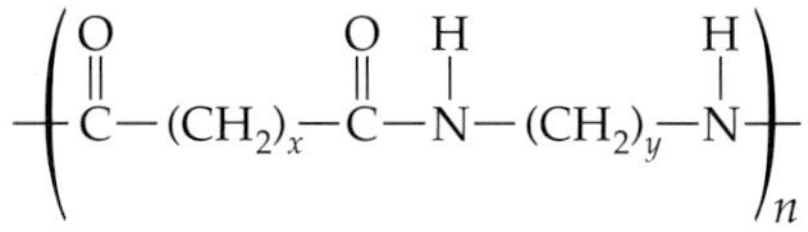

Formaldehyde-Based Network Polymers

A formaldehyde-amine resin is used to make almost indestructible dinnerware (Melmac) and counter tops (Formica). A formaldehyde-urea resin is used to make particle-board binder.

Most of the polymers we have considered so far—including polyethylenes, substituted polyethylenes, polyesters, and nylons—are thermoplastics (plastics that can be softened by heat and remolded). Formaldehyde-based polymers, however, are thermosetting plastics and cannot be remolded once hardened. They form as sticky, viscous liquids called **resins** that become hard on exposure to air.

The first thermosetting plastic was synthesized in 1909 by Leo Baekeland and marketed under the trade name Bakelite. Bakelite is produced from formaldehyde and phenol monomers by means of a two-step polymerization reaction (**Fig. 9.12a**). In the first step, formaldehyde is added to the *ortho* and *para* positions of phenol.

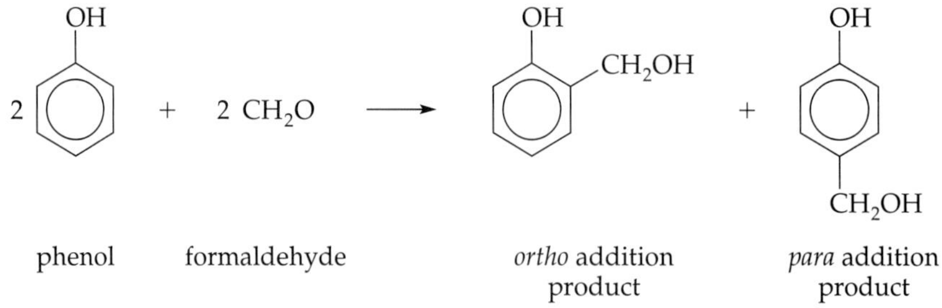

$$2\, C_6H_5OH + 2\, CH_2O \longrightarrow o\text{-}HOC_6H_4CH_2OH + p\text{-}HOC_6H_4CH_2OH$$

phenol — formaldehyde — *ortho* addition product — *para* addition product

FIGURE 9.12

(a) Preparation of phenol-formaldehyde polymer. (b) The resin formed from phenol and formaldehyde has a three-dimensional network structure.

In the second step, the substituted monomers undergo a condensation reaction, with the elimination of water. (Recall from Chapter 8 that a hydrogen atom, which is not usually shown, is present at each of the unsubstituted positions on the benzene ring.)

$$\text{OH-C}_6\text{H}_4\text{-CH}_2\text{OH} + \text{H-C}_6\text{H}_3(\text{OH})\text{-CH}_2\text{OH} \longrightarrow \text{OH-C}_6\text{H}_4\text{-CH}_2\text{-C}_6\text{H}_3(\text{OH})\text{-CH}_2\text{OH} + H_2O$$

Unlike the bifunctional monomers used to make polyesters and polyamides, each phenol monomer can be linked to *three* other monomers to form a complex, three-dimensional, cross-linked network (**Fig. 9.12b**). The benzene rings are joined through $-CH_2-$ linkages derived from formaldehyde. The numerous cross-links make the polymer very hard and rigid.

Bakelite has excellent insulating properties and is used to make housings for radios and other electric circuits. It is also used for buttons and bottle caps, but its primary use is as an adhesive in the production of plywood and particle board.

9.6 Rearrangement Polymers: Polyurethanes

The most important rearrangement polymers are the **polyurethanes,** which are made from dialcohol and diisocyanate monomers. Isocyanates are compounds that contain the functional group $O{=}C{=}N-$. The rearrangement reaction that leads to the formation of a urethane linkage is shown below, using general formulas for the dialcohol and the diisocyanate monomers.

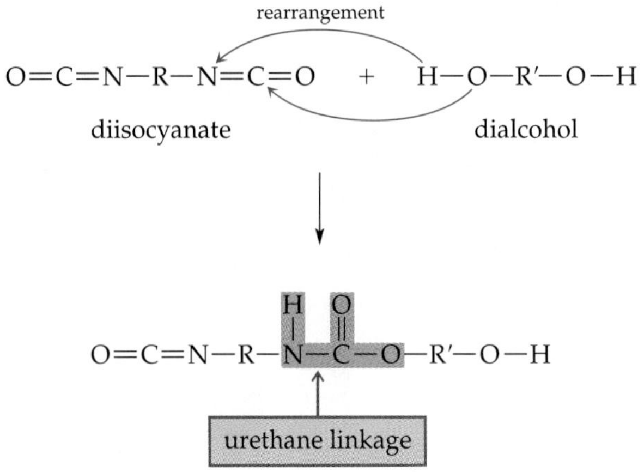

rearrangement

O=C=N—R—N=C=O + H—O—R′—O—H

diisocyanate dialcohol

↓

O=C=N—R—N(H)—C(=O)—O—R′—O—H

urethane linkage

The reaction is not a condensation reaction; there is a *rearrangement* of atoms, but no atoms are lost. As the urethane linkage forms, a hydrogen atom on the alcohol moves to a nitrogen atom on the isocyanate, and the nitrogen-carbon double bond is converted to a nitrogen-carbon single bond. The carbon atom in the isocyanate bonds to the alcohol through an oxygen atom.

Further rearrangement reactions lead to the formation of a long-chain polyurethane polymer.

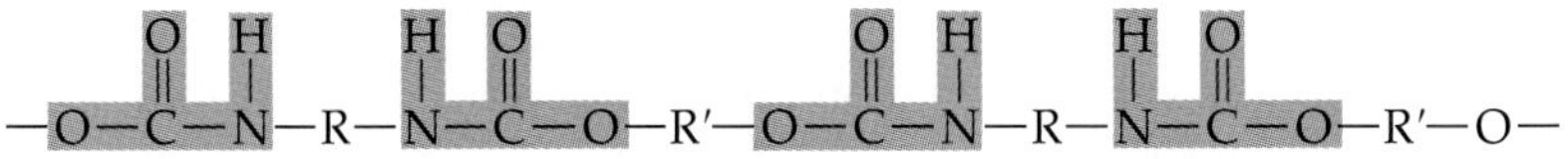

—O—C(=O)—N(H)—R—N(H)—C(=O)—O—R′—O—C(=O)—N(H)—R—N(H)—C(=O)—O—R′—O—

polyurethane

Soft polyurethane foam is often called *foam rubber*.

A urethane linkage is similar (although not identical) to an amide linkage. As a result, polyurethanes are similar in many ways to nylons. They can be drawn into fibers, which are used primarily for upholstery, and they can be processed into film. Some polyurethanes are rubbery materials that have found uses as skin replacements for burn victims and as valves in artificial hearts (**Fig. 9.13**). Soft polyurethane foam is used as padding in furniture and mattresses. Rigid polyurethane foam is used in the building industry.

9.7 Silicones

All the polymers we have discussed so far have been organic polymers with chains of carbon atoms as their backbones. Chemists have also made very useful polymers with silicon backbones.

Silicon is in the same group of the periodic table as carbon, and it forms a few low-molecular-weight compounds with hydrogen, including SiH_4, Si_2H_6, and Si_3H_8; these are called *silanes* and are analogous to simple hydrocarbons. Si—Si and Si—H covalent bonds are not as stable as C—C and C—H covalent bonds, and silicon does not form stable long-chain molecules analogous to long-chain hydrocarbons. But the hydrogen atoms in silanes can be replaced quite easily by other atoms, or groups of atoms, to give compounds that can readily polymerize. For example, dihydroxy silicon monomers with the general formula $R_2Si(OH)_2$ (where R is CH_3, C_2H_5, C_3H_7, or some

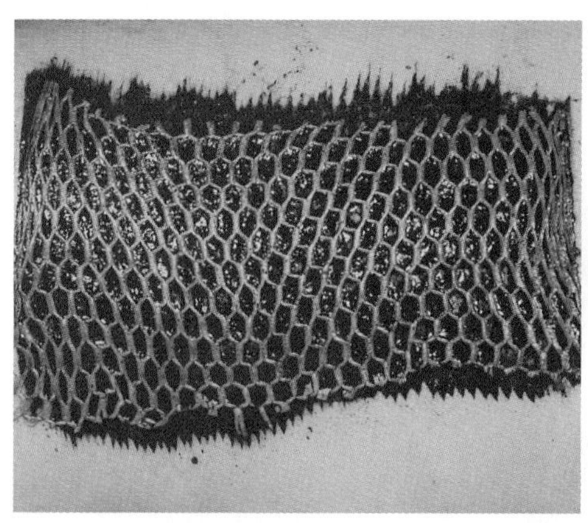

a.

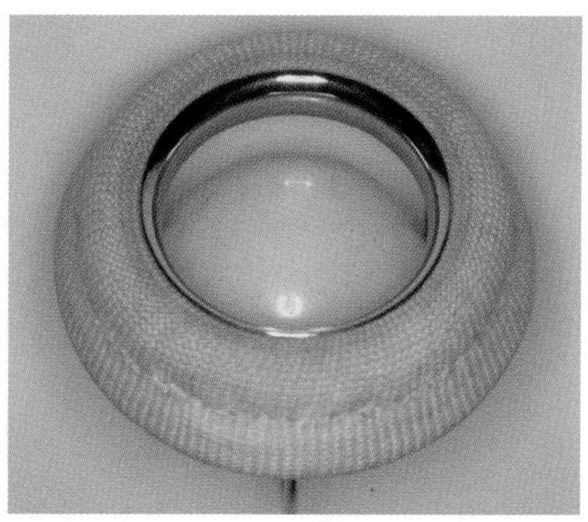

b.

FIGURE 9.13

(a) Thin sheets of polyurethane are used as skin replacements for burn patients. (b) Rubbery polyurethanes are used in artificial heart valves.

other hydrocarbon group) can undergo condensation reactions to form both linear and cross-linked network polymers known as **silicones.**

$$\mathrm{H{-}O{-}Si(CH_3)_2{-}O{-}H + H{-}O{-}Si(CH_3)_2{-}O{-}H + H{-}O{-}Si(CH_3)_2{-}O{-}H + H{-}O{-}Si(CH_3)_2{-}O{-}H}$$

alcohol monomers

↓ polymerization

$$\mathrm{{-}O{-}Si(CH_3)_2{-}O{-}Si(CH_3)_2{-}O{-}Si(CH_3)_2{-}O{-}Si(CH_3)_2{-}O{-}} + 3\,H_2O$$

↓ chemicals or heat

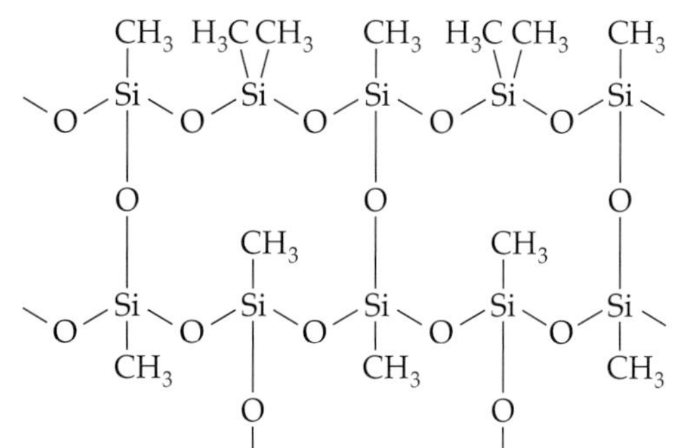

methyl silicone cross-linked polymer

Silicones are water-repellent, heat-stable, and very resistant to chemical attack. Depending on the nature of the $R_2Si(OH)_2$ monomer and the degree of polymerization, the silicones obtained may be oils, greases, rubberlike materials, or resins **(Fig. 9.14a)**. Silicone oils, unlike hydrocarbon oils, do not decompose at high temperatures and do not become viscous. They are particularly valuable as lubricants for machinery and

a.

b.

FIGURE 9.14

(a) Items containing silicone polymers. (b) Tubing in automobile engines is made of silicone rubber.

engines that must operate at extremes of temperature. Other silicones are used as hydraulic fluids, electrical insulators, and moisture-proofing agents in fabrics.

Silicone rubbers (**Fig. 9.14b**), which are produced when long polymer chains are extensively cross-linked with —Si—O—Si— links, are less subject to oxidation and attack by chemicals than either natural or synthetic carbon-based rubber. Another advantage is that silicone rubbers retain their elasticity at very low temperatures, which makes them valuable as sealants for aircraft and space vehicles. Silicone rubber is easily molded, and, because it is so unreactive, it is an excellent replacement material for body parts. It is used for repairing noses and ears and in the construction of artificial joints.

Silly Putty owes its dual elastic and liquid properties to its structure, which is between that of a silicone rubber and a silicone oil.

Until recently, liquid silicone contained in urethane bags was used extensively for breast augmentation. In 1992, reports that implants increased the risk of developing cancer and that leaking silicone might cause a severe immune response prompted the Food and Drug Administration to limit the use of silicone to breast augmentation for mastectomy patients. The controversial decision has been criticized by researchers who oppose the ban.

9.8 Polymers for the Future

Polymer research is a fertile field. Many of the newest polymers are *composites*, exceptionally strong materials in which various fibers are embedded in a polymer matrix. The fibers give support to the polymer, and the polymer prevents the fibers from bending or breaking. Composites are extremely lightweight but stronger than steel, and they are very unreactive. The most common composites are made of polyester reinforced with glass fibers. They are used to make automobile body parts and boat hulls. Polymers reinforced with graphite are used in aircraft construction and to make tennis raquets, golf clubs, and other sports equipment.

Although polymers usually act as electrical insulators, stable polymers that conduct electricity have been developed. These materials, which in most cases are types of acetylene polymers, are lightweight and flexible. They are expected to have applications in batteries, semiconductor circuits, radar dishes, and electrical wiring. Thermoplastic elastomers (TPEs) may soon replace natural and synthetic rubbers in many applications. TPEs are easier and less expensive to produce, and they present fewer disposal problems because they can be ground up and reused. TPEs are not suitable for tires, but they are ideal for most other rubber automotive parts, including bumpers, belts, drive boots, and air-conditioning hoses.

9.9 Problems with Polymers

The major problem posed by synthetic polymers is how to dispose of the discarded items made from them. There is also concern about the hazardous chemicals that some polymer materials release into the environment. Another problem is the dwindling supply of petroleum, the primary source of the raw materials from which synthetic polymers are made.

FIGURE 9.15

The disposal of plastic waste has become a serious problem.

Disposal of Plastics

The very properties that make plastics so useful—their stability and resistance to oxidation and attack by chemicals and bacteria—make them almost indestructible. They have become unsightly litter along roadsides and wherever else they are discarded, and they are found floating on the surfaces of all the oceans of the world.

Plastics account for only 8% of the total weight of solid waste generated in the United States, but they make up more than 20% of its volume (**Fig. 9.15**). Because they are not biodegradable, when buried in a landfill, they remain there unchanged almost indefinitely.

There is no single way to solve this problem. It will have to be tackled using a combination of tactics, including recycling, incineration, making plastics more degradable, and reducing consumption.

Recycling About 87% of all plastics are thermoplastic and can be recycled by melting and remolding. At present, however, only about 5% of plastic is being recycled, compared with 20% of paper and more than 60% of aluminum cans (Chapter 19). A major problem is that different types of plastics have different properties, and, to be of value to industry, recycled plastics must be separated according to type.

To assist in sorting, most plastic containers and other items are stamped with a code consisting of three triangulated arrows surrounding a single-digit code number, with the abbreviation of the name of the plastic underneath (**Table 9.2**). At present, most municipalities that have recycling programs collect only HDPE (#2) milk and juice jugs and PET (#1) soft drink bottles.

Recycled HDPE is used to make drainage pipes and containers for a variety of liquid products. Recycled PET is used to make carpeting and insulation for ski jackets and parkas. Used polystyrene (PS, #6) from coffee cups and throw-away plates and food trays, is being converted to plastic "lumber." This lumber can be drilled, cut, or nailed, just like real lumber, and it has the advantage that it is not eaten by termites or rotted by water. Recycled PVC (V, #3) from automobile seats is being converted into sewer drainpipes. Polyethylene grocery bags can be made into new bags.

Incineration Incineration of plastics to produce energy is another approach to disposal, but it has the drawback that certain plastics release toxic gases when burned. Vinyl chloride polymers, for example, release toxic hydrogen chloride gas; polyurethanes and polyacrylonitriles produce deadly hydrogen cyanide. Even if a polymer does not release a toxic gas when burned, it will produce carbon dioxide, which contributes to the greenhouse effect (Chapter 11). The formation of toxic gases during burning is one of the dangers associated with accidental fires in homes, other buildings, and aircraft, where many furnishings and other fittings are made of plastics.

Table 9.2 Plastics That Are Recycled and Their Codes

Code	Name of Plastic	Products Commonly Used In	Uses after Recycling
1 PETE	Polyethylene terephthalate (PET)	2-liter soda bottles and other bottles, other containers such as peanut butter jars	Strapping (for packaging), fiberfill for winter clothing, carpets, surfboards, sailboat hulls
2 HDPE	High-density polyethylene (HDPE)	Milk jugs, bleach and detergent bottles, motor-oil bottles, other containers, toys, pipe	Trash cans, detergent bottles, drainage pipes, base cups for soda bottles
3 V	Poly(vinyl chloride) (PVC)	Vinyl siding, plastic pipes and hoses, shower curtains, some cooking oil and shampoo bottles and bottles for household chemicals, imitation leather	Fencing, siding, handrails, pipes
4 LDPE	Low-density polyethylene (LDPE)	Plastic wrap, bread bags, trash bags, other kinds of containers	Grocery and trash bags
5 PP	Polypropylene	Margarine and yogurt containers, some lids and caps	Auto battery casings, bird feeders, water buckets
6 PS	Polystyrene	Styrofoam: coffee cups, "peanuts" for packing, egg cartons, meat trays Nonfoam: glasses, toys, plastic utensils, videocassettes	Tape dispensers, cafeteria trays, plastic lumber
7 OTHER	Plastic other than the six listed above and objects produced from mixtures of plastics		Plastic lumber (used to manufacture benches, picnic tables and other outdoor furniture, marine pilings)

Degradable Plastics Manufacturers are increasingly seeking ways to make plastics more degradable. Photodegradable plastics, which are broken down by ultraviolet light, can be made by incorporating light-sensitive molecules containing a carbonyl group (—C=O) into the polymer chains. Biodegradable plastics that can be digested by microorganisms are made by incorporating starch or cellulose into the polymer during production. As bacteria consume the starch or cellulose, the plastic is broken down into small pieces. This approach is particularly useful for thermosetting plastics, which are not readily recycled.

The major value of both types of degradable plastics is that they reduce litter. When buried in a landfill, they degrade very slowly, if at all, because of lack of light, oxygen, and moisture. Even naturally biodegradable materials such as newspapers have been found to be almost unchanged after 20 years in landfills. Degradable plastics are used primarily for six-pack beverage rings, trash bags, and disposable diapers. The decision to use degradable plastic in these products was made because they entangle wildlife, make up a high percentage of litter, or cause pollution.

Before plastic six-pack yokes are discarded, each ring should be cut to prevent animals from becoming caught in them.

In a development that holds promise for the future, biodegradable polymers based on carbohydrates, proteins, and carboxylic acids have been made by bacteria in fermentation processes. The same bacteria that produce the polymers will decompose them under the anaerobic conditions of a landfill. Scientists have produced a cornstarch-based polymer that can be used as an alternative to Styrofoam in packing materials. At present, these biodegradable materials have a limited market because they are considerably more expensive to produce than nondegradable alternatives.

Reducing Plastic Usage Nearly one-third of all plastic produced goes into making packaging materials and other quickly discarded items. Plastic cups, plates, forks, carrying bags, and numerous other products are discarded after one use. In many instances, washable, reusable utensils and reusable string or cloth bags could be used instead of the throw-away items. Fast-food establishments are discovering that it is good public relations to cut back on the amount of plastic materials they use for food packaging. Some local governments have banned the use of certain types of plastic containers. However, to have any significant impact on the plastic waste problem, such measures will have to be adopted much more widely.

According to the Environmental Defense Fund, using 1000 disposable plastic teaspoons consumes more than ten times the energy and natural resources used in making one stainless steel teaspoon and washing it 1000 times.

Polymer Additives: Plasticizers

Another problem associated with polymers is the chemicals that are frequently added to them during production to improve or alter their properties; these chemicals may contaminate the environment when the plastic is discarded. Polymers that are brittle and hard can often be made soft and pliable by the addition of compounds called **plasticizers.** For example, PVC treated in this way can be used to make raincoats, seat coverings, and other similar materials. For many years until they were banned, PCBs (Chapter 10) were widely used as plasticizers. Today, the most commonly used plasticizers are phthalates, esters of phthalic acid.

Over time, plasticizers evaporate, and the material loses its flexibility and tends to crack.

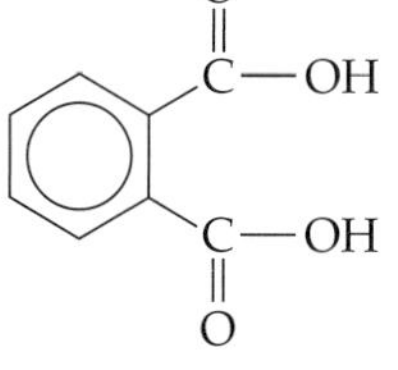

$C_6H_4(COOH)_2$

phthalic acid

$C_6H_4(COOCH_2CH(CH_2CH_3)CH_2CH_2CH_2CH_3)_2$

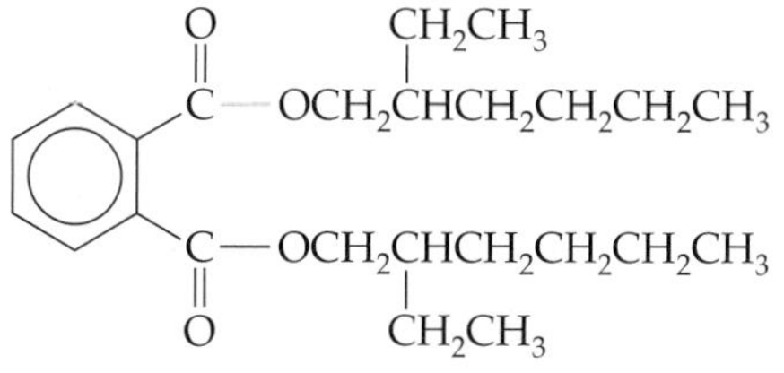

dioctyl phthalate (DOP)

EXPLORATIONS

The Electric Car—Upscale Vehicle for the 21st Century

The gasoline engine has dominated road transportation for almost a hundred years. However, it may soon be facing stiff competition from a battery-powered motor. As petroleum supplies inevitably dwindle, gasoline prices rise, and the demand for vehicles emitting fewer pollutants increases, the motoring public may be ready for an electric car. To date, electric vehicles (EVs) have had little success because of their relatively poor performance, limited range between battery charges, and high cost. This situation may soon change. The big automakers in the United States and abroad are devoting large sums of money on research to develop a viable electric car.

Early Versions of EVs

At the end of the 19th century, when the production of a horseless carriage was still little more than a dream, three technologies were being explored: gasoline engines, steam engines, and battery-powered motors. By the early 1900s, 40% of the approximately 8000 automobiles in the United States were powered by steam, 38% by batteries, and only 22% by gasoline. At that time, gasoline engines were difficult to start, noisy, and unreliable. Steam-powered cars were easier to operate, but they needed a constant supply of water and high steam pressure, which made them expensive and difficult to maintain; most were gone by 1910. Early EVs had many advantages over their competitors: They started up instantly, were easy to operate, required minimum maintenance and ran silently—so silently that they were judged dangerous because pedestrians could not hear them coming! At the height of their popularity in 1912, EVs had a cruising speed between 15 and 20 miles per hour and could travel 30 to 40 miles between charges.

The popularity of EVs was short-lived, however; by the 1920s, gasoline-powered vehicles were firmly in the lead. Two factors doomed the electric car: the limited capacity of the battery, which restricted speed and range, and, ironically, the replacement of the gasoline motor's dangerous hand crank with a battery-powered electric starter.

Despite their limitations, EVs never completely disappeared. In the United Kingdom, in particular, they have been in continuous use for over 70 years as neighborhood delivery vehicles. In this role, their modest speed and range are not a handicap, and their low operating costs and few repair needs make them more economical than gas-powered vehicles. And we are all familiar with electric golf carts and the similar vehicles that are used at airports. However, it was not until the oil crisis in the 1970s, when gasoline prices soared and long lines formed at gas stations, that automakers made a serious commitment to market electric cars for general use. By 1979, large and small companies in the United States and abroad were producing an assortment of often strange-looking vehicles, most with cruising speeds between 30 and 50 miles per hour. Once the oil crisis ended, interest in EVs again waned.

Environmental Benefits of EVs

Today, as concern for the environment increases, EVs look attractive. Many automakers around the world are already marketing, or planning to market, electric cars. In the United States, a significant spur to their development is legislation in California and several other states requiring that by 2003, 10% of new cars sold in the state must meet zero-emission standards. Currently, only EVs can meet this standard. Even when the power-plant emissions associated with generating the electricity needed to power the batteries are taken into account, EVs emit 99% less carbon monoxide, 98% less hydrocarbons, 89% less nitrogen oxides, and more than 50% less carbon dioxide per mile than conventional automobiles.

General Motors, Ford, and Chrysler, which all have EV programs, are well aware that to be acceptable to the general public an EV must be comparable in cost and performance to a gasoline-powered car. In 1990, when GM's test model, the Impact, was first displayed, it created a sensation. The vehicle had a top speed of 110 miles per hour and a range of 120 miles between charges. Electricity was generated by a row of advanced-design lead-acid batteries, and an inverter converted the direct current to alternating current and transferred it to two induction motors—one for each front wheel. Despite its enormous potential, however, the Impact was not a prototype for a marketable vehicle because it was prohibitively expensive and it also had numerous safety and design problems.

The main difficulty with EVs has always been the battery. In 1991, the three big U.S.

automakers, the Department of Energy, and the electric utilities began a joint effort to develop promising battery technologies for future EVs. Nickel-iron, nickel-cadmium, nickel-metal hydride, lithium-polymer, and a number of other systems are being investigated. Fuel cells are another promising option (see Chapter 13). But no other battery system can yet compete with an optimally efficient lead-acid battery. Lead-acid batteries are heavy, and since every pound reduces efficiency, the challenge is to produce a lightweight, but strong, auto body with the least possible resistance to air.

In 1996, GM introduced its two-door, two-seat electric sports car, the EV1. Its frame, which is made of aluminum instead of steel, is extremely light, and the vehicle's sleek shape gives it a very low drag coefficient. Body parts are made of the most advanced lightweight but strong plastics; new synthetic rubber that has less rolling resistance than conventional rubber is used to make the tires. The EV1 runs on electricity generated by 26 lead-acid batteries, which together weigh over 1000 pounds. It is almost noiseless and accelerates to 60 miles per hour in less than 9 seconds. It has a top speed of 80 miles per hour and a range of 90 miles between recharges. Recharging takes from 3 to 12 hours. In California, over 4000 public recharging stations are already in operation. Some stations run on solar cells, thus eliminating the power-plant pollution typically associated with generating electricity.

The big automakers in the United States and abroad are devoting large sums of money on research to develop a viable electric car.

At $35,000, the EV1 is considerably more expensive than the average conventional automobile. It is estimated, however, that the EV1 would be more economical than a gas-powered car over the lifetime of the vehicle. There are few parts to wear out, no oil to change, and maintenance costs would be much lower. GM is promoting the EV1 as an ideal car for the daily commute to work, which for most people rarely exceeds 35 miles.

Hybrid Cars

Despite the environmental and potential cost advantages of EVs, the average motorist is not ready to invest in one. At present, the most attractive of the environmentally-friendly cars are Toyota's roomy, four-door Prius and Honda's two-door Insight. These cars, which are gas-electric hybrids, were introduced to the American market in 2000. Although not zero-emission vehicles, they get over 60 miles per gallon—nearly five times as much as the average sport utility vehicle—and emit one-tenth the pollution of the average new sedan. Using a combination of a small standard gasoline engine, an electric motor, and a battery, hybrids perform as well as the average new car. The battery powers the car at start-up and at very low speeds. Then, at about 25 miles per hour, the gas engine automatically takes over. Unlike the battery of all-electric cars, the battery in these hybrids never needs recharging between trips because it automatically recharges when the brakes are applied and the car slows down. If the savings at the gas pump are factored in, the $20,000 cost of a Prius is no more than Americans are used to paying for a new sedan.

Honda's hybrid car, the Insight

Although many problems remain to be solved before an all-electric zero-emission car is standard on American highways, it is likely that by the end of the 21st century, gas-guzzlers will seem as old-fashioned as the phonograph record does to today's teenagers.

References: M. Fischetti, "Here Comes the Electric Car—It's sporty, aggressive, and clean," *Smithsonian* (April, 1992), p. 34; *World Guide to Battery-Powered Road Transportation. Comparative Technical and Performance Specifications*, compiled by Jeffrey M. Christian (New York: McGraw-Hill, 1980).

Phthalates, as a result of their widespread use in plastics, have been detected in virtually all soil and water ecosystems, including the open ocean. The levels found are very low, and the toxicity of phthalates is also low. Because of their widespread distribution, however, some scientists believe that they may cause adverse health effects (see the section on endocrine disruptors in Chapter 14).

Phthalates are present in soft PVC toys, and there is concern that when small children chew on these toys, they may ingest sufficient phthalates to cause kidney and liver damage. A number of scientists and environmental groups are working to have these toys removed from the market. Another concern is that phthalates in PVC intravenous (IV) bags and tubing may leach into the fluids being delivered to patients. Significant amounts of phthalates have been found in PVC bags used to store blood.

The Raw Materials for Polymer Production

Ethylene, propylene, butylene, styrene, and phenol, the monomers from which many of the giant polymer molecules are made, are derived mainly from petroleum (Chapter 12). As petroleum reserves are depleted, more of the raw materials needed to produce synthetic polymers are expected to be obtained from coal. Another possibility is using biomass conversion (Chapter 13) to produce small molecules such as methyl and ethyl alcohol and methane, from which the necessary monomers could be synthesized.

Chapter Summary

1. Polymers are giant molecules composed of thousands of small repeating units called monomers. Monomers can be linked together to form long chains or three-dimensional networks.
2. Thermoplastic polymers can be softened by heat and remolded; thermosetting polymers cannot.
3. Naturally occurring polymers include silicates, diamond, starches, cellulose, proteins, DNA, and natural rubber.
4. There are three general types of polymers: (a) addition polymers, (b) condensation polymers, and (c) rearrangement polymers.
5. The simplest addition polymer is polyethylene. No atoms are lost from the monomer units in forming an addition polymer.
6. High-density polyethylene (HDPE) has unbranched chains and is used to make beverage containers and toys. Low-density polyethylene (LDPE) has branched chains and is used to make shopping bags.
7. Substituted polyethylenes are made by replacing one or more of the hydrogen atoms in ethylene with another atom or group of atoms (see Table 9.1). These polymers include poly(vinyl chloride) (PVC), polypropylene, polystyrene, polytetrafluoroethylene (Teflon), and polyacrylonitrile.
8. The addition of sulfur to rubber, a process called vulcanization, improves its qualities. Rubber owes its elasticity to its flexible, randomly intertwined chains.
9. Both latex and acrylic paints contain polymers, called binders, which harden as the paint dries.

10. In a condensation reaction, two molecules combine to form a larger molecule with the elimination of a small molecule, often water.
11. Condensation polymers include polyesters (e.g., PET), which are copolymers of a diol and a dicarboxylic acid; polyamides (nylons), which are copolymers of a diamine and a dicarboxylic acid; and Bakelite, which is a phenol-formaldehyde polymer.
12. Bakelite, the first completely synthetic polymer to be produced, is thermosetting. Nearly all other synthetic polymers are thermoplastic.
13. When a polymer is formed in a rearrangement reaction, atoms of the monomer units are rearranged, but none are lost. Rearrangement polymers include polyurethanes, which are made from a diol and a diisocyanate.
14. Some silicon compounds can form useful polymers called silicones.
15. Because synthetic polymers are not biodegradable, their disposal creates a serious problem. Solutions include recycling, incineration, reducing consumption, and making polymers more degradable.
16. Plasticizers and phthalates, which are added to polymers to improve their properties, are suspected of causing health and environmental problems.
17. Most of the monomers from which polymers are made are derived from petroleum, which is a diminishing and nonrenewable natural resource.

Key Terms

addition polymer (p. 254)
condensation polymer (p. 254)
copolymer (p. 252)
elastomer (p. 260)
high-density polyethylene (HDPE) (p. 255)
homopolymer (p. 252)
low-density polyethylene (LDPE) (p. 256)
monomer (p. 252)
neoprene (p. 262)
plastic (p. 252)
plasticizer (p. 275)
polyacrylonitrile (p. 259)
polyamide (p. 264)
polyester (p. 263)
polyethylene (p. 254)
polymer (p. 252)
polypropylene (p. 257)
polystyrene (p. 257)
polytetrafluoroethylene (p. 258)
polyurethanes (p. 269)
poly(vinyl chloride) (PVC) (p. 256)
rearrangement polymer (p. 254)
resin (p. 266)
silicones (p. 269)
thermoplastic (p. 253)
thermosetting plastic (p. 253)
vulcanization (p. 261)

Questions and Problems

1. Define the following:
   a. HDPE
   b. thermoplastic
   c. copolymer
2. Define the following:
   a. monomer
   b. vulcanization
   c. addition polymer
3. What general name is given to a polymer made from two or more different monomers?
4. Nylon is a copolymer. Orlon and polyethylene are not. Explain.
5. What is the difference between a thermoplastic and a thermosetting plastic?

6. The first synthetic polymer was celluloid. What was it made from?
7. Write a chemical equation that shows how polyethylene is formed from ethylene.
8. Write a chemical equation that shows how poly(vinyl chloride) is formed from vinyl chloride.
9. Describe the process of addition polymerization.
10. Give the structure and name of the monomer used to make the polymer $\left(C_2F_4 \right)_n$. Is this an addition or condensation polymer?
11. Polyethylene can be manufactured in two different forms: high-density polyethylene (HDPE) and low-density polyethylene (LDPE).
    a. Is there a difference in molecular structure between HDPE and LDPE?
    b. List the properties and uses of HDPE and LDPE.
12. Explain what is meant by *vulcanization.*
13. Why is rubber elastic?
14. How is latex obtained?
15. Because of its low reactivity, Teflon is used to line cookware. Explain why this polymer is unreactive. Why is polystyrene unsuitable for lining cookware?
16. How is a water molecule involved in the reaction that produces condensation polymers?
17. The very strong condensation polymer Kevlar, which is used to make bullet-proof vests, has the following structure.

$$-NH\left(\overset{O}{\overset{\|}{C}}-C_6H_4-\overset{O}{\overset{\|}{C}}-NH-C_6H_4-NH\right)_n\overset{O}{\overset{\|}{C}}-C_6H_4-$$

    What monomers are used to make Kevlar?
18. Burning plastics can give off choking and even poisonous gases. Name the plastic that would give off each of the following if burned:
    a. HCl b. HCN c. HCHO d. C_2H_3Cl
19. Draw the structure of a polymer made from styrene monomers (show five repeating units).
20. Draw the structure of a polymer made from tetrafluoroethylene monomers (show five repeating units).
21. Draw the structure of the polymer made from vinyl chloride monomers (show five repeating units).
22. Name the polymer that is formed from ethylene molecules that have all of their hydrogens replaced with fluorine atoms.
23. The polymer PET is a copolymer. From what two monomers is it made?
24. Name a commercial plastic that contains each of the following functional groups:
    a. chlorine b. fluorine
    c. nitrogen d. benzene ring
25. Natural rubber is made from isoprene.
    a. Draw the structure of isoprene.
    b. Draw the polymer made from isoprene monomers polymerized in a *cis* arrangement.
    c. On the basis of your drawing for part (b), would you expect this polymer to be easily stretched?
26. How does the addition of sulfur to natural rubber allow the rubber to be used for making tires?
27. Besides natural or synthetic rubber, what other material is used to make tires?
28. Draw the structures of phenol and formaldehyde and the polymer they form when reacted together.
29. List five paint pigments and the color each produces. Which white pigment has been banned?
30. What does the binder do as paint dries? What compounds are considered the best binders?
31. What feature do all condensation reactions have in common?
32. What are the main components of paint? What is the purpose of each ingredient?
33. Describe the changes that take place when an oil-based paint dries. Does the wet paint react with any atmospheric gases? Does it add anything to the atmosphere?
34. Nylon 46 is made from a diamine with four carbon atoms and a diacid with six carbon atoms. Draw the structure of nylon 46.
35. What raw material is the source for the monomers used in the synthesis of most polymers?

36. What properties make polyester a good material for clothing? Would HDPE be suitable for making clothing? Explain.

37. Polyesters are formed by the reaction of an acid with a ________.

38. Draw the structure and give a brand name for each of the following:
   a. a thermosetting polymer
   b. a thermoplastic polymer
   c. a silicone polymer

39. List three factors that make silicone rubbers superior to some carbon-based rubber.

40. What type of polymerization reaction produces polyurethanes?

41. Polyurethanes are held together by urethane linkages, which are similar to amide linkages. What other kind of polymer has amide linkages?

42. Draw the structure of a methyl silicone polymer containing five silicon atoms.

43. How do composites differ from polymers?

44. What is the main environmental issue that needs to be addressed concerning the continued use of plastics?

45. What percentage of plastic waste is recycled? How does this amount compare with the amounts of aluminum and paper that are recycled?

46. Name two substances that can be added to a plastic to make it biodegradable.

47. Why is a plasticizer added to a polymer? Name a commonly used plasticizer.

48. List three ways to dispose of used plastics. Give two problems associated with each.

49. Why are plasticizers added to polymers? Give two examples of polymers you have used that are plasticized.

50. Would you support a national container deposit law? Explain.

51. Why is incineration a poor choice for disposal of plastics?

52. If you were going to recycle discarded plastics in your hometown, describe how you would separate the plastics from other trash and what you would do with them once separated.

Answers to Practice Exercises

9.1 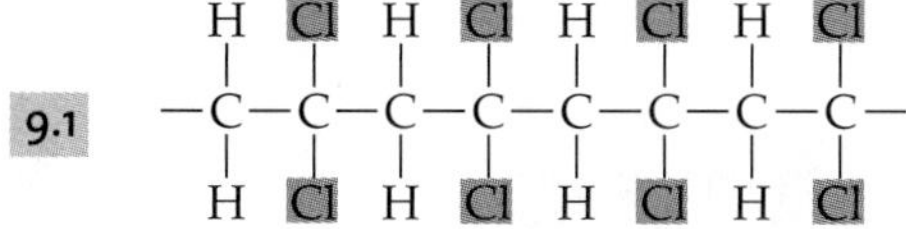

9.2 $HOOC-(CH_2)_x-COOH$ and $H_2N-(CH_2)_y-NH_2$

Chapter 10 Water Resources and Water Pollution

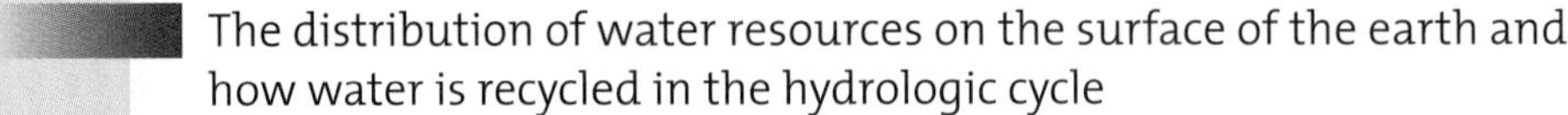

Chapter Objectives

In this chapter, you should gain an understanding of:

- The distribution of water resources on the surface of the earth and how water is recycled in the hydrologic cycle
- The unique physical properties of water and why these properties are so important
- Water management and conservation
- The different types and sources of water pollutants
- Sewage treatment

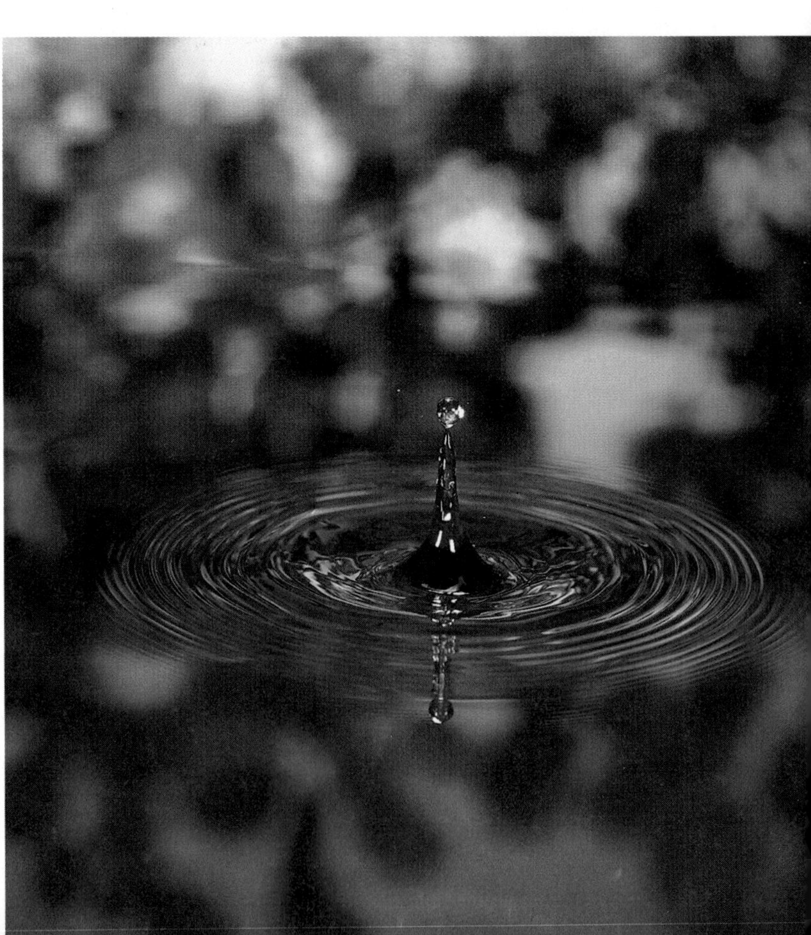

Without water, life on earth would not exist. Although a very common substance, water, has many unique properties that result from its molecular structure.

WATER IS ONE OF THE WORLD'S most precious resources; without it, there would be no life on earth. We use water in the home and for recreation, and it is crucial for agriculture and industry. Although the total amount of water on earth is fixed and cannot be increased, we are in no danger of running out. Water is constantly being recycled and replenished by rainfall; there is plenty of freshwater to meet the needs of everyone on earth. However, because of the uneven distribution of rainfall and the heavy use of water in certain areas, many regions in the United States and other parts of the world are experiencing severe water shortages.

Water in all its forms—ice, liquid water, and water vapor—is very familiar, and yet, water has many unusual properties. For example, water's exceptional ability to store heat modifies the earth's climate, and ice's ability to float on water allows aquatic creatures to survive in winter.

Water is an excellent solvent that can dissolve a wide variety of ionic and polar substances. Thus, it is an effective medium for carrying nutrients to plants and animals. It is also a good medium for carrying toxic substances and other pollutants.

In this chapter, we examine the composition of natural waters, the way water is distributed on the earth, and how it is recycled. We see how the molecular structure of water accounts for its extraordinary properties, and we consider the implications of those properties for life on earth. We discuss the ways water supplies are used and the measures that can be taken to better manage and conserve them. Last, we study water pollution and examine its main sources, the effects of pollutants on human health and the environment, and the steps that can be taken to preserve water quality.

ONLINE FEATURES

www.jbpub.com/chemistry

- Chemistry and the Environment
- eLearning
- Chemistry in the News
- Special Topics
- Research & Reference Links
- Ask the Authors

10.1 Distribution of Water on the Earth

Water is the most abundant compound on earth, covering nearly three-quarters of the planet's surface. Although the total amount of water on earth is enormous, only a small percentage is freshwater—and much of that is not readily available for human use. Over 97% of the world's total supply of water is found in oceans and is too salty for drinking, irrigation, and most industrial and household needs. Of the remaining approximately 3%, about 2% is in the form of glaciers and polar ice caps. Most of the rest is found underground (groundwater), and much of that is too difficult or too expensive to tap. Lakes and rivers, which are the major sources of the world's drinking water, account for less than 0.01% of the earth's total water **(Fig. 10.1)**.

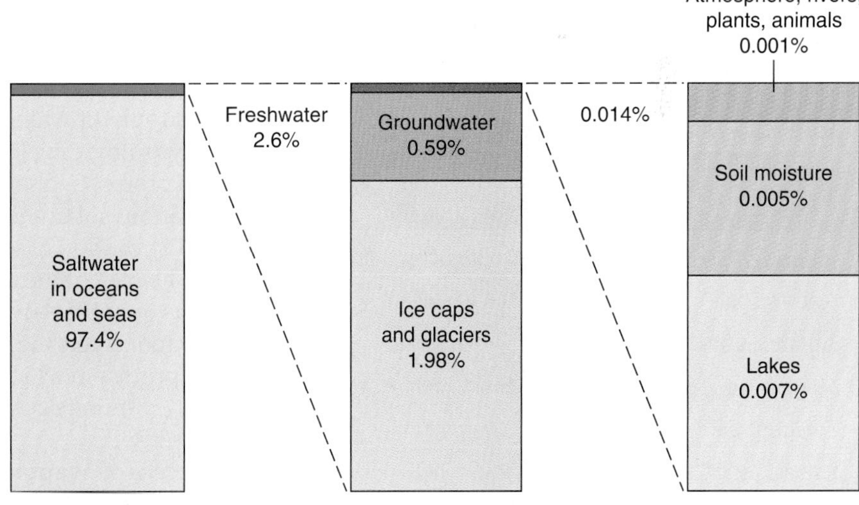

FIGURE 10.1

Most of the water on earth is saltwater in the oceans. Less than 3% is freshwater, and most of that exists as ice in glaciers and polar ice caps. The freshwater needs of humans must be supplied by the approximately 0.6% of the earth's water that is found underground and in lakes and rivers.

Water is the major component of all living things **(Table 10.1)**. It constitutes 70% of every adult human body and from 50% to 90% of all plants and animals. Water enters our bodies in the liquids we drink and the foods we

Table 10.1 Water Content of Selected Organisms and Foods

	Percentage of Mass
Organisms	
Marine invertebrates	97
Human fetus (1 month)	93
Fish	82
Human adult	70
Foods	
Broccoli	90
Milk	88
Apples	85
Grapes	80
Eggs	75
Potatoes	75
Steak	73
Cheese	35

eat; it leaves in urine, feces, sweat, and the air we exhale. Humans can survive only a few days without water.

10.2 The Composition of Natural Waters

All bodies of natural water are solutions that contain varying concentrations of dissolved solids. Water containing up to 0.1% (1000 ppm) of dissolved solids is generally termed **freshwater.** However, such water is not necessarily suitable for drinking. For drinking water, the U.S. Public Health Service recommends a solid content of no more than 0.05% (500 ppm). The concentration of dissolved solids in seawater averages 3.5% (35,000 ppm). Water is termed *brackish* when the solid content lies somewhere between that of freshwater and seawater.

The ions present in seawater at a concentration of 1 ppm or more are shown in **Table 10.2**. Present at lower concentrations are at least 50 other elements. The high concentration of salts in seawater comes primarily from the constant evaporation of water from ocean surfaces.

The concentration of ions in freshwater in streams, rivers, and lakes is much lower than that in seawater, and the distribution of ions is quite different **(Table 10.3)**. In seawater, the main cation is sodium (Na^+), and the main anion is chloride (Cl^-). In freshwater, on the other hand, calcium (Ca^{2+}) and magnesium (Mg^{2+}) are the dominant cations, and bicarbonate (HCO_3^-) is the dominant anion. The ions in freshwater come from the weathering of rocks and soil. A main source of the Na^+ and Cl^- ions in rivers, particularly those near a seacoast, is salt spray that is thrown up into the atmosphere from the ocean and then deposited on land and carried, by runoff, into the rivers.

In the Great Salt Lake in Utah and the Dead Sea in Israel, the concentration of dissolved solids is as high as 25%.

10.3 The Hydrologic Cycle: Recycling and Purification

Table 10.2 Major Constituents of Seawater

Ion	Concentration (ppm)
Chloride, Cl^-	19,000
Sodium, Na^+	10,600
Sulfate, SO_4^{2-}	2,600
Magnesium, Mg^{2+}	1,300
Calcium, Ca^{2+}	400
Potassium, K^+	380
Bicarbonate, HCO_3^-	140
Bromide, Br^-	65
Other substances	34
Total	34,519

The earth's supply of water is continually being purified and recycled in the **hydrologic cycle,** or *water cycle,* illustrated in **Figure 10.2**. Through the processes of evaporation, transpiration, condensation, and precipitation, solar energy and gravity are responsible for the ceaseless redistribution of water among oceans, land, air, and living organisms. The heat of the sun warms the surface of the earth, causing enormous quantities of water to evaporate from the oceans. On land, water vapor enters the atmosphere from plants as a result of *transpiration,* the process in which water escapes through the pores on leaf surfaces and evaporates. Additional water evaporates from lakes, rivers, and wet soil.

In the process of evaporation, water is purified. Dissolved substances are retained in the oceans, soil, and plants, and purified water vapor enters the atmosphere. As the water vapor rises, it cools and condenses into fine droplets that form into clouds. Prevailing winds carry moist air and clouds across the surface of the earth. If the moist air cools sufficiently, water droplets or ice crystals fall to earth as precipitation (rain, sleet, hail, or snow). Precipitation

Table 10.3 Comparison of the Concentrations of Major Ions in Freshwater and Seawater (percent of total ionic concentration)

Ion	Freshwater	Seawater
HCO_3^-	41.0	0.2
Ca^{2+}	16.0	0.9
Mg^{2+}	14.0	4.9
Na^+	11.0	41.0
Cl^-	8.5	49.0

that falls through clean air is pure and contains only gases dissolved from the atmosphere and traces of dissolved salts. Some precipitation becomes locked in glaciers, but most of it either sinks into the soil or flows downhill, as runoff, into nearby streams and lakes, eventually making its way through rivers and wetlands to the ocean.

The term *vapor* is generally used rather than *gas* to describe a substance that is in a gaseous form below its normal boiling point.

Precipitation that seeps into the soil either is taken up by plant roots or continues to percolate into the ground until it reaches an impermeable layer of rock that stops its downward progress. Water collects in the porous rock above this impermeable layer, forming a reservoir of groundwater termed an **aquifer** (refer to Fig. 10.2). Groundwater is generally of high purity because the porous rock acts as a filter and retains suspended particles and bacteria. Groundwater is replenished much more slowly than is water at the surface of the earth; it flows slowly (a speed of 15 meters, or 50 feet, per year is typical) through an aquifer until it reaches an exit to the surface where it either emerges as a spring or seeps out over a relatively wide area. These springs and seeps feed lakes, streams, and rivers—and, ultimately, the oceans.

On land, more water vapor typically enters the atmosphere through transpiration than through direct evaporation from lakes, rivers, and soil. Worldwide, direct evaporation from the oceans is the major source of atmospheric water vapor.

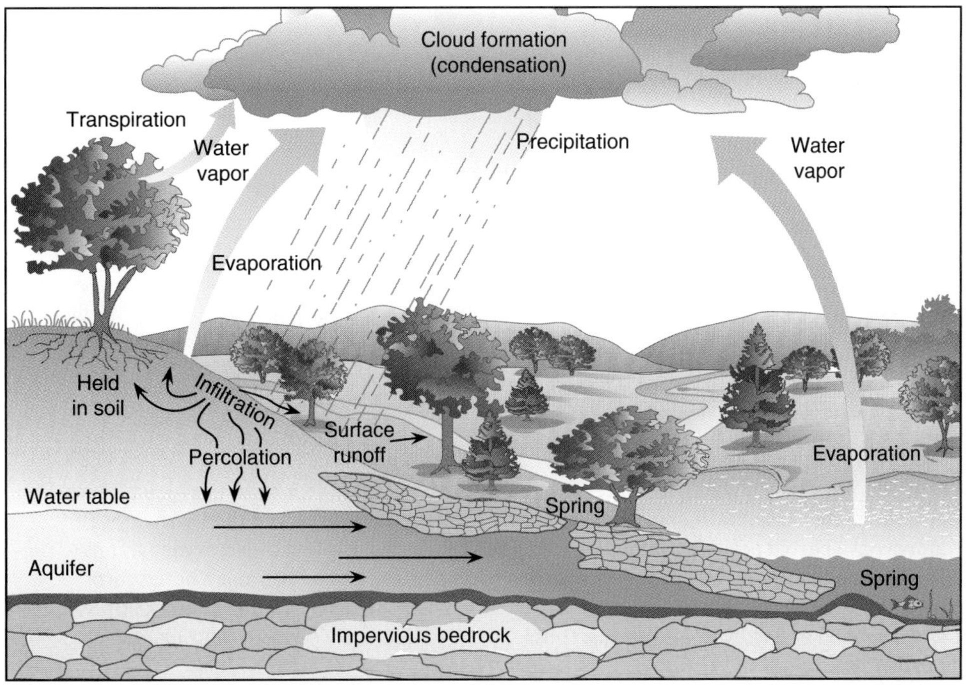

FIGURE 10.2

The hydrologic cycle through which the water on earth is continuously recycled. Powered by the sun, evaporation, condensation, and precipitation redistribute water among the oceans, the land, the atmosphere, and living organisms.

10.4 The Unique Properties of Water

Water is the only common liquid on earth. Water also possesses many unusual—even unique—properties. If it behaved like other chemical compounds of similar molecular weight and structure, life on earth could not exist. The reason for most of water's extraordinary properties is hydrogen bonding (Chapter 4). Although hydrogen bonds are much weaker than ionic or covalent bonds, they have a profound effect on the physical properties of water in both its liquid and solid states.

The Water Molecule and Hydrogen Bonding

The bond angle in the water molecule is 104.5°.

A water molecule can form four hydrogen bonds **(Fig. 10.3)**. The oxygen atom in the molecule can bond to two hydrogen atoms in other molecules because it has two pairs of unshared electrons. Each hydrogen atom in a water molecule can bond to an oxygen atom in another molecule. These bonds are directional, which means they can form only when the molecules are correctly oriented relative to each other.

In the liquid state, water molecules are in constant motion, and hydrogen bonds are continually being formed and broken. The arrangement of the molecules is random, and not all possible hydrogen bonds are formed. In solid water (ice), molecular motion is at a minimum, and molecules become oriented so that the maximum number of hydrogen bonds are formed. This results in an ordered, strong, extended, three-dimensional, open lattice structure **(Fig. 10.4a)**. The size of the "holes" in the lattice is dictated by the bond angle in the water molecule, which determines how close adjacent molecules can be. As a result, adjacent water molecules in ice are not as close to each other as they can

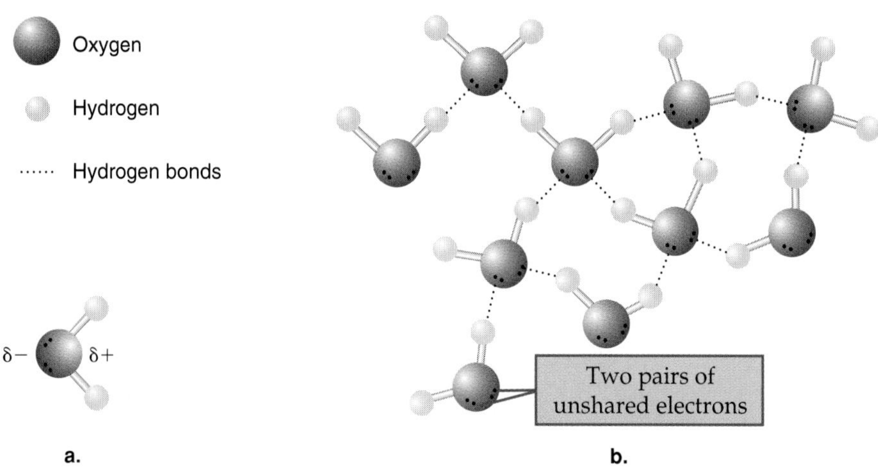

FIGURE 10.3

Hydrogen bonding in water: (a) Water is an angular polar molecule. Because oxygen is more electronegative than hydrogen, it has a partial negative charge (δ−), and the two hydrogen atoms have partial positive charges (δ+). (b) The slightly negatively charged oxygen atoms are attracted to the slightly positively charged hydrogen atoms, and hydrogen bonds form between adjacent water molecules. Each oxygen atom has two pairs of unshared electrons and can bond to two hydrogen atoms. Thus, each water molecule can form four hydrogen bonds.

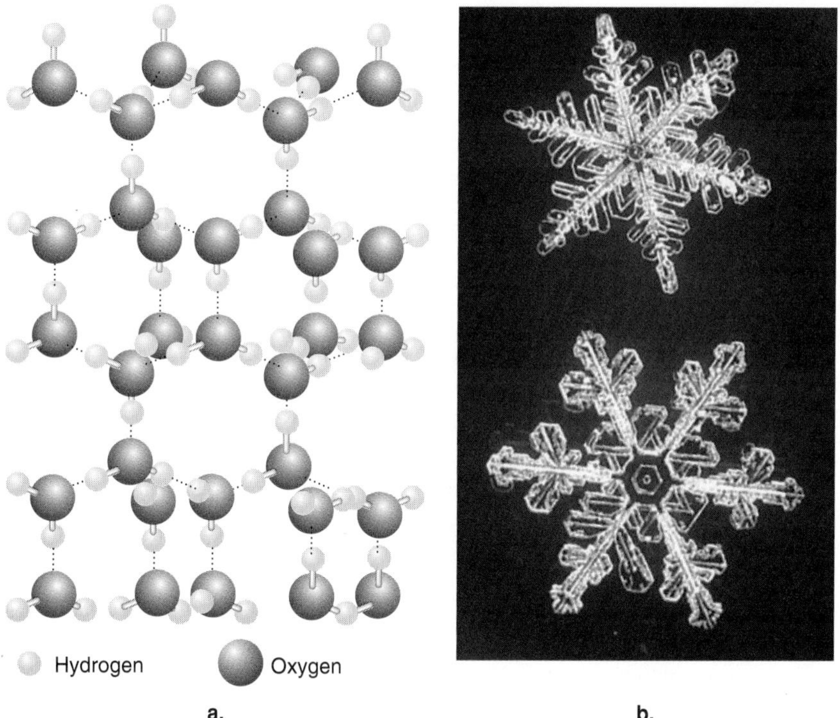

FIGURE 10.4

(a) In the well-ordered three-dimensional structure of ice, each water molecule is hydrogen-bonded to four other water molecules. (b) The ordered arrangement of water molecules in ice accounts for the intricate hexagonal shapes of snowflakes.

be in liquid water. The ordered arrangement of atoms in ice accounts for the symmetry of ice crystals in snowflakes **(Fig. 10.4b)**.

Boiling Point and Melting Point

Compared with hydrogen compounds of other elements in Group VIA of the periodic table (H_2S, H_2Se, and H_2Te), water (H_2O) has an unexpectedly high boiling point **(Fig. 10.5)**. Normally, boiling points in a series of compounds of elements in the same group increase regularly with increasing molecular weight; this does occur for H_2S, H_2Se, and H_2Te. The unexpectedly high boiling point of H_2O is caused by the hydrogen bonding that water molecules engage in; the molecules of the other hydrogen compounds of Group VIA elements do not hydrogen bond to any significant extent. When water is converted to vapor, additional energy in the form of heat is required to break these hydrogen bonds; consequently, water's boiling point is higher than would be expected. If water boiled at the predicted temperature of −80°C (−112°F), it would be a gas at the temperatures found on earth, and life as we know it would not be possible.

Water also has an exceptionally high melting point because of the large quantity of heat energy required to break its hydrogen bonds. When ice melts, about 15% of its hydrogen bonds are broken. The three-dimensional lattice structure collapses, and water forms.

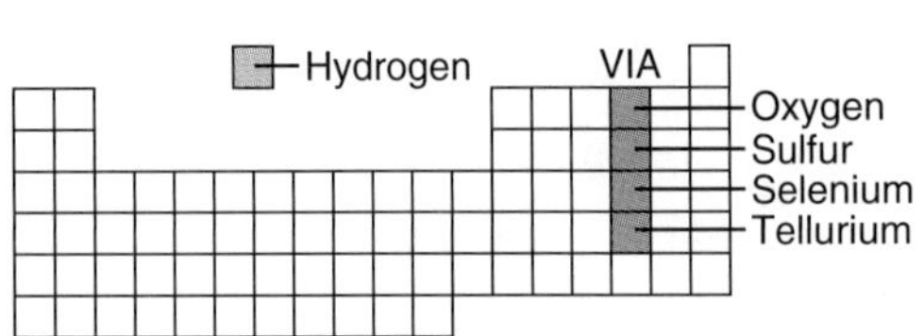

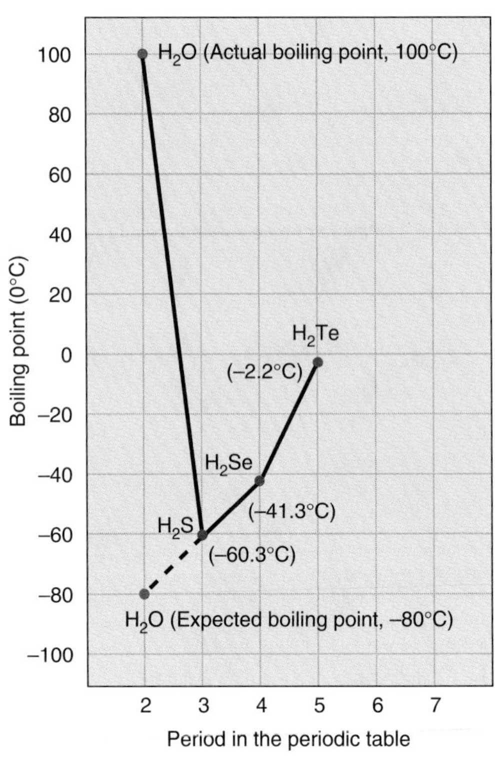

FIGURE 10.5

With the exception of oxygen, the boiling points of the hydrogen compounds of the Group VIA elements increase regularly going down the group. The unexpectedly high boiling point of water is due to hydrogen bonding, which occurs in H_2O but not to any extent in H_2S, H_2Se, and H_2Te.

Heat Capacity

A calorie is defined as the amount of heat required to raise the temperature of 1 gram of water by 1°C.

One food Calorie is equivalent to 1000 calories.

The high heat capacity of water helps organisms maintain a constant internal temperature when outside temperature fluctuates.

Heat capacity is the quantity of heat required to raise the temperature of a given mass of substance by 1°C. It takes 1 calorie of heat to raise the temperature of 1 gram of liquid water by 1°C. Water has the highest heat capacity of any common liquid or solid. The higher the heat capacity of a substance, the less its temperature will rise when it absorbs a given amount of heat, and, conversely, the less its temperature will fall when the same amount of heat is released.

The high heat capacity of water has enormous implications for the earth's climate. Because of water's heat capacity, the oceans can absorb very large amounts of heat without showing a corresponding rise in temperature. The oceans absorb heat from the sun on warm days, primarily in the summer, and release it in the winter. If there were no liquid water to absorb and release heat, temperatures on the earth would fluctuate as drastically as they do on the waterless moon and the planet Mercury, varying by hundreds of degrees during each light-dark cycle.

Heat of Fusion and Heat of Vaporization

Heat of fusion is the amount of heat required to convert 1 gram of a solid to a liquid at its melting point; the same amount of heat is released when 1 gram of the liquid is converted to the solid. **Heat of vaporization** is the amount of heat required to convert 1 gram

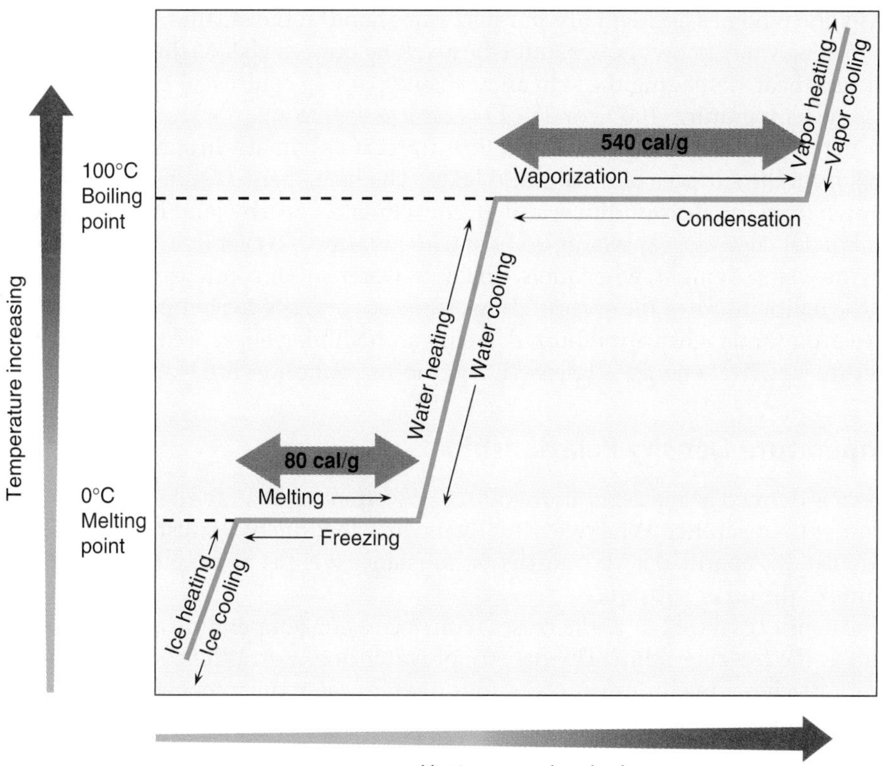

FIGURE 10.6

Changes in state as water is heated and cooled. There is no change in temperature as water melts, vaporizes, condenses, or freezes until the process has been completed. During melting and vaporization of water, 80 and 540 cal/g, respectively, are absorbed; during condensation and freezing, 540 and 80 cal/g, respectively, are released.

of a liquid to a vapor at its boiling point; the same amount of heat is released when 1 gram of the vapor condenses to its liquid.

During the process of fusion (melting), heat energy is absorbed but the temperature of the solid/liquid mixture does not begin to rise above the melting point until melting is complete. During the reverse process of freezing, heat energy is released to the surroundings; the temperature does not begin to fall until freezing is complete. Similarly, there is no change in the temperature of a substance during vaporization and condensation, until each process is complete. The changes in energy and temperature that occur when water changes successively from a solid to a liquid and to a gas, and then back again, are shown diagrammatically in **Figure 10.6**.

Since heat of fusion and heat of vaporization are related to heat capacity, it is not surprising that their values are higher for water than for practically any other substance. Again, the explanation is hydrogen bonding: For ice to melt and for water to vaporize, hydrogen bonds must be broken, and breaking them requires a considerable input of energy in the form of heat.

The fact that a relatively large amount of heat is required to evaporate a small volume of water has important consequences. It means that the human body can be cooled efficiently by the evaporation of a small amount of water (perspiration) from the skin. Extensive water loss from the body, which could upset the internal fluid balance, is thus kept at a minimum.

Anyone who has accidentally put his or her hand in the steam coming from a kettle of boiling water knows how painful the resulting burn can be. As the steam condenses, it releases heat, damaging the skin and causing pain. In contrast, a burn from water at the same temperature (100°C or 212°F) is far less severe.

Water's high heat of vaporization affects the earth's climate. In summer, water evaporates from the surfaces of oceans and lakes. The heat energy needed for evaporation is drawn from the surroundings, and, in consequence, nearby land masses are cooled. On a hot day, land close to a large body of water is always cooler than land farther away from the water. At night, when moist air cools, water vapor condenses, heat is released, and the temperature of the surroundings is raised. In this way, temperature variations between day and night are minimized. A similar modifying effect occurs in winter. When water freezes, heat energy is released, and the surroundings are warmed.

It takes 540 calories to vaporize 1 gram of water at its boiling point, compared to 73 cal/g for mercury and 204 cal/g for ethyl alcohol at their boiling points.

Temperature-Density Relationship

Density is defined as mass per unit volume. We often say that one substance is heavier or lighter than another. What we actually mean is that the two substances have different densities: A particular volume of one substance weighs more or less than the same volume of the other substance.

Wood floats on water because its density is less than that of water; lead sinks because it is more dense than water.

The density of most liquids increases with decreasing temperature and reaches a maximum at the freezing point. The density of water does not. When water is cooled, its density reaches a maximum at 4°C—four degrees *above* the freezing point—and then decreases until the freezing point is reached at 0°C (32°F). The fortunate consequence of this property is that ice floats on the surface of water. This behavior is so familiar that we tend to forget that it is not typical of most liquids. For example, if a piece of solid paraffin is put into a container of liquid paraffin, it sinks to the bottom of the container because it is more dense than the liquid **(Fig. 10.7)**.

FIGURE 10.7

Left: A solid piece of paraffin sinks in liquid paraffin. Right: Ice cubes float in water. This behavior of water is exceptional; most chemical compounds behave like paraffin.

The unusual behavior of ice is the result of the open lattice structure of hydrogen-bonded molecules that forms when water freezes (refer to Fig. 10.4). As noted earlier, molecules of water are farther apart in ice than they are in liquid water. As a result, when water starts to freeze, the number of molecules per unit volume (and thus the mass per unit volume, or the density) decreases.

The fact that ice is less dense than water has important consequences for aquatic life. When the air temperature falls below freezing in winter, water at the surface of lakes and ponds begins to freeze, and a layer of ice forms and floats on the water below it. The ice covering the surface acts as an insulating layer, reducing heat loss from the water under it. As a result, most lakes and large ponds in temperate climates never freeze to a depth of more than a few feet, and fish and other aquatic organisms are able to survive the winter in the water under the ice. If water behaved like most other liquids, ice would sink to the bottom as it formed, and lakes and ponds would freeze from the bottom up. Even deep lakes would freeze solid in winter, and aquatic life would be killed.

The unusual behavior of water when freezing has other consequences for the environment. When water trapped in cracks in rocks freezes, it expands. The force of its expansion is so powerful that the rock may split, an important factor in its weathering.

10.5 Water Use and Water Shortages

As human populations throughout the world have grown, the demand for water has increased tremendously. Water tables are falling on all continents, and many regions, including parts of the United States, are experiencing serious water shortages. In some countries, water scarcity is threatening food production; water pollution has added to the problem by reducing the proportion of water that is drinkable (potable).

There are two sources of usable water: (1) *surface water* in lakes and rivers, and (2) **groundwater** from wells drilled into aquifers (see Fig. 10.2). At the present time, only groundwater that lies within about 1000 meters (about 3000 feet) of the earth's surface can be tapped economically.

Worldwide, 70% of the water drawn from rivers, lakes, and aquifers is used for irrigation and 20% is used by industry. In the United States, industry uses a much higher share, about 60%; irrigation accounts for 30%, and domestic and municipal use accounts for the remaining 10%. Oil refining, steel making, and paper manufacturing all require water; the major U.S. user, however, is the electric power industry, which uses large quantities of water for cooling purposes. This water is recycled at power plant sites and reused. On the other hand, water used for irrigation, which accounts for 70% of all groundwater withdrawals, is consumed and must be continually resupplied.

The arid Plains States depend on the water in the vast Ogallala aquifer to irrigate over 10 million acres of cropland **(Fig. 10.8)**. Although the aquifer is enormous, the current rate at which water is being withdrawn greatly exceeds the rate of recharge, which is extremely slow because the aquifer underlies a region of low rainfall. It has been estimated that, at the present rate of withdrawal, much of the Ogallala aquifer could be dry by 2020.

When withdrawal of water from an aquifer is excessive, the land above it may collapse, producing a large sinkhole **(Fig. 10.9)**. Depletion of groundwater near coastal areas can cause seawater to flow into an aquifer, making its water unfit for drinking. Regions of the United States where groundwater has been depleted are shown in **Figure 10.10**.

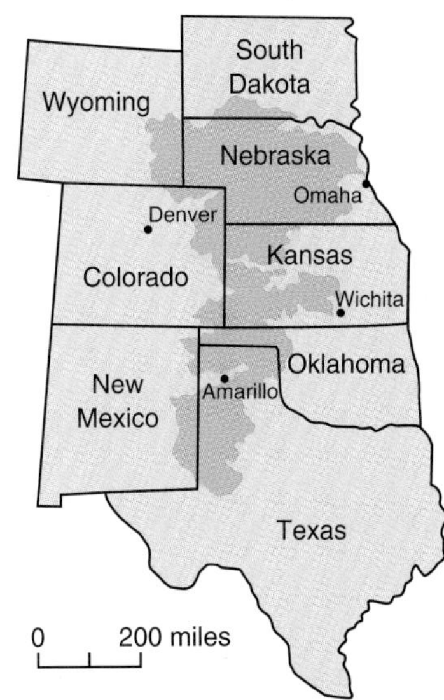

FIGURE 10.8

The vast Ogallala aquifer lies beneath the Great Plains, stretching from South Dakota to Texas. It was formed more than 2 million years ago from melting glaciers. Today, water is being withdrawn from the aquifer much more rapidly than it is being returned to it.

FIGURE 10.9

Excessive withdrawal of groundwater caused collapse of the land above it, forming this sinkhole in Winter Park, Florida.

EXPLORATIONS describes a place where overuse of water for irrigation led to an environmental disaster and the disappearance of a lake (see pp. 310–311).

Potable water for domestic purposes, which is obtained almost equally from surface water and groundwater, comprises approximately 10% of all water used in the United States. Although each person needs to drink only about one-half gallon of water per day, average per capita daily water use in the United States is 90 gallons, divided up as shown in **Table 10.4**. This is nearly 3 times the average per capita daily use worldwide and up to 20 times the average per capita daily use in less developed countries. In some arid areas of the western United States, millions of gallons of water are used

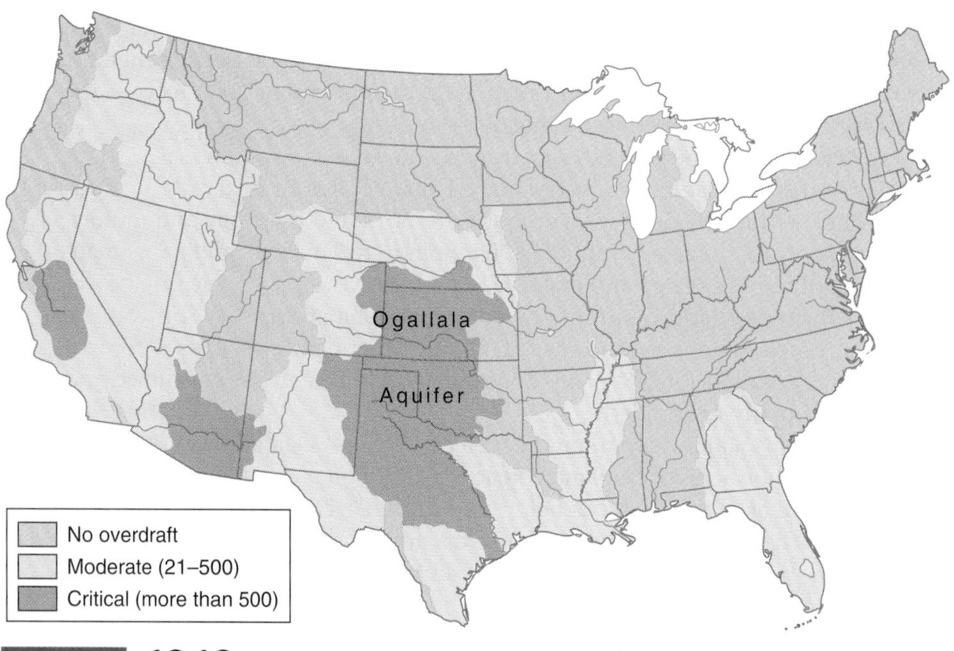

FIGURE 10.10

In many parts of the United States, groundwater is being withdrawn faster than the recharge rate, creating a groundwater overdraft. This map shows the groundwater overdraft in millions of gallons per day.

daily to water golf courses, and average per capita water consumption is approaching 400 gallons per day.

Such extravagant use of water has led to a serious water crisis. While we cannot increase the earth's supply of water, we can take steps to manage the water that we have more efficiently.

Table 10.4 Average Per Capita Daily Water Use in the United States

Use	Rate (gallons)
Flushing toilets	30
Bathing	23
Laundry	11
Drinking and cooking	10
Washing dishes	6
Miscellaneous	10
Total	90

10.6 Water Management and Conservation

Water Transfer

Limited water supplies in one region can often be increased by diverting river water to the depleted area or by building dams along rivers to collect water in large reservoirs to serve the area. However, these undertakings are always controversial. When dams are built, agricultural land and wildlife habitat are lost, and nutrient levels fall in water downriver from the dam. Because it has been subjected to ten major dams and many diversions, the Colorado River, for example, frequently runs dry before reaching the ocean.

River diversion often leads to acrimonious debate between residents of neighboring regions, who are competing for the same water supplies. There have been frequent disputes about allocations from the Colorado River, which supplies water to seven states and is the main source of water for the arid Southwest. For years, California was able to withdraw more than its allotted share because less populous states were not taking their full shares. Now, as their populations have grown, Arizona and Nevada are claiming their allotments. Faced with a reduction in its water supply, California must decide how best to divide it between the needs of the cities and agriculture.

Worldwide, many nations that depend on the flow of water from other countries face similar problems, which in some cases could lead to hostilities. Egypt, which is entirely dependent on the Nile River for water, is concerned about diversions that are planned by Ethiopia, which controls the river's headwaters. In Turkey, dams on the Tigris and Euphrates Rivers will significantly reduce water flow into Syria and Iraq. The dry Middle East relies primarily on the Jordan River for water, and many believe that another war in that region could arise not over oil but over water, for which there is no substitute.

Conservation

Conservation of water is essential if serious shortages are to be avoided. Farmers could reduce their use of water substantially by lining irrigation canals with plastic to reduce loss by seepage and by burying pipes to irrigate from below ground. The commonly used method of spraying water over the land is inefficient because so much water evaporates before it can sink into the soil and reach plant roots.

In most industrialized countries, significant water usage reductions could also be made in the home. The average toilet converts 19 liters (5 gallons) of drinkable water into wastewater with each flush and could be replaced with a water-saving model that works effectively with as little as 2 liters (0.5 gallon) of water. The excessive amount of water released by a conventional shower (almost 40 liters, or 10 gallons, per minute) can be reduced very cheaply by switching to a water-saving showerhead.

Water can also be saved by recycling wastewater from kitchen sinks and laundry tubs and using it for such purposes as flushing toilets and watering lawns (**Fig. 10.11**). Another

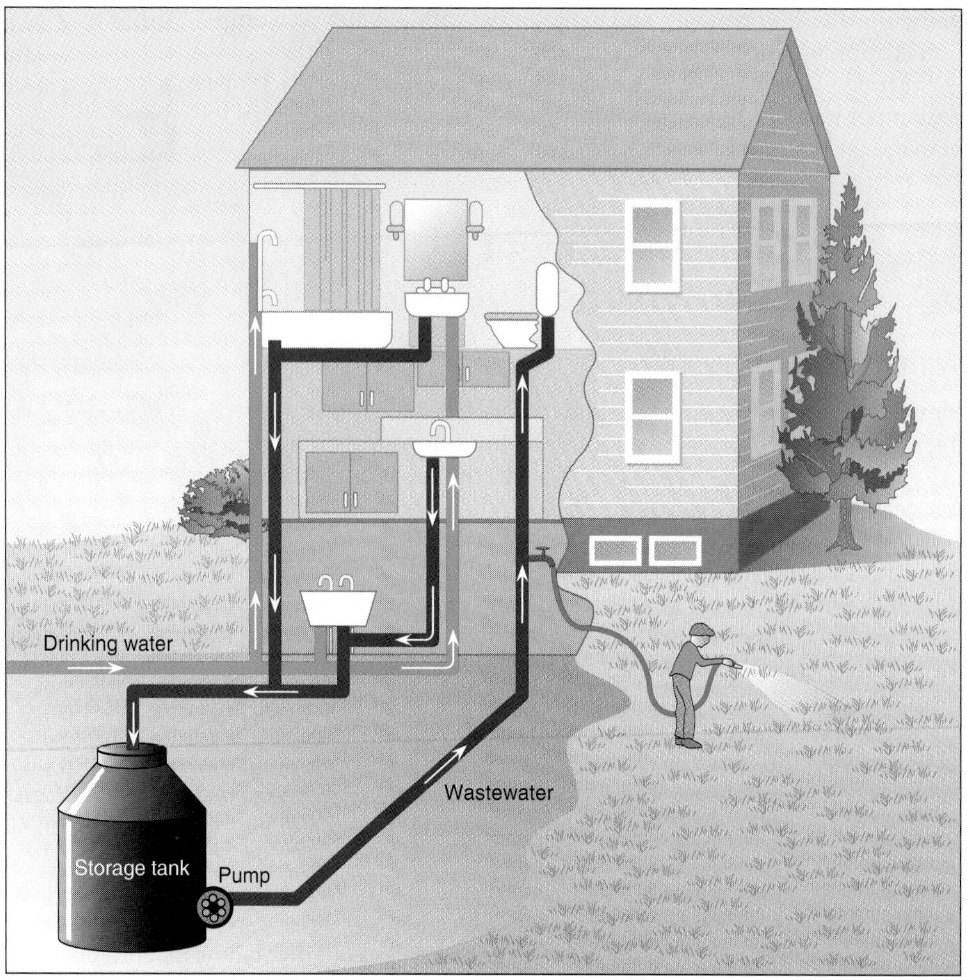

FIGURE 10.11

Water could be saved in homes if wastewater from sinks and bathtubs were piped into a storage tank and reused for flushing toilets, washing cars, and other activities that do not require drinking-quality water.

idea is for public water systems to provide two kinds of water: one of drinking quality and a second that would not meet all drinking-water standards but would be free of bacteria and suitable for many other household purposes.

In the United States and many other countries, there is currently little incentive to economize because water is so cheap. Farmers would be more likely to use efficient irrigation methods if they were not so heavily subsidized by their governments. And if water prices were higher, people in the dry Western states of the United States might exchange their green golf courses and lawns for more suitable desert vegetation.

Potable Water from Wastewater

In most urban areas of developed countries, wastewater from homes, businesses, and industrial sites flows through a network of sewer pipes to sewage treatment plants. After it has been treated, this water is discharged into rivers or oceans. Water could be conserved—and money saved—if the purified water from sewage plants were recycled into public water systems. Although the idea may not be attractive to many people, it is technologically feasible to obtain drinking-quality water from sewage effluent. We will consider wastewater treatment in detail later in the chapter.

Desalination

One way to increase the supply of freshwater is to employ **desalination,** the process of removing salts from seawater and brackish water. Although the technology exists, desalination is very expensive because it requires a large input of energy. Consequently, this process is used only in regions where there is no alternative source of freshwater. Another problem with desalination, besides its high cost, is disposing of the mountains of salt it produces. If the salt is returned to the ocean near a coast, the increased salt content could have an adverse effect on wildlife in the coastal waters.

Distillation and reverse osmosis are the most widely used methods for converting salt water to freshwater. In **distillation,** salt water is heated to evaporate off the water; the water vapor is condensed to pure water, leaving the salts behind. Since the need for water is greatest in hot arid lands where sunlight is plentiful, energy costs can be reduced by using solar radiation to heat the salt water.

In the process of **osmosis,** two compartments are separated by a *semipermeable membrane,* a thin sheet of material that contains minute pores through which water molecules, but not larger ions or molecules, can pass. When pure water is placed on one side of the membrane and salt water on the other **(Fig. 10.12a)**, water molecules pass through the membrane from the water compartment to the salt compartment by the process of *osmosis.* The level of liquid in the salt compartment rises and reaches a maximum at equilibrium (when the movement of water molecules back and forth between the two compartments is equal) **(Fig. 10.12b)**. The pressure that must be exerted to prevent this osmosis is termed the *osmotic pressure.* In the process of **reverse osmosis,** pressure in excess of the osmotic pressure is exerted on the liquid in the salt compartment, and pure water flows from the salt compartment through the membrane into the water compartment **(Fig. 10.12c)**. Reverse osmosis plants are a major source of freshwater in such arid places as Israel, Saudi Arabia, and the Mediterranean island of Malta, where freshwater is in very short supply.

10.7 Water Pollution: A Historical Perspective

Human waste was the first pollution problem. In ancient times, people naturally settled near sources of water, so communities grew up beside lakes, along rivers, and in areas where spring or well water was available. People often drank water from the same river in which they disposed of their wastes and in which they washed themselves and their clothes. As a result, they often became sick because their drinking water was contaminated with disease-causing microorganisms from human and animal wastes.

During the Industrial Revolution of the 19th century, cities in the United States and Western Europe grew at a tremendous rate. Refuse of all kinds—including great quantities of horse manure—ended up in the streets, in open sewers, and in nearby rivers. Devastating epidemics of water-borne diseases such as cholera, typhoid, and dysentery were common in large cities.

By the early part of the 20th century, the connection between diseases and sewage-borne microorganisms had been recognized, and safe water supplies were established in most industrialized nations. As a result, water-borne diseases have been virtually elim-

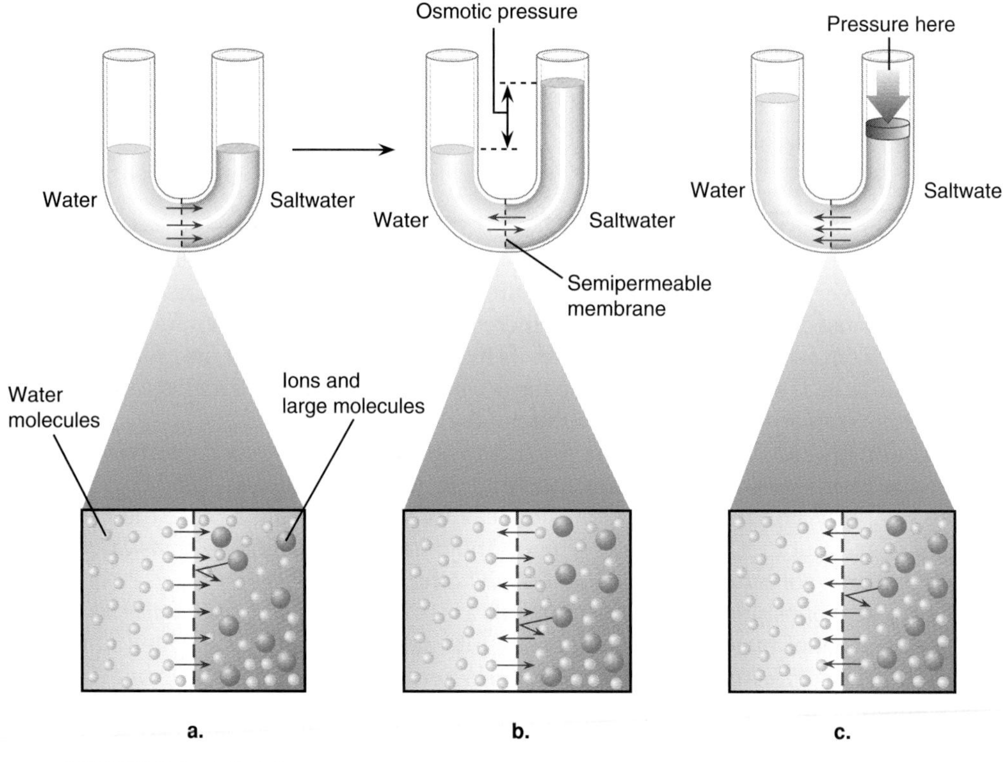

FIGURE 10.12

The processes of osmosis and reverse osmosis: (a) Water and a salt solution are separated by a semipermeable membrane. Osmosis causes water molecules to pass through the membrane into the salt solution, raising the level of liquid in that compartment. (b) The difference in levels between the two compartments represents the osmotic pressure. (c) In reverse osmosis, pressure in excess of the osmotic pressure is applied to force water molecules from the saltwater compartment into the other compartment.

inated in those countries. However, they are still very common in less developed countries, where waste disposal systems are often inadequate or nonexistent. In industrialized nations, contamination with hazardous chemicals has become the main threat to water supplies.

10.8 Types of Water Pollutants

Water pollutants can be divided into these broad categories: (1) disease-causing agents, (2) oxygen-consuming wastes, (3) plant nutrients, (4) suspended solids and sediments, (5) dissolved solids, (6) toxic substances, (7) heat (thermal pollution), (8) radioactive substances, (9) oil, and (10) acids. In this chapter, we consider the sources of the first seven types of pollutants, as well as their effects on the environment and the steps that can be taken to control them. Radioactive substances will be discussed in Chapter 19, and oil will be considered in Chapter 12. Acids were covered in Chapter 7.

Point and Nonpoint Sources of Water Pollutants

Pollutants enter waterways from both point and nonpoint sources (Fig. 10.13). **Point sources** include sewage treatment plants, factories, electric power plants, mines, and offshore oil-drilling rigs—in other words, sources that discharge pollutants at specific locations, usually through pipes. **Nonpoint sources** discharge pollutants over a wide area through runoff; these sources include feedlots, cultivated land, forests that have been clear-cut (i.e., every tree has been cut down), cities, and construction sites. Discharges from nonpoint sources are generally more dilute than those from point sources, but they are more difficult to identify and regulate. They occur irregularly, often as the result of severe storms.

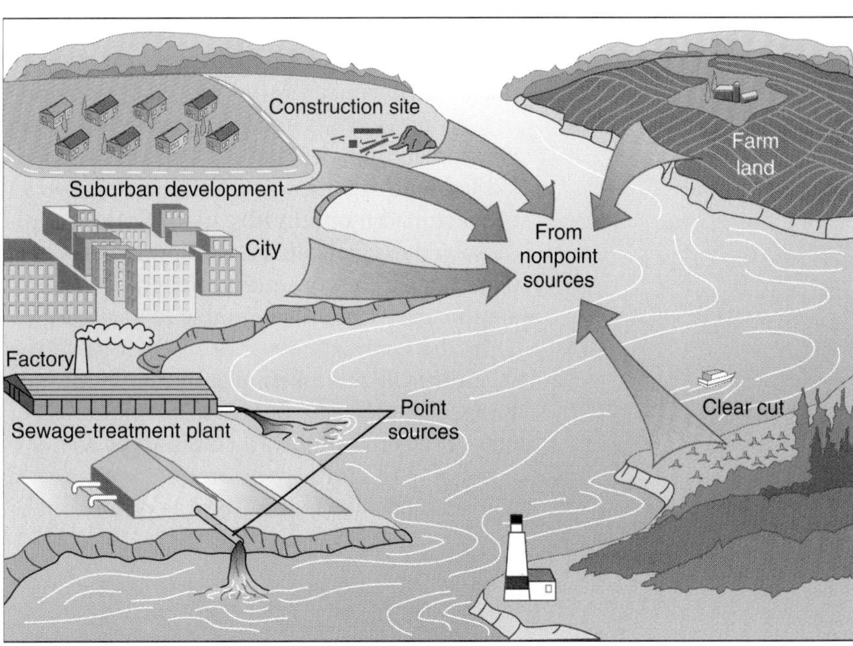

FIGURE 10.13

Point and nonpoint sources of water pollutants.

10.9 Disease-Causing Agents

Outbreaks of disease, sometimes of epidemic proportions, can occur if feces from people infected with **pathogens** (disease-causing agents) contaminate water supplies. Diseases that are transmitted when people drink contaminated water or swim in it include cholera, typhoid fever, dysentery, infectious hepatitis, and polio. According to the United Nations, in the developing countries, where clean water is scarce, as many as 10 million people, half of them children, die each year from drinking pathogen-contaminated water.

The coliform bacteria count is used to determine whether water has been contaminated by microorganisms. Coliform bacteria live naturally in the human intestinal tract, and the average person excretes billions of them in feces each day. Coliform bacteria are harmless and cause no diseases, but their presence in water is an indication of fecal contamination. If none are found, the water is free from fecal contamination and can be assumed to be free from pathogens.

10.10 Oxygen-Consuming Wastes

Animals and plants that live in aquatic habitats depend on oxygen dissolved in the water for their survival. Oxygen is not very soluble in water, and the amount that does dissolve depends on both the temperature and the altitude of the water. As shown in **Figure 10.14**, the solubility decreases with increasing temperature and with increasing altitude. At 20°C (68°F), at sea level, the concentration in oxygen-saturated water (water that contains the maximum amount of oxygen it can dissolve) is about 9 ppm, compared to only 6 ppm at 30°C (86°F) at 2000 meters (6600 feet).

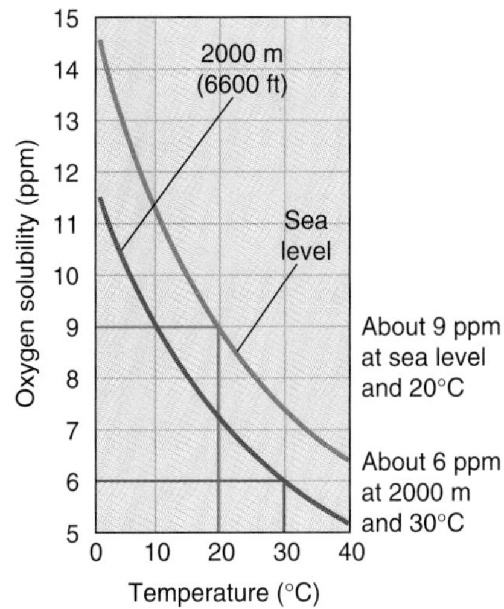

FIGURE 10.14

The solubility of oxygen in water at different temperatures and altitudes. The solubility decreases as temperature rises and as altitude increases.

Dissolved oxygen (often abbreviated as DO) in lakes and rivers is rapidly depleted if organic waste materials are released into the water. Typical examples of such wastes—collectively called **oxygen-consuming wastes**—

are human and animal feces and industrial wastes from paper mills, tanneries, and food-processing plants. Wastes from slaughterhouses and meat-packing plants are a particularly concentrated source of oxygen-consuming wastes.

Organic detritus in aquatic ecosystems is ordinarily broken down by aerobic (oxygen-consuming) decomposers, primarily bacteria and fungi (Chapter 2). If water is overloaded with organic wastes, the aerobic decomposers proliferate, and dissolved oxygen is consumed more rapidly than it can be replaced from the atmosphere. If the level of dissolved oxygen falls below 5 ppm, fish—particularly desirable game fish—start to die **(Fig. 10.15)**. If the concentration of dissolved oxygen continues to fall, invertebrates and aerobic bacteria will be unable to survive.

In the complete absence of dissolved oxygen, decomposition of organic detritus continues but it is taken over by anaerobic (non–oxygen-requiring) bacteria. If this stage is reached, the water begins to smell unpleasant because different decomposition pathways are being followed **(Table 10.5)**. Many of the end products of anaerobic decomposition—including hydrogen sulfide, ammonia, amines, and phosphorus compounds—have disagreeable odors.

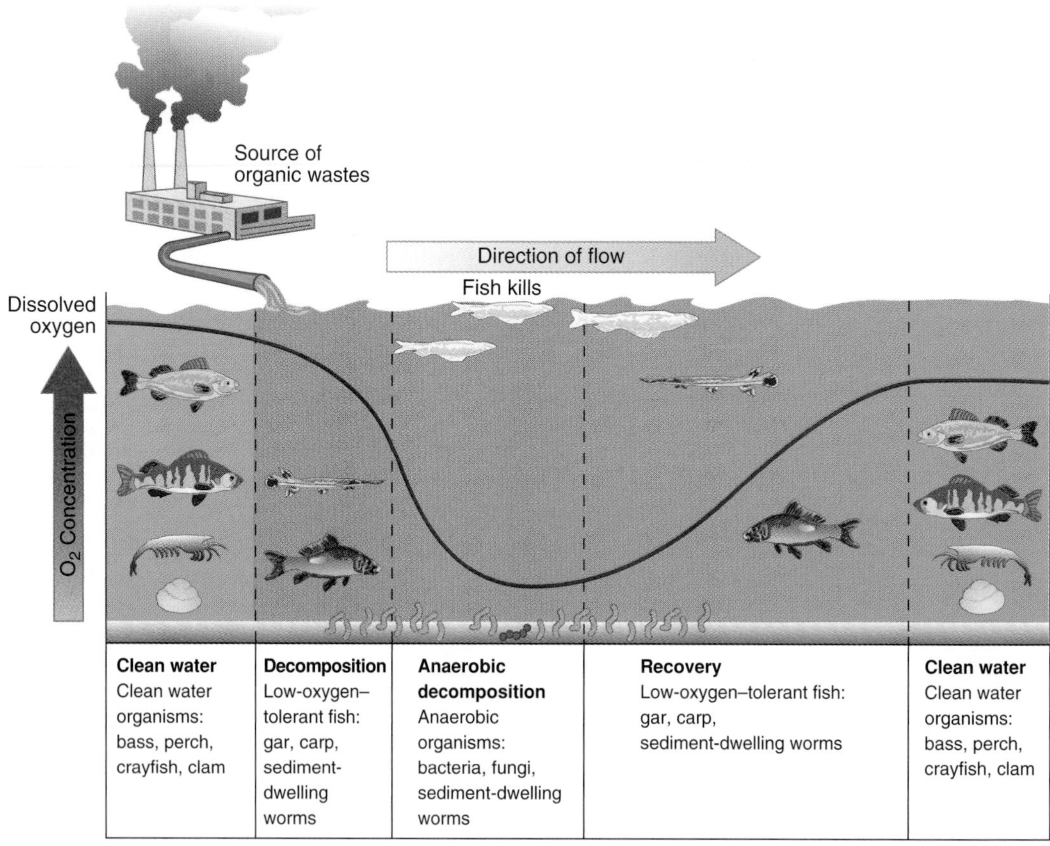

FIGURE 10.15

If organic wastes are discharged into a river, the level of dissolved oxygen in the water falls, and aquatic organisms begin to die. First to be affected by the decrease in dissolved oxygen are crayfish, clams, and game fish such as bass and perch. If the level of dissolved oxygen continues to fall, aerobic bacteria disappear, other fish such as gar and carp die, and decomposition is taken over by anaerobic bacteria. Once the waste discharge ceases, the river recovers.

Table 10.5 End Products of Decomposition of Organic Compounds under Aerobic and Anaerobic Conditions

Element in Organic Compound	End Product(s) of Decomposition	
	Aerobic conditions	Anaerobic conditions
Carbon (C)	CO_2	CH_4
Nitrogen (N)	NO_3^-	NH_3 and amines
Sulfur (S)	SO_4^{2-}	H_2S
Phosphorus (P)	PO_4^{3-}	PH_3 and other phosphorus compounds

In an undisturbed environment, eutrophication occurs naturally over thousands of years as lakes age and nutrients are slowly enriched. When human activities add nutrients, the process is greatly accelerated.

In a river, water downstream recovers quite quickly once the discharge of organic wastes ceases, particularly if the water is free-flowing and turbulent. Organic material is diluted, and the water is reoxygenated as it is brought to the surface. Lakes, which have little flow of water, take much longer to recover.

The effects of eutrophication are most severe in reservoirs, lakes, and estuaries, or near shorelines where there is little or no flow of water.

10.11 Plant Nutrients

Aquatic plants need the elements nitrogen (as nitrates) and phosphorus (as phosphates) for proper growth. These elements are usually scarce in most natural bodies of water, and this shortage normally keeps the spread of vegetation under control. However, if the amount of nitrogen or phosphorus in the water is increased, excessive plant growth occurs, and the algae population explodes.

As a result of this phenomenon, which is termed **eutrophication,** dense mats of rooted and floating plants are formed. Waterways become clogged, and boat propellers are fouled. The cost of removing the unwanted vegetation is usually prohibitive. The carpets, or *blooms,* of green algae and cyanobacteria that appear on the water's surface **(Fig. 10.16)** release substances that are unpleasant smelling. The water becomes turbid, and once-attractive recreation areas are spoiled. As the vegetation and algae decay, they consume oxygen dissolved in the water, producing the results described in the preceding section. The effects of eutrophication are most severe in reservoirs, lakes, and estuaries, or anywhere there is little or no flow of water.

A main cause of eutrophication is domestic sewage, which is made up largely of nitrogen-containing human wastes and, except where their use is prohibited by law, phosphorus-containing detergents. In the United States, 20 states now ban or limit phosphates in laundry detergents. Other causes are animal wastes from feedlots, runoff from agricultural land that has been treated with nitrate and phosphate fertilizers, discharges from industries, and phosphate mining.

Excess nutrients in streams and rivers often become mixed with sediments (see the next section). When these nutrient-rich sediments reach the ocean, they cause eutrophication. The resulting blooms of ocean algae produce toxic substances that kill aquatic life, and as a result many coastal regions, including a considerable area in the Gulf of Mexico, are now dead zones. Wastes from coastal fish farms add to the problem. Ocean algal blooms (often known as *red tides* because many of the algae are red) occur naturally and have been known since Biblical times, but they are occurring with increasing frequency and severity.

FIGURE 10.16

If wastewater that contains plant nutrients is discharged into a natural body of water, excessive growth of algae and other aquatic plants occurs.

Control of Eutrophication

Action to control eutrophication has been focused on limiting discharges of phosphorus-containing wastes, which come mainly from point sources: sewage treatment plants, industrial plants that use large amounts of phosphorus-containing cleaning agents, and phosphate mines. Nitrogen-containing wastes are discharged primarily from nonpoint sources: agricultural land treated with manure or nitrate fertilizers, slaughterhouses, and stockyards. Nitrogen is usually present in wastes as nitrates, compounds that are very soluble in water and therefore difficult to remove. Phosphates, on the other hand, are much less soluble and can be removed by precipitation before the water is discharged (see the section on sewage treatment later in this chapter).

Limiting phosphate discharges is effective in controlling eutrophication. Without an adequate supply of phosphorus, excessive plant growth is greatly reduced even if sources of nitrogen remain abundant.

10.12 Suspended Solids and Sediments

As a result of the natural erosion of rock and soil, all bodies of water contain undissolved particles termed *sediments.* Fine particles of clays remain suspended in water for months; coarser particles, like sand and silt, settle out quite rapidly.

Many human activities increase the formation of sediments. For example, bulldozing for housing developments, clear-cutting of timber, strip mining, overgrazing, and plowing are all practices that remove natural ground cover and accelerate soil erosion. Soil loss is greatest where the ground slopes steeply. It has been estimated that construction sites contribute 10 times as much sediment per unit area as farmland, 200 times as much as grassland, and 2000 times as much as undisturbed forest.

Increased loads of sediments in streams, rivers, and lakes can cause many problems. Suspended particles make water turbid, thereby reducing light penetration and slowing the rate of photosynthesis. As sediments settle, they can bury bottom-dwelling organisms and cover fish spawning grounds, generally disrupting aquatic habitats. The effects of suspended solids can be especially damaging where rivers meet the sea in estuaries and bays—regions that are particularly important as breeding grounds for fish and shellfish and as feeding grounds for birds and other wildlife.

Sediments cause problems by filling irrigation ditches and clogging harbors and lakes. Also, when toxic substances such as metals and pesticides are released into turbid water, the toxins adhere to the suspended particles and become concentrated in sediments. Then, a subsequent disturbance in the water, such as an increase in acidity, may release the toxins. In clear water, where there are few suspended particles, toxic substances cause fewer problems because they are more likely to remain in solution and gradually become diluted to insignificant concentrations by the flow of water.

10.13 Dissolved Solids

Freshwater always contains some dissolved solids. The ions in solution (see Table 10.3) are essential for the normal growth of most life forms, but at concentrations above certain levels, they become harmful and may destroy freshwater fish and other organisms. Freshwater that contains an abnormally high concentration of dissolved ionic salts is termed *saline water.* If water contains over 500 ppm of such salts, the U.S. Public Health Service does not recommend it for drinking. The salts in saline water may not be toxic, but they are considered pollutants if they are present in concentrations high enough to render the water unsuitable for normal purposes.

Irrigation is the main cause of increased salinity of natural waters. All over the world, food production has been increased dramatically by irrigating once-arid land. Unfortunately, this practice can lead to serious problems.

Agricultural land that is irrigated is located in regions where rainfall is sparse and the climate is generally hot and dry. Consequently, the rate of evaporation of water from the soil is high, and salts tend to accumulate on the soil surface. Any runoff from the land and any subsurface drainage, if returned to the land, carry an increased load of dissolved salts.

Because rainfall is scarce on irrigated land, the only way to remove excess salts is by further irrigation. Frequently, irrigation water is recycled, and it becomes increasingly saline with each recycling. It is estimated that 25–30% of the irrigated cropland in the United States suffers from increased salinity, and that thousands of acres have ceased to be productive. Crops that are particularly susceptible to increased salinity include beans, carrots, and onions; cabbage, broccoli, and tomatoes are more tolerant.

Other causes of increased salinity of natural waters are discharges from industrial and municipal waste-treatment plants and runoff from fertilized agricultural land and urban areas. Salt spread on roads to de-ice them can be a factor in regions where winters are severe.

By the natural processes of erosion, solids are continually added to a river as it flows to the sea. Thus, water near the mouth of a river always contains more dissolved solids than does water near the source. It has been estimated that in many rivers in the United States, including the Colorado, as much as half the load of dissolved solids in the river water at the mouth is due to human activities.

10.14 Toxic Substances

A **toxic substance** is one that, when ingested by an organism, causes harm by interfering with the organism's normal metabolic processes. In the context of water pollution, only substances that cause harm when present at very low concentrations—parts per trillion to parts per million—are generally labeled toxic. Many substances, including common table salt, are harmful at high concentrations but are not termed toxic substances.

Of particular concern in aquatic ecosystems is **bioaccumulation,** the process by which a toxic substance becomes more concentrated as it moves upward through a food chain (Chapter 2). For example, lake water may contain an insignificant concentration of a toxic substance, but as a result of bioaccumulation, the substance can become increasingly concentrated as it passes from the water to microorganisms, then to small fish, and finally to large fish (**Fig. 10.17**). The concentration in large fish may be high enough to cause harm to birds of prey or humans who eat the fish. The concentration of a toxic substance in large fish is a useful measure of toxic pollution in a body of water.

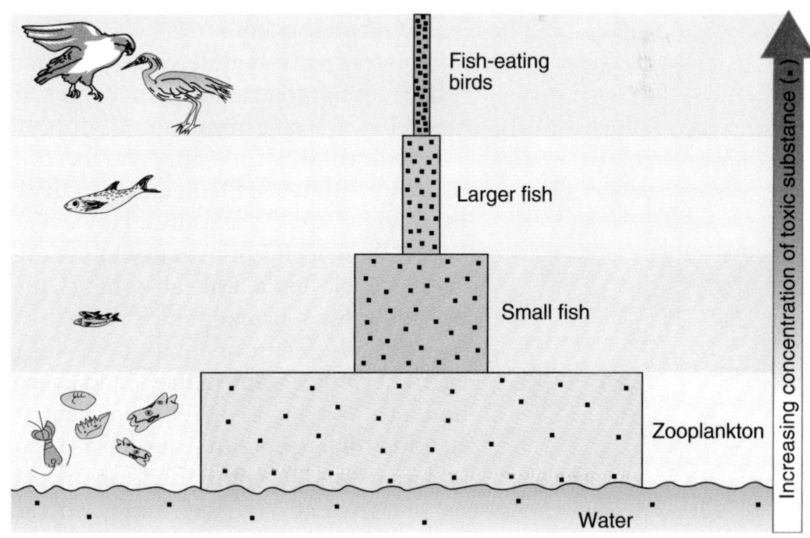

FIGURE 10.17

As a result of bioaccumulation of a toxic substance in an aquatic food chain, the concentration of the substance in a fish-eating bird may be more than a million times as great as the concentration in the water.

Table 10.6 Maximum Permissible Levels for Metals in Drinking Water

Metal	Maximum Level (ppb)
Antimony	6
Arsenic*	50
Barium	2000
Beryllium	4
Cadmium	5
Chromium (total)	100
Copper	1300
Lead	15
Mercury (inorganic)	2
Selenium	50
Thallium	2

Source: EPA, June 1999.

*In 2001, the maximum permissible level for arsenic was lowered to 10 ppb.

Because aquatic ecosystems have from four to six trophic levels (Chapter 2) bioaccumulation is a more serious problem in these ecosystems than in terrestrial ones, which usually have only two or three trophic levels. Toxic substances that are of particular concern are metals and organic chemicals. A variety of industrial activities and farming practices can result in the release of these substances into water supplies.

Toxic Metals

Many metals—including sodium, potassium, calcium, magnesium, copper, and zinc—are essential for the normal development and well-being of humans and other animals. Other metals, even if present in an organism at very low concentrations, are toxic. The EPA has established maximum permissible levels in drinking water for the following toxic metals: antimony, arsenic (which is not a metal but is usually listed with them), barium, beryllium, cadmium, chromium, copper, lead, mercury, selenium, and thallium (**Table 10.6**). Municipal water supplies are monitored constantly to check that these limits are not exceeded.

Because metals are elements, they cannot be broken down into simpler, less toxic forms. They persist unchanged in the environment for many years and bioaccumulate at the tops of food chains. Toxic metals can cause brain damage, kidney and liver disorders, and bone damage. Many are carcinogens (Chapter 18). Toxic metals enter waterways from two main sources: industrial waste discharges and particulates in the atmosphere that settle out and are carried in runoff. We will consider three metals—mercury, lead, and cadmium—that are particularly dangerous because of their widespread use and high toxicity.

MERCURY Mercury is the only common metal that is a liquid at room temperature. It is a component of many rocks and is released continually, but very slowly, into natural waters by normal chemical weathering processes. Small quantities of mercury also vaporize into the atmosphere from mercury-containing rocks in the earth's crust. Concentrations of mercury in water from natural sources generally amount to a few parts per billion.

In bodies of water, mercury and mercury compounds tend to settle at the bottom and adhere to sediments. Elemental mercury and its inorganic compounds are not readily absorbed by aquatic creatures. However, anaerobic bacteria in bottom sediments can convert these forms of mercury to the organic methyl mercury ion (CH_3Hg^+), which is very soluble and readily absorbed. Unless swallowed in considerable quantity, mercury and most of its inorganic salts cause little harm because they are rapidly excreted from the body. Methyl mercury, however, remains in the body for months and can damage the nervous system, kidneys, and liver and cause birth defects. Methyl mercury bioaccumulates as it moves up the food chain; for this reason, it can be dangerous to eat large quantities of large ocean fish such as tuna and swordfish.

Mercury and its compounds are used in the production of chemicals, paints, plastics, pharmaceuticals, steel, and electrical equipment. They are also used as algicides in paper manufacture and as fungicides to treat seeds. Mercury and its compounds can thus reach natural bodies of water from manufacturing or farming sources as well as from sites where mercury-rich zinc and silver ores are mined. The major source of mercury emissions into the atmosphere is the burning of coal, which always contains some mercury, by electric power plants. Another source is waste incineration. In both cases, mercury-containing fly ash is released from smokestacks and can enter bodies of water as it falls back to earth in rainfall.

Inhalation of mercury vapor or absorption of mercury salts through the skin causes serious neurological problems. In the 19th century, workers making felt hats from animal pelts suffered brain damage as a result of absorbing mercury while they washed skins in solutions of mercury(II) nitrate. This is the origin of the expression "mad as a hatter" and the explanation for the name and behavior of the Mad Hatter in Lewis Carroll's *Alice in Wonderland.*

In the 1950s, the toxicity of mercury was tragically demonstrated when a plastics factory in Japan released large quantities of mercury compounds into Minimata Bay. Fish and shellfish became contaminated with methyl mercury, and many people who relied heavily on fish from the bay for their diet became sick. Forty-four people died, and others suffered severe neurological symptoms, including numbness, impaired vision, paralysis, and brain damage.

A worse tragedy occurred in Iraq in 1972. Over 6000 people became sick and nearly 500 died after eating bread made from wheat grain that had been treated with a methyl mercury fungicide. The imported grain was intended for planting, but the warning label was written in a language the Iraqis did not understand.

In the United States, the main concern is the mercury content of fish, which is so high in fish caught in the Great Lakes that people are advised not to eat them. Because the conversion of mercury to methyl mercury is accelerated in acidic water, the acidification of lakes (Chapter 7) can increase concentrations of this toxic form of mercury in fish.

The biological mechanism that accounts for mercury's toxicity will be considered in Chapter 18.

LEAD Lead has many useful properties. It has a low melting point, does not corrode, and is malleable and dense. Its major use in the United States is for the manufacture of lead-acid storage batteries (Chapter 7). It is used as solder, is a component of many alloys, and is present in glazes used on some ceramic ware. Until restricted by law, lead compounds were ingredients in paint (Chapter 9).

In the past, plumbing was a source of unacceptably high lead levels in drinking water. Today, although plastic piping has replaced metal piping in many buildings and is used in new construction, some older buildings still receive water through lead pipes or through copper pipes sealed with lead solder. The amount of lead that dissolves is highest in acidic water that has a low mineral content, particularly if the water has been in contact with lead for a long time. The current limit for lead in drinking water is 15 ppb (see Table 10.6). Until quite recently, canned foods and beverages, particularly if they were acidic, were another source of lead because can seams were soldered with lead.

The word *plumbing* is derived from *plumbum*, the Latin name for lead.

In 1990, many older lead-lined water coolers had to be replaced when it was discovered that water taken from them had unacceptable levels of lead—in some cases as high as 200 ppb.

Although lead has been virtually eliminated from gasoline (Chapter 12), lead-containing emissions from automobiles remain a problem in traffic-congested urban areas. Lead in auto emissions adheres to dust particles in the air, the particles settle out onto streets, and the lead enters water supplies in runoff. Lead is present in the earth's crust in many ores, primarily lead sulfide (PbS); mining and smelting of these ores also adds lead compounds to waterways.

Lead is particularly toxic to young children. Even at very low concentrations in the blood, it can retard development and cause brain damage. In adults, it causes neurological disorders and complicates pregnancy. Lead is a cumulative poison and tends to concentrate in bones. From there, it can be remobilized and enter the bloodstream. Lead disrupts metabolic processes primarily by inhibiting the activity of vital enzymes (Chapters 14 and 18).

In the past, paints were often colored red with mercury sulfide and cadmium selenide or yellow with cadmium sulfide.

Cadmium Cadmium is used in electroplating and in the manufacture of paints, plastics, nickel-cadmium batteries, and control rods used in the nuclear power industry (Chapter 13). These products are potential sources of cadmium that contaminates water, but the main cause for concern is cadmium in zinc products. Most zinc ores contain small amounts of cadmium; as a result, some cadmium is usually present as an impurity in zinc metal and zinc compounds produced from the ores. Cadium is chemically very similar to zinc (both are in Group IIB of the periodic table; see Chapter 3), and it is thought that the mechanism of cadmium poisoning may involve the substitution of cadium for zinc in certain enzymes. When cadium enters the body, it accumulates in the liver. Cadmium is a carcinogen; long-term effects include increased blood pressure and liver, kidney, and lung disease.

Zinc and its compounds have many commercial uses. Large quantities of zinc are used to galvanize iron—a process in which iron is coated with zinc to protect it from corrosion. Because zinc is an essential element for plants, it is included in most inorganic fertilizers. Leafy vegetables and tobacco leaves absorb zinc—along with any cadmium associated with it—from water in the soil. Smokers run a much greater risk of cadmium poisoning than nonsmokers do.

In the 1950s, numerous cases of cadmium poisoning occurred when effluents from a zinc mine were discharged into the Zintsu River in northern Japan. Water contaminated with the discharge was used to irrigate rice fields. People who ate the rice developed an extremely painful skeletal disorder known as *itai-itai byo* (or "ouch-ouch" disease). Their bones became brittle from loss of calcium and broke very easily. Many also suffered from abdominal pain, diarrhea, vomiting, liver damage, and kidney failure.

Synthetic Organic Chemicals

Thousands of organic chemicals are synthesized each year for use in insecticides, herbicides, detergents, insulating materials, and many other products, and most of them are not adequately tested for toxicity before being put on the market. Many organic chemicals, known as **persistent organic pollutants (POPs),** persist in the environment for long periods of time and, if they enter waterways, can cause serious health and environmental problems. There is growing concern that POPs may act as hormone disrupters. *Hormones* are the chemical messengers that are produced by specialized glands in the body and control reproduction, growth, and development. They will be considered in more detail in Chapter 14. POPs are also suspected of causing neurological disorders, suppressing the immune system, and increasing the risk of cancer.

The most notorious of the POPs are certain organic chlorine compounds (Chapter 8), called **polychlorinated hydrocarbons.** These compounds pose a threat to aquatic life because they are stable and do not readily break down into simpler, less toxic forms. They persist in the environment for long periods of time and, like toxic metals, bioaccumulate at the tops of food chains. Polychlorinated hydrocarbons are insoluble in water but soluble in fats. They become concentrated in the fatty tissues of fish and of birds and humans who eat the fish.

We will consider polychlorinated hydrocarbons that are of particular concern as water pollutants: DDT, dioxins, and PCBs.

Scientists have observed that DDT concentration may increase as much as 7 million times as it passes through food chains from water to fish-eating birds.

DDT When DDT (dichlorodiphenyltrichlorethane) was introduced in the late 1940s, it appeared to be an ideal insecticide. It was cheap to produce and apparently nontoxic to humans and other mammals. Because it did not break down easily, it continued to kill insects for a long time after it had been applied. Later, however, it was discovered that water contaminated with DDT from aerial spraying and from runoff from treated land was having a devastating effect on fish-eating birds. By the 1950s and 1960s, populations of bald eagles, peregrine falcons, and brown pelicans had shrunk to disturbingly low levels. Also, insects became increasingly resistant to the pesticide, and larger and larger amounts had to be applied to achieve the desired results (Chapter 16).

The first person to draw public attention to the dangers of pesticides was biologist Rachel Carson. In her classic bestseller, *Silent Spring*, published in 1962, she warned that indiscriminate reliance on synthetic organic chemicals to control insects could lead to a spring without songbirds and, eventually, to the disruption of all life. She was vigorously attacked by the chemical and agricultural industries but supported by other scientists and the general public. Rachel Carson is credited with inspiring the environmental movement in the 1970s, which led directly to the creation of the Environmental Protection Agency (EPA).

FIGURE 10.18

As a result of ingestion of DDT, many birds laid eggs whose shells were so fragile that they cracked.

Research has shown that DDT interferes with calcium metabolism in birds; as a result, eggshells (which are composed primarily of calcium compounds) become thin and break when parent birds attempt to incubate the eggs **(Fig. 10.18)**. The use of DDT was banned in the United States in 1973. Since that time, bald eagles and other fish-eating birds have made a dramatic recovery. DDT continues to be produced in the United States for export to developing countries, where its success in controlling the scourge of mosquito-borne malaria outweighs its disadvantages.

DIOXINS *Dioxins* are a family of chlorinated hydrocarbons, but the name has come to be associated with one particular compound, 2,3,7,8-tetrachlorodibenzo-*p*-dioxin, usually abbreviated TCDD. Like DDT, TCDD persists in the environment for a very long time and bioaccumulates upward in food chains.

TCDD is not produced intentionally. Its major source is the burning of chlorine-containing medical and municipal wastes. With the introduction of efficient high-temperature waste incinerators, which convert dioxins to harmless products, emissions from this source have decreased substantially. In the past, TCDD also reached waterways in the effluent from paper mills, which used chlorine to bleach paper pulp. Most mills have switched to a bleaching agent that does not produce TCDD. TCDD is also formed as a by-product of the manufacture of trichlorophenol, a chemical that is used in the manufacture of a variety of herbicides, including 2,4-D (Silvex). Silvex is widely used on croplands and suburban lawns to control broad-leaved weeds.

During the Vietnam War, 2,4-D was mixed in equal quantities with the related herbicide, 2,4,5-T, to make the defoliant Agent Orange. Present in the 2,4,5-T as an impurity was TCDD. Because TCDD is extraordinarily toxic to guinea pigs and is a known carcinogen for many animals, there was concern that humans could also be affected. As a result of these concerns, the EPA banned the use of 2,4,5-T in 1985.

The effects on humans of exposure to Agent Orange have been debated for many years. In 1993, a comprehensive study linked Agent Orange to three types of cancer and two skin conditions, but failed to support Vietnam veterans' claims that the chemical caused birth defects, infertility, and other disorders. However, more recently, the EPA concluded that exposure to TCDD, in addition to increasing the risk of cancer, may also disrupt some reproductive mechanisms and suppress the immune system.

PCBs PCBs (polychlorinated biphenyls) are structurally similar to DDT and, like DDT and dioxins, are fat-soluble and bioaccumulate at the upper levels of food chains. PCBs present in water in negligible concentrations can become over a million times more concentrated in fish. In 1977, production of PCBs was halted in the United States, and disposal of PCB-containing products is strictly regulated. As a result, there has been a dramatic decrease in PCB contamination of fish. Because PCBs are stable, however, they persist in the environment for many years, and fish in many lakes still contain significant levels of these compounds. PCBs continue to be produced and used in Russia and in many developing countries.

PCBs are fire-resistant, stable at high temperatures, and have high electrical resistance. They were widely used as insulating materials in transformers, electrical capacitors, and condensers and are still present in older equipment. PCBs were also used extensively in the plastics industry as plasticizers (Chapter 9). Plastics tend to be brittle; the addition of PCBs makes them more flexible and resistant to cracking.

PCBs were spread widely in the environment and entered surface waters in industrial discharges and in particulates from incinerators. When plastic wastes and other PCB-containing materials are burned, PCB vapors condense on airborne particles that then fall directly onto open water or reach waterways in runoff from the land. PCBs are now found in the body fat of animals living in the farthest corners of the earth, including polar bears in the Arctic and albatrosses on remote Pacific islands.

PCBs cause eggshell thinning and neurological damage in birds, and they impair the reproduction of aquatic species. In humans, they cause chloracne (a serious form of acne) and liver damage. Most important, they can be transferred from mother to fetus through the placenta and from mother to infant through breast milk. PCBs cause stillbirths and retard growth and, like other POPs, have been linked to reproductive disorders, birth defects, and cancer.

In response to growing concerns about POPs, an international conference to plan for the phasing out of their production was held in Montreal in 1998. A major problem is the need to find a cheap pesticide to replace DDT, which is so effective in controlling insect-borne diseases in developing countries.

The chemical formulas of DDT, TCDD and related dioxins, and representative PCBs are shown in **Figure 10.19**. The mechanisms by which POPs disrupt reproductive systems will be considered in Chapter 14.

10.15 Thermal Pollution

A different type of water pollution involves heat and is called **thermal pollution.** The electric power industry and many other industries draw huge volumes of water from rivers and lakes for cooling purposes. During operation, electric power plants produce enormous quantities of waste heat, which is removed in water that is circulated through the plant. As the water circulates, its temperature may rise by as much as 10–20°C. If the warm water is returned directly into a waterway, the temperature of the water at the point of discharge increases.

A rise in water temperature can adversely affect aquatic life by increasing the body temperature of the organisms. High body temperatures can be fatal to some fish, particularly game fish. Warm water also raises respiration rates, which in turn increase oxygen consumption. At the same time, because the solubility of oxygen in water decreases with rising temperature (refer back to Fig. 10.14), the dissolved oxygen content of the water decreases. Fish and other aquatic organisms may die from lack of oxygen if they are not killed directly by increasing temperatures. The life cycles of many aquatic

DDT
(dichloridiphenyltrichloroethane)

a.

2,4-dichlorophenoxyacetic acid
(2,4-D)

2,4,5-trichlorophenoxyacetic acid
(2,4,5-T)

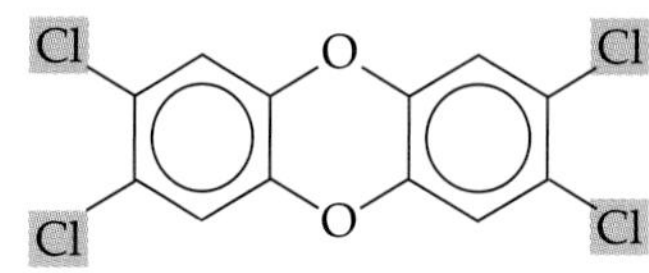

2,3,7,8-tetrachlorodibenzo-*p*-dioxin
(a dioxin)

b.

biphenyl

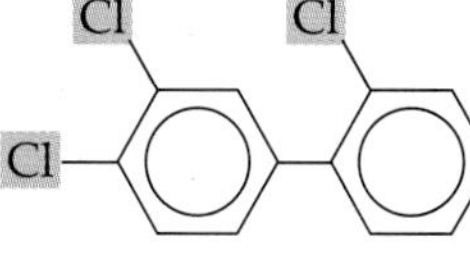

PCB_1

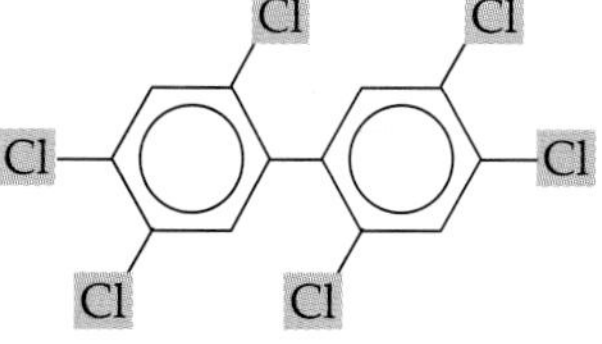

PCB_2

c.

FIGURE 10.19

Chemical structures of (a) DDT, (b) TCDD and other dioxins, (c) biphenyl and two PCBs that can be derived from it.

organisms are controlled by temperature. For example, the number of days it takes for trout eggs to hatch depends on the temperature, and some fish migrate and spawn in response to slight changes in temperature. Any unusual changes in water temperature can disrupt their normal development.

Thermal pollution can be prevented in several ways. Heated water can be made to flow through a cooling pond, or it can be sprayed into a cooling tower and cooled by evaporation. More efficient, but more costly, is a dry tower, in which heated water transfers heat to the surrounding air as it circulates through pipes.

10.16 Pollution of Groundwater

So far in this chapter, we have concentrated on pollution of surface water—lakes, rivers, and streams. Of even greater concern is pollution of groundwater. Because groundwater flows slowly and is renewed slowly, contaminants are not diluted and washed away as they are in a swiftly flowing river. Instead they remain in the water for a very long time. Normally, groundwater is of such high quality that it meets safe drinking water standards without the need for purification or treatment, and it is usually pumped directly from the ground into homes. However, in a number of states, including New Jersey and Florida, formerly pure groundwater has been severely contaminated with hazardous substances, and hundreds of wells have had to be closed.

Not all hazardous substances are toxic, but all toxic substances are hazardous.

The main sources of hazardous materials that pollute groundwater are dump sites where waste chemicals produced by industry are leaking from corroded metal drums. In the past, many industries disposed of their hazardous wastes carelessly—and often illegally—without considering the consequences. Chemical wastes generated in the production of paints, metals, textiles, fertilizers, pesticides, plastics, and petroleum products were often discarded in open dumps and landfills or buried. The problem of cleaning up these dump sites will be discussed in Chapter 19.

Septic tanks and cesspools can pollute groundwater if they are not operated properly or if they are located where the soil's natural filtering capacity is inadequate.

Other types of groundwater pollutants include pesticides and fertilizers from farmland, sewage from septic tanks and leaking sewer pipes, and gasoline leaking from service station storage tanks.

As hazardous chemicals from dump sites and other sources seep down through the ground, some pollutants are filtered out by the soil and travel only short distances. Soluble substances such as nitrates, however, are not filtered out and can percolate downward into groundwater from septic tanks, feed lots, and fertilized land. Consumption of very dilute solutions of nitrates can cause abortions in cattle and methemoglobinemia (blue baby disease) in human infants. Microorganisms in the human digestive tract convert nitrate ions to nitrite ions. Nitrite ions oxidize the iron in hemoglobin from Fe(II) to Fe(III), forming an abnormal hemoglobin called *methemoglobin*. Because methemoglobin is incapable of combining with oxygen, the individual's blood becomes oxygen-deficient.

10.17 Sewage Treatment

Pollution problems can result if cities have a single line for both sewage and storm-water drainage. If the volume of water reaching the sewage plant during heavy rains exceeds the plant's capacity, untreated sewage enters waterways.

Raw sewage is 99.9% water. The water comes primarily from flush toilets, showers and bathtubs, kitchen sinks, laundry facilities, car washes, and storm drains. In the United States, sewage is usually purified in two processes: primary and secondary treatment. Tertiary treatment, a process designed to take care of potentially harmful substances not removed in the first two stages, is currently used by few municipal treatment plants.

Primary Treatment

As the first step in **primary treatment** (Fig. 10.20), the sewage is passed through a screen to remove large pieces of debris, such as sticks, stones, rags, and plastic bags, that have washed in through storm drains. The sewage then enters a grit chamber, where flow rate is slowed just enough to allow coarse sand and gravel to settle out on the bottom. The grit is collected and disposed of in landfills. As water enters the sedimentation tank, its flow rate is further decreased to permit suspended solids—which account for about 30% of the total organic wastes—to settle out as **sludge.** Any oily material floats to the surface and is skimmed off.

Calcium hydroxide and aluminum sulfate are often added to speed up the sedimentation process. These two chemicals react to produce a gelatinous precipitate of aluminum hydroxide, which settles out slowly, carrying suspended material and bacteria with it.

$$3\ Ca(OH)_2 + Al_2(SO_4)_3 \longrightarrow 2\ Al(OH)_3 + 3\ CaSO_4$$

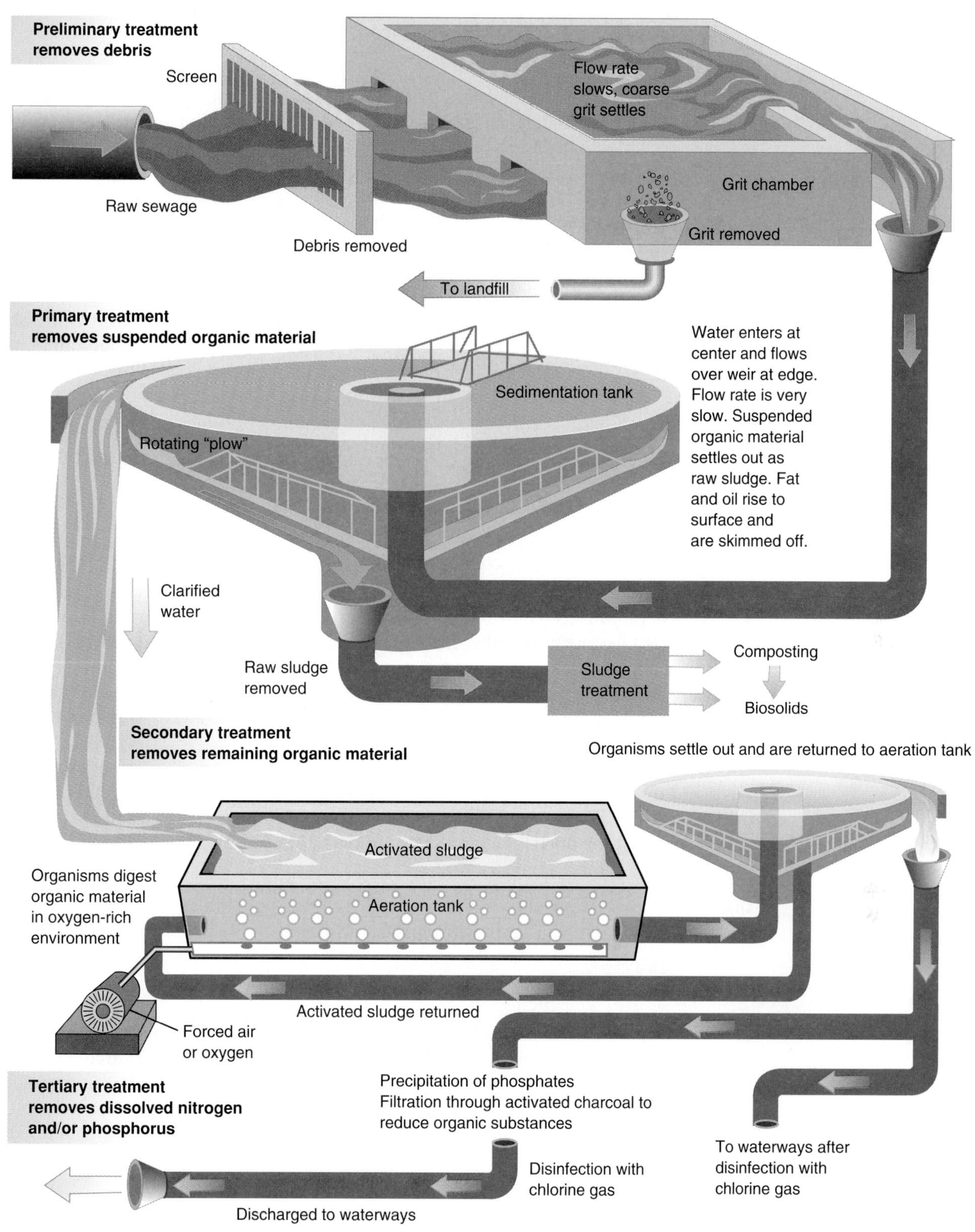
Preliminary treatment
removes debris
Screen
Flow rate
slows, coarse
grit settles
Raw sewage
Grit chamber
Grit removed
Debris removed
To landfill
Primary treatment
removes suspended organic material
Water enters at
center and flows
over weir at edge.
Flow rate is very
slow. Suspended
organic material
settles out as
raw sludge. Fat
and oil rise to
surface and
are skimmed off.
Sedimentation tank
Rotating "plow"
Clarified
water
Raw sludge
removed
Sludge
treatment
Composting
Biosolids
Secondary treatment
removes remaining organic material
Organisms settle out and are returned to aeration tank
Activated sludge
Organisms digest
organic material
in oxygen-rich
environment
Aeration tank
Activated sludge returned
Forced air
or oxygen
Tertiary treatment
removes dissolved nitrogen
and/or phosphorus
Precipitation of phosphates
Filtration through activated charcoal to
reduce organic substances
To waterways after
disinfection with
chlorine gas
Disinfection with
chlorine gas
Discharged to waterways

FIGURE 10.20

The main steps in the processes used for sewage treatment.

In the past, the raw sludge that settled out was incinerated, put in landfills, or dumped at sea. Now, much of it is composted, or otherwise treated, to produce a product known as **biosolids,** a nutrient-rich, bacteria-free, humuslike material that is used as fertilizer. However, there is concern that primary treatment does not remove toxic metals and nonbiodegradable hazardous chemicals that may be present in sewage. These pollutants get into sewage if metal-containing products, pesticides, and other dangerous compounds are carelessly flushed down drains.

In 1972—the year the Clean Water Act was passed—one-third of all sewage treatment plants in the United States chlorinated the water remaining after primary treatment to kill pathogens and then discharged it into waterways without further treatment. This practice caused problems because the discharged water still contained a large amount of oxygen-consuming wastes, which often depleted dissolved oxygen in the waterways and caused eutrophication. Most sewage in the United States now receives both primary and secondary treatment, and, as a result, water quality has improved in many waterways that were once seriously polluted.

Secondary Treatment

Secondary treatment (see Fig. 10.20) is a biological process that relies on aerobic bacteria and other detritus feeders to break down nearly all the oxygen-consuming wastes remaining in the water after primary treatment. In the most frequently used method, a mixture of organisms—termed *activated sludge*—is added to the sewage effluent. Air or oxygen is vigorously bubbled through pipes into the effluent as it moves slowly through a tank. In this oxygen-rich environment, the organisms digest the organic material and break it down into carbon dioxide and water. After settling out in a sedimentation tank, the organisms, together with any remaining undecomposed material, are returned to the aeration tank.

Effluent released from sewage treatment plants after secondary treatment is a major cause of eutrophication because it contains phosphates.

Most municipal sewage treatment plants chlorinate the water after secondary treatment and then release it into waterways. Although about 90% of the original organic matter is removed by primary and secondary treatment, many other substances (including phosphates) are much less completely removed. Also, disinfection with chlorine can introduce hazardous chemicals.

Disinfection with Chlorine Chlorine gas (Cl_2) reacts with water to form hypochlorous acid (HClO), which is a powerful oxidizing agent.

$$Cl_2 + H_2O \longrightarrow HClO + H^+ + Cl^-$$

Hypochlorous acid kills disease-causing bacteria, destroys some (but not all) viruses, and removes color from the water. Disinfection with chlorine gas is relatively cheap, and it has almost completely eliminated the risk of contracting water-borne diseases from municipal water supplies. But chlorine can cause a number of environmental problems. It is very toxic, and there is always the danger of accidental release from tanks during transportation or at the plant. Further, low concentrations of chlorine in water are toxic to fish. Perhaps the most serious problem is the finding that chlorine reacts with residual organic and inorganic substances in the water to form by-products known as **disinfectant by-products (DBPs),** which include chloroform ($CHCl_3$), bromodichloromethane ($CHBrCl_2$), dibromochloromethane ($CHBr_2Cl$), and bromoform (CH_3Br), and are suspected of causing miscarriages, birth defects, and cancer. The bromine-containing compounds are produced as a result of the interaction of chlorine with bromine ions present in the water. The dangers associated with DBPs are minimized if tertiary treatment is used to remove organic substances.

Alternatives to chlorine that are sometimes used for disinfection include chloramine, chlorine dioxide, ozone, and UV radiation. Chloramine and chlorine dioxide produce lower levels of DBPs than chlorine gas does, but they are less effective in killing bacteria.

Ozone (O_3) is a powerful oxidizing agent that is very effective in killing bacteria and, in the process, is converted to oxygen, which improves water quality. When ozone is used, it is manufactured on site by passing oxygen or air through an electric discharge. Ozone disinfection, which is relatively expensive, is used quite widely in Europe but less frequently in the United States. Unfortunately, ozone can also generate by-products that may be associated with health risks.

UV radiation is a relatively new disinfection technique. It is effective in killing microorganisms and is becoming increasingly popular in the United States as an alternative to chlorination. The UV source is a low-pressure mercury arc lamp.

At most sewage treatment plants in the United States, DBPs can be kept below the level allowed by the Safe Drinking Water Act by adjusting the amount of chlorine used and improving the sedimentation process. Where this is not possible, tertiary treatment may be necessary.

Tertiary Treatment

Tertiary treatment (see Fig. 10.20) removes DBPs by passing the chlorinated water that has undergone secondary treatment through granular charcoal that has been activated by heat. DBPs and other organic substances become attached to the surface of the carbon particles and are thus removed from the water. This procedure is effective but expensive. If necessary, reverse osmosis can be used to remove remaining dissolved organic and inorganic substances, including toxic metal ions.

Excess phosphates are removed in one of two ways. One method involves the addition of aluminum sulfate or lime (CaO), which results in the precipitation of phosphate as insoluble aluminum or calcium phosphates:

$$3\ PO_4^{3-} + Al^{3+} + 3\ Ca^{2+} \longrightarrow AlPO_4 + Ca_3(PO_4)_2$$

The second method is similar to the use of activated sludge during secondary treatment. Microorganisms that absorb phosphate are added and then removed, taking the phosphates with them. The procedure also removes some remaining organic materials.

Because tertiary treatment is still experimental and expensive, it is not used by many municipal sewage treatment plants.

Regulation of Water Quality

By the 1970s, many rivers and lakes in the United States had become so polluted that they were unsuitable for recreational activities, and it was often dangerous to eat fish taken from them. Some communities lacked safe drinking water. Since then, water quality has improved dramatically as a result of federal regulations—primarily the Clean Water Act of 1972 and amendments to it in 1977 and the Safe Drinking Water Act of 1974 and amendments to it in 1986. Also, water discharged from industrial plants and municipal sewage treatment plants must meet certain minimum standards before it can be released into natural waterways, and states are required to find ways to control pollution from nonpoint sources.

The Safe Drinking Water Act requires that all communities regularly monitor their drinking water supplies for toxic substances, including certain DBPs, to be sure that they do not exceed the EPA standards for drinking water. Because of the possible risks associated with drinking tap water, many people have switched to bottled water. However,

EXPLORATIONS

The Aral Sea—The Great Lake That Is Disappearing

In 1960, the Aral Sea, which is bordered by Kazakhstan and Uzbekistan in Central Asia, had a surface area of approximately 26,000 square miles and was the fourth largest lake in the world. Only the Caspian Sea, Lake Superior, and Lake Victoria were larger. Today, the Aral Sea is in imminent danger of completely disappearing. In the last 40 years, so much water has been diverted from the rivers that normally flow into it that the lake has lost 75% of its volume, and its area has shrunk by 50%, to 13,000 square miles.

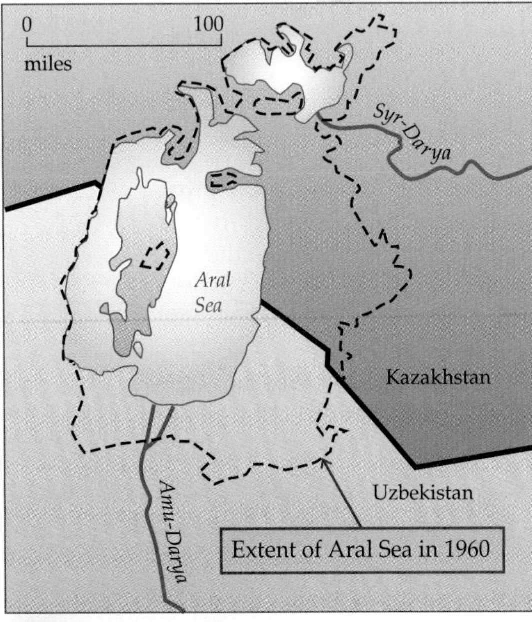

Shrinkage of the Aral Sea during the period from 1960 to 1998.

As the lake has shrunk, the water has become increasingly salty. Since 1960, salinity has risen from 10% to 23%; fish that once were plentiful have disappeared, and the lake is lifeless. The receding water has left behind an 11,000-square-mile desert of salt and sand. Wetlands in the river deltas that once teemed with deer, boar, muskrats, and egrets have dried up, and the animals have vanished.

Causes of the Problem

For thousands of years, the Aral Sea was replenished by two rivers: the fabled 1578-mile-long Amu-Darya (known to Alexander the Great as the Oxus) and the 1370-mile-long Syr-Darya. These two great rivers, which arise in mountains to the southeast, flowed north to the Aral Sea, bringing life to the hot arid steppes they traversed. Today, the rivers frequently run dry before they reach the Aral Sea, because their water has been diverted into immense canals and drawn off into reservoirs to irrigate the world's largest cotton belt.

The seeds of the Aral Sea's destruction were sown almost as soon as the communists came to power in 1917. In the following year, the Soviet government ordered the cultivation of millions of acres of land south of the Aral Sea for the production of cotton. Water for the project was drawn from the Amu-Darya and the Syr-Darya. The original aim was to make the Soviet Union self-sufficient in cotton, but officials soon realized that cotton could become a valuable source of revenue. In 1945, after World War II, cotton became a vital export, and to fulfill quotas set in Moscow, more and more land was brought under cultivation. It became a patriotic duty to increase production, and much of the land previously used for pasture, orchards, or vegetable production was turned over to cotton.

Cotton requires a great deal of water, and the Kara Kum Canal—the longest canal in the world—was built in the 1950s to meet the demand for water to grow cotton. Water for the canal, which runs for 850 miles along a course parallel to the borders of Afghanistan and Iran, was siphoned from the Amu-Darya. In the 30 years following the opening of the canal in 1956, cotton production doubled.

With the completion of the immense canal, the continual withdrawal of water from the rivers began to have a significant effect on the water level in the Aral Sea. As the flow of water into the lake slowed and then almost ceased, the delicate balance between inflow and evaporation that had maintained the water at a constant level was upset. Between 1957 and 1984, the water level in the lake fell 6 feet. It is estimated that if depletion continues at the present rate, the Aral Sea will completely disappear by 2010.

Effects on Local Industry and Climate

The Aral Sea was once the center of a thriving fishing industry. In the early 1960s, 24 native species inhabited the lake, and more than 150 tons of fish were caught daily. Approximately 10,000 fishermen in and around the lakeside town of Muynak made a living catching pike, perch, bream, and other edible species. Thousands of other men and women had jobs at a nearby cannery. Today, the fish are gone. Muynak is 20 miles inland from the lake shore, and fishing boats that once sailed the waters of the lake lie rusting on a salty desert wasteland.

Stranded fishing boat near the Aral Sea.

The decrease in size of the Aral Sea has had a profound effect on the region's climate. In the past, the huge body of water moderated the temperature, having a cooling effect in summer and a warming effect in winter. As the lake has shrunk, temperatures have become more extreme. The region was always dry, but now summers are hotter and dryer than ever before; winters are longer, more severe, and snowless—the wind blows almost constantly. Sand and salt from the drying lake are swept up in frequent dust storms that contaminate the land and threaten the health of the people. Much of the land is so encrusted with sodium chloride and sodium sulfate that nothing will grow. When the wind blows, people taste salt on their lips and salt stings their eyes. Cases of eye disease, respiratory disease, and throat cancer have risen dramatically in the last 25 years.

By the 1980s, the overworked land had begun to lose productivity and immense quantities of fertilizers and pesticides were applied to meet cotton quotas. It has been estimated that each acre of cotton-growing land received 48 pounds of hazardous chemicals annually. Often the pesticides were sprayed when the workers were in the fields. Residues of the chemicals are now found in rivers and groundwater; in 1989, over 80% of the region's drinking water was declared unsafe for human consumption. Trace quantities of pesticides and heavy metals have been detected in breast milk; infant mortality in the area is higher than in any other region in the former Soviet Union, and there has been a dramatic increase in the number of cases of cancer, hepatitis, typhoid, and other illnesses. Women who pick the cotton are glad to have jobs and have little understanding of the cause of their plight.

Barriers to a Solution

Before the collapse of the Soviet Union, no one dared question the directives from Moscow. Since then, people have been speaking out indignantly and calling for action to prevent further loss of water from the lake. Little has been done, however, because of lack of resources and disagreements among the newly independent republics bordering the sea, which are competing for water. In 1998, authorities in Kazakhastan, where the northern basin of the Aral Sea (the Little Sea) is located, began building a dike, made of sand, to completely separate the Little Sea, which is fed by the Syr-Darya, from the much larger southern section (the Big Sea), which is located in Uzbekistan. This action will prevent any water from the Syr-Darya reaching the Big Sea, which was fed primarily by the now dried up Amu-Darya. Evaporation will accelerate, and there is little hope that the whole sea can be saved. Those responsible for building the dam are determined to try to save a small part of it. A major problem is that the dike is being made with sand, not with stone and clay, and eventually it is sure to leak.

In the span of a single lifetime, the obsession for cotton has devastated practically all of formerly Soviet Central Asia.

In the span of a single lifetime, the obsession for cotton has devastated practically all of formerly Soviet Central Asia. It has polluted the land, altered the climate, destroyed the wildlife, impoverished the people, and ruined their health. No matter what is done, the Aral Sea will never again be the immense and beautiful inland sea, teeming with life, that was extolled by poets and travelers throughout the centuries. Its disappearance must surely rank as one of the worst environmental disasters of all time.

References: W. S. Ellis, "The Aral: A Soviet Sea Lies Dying," *National Geographic*, 177 (1990): p. 73; J. Rupert, "A Death in the Desert: Soviet Plan Kills a Sea," *Washington Post* (June 20, 1992), p. A1; D. Williams, "After the Empire: Life in the Former Soviet Republics. The Sinking Sea," *Washington Post* (November 12, 1998), p. A23.

this does not eliminate all risk. Because DBPs are volatile and vaporize in hot water, there is a risk of inhaling them when showering. In 1998, because of growing concern about health problems, the EPA lowered the maximum allowable level for certain DBPs by 20%.

Chapter Summary

1. Water covers nearly three-quarters of the earth's surface, but less than 1% of the total is usable freshwater.
2. Water is recycled and purified in the hydrologic cycle.
3. Water, the only common liquid on earth, owes most of its unique properties to hydrogen bonding.
4. Water's high heat capacity, heat of fusion, and heat of vaporization serve to moderate the earth's climate.
5. Ice is less dense than water. It floats on water, which allows aquatic life to survive winter under the ice.
6. Drinking water is obtained from lakes and rivers and from groundwater.
7. Many areas of the world are suffering serious water shortages.
8. Freshwater can be obtained from seawater by distillation and reverse osmosis.
9. Water is an excellent solvent, not only for nutrients needed by plants and animals, but also for many pollutants.
10. Major causes of water pollution are
    - **a.** disease-causing agents
    - **b.** oxygen-consuming wastes
    - **c.** plant nutrients
    - **d.** suspended solids and sediments
    - **e.** dissolved solids
    - **f.** toxic substances
    - **g.** thermal pollution
11. Pollutants enter waterways from point and nonpoint sources.
12. Disposal of organic wastes into waterways depletes dissolved oxygen and kills aquatic life.
13. High levels of nitrates and phosphates in water cause eutrophication.
14. Increased loads of suspended solids and sediments cover bottom-dwelling creatures and increase the risk of contamination by toxic substances.
15. Irrigation is a major cause of increased salinity in water.
16. If present in water, many toxic substances, including certain metals and synthetic organic compounds, bioaccumulate in fish at the top of aquatic food chains.
17. The environmental movement was started in the 1970s by Rachel Carson, who first warned of the dangers of DDT.

18. Hot water, if released into waterways by industry, disrupts aquatic life by decreasing the solubility of oxygen.

19. Groundwater in the United States is threatened by chemicals leaking from dumpsites.

20. Sewage is usually treated in two processes: removal of organic matter, and disinfection with chlorine.

21. In the United States, the purity of water is regulated by the 1972 Clean Water Act and subsequent amendments to it.

Key Terms

aquifer (p. 283)
bioaccumulation (p. 299)
biosolids (p. 308)
desalination (p. 293)
disinfectant by-products (DBPs) (p. 308)
distillation (p. 293)
eutrophication (p. 297)
freshwater (p. 282)
groundwater (p. 289)
heat capacity (p. 286)
heat of fusion (p. 282)
heat of vaporization (p. 286)
hydrologic cycle (p. 282)
nonpoint source (p. 295)
osmosis (p. 293)
oxygen-consuming wastes (p. 295)
pathogen (p. 295)
persistent organic pollutant (POP) (p. 302)
point source (p. 294)
polychlorinated hydrocarbons (p. 302)
primary treatment (p. 306)
reverse osmosis (p. 293)
secondary treatment (p. 308)
sludge (p. 306)
tertiary treatment (p. 309)
thermal pollution (p. 304)
toxic substance (p. 299)

Questions and Problems

1. Explain why:
   a. Water is a liquid at room temperature.
   b. Ice floats on liquid water.
   c. Water has such an exceptionally high boiling point.

2. a. Define heat of vaporization.
   b. Why is it cooler along the coast than inland?
   c. Is water, which has a large heat of vaporization, a more or less effective coolant for our bodies than another liquid with a smaller heat of vaporization?

3. a. What percentage of the earth's water is seawater?
   b. List four metal ions present in seawater.
   c. What is the approximate concentration (in ppm) of dissolved solids in seawater?

4. Why is seawater salty?

5. What is the dominant cation in each of the following?
   a. seawater b. freshwater

6. If the water molecule were linear rather than bent, would you expect ice to be less dense than liquid water? Explain.

7. At what temperature does water reach its maximum density? What are the implications of this property for life in a pond?

8. In which of the following steps of the hydrologic cycle is water purified?
   a. condensation b. precipitation
   c. evaporation d. transpiration

9. During the Persian Gulf War, desalination plants in Saudi Arabia and other Middle Eastern countries were contaminated with crude oil intentionally released by the Iraqis. Describe the following

desalination processes. Do you think oil would interfere with these processes?

a. distillation b. reverse osmosis

10. Describe the natural process that replenishes supplies of freshwater.

11. How close to the surface does groundwater need to be in order to be considered accessible?

12. Describe ways in which the average global temperature could affect the hydrologic cycle.

13. What is the primary atmospheric gas that dissolves in falling rain and causes its pH to be 5.6?

14. It has been proposed that water from the Great Lakes be diverted to raise the level of the Mississippi River. List possible benefits and problems that need to be considered before such diversion begins.

15. How long does groundwater take to travel 15 meters, or 50 feet, through an aquifer?

16. Explain why land subsides when groundwater is depleted. If groundwater removal is discontinued, will the land rise up to its previous level?

17. Many urban areas, such as Washington, DC, draw their drinking water from a nearby river. Much of this water is then returned to the river in the form of treated sewage effluent. Does this affect the hydrologic cycle of the locality? How does this practice affect water quality?

18. San Diego, California, draws its water supply from the Colorado River, which is several hundred miles away. The water flows in open aqueducts from the river to the city. Would you expect this to have any effect on the quality of the water?

19. Where does the drinking water used in your community come from? What alternative sources does your community have?

20. Assume you are a homeowner and you have a limited amount of water available from the following sources: (a) water from your bath, (b) good well water, and (c) rainwater drained from the roof. You want to cook dinner, give your little dog a bath, wash the car, and water your vegetable garden. Which source of water would you use for each task?

21. List five ways you can conserve water. In your opinion, should the price of water for all uses be raised to encourage conservation in the United States?

22. List eight types of pollutants that can be found in water supplies.

23. In the middle of the 19th century, what was the main cause of water pollution in the United States and Europe? Explain briefly how this occurred.

24. Name four water-borne diseases that are caused by human wastes. Are these diseases common in the United States today? Explain.

25. What are pathogens? Describe the test that is generally used to determine if water is free from pathogens.

26. How is the solubility of oxygen in water affected by each of the following?

    a. an increase in water temperature

    b. a decrease in pressure caused by increasing altitude

27. What are oxygen-consuming wastes? Name typical sources.

28. Describe the sequence of events that leads to lowered dissolved oxygen (DO) in a river that is polluted with animal wastes from a meat packing plant.

29. What are the end products of the decomposition of organic materials containing the elements carbon, sulfur, and nitrogen under the following conditions?

    a. aerobic conditions b. anaerobic conditions

30. What is meant by point and nonpoint sources of pollutants? Give an example of each.

31. What is eutrophication? How is it caused? Before regulations were introduced to limit its discharge, which consumer product was a major cause of eutrophication?

32. Describe the effects of eutrophication on the aquatic life in an estuary.

33. a. Describe human activities that cause erosion and the deposition of sediments in waterways.

    b. List types of land use in the order in which they cause the most erosion.

34. What are the effects on aquatic life of increasing the amount of sediments in rivers? What are the economic consequences of increased sediments?

35. Water is drawn from a river to irrigate agricultural land. Explain how this can lead to increased salinity of the river water.

36. Name the process in which the concentration of a toxic substance increases toward the top of a food chain. Is this process a more serious problem in a lake than in a woodland ecosystem? Explain.

37. Explain how chlorinated hydrocarbons, such as DDT, that are present in very low concentrations in a body of water can become a serious pollution problem.

38. Why are toxic metals a threat to aquatic environments? List sources of mercury, lead, and cadmium.

39. Is the water obtained from the water fountains in your school or office safe to drink? What are the most likely pollutants, and how might they get into the water?

40. Describe how mercury compounds discharged into a lake can be a threat to humans who eat fish taken from the lake.

41. Which organ in a person's body is most harmed by lead poisoning?

42. What is the major source of mercury poisoning in the United States?

43. Why is DDT so dangerous?

44. The lethal dose of DDT for a human has been estimated to be 0.5 g/kg of body weight. How much DDT would be fatal to a person weighing 55 kg?

45. List the properties of PCBs that make them valuable insulating materials. If discharged into a river, how do PCBs threaten wildlife and humans?

46. What was the purpose of spraying Agent Orange over the Vietnamese countryside during the Vietnam War? Discuss the risk Agent Orange may pose to humans.

47. Explain how thermal pollution of natural waters has an adverse effect on fish. What effects does thermal pollution have on the metabolism and reproduction of aquatic organisms?

48. Discuss ways in which thermal pollution by industry can be prevented.

49. What effects does thermal pollution have on aquatic organisms?

50. In arid regions, how should limited water resources be used? Should water be used for agriculture, industry, city water supplies, or other uses?

51. Using diagrams to show the flow of material, describe primary, secondary, and tertiary treatments of sewage.

52. Chlorine gas is used for disinfection of drinking water.

a. What is the active oxidizing agent that is formed when chlorine is added to water?

b. Describe any health risks associated with chlorination.

c. What alternate disinfectant is used in Europe?

53. What do primary and secondary treatment in a sewage plant remove from the waste stream?

54. How are phosphates removed in tertiary treatment of sewage?

55. How do sewage plants deal with raw sludge obtained during the primary treatment of sewage?

The Air We Breathe

Chapter Objectives

In this chapter, you should gain an understanding of:

- The gas laws
- The composition of the atmosphere
- The five primary air pollutants (their sources and effects on human health and the environment) and the secondary air pollutants
- Ozone as a pollutant
- The importance of the ozone layer and how it is being depleted
- Sources of indoor air pollution
- Greenhouse gases and their potential for causing global warming

In this view of a sunset taken from outer space, the atmosphere is seen as a thin gaseous blanket around the earth.

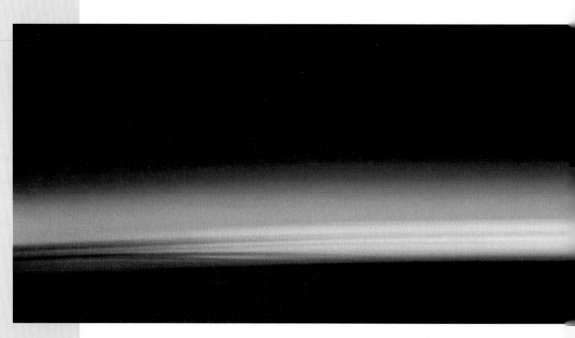

ONLINE FEATURES

www.jbpub.com/chemistry

- Chemistry and the Environment
- eLearning
- Chemistry in the News
- Special Topics
- Research & Reference Links
- Ask the Authors

THE ATMOSPHERE IS a thin blanket of gas that envelops the earth. It provides the carbon dioxide plants need for photosynthesis and the oxygen animals need for respiration. It is also the ultimate source of nitrogen for plant growth. Freshwater reaches the earth from the atmosphere as dew, rain, and snow. The atmosphere shields us from the sun's cancer-causing ultraviolet (UV) radiation, and it also moderates the earth's climate. Without it, the earth would experience the extremes of hot and cold found on planets that have little or no atmosphere.

The atmosphere is obviously vital for human existence, and yet we have been polluting it for years. By the end of the 19th century, huge quantities of coal were being burned to fuel the Industrial Revolution, and smokestacks belching great brown clouds into the atmosphere became a sign of prosperity. By the middle of the 20th century, the automobile had become another significant source of air pollution. Today, in the United States, pollution from these sources has been greatly reduced as a result of legislation, but we face other problems. There is growing concern that the earth's protective ozone layer is being destroyed by certain chemicals produced by industry. Also, our continued dependence on fossil fuels for energy is introducing increasingly large quantities of carbon dioxide into the atmosphere. Many believe this practice will cause the atmosphere to become warmer, a trend that could have disastrous consequences for the world's climate.

In this chapter, we first examine the laws that govern the physical behavior of gases and study the composition and properties of the atmosphere. We then study each of the major air pollutants and examine their sources, their effects on health and the environment, and the steps that are being taken to lower emissions. We review the evidence that strongly suggests that chlorofluorocarbons (CFCs) are the cause of ozone depletion, and we discuss why carbon dioxide and certain other gases may be causing a global warming trend. Lastly, we consider federal legislation that has been passed to control pollutants.

11.1 The Gas Laws

Before we consider the gases that make up the atmosphere, let's consider briefly some of the properties relating to pressure, volume, and temperature that are common to all gases and are expressed in the *gas laws.*

Boyle's Law

The relationship between the pressure and the volume of a gas was discovered by the British scientist Robert Boyle in 1662. He found that if the pressure on a sample of gas is increased, the volume occupied by the gas will decrease proportionally. If the pressure is doubled, the volume will be halved **(Fig. 11.1a)**; if the pressure is halved, the volume will be doubled. **Boyle's law** states that *the volume of a fixed amount of gas maintained at constant temperature is inversely proportional to the gas pressure.*

A large volume of a gas—for example, oxygen, which is frequently required by patients in hospitals—can be compressed into a small tank.

For a given amount of gas at a constant temperature, Boyle's law can be expressed mathematically as:

$$\text{pressure} \times \text{volume} = \text{a constant}$$

or

$$PV = k$$

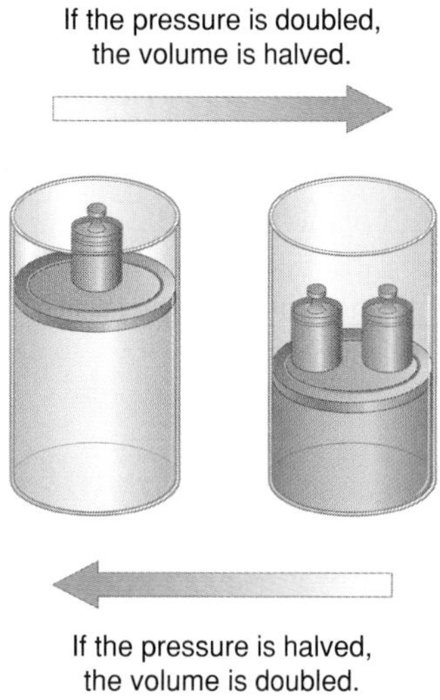

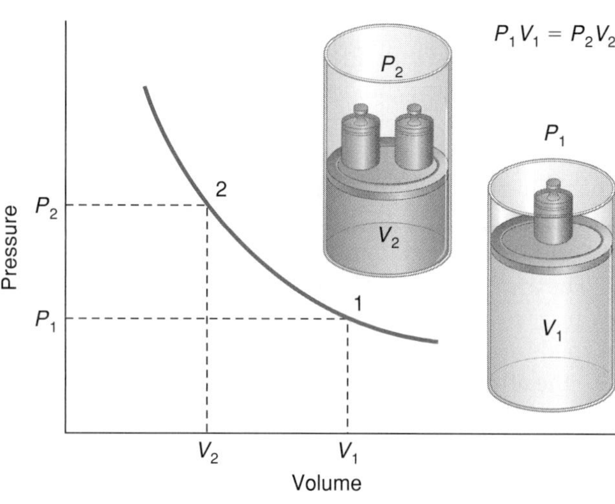

FIGURE 11.1

(a) Boyle's law states that at constant temperature, if the pressure on a fixed volume of gas is increased, the volume of the gas decreases proportionately. (b) Graphical representation of Boyle's law: If the initial pressure (P_1) on a volume of gas (V_1) is doubled (P_2), the volume of gas is halved (V_2) and $P_1V_1 = P_2V_2$.

The value of the constant, k, depends on the temperature and volume of the gas. The relationship between pressure and volume is shown graphically in **Figure 11.1b**.

Boyle's law may also be expressed as:

$$P_1V_1 = P_2V_2$$

where P_1 and V_1 are the initial pressure and volume of a gas sample and P_2 and V_2 are the values after either the pressure or the volume has been changed. The equation assumes that there has been no change in temperature.

Weather balloons **(Fig. 11.2)** are only partly filled with helium when they are launched because as they ascend into the atmosphere, the atmospheric pressure on the outside of the balloon decreases. This causes the balloon to expand as the helium inside it expands.

FIGURE 11.2

As this weather balloon ascends into the upper atmosphere, where pressure is lower than at the earth's surface, the volume will increase.

EXAMPLE 11.1 Using Boyle's law

The volume of gas in a weather balloon is 9.4 liters (L) on a day when the atmospheric pressure is 665 mm Hg. What will be the volume of the gas if the atmospheric pressure falls to 630 mm Hg, assuming that the temperature remains constant?

Solution Begin by listing the information you are given:

Initial	Final
$P_1 =$ 665 mm Hg	$P_2 =$ 630 mm Hg
$V_1 =$ 9.4 L	$V_2 =$?

From Boyle's law, you know that

$$P_1V_1 = P_2V_2$$

Solve for the final volume (V_2):

$$V_2 = \frac{P_1V_1}{P_2}$$

Enter the given values:

$$V_2 = \frac{665 \text{ mm Hg} \times 9.4 \text{ L}}{630 \text{ mm Hg}} = 9.9 \text{ L}$$

The volume of the gas at 630 mm Hg will be 9.9 L.

Practice Exercise 11.1

A sample of oxygen gas occupies a volume of 300 milliliters (mL) at 25°C and 47 lb/in^2. What pressure must be applied to reduce the volume to 25 mL at the same temperature?

Boyle's law can be explained by the **kinetic-molecular theory.** According to this theory, the individual molecules in a gas have negligible volume and are widely separated from each other. A volume of gas is therefore mostly empty space. The molecules are in constant, rapid, random motion, and they exert no attraction on each other. Molecules possess kinetic energy of motion, and if the temperature is increased, the molecules move more rapidly. Their faster motion increases their kinetic energy. Also, the molecules are completely elastic: When they collide with each other or with the walls of their container, they bounce off without losing any energy. As the molecules collide with the walls of the container, they exert a pressure on the walls.

It therefore follows from this theory that if the volume occupied by a fixed amount of gas kept at constant temperature is decreased, the rate per unit area at which molecules collide with the walls increases and the gas pressure increases (**Fig. 11.3**). Conversely, if the volume is increased, the number of collisions decreases and the gas pressure decreases.

The kinetic-molecular theory cannot completely explain the behavior of gases at very low temperatures and very high pressures, but it explains their behavior under moderate conditions very satisfactorily.

When volume decreases, rate of collisions increases and pressure increases.

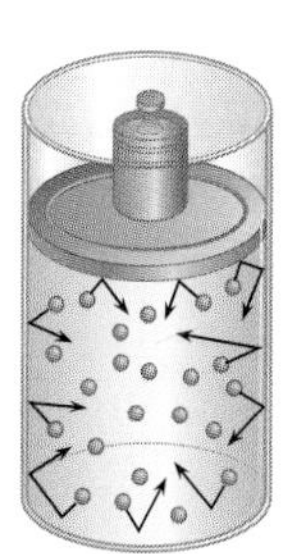

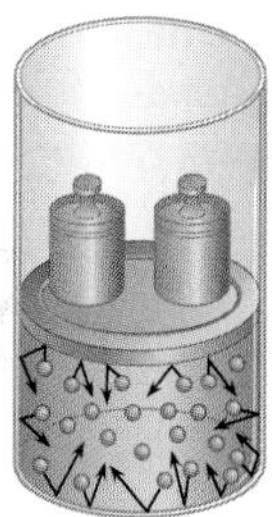

When volume increases, rate of collisions decreases and pressure decreases.

FIGURE 11.3

Boyle's law is supported by kinetic-molecular theory. When the volume of a gas decreases, the rate at which gas molecules collide increases and pressure increases. Conversely, when the volume increases, the collision rate decreases and the pressure decreases.

Charles's Law

In 1787, more than 100 years after the discovery of Boyle's law, the French scientist Jacques Charles investigated the relationship between the temperature of a gas and its volume. He found that if a gas is heated while the pressure is held constant, the gas expands. **Charles's law** states that *at a constant pressure, the volume of a fixed amount of gas is directly proportional to its temperature.*

The relationship between the volume and the temperature of a gas held at constant pressure is shown graphically in **Figure 11.4**. When the volume is plotted against the temperature, a straight line is obtained. If the line is extended toward the left (extrapolated), it intersects the temperature axis at −273.15°C. This is the temperature at which all gases, if they did not condense, would have a volume of zero. It is the lowest temperature that can be attained, and it is called **absolute zero.**

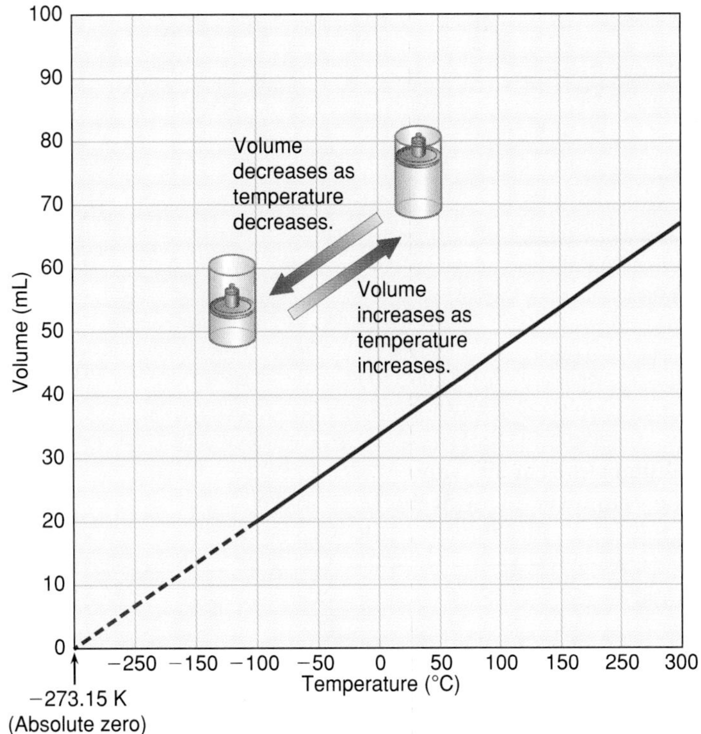

FIGURE 11.4

Charles's law concerning the relationship between the volume and temperature of a gas. If pressure is kept constant, the volume occupied by a fixed amount of any gas decreases with decreasing temperature and increases with increasing temperature. If the line in this graph is extended to zero volume, it intersects the temperature axis at −273.15°C, or absolute zero.

Absolute zero represents the zero point on the Kelvin temperature scale (Appendix B). To convert from the Celsius to the Kelvin scale, it is necessary to add 273 to the Celsius temperature:

$$T(\text{K}) = T(°\text{C}) + 273$$

Charles's law can be expressed mathematically as:

$$V = T \times (\text{a constant})$$

or

$$\frac{V}{T} = k$$

where T is the temperature on the Kelvin scale.

Charles's law can also be expressed as:

$$\frac{V_1}{T_1} = \frac{V_2}{T_2}$$

where V_1 and T_1 are the intial volume and temperature of a gas sample, and V_2 and T_2 are the values after either the volume or temperature of the sample has been changed with the pressure remaining the same.

The effect of temperature on the volume of a gas can be illustrated dramatically by placing a balloon filled at room temperature in contact with liquid nitrogen (−196°C) (**Fig.11.5**). The balloon collapses as the gas inside cools rapidly and occupies a smaller volume.

EXAMPLE 11.2 Using Charles's law

A sample of gas occupies 200 mL at 25°C. What volume will it occupy at 50°C if there is no change in pressure?

Solution

1. Convert the given temperatures (T) from Celsius to Kelvin:

$$T(°\text{C}) + 273 = T(\text{K})$$
$$T_1 = 25 + 273 = 298 \text{ K}$$
$$T_2 = 50 + 273 = 323 \text{ K}$$

2. List the information you are given:

Initial	Final
$V_1 = 200$ mL	$V_2 = ?$
$T_1 = 298$ K	$T_2 = 323$ K

FIGURE 11.5

When a balloon filled with gas at room temperature is placed in contact with liquid nitrogen (−196°C), its volume decreases and it collapses.

3. From Charles's law, you know that

$$\frac{V_1}{T_1} = \frac{V_2}{T_2}$$

Solve for the final volume (V_2):

$$V_2 = \frac{V_1 \times T_2}{T_1}$$

Enter the given values:

$$V_2 = \frac{200\text{ mL} \times 323\text{ K}}{298\text{ K}} = 217\text{ mL}$$

The gas will occupy a volume of 217 mL at 50°C.

Practice Exercise 11.2

At a temperature of 80°C, a balloon has a volume of 3.4 L. What is the volume of the balloon if the temperature is lowered to −10°C, assuming that there is no change in pressure?

Charles's law is easily explained by the kinetic-molecular theory, which states that, when the temperature of a gas is increased, the molecules move more rapidly and their kinetic energy increases. These faster molecules collide more frequently and with greater energy with each other and the walls of the container. For the pressure to remain constant, the volume of the gas must increase so that the molecules hit the walls of the container less frequently. Conversely, if the temperature is lowered, the frequency and energy of collisions decrease, and the volume must decrease in order for the pressure to remain constant.

The Combined Gas Laws

Boyle's and Charles's laws can be conveniently combined into a single law called the **ideal gas law,** which shows the relationships between the pressure, volume, and temperature of a fixed mass of a gas. The ideal gas law can be expressed mathematically as follows:

$$\frac{P_1V_1}{T_1} = \frac{P_2V_2}{T_2}$$

This expression can be used to determine the change in one of the variables (P, V, or T) if *both* the other variables are changed.

EXAMPLE 11.3 Using the ideal gas law

A weather balloon has a volume of 3.0 L at sea level, where the pressure is 760 mm Hg and the temperature is 26°C. What will the volume of the balloon be if it ascends to 10 km above the earth's surface, where the pressure is reduced to 250 mm Hg and the temperature is −60°C?

Solution

1. Convert the given temperatures (T) from Celsius to Kelvin:

$$T(°\text{C}) + 273 = T(\text{K})$$
$$T_1 = 26 + 273 = 299\ \text{K}$$
$$T_2 = -60 + 273 = 213\ \text{K}$$

2. List the information you are given:

Initial	Final
$V_1 = 3.0$ L	$V_2 = ?$
$P_1 = 760$ mm Hg	$P_2 = 250$ mm Hg
$T_1 = 299$ K	$T_2 = 213$ K

3. From the ideal gas law, you know that

$$\frac{P_1V_1}{T_1} = \frac{P_2V_2}{T_2}$$

Solve for the final volume (V_2):

$$V_2 = \frac{P_1V_1}{T_1} \times \frac{T_2}{P_2}$$

4. Enter the given values:

$$V_2 = \frac{760\ \text{mm Hg} \times 3.0\ \text{L}}{299\ \text{K}} \times \frac{213\ \text{K}}{250\ \text{mm Hg}} = 6.5\ \text{L}$$

The volume of the weather balloon at 10 km above the earth is 6.5 L.

Practice Exercise 11.3

A gas occupies 10 L at 40°C and 1.2 atm. What will the pressure be when the gas occupies a volume of 25 L at 0°C?

11.2 The Major Layers in the Atmosphere

The gases that make up the atmosphere are held close to the earth by the pull of gravity. With increasing distance from the earth's surface, the temperature, density, and composition of the atmosphere gradually change. On the basis of air temperature, the atmosphere can be divided vertically into four major layers: troposphere, stratosphere, mesosphere, and thermosphere (Fig. 11.6).

Temperature Changes in the Atmosphere

The **troposphere** extends above the earth to a distance of 10–16 kilometers (6–10 miles). The lower part of the troposphere, which interacts directly with the surface of the earth, is the part of the atmosphere we generally call air. Most weather occurs in the troposphere. Temperature decreases steadily as the distance from the earth's warm surface increases, until it reaches about −57°C (−70°F) at the top of the troposphere, in the region called the *tropopause*.

The location of the upper limit of the troposphere varies with the temperature at the earth's surface, the nature of the underlying terrain, and other factors. It is generally higher above the equator.

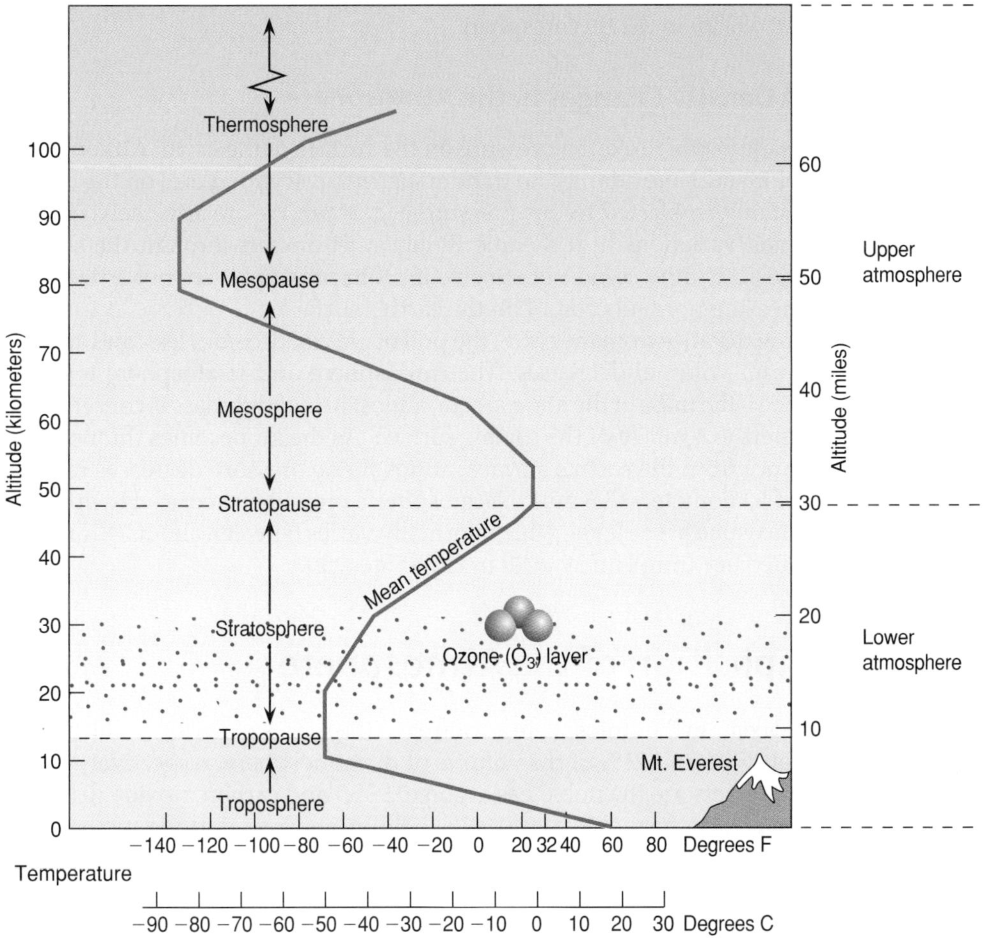

FIGURE 11.6

The earth's atmosphere is subdivided vertically into four major regions based on the air-temperature profile. The ozone layer that protects us from the sun's ultraviolet radiation is in the stratosphere.

Because there is little exchange of air between the stratosphere and the troposphere, pollutants that enter the stratosphere tend to remain there for long periods of time.

Above the troposphere is the **stratosphere,** which extends to about 50 kilometers (30 miles) and includes the ozone layer. To avoid going through bad weather in the troposphere, jet aircraft fly in the lower stratosphere where the atmosphere is calm. Temperature remains constant in the lower part of the stratosphere but begins to rise with increasing altitude, reaching a maximum of approximately −1°C (30°F) at the *stratopause,* which is the boundary between the stratosphere and the mesosphere. This rise in temperature is caused by the ozone in the stratosphere, which absorbs ultraviolet (UV) radiation and converts the radiant energy into heat. It might be expected that the maximum temperature would be reached in the region where ozone concentration is at a maximum (refer to Fig. 11.6). However, because ozone is such an efficient absorber of UV radiation, the relatively few ozone molecules in the upper stratosphere are sufficient to absorb most of this radiation before it reaches the region of maximum ozone concentration.

There is no precise altitude at which the earth's atmosphere ends and space begins.

Together, the troposphere and stratosphere are called the *lower atmosphere.* The *upper atmosphere* extends out beyond the stratosphere and is divided into the **mesosphere** and the **thermosphere.** Continuing outward through the mesosphere, the temperature again falls. At an altitude between about 80 and 90 kilometers (50 and 56 miles), the lowest temperature in the atmosphere, approximately −90°C (−130°F), is reached. Above the mesosphere, the temperature rises once more and reaches a maximum of approximately 1200°C (2192°F) in the thermosphere.

Pressure and Density Changes in the Atmosphere

The average pressure at sea level is approximately equal to 1 kilogram per square centimeter (14.7 pounds per square inch).

The gases in the atmosphere exert a pressure on the surface of the earth. Although we are not aware of it and have adapted to it, humans and everything else on the earth's surface are constantly subjected to this pressure **(Fig. 11.7a)**. We are adversely affected by relatively small variations in it. People flying in jet aircraft through the stratosphere, where the air is thin, could not survive if cabin pressure were not adjusted to match the air pressure normally found on the earth's surface.

The Italian physicist, Evangelista Torricelli (1608–1647) was the first to suggest that the atmosphere exerted a pressure on the earth's surface. He invented an instrument for measuring atmospheric pressure, which he called a *barometer* (from the Greek words *baros,* meaning "weight," and *metros,* meaning "measure").

With increasing distance from the earth, the pull of gravity becomes less, and air density (mass per unit volume) decreases. The troposphere and stratosphere together account for 99.9% of the mass of the atmosphere; almost half of this mass is concentrated within 6 kilometers (3.6 miles) of the earth's surface. As the air becomes thinner with increasing distance from the earth's surface, atmospheric pressure decreases rapidly. At an altitude of 6 kilometers (3.6 miles), atmospheric pressure is reduced to approximately 50% of the value at sea level, which normally varies between about 740 and 770 millimeters of mercury (mm Hg; 29 to 30 in Hg) **(Fig. 11.7b)**.

11.3 Composition of the Atmosphere

The major components in the atmosphere are nitrogen (N_2) and oxygen (O_2), which make up approximately 78% and 21% of the volume of the atmosphere, respectively **(Table 11.1)**. Minor components are the noble gas argon (0.93%) and carbon dioxide (0.037%). Smaller amounts of the other noble gases (neon, helium, and krypton) and methane are present. The percentage of carbon dioxide in the atmosphere is extremely small; but carbon dioxide is the essential raw material for photosynthesis, so this very small amount is vital for life on earth. As we shall see later, carbon dioxide also plays an important role in maintaining the earth's heat balance.

Water vapor is not included in Table 11.1 because its concentration in air is variable. Depending on temperature, precipitation, rate of evaporation, and other factors at a particular location, the percentage of water vapor in the atmosphere may be as low as 0.1%

Table 11.1 Composition of Pure Dry Air at Ground Level

Gas	Percent by Volume	Parts per Million
Nitrogen (N_2)	78.08	780,840
Oxygen (O_2)	20.94	209,440
Argon (Ar)	0.93	9,340
Carbon dioxide (CO_2)	0.04	370
All other gases	0.01	10

or as high as 5%. It generally lies between 1% and 3%, making water the third most abundant constituent of the air.

If water vapor is excluded, the composition of the air is remarkably constant. In the absence of pollution, no matter where you may be on the surface of the earth, the air you breathe is the same. This homogeneity results from the mixing that is brought about by the continuous circulation of the air in the troposphere.

In addition to water vapor and gases, the atmosphere contains many airborne particles. These particles are the nuclei around which ice crystals and water droplets form. Under appropriate conditions, the droplets coalesce to produce clouds and ultimately rain. Airborne particles range in size from those that are visible, such as dust, to others that can be seen only with a high-powered microscope. Minute particles with diameters less than about 10 microns (μm) are termed **aerosols;** larger particles are called **particulates.** Both types of particles can be either liquid or solid.

Relative to their size, small particles have very large surface areas that act as sites for chemical interactions. Depending on the nature of the particle and of the impacting molecule (or other species), chemical reactions may occur at the surface of a particle or within it. If impacting molecules become attached to the particle's surface, the process is termed **adsorption.** If, in the case of liquid particles, molecules are drawn inside and dissolved, the process is termed **absorption.**

Particulates will be studied in more detail when we consider their role as pollutants.

Ozone in the Stratosphere

In the absence of pollution, ozone (O_3) is not present to any appreciable extent in the troposphere, but it occurs naturally in the stratosphere. Its concentration is greatest at an altitude of 20–30 kilometers (12–19 miles) from the earth's surface (see Fig. 11.6). The **ozone layer** is formed when ordinary molecules of oxygen gas (O_2) in the stratosphere absorb UV radiation from the sun with wavelengths less than 242 nm (refer to Fig. 1.13), which causes them to dissociate into single oxygen atoms (O). Single oxygen atoms are very reactive and immediately combine with O_2 to form O_3:

$$O_2 \xrightarrow[(<242\text{ nm})]{\text{UV}} O + O$$

$$O + O_2 \longrightarrow O_3$$

The air above each square inch of surface at sea level exerts a pressure of 14.7 pounds.

FIGURE 11.7

(a) At sea level, the air above each inch of the earth's surface exerts a pressure of 14.7 pounds. (b) At sea level and 0°C, the average pressure of the atmosphere supports a column of mercury 740 to 770 mm in height.

At the same time, ozone, which very strongly absorbs UV radiation with wavelengths of 220–330 nm, is being broken down to form an oxygen molecule and an oxygen atom:

$$O_3 \xrightarrow[\text{(<330 nm)}]{\text{UV}} O_2 + O$$

Because few molecules in the atmosphere absorb UV radiation as strongly as ozone does, any reduction in the ozone layer increases the amount of UV radiation reaching the earth's surface.

A dynamic equilibrium (Chapter 6) is established, and as a result, a fairly constant concentration of ozone is maintained in the stratosphere. The concentration varies with season and latitude but averages 10 ppm. Although it is low, this concentration of ozone is sufficient to screen out 95–99% of the sun's dangerous UV radiation.

11.4 Types and Sources of Air Pollutants

Although the emission of carbon dioxide into the atmosphere during the combustion of fossil fuels may be causing global warming, carbon dioxide is not regarded as a pollutant.

An *air pollutant* is defined as a substance that is present in the atmosphere at a concentration sufficient to cause harm to humans, other animals, vegetation, or materials. Each day humans inhale about 20,000 liters (5,300 gallons) of air. If harmful gases or fine toxic particles are present in the air, they are also drawn into the lungs, where they may cause serious respiratory diseases and other health problems.

Approximately 90% of all air pollution in the United States is caused by five **primary air pollutants:** carbon monoxide, sulfur dioxide, nitrogen oxides, volatile organic compounds (mostly hydrocarbons), and suspended particles. Their major sources are shown in **Figure 11.8**. Emissions of all five are now regulated in the United States.

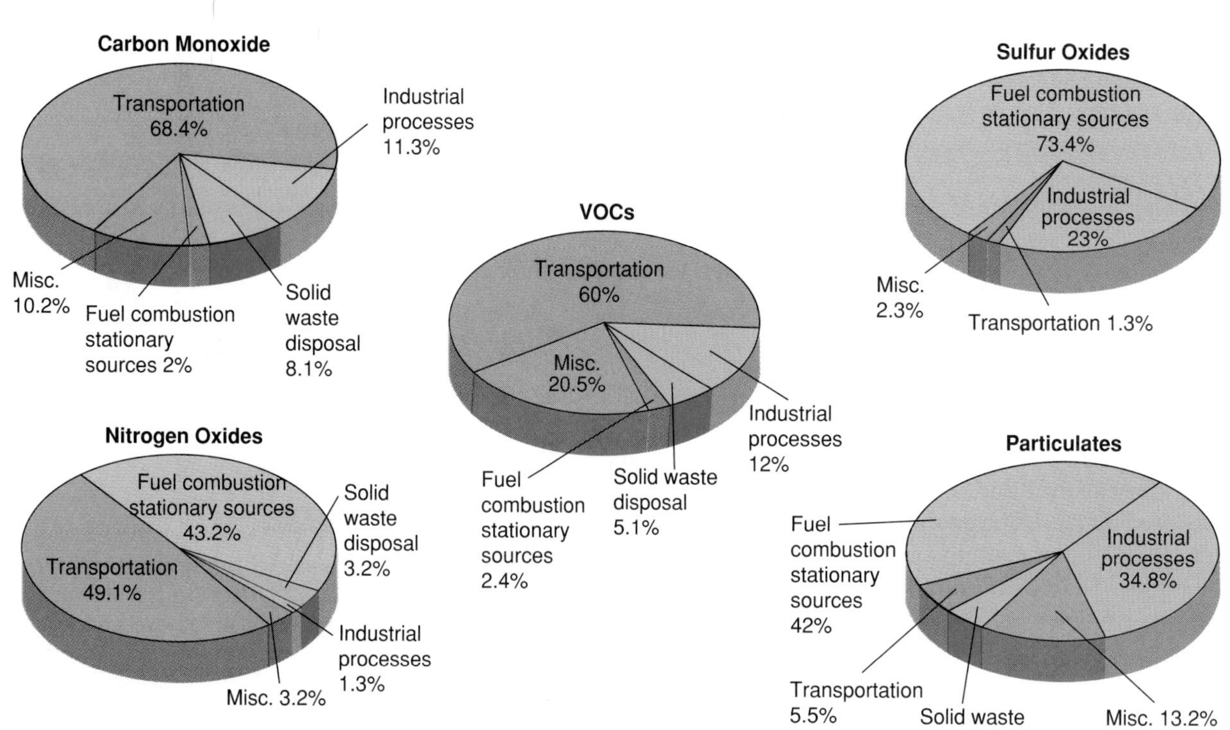

FIGURE 11.8

Nationwide emissions of primary air pollutants according to source.

The transportation industry is responsible for nearly 50% of all air pollution from *anthropogenic sources* (caused by human activities). In addition to carbon monoxide, automobiles emit nitrogen oxides and hydrocarbons. The burning of fossil fuels by stationary sources (power plants and industrial plants) accounts for approximately one-third of air pollutants, mainly in the form of sulfur oxides. Other industrial activities, along with a variety of processes including incineration of solid wastes, contribute smaller amounts.

Approximately 50% of the electric power generated in the United States is produced by burning coal.

If air pollutants were distributed evenly over the entire country, their harmful effects would be greatly reduced. But because the pollutants tend to be concentrated in urban areas, where industry is more common and automobile traffic is congested, large segments of the population are exposed to their harmful effects, particularly during daily rush hours.

In addition to the five primary air pollutants, the atmosphere is contaminated with **secondary air pollutants,** which are harmful substances produced by chemical reactions between primary pollutants and other constituents of the atmosphere. Secondary pollutants include sulfuric acid, nitric acid, sulfates and nitrates (which contribute to acid deposition), and ozone and other photochemical oxidants (which contribute to photochemical smog).

Before we study the individual pollutants, we next examine the influence of solar radiation on chemical reactions in the atmosphere. Many reactions, particularly those occurring in the upper atmosphere, are quite unlike reactions we have studied so far.

11.5 Chemical Reactions in the Atmosphere

In Chapter 4, you learned that reactions between chemical species (atoms, molecules, and ions) usually involve the loss and gain of electrons in such a way that the products achieve stable electron configurations, with electrons arranged in pairs. In the atmosphere, species called **free radicals** (often referred to simply as *radicals*) are often formed under the influence of solar radiation. Radicals are uncharged fragments of molecules that, unlike ordinary chemical species, have an unpaired electron. As a result, radicals are highly reactive and very short-lived. They are responsible for many of the complex, often poorly understood, reactions that occur in the normal and polluted atmosphere.

Central to the chemistry of the troposphere is the **hydroxyl radical.** The hydroxyl radical (•OH) is uncharged and thus quite different from the negatively charged hydroxide ion (OH^-). It is written with a dot (•) beside it to indicate an unpaired electron.

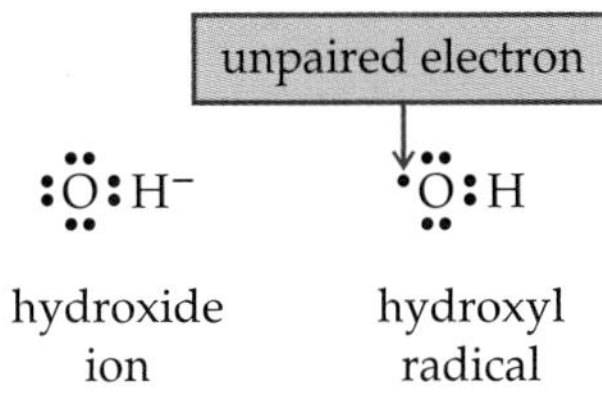

Lewis dot structures

Gamma rays cause harm to humans if they enter the body because they can release hydroxyl radicals from water in the tissues. These radicals attack other compounds and produce more radicals, which then disrupt the normal functioning of the cells.

Hydroxyl radicals are continually formed and consumed in the troposphere. They are produced as the result of a series of complex reactions involving, primarily, ozone, water, and nitrogen dioxide. They play a role in the removal of carbon monoxide from the atmosphere and in the formation of nitric acid, sulfuric acid, and photochemical smog from atmospheric gases.

11.6 Carbon Monoxide

Sources of Carbon Monoxide

The main anthropogenic source of **carbon monoxide** is the combustion of gasoline in automobile engines (refer to Fig. 11.8). Gasoline is a complex mixture of hydrocarbons (Chapter 8). If it is ignited in an adequate supply of oxygen, the products are carbon dioxide and water, as shown below for octane (C_8H_{18}), a representative gasoline hydrocarbon.

$$2\ C_8H_{18} + 25\ O_2 \longrightarrow 18\ H_2O + 16\ CO_2$$

For every 100 gallons of gasoline burned, 230 pounds of carbon monoxide are produced.

In the confined space of the internal combustion engine, however, atmospheric oxygen is in limited supply, and combustion is incomplete. Carbon monoxide (CO) is formed and released to the atmosphere in automobile exhaust.

$$2\ C_8H_{18} + 17\ O_2 \longrightarrow 18\ H_2O + 16\ CO$$

EXAMPLE 11.4 Understanding combustion reactions

Write balanced equations for the combustion of $C_{10}H_{22}$, a component of gasoline, to yield each of the following sets of products.

a. water and carbon dioxide **b.** water and carbon monoxide

Solution

a. Write an equation showing the reactants and products:

$$C_{10}H_{22} + O_2 \longrightarrow H_2O + CO_2$$

Balance the C and H atoms:

$$C_{10}H_{22} + O_2 \longrightarrow 11\ H_2O + 10\ CO_2$$

To balance the 31 (11 + 20) O atoms on the product side of the equation, we need 31 O atoms on the reactant side. But, because oxygen is diatomic, the number of O atoms must be *even*. Therefore, we need 2 molecules of $C_{10}H_{22}$ and 31 molecules of O_2 (62 O atoms).

$$2\ C_{10}H_{22} + 31\ O_2 \longrightarrow 22\ H_2O + 20\ CO_2$$

44 H 20 C 62 O   44 H 20 C 62 (22 + 40) O

b. Use the same procedure as in part a:

$$C_{10}H_{22} + O_2 \longrightarrow H_2O + CO$$

$$C_{10}H_{22} + O_2 \longrightarrow 11\ H_2O + 10\ CO$$

To balance the 21 (11 + 10) O atoms on the product side, we need 21 O atoms on the reactant side. But there must be an even number of O atoms. The equation is as follows:

$$2\ C_{10}H_{22} + 21\ O_2 \longrightarrow 22\ H_2O + 20\ CO$$

44 H 20 C 42 O   44 H 20 C 42 (22 + 20) O

Practice Exercise 11.4

Write balanced equations for the combustion of the hydrocarbon C_6H_{14} to yield:

a. H_2O and CO_2 **b.** H_2O and CO

Since the introduction of the catalytic converter (described later in this chapter), carbon monoxide emissions have been greatly reduced. In addition to the automobile, other anthropogenic sources of carbon monoxide are combustion processes used by the electric power industry, various industrial processes, and solid-waste disposal.

It is perhaps surprising to discover that natural sources release approximately ten times more carbon monoxide into the atmosphere than all the anthropogenic sources combined. The main natural source is methane gas, which is released during the anaerobic decay of plant materials (refer to Table 10.5) in swamps, rice paddies, and other wetlands, where vegetation is submerged in oxygen-depleted water (**Fig. 11.9**). Methane is also produced in the stomachs of ruminants (cattle and sheep) and the intestines of termites. Cattle, as they digest food, produce methane in their intestines. The gas then enters the bloodstream, and when the blood reaches the lungs, the methane is released and exhaled in normal breathing. Oxygen in the atmosphere oxidizes the methane to carbon monoxide.

FIGURE 11.9

Anaerobic bacteria, which thrive in rice paddies like this one, are a major source of methane gas in the atmosphere.

$$2\ CH_4 + 3\ O_2 \longrightarrow 2\ CO + 4\ H_2O$$

Unlike anthropogenic sources, natural emissions of carbon monoxide are dispersed over the entire surface of the earth. Two mechanisms are believed to be at work to maintain the average global level constant at about 0.1 ppm: the conversion of carbon monoxide to carbon dioxide in reactions involving hydroxyl radicals, and the removal of carbon monoxide from the atmosphere by microorganisms in soil. In cities, where soil has been largely replaced with asphalt and concrete and where emissions are very concentrated, nature's natural defense mechanism is overwhelmed, and atmospheric carbon monoxide levels increase.

An estimated 170 million tons of methane are produced annually by termites.

Effects of Carbon Monoxide on Human Health

Although carbon monoxide is the most abundant air pollutant, it is not very toxic at the levels usually found in the atmosphere. However, if it builds up in a confined space, it can cause serious health problems.

Carbon monoxide interferes with the oxygen-carrying capacity of blood. Normally, hemoglobin (Hb) in red blood cells combines with oxygen in the lungs to form oxyhemoglobin (HbO_2). The oxyhemoglobin is carried in the bloodstream to the various parts of the body, where the oxygen is released to the tissues.

Carbon monoxide binds much more strongly to hemoglobin than oxygen does. If carbon monoxide is present in the lungs, it displaces oxygen from hemoglobin and thus reduces the amount of oxygen that can be delivered to the tissues.

$$\underset{\text{oxyhemoglobin}}{HbO_2} + CO \longrightarrow \underset{\text{carboxyhemoglobin}}{HbCO} + O_2$$

The treatment for carbon monoxide poisoning is inhalation of pure oxygen, which reverses the direction of the above reaction.

The symptoms of carbon monoxide poisoning are those of oxygen deprivation: headache, dizziness, impaired judgment, drowsiness, slowed reflexes, respiratory fail-

Exposure to air containing over 1000 ppm of carbon monoxide for about 4 hours converts approximately 60% of the body's hemoglobin to carboxyhemoglobin, and usually results in death.

ure, and eventually loss of consciousness and death. Prolonged exposure to carbon monoxide levels as low as 10 ppm can be harmful. The danger from carbon monoxide is heightened by the fact that the gas is colorless, tasteless, and odorless; people succumb to its effects before they are aware of its presence. On busy city streets, carbon monoxide concentration may reach 50 ppm and may be much higher in underground garages and traffic jams.

11.7 Nitrogen Oxides (NO_x)

Nitrogen dioxide (NO_2) is the major nitrogen oxide pollutant in the atmosphere. It is formed from nitric oxide (NO). Collectively, these related nitrogen oxides are designated as NO_x.

Sources of Nitrogen Oxides

Practically all anthropogenic nitrogen oxides enter the atmosphere from the combustion of fossil fuels by automobiles, aircraft, and power plants (refer to Fig. 11.8). At normal atmospheric temperatures, nitrogen and oxygen—the two main components of air—do not react with each other. However, at the very high temperatures that exist in the internal combustion engine and in industrial furnaces, normally unreactive atmospheric nitrogen reacts with oxygen. In a series of complex reactions, the two gases combine to form nitric oxide:

$$N_2 + O_2 \longrightarrow 2\ NO$$

When released to the atmosphere, nitric oxide combines rapidly with atmospheric oxygen to form nitrogen dioxide:

$$2\ NO + O_2 \longrightarrow 2\ NO_2$$

A third nitrogen oxide, nitrous oxide (N_2O), is formed naturally in soil during nitrogen fixation. This oxide, which is known as "laughing gas," is used as an anesthetic for minor surgery, particularly in dentistry.

As is the case with carbon monoxide, far more nitrogen oxides are released to the atmosphere by natural processes than by human activities. Recall from Chapter 2 that during electrical storms atmospheric nitrogen and oxygen react to form nitric oxide, which then rapidly combines with more atmospheric oxygen to form nitrogen dioxide, as shown in the above equations. Bacterial decomposition of nitrogen-containing organic matter in soil is another natural source of nitrogen oxides. Because emissions from natural processes are widely dispersed, they do not have an adverse effect on the environment.

The Fate of Atmospheric Nitrogen Oxides

Nitrogen dioxide, regardless of its source, is ultimately removed from the atmosphere as nitric acid and nitrates in dust and rainfall. In a complex series of reactions involving hydroxyl radicals, nitrogen dioxide combines with water vapor to form nitric acid. The overall reaction can be written as

$$4\ NO_2 + 2\ H_2O + O_2 \longrightarrow 4\ HNO_3$$

Much of the nitric acid in the atmosphere is formed within aqueous aerosols. If weather conditions are right, the aerosols coalesce into larger droplets in clouds, and the result is acid rain. Some of the nitric acid formed reacts with ammonia and metallic particles in the atmosphere to form nitrates. Ammonium nitrate is formed as follows:

$$HNO_3 + NH_3 \longrightarrow NH_4NO_3$$

Nitrates dissolve in rain and snow or settle out as particles. The combined fallout contributes to acid deposition.

Effects of Nitrogen Oxides on Human Health and the Environment

Nitrogen dioxide is a red-brown toxic gas with a very unpleasant acrid odor. It can cause irritation of the eyes, inflammation of lung tissue, and emphysema. Even in badly polluted areas, however, its concentration in the atmosphere is rarely high enough to produce these symptoms. Nitrogen oxides are a serious health problem because of their role in the formation of the secondary pollutants associated with photochemical smog (discussed later in this chapter).

Emissions from stationary fuel combustion sources are difficult to control. Lowering the combustion temperature of the furnace decreases formation of nitric oxide, but it decreases efficiency at the same time. Most research has concentrated on reducing automobile emissions by means of the catalytic converter.

11.8 Volatile Organic Compounds (VOCs)

A great variety of **volatile organic compounds (VOCs),** including many hydrocarbons, enter the atmosphere from both natural and anthropogenic sources (refer to Fig. 11.8). Most are not pollutants themselves, but they create problems when they react with other substances in the atmosphere to form the secondary air pollutants associated with photochemical smog.

The petroleum industry is the main anthropogenic source of hydrocarbons in the atmosphere. Gasoline is a complex mixture of many volatile hydrocarbons (Chapters 8 and 12), and in urban areas, gasoline vapors can escape into the atmosphere in several ways: when gas is pumped at gas stations, during filling of storage tanks, and as unburned gasoline in automobile exhaust.

In the natural world, the pleasant aroma of pine, eucalyptus, and sandalwood trees is caused by the evaporation of VOCs called **terpenes** from their leaves. Natural sources account for about 85% of total emissions of volatile hydrocarbons. The remaining 15% come from anthropogenic sources, which are of concern because, unlike natural sources, they are not evenly distributed but are concentrated in urban areas.

11.9 Automobile Pollutants and the Catalytic Converter

Motor vehicles are a major source of carbon monoxide, nitrogen oxides, and volatile hydrocarbons (see Fig. 11.8). Since 1975, when all new cars in the United States were required by law to be equipped with a catalytic converter, emissions of those pollutants have been reduced significantly. Today's cars emit 95% less pollutants than pre-1970 vehicles, despite the fact that the number of miles traveled almost doubled in the last 20 years.

In the three-way **catalytic converter** (in use since 1981), two opposing chemical reactions take place: an oxidation and a reduction. Hot exhaust gases from the engine pass through the converter before they enter the muffler. The converter is a very fine honeycomb structure made of a ceramic coated with the precious metals platinum (Pt), palladium (Pd), and rhodium (Rh), which act as catalysts (**Fig. 11.10**). As the gases enter, rhodium catalyzes the reduction of nitrogen oxides to nitrogen gas. Then air is injected into the exhaust stream to provide oxygen, which, in the presence of the platinum and palladium catalysts, oxidizes carbon monoxide to carbon dioxide and hydrocarbons to water and carbon dioxide.

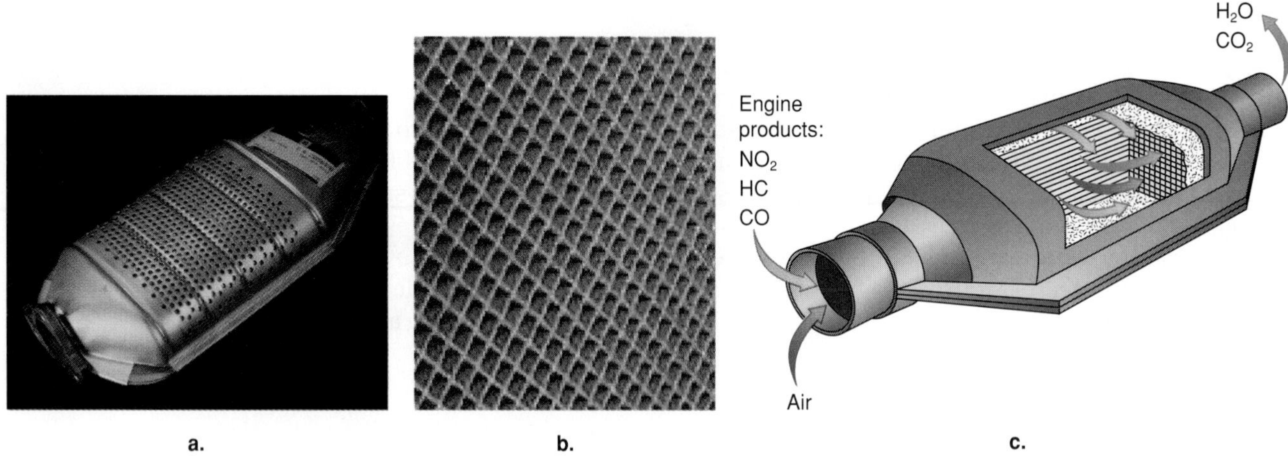

FIGURE 11.10

(a) The catalytic converter, which has been standard equipment in automobiles sold in the United States since 1975, reduces engine emissions. (b) A three-way catalytic converter consists of a ceramic honeycomb coated with the precious metals platinum, palladium, and rhodium. (c) The metals catalyze the conversion of nitrogen oxides and carbon monoxide to nitrogen and carbon dioxide and the conversions of hydrocarbons to carbon dioxide and water.

Catalytic converters currently in use do not reach their minimum operating temperature until about 90 seconds after the cold start of the automobile. During this time 50% of the auto's emissions pass unreacted from the tail pipe. Electrically heated catalysts are being designed to reduce this cold-start time.

The overall reaction for the reduction of nitric oxide and the oxidation of carbon monoxide can be written as

$$2\ NO + 2\ CO \xrightarrow[\text{catalysts}]{\text{Rh, Pt, Pd}} N_2 + 2\ CO_2$$

The oxidation of a typical gasoline hydrocarbon occurs as follows:

$$2\ C_8H_{18} + 25\ O_2 \xrightarrow[\text{catalysts}]{\text{Pt, Pd}} 16\ CO_2 + 18\ H_2O$$

The formation of carbon dioxide in these two reactions results in an increased level of carbon dioxide in the atmosphere, which, as we'll discuss later, may be linked to global warming.

Recall from Chapter 6 that a catalyst is a substance that increases the rate of a chemical reaction without itself being used up in the reaction. Platinum and rhodium are very expensive, but since they are not used up, the small amounts needed last a long time. Lead-free gasoline must be used in cars fitted with catalytic converters because lead inactivates the catalysts.

Today's catalytic converters remove 96% of carbon monoxide, 98% of hydrocarbons, and 95% of nitrogen oxides from auto exhausts.

A 1000-megawatt coal-fired electric power plant burns approximately 700 tons of coal per hour. If the coal is 3% sulfur by weight, 42 tons of sulfur dioxide per hour (nearly 370,000 tons per year) will be produced.

11.10 Sulfur Dioxide (SO_2)

Sources of Sulfur Dioxide

Release of sulfur dioxide (SO_2) to the atmosphere is the primary cause of acid rain (Chapter 7) in the United States. Fossil fuel combustion at electric power generating plants accounts for nearly 75% of the emissions; industrial sources contribute approximately 25% (see Fig. 11.8).

Coal, oil, and all other fossil fuels naturally contain some sulfur because the plant materials from which they were formed included sulfur-containing compounds. Coal frequently contains additional sulfur in the form of the mineral pyrite (FeS_2). Coal mined in the United States is, typically, between 1% and 4%, by weight, sulfur. The percentage is lower in coal from the western states than in coal mined in the central and eastern states.

When sulfur-containing coal is burned, the sulfur is oxidized to sulfur dioxide:

$$S + O_2 \longrightarrow SO_2$$

Natural sources account for approximately half of all sulfur dioxide emissions. Hydrogen sulfide produced as an end product of the anaerobic decomposition of sulfur-containing organic matter by microorganisms (refer to Table 10.5) is the main source. On entering the atmosphere, hydrogen sulfide is oxidized to sulfur dioxide:

$$2\ H_2S + 3\ O_2 \longrightarrow 2\ SO_2 + 2\ H_2O$$

Volcanic eruptions are another, more localized natural source of sulfur dioxide. It has been estimated that the eruption of Mt. Pinatubo in the Philippines in June 1991 (**Fig. 11.11**) injected as much as 25 million tons of sulfur dioxide into the stratosphere, where it was converted into sulfuric acid aerosols.

FIGURE 11.11

The eruption of Mt. Pinatubo in the Philippines in June 1991 ejected huge quantities of sulfur dioxide and ash into the atmosphere.

The Fate of Atmospheric Sulfur Dioxide: Acid Rain

Sulfur dioxide in the atmosphere reacts with oxygen to form sulfur trioxide (SO_3), which then reacts readily with water vapor or water droplets to form sulfuric acid. The overall reactions can be written as follows:

$$2\ SO_2 + O_2 \longrightarrow 2\ SO_3$$

$$SO_3 + H_2O \longrightarrow H_2SO_4$$

The actual pathways are more complex and involve hydroxyl radicals.

Sulfuric acid in the atmosphere becomes concentrated near the base of clouds, where pH levels as low as 3 (about the same pH as orange juice) have been recorded. Thus, cloud-enshrouded, high-altitude trees and vegetation may be exposed to unusually high acidity. Since rain is made up of moisture from all cloud levels, it is less acidic than moisture at the lower cloud levels.

Ash particles are usually emitted together with sulfur dioxide from electric power generating plants. Sulfur oxides become adsorbed onto particle surfaces and may be carried many miles from their source before settling out or being washed out by precipitation. Like nitric acid and nitrates formed from nitrogen oxides, sulfuric acid, sulfates, and particulates all contribute to acid deposition.

The formation of SO_3 from SO_2 is influenced by prevailing atmospheric conditions such as sunlight, temperature and humidity. During episodes of photochemical smog, when hydrocarbons, nitrogen oxides, and particulates are present in the atmosphere, the formation of SO_3 is accelerated.

Effects of Sulfur Dioxide on Human Health and the Environment

Sulfur dioxide is a colorless, toxic gas with a sharp, acrid odor. Exposure to it causes irritation of the eyes and respiratory passages and aggravates symptoms of respiratory disease. Children and the elderly are especially susceptible to its effects.

Sulfur dioxide is also harmful to plants. Crops such as barley, alfalfa, cotton, and wheat are particularly likely to be adversely affected. Other environmental effects of acid deposition associated with sulfur dioxide were considered in Chapter 7.

Plants are even more susceptible to sulfur dioxide than humans are. Before emissions controls were introduced, land around smelters and coal-burning power plants was often completely devoid of vegetation.

Methods for Controlling Emissions of Sulfur Dioxide

As a result of the Clean Air Act of 1970 and amendments made to it in 1990, coal-fired electric power plants were required to make significant reductions in their emissions of sulfur dioxide. Reductions can be achieved in two ways: (1) Sulfur can be removed from coal before combustion, or (2) sulfur dioxide can be removed from the smokestack after combustion—but before it reaches the atmosphere. The second, cheaper approach is generally chosen.

The most commonly used method is **flue-gas desulfurization (FGD)** in which sulfur-containing compounds are washed out (or *scrubbed*) by passing the chimney (flue) gases through a slurry of water mixed with finely ground limestone ($CaCO_3$) or dolomite [$CaMg(CO_3)_2$] or both. On heating, the basic calcium carbonate reacts with acidic sulfur dioxide and oxygen to form calcium sulfate ($CaSO_4$).

$$2\ SO_2 + 2\ CaCO_3 + O_2 \longrightarrow 2\ CaSO_4 + 2\ CO_2$$

Scrubbers, which remove up to 90% of the sulfur dioxide in the flue gas, can be quite easily, and inexpensively, retrofitted onto existing power plants.

A promising newer method is **fluidized bed combustion (FBC),** a process in which a mixture of pulverized coal and powdered limestone is burned, with air being introduced to keep the mixture in a semifluid state. The limestone is converted to calcium sulfate according to the above equation. Unlike FGD, this method does not require water, which makes it particularly useful in arid locations. The disadvantage of FBC is that it cannot be added to existing power plants, but it is the preferred technology for installing in a new power plant. Both FGD and FBC have the problem of disposing of large quantities of calcium sulfate.

The world's tallest stack is the 1250-foot stack at the Copper Cliff Nickel Smelter in Sudbury, Ontario, Canada.

Another, less desirable approach is the installation of very tall stacks, which reduce pollutants in the immediate neighborhood but disperse them more widely. Communities distant from the source of pollution then suffer most of the adverse effects.

Legislation to Control Emissions of Sulfur Dioxide

Prior to the passage of the Clean Air Act Amendments of 1990, the electric utility industry was unwilling to take steps to reduce its sulfur dioxide emissions because of cost considerations. The law required that total emissions of sulfur dioxide be reduced to 10 million tons per year by 2000. Power companies could meet this requirement by installing efficient scrubbers or by switching to low-sulfur western coal, a move that would mean a loss of jobs and great economic hardship for the high-sulfur coal miners of the Ohio Valley and West Virginia. The law gives each power plant an emissions allowance that permits it to release a certain amount of sulfur dioxide per year. To make it easier for the industry to reach the required overall reduction, the EPA introduced a free-market system of emissions trading, which allows a plant emitting less sulfur dioxide than its allowance to sell the difference to a plant that is emitting more than its allowance. This trading of pollution reduction credits allows the lower overall emission levels required by law to be achieved, but the disadvantage is that the oldest and dirtiest plants can continue to release unacceptably high levels of pollution by buying credits from newer, cleaner plants.

Between 1980 and 1995, emissions of sulfur dioxide from burning fossil fuels fell by almost 50% in Europe and 30% in the United States and are expected to continue to decrease. However, during this same period, emissions in the developing countries more than doubled and are expected to rise further as populations in these countries increase and they become more industrialized.

11.11 Suspended Particulate Matter

Natural sources of airborne particles, which may be solid or liquid, include smoke and ash from forest fires and volcanic eruptions, dust, sea salt spray, pollen grains, bacteria, and fungal spores. Aerosols and particulates that are liquid are generally called *mist*, which includes fog and raindrops. A significant quantity of harmful particulate matter is also emitted as a result of human activities **(Fig. 11.12)**.

Eventually, all particulate matter is deposited on the earth's surface. Relatively large particles (diameters greater than 10 microns) settle out under the influence of gravity within 1–2 days; medium-sized particles (diameters of 1–10 microns) remain suspended for several days; fine particles (diameters less than 1 micron) may remain in the troposphere for several weeks and in the stratosphere for up to 5 years. Aerosols, acting as nuclei for the formation of droplets in clouds, reach the ground when the droplets condense and fall as rain or snow. Fine particles can be transported considerable distances by winds before settling to the ground or being washed out by falling rain or snow.

In cloud seeding, a powdered chemical, usually silver iodide (AgI), is injected into a cloud. The particles of powder act like aerosols, becoming nuclei, or centers, around which moisture collects to form water droplets, which then fall as rain.

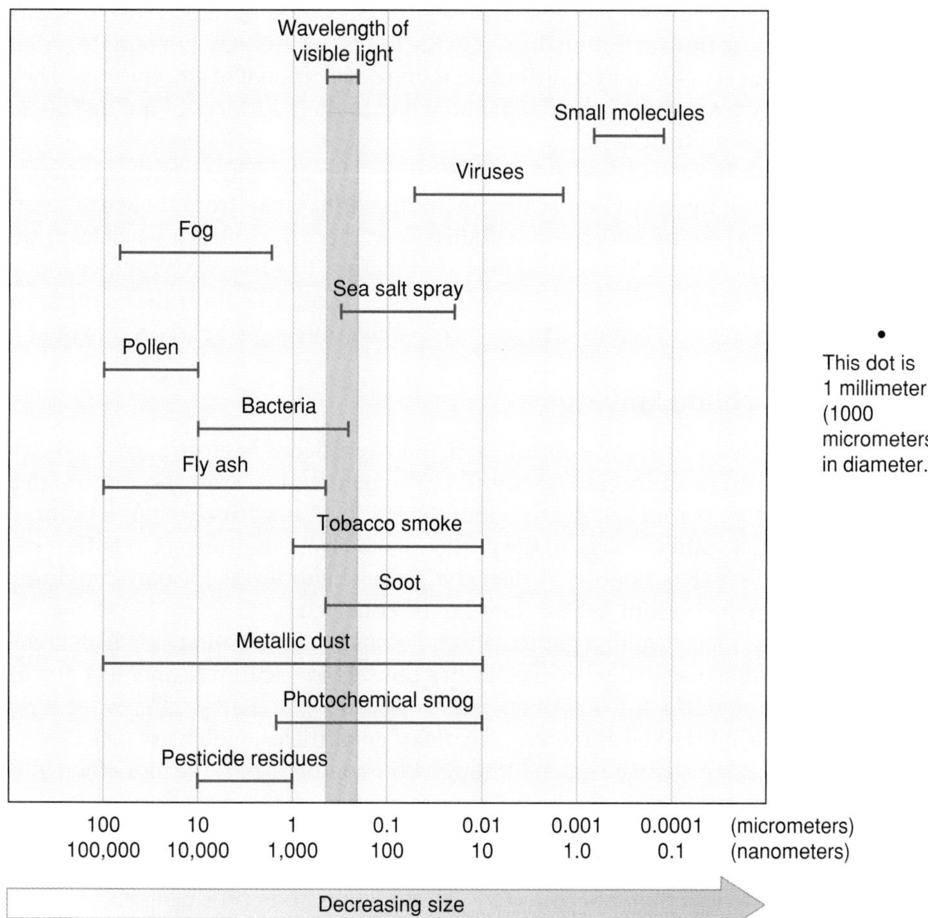

FIGURE 11.12

Suspended particulate matter of many types and sizes enters the atmosphere from both natural and anthropogenic sources.

Anthropogenic Sources of Particulate Matter

Suspended particulate matter is the most visible form of air pollution. The major sources of such particles are the combustion of coal by electric power plants and industrial processes (see Fig. 11.8). One product of the incomplete combustion of coal is *soot*, a finely divided impure form of carbon. Other products include metallic and nonmetallic oxides formed from minerals present in coal. These materials, called *fly ash*, are swept up smokestacks in drafts from roaring furnaces and emitted into the atmosphere.

Other anthropogenic sources of particulate emissions are solid-waste incineration, diesel engine and automobile exhaust, mining and ore processing, construction sites, and agricultural activities. Toxic metal particles, cement dust, and pesticide and fertilizer residues that are released in these activities are all potentially hazardous.

Effects of Suspended Particulate Matter on Human Health and the Environment

The effects of airborne particulates on human health and the environment depend on both the size and the nature of the particles. Very fine particles (diameters less than about 1 micron) are the most hazardous to human health. They are not filtered out by hairs and mucus in the nose but are drawn deep into the lungs, where they can remain indefinitely, causing tissue damage and contributing to the development of emphysema. They take with them any toxic chemicals that are attached to their surfaces or are dissolved within them, such as sulfuric acid.

Particulates and aerosols have a direct effect on climate. Those with diameters greater than 1 micron intercept and scatter incoming light away from the earth's surface and thus exert a cooling effect. As geologists predicted, the fine ash particles emitted into the stratosphere in the 1991 eruption of Mt. Pinatubo (the largest volcanic eruption since that of Krakatoa, described in Chapter 2) caused a cooling effect that temporarily offset any warming trend caused by an increase in atmospheric carbon dioxide.

Control of Particulate Emissions

One method used by industry to control particulate emissions is **electrostatic precipitation** (Fig. 11.13a). In this process, gases and particulate matter are passed through a high-voltage chamber before leaving the chimney stacks. A negatively charged central electrode imparts a negative charge to the particles, which are then attracted to the positively charged walls of the chamber. As their charges are neutralized, the particles clump together and fall to the bottom, where they can be collected.

A second method for controlling particulate emissions is **bag filtration,** in which lined-up fabric bags function essentially like the bag in a household vacuum cleaner (Fig. 11.13b). Gases can pass through the finely woven bags but particulate matter cannot—it is filtered out. The bags are shaken at intervals to dislodge particles, which are collected in a hopper located underneath the bags. Although both methods remove more than 98% of particulates, they fail to remove the finest ones, which are the most dangerous.

Another, less efficient method is **cyclone separation,** in which the particle-laden gases are subjected to a centrifugal force. The spinning particles hit the walls, settle out, and fall to the bottom, where they are collected. All three methods produce solid wastes, which must be disposed of and which mostly end up in landfills (Chapter 19).

Although the 1970 Clean Air Act mandated standards for particulate emissions, many areas failed to meet the requirements because of difficulties in enforcing the law. Under the 1990 amendments to the act, areas not in compliance must take steps to meet the standards.

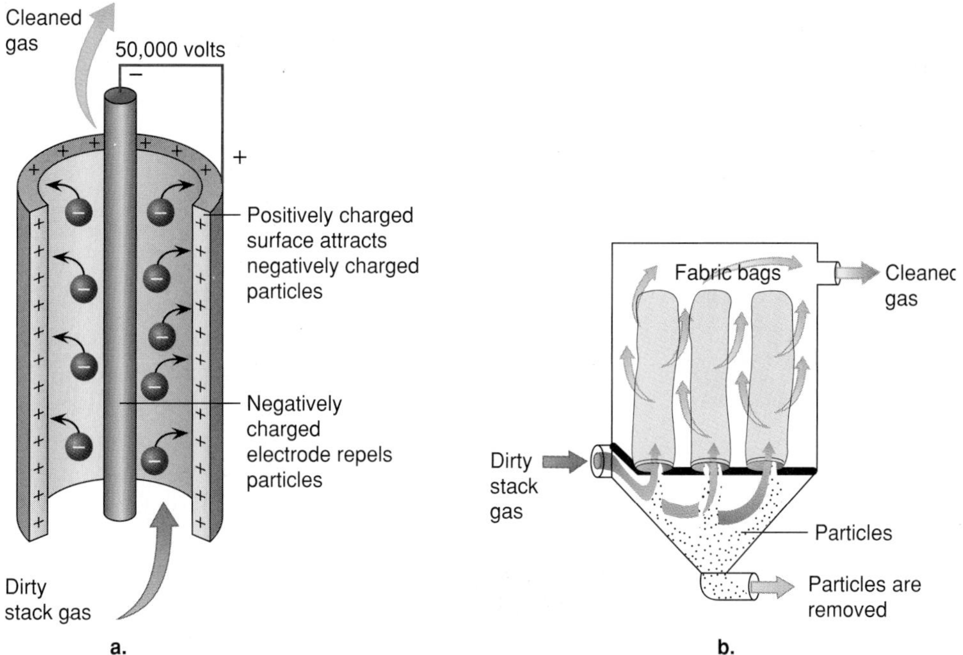

FIGURE 11.13

Two methods for controlling particulate emissions: (a) In electrostatic precipitation, the dirty gas is passed through a chamber, where particles acquire a negative charge from a negatively charged central electrode. The charged particles are attracted to the positively charged wall of the chamber, where they are deposited. (b) In bag filtration, the dirty gas is forced through fabric bags, which trap particles. The bags are shaken at intervals, causing the particles to fall into a hopper.

The word *smog* is thought to be derived from a combination of *sm*oke and f*og*.

11.12 Industrial Smog

Particulate matter and sulfur dioxide can be a deadly combination. Released into the atmosphere together when coal is burned, they can form **industrial smog** (sometimes called *London smog*), a mixture of fly ash, soot, sulfur dioxide, and some volatile organic compounds.

In the 19th and 20th centuries, industrial smog was common in the industrial centers of Europe and the United States. It formed in winter, typically in cities where the weather was cold and wet. Visibility was often reduced to a few yards, and people in factory towns lived under a pall of black smoke (**Fig. 11.14**).

FIGURE 11.14

Industrial smog is created by the burning of coal and is made up primarily of fly ash, soot, and sulfur dioxide.

11.13 Photochemical Smog

The origin of **photochemical smog** is quite different from that of industrial smog. Typically, photochemical smog develops as a yellow-brown haze in hot sunny weather in cities, such as Los Angeles, where automobile traffic is congested (**Fig. 11.15**). The reactions that lead to its formation are initiated by sunlight and involve the hydrocarbons and nitrogen oxides emitted in automobile exhaust. Nitrogen dioxide is responsible for the brownish color of the haze.

FIGURE 11.15

Photochemical smog forms in cities such as Los Angeles when nitrogen oxides and hydrocarbons in the air interact in the presence of sunlight to form a yellow-brown haze.

The reactions that produce photochemical smog are complex and not fully understood. A simplified reaction sequence is shown in the following equations.

$$NO_2 \xrightarrow{UV} NO + O$$

$$O + NO \longrightarrow NO_2$$

$$O + O_2 \longrightarrow O_3$$

$$O + H_2O \longrightarrow 2\ \bullet OH$$

In the initiating step, absorption of sunlight causes nitrogen dioxide to split into nitric oxide and atomic oxygen. The very reactive atomic oxygen is involved in several more steps: It reacts with nitric oxide to reform nitrogen dioxide in a continuing cycle; it combines with ordinary molecular oxygen to form ozone; and it reacts with water to form hydroxyl radicals, which react readily with organic compounds.

Unburned hydrocarbons in automobile exhaust (represented by RH in the following equation) react with hydroxyl radicals to form a number of secondary pollutants, including the hydrocarbon radical ($RO_2\bullet$). This radical then reacts with nitric oxide to form aldehydes and ketones (Chapter 8) and the hydroperoxide radical ($HO_2\bullet$).

$$RH + \bullet OH + O_2 \longrightarrow RO_2\bullet + H_2O$$

$$RO_2\bullet + NO \longrightarrow \text{(aldehydes and ketones)} + NO_2 + HO_2\bullet$$

Further reactions produce peroxyacetyl nitrate (PAN):

$$\underbrace{CH_3-\overset{\overset{\displaystyle O}{\|}}{C}}_{\text{acetyl}}-\overbrace{O-O}^{\text{peroxy}}\!\!\!\!\!\!\underbrace{\;-NO_2}_{\text{nitrate}}$$

Ozone, aldehydes, and PAN all contribute to the harmful effects of photochemical smog, but ozone—the pollutant produced in greatest quantity—causes the most serious problems.

Ozone: A Pollutant in the Troposphere

Ozone in the stratosphere protects us from damaging UV radiation from the sun, but ozone in the troposphere is a dangerous pollutant. Ozone is a powerful oxidizing agent. It is a colorless, pungent, very reactive gas, which irritates the eyes and nasal passages. People with asthma or heart disease are particularly susceptible to its harmful effects. Exposure to ozone levels as low as 0.3 ppm for 1–2 hours can cause fatigue and respiratory difficulties. In Los Angeles, schoolchildren are kept indoors if ozone levels reach 0.35 ppm. Fortunately, because of pollution controls, these levels are rarely reached today.

Tomatoes, tobacco, and other leafy plants are very susceptible to damage by ozone. Brief exposure to concentrations as low as 0.1 ppm reduces photosynthesis, causing leaves to turn yellow.

Ozone is also very toxic to plants. In California, crop damage caused by ozone and other photochemical pollutants costs the state millions of dollars a year. Ozone also damages fabrics and the rubber in tires and windshield wiper blades.

11.14 Temperature Inversions and Smog

Certain meteorological and geographical conditions favor the formation of both industrial and photochemical smog. Normally, air temperature in the troposphere decreases with increasing altitude (refer back to Fig. 11.6). Warm air at the earth's surface expands, becomes less dense, and rises. As it does so, cooler air from above flows in to replace it. In turn, the cooler air is warmed and rises. Through this process, the air is continually renewed, and pollutants are dispersed by vertical currents and prevailing winds **(Fig. 11.16a)**.

A reversal of the usual temperature pattern, called a **temperature inversion,** sometimes occurs; after an initial decrease, the air temperature, instead of continuing to decrease with increasing altitude, begins to increase. A lid of warm air forms over cooler air near the earth's surface **(Fig. 11.16b)**. The cooler, denser layer cannot rise through the warm lid of air above it and becomes trapped, sometimes for days. There is no vertical circulation, and pollutants accumulate.

A particularly serious incident occurred in 1948 in the Donora Valley, an industrial area in Pennsylvania. As a result of a temperature inversion, industrial smog settled over the valley for 5 days. Many people died, and almost half of the population suffered from respiratory ailments. In 1952, a similar incident in London resulted in 4000 deaths. In that case, smoke from coal burned for heat in homes and workplaces was the main cause of the air pollution.

If a temperature inversion occurs in an area partly surrounded by mountains—for example, Los Angeles, Denver, or Salt Lake City—photochemical smog buildup is particularly serious. Because of the encircling mountains, the pollutants cannot be dispersed horizontally. They remain in a blanket over the city until the weather changes and the wind disperses the polluted air.

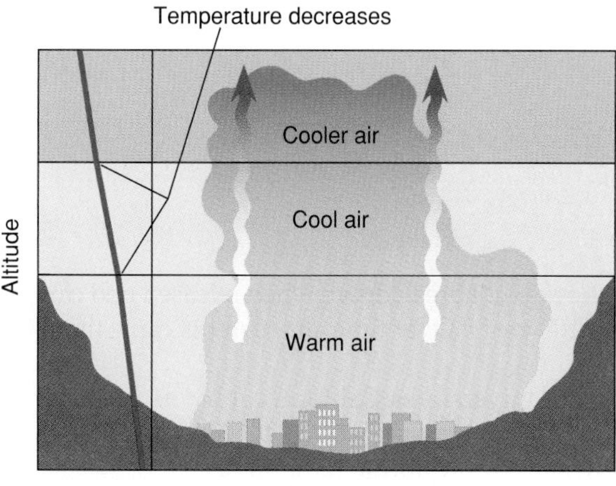

a. Normal conditions

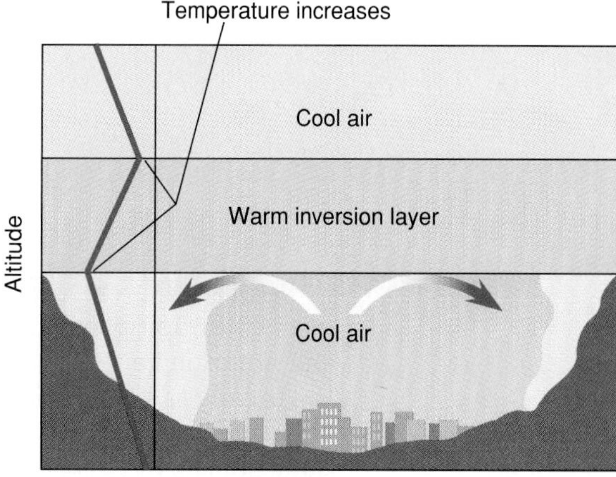

b. Temperature inversion

FIGURE 11.16

(a) Under normal atmospheric conditions, air warmed at the earth's surface rises and mixes with cooler air above. Any pollutants are dispersed upward. (b) When a temperature inversion occurs, a layer of warm air settles over cooler air, preventing it from rising. Any pollutants become trapped under the warm lid until atmospheric conditions change.

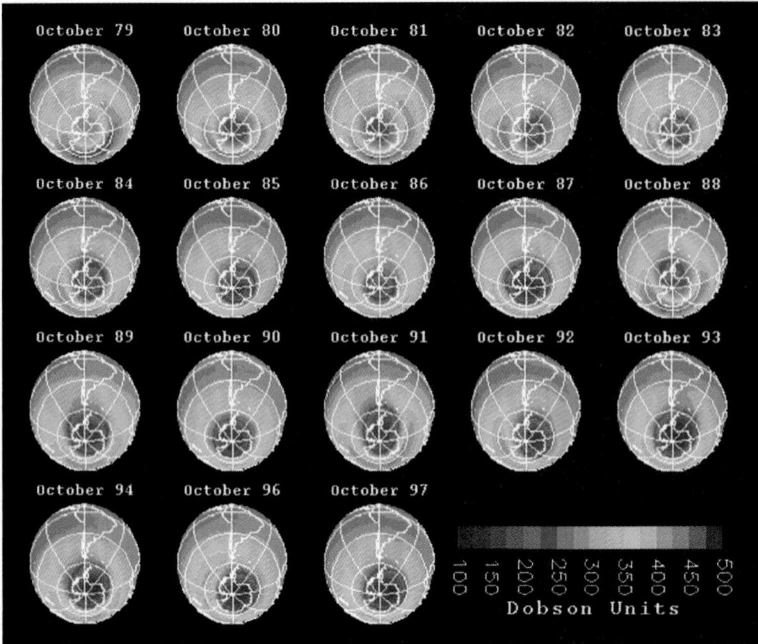

FIGURE 11.17

The ozone hole over the South Pole (shown in purple) was the largest ever recorded in September 1998.

11.15 Depletion of the Protective Ozone Layer in the Stratosphere

Since 1985, satellite images have revealed a "hole" in the ozone layer each spring over the South Pole (remember that spring in the Southern Hemisphere is autumn in the Northern Hemisphere) **(Fig. 11.17)**. In September 1998, the **ozone hole** was the largest ever recorded, and one of the deepest. Measuring 10.5 million square miles, it covered an area greater than all of North America. There is convincing evidence that **chlorofluorocarbons (CFCs),** a class of synthetic organic compounds, are responsible for this destruction of the ozone layer.

CFCs are inert, nontoxic, nonflammable compounds that were first produced in the 1930s. They are very useful as coolants for refrigerators and air conditioners (including those in automobiles) and as blowing agents in the production of polymer foam insulation. They are unsurpassed as solvents for cleaning electronic microcircuits and were used for many years as propellants for aerosol spray cans. In all these uses, except refrigeration, CFCs are released directly into the atmosphere. Commercially, the most important CFCs are the halogenated methanes, Freon-11 and Freon-12 (Freon is a DuPont trade name).

```
     Cl              Cl
     |               |
Cl—C—F          Cl—C—F
     |               |
     Cl              F

 Freon-11        Freon-12
```

Because CFCs are so unreactive, they do not break down when released and can persist in the troposphere for more than 100 years. Over time, air currents carry them into the stratosphere.

As early as 1974, two chemists at the University of California at Irvine, F. Sherwood Rowland and Mario Molina, predicted that when exposed to UV radiation in the stratosphere, CFCs would break down to form chlorine radicals (Cl•):

$$\textbf{Step 1}\quad CF_2Cl_2 \xrightarrow{UV} CF_2Cl + Cl\bullet$$

Each chlorine radical would then destroy a molecule of ozone with the formation of a chlorine monoxide radical (ClO•) and a molecule of oxygen (Step 2). The chlorine monoxide radical in turn would react with an oxygen atom to yield another molecule of oxygen and another chlorine radical—which could then start the chain reaction over again (Step 3). The net result would be the destruction of a molecule of ozone.

Step 2	$Cl\bullet$	+	O_3	$\xrightarrow{UV}$	$ClO\bullet$	+	O_2
Step 3	$ClO\bullet$	+	O	$\longrightarrow$	$Cl\bullet$	+	O_2
Net result:	O_3	+	O	$\longrightarrow$	$2\ O_2$		

Convincing field evidence for the involvement of CFCs in ozone depletion came in 1987. Between August and September of that year, a NASA research plane, equipped with sophisticated analytical instruments, flew 25 missions in the region of the ozone hole in the stratosphere over Antarctica. Data collected showed conclusively that as ozone concentration decreased, the concentration of the chlorine monoxide radical ($ClO\bullet$) rose **(Fig. 11.18)**. The chlorine monoxide radical has been dubbed the "smoking gun" of ozone depletion. The above chain reaction is now thought to account for about 80% of the ozone loss in the stratosphere. It has been estimated that each chlorine radical involved in a chain reaction has the potential to destroy 100,000 molecules of ozone before winds carry it back to the troposphere.

Destruction of ozone is greatest over Antarctica because of unusual meteorological conditions that exist there. During the Antarctic winter (June to September) when there is little or no sunshine, the temperature in the stratosphere over the South Pole becomes colder than at any other place on earth. Frigid temperatures and winds set up a rapidly rotating polar vortex of air, in which stratospheric ice clouds containing nitric acid (HNO_3) crystals form. Active chlorine species condense on the cloud surfaces. In the spring, the returning sunlight initiates reactions leading to the release of chlorine and chlorine monoxide radicals and the destruction of ozone. Warmer weather eventually breaks up the vortex, and ozone-poor air masses spread out toward Australia, New Zealand, and the southern tips of Africa and South America where they may remain for several weeks. If the winter temperature over the South Pole were not so low, nitric acid crystals would not be formed. Instead, gaseous nitrogen oxides would be present and would react with chlorine monoxide radicals to form chlorine nitrate ($ClONO_2$), thus limiting the formation of ozone-destroying chlorine radicals (see Step 3 above).

A number of related bromine-containing compounds called *halons* have also been implicated in ozone destruction. Halons, which include CF_2ClBr and CF_3Br, are very effective as fire-extinguishing agents. Like CFCs, they break down in the stratosphere but form bromine radicals ($Br\bullet$), which initiate the same type of chain reaction as shown for chlorine radicals (see Steps 1–3 above).

Another cause for concern is the release of nitrogen oxides (NO_x) in the exhaust of aircraft flying in the stratosphere. Normally, nitrogen oxides, which are derived from nitrous oxide (N_2O) emitted by microorganisms in the soil and oceans, occur naturally in the stratosphere, where they are responsible for some of the normal destruction of ozone. Because adding to stratospheric nitrogen oxides could further increase ozone depletion, plans by the United States to build a fleet of supersonic transport planes was halted.

The importance of the work of Rowland and Molina in elucidating the role of CFCs in ozone destruction was rec-

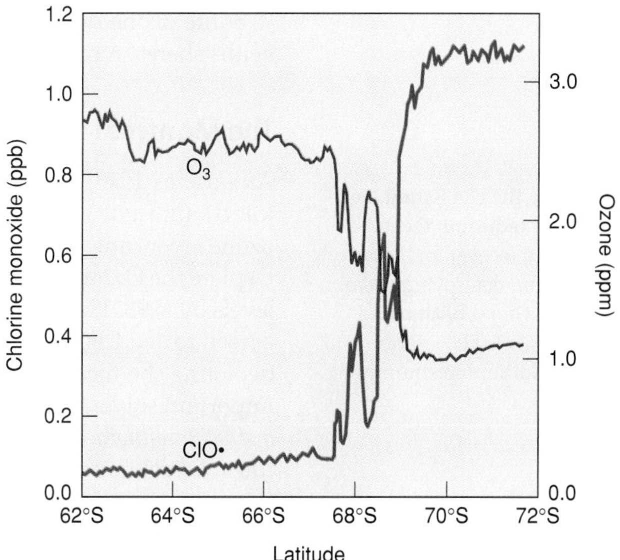

FIGURE 11.18

In September 1987, instruments aboard a NASA research plane simultaneously measured ozone and chlorine monoxide concentrations as the plane flew southward from Chile toward Antarctica. As the plane entered the ozone hole, chlorine monoxide concentration increased rapidly to about 500 times its normal atmospheric level, while ozone concentration fell dramatically.

ognized in 1995: Along with Paul Crutzen, a Dutch scientist who had demonstrated the influence of nitrogen oxides in maintaining normal stratospheric ozone concentration, they were awarded that year's Nobel Prize in chemistry.

Effects of Ozone Depletion on Human Health and the Environment

Depletion of the ozone layer leaves the earth vulnerable to the damaging effects of UV radiation. UVB radiation (wavelengths of 280–320 nm), because it transmits more energy, is more damaging to living organisms than longer-wavelength UVA radiation (320–400 nm). In humans, exposure to UVB radiation causes cataracts, tanning and sunburn, and suppression of the immune system. By damaging DNA, UVB radiation can cause several types of skin cancer, including deadly melanoma. It also harms crops and kills phytoplankton, the microscopic photosynthetic organisms that are at the lowest level of ocean food chains. It has been calculated that for every 1% decrease in stratospheric ozone, there is a 2% increase in the amount of UVB radiation reaching the earth's surface.

Ozone Loss over the Arctic and the Middle to High Latitudes

A springtime depletion of ozone over the Arctic was first observed in 1995. However, because it does not get as cold at the North Pole as at the South Pole and because the northern polar vortex is weaker, ozone loss over the Arctic has not been as great as over the Antarctic. However, since three-quarters of the world's population lives in the Northern Hemisphere, continued depletion of ozone over the Arctic could have even more serious consequences than depletion over the Antarctic.

Some ozone depletion has also occurred over the middle to high latitudes in both hemispheres, record losses being observed in 1998.

The Montreal Protocol

In 1993, the EPA issued regulations requiring technicians to recover and recycle all ozone-destroying chemicals during servicing and disposal of refrigeration and air conditioning equipment.

As early as 1978, the use of CFCs as aerosol propellants was banned in North America, though not in most other countries. The first international effort to protect the ozone layer came in 1987 with the signing of the Montreal Protocol on Substances That Deplete the Ozone Layer, which called for CFC production to be cut back to 5% of 1986 levels by 1998. This protocol was amended in 1990 and again in 1992, when 140 nations agreed to end CFC production by 1995 and to speed up the phase-out of all other ozone-depleting chemicals. Targeted chemicals include halons, carbon tetrachloride (CCl_4), an important solvent; methyl bromide (CH_3Br), a widely used agricultural fumigant; and *hydrochlorofluorocarbons (HCFCs)*, the compounds presently being used as CFC substitutes. Developing nations were given until 2010 to halt their production of CFCs and their use of methyl bromide and HCFCs was not restricted.

Global production of CFCs has plummeted, but the drop has been less than hoped for because production is continuing to rise in China, India, and other developing nations (**Fig. 11.19**), and because Russia failed to meet its 1996 deadline. Further, there is a flourishing black market in CFCs; they are being smuggled into the United States and other industrialized countries, mainly from Mexico and the former Eastern Bloc countries. In the United States, only the drug trade is more profitable than the traffic in CFCs.

Because CFCs are so long-lived in the atmosphere, even if all nations abide by their commitments, the ozone hole is expected to continue to grow until about 2005 before very slowly beginning to mend.

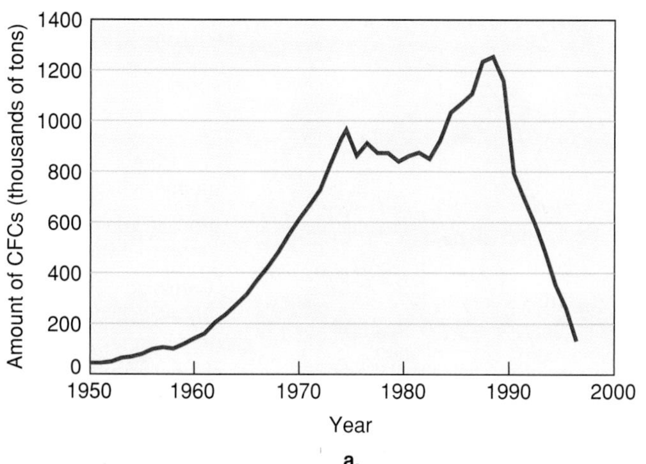

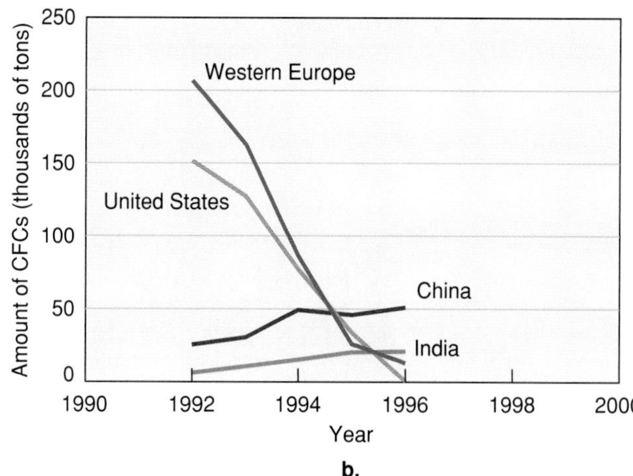

FIGURE 11.19

(a) Worldwide production of CFCs has fallen dramatically since the signing of the Montreal Protocol in 1987. (b) CFC production continues to decrease in the United States and Western Europe (France, Germany, Italy, the Netherlands, Spain, United Kingdom), but is increasing in China and India.

The Montreal Protocol, despite its limitations, is an encouraging example of international cooperation in the interest of solving a major global environmental problem. It can be credited with preventing numerous cases of skin cancer and halting widespread damage to the environment.

Alternatives to CFCs

The first CFC substitutes to be introduced were HCFCs, such as CF_3CHCl_2 and CHF_2Cl, compounds that have fewer chlorine atoms than do CFCs. HCFCs break down more readily in the troposphere than CFCs and so are less likely to reach the stratosphere. However, because they can cause some ozone destruction, they are scheduled to be phased out by 2030. Much better substitutes for CFCs are the *hydrofluorocarbons (HFCs)*, which contain no chlorine. One of these, CF_3CH_2F, has been used successfully as a refrigerant and, since 1994, has replaced Freon in nearly all car air conditioners. In the electronics industry, soapy water followed by rinsing and air drying is now used instead of CFCs to clean microcircuits.

Unfortunately, there is a serious problem associated with the long-term use of HFCs. Like CFCs and HCFCs, they contribute to climate change (see Section 11.17). Research is ongoing to find chemicals that are both efficient refrigerants and environmentally friendly.

11.16 Indoor Air Pollution

You might expect to be safer from air pollutants indoors, but in today's well-sealed homes and offices, this is often not the case. In buildings where there is little or no circulation of fresh air, pollutants may accumulate to dangerous levels. **Figure 11.20** shows sources of the major indoor pollutants.

Smoking is a particularly dangerous cause of indoor air pollution. In addition to tars and nicotine, tobacco smoke contains high levels of all the primary pollutants associ-

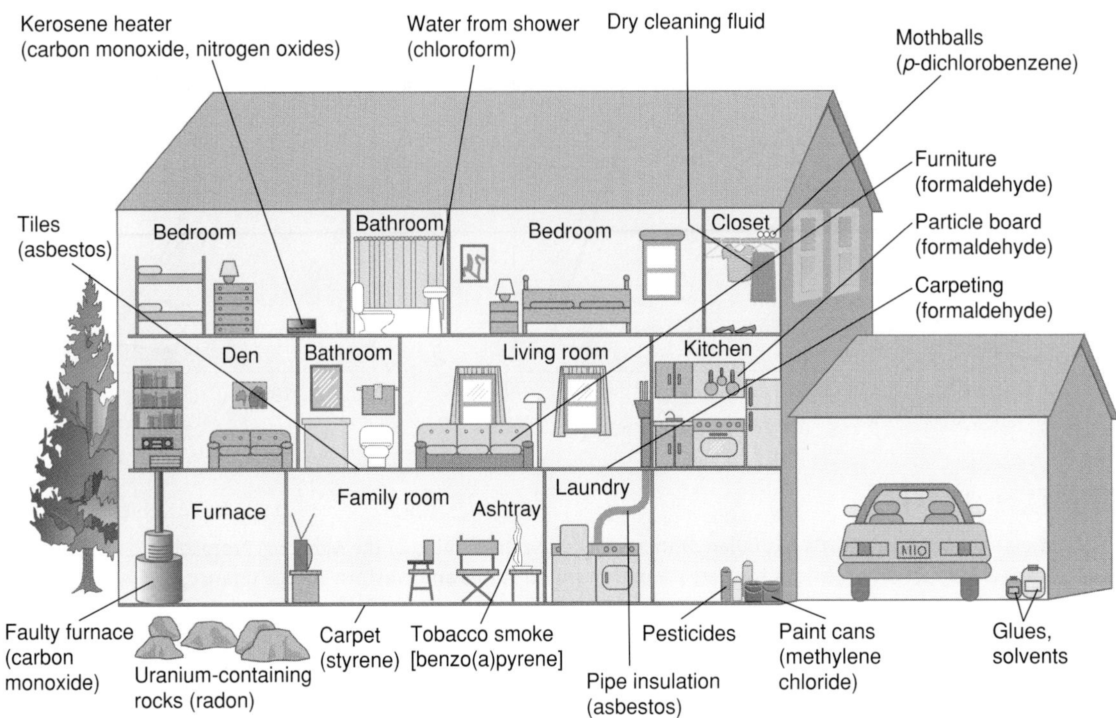

FIGURE 11.20

Sources of major indoor air pollutants.

Cigarette smokers inhale much higher concentations of pollutants than are found in even severely polluted air.

ated with combustion. Smoking causes emphysema, lung cancer, and coronary heart disease. Of increasing concern is the risk of smoking to nonsmokers. Young children of smokers, for example, are more likely to suffer from asthma and bronchitis than are children of nonsmokers.

Gas stoves, kerosene heaters, wood stoves, and faulty furnaces are potential sources of nitrogen oxides and carbon monoxide. Paint, paint strippers and thinners, gasoline, and pesticides, which many people store in their basements, release harmful vapors and dust particles. Formaldehyde, a toxic, irritating gas, is released from the polymers used to manufacture certain types of insulation foam and furniture stuffing and from newly installed carpeting and paneling. Clothes brought home from the dry cleaners may also cause a problem: Traces of harmful volatile solvents used in the cleaning process are retained by the garments and later released into the atmosphere. Even taking a hot shower or bath may be harmful since chloroform can be released from chlorine-treated water (Chapter 10). A particularly insidious indoor air pollutant is radon (Chapter 5).

Apart from preventing pollutants from entering a building in the first place, one way to control indoor air pollution is to install air-to-air heat exchangers that circulate fresh air without adversely upsetting the temperature of the indoor air. Air conditioners, smoke removers, and vacuum cleaners all help to reduce indoor air pollutants.

11.17 Global Warming and Climate Change

There is growing concern that increases in concentration of certain trace gases in the atmosphere, primarily carbon dioxide, are causing a general warming trend, referred to as **global warming.** To understand how these gases may be causing a rise in temperature, you need to know how the earth maintains its heat balance.

EXPLORATIONS looks at how the processes related to global warming might be used to alter the atmosphere on Mars (see pp. 352–353).

The Earth's Heat Balance

Solar energy is transmitted through space as electromagnetic radiation, which, proceeding from longer to shorter wavelengths, includes radio waves, microwaves, infrared radiation, visible light, UV radiation, and X-rays (see Fig. 1.13). The amount of solar energy that reaches the outer limits of the earth's upper atmosphere is enormous; if all of it penetrated to the earth's surface and were retained, the very high temperature would have prevented the development of any life on earth. Fortunately, as solar radiation travels through the atmosphere, interactions with gases and particulates prevent about half of it from penetrating to the earth's surface. The 50% of solar energy that eventually reaches the earth includes the entire visible region of the spectrum together with smaller portions of the adjacent UV and infrared (IR) regions. This incoming radiation is largely absorbed at the surface and then reradiated back to space. If this radiation away from the surface did not occur, the earth would become increasingly warm as solar energy continued to flow in. Outgoing radiation from the earth is in the longer-wavelength IR region.

A small amount of incoming radiation is scattered when it encounters gases and particulates. Scattering of the blue part of visible light by nitrogen and oxygen molecules accounts for the blue color of the sky.

A considerable amount of the outgoing IR radiation does not escape into space but is reabsorbed by gases in the atmosphere called **greenhouse gases** and then reradiated back to earth. As a result of this absorption and reradiation, the atmosphere is warmed (**Fig. 11.21**). Greenhouse gases include carbon dioxide, water vapor, methane, nitrous oxide (N_2O), CFCs, and ozone. The most important is carbon dioxide. Although water vapor is present in the atmosphere at a much higher concentration than carbon dioxide, it absorbs IR radiation less strongly than carbon dioxide does, and so its effect is much less.

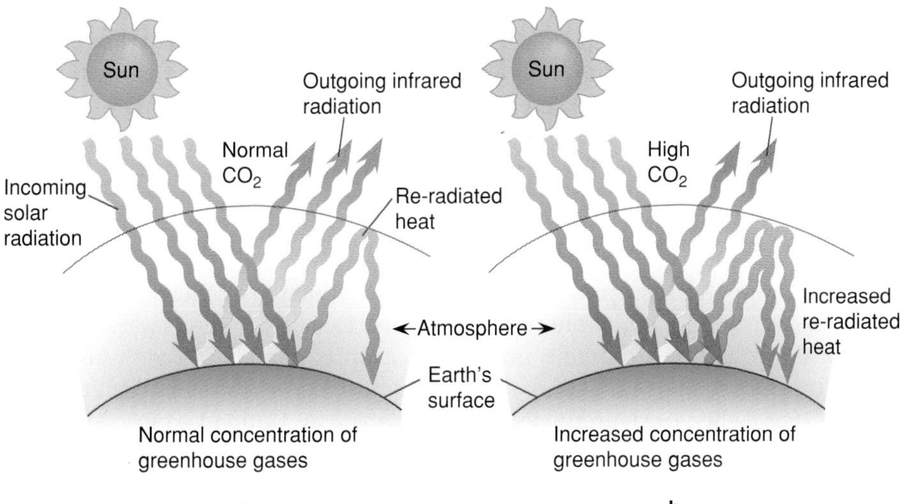

FIGURE 11.21

(a) Light from the sun penetrates the atmosphere and warms the earth's surface. Longer-wavelength infrared radiation is radiated away from the surface; some of this outgoing radiation is absorbed by carbon dioxide and other greenhouse gases and then reradiated back to the surface. (b) When concentrations of greenhouse gases increase, more infrared radiation is returned to the earth and the surface temperature rises.

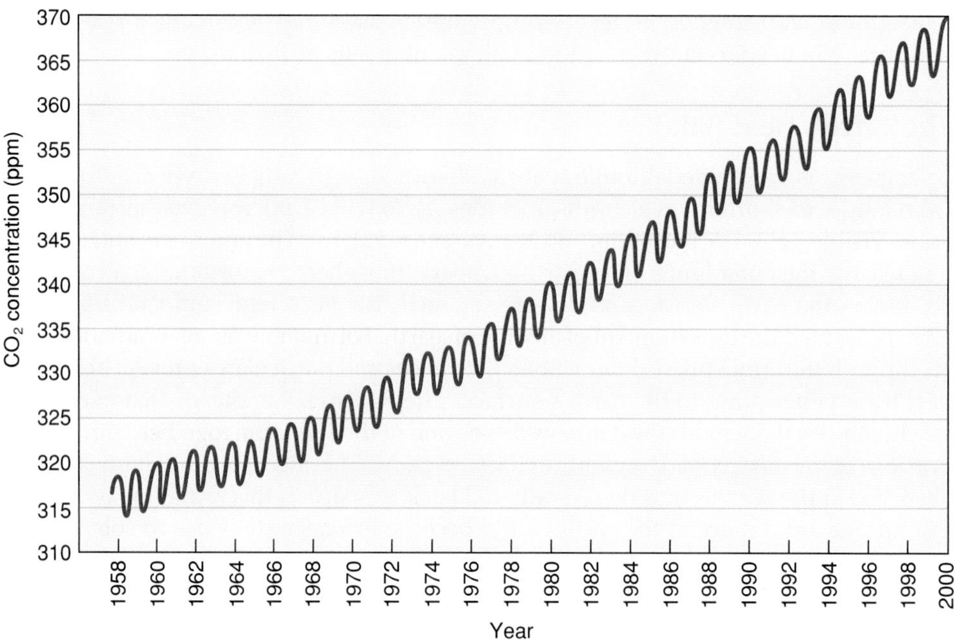

FIGURE 11.22

Since the first measurements were made at Mauna Loa, Hawaii, in 1958, carbon dioxide concentration in the atmosphere has risen dramatically. The yearly seasonal variations are caused by the removal of carbon dioxide from the atmosphere by growing plants in the summer and the return of carbon dioxide to the atmosphere in the winter when plants decay.

The warming effect caused by the absorption and reradiation of IR radiation by the greenhouse gases is commonly called the **greenhouse effect.** This name arose because the gases act somewhat like a pane of glass in a greenhouse: Visible light passes through the glass and is absorbed by the objects inside the greenhouse. The objects are warmed and emit heat energy in the form of IR radiation, which cannot pass through glass. It is trapped and the temperature rises inside the greenhouse.

The Increase in Atmospheric Carbon Dioxide

Analysis of air trapped in samples of ancient ice shows that the average concentration of carbon dioxide in the atmosphere has been increasing for thousands of years. Twenty thousand years ago, the concentration was approximately 200 ppm; before the start of the Industrial Revolution at the end of the 19th century, it had risen to about 280 ppm. In 1958, when measurements were first made at Mauna Loa in Hawaii, the atmospheric carbon dioxide concentration was 315 ppm; by 2000, it had reached 370 ppm (Fig. 11.22). Thus, there has been a 30% increase in just 100 years. The major cause of the increase in atmospheric carbon dioxide is the burning of fossil fuels by electric utilities, automobiles, and industry.

Tree planting is encouraged as a way to combat carbon dioxide buildup in the atmosphere. It has been estimated that a tree farm the size of Australia would be needed to absorb the excess carbon dioxide produced each year.

A secondary cause of the increase is deforestation. As trees grow, they take carbon dioxide from the atmosphere in the process of photosynthesis; when they decay or are burned in forest fires, carbon dioxide is released back into the atmosphere. Any new growth in a cleared area, whether crops or grasses, takes much less carbon dioxide from the atmosphere than a mature forest does. By 2000, 2% of all rain forests were being cut

Table 11.2 Concentrations of Greenhouse Gases in the Atmosphere Today and 100 Years Ago

Gas	Average Concentration (ppb)	
	100 Years Ago	Today
Carbon dioxide (CO_2)	280,000	370,000
Methane (CH_4)	900	1,700
Nitrous oxide (N_2O)	285	310
Chlorofluorocarbons (CFCs)	0	3

and burned each year. Destruction is particularly severe in tropical rain forests, especially in Brazil.

The atmospheric concentrations of other greenhouse gases have also risen in recent years (**Table 11.2**). Although their combined total concentration is considerably less than that of carbon dioxide, they absorb and re-emit more IR radiation, molecule for molecule, than carbon dioxide does. Methane, for example, is approximately 25 times more effective than carbon dioxide at trapping heat. If the levels of the other greenhouse gases continue to rise at the present rate, it is estimated that their warming effect will soon equal that of carbon dioxide.

The steady rise in methane is attributed primarily to a worldwide increase in the number of cattle and the number of rice paddies, both of which are sources of methane. Other anthropogenic sources of methane include landfills (Chapter 19) and coal mines. Nitrous oxide is formed naturally in the soil during microbiological processes associated with nitrogen fixation (Chapter 2). Its increase in the atmosphere is believed to be caused primarily by the increased use of nitrogen-containing fertilizers.

Evidence for Global Warming

The evidence for global warming is considerable. Glaciers all over the world have been retreating steadily for many years and unusually large icebergs have been breaking off Antarctic ice shelves. The average temperature at the earth's surface has increased by 0.3–0.6°C (0.5–1.1°F) since the last half of the 19th century. During the same period, sea levels around the world have risen 10–25 centimeters (4–10 inches) as water has expanded and ice melted. The 14 warmest years since 1860 (when reliable record-keeping began) have all occurred since 1980; 1998 was the hottest year on record. There was a slight drop in average temperature in 1999 and no further change in 2000 (**Fig. 11.23**). The rate of temperature increase in the 20th century was greater than at any time since 1200.

Recent studies of ancient ice have provided convincing evidence that the warming trend is likely to continue if carbon dioxide emissions continue to rise. Scientists took mile-long ice cores, dating back 160,000 years, from Antarctic glaciers and analyzed the air trapped in pockets in the ice for carbon dioxide. They

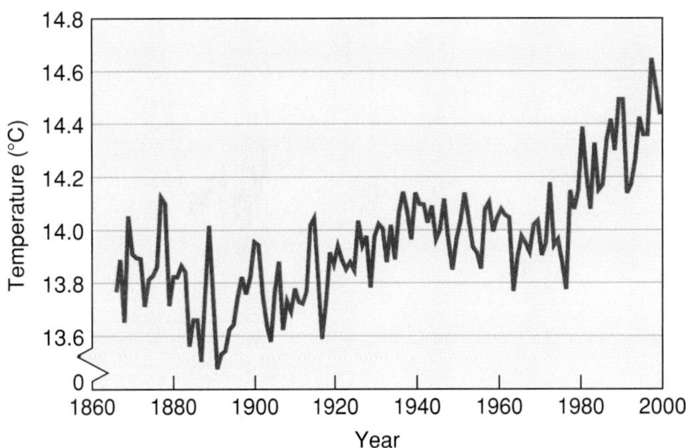

FIGURE 11.23
Average temperature at the earth's surface, 1866–2000.

also estimated the air temperature at the time the air became locked in the ice, using a technique based on measurements of the ratio of deuterium (heavy hydrogen, ${}^{2}_{1}H$) to ordinary hydrogen (${}^{1}_{1}H$) in the ice surrounding the air pockets (Chapter 5). Lighter water molecules evaporate more readily than heavy water molecules. Therefore, the higher the ${}^{2}_{1}H/{}^{1}_{1}H$ ratio in the frozen water, the higher the temperature. From their findings, the scientists were able to estimate temperatures at the time the different layers of ice were formed. As **Figure 11.24** shows, there is a remarkable correlation between carbon dioxide concentration and air temperature. The air temperature rises and falls in step with increases and decreases in the carbon dioxide concentration.

Computer-generated climate models also predict a warming trend. These models are based on numerous complex mathematical equations representing the many variables that affect climate. The information is fed into supercomputers, which then project temperature changes that are likely to occur if atmospheric carbon dioxide increases. Because many factors, such as the influence of clouds and the role of deep ocean currents, are not well understood and are not included in the computer models, some scientists have been skeptical about global warming. However, in 1995, the Intergov-

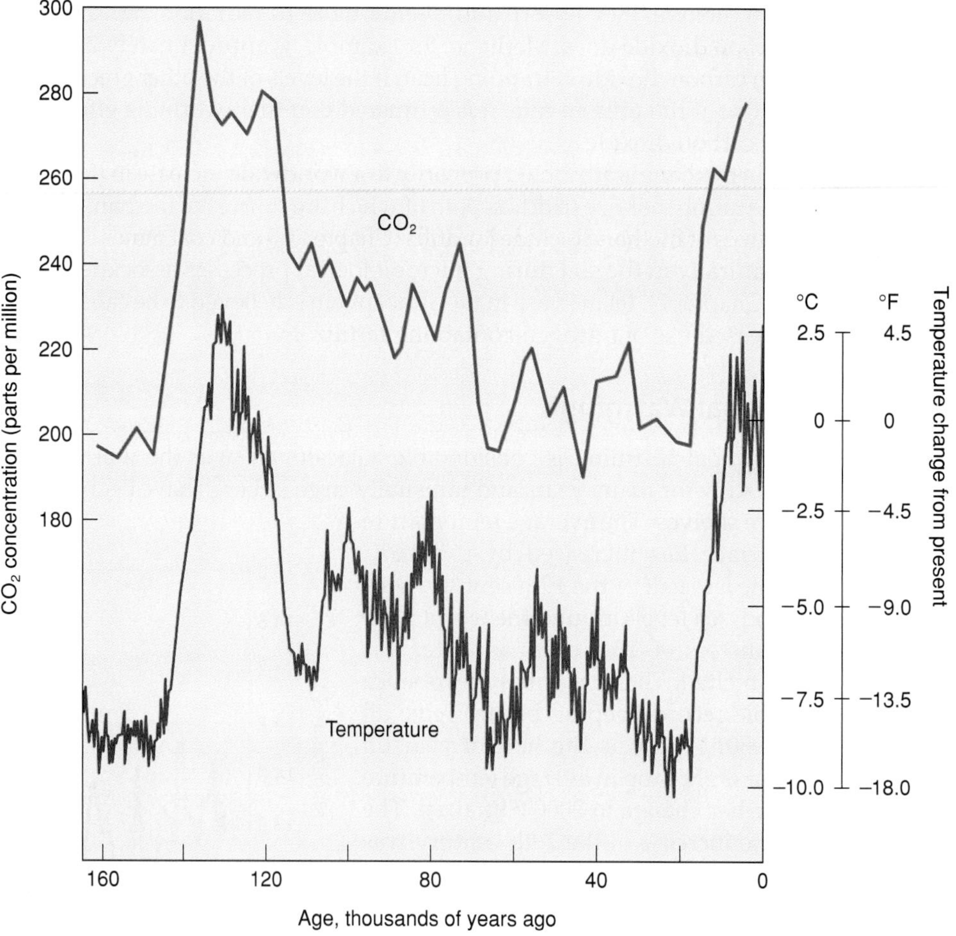

FIGURE 11.24

Variations in atmospheric carbon dioxide concentration and air temperature over the last 160,000 years were revealed by analysis of Antarctic ice cores. Air temperature has risen and fallen in step with increases and decreases in carbon dioxide concentration.

ernmental Panel on Climate Change (IPCC), a group of 2500 scientists from 100 countries, concluded that "the balance of evidence suggests that there is a discernible human influence on global climate."

Effects of Global Warming

The effects of global warming could be devastating. The IPCC predicts that if nothing is done to lower emissions of greenhouse gases, by 2100, average global surface air temperature will increase 1–3.5°C (1.8–6.3°F) and world sea levels will rise 15–90 centimeters (6–36 inches). These may seem like small increases, but it should be remembered that at the time of the last Ice Age, the earth's temperature was only 5°C lower than it is today. A sea level rise of only 15 centimeters (6 inches) would cause flooding in many coastal areas.

With rising temperatures, air circulation, ocean currents, and rainfall patterns would change, causing generally violent weather. As a result, some regions of the world, including much of the United States, would experience droughts; other regions would become much wetter. Ecosystems all over the world would be disrupted, and some species might face extinction. Climate-related diseases such as malaria would invade areas where they are currently unknown.

Controlling Global Warming

Because of its potential to change world climate, scientists in many nations view global warming as the major environmental problem of the 21st century. The first treaty to address this problem was signed at the United Nations Conference on the Environment (the Earth Summit) in Rio de Janeiro, Brazil, in June 1992. The treaty set goals to reduce emissions of carbon dioxide and other greenhouse gases by industrialized nations to 1990 levels by 2000. However, because the proposed reductions were not binding, they were not met.

At a conference on climate change held in Buenos Aires, Argentina, in November 1998, 106 nations, including the United States, agreed to begin implementing the agreement reached the year before at Kyoto, Japan. The market-based Kyoto Protocol commits the 38 industrialized nations, including the republics of the former Soviet Union, to reducing global greenhouse gas emissions to at least 5% below 1990 levels by 2012. Different nations have different targets and timetables. Each has an emission allowance, which, as in the United States program for reducing emissions of sulfur dioxide (discussed earlier in this chapter) can be traded.

The major flaw in the Kyoto Protocol is that it does not require cuts by the developing nations, most of which, including India and China, are opposed to making any reductions in the near future. An encouraging sign, however, was the voluntary agreement by Argentina and Kazakhstan to reduce their greenhouse gas emissions. Although, at present, developing countries produce one-tenth as much carbon dioxide per person as industrialized countries, in the last ten years, as their economies have grown, their emissions have increased 75%. It is predicted that even if the industrialized nations reach their targets, the total concentration of greenhouse gases in the atmosphere will continue to rise.

Taking steps to reduce emissions will be both difficult and costly. In the United States, there is opposition from the oil and coal industries and their supporters in Congress, many of whom continue to deny the evidence for global warming. As a result, in 2001, the U.S. government withdrew its support of the Kyoto Protocol. However, because of unprecedented losses from recent weather-related disasters, the banking and insurance industries are taking the problem seriously. And although most oil companies oppose a global warming treaty, several major European corporations, including British Petroleum and Royal Dutch Shell, are now in favor of it.

11.18 Regulating Air Pollution

The first person to enact legislation to prevent air pollution was probably King Edward I (1239–1307) of England, who banned the burning of a high-sulfur coal called "seale" coal, because it produced noxious fumes.

The Clean Air Act of 1970 mandated air quality standards for five air pollutants: suspended particles, sulfur dioxide, carbon monoxide, nitrogen oxides, and ozone. A few years later, lead was added to the list. Until a phase-out began in 1975, lead was added to gasoline to prevent knocking (Chapter 12), and large quantities of lead were released to the atmosphere in engine exhaust, causing serious problems in urban areas. Despite great improvements in air quality, including a dramatic drop in lead emissions (**Fig. 11.25**), many urban areas were not meeting the desired standards by the late 1980s. Thus, in 1990, tougher standards were established in the Clean Air Act Amendments.

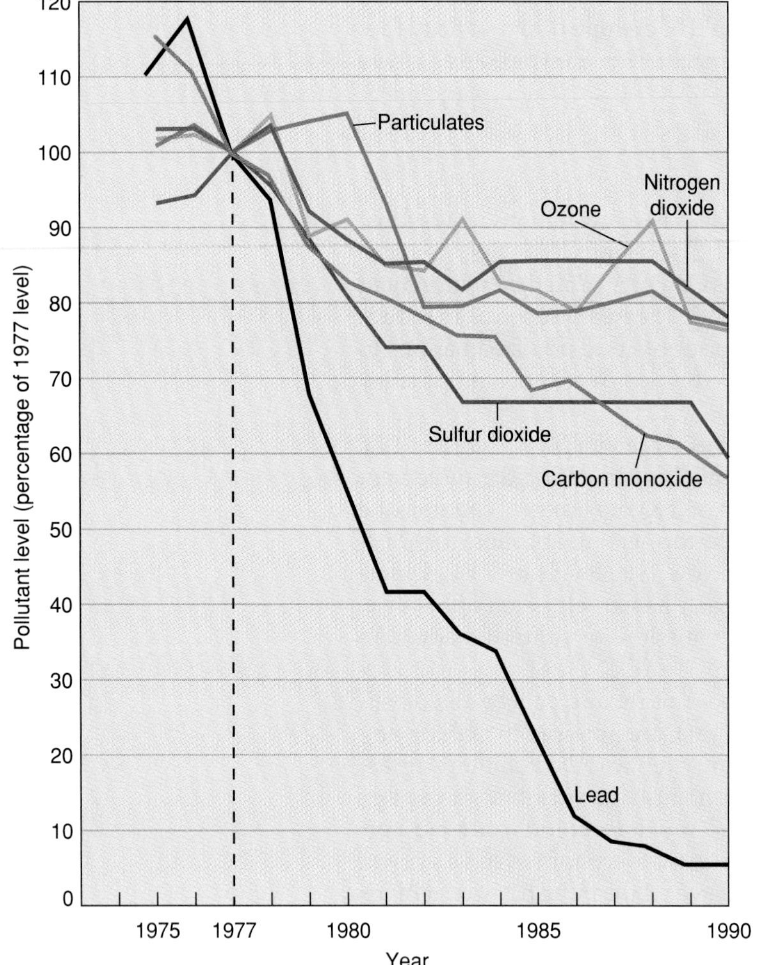

FIGURE 11.25

Reductions in six air pollutants, using 1977 values as 100% levels. The most dramatic reduction was in lead emissions.

A significant problem not addressed by the original act was the emission by industry of thousands of tons of unregulated hazardous chemicals, many of which are suspected of being carcinogens. The 1990 amendments require that industries emitting any of 189 specifically named toxic chemicals install control devices capable of reducing those emissions by at least 90%. All industries must be in compliance by 2003.

As noted earlier, the 1990 amendments include other important provisions: Coal-burning power plants must reduce annual sulfur dioxide emissions, and new cars will have to meet increasingly strict emission standards. At present, light trucks (sport-utility vehicles, minivans, and pickups) emit three times as much pollution as the average passenger car because they are not required to meet the same emissions standards, but this may soon change. The EPA recently proposed new rules that would require all passenger vehicles to meet the same standards by 2004. As you will see in Chapter 13, to meet new stricter emission standards, automakers are developing cleaner and more fuel-efficient cars, including cars powered by fuel cells, electric cars, and hybrid cars in which a battery and small gasoline engine work together.

Since 1992, cities with unacceptably high emissions of carbon monoxide have been required to sell gasoline containing 2.7% oxygen during the winter months to reduce these

Table 11.3 National Ambient Air Quality Standards Established by the EPA in 1992

Pollutant	Time Period*	Limit in 1992	Proposed Limit in 1996
Carbon monoxide	8 hours 1 hour	9 ppm 35 ppm	No change
Sulfur dioxide	1 year 24 hours 3 hours	0.03 ppm 0.14 ppm 0.5 ppm	No change
Nitrogen oxides	1 year	0.05 ppm	No change
Ozone	1 hour	0.12 ppm	0.018 ppm
Particulates of diameter <10 microns	1 year 24 hours	50 $\mu g/m^3$ 150 $\mu g/m^3$	Diameter reduced to 2.5 microns

*Period over which the concentrations are measured and averaged.

emissions. The oxygen content can be increased by adding ethanol or methanol, but MTBE (methyl-*t*-butyl ether) is preferred because of its higher octane rating (see Chapter 12). As a result of its increased oxygen content, this reformulated gasoline emits less carbon monoxide when it is burned than does regular gasoline. However, MTBE, a known animal carcinogen, has been detected in groundwater in several states following gasoline spills and leaks from underground storage tanks. The EPA has recommended a nationwide ban on its use because of the threat to human health.

The Clean Air Act and amendments required the EPA to establish National Ambient Air Quality Standards (NAAQS) for the major pollutants (**Table 11.3**). For each pollutant, these standards specified a concentration, averaged over a specified time period, which must not be exceeded. At the end of 1996, the EPA proposed even more stringent standards for ozone and particulates: The allowable level for ozone was lowered from 0.12 ppm to 0.08 ppm, and the size (diameter) of particulates covered by the standards was reduced from 10 microns to 2.5 microns. Because of the costs involved and questions about the potential health benefits, the tougher standards are opposed by industry, state governors, and cities and have not been adopted.

In 1998, the EPA announced further new rules, this time aimed at reducing the interstate flow of smog-causing emissions from 22 industrial eastern states. Power plants in the targeted states have to limit emissions of nitrogen oxides, which are precursors of both ozone and photochemical smog.

Air pollution is a worldwide problem that is far more serious in many other countries than it is in the United States. For example, in Mexico City, Sao Paulo (Brazil), and many industrial cities in China, Eastern Europe, and the former Soviet Union, people live under a pall of toxic smog. The sun is obscured, severe respiratory ailments are common, and infant mortality rates are high; often, there are few complaints because the smog and pollution mean jobs.

Cleaning up the atmosphere will be costly and will bring economic hardship, particularly to the developing countries of the world. But the alternative of continuing to pour pollutants into the air will have far more serious consequences. As we have seen, pollutants are already destroying the protective ozone shield, and if, as many fear, human activities are warming the earth, the resulting alterations in the world's climate could change, if not destroy, our entire way of life.

EXPLORATIONS

Setting Up House on Mars, the Red Planet

All the planets in our solar system were born from common cosmic material some 4.6 billion years ago, but only the Earth is known to support life. Mercury and Venus, the two planets closer than the Earth to the Sun, are too hot to sustain life; Mars, the next planet outward from the sun, is too cold. But would it not be possible to warm Mars and make it habitable? Then, if the Earth became too crowded or its environment deteriorated beyond repair, there would be another planet to escape to in the future. We have already learned, at a cost, that introducing greenhouse gases into Earth's atmosphere can raise temperatures and alter climate on a global scale. Why not use the same approach on Mars?

This notion sounds like the stuff of science fiction, but space scientists believe it could be done. They point out that data from the Viking (1976), Pathfinder (1997), and Global Surveyor (1999) missions to Mars indicate that many of the ingredients necessary for life, including carbon dioxide and water (potential sources of oxygen) and nitrogen, are already present on Mars, frozen into its surface. Mars's gravity (about two-fifths as strong as that of Earth) is adequate for life, and, conveniently, a day on Mars is almost the same length as a day on Earth, although the martian year extends to 687 days.

Present and Past Climate on Mars

At present, Mars is very inhospitable. It is dry and barren, its surface frequently swept by windblown clouds of the iron-rich dust that give it its characteristic red color. Located 1.5 times as far from the sun as the Earth, Mars receives 57% less sunlight. Temperature at its equator may reach 20°C (68°F), but average surface temperature is a frigid −95°C (−139°F), very much colder than the earth's average of 15°C (59°F). The sparse martian atmosphere is 150 times thinner than earth's and consists primarily of carbon dioxide (95%), with some nitrogen (2–3%), and smaller amounts of other gases.

Early in its history, Mars was very different. Approximately 3.5 billion years ago, when the planet was still very hot, volcanoes were erupting on its surface and emitting carbon dioxide, water vapor, and nitrogen into the martian atmosphere. As Mars cooled, liquid water is

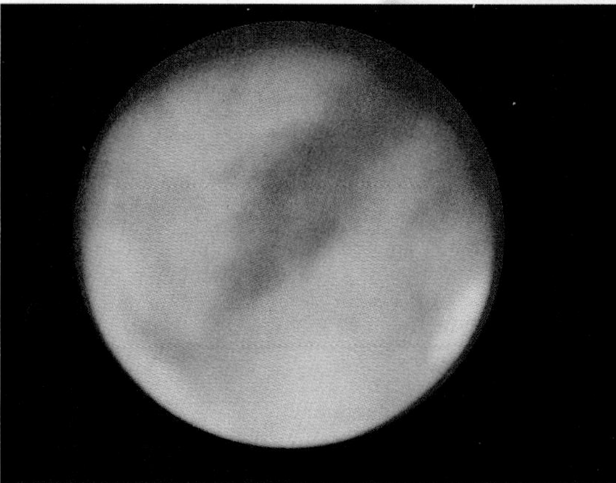

Mars, the red planet

believed to have flowed over its surface, carving out the riverlike channels that can be seen on the red planet today. In time, most of the atmosphere disappeared. Some was lost to space, but evidence suggests that huge quantities of carbon dioxide were absorbed into the ground and froze in subsurface reservoirs as the planet continued to cool. Some carbon dioxide may have reacted with minerals to form carbonates when water was present on Mars. The extensive polar ice caps visible year-round are composed of carbon dioxide ice in the south and water ice in the north. Large amounts of water are also probably tied up in permafrost.

Adding CFCs to the Martian Atmosphere

The key to making Mars habitable is to warm it up and thus unfreeze these stores of frozen carbon dioxide and water. Without liquid water on its surface, Mars could not support life. Scientists suggest that the simplest way to warm the planet is to release greenhouse gases into its atmosphere. These gases, by absorbing heat-producing infrared (IR) radiation from the sun and reradiating it back to the planet's surface, would raise the temperature. Chlorofluorocarbons (CFCs), the compounds that are of concern on earth because of their destructive effect on our ozone layer, would be suitable warming agents. Since the mass of CFCs required to cause a significant rise in temperature would far exceed the amount that could be lifted by a space shuttle, the gases would have to be manufactured on Mars. Adequate supplies of chlorine, fluorine, carbon, and other necessary raw materials are believed to be present in the martian soil. All that is needed is the development of technology to manufacture CFCs from these materials in the harsh martian environment.

It is estimated that, allowing for some breakdown of CFCs by the sun's ultraviolet (UV) radiation, several million tons of these compounds would be needed to raise the temperature of the martian surface above −30°C (approximately −20°F). At that point, carbon dioxide frozen in the ice cap and absorbed in the martian soil should begin to be released. Since carbon dioxide is also a greenhouse gas, it would augment the warming effect of the CFCs. More and more carbon dioxide would be released, and temperature would rise in a continuing cycle until it eventually climbed above the freezing point of water. Ice in the polar cap, in permafrost, and in underground reservoirs would then begin to melt. Water would flow on the martian surface, and water vapor (another greenhouse gas) would enter the atmosphere.

Introducing Plants to the Planet

At that stage, Mars should have an atmosphere and a surface environment capable of sustaining simple terrestrial plants. An adequate supply of nitrogen, an essential element for all organisms, is thought to be present in the martian soil in the form of nitrites and nitrates. Other necessary elements, including sulfur, phosphorus, and various trace metals, were detected in the martian rocks and soil by the Viking landers. Scientists estimate that, if the frozen stores of carbon dioxide are close to the surface, where they could be warmed up and the carbon dioxide released relatively rapidly, it would require 100 to 200 years to reach a time when plants could survive. If, however, the stores are buried deeply, where it would take much longer for heat to penetrate, this time frame could be as long as 100,000 years.

We have already learned, at a cost, that introducing greenhouse gases into Earth's atmosphere can raise temperatures and alter climate on a global scale.

A factor that would have to be considered is that once ponds and lakes began to form, the newly discharged carbon dioxide would dissolve in the water and gradually become tied up in carbonate deposits. Some means might have to be devised to release carbon dioxide from carbonate rocks to prevent depletion of the atmosphere.

An Atmosphere Fit for Humans

If green plants could be introduced on Mars, they would immediately start to photosynthesize, converting carbon dioxide and water into carbohydrates and oxygen. If the oxygen content of the atmosphere could be raised sufficiently, animals and humans might be able to take up residence on Mars. Unfortunately, plants are very inefficient producers of oxygen. Unless genetic engineering could improve their functioning, it could take up to 100,000 years to reach the required oxygen level.

Assuming it was possible to raise the oxygen level high enough to enable humans to breathe, nitrogen would have to be added to the atmosphere to act as a buffer gas, because at high concentration oxygen is toxic. The small amount of nitrogen known to be present in the martian atmosphere would need to be supplemented by introducing microorganisms capable of releasing nitrogen from compounds in the soil.

Once oxygen entered the martian atmosphere, some of it would be converted into a protective ozone layer by the action of UV radiation from the sun. To prevent destruction of the ozone by chlorine compounds released from the CFCs, alternatives to these greenhouse gases (probably related compounds containing no chlorine) would have to be manufactured at that point.

At present, of course, this plan for colonizing Mars is no more than scientific speculation. Before any action could even be considered, space scientists would have to determine if, in fact, Mars contains hidden reservoirs of carbon dioxide, water, and nitrogen. And there is the moral issue of whether humans have any right to attempt to alter the natural world on such a vast scale. Certainly, we should not do so until the possibility of life already existing on Mars has been completely ruled out. We must conclude that for the foreseeable future we shall have to accommodate our burgeoning population on our own planet.

References: J. Kluger, "Mars: in Earth's Image," *Discovery* (September 1992), p. 70; C. P. McKay, O. B. Toon, and J. F. Kasting, "Making Mars Habitable," review article, *Nature* (August 1991), p. 489; W. Newcott, "Return to Mars," *National Geographic* (August, 1998), p. 8.

Chapter Summary

1. Boyle's law states that at constant temperature, the volume of a fixed amount of gas is inversely proportional to the gas pressure ($P_1V_1 = P_2V_2$).
2. Charles's law states that at constant pressure, the volume of a fixed amount of gas is directly proportional to its temperature ($V_1/T_1 = V_2/T_2$).
3. The Ideal Gas Law can be expressed mathematically as $P_1V_1/T_1 = P_2V_2/T_2$.
4. Boyle's and Charles's laws are explained by the kinetic molecular theory.
5. Moving outward from the earth, the atmosphere is divided into four layers: troposphere, stratosphere, mesosphere, and thermosphere.
6. Atmospheric pressure decreases with increasing distance from the earth's surface. At sea level, the average pressure is 760 mm of mercury.
7. Dry air is approximately 78% nitrogen, 21% oxygen, 1% argon, and smaller amounts of other gases. Carbon dioxide contributes 0.04%.
8. The ozone (O_3) layer located in the stratosphere protects life on earth from the sun's harmful UV radiation.
9. The five primary air pollutants are carbon monoxide, sulfur dioxide, nitrogen oxides, volatile organic compounds, and particulates.
10. Secondary pollutants include sulfuric and nitric acids, ozone, and other photochemical oxidants.
11. Automobile emissions and the burning of fossil fuels by stationary sources are the major anthropogenic causes of air pollution.
12. The main anthropogenic source of carbon monoxide is the incomplete combustion of gasoline in automobile engines. Carbon monoxide interferes with the oxygen-carrying capacity of blood by displacing oxygen from hemoglobin.
13. The main anthropogenic source of nitrogen oxides is combustion of fossil fuels by automobiles, aircraft, and power plants.
14. Catalytic converters in automobiles have significantly reduced emissions of carbon monoxide, nitrogen oxides, and hydrocarbons.
15. Sulfur dioxide emitted by electric power generating plants is the main cause of acid deposition. Emissions can be controlled by flue-gas desulfurization (FGD) and fluidized bed combustion (FBC).
16. Natural sources of sulfur dioxide include volcanic eruptions and the anaerobic decomposition of sulfur-containing organic matter.
17. The major source of suspended particulate matter is coal combustion by electric power generating plants. Emissions can be controlled by electrostatic precipitation, bag filtration, and cyclone separation.
18. Industrial smog, common in coal-burning industrial cities in colder climates in the 19th century, is made up of fly ash, soot, sulfur dioxide, and volatile hydrocarbons.
19. Photochemical smog, which is initiated by sunlight, occurs in hot, sunny cities where levels of nitrogen oxides and hydrocarbons from automobile emissions are high. Ozone in photochemical smog causes respiratory difficulties.

20. In a temperature inversion, a layer of warm air traps a layer of cool air near the earth's surface.

21. Chlorofluorocarbons (CFCs) are responsible for the depletion of the ozone layer. As a result of the Montreal Protocol, signed by many nations, production of CFCs and related compounds has been severely cut back.

22. The increase in atmospheric carbon dioxide and other greenhouse gases (methane, CFCs, and nitrous oxides) in the last 100 years is causing global warming and climate change.

23. The 1970 Clean Air Act and subsequent amendments have established air quality standards for the major air pollutants and banned the use of lead in gasoline.

Key Terms

absolute zero (p. 319)
absorption (p. 325)
adsorption (p. 325)
aerosols (p. 325)
bag filtration (p. 336)
Boyle's law (p. 317)
carbon monoxide (p. 328)
catalytic converter (p. 331)
Charles's law (p. 319)
chlorofluorocarbons (CFCs) (p. 340)
cyclone separation (p. 336)
electrostatic precipitation (p. 336)
flue-gas desulfurization (FGD) (p. 334)
fluidized bed combustion (FBC) (p. 334)
free radicals (p. 327)
global warming (p. 345)
greenhouse effect (p. 346)
greenhouse gases (p. 345)
hydroxyl radical (p. 327)
ideal gas law (p. 322)
industrial smog (p. 337)
kinetic-molecular theory (p. 319)
mesosphere (p. 324)
ozone hole (p. 340)
ozone layer (p. 325)
particulates (p. 325)
photochemical smog (p. 337)
primary air pollutants (p. 326)
secondary air pollutants (p. 327)
stratosphere (p. 324)
temperature inversion (p. 339)
terpenes (p. 331)
thermosphere (p. 324)
troposphere (p. 323)
volatile organic compounds (VOCs) (p. 331)

Questions and Problems

1. What is the concentration of each of the following gases in clean, dry air? Express the concentration as a percentage and in parts per million (ppm).
   a. O_2 **b.** CO_2 **c.** N_2 **d.** Ar

2. Draw a diagram showing the four main layers into which the earth's atmosphere can be divided. Add labels showing:
   a. The distance the two layers closest to the earth extend out from the earth's surface
   b. The changes in temperature with increasing distance from the surface of the earth
   c. The location of the ozone layer

3. Explain why the stratosphere, which is more than 20 miles thick, contains a smaller total mass than the troposphere, which is less than 10 miles thick.

4. What will happen to the diameter of a balloon filled with nitrogen gas in each of the following cases?
   a. The pressure outside the balloon is increased.
   b. The temperature of the gas inside the balloon is lowered.

5. It is very dangerous to throw an aerosol can into a fire. Given what you know about the gas laws, what do you expect to happen as the can heats up? Explain.

6. When the brake pads on your car are changed, it is very important to remove any air from the brake lines that bring the liquid brake fluid to the wheels. Why is it so important for the mechanic to "bleed" the air out of the brake lines?

7. The volume of helium in a balloon is 4.0 L in a gift shop that is at 25°C. The balloon is taken outside on a cold winter day when the air temperature is −10°C. Assuming the barometric pressure to be the same inside and outside, what will be the volume of the balloon when it is outside?

8. A sample of argon gas occupies a volume of 500 mL at 25°C and a pressure of 760 mm Hg. What pressure must be applied to reduce the volume to 250 mL at the same temperature?

9. Methane from a leaking gas stove is filling the kitchen with gas. The kitchen has a volume of 5000 L, a pressure of 1 atm, and a temperature of 25°C. Assume that an explosive mixture is one with 25% methane and 75% air. How long will it be before an explosion can occur if the methane is leaking at 100 mL per minute?

10. A balloon occupies a volume of 10 L at 25°C and 760 mm Hg. To what temperature must the balloon be cooled to reduce the volume to 9 L at the same pressure?

11. At a temperature of 25°C and a pressure of 1 atm, a balloon has a volume of 3500 mL. What is the volume of the balloon if the temperature is raised to 100°C, assuming there is no change in pressure?

12. A balloon has a volume of 2500 mL at 1 atm and 25°C. It is allowed to rise through the atmosphere to a point where the temperature is −20°C and the pressure is 500 mm Hg. What is the new volume of the balloon?

13. A gas occupies a volume of 5.0 L at 50°C and 1.5 atm. What will the pressure be when the gas occupies a volume of 4.0 L at 0°C?

14. Write equations to show how ozone is formed naturally from O_2 in the ozone layer.

15. The percentage of water vapor in the atmosphere is variable. In which range does it generally lie?
    a. 0.05–1% b. 1–3%
    c. 5–10% d. 10–20%

16. State briefly why the ozone layer is so important to all living things on earth.

17. Draw a diagram showing how atmospheric pressure varies with altitude.

18. Define *air pollutant.* What are primary air pollutants? What are secondary air pollutants?

19. List the five types of substances that are responsible for 90% of the air pollution in the United States. What are the major sources of these pollutants?

20. What are the major air pollutants produced by the following industries?
    a. trucking
    b. electric power generation

21. Write chemical equations to show how carbon monoxide and carbon dioxide are formed during the burning of fossil fuels. Which gas (CO or CO_2) is the main product in each of the following situations?
    a. Gasoline is burned in an automobile engine.
    b. Coal is burned in an open fireplace.

22. a. Draw a diagram of a catalytic converter and describe how it reduces pollutants in automobile exhaust.
    b. Catalytic converters reduce automobile emissions by 95%. Why are there calls for devices with greater efficiency?

23. Why will the use of leaded gasoline in an automobile destroy the effectiveness of its catalytic converter?

24. Are the following statements true or false?
    a. Anthropogenic sources release more carbon monoxide into the atmosphere than do natural sources.
    b. The electric power industry is a major source of VOCs.
    c. CFCs are very reactive, unstable compounds.

25. What are the health effects of carbon monoxide inhalation?

26. Write the formula for the nitrogen oxide that is the major air pollutant. How is it produced, and how is it ultimately removed from the atmosphere?

27. a. What are the main anthropogenic sources of sulfur dioxide?
    b. Why does the burning of coal release sulfur dioxide?
    c. Write equations to show how sulfur dioxide reacts in the atmosphere.

28. What adverse effects does sulfur dioxide have on human health?

29. Describe two methods for removing sulfur dioxide from smokestack emissions.

30. In your opinion, is the installation of very tall smokestacks a good way to deal with hazardous industrial emissions? Explain.

31. What does VOC stand for? What are the main anthropogenic and natural sources of VOCs?

32. Using chemical equations, show how photochemical smog is formed.

33. What is a photochemical oxidant? Give equations to show how they are formed. Which photochemical oxidant is produced in greatest quantity?

34. What adverse effects does ozone have on plants and human health?

35. Describe the difference between an aerosol and a particulate.

36. Define the following:
    a. mist **b.** fog **c.** soot **d.** fly ash **e.** smog

37. Give four examples of particulate matter that arises from:
    a. natural sources **b.** anthropogenic sources

38. Explain how the surface area of particulates and aerosols is related to their adverse effects on human health.

39. How is industrial smog formed?

40. In what way does the size of a particle influence its effect on human health?

41. Describe two different methods for removing particulate matter from industrial emissions.

42. Describe the two major effects particulate matter has on the weather.

43. Use a diagram to show how a temperature inversion occurs. What geographical features promote the formation of an inversion?

44. How does a free radical differ from an ordinary molecule?

45. What does CFC stand for? Write the chemical formula of a representative CFC. List three uses for CFCs.

46. Describe the evidence that implicates CFCs in the destruction of the ozone layer.

47. Write the overall series of reactions that chemists believe account for the destruction of the ozone layer by CFCs.

48. Where is the ozone hole located? Why is its formation cause for alarm? In your opinion, what steps should be taken to prevent destruction of the ozone layer?

49. What is the fate of outgoing infrared (IR) radiation emitted from the earth's surface? How does this infrared radiation affect the temperature of the earth's atmosphere?

50. List the greenhouse gases. Which of these gases occurs in the atmosphere in greatest concentration? How do the effects of other greenhouse gases compare with those of carbon dioxide?

51. With the aid of a diagram, explain how an increase in the level of carbon dioxide in the atmosphere can produce a warming trend. What are the main causes of the increase in carbon dioxide in the atmosphere?

52. Make a list of indoor air pollutants. Give the source of each one.

53. Why is tobacco smoke such a dangerous air pollutant? In your opinion, should cigarette smoking be permitted in public buildings?

54. In your opinion, is some level of air pollution an inevitable consequence of living in an industrialized society?

55. What are the major types and sources of air pollution in developing countries?

Answers to Practice Exercises

11.1 564 lb/in^2

11.2 2.5 L

11.3 0.4 atm

11.4 **a.** $2\ C_6H_{14} + 19\ O_2 \longrightarrow 14\ H_2O + 12\ CO_2$

b. $2\ C_6H_{14} + 13\ O_2 \longrightarrow 14\ H_2O + 12\ CO$

Chapter 12

Fossil Fuels

Our Major Source of Energy

Chapter Objectives

In this chapter, you should gain an understanding of:

- How energy use by humans has changed historically
- The difference between power and energy
- The production of energy from fossil fuels
- The formation, composition, uses, and environmental impact of the three fossil fuels: petroleum, natural gas, and coal
- The production of electricity

Coal-fired power plants supply much of our electrical energy. They are also a major contributor to air pollution, acid deposition, and global warming.

ONLINE FEATURES

www.jbpub.com/chemistry

- Chemistry and the Environment
- eLearning
- Chemistry in the News
- Special Topics
- Research & Reference Links
- Ask the Authors

ENERGY EXISTS IN MANY FORMS, including solar energy, heat energy, mechanical energy, electrical energy, and chemical energy. These different forms of energy are continually being converted from one form to another. For example, when animals run, their bodies convert chemical energy stored in carbohydrates into mechanical energy to contract muscles. In a power plant, combustion of coal produces heat energy that is used to change water into steam. The steam turns a turbine, and the mechanical energy of the turning turbine produces electricity (electrical energy).

As determined by the second law of thermodynamics, some energy is wasted in these and all other energy conversions. For example, when you drive a car, only about 10% of the chemical energy in the gasoline is converted to mechanical energy of motion. As much as 90% is converted to heat, which is lost to the surroundings.

All life forms and all societies require a constant input of energy. The more industrialized and technologically advanced the society is, the greater is its need for energy. Today, the United States and other industrialized nations obtain most of their energy from fossil fuels: coal, petroleum, and natural gas. Fossil fuels are the remains of plants, animals, and microorganisms that lived millions of years ago and, over time, were buried and converted to the three fossil fuels. Thus, for energy, modern societies are almost entirely dependent on photosynthesis that occurred billions of years ago.

Fossil fuels, like all living and once-living materials, are composed of carbon compounds. The majority of these compounds are hydrocarbons, which contain only carbon and hydrogen. As we saw in Chapter 8, these two elements can combine in many proportions and ways to form thousands of different hydrocarbons. The differences in properties of the three fossil fuels are due primarily to differences in their hydrocarbon composition. Petroleum, natural gas, and coal are all excellent fuels, because when they are burned, some of the chemical energy stored in their hydrocarbon bonds is released as heat energy, which can perform useful work.

In this chapter, we review the development of energy sources and discuss why fossil fuels are such a good source of energy. We study the formation, extraction, composition, and uses of each of the three fossil fuels, and we examine the environmental impact of modern society's reliance on fossil fuels.

12.1 Energy Use: A Historical Overview

Throughout human history, as societies have become more advanced, their energy use has increased (Fig. 12.1). In primitive cultures, people relied on their own physical labor to supply their energy requirements. They obtained the food they needed by hunting animals and gathering wild plants. With the discovery of fire, humans began to use energy that was not obtained from food. Heat obtained from the burning of wood was used to heat dwellings and cook food.

As civilization advanced, the horse and ox were domesticated and put into service to supply energy needed for farming, transportation, and other activities. Water power and wind power were harnessed to do other useful work. As early as the 1st century A.D., Egyptians were using waterwheels to grind grain. By the Middle Ages, windmills were common in Europe, furnishing the energy needed to grind grain and pump water.

As late as the first quarter of the 19th century, the main sources of energy, apart from human and animal muscle, were wood fires, windmills, and waterwheels (which could be used to power lumber mills, textile manufacturing mills, and other small industries.) A great surge in the use of energy came with the development of the steam engine in

FIGURE 12.1

Throughout history, the discovery of new sources of energy has paralleled the development of new technologies and changed the pattern of human life.

In a steam engine, water is boiled to produce high-pressure steam that thrusts a piston back and forth in a cylinder. The piston is attached to a crankshaft that transmits the power to the drive wheels of the machinery.

From the 16th century until the discovery of petroleum in the 19th century, whale oil obtained principally from whale blubber, was the main source of fuel for lamps.

the late 18th century. The steam engine, for the first time, enabled heat energy from fuel to be converted directly into useful work. This development was the beginning of the Industrial Revolution, the period when machines began to replace human and animal labor. In the United States, an essentially agrarian economy changed to one based on manufacturing. Because steam engines could be located almost anywhere and provided unprecedented amounts of energy, factory sites were no longer dictated by the availability of water. Steam engines could be mounted on wheels, and the development of the steam locomotive marked the beginning of mass transportation. Enormous quantities of fuel were required to supply energy to operate the new machinery; in the 200 years between 1725 and 1925, per capita energy consumption in industrialized nations increased tenfold.

The changing pattern of energy use in the United States since 1860 is shown in **Figure 12.2**. Steam engines—including those in railroad locomotives and riverboats—were initially fueled by wood; in the 1830s, however, wood became scarce, and coal was gradually substituted. By 1910, coal was the source of 70% of the energy used in the United States. Petroleum did not become a significant energy source until the end of the 19th century. It was first discovered in the United States, near Titusville, Pennsylvania, in 1859 (**Fig. 12.3**) and in the Middle East, in Iran, in 1908. For the first 60 years after its discovery in the United States, petroleum was used principally to produce kerosene for lighting and heating.

The development of petroleum as a major energy source paralleled the development of the internal combustion engine and the growth of the automobile industry (**Fig. 12.4**). As a fuel for transportation, petroleum has a tremendous advantage over coal: Gasoline-powered engines are many times smaller and lighter than coal-powered steam engines.

Natural gas, which is often found with petroleum, was not used as a fuel until an economical means for transporting it from the well to the consumer was established.

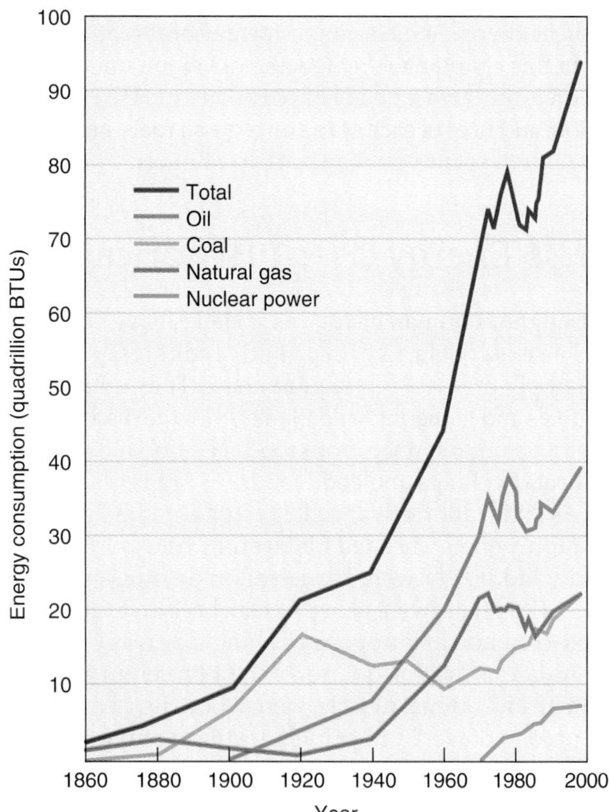

FIGURE 12.2

The contribution of different energy sources to total U.S. energy consumption. Before 1880, wood was the major source of energy, but then coal consumption began to increase sharply. Petroleum consumption rose dramatically throughout the first three-quarters of the 20th century as the number of automobiles and trucks on U.S. highways multiplied. At the same time, the use of natural gas for home heating increased. The brief decline in fuel consumption in the 1970s was the result of an oil shortage engineered by the major oil-producing nations. Nuclear power continues to be a relatively minor source of energy.

In the early part of the 20th century, as much as 90% of the natural gas found with petroleum was burned—or "flared"—at the well site as a waste product. By 1945, a network of pipelines had been constructed throughout the United States to link the gas fields, located mainly in the Southwest, with the states in the Midwest and Northeast, where gas for heating was in greatest demand.

FIGURE 12.3

The first commercial oil well in the United States, near Titusville, Pennsylvania, began producing oil in 1859.

Current Use of Energy

In the industrialized nations of the world, energy is used for four main purposes: (1) industrial processes, (2) transportation, (3) residential use, and (4) commercial use.

The increase in the consumption of energy in the United States between 1949 and 2000 is shown in **Figure 12.5**. Transportation, which includes road vehicles, trains, boats, and airplanes, consumes approximately 27% of all the energy used in the United States. Energy for transportation comes almost exclusively from gasoline and diesel oil, both of which are derived from crude oil (the word *oil* is frequently used as a synonym for *petroleum*). The industrial sector accounts for approximately 38% of energy use. Approximately 25% of this is used to manufacture the goods Americans depend on for a high standard of living. The residential and commercial sectors together account for another 36% of total U.S. energy use. Over one-third of the energy consumed in the United States is used to produce electricity. Electricity, unlike fossil fuels, nuclear power, and water power, is not a primary source of energy. It is called a secondary source because it is produced from a primary source.

Primitive people consumed about 2,000 kcal each per day—the minimum amount of energy needed for survival. Today, the average person in the United States consumes about 230,000 kcal per day.

As Figure 12.2 shows, fossil fuels—petroleum, natural gas, and coal—are used to meet almost all the energy requirements of the United States. Nuclear energy, water power, solar energy, and other less conventional sources of energy fulfill a minute fraction of U.S. energy needs.

12.2 Energy and Power

It is important to understand the distinction between energy and power. **Energy** is the ability to do work, and it is measured in units such as calories, joules, and British thermal units. **Power** is a measure of the *rate* at which energy is used. It is expressed in terms of units of energy per unit of time, such as calories per second or joules per second (watts). Units in which energy and power are measured are explained in **Table 12.1**.

FIGURE 12.4

Motor vehicles clog U.S. highways and consume more than 60% of the oil burned in this country.

12.3 Energy from Fuels

Fuels are defined as substances that burn readily in air and release significant amounts of heat energy. The process of burning (or combustion) is an oxidation process, and it is exothermic (Chapter 6). Most materials on the earth—minerals in rocks and soil, and water, for example—are already in an oxidized state. They do not burn and cannot be used as a source of heat energy.

What makes petroleum, natural gas, and coal such excellent fuels? The answer lies in their chemical composition. Fossil fuels are composed primarily of hydrocarbons (Chapter 8); when hydrocarbons are burned in an adequate supply of air, they are oxidized, and carbon dioxide and water are formed. Chemical bonds are broken, new bonds are formed, and, as a result, energy is released as heat.

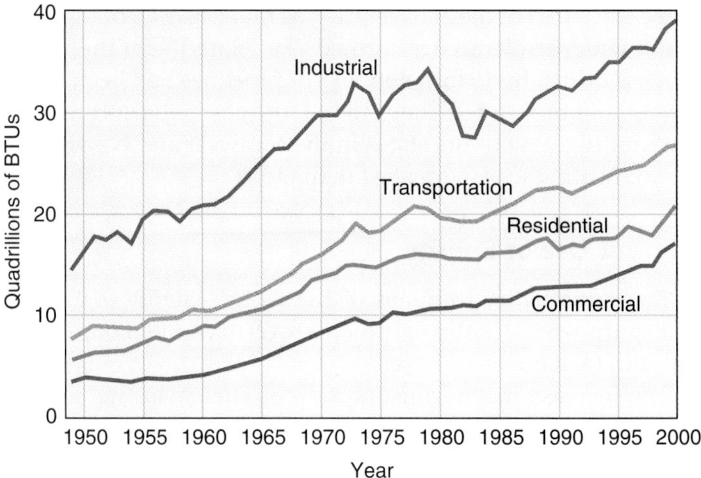

FIGURE 12.5

Energy consumption in the United States has continued to grow during the last 50 years. Energy is consumed in four main sectors: industrial, transportation, residential, and commercial.

The value of 1 horsepower (the power needed to raise 33,000 pounds 1 foot in 1 minute) was established in the late 18th century, based on results obtained with strong dray horses. It is about 50% greater than the rate that can be sustained by the average horse over a working day.

Equations for the combustion of methane (CH_4, the major component of natural gas) and octane (C_8H_{18}, a component of gasoline) are shown below:

$$CH_4 + 2\,O_2 \longrightarrow CO_2 + 2\,H_2O + \text{heat}$$

$$2\,C_8H_{18} + 25\,O_2 \longrightarrow 16\,CO_2 + 18\,H_2O + \text{heat}$$

When these hydrocarbons are burned in air, their carbon-hydrogen bonds are broken, as are bonds between oxygen atoms in O_2 molecules. New bonds form between carbon and oxygen atoms and between hydrogen and oxygen atoms, as CO_2 and H_2O molecules are produced. Heat is produced in these reactions because the energy released in making the new bonds in carbon dioxide and water is greater than the energy

Table 12.1 Glossary of Energy Terms

Term	Meaning
Calorie (cal)	The amount of heat needed to raise the temperature of 1 g of water 1°C*
Kilocalorie (kcal)	One thousand calories (1000 × 1 cal)
Joule (J)	The joule is the standard unit of energy: 4.18 J = 1 cal
British thermal unit (BTU)	The amount of heat needed to raise the temperature of 1 lb of water 1°F
Quad	One quadrillion (1000 trillion) British thermal units (BTU): one quad = 10^{15} BTU
Horsepower	One horsepower = 33,000 ft lb of work/min
Watt (W)	An amount of power available from an electric current of 1 ampere (A) at the potential difference of one volt (V): 1 W = 1 V × 1 A; 1 W = 1 J/s
Kilowatt (kW)	One thousand watts (1000 × 1 W)
Megawatt (MW)	One million watts, or 1000 kW
Kilowatt-hour (kWh)	Unit of electrical energy equivalent to the energy delivered by the flow of 1 kW of electrical power for 1 hour

*Energy derived from food is measured in Calories (Cal): 1 Cal = 1 kcal = 1000 cal.

required to break the bonds in the hydrocarbons and in oxygen. (If the energy released in making the new bonds were *less* than the energy required to break the old bonds, heat would be consumed, and the reaction would be endothermic.)

The amount of energy required to break 1 mole (Chapter 6) of a particular kind of bond is the **bond energy.** Average values for bond energies are given in **Table 12.2**. The same amount of energy is *released* if the bond is reformed from individual atoms. Bond energy is measured in kilocalories (kcal) (see Table 12.1).

We can use the values given in Table 12.2 to calculate the heat energy that will be released when 1 mole of methane (CH_4) is oxidized to yield carbon dioxide and water, as shown in the preceding equation.

Bond	Bond energy (kcal)	Number of bonds broken or formed	Bond energy/ mole (kcal)
C—H	99	4	4 × 99 = 396
O—O	118	2	2 × 118 = 236
		Energy required to break bonds	= 632
C=O	192	2	2 × 192 = 384
O—H	111	4	4 × 111 = 444
		Energy released in forming bonds	= 828

CH_4 CO_2 O_2 H_2O

Thus, more energy is released in making new bonds than is consumed in breaking the bonds in the hydrocarbon and oxygen molecules. The excess energy (828 − 632 = 196) is released as heat.

This calculation explains why fossil fuels are valuable as energy sources. When they are burned, numerous carbon-hydrogen bonds are broken, and as carbon dioxide and water are formed, a relatively large quantity of energy is produced. The energy output (fuel value) of various common fuels is given in **Table 12.3**.

> Energy is obtained from a chemical reaction when the energy produced from making bonds to form products is greater than the energy required to break the bonds in the reactants.

Table 12.2 Approximate Bond Energies for Selected Bonds

Type of Bond	Bond Energy (kcal/mole)
C—H	99
C—C	83
C=C	147
C—O	85
C=O	173
C=O in CO_2	192
C—Cl	78
H—H	104
H—Cl	103
O—O	35
O—O in O_2	118
O—H	111
Cl—Cl	58

Table 12.3 Fuel Values of Some Common Fuels

Fuel	Fuel Value (kcal/g)
Wood (pine)	4.3
Bituminous coal	7.4
Anthracite	7.6
Crude oil	10.8
Gasoline	11.5
Natural gas	11.7
Hydrogen	33.9

EXAMPLE 12.1 Calculating the energy from burning fossil fuels

How much energy is released when 1 mole of propane (C_3H_8) is burned?

Solution

1. Write the balanced equation for the reaction:

$$C_3H_8 + 5\,O_2 \longrightarrow 3\,CO_2 + 4\,H_2O$$

2. Determine the type and number of bonds broken by drawing the structural formula of C_3H_8:

```
   H  H  H
   |  |  |
H—C—C—C—H
   |  |  |
   H  H  H
```

Clearly, 8 C—H and 2 C—C bonds are broken; 5 O—O bonds are also broken.

3. Use Table 12.2 to obtain the bond energy of each type of bond. Calculate the total energy required to break all the bonds:

Bond	Bond energy (kcal)	Number of bonds broken	Bond energy/ mole (kcal)
C—H	99	8	8 × 99 = 792
C—C	83	2	2 × 83 = 166
O—O	118	5	5 × 118 = 590
		Energy required to break bonds	= 1548

4. Similarly, calculate the energy released when the new bonds are formed:

Bond	Bond energy (kcal)	Number of bonds formed	Bond energy/ mole (kcal)
C=O	192	6	6 × 192 = 1152
O—H	111	8	8 × 111 = 888
		Energy released in forming bonds	= 2040

5. Find the difference between the energy consumed and the energy released:

$$2040 - 1548 = 492$$

Thus, 492 kcal/mole of energy are released when 1 mole of propane is burned.

Practice Exercise 12.1

How much energy is released when 1 mole of ethane (C_2H_6) is burned?

12.4 Petroleum

The Formation of Oil Fields

Petroleum probably originated from microscopic marine organisms that once lived in great numbers in shallow coastal waters. When the organisms died, their remains collected in bottom sediments, where the supply of oxygen was insufficient to oxidize all the organic material. Over millions of years, the organic matter was buried by sediments, and, in some locations, the seas dried up, leaving behind a new land mass. The buried material was subjected to high temperature and pressure; chemical reactions converted some of the organic matter to liquid hydrocarbons (crude oil or petroleum) and gaseous hydrocarbons (natural gas).

Based on their understanding of the earth's history, geologists are able to recognize the major geological features that indicate the presence of large accumulations of oil (**Fig. 12.6**). They then assess the probable size of the oil-bearing formation, its location, and other factors to determine the economic feasibility of extracting the petroleum.

Petroleum and natural gas are generally found together. They accumulate in porous, permeable rock called **reservoir rock.** Rock is described as *porous* if, like a sponge, it contains open spaces in its structure where gases and liquids can collect. It is *permeable* if the spaces are interconnected so that fluids—gases, water, and oil—can migrate through the pores. The porosity of reservoir rock, which is usually sandstone or limestone, ranges from 10–30%; usually, one-half of the pore space is occupied by water.

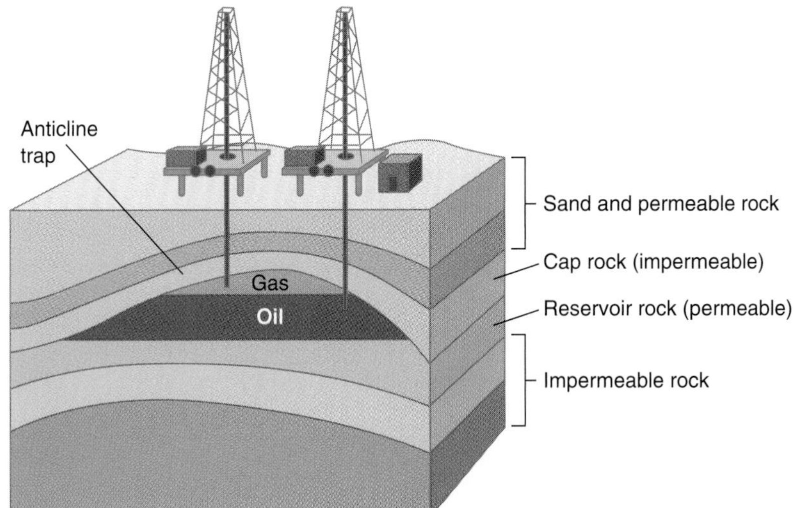

FIGURE 12.6

Geological features associated with petroleum and natural gas deposits: As a result of folding and faulting of rock layers, oil and gas become trapped under a dome of impermeable cap rock.

For petroleum to accumulate in quantity, there must be a layer of impermeable rock (known as *cap rock*) above the reservoir rock to prevent the fluids in the oil-bearing rock from escaping by moving upward. There must also be a folding or faulting of the rock strata to create a dome, or *anticline trap,* under which the petroleum and natural gas are held. The trap, which may extend over many miles, prevents lateral movement of fluids in the reservoir rock and allows oil and gas to collect. In time, the fluids in the reservoir rock separate out according to density. Because petroleum and natural gas are less dense than water, they rise above the water and collect directly under the anticline trap, with the gas above the oil (refer to Fig. 12.6).

Oil cannot be withdrawn from an oil field easily because it does not accumulate in liquid pools. Instead, it is dispersed within the pores of the reservoir rock and must be forced out. The oil and gas are under pressure. Drilling into the reservoir rock releases the pressure, and the crude oil is forced up the bore hole. If the pressure is sufficiently high, the oil may be released as a gusher. As the oil flows out, the pressure gradually decreases, and the rate of flow declines. The average oil field ceases to be productive after only about one-third of the oil in the formation has been recovered. That is, two barrels of oil are left in the ground for every barrel extracted. A further one-third can be obtained by repressurizing the oil field. In repressurizing, auxiliary wells are drilled, and water, steam, or a gas such as carbon dioxide or nitrogen is injected into the reservoir rock to force more oil out. Repressurizing is costly; whether it is worthwhile depends primarily on the prevailing price of oil.

A barrel of oil is equal to about 42 gallons.

A drill bit can cut through about 30 meters (100 feet) of soft shale per hour and lasts about 2 days before it dulls. If the rock is hard, the bit may dull after cutting through just a few feet.

The Composition of Petroleum

Crude petroleum is a complex mixture of thousands of organic compounds. The majority of these compounds are hydrocarbons, including alkanes, cycloalkanes, alkenes, and aromatic hydrocarbons. Small amounts of sulfur-, nitrogen-, and oxygen-containing compounds are also present. The proportions of different hydrocarbons in a sample of crude petroleum vary with location. For example, the percentage of straight-chain and branched-chain hydrocarbons is much higher in crude oil from Pennsylvania than in crude oil from California, which has a high content of aromatic hydrocarbons.

a.

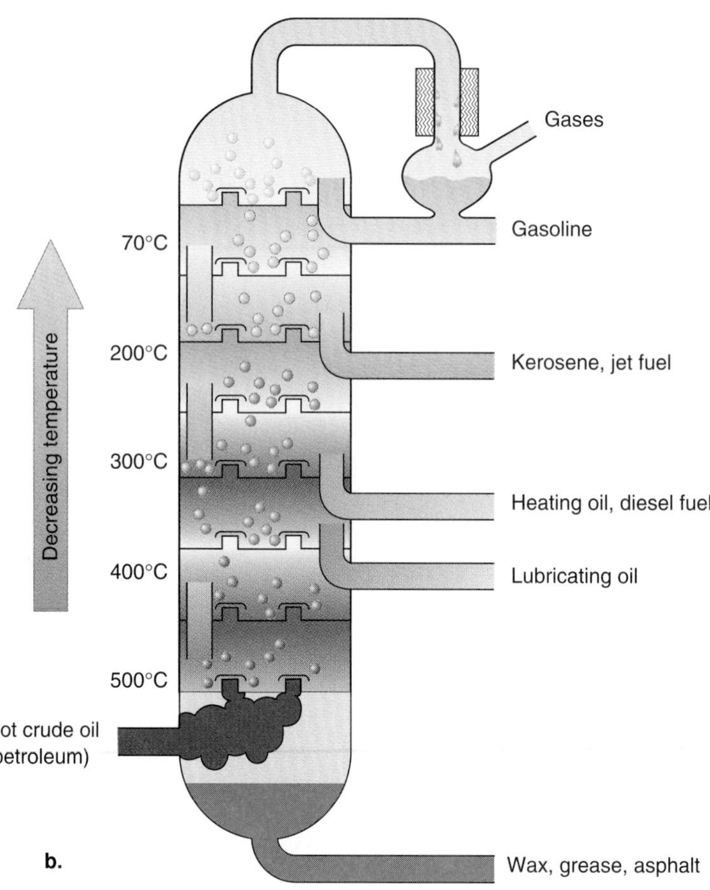

FIGURE 12.7

(a) Distillation towers at a refinery. (b) Crude petroleum is heated at the bottom of a distillation tower, and most of the hydrocarbon constituents vaporize. Inside the tower, temperature decreases with increasing height. High-boiling components, or fractions, condense in the lower part of the tower; lower-boiling components (including the gasoline fraction) condense near the top. Fractions are removed at various heights as they condense.

Petroleum Refining

Crude petroleum in the form in which it is recovered has few immediate uses. To be useful, it must be sent to a refinery and separated into fractions by **fractional distillation** (Fig. 12.7a), a process that separates the hydrocarbons according to their boiling points.

Figure 12.7b shows the essential parts of a fractional distillation tower. The tower may be as tall as 30 meters (100 feet). Crude oil at the bottom of the tower is heated to about 500°C (930°F), and the majority of the hydrocarbons vaporize. The mixture of hot vapors rises in the fractionating tower; components in the mixture condense at various levels in the tower and are collected. The temperature in the tower decreases with increasing height. The more volatile fractions with lower boiling points condense near the top of the tower; high boiling, less volatile fractions condense near the bottom. Any uncondensed gases are drawn off at the top of the tower. Residual hydrocarbons that do not vaporize collect at the bottom of the tower and are transferred to a vacuum distillation tower, where they are vaporized under reduced pressure to yield further fractions. This process makes the full range of hydrocarbons in crude oil available for useful purposes. The properties and uses of typical petroleum fractions are listed in Table 12.4.

Reducing the external pressure lowers the boiling point of a liquid.

The relative amount of each fraction produced by fractional distillation varies with the composition of the original crude petroleum; it rarely reflects consumer needs. Before the advent of the automobile, the fraction in greatest demand was kerosene, which was used for heating and lighting. Today, the greatest need is for the more volatile frac-

Table 12.4 Typical Petroleum Fractions

Fraction	Boiling Point (°C)	Composition	Uses
Gas	Up to 20	Alkanes from CH_4 to C_4H_{10}	Synthesis of other carbon compounds; fuel
Petroleum ether	20–70	C_5H_{12}, C_6H_{14}	Solvent; gasoline additive for cold weather
Gasoline	70–180	Alkanes from C_6H_{14} to $C_{10}H_{22}$	Fuel for gasoline engines
Kerosene	180–230	$C_{11}H_{24}$, $C_{12}H_{26}$	Fuel for jet engines
Light gas oil	230–305	$C_{13}H_{28}$ to $C_{17}H_{36}$	Fuel for furnaces and diesel engines
Heavy gas oil and light lubricating distillate	305–405	$C_{18}H_{38}$ to $C_{25}H_{52}$	Fuel for generating stations; lubricating oil
Lubricants	405–515	Higher alkanes	Thick oils, greases, and waxy solids; lubricating grease; petroleum jelly
Solid residue			Pitch or asphalt for roofing and road material

tion known as *straight-run gasoline,* which provides the fuel for automobiles (see Table 12.4).

To meet the increasing need for gasoline, refineries must convert higher-boiling fractions to gasoline by the process of **cracking.** The higher-boiling fractions contain long-chain hydrocarbons, for which there is little demand. When heated under pressure in the absence of air, the long-chain hydrocarbons break (or crack) into shorter-chain hydrocarbons, including both alkanes and alkenes, some of which are in the desired gasoline range (their molecules have six to ten carbons).

$$C_{14}H_{30} \xrightarrow[\text{absence of air}]{\text{heat, pressure}} \underset{\text{alkane}}{C_7H_{16}} + \underset{\text{alkene}}{C_7H_{14}}$$

If a suitable catalyst is added, the reaction proceeds at lower pressure, and the yield of hydrocarbons in the gasoline range is increased.

To meet the increased demand for heating oil in the winter months, a refinery can switch to the reverse process, in which smaller hydrocarbons are combined to produce larger ones.

Octane Rating

Straight-run gasoline consists primarily of straight-chain hydrocarbons and is of limited value as an automobile fuel. A high content of straight-chain hydrocarbons causes a fuel to begin burning before it is ignited by the spark plug. This premature ignition produces a knocking sound and leads to loss of engine power and eventual damage to the engine. The tendency of gasoline to cause knocking is rated according to an arbitrary scale known as the **octane rating.**

The compound 2,2,4-trimethylpentane is just one of several isomers of octane; therefore, it is not strictly correct to call it isooctane.

The prefix *n*- in *n*-heptane stands for *normal*. It is used to signify a straight-chain hydrocarbon.

The octane rating was established in 1927, when a large number of straight- and branched-chain hydrocarbons in the gasoline range were tested separately for performance in a standard engine. The branched hydrocarbon 2,2,4-trimethylpentane (isooctane) was found to be a superior fuel that burned without causing knocking; it was assigned an octane rating of 100. The straight-chain hydrocarbon *n*-heptane, in contrast, caused serious knocking and was given a rating of 0.

$$CH_3-CH(CH_3)-CH_2-C(CH_3)_2-CH_3$$

2,2,4-trimethylpentane (isooctane)

$$CH_3-CH_2-CH_2-CH_2-CH_2-CH_2-CH_3$$

n-heptane

The octane rating of a particular gasoline is determined by burning it in a standard engine and comparing its knocking properties with those of standard mixtures of isooctane and *n*-heptane. A gasoline that performs in the same way as a mixture containing 90% isooctane and 10% *n*-heptane is assigned an octane rating of 90. **Table 12.5** lists the octane ratings for pure samples of some of the hydrocarbons that are present in gasoline.

Most gas stations offer customers a choice of three octane ratings: 87, 89, and 93.

The straight-run gasoline fraction from the distillation tower has an octane rating between 50 and 55, much too low for today's automobile engines, which require gasoline with a rating between 87 and 93. The octane rating of gasoline can be increased in three main ways: (1) cracking, (2) catalytic reforming, and (3) the addition of octane enhancers.

Table 12.5 Octane Ratings of Selected Hydrocarbons

Name	Formula	Octane Rating
Straight-chain and branched hydrocarbons		
n-Butane	C_4H_{10}	94
n-Pentane	C_5H_{12}	62
2-Methylbutane		94
n-Hexane	C_6H_{14}	25
2-Methylpentane		73
2,2-Dimethylbutane		92
n-Heptane	C_7H_{16}	0
2-Methylhexane		42
2,3-Dimethylpentane		90
2-Methylheptane	C_8H_{18}	22
2,3-Dimethylhexane		71
2,2,4-Trimethylpentane (isooctane)		100
Aromatic hydrocarbons		
Benzene		106
Toluene		118
o-Xylene		107
p-Xylene		116

Cracking increases the octane rating by increasing the percentage of short-chain hydrocarbons, which, in general, have higher octane numbers than longer-chain hydrocarbons (see Table 12.5). For example, as the length of the hydrocarbon chain decreases from heptane (C_7H_{16}) to butane (C_4H_{10}), the octane rating increases from 0 to 94.

Branched-chain hydrocarbons have higher octane numbers than the corresponding straight-chain isomers. It can be seen in Table 12.5 that as the degree of branching increases in the three isomeric hexanes (C_6H_{14})—*n*-hexane, 2-methylpentane, and 2,2-dimethyl butane—the octane rating increases from 25 to 92. Table 12.5 also shows that aromatic hydrocarbons have higher octane ratings than nonaromatic hydrocarbons do. Benzene, toluene, and the xylenes, for example, have octane ratings above 100. Octane ratings above 100 are obtained by comparing a fuel with reference samples containing isooctane (octane number, 100) and known amounts of an octane enhancer. A conversion chart is used to obtain the octane number.

It therefore follows that if the percentages of branched-chain hydrocarbons and aromatic hydrocarbons can be increased, the octane rating of a gasoline will be improved. This improvement can be achieved by **catalytic reforming,** a process in which hydrocarbon vapors from straight-run gasoline are heated in the presence of suitable catalysts such as platinum. By this means, *n*-hexane, for example, is converted to 2,2-dimethyl-butane:

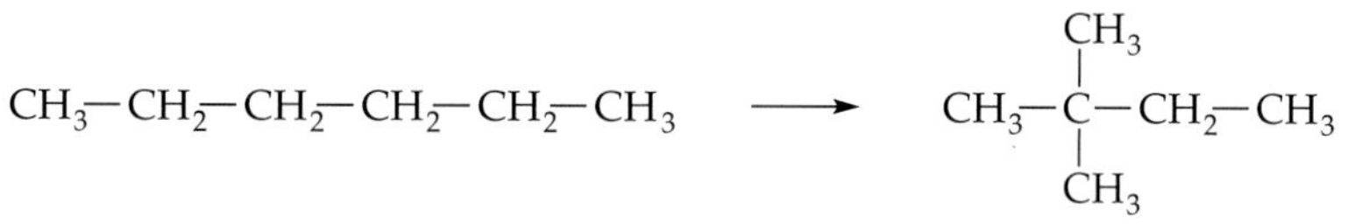

$$CH_3{-}CH_2{-}CH_2{-}CH_2{-}CH_2{-}CH_3 \longrightarrow CH_3{-}\underset{\displaystyle CH_3}{\overset{\displaystyle CH_3}{\underset{|}{\overset{|}{C}}}}{-}CH_2{-}CH_3$$

n-hexane 2,2-dimethylbutane

Also, *n*-heptane is converted to toluene:

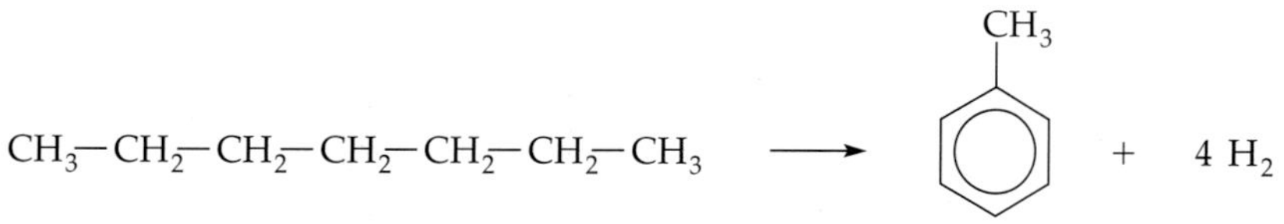

$$CH_3{-}CH_2{-}CH_2{-}CH_2{-}CH_2{-}CH_2{-}CH_3 \longrightarrow C_6H_5CH_3 + 4\ H_2$$

n-heptane toluene

The octane rating of gasoline can also be increased by adding *octane enhancers*, or antiknock agents. Before 1975, the most widely used octane enhancer was tetraethyl lead (TEL), $(C_2H_5)_4Pb$, which was both cheap and effective.

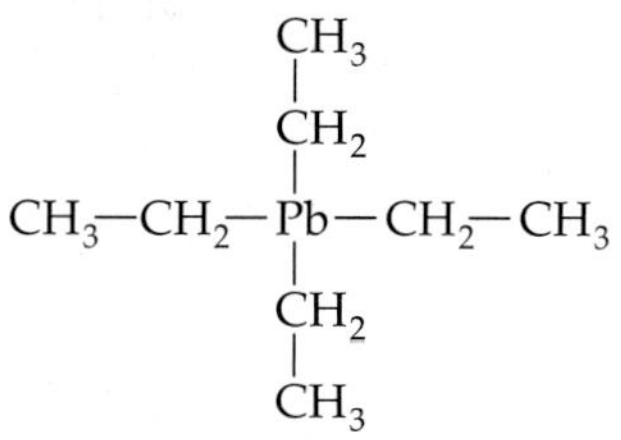

$$CH_3{-}CH_2{-}\underset{\displaystyle \underset{\displaystyle CH_3}{\underset{|}{CH_2}}}{\overset{\displaystyle \overset{\displaystyle CH_3}{\overset{|}{CH_2}}}{\underset{|}{\overset{|}{Pb}}}}{-}CH_2{-}CH_3$$

tetraethyl lead (TEL)

Adding as little as 0.1% of TEL to gasoline can increase the octane rating by 10–15 points. However, lead is toxic (Chapters 10 and 18), and recognition of the health hazards asso-

Table 12.6 Octane Number of Gasoline Additives

Additive	Octane Number
Methanol	107
Ethanol	108
Methyl-*t*-butyl ether (MTBE)	116
Ethyl-*t*-butyl ether (ETBE)	118

ciated with its release into the atmosphere from automobile exhausts led to the mandatory phasing out of TEL as a gasoline additive. The 1975 requirement that all new cars be fitted with a catalytic converter to control pollutants (Chapter 11) was a further factor in reducing the use of TEL. Only unleaded gasoline can be used in automobiles fitted with catalytic converters because lead inactivates the catalysts.

Octane enhancers that have replaced TEL include methyl-*t*-butyl ether (MTBE), ethyl-*t*-butyl ether (ETBE), methanol, and ethanol, all of which have high octane ratings (Table 12.6). The most popular octane enhancer is MTBE. However, it has been detected in water supplies and, for health reasons, the EPA has recommended a ban on its use (Chapter 11).

Methanol and ethanol, which are both fuels, have been added to gasoline in amounts up to 10% to produce gasoline blends known as **gasohol.** The use of methanol as an alternative fuel will be considered in Chapter 13.

Oil Shale and Tar Sands

Most oil shales yield approximately 25–50 gallons of oil per ton; a few yield as much as 150 gal/ton.

A largely untapped source of petroleum is **oil shale** (Fig. 12.8a), a common sedimentary rock that contains a solid organic material called kerogen. **Kerogen** consists primarily of heavy hydrocarbons together with small quantities of sulfur-, nitrogen-, and oxygen-containing compounds. Oil shale is found close to the earth's surface and can be mined like coal. When oil shale is crushed and heated in the absence of air at about 48°C (118°F), a thick brown liquid called **shale oil** is produced. After it has been treated to remove sulfur and nitrogen impurities, shale oil can be refined like petroleum. Fractional distillation yields mainly high-molecular-weight hydrocarbons, which can then be cracked to yield the more desirable hydrocarbons in the gasoline range.

The United States has immense deposits of oil shale, mainly in Utah and Wyoming. Although the technology is available to exploit this resource, there are environmental and economic problems that must first be overcome. For example, very large quantities of waste rock are produced when oil shale is mined, and because this waste is alka-

a.

b.

FIGURE 12.8

Oil can be extracted from (a) oil shale, a sedimentary rock rich in the hydrocarbon material called *kerogen*, and from (b) tar sands, a mixture of sandstone and a black, high-sulfur, tarry substance called *bitumen*.

line and poor in nutrients, very little will grow on it. Furthermore, for every gallon of shale oil produced, 2–4 gallons of water are required. When oil prices are relatively low, mining of oil shale for fuel is not economically attractive.

Tar sands are another potential source of petroleum. Tar sands **(Fig. 12.8b)** consist of sandstone mixed with a black, high-sulfur, tarlike oil known as **bitumen.** Large deposits exist in the United States, particularly in Utah, but the most extensive known deposits are found in Northern Alberta, Canada, where they are nearer the surface than in the United States and much easier to mine. In 2000, 15% of Canada's oil was obtained from tar sands. The extraction process involves injecting steam under pressure into the tar sand to liquefy the bitumen so that it can float to the surface and be collected. After it is purified, the oil can be refined in the usual manner.

It has been estimated that 300 billion barrels of oil are recoverable from tar sands in Canada.

Producing oil from tar sands is expensive and requires a great deal of energy. As with oil shale, environmental and economic considerations will determine the future of tar sands as a viable energy source. As petroleum reserves are depleted, interest in both oil shale and tar sands will undoubtedly increase.

The Petrochemical Industry

As we saw in Chapter 8, all organic compounds are either hydrocarbons or derivatives of hydrocarbons. Petroleum is the primary source of the hydrocarbons that are needed as starting materials for the synthesis of organic compounds used in the manufacture of plastics, synthetic fibers, synthetic rubber, pesticides, pharmaceuticals, and numerous other consumer products **(Fig. 12.9)**. Approximately 10% of the petroleum refined today is used for this purpose. The organic compounds synthesized from petroleum hydrocarbons are called *petrochemicals*. In the future, as petroleum reserves decline, more organic compounds are expected to be produced using synthesis gas derived from coal (discussed later in this chapter) as the starting material.

Oil Pollution

The environmental cost of producing, transporting, and using oil is considerable. The worst oil spill in U.S. history occurred in March of 1989, when the supertanker *Exxon Valdez* ran aground 28 miles south of the trans-Alaska pipeline terminal at Valdez and spilled 260,000 barrels (11 million gallons) of crude oil into Prince William Sound. More than 1000 miles of shoreline were contaminated with oil, and tens of thousands of oil-coated seabirds **(Fig. 12.10)** and at least a thousand sea otters died.

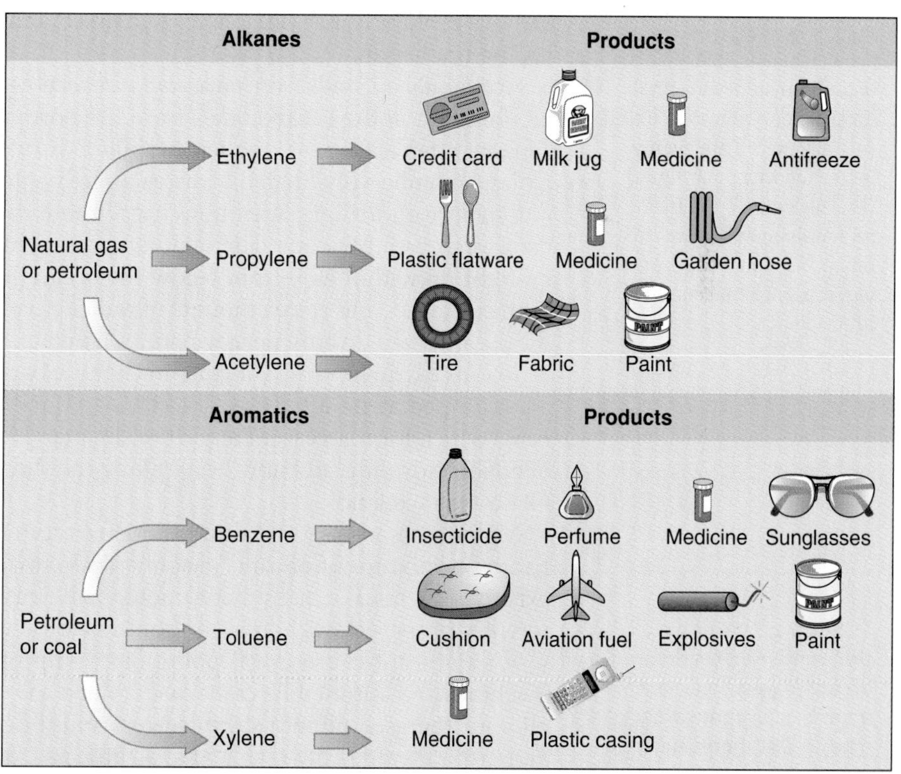

FIGURE 12.9

Some products made from petrochemicals (chemicals derived from petroleum).

FIGURE 12.10

When seabirds become coated with oil, they are unable to fly. Their feathers no longer insulate them from the cold, and most die of exposure.

When crude oil is spilled into the ocean, it is dispersed and changed by many physical and chemical processes. Nearly all the components in crude oil are less dense than water and insoluble in it; they float on the ocean's surface as a layer that gradually thins as it spreads outward. The most volatile components in the layer (usually about 25% of the oil) evaporate into the atmosphere. Less volatile components are broken up by wave action into fine droplets that become dispersed in the water. Near shore, oil droplets tend to adsorb onto any suspended sand and silt particles and sink. At the ocean surface, the oil is whipped by wind and wave action into an oil-in-water emulsion (or mousse). Oil that washes ashore and oil near the ocean surface is gradually decomposed by bacteria and sunlight, but these processes occur very slowly in cold seawater. As the emulsion breaks up, tar balls form and are washed ashore. The small fraction of crude oil that is heavier than water sinks to the ocean floor, where it can coat and destroy bottom-dwelling organisms.

EXPLORATIONS describes an environmental catastrophe involving oil that was a deliberate act of terrorism (see pp. 378–379).

In an emulsion, very small droplets of a nonpolar liquid (usually an oil) are evenly dispersed in a polar liquid (usually water). We make an emulsion when we mix oil and vinegar together vigorously to make a salad dressing.

The long-term environmental effects of oil spills are difficult to assess. Where strong waves pound exposed rocky shores, cleanup is rapid, but on sheltered sandy beaches and in salt marshes and tidal flats, oil may become buried and remain without decomposing for 2–3 years. Hydrocarbons are insoluble in water; when ingested by marine creatures, they dissolve in the animals' fatty tissues. Fish metabolize and excrete hydrocarbons rapidly, but shellfish may remain contaminated for a long time, smelling and tasting unpleasant.

Accidental oil spills from supertankers receive the most publicity, but nearly 70% of oil pollution in the ocean comes from the normal routine operation of these tankers. Tankers often leak, oil is frequently spilled during unloading and refueling, and most tankers intentionally flush oil wastes directly into the sea.

Offshore oil-drilling operations are a minor cause of oil pollution, mainly from the occasional well blowout. The main environmental concern is the wastewater that is brought up with the oil from below the ocean floor. This ancient water contains barium, zinc, lead, and trace amounts of radioactive materials, and when it is returned to the ocean, it can destroy nearby natural habitats. Another concern is the clay mud that is used as a drill lubricant; it contains lignite, barium, and other chemicals and is usually dumped at sea after use. The clay can bury and destroy bottom-dwelling organisms, which form an essential part of the marine food chain. The most serious effects of offshore drilling probably occur along the shore, where facilities built to receive the oil destroy wetlands.

Motor boats are a source of oil pollution on recreational waters. Many outboard motors run on a mixture of oil and gasoline and emit this fuel half-burned through their exhaust systems. For many years, motor boat emissions were not regulated, but the EPA recently proposed strict emission controls for new outboard motors.

As much as 50% of the oil that contaminates the oceans is waste oil that is generated by industries, cities, gas stations, and individuals and flows into the oceans from sewers, streams, and rivers.

A significant cause of oil pollution is improper disposal of lubricating oil from machines and automobile crankcases. Many car owners who change their own motor oil pour the used oil into storm sewers and other areas that drain into creeks, rivers, and groundwater. Few realize that as little as 1 quart of oil can contaminate 2 million gallons of drinking water; 1 gallon of oil can form an oil slick measuring nearly 8 acres. Because of this danger, many communities have set up collection centers where used oil can be brought for recycling.

12.5 Natural Gas

The geological conditions necessary for the formation of natural gas are similar to those required for the formation of oil (see Fig. 12.6). As we saw earlier, oil and gas are often found together, but natural gas is also found by itself. In the formation of natural gas, part of the original, buried organic material was converted to light gaseous hydrocarbons; in the formation of oil, heavier liquid hydrocarbons were formed.

Natural gas is composed chiefly of methane and small amounts of ethane, propane, and butane. Typically, it contains 60–80% methane, but the exact composition varies with the source. Crude natural gas also contains small quantities of larger alkanes, carbon dioxide, nitrogen, hydrogen sulfide (H_2S), and helium. The major natural gas–producing states are Texas, Louisiana, Oklahoma, and New Mexico.

The household gas that has often been used to commit suicide and is mentioned in detective novels by Agatha Christie and others as a cause of death is coal gas, not natural gas.

Before crude natural gas can be used as a commercial fuel, it must be treated to remove carbon dioxide, sulfur compounds, water vapor, and most of the hydrocarbons with molecular weights greater than that of ethane (C_2H_6). Carbon dioxide is undesirable because it reduces the heat output of the gas; sulfur compounds cause corrosion and release unpleasant odors. Water vapor can cause problems if it condenses in a pipeline.

People who are killed by breathing natural gas die from oxygen deprivation.

Propane (C_3H_8) and butane (C_4H_{10}) are useful by-products. After they have been removed from crude natural gas, these by-products are converted to a liquid known as **liquefied petroleum gas (LPG)** and marketed as heating fuel. Helium, which is often present in crude natural gas, is another important by-product. It is used to provide an inert atmosphere for welding and as a lighter-than-air gas for balloons and other devices.

Natural gas is nontoxic and safe, but many people mistakenly believe that it is dangerous. This misconception dates from the early part of the 20th century, when coal gas (discussed in the next section), a toxic mixture of carbon monoxide and hydrogen, was piped into homes. Of course, if natural gas accumulates in a confined space, a spark will ignite it explosively.

Because natural gas is odorless, a small quantity of an unpleasant smelling organic compound (usually a mercaptan) that can be detected when it is present in concentrations of parts per million is added to it. If a leak occurs, the escaping gas can be detected before it can build up to form an explosive atmosphere.

Natural gas is relatively inexpensive. It is a superior fuel and has many advantages over coal and petroleum. It burns cleanly, leaves no residue, and, weight-for-weight, provides a higher heat output than any other common fuel (see Table 12.3). It emits less carbon dioxide per unit of energy than other fossil fuels and generally produces no sulfur oxides. The extensive system of pipelines that crisscrosses the United States, taking gas directly to the consumer, reduces the need for expensive storage facilities.

12.6 Coal

The Formation of Coal Deposits

Three hundred million years ago, when the earth was much warmer than it is now, plants grew in great abundance in the widespread tropical freshwater swamps and bogs (**Fig. 12.11**). When the luxuriant growth died, some of the plant material sank under water before it could be oxidized by atmospheric oxygen in the usual way. In the absence of air, little decomposition occurred. The material accumulated and became buried under sediments, falling leaves, and other vegetation. In time, it was compressed and converted to a porous brown organic material now known as *peat*. In some locations, because of geologi-

FIGURE 12.11

Approximately 300 million years ago, huge ferns and other plants grew in great abundance on the earth. Over time, dead plant material became buried under layers of sediments, and some of the material was ultimately converted to coal.

Table 12.7 Characteristics of Different Types of Coal

Type of Coal	Carbon (%)	Water (%)	Fuel Value
Peat	5	90	Very low
Lignite	30	40	Low
Subbituminous coal	40	9	Medium
Bituminous coal	65	3	High
Anthracite	90	3	High

Note: Values may vary considerably with the source of coal.

cal changes, the peat became more deeply buried; increasing pressure compressed it and changed it to a harder material called *lignite.*

Over thousands of years, deeper burial and the resulting increase in temperature and pressure transformed the lignite into various grades of **bituminous coal** (also known as *soft coal*). In areas where mountains formed as a result of deformation and uplifting of rock formations, further changes occurred. The very high pressure and temperature associated with these geological processes converted bituminous coal to **anthracite** (or *hard coal*).

With each step in the transformation from peat to anthracite, chemical reactions occur. Volatile compounds containing carbon, oxygen, and hydrogen are released; the water content of the material decreases; and the carbon content increases (**Table 12.7**). At the same time, the material becomes hard and shiny. The quality of coal as a fuel increases with its carbon content. Anthracite, which is approximately 90% carbon, is the most desirable form of coal. Peat is not a true coal; it is considered a low-grade fuel because of its very high water content (90%) and low carbon content (5%). Dried peat, however, is used as a fuel in some parts of the world.

The locations of the principal coal deposits in the United States are shown in **Figure 12.12**. Most of the anthracite has already been mined.

The Composition of Coal

Coal is a complex mixture of organic compounds. It is composed primarily of hydrocarbons and small amounts of oxygen-, nitrogen-, and sulfur-containing compounds. Compared with petroleum, coal contains a much higher percentage of aromatic hydrocarbons, including many polycyclic aromatic compounds (Chapter 8). The composition of coal varies with its type and the location of the deposit.

Problems with Coal

Most coal is extracted from the earth by strip mining, which devastates the land unless adequate steps are taken for reclamation (**Fig. 12.13**). Both strip mining and underground mining produce large quantities of waste rock, and acid leaching from the wastes can contaminate nearby streams and rivers (Chapter 7). Coal is dirty, bulky, and expensive to transport. When coal is burned, large quantities of ash are produced, presenting a disposal problem. Some ash is used to make cement and some for fill. All coal contains some sulfur, and sulfur dioxide emitted during the combustion of coal is the primary cause of acid rain (Chapter 11). Also, the burning of coal produces carbon dioxide, and, as we saw earlier (Chapter 11), the constantly increasing concentration of car-

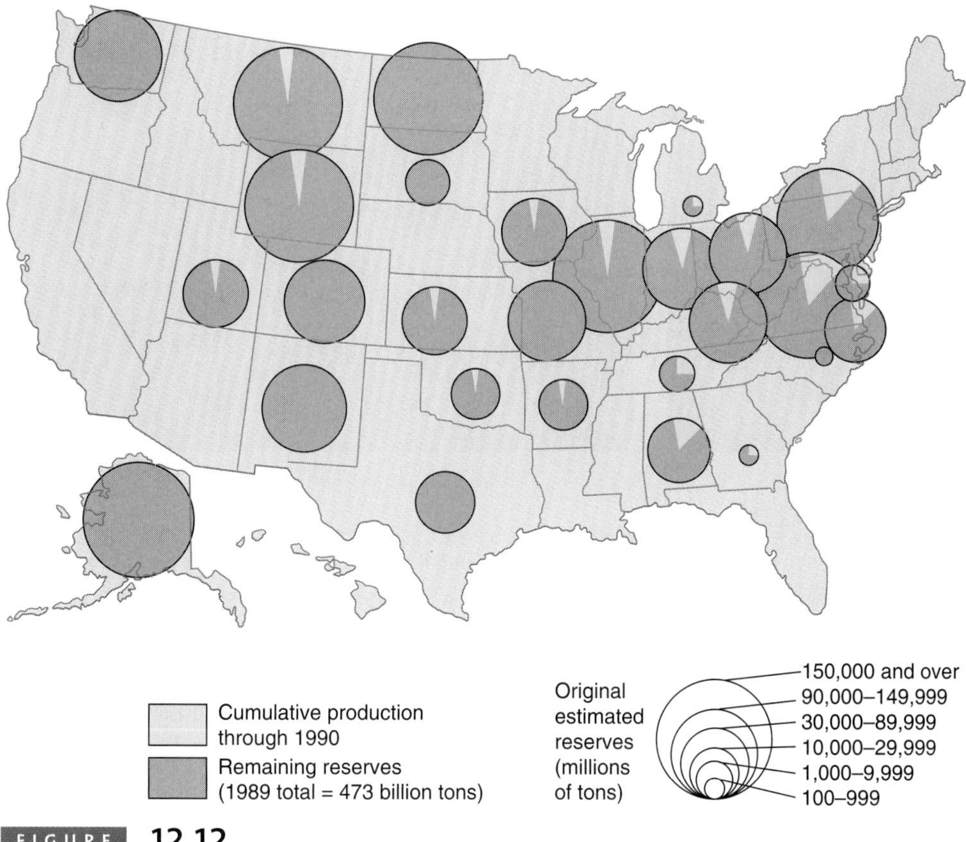

FIGURE 12.12

Major coal deposits in the United States.

bon dioxide in the atmosphere has the potential to change the earth's climate drastically. Coal accounts for 43% of annual global carbon emissions (in the form of carbon dioxide); and it produces 29% more carbon per unit of energy than oil and 80% more than natural gas. The environmental effects of coal use are pictured in **Figure 12.14**.

Apart from the environmental problems associated with its use, solid coal is not a versatile fuel. It is not suitable for today's home and office heating systems or as a fuel for automobiles or airplanes. The main consumer of coal today is the electric power industry. Many of the disadvantages of coal as a fuel can be overcome if it is converted to a gas or liquid before it is burned. In a gaseous or liquid form, coal is easier to transport and burns more cleanly.

The gaseous and liquid fuels obtained from coal are called *syn fuels.*

Coal Gasification

Coal gasification (the production of a gas from coal) is not a new idea. As early as 1807, a municipal system was used to light streets in London. Coal was heated in the absence of air—by a process known as *pyrolysis*—and yielded *coal gas,* a mixture of hydrocarbons, hydrogen, and carbon monoxide, that could be burned to produce light. As late as 1932, gas supplied to homes in the eastern United States was derived from coal, but, with the construction of the gas pipeline network in the 1940s, less expensive and safer natural gas replaced it. Coal makes up the largest reserve of a fossil fuel in the United

a.

b.

FIGURE 12.13

(a) Strip mining of coal can devastate the land, leaving it bare and subject to erosion. (b) Land destroyed by strip mining can be reclaimed by regrading and planting trees and grass.

States, and as concerns over U.S. dependence on imported oil have grown, interest in both coal gasification and coal liquefication has been renewed.

In the first step in a modern coal gasification process, heated crushed coal is treated with superheated steam under carefully controlled conditions. Carbon monoxide and hydrogen are formed as follows:

$$\underset{\text{coal}}{C} + \underset{\text{steam}}{H_2O} \longrightarrow CO + H_2$$

Carbon monoxide then reacts with hydrogen to produce methane and water and with steam to produce carbon dioxide and more hydrogen:

$$CO + 3\ H_2 \longrightarrow CH_4 + H_2O$$

$$CO + H_2O \longrightarrow CO_2 + H_2$$

Conditions are adjusted to give the maximum possible yield of methane. Unreacted carbon monoxide and hydrogen are recycled through the system, and sulfur compounds and other impurities are removed. The final product, which is predominantly methane, is called **synthesis gas** (or *syn gas*). The process, which is relatively inexpensive, can also be used to produce methanol, another useful fuel.

Coal Liquefication

Petroleum-like liquids can be obtained from coal if the complex organic molecules in coal are broken down into smaller molecules and if the hydrogen content of the molecules is increased. This result can be achieved by subjecting a mixture of coal and hydrogen to high pressure in the presence of a metal catalyst. A material similar to crude oil is produced; it can be fractionally distilled to yield products similar to those obtained from petroleum: kerosene, diesel fuel, gasoline, and lubricants.

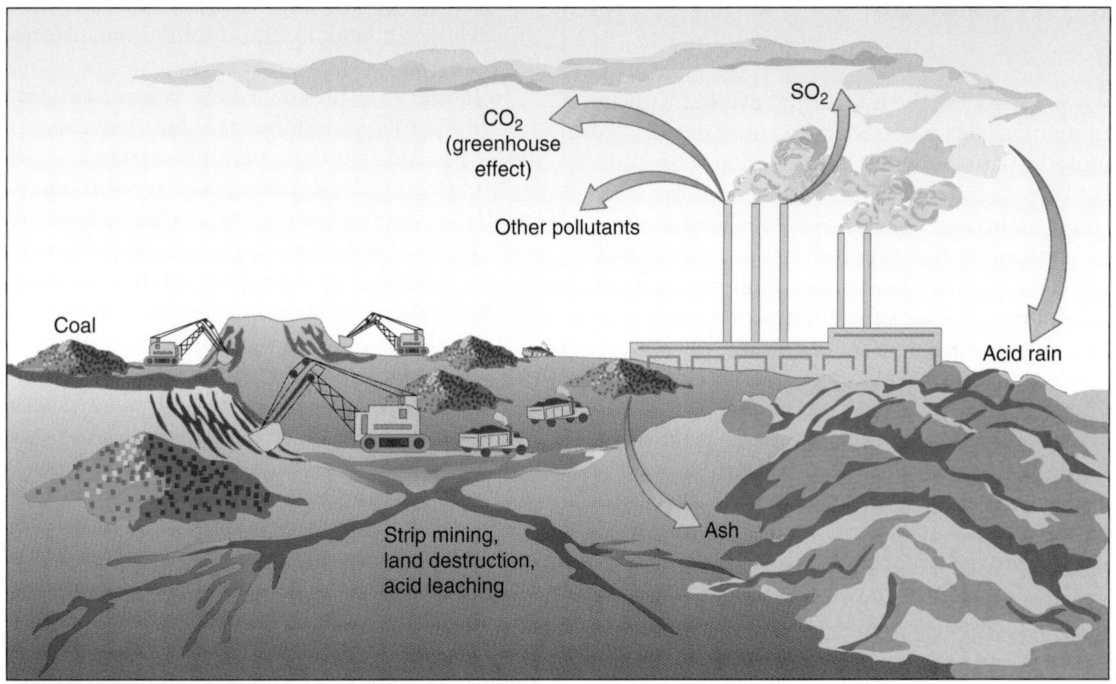

FIGURE 12.14

Environmental effects of the use of coal as an energy source: Combustion of sulfur-containing coal is the major cause of acid rain. When coal is burned, the greenhouse gas carbon dioxide is added to the atmosphere, increasing the possibility of global warming. Coal mining can result in acid drainage and destruction of land.

The future of coal liquefication will depend on the world price of oil and the degree of U.S. reliance on imported oil. It should be remembered that the conversion of coal to a liquid or a gas in no way reduces the amount of carbon dioxide released to the atmosphere during combustion.

12.7 Electricity

Electricity, the most important secondary source of energy in the United States, plays a critical role in modern society. When the first electrical generating plants began operating at the end of the 19th century, electricity was used only for lighting. It was not until the beginning of the 20th century, when the electrification of industry began, that electrical appliances were introduced in the marketplace. The first clothes washers and vacuum cleaners were produced in 1907. Today, more than one-third of all the energy consumed in the United States is used to generate electricity. Approximately two-thirds of this production powers electric motors, which are vital to the industrial, domestic, and military sectors of the U.S. economy. Elevators, air conditioners, communications systems, TV sets, microwave ovens, toasters, and numerous other systems and appliances that depend on electricity have become commonplace, taken for granted as necessities for the American way of life. In the United States, coal provides more than half of the energy used to generate electricity. Nuclear power provides 20%, natural gas

EXPLORATIONS

Environmental Terrorism in the Persian Gulf

In August 1990, the Iraqi army invaded Kuwait. Six months later, when Saddam Hussein still refused to comply with United Nations demands that he withdraw his forces, troops from the United States and its allies drove the invaders out. In February 1991, during the final days of the war, the retreating Iraqis committed an unprecedented and unforgivable act of environmental terrorism. They fired more than 600 of Kuwait's oil wells.

The dynamited wells exploded in flaming fountains of burning oil that roared 200 feet into the air. Huge plumes of black smoke from the fires were blown by the wind, creating immense black clouds that blotted out the sun. A steady drizzle of soot and oil droplets fell from the plumes and coated every surface below with a black film. Mixed with sand, the oily soot formed a black crust over the desert. Soot particles, rising as high as 20,000 feet into the air, were carried by the prevailing winds mostly southward over Saudi Arabia, although some reached India, 1500 miles to the east, before falling to the ground or being washed out in rain or snow. Each day the raging fires consumed about 5 million barrels of oil and released approximately 0.5 billion tons of pollutants into the atmosphere.

At some of the dynamited wells, the oil did not catch fire but surged into the air in great dark brown jets. Millions of barrels of crude oil spilled onto the desert, creating huge oil lakes that became a threat to groundwater. Desert plants, insects, lizards, and small mammals drowned in the oil. Thousands of birds became mired in the thick crude when they landed on the oil lakes, mistaking the shimmering surfaces for water.

In January 1991, a few weeks before Saddam Hussein's troops fired the oil wells, they callously poured as much as 6 million barrels (approximately 250 million gallons) of oil into the Persian Gulf, creating the world's largest oil spill. The Iraqis deliberately ruptured pipelines and storage tanks and emptied oil from as many as seven loaded tankers. A 600-square-mile area of the gulf was covered with a layer of oil, and black oil washed up along 300 miles of the Saudi Arabian coast. The spill could have been much worse if employees at the Kuwait Oil Company had not secretly closed a vital valve. Their courageous action prevented 8.5 million barrels of oil in storage tanks from pouring through damaged pipelines to the dynamited offshore loading docks on Sea Island.

Rich Ecology of the Gulf

The Persian Gulf is a shallow sea, rarely more than 300 feet deep, that opens into the Arabian Sea through the narrow Strait of Hormuz. The environment in the gulf is unique. Although the water is one and a half times saltier than typical seawater, it is rich in nutrients that flow in from the Tigris and Euphrates rivers, and the gulf teems with life. Blue-green algae and sea grasses flourish in the broad tidal shallows that stretch along the Saudi Arabian coast. Minute organisms living in the algae and plankton in the water are

Oil wells burning in Kuwait in 1991

the first links in the food chain that supports the gulf's wildlife. The sea grass meadows are habitat for juvenile shrimp, fry, and other small creatures. The waters, rich in shrimp and commercially valuable fish, including mackerel, mullet, snapper, and grouper, have been fished for millennia. The marine life supports large flocks of migratory and resident birds. Ospreys, flamingos, terns, cormorants, other waterfowl, and shorebirds such as plover and sandpipers are abundant along the intertidal sandflats. Small islands in the gulf, many ringed with fragile coral, provide nesting sites for the endangered hawksbill turtle and for many seabirds. The southern end of the gulf, below Bahrain, is home to the dugong, a relative of the manatee.

Oil Spills Common

As the main shipping route through which Middle East oil reaches the outside world, the Persian Gulf is constantly subjected to oil spills. Annually, in the normal course of operations, about a quarter of a million barrels of oil—an amount equal to the 1989 *Exxon Valdez* spill in Prince William Sound in Alaska—spill into the gulf. The Iraqi's act of sabotage in January 1991 released more than 20 times that amount in a matter of days. Prevailing northwesterly winds, combined with the counterclockwise currents in the water, dispersed the oil southward along the Saudi coastline.

Saudi and Kuwaiti authorities were overwhelmed by the enormity of the spill and were completely unprepared to deal with it. Booms were deployed, and cleanup crews engaged to protect refineries, petrochemical facilities, and desalination plants, but little was done to protect natural habitats. Compared with the remedial actions taken after the much smaller *Exxon Valdez* spill, the response to the gulf tragedy was slow and inadequate.

Despite cleanup efforts and encouraging signs of healing, it is likely to be decades before the gulf fully recovers.

Extensive Environmental Damage

A black ribbon of oil half a mile wide soon covered the tidal zone, and with each incoming tide, more oil seeped into small estuaries and creeks. As the oily tides retreated, they left behind thousands of dead, oil-soaked crabs, fish, and dead and dying birds. A bird rehabilitation center treated hundreds of oil-soaked birds, but it is estimated that at least 20,000 birds died. Unique stands of black mangrove trees were other victims of the oil. Much of the oil in the shallows mixed with sand and sank, smothering numerous bottom-dwelling creatures. Farther out from the shore, the shimmering film of oil on the water gradually thinned. It is estimated that 40–50% of the slick evaporated during the first week.

The damage to the environment from the oil fires and the spills was enormous, but a year after the disaster, the damage appeared to be less severe than originally feared. Although crabs were still dying in the shallows where oil remained on the bottom, fresh green sea grass was growing vigorously, and small fish were swimming around it. Thousands of terns had bred successfully on Karan Island despite a hard layer of tar that contaminated their rocky nesting sites.

It had been predicted that the fired wells might burn for at least 5 years and have serious effects on climate and human health. In fact, all the damaged wells were capped within 9 months. Even when the fires were at their height, concentrations of sulfur dioxide and other toxic gases in the plumes, although above prewar levels, were within U.S. air quality limits. High levels of fine particulates in the atmosphere accounted for the increase in respiratory ailments that was observed. After the fires were extinguished, there was still some concern that unidentified carcinogens in the smoke might cause health problems in the future.

Despite cleanup efforts and encouraging signs of healing, it is likely to be decades before the gulf fully recovers. Accidental oil spills and other insults to the environment are, unfortunately, likely to occur in the future but we must hope that no nation will ever again intentionally devastate the natural world.

References: T. Y. Canby, *National Geographic,* 180, No. 2 (August 1991), p. 2; S. A. Early, *National Geographic,* 181, No. 2 (February 1992), p. 122.

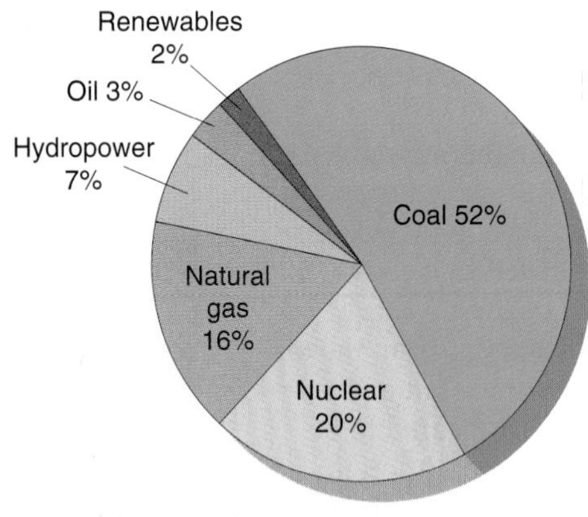

FIGURE 12.15

In 2000, fossil fuels (oil, natural gas, and coal) comprised more than 70% of the fuel sources for electric power generation in the United States.

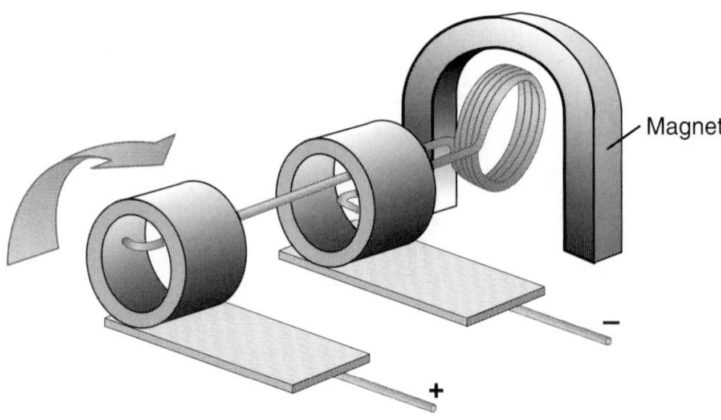

FIGURE 12.16

In an electric generator, when a coil of wire is rotated in a magnetic field, an electric current flows in the wire.

supplies 16% and hydropower, oil, and renewable sources supply smaller amounts. **(Fig. 12.15)**

> Many people are concerned that living close to power lines may increase the risk of developing cancer, but there is as yet no convincing evidence to indicate that this is the case.

Electricity is produced whenever a coil of wire is moved through a magnetic field—or whenever a magnetic field is moved within a coil of wire. The movement induces a flow of electrons (electricity) in the wire **(Fig. 12.16)**. Most modern electric generators are operated by rotating a coil of wire within a circular arrangement of magnets. Energy, of course, is needed to rotate the coil. In the United States, this energy is obtained primarily from coal. Heat from the combustion of coal converts water in a boiler into high-pressure steam. The steam strikes the blades of a turbine, which is coupled to a generator. The turbine turns, causing the coil of wire to rotate in a magnetic field, and electricity is generated **(Fig. 12.17a)**. After the steam has passed through the turbine, it is cooled, condensed back to water, and returned to the boiler to be reused.

Generation of electricity is not an efficient process. Inevitably, as a consequence of the second law of thermodynamics, energy is wasted. In the average coal-fired electric power plant, 60–70% of the energy used to produce steam is lost in the form of heat as the steam is cooled. Only 30–40% of the energy derived from coal is converted into electricity. The waste heat is removed by circulating large quantities of water, usually taken

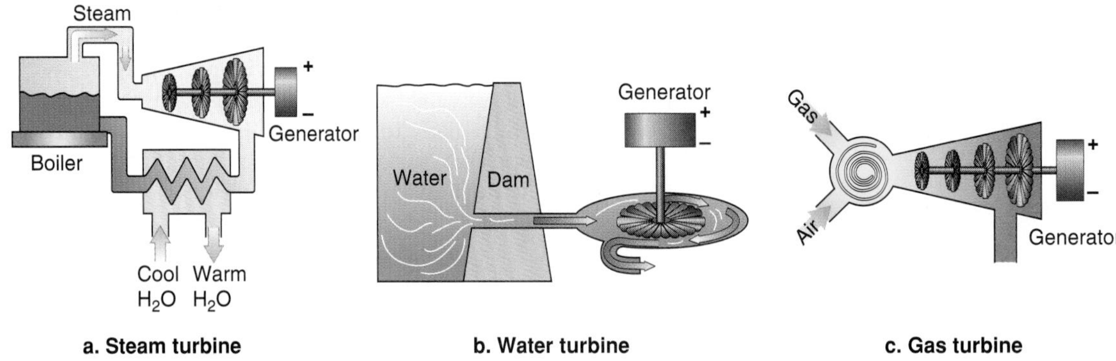

FIGURE 12.17

(a) Steam turbines, (b) water turbines, or (c) gas turbines can be used to generate electricity.

directly from a nearby river or lake, around the condenser. As we saw in Chapter 10, the increased temperature of the water as it is returned to its natural source can adversely affect fish and other marine organisms.

Steam turbines generate approximately 80% of the electricity produced in the United States. Coal is the major heat source used to convert water to steam, but oil and nuclear energy are also used. Approximately 10% of the electricity produced in the United States is generated with hydroelectric power **(Fig. 12.17b)**. In this process, the kinetic energy released when water held behind a dam or at the top of a waterfall drops to a lower level is used to turn the blades of a turbine. Another 10% of U.S. electricity is generated by gas turbines **(Fig. 12.17c)**. High-pressure gases produced in the combustion of natural gas turn the turbine blades.

Electricity is a very clean and convenient form of energy at the consumer level, but its production creates many environmental problems. These problems are related to the primary energy source, coal, which provides 52% of the energy used to generate electricity; its combustion by power plants produces the adverse environmental effects described earlier in this chapter. As demand for electricity has escalated, the world has become increasingly dependent on coal and nuclear energy to meet energy needs.

An electric bill charges for the kilowatt hours of electrical energy consumed in a stated period of time.

Nuclear energy and other nonfossil fuel energy sources are considered in Chapter 13.

Chapter Summary

1. Fossil fuels (petroleum, natural gas, and coal) are the remains of plants, animals, and microorganisms that lived millions of years ago.
2. Energy, which is the ability to do work, is measured in calories, joules, or British thermal units. Power, which is a measure of the rate at which energy is used, is measured in calories per second or watts.
3. Fossil fuels are composed primarily of hydrocarbons.
4. When fossil fuels are burned, they release energy in the form of heat. The amount of energy released is determined by the number and kinds of bonds that are broken and that are reformed.
5. The amount of energy in kilocalories required to break, or make, 1 mole of a particular kind of bond is its bond energy.
6. Petroleum and natural gas accumulate together in permeable rock (reservoir rock) under a layer of impermeable cap rock.
7. Petroleum is purified by fractional distillation, which separates hydrocarbons according to their boiling points.
8. A process called cracking is used to convert higher-boiling hydrocarbons to the lower-boiling, straight-chain hydrocarbons that are suitable as fuel for automobiles.
9. The tendency of gasoline to cause knocking is measured according to an arbitrary scale called the octane rating.
10. The octane rating of a gasoline can be improved by catalytic reforming, a process in which straight-chain hydrocarbons are converted to branched hydrocarbons and aromatic hydrocarbons.
11. Tetraethyl lead (TEL), used for many years as an octane enhancer, was phased out because it is a health hazard. Its replacement, methy-*t*-butyl ether (MTBE), has also been shown to be a health hazard.

12. Shale oil and tar sands are potential sources of petroleum.
13. Petroleum is the primary source of the hydrocarbons needed as starting materials for the manufacture of plastics, synthetic fibers, synthetic rubber, pesticides, pharmaceuticals, and other consumer products.
14. Oil spilled from tankers during normal operations and in accidents fouls coastlines, kills aquatic birds and mammals, and destroys wetlands.
15. Natural gas is composed mainly of methane, with small amounts of ethane, propane, and butane. It is cleaner burning and emits less carbon dioxide per unit of energy than either oil or coal.
16. Environmental problems associated with the use of coal include land devastation by strip mining, acid leaching from waste rock into waterways, and sulfur dioxide and carbon dioxide emissions when coal is burned.
17. Coal gasification and liquefication convert coal into a cleaner burning fuel.
18. Electricity is produced when a coil of wire is moved through a magnetic field or when a magnetic field is moved through a coil of wire.
19. In a typical coal-burning electric power plant, heat from burning coal converts water in a boiler into high-pressure steam, which turns the blades of a turbine coupled to a generator.

Key Terms

anthracite (p. 374)
bitumen (p. 371
bituminous coal (p. 374)
bond energy (p. 363)
catalytic reforming (p. 369)
cracking (p. 367)
energy (p. 361)
fractional distillation (p. 365)
fuel (p. 361)
gasohol (p. 370)
kerogen (p. 370)
liquefied petroleum gas (LPG) (p. 373)
octane rating (p. 368)
oil shale (p. 370)
power (p. 361)
reservoir rock (p. 364)
shale oil (p. 370)
synthesis gas (p. 376)
tar sands (p. 371)

Questions and Problems

1. Describe two ways in which the steam engine changed society.
2. Prior to the discovery of petroleum, what was the major source of energy in the United States?
3. Describe the difference between energy and power. What units are used to measure each?
4. What is the origin of petroleum?
5. What are the geological features that are required for the formation of petroleum in amounts that can be extracted from the earth?
6. Besides crude oil and natural gas, what other substance is often found in reservoir rock? What problem does this substance create?
7. Why was natural gas once treated as a waste product of petroleum production?
8. Give two reasons why natural gas is considered to be a good fuel.
9. Why is the fuel value for crude oil higher than the value for coal or wood (see Table 12.3)?
10. Why does sulfur pose a problem when it is a component of crude oil?
11. Write a chemical equation for the complete combustion of butane.
12. When burned, methane (CH_4) releases 192 kcal/mole. How much energy is released when 10.0 moles of methane are burned?

13. Which of the following are fuels?
   a. CH_4 **b.** H_2O **c.** C_4H_{10}

14. How much energy is released when butane (C_4H_{10}) is burned?

15. Describe what is meant by *fractional distillation.*

16. What is the difference between gasoline and kerosene?

17. What is straight-run gasoline?

18. Describe what is meant by *catalytic reforming.*

19. Draw all hydrocarbon isomers with the formula C_5H_{12}. Using Table 12.5, list them according to increasing octane number.

20. Which has a higher octane rating, toluene (C_7H_8) or heptane (C_7H_{16})?

21. For each of the following pairs of hydrocarbons, indicate which has the higher octane rating.
   a. pentane or octane
   b. 2-methylpentane or 2,2-dimethylbutane
   c. octane or 2-methylheptane

22. Why was tetraethyl lead (TEL) used as an octane enhancer?

23. Give two reasons why tetraethyl lead (TEL) was banned as a gasoline additive.

24. Explain the octane rating system.

25. Describe three ways for increasing the octane rating of gasoline.

26. What is the purpose of an antiknock agent? Why does knocking occur in an automobile engine?

27. What is gasohol? List two benefits of using gasohol rather than gasoline.

28. What is oil shale? What organic material does it contain?

29. Describe the problems associated with recovering oil from oil shale.

30. How does the current price of petroleum affect the production of shale oil?

31. What are tar sands? What organic material do they contain?

32. What is the difference in chemical composition between commercial natural gas and crude natural gas?

33. What is the difference between bituminous and anthracite coal? Why is anthracite coal a better fuel?

34. Why is peat not considered a good fuel?

35. What are the two main methods for mining coal, and which one is used more extensively? What is the environmental impact of each method?

36. Identify two environmental problems that are associated with strip mining.

37. Write a chemical equation for the complete combustion of coal.

38. Why is coal, which was a major fuel for residential heating in the 1930s, considered unsuitable for home heating in the 1990s?

39. Is coal gasification a way to convert coal to a more suitable fuel for home use?

40. What is *synthesis gas*? How is it formed in coal gasification? Is it a high-energy fuel?

41. Make two lists: one of the environmental advantages of coal gasification and the other of the environmental disadvantages.

42. A recently developed technique for coal gasification produces methane from coal. List the reactants needed for this process, and write a chemical equation that describes it.

43. The generation of electricity produces significant quantities of various pollutants. List the types of fuels used to generate electricity, including the percentage that each contributes to total U.S. production of electricity and the pollutants each produces.

44. Producing electric power from fossil fuels causes two major environmental problems. What are they?

45. What is the average efficiency of a coal-burning electric power generating plant?

46. Describe the relationship between the demand for a fossil fuel and growth in each of the following U.S. industries (indicate increasing or decreasing demand and which fossil fuel):
   a. farming
   b. food processing
   c. manufacturing
   d. transportation

47. Describe the principle of an electric generator.

48. Describe how energy produced by burning coal is used to turn a turbine and generate electricity.

49. Why are electric power plants often located near large bodies of water?

50. Every day, 100 railroad cars of coal enter a medium sized coal-burning electric power plant. By the end of the day, 30 railroad cars of ash are removed from the plant. Describe what happened to the contents of the railroad cars.

51. Describe how a gas turbine makes electricity.

Answer to Practice Exercise

12.1 344 kcal/mole

Chapter 13 Energy Sources for the Future

Chapter Objectives

In this chapter, you should gain an understanding of:

- Energy conservation's crucial role for the future
- Nuclear power for generating electricity, the nuclear fuel cycle, and nuclear fusion
- Active and passive solar energy
- Biomass as a source of energy
- Wind, geothermal resources, tides, dams, fuel cells, and hydrogen as sources of energy
- Alternative fuels for automobiles

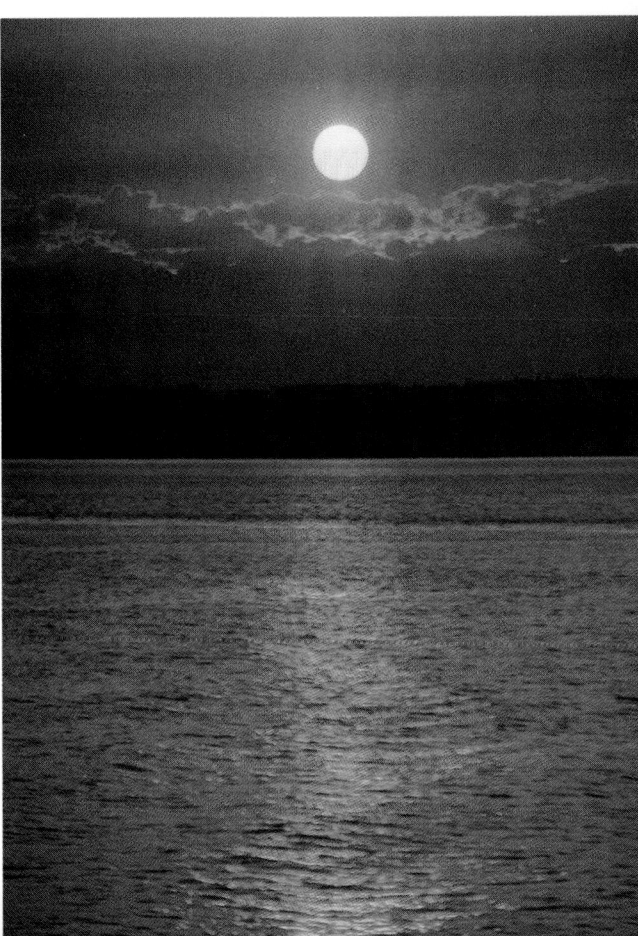

The sun is our ultimate source of energy. If we can harness this energy for society's needs, we will have a source that is inexhaustible.

ONLINE FEATURES

www.jbpub.com/chemistry

- Chemistry and the Environment
- eLearning
- Chemistry in the News
- Special Topics
- Research & Reference Links
- Ask the Authors

THROUGHOUT HUMAN HISTORY, improvements in living standards have been linked to the availability of energy. Starting at the end of the 18th century, material well-being in the industrialized nations has become almost entirely dependent on fossil fuels as the source of energy. These finite, nonrenewable resources are being rapidly depleted, and other sources of energy will eventually have to be developed or discovered. In terms of human history, this period of dependence on fossil fuels for energy will be seen as a very brief episode (**Fig. 13.1**).

In the United States, the most immediate concern focuses on the availability of petroleum, which currently supplies approximately 40% of U.S. energy needs. If world consumption of petroleum continues at its present rate, there could be shortfalls between world production and world demand by about 2050, and economically recoverable reserves could be gone by the end of the 21st century. However, because world consumption, particularly in the developing countries, is expected to increase rapidly in the coming years, reserves are likely to be depleted even sooner.

To avoid serious economic and social disruptions, Americans need to begin making a gradual transition to other sources of energy before oil becomes scarce and to use energy more efficiently and less extravagantly. Although the United States has only 5% of the world's population, it consumes 30% of the world's supply of energy. If other countries used as much energy per capita as the United States does, fossil fuels would be depleted within the next decade.

In this chapter, we consider the extent of the world's known reserves of fossil fuels and current dependence on this energy source. We examine ways in which energy can be conserved, and we study the advantages and disadvantages of alternative sources of energy.

13.1 Fossil Fuel Resources and the Energy Crisis

Eighty percent of the world's energy is supplied by nonrenewable fossil fuels. How long these rapidly dwindling resources will last depends on several factors: the total amount of fossil fuels that exist in the earth, how much of this total is economically recoverable, the probability of discovering new deposits, and the rate at which these resources are consumed. One thing is certain: Sooner or later, the earth's store of fossil fuels will be used up.

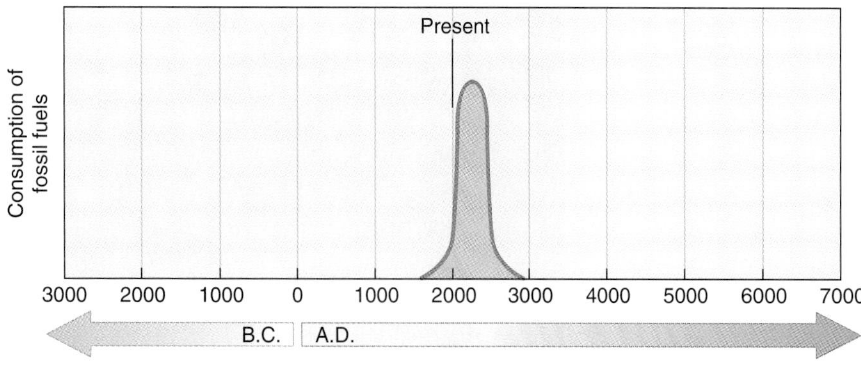

FIGURE 13.1

Fossil fuel use over the past 5000 years and projected use for the next 5000 years. Fossil fuel reserves are expected to be used up during the next few hundred years. From a historical perspective, the period of dependence on fossil fuels will be very brief.

Oil Reserves

Oil is very unevenly distributed around the world (**Fig. 13.2**). Over half of the world's proven reserves of oil are found in the Middle East.

As we saw in Chapter 12, oil has replaced coal as the major source of energy in the United States (refer back to Fig. 12.2), and now supplies approximately 40% of U.S. energy needs. From 1950 to the mid-1970s, oil was cheap and consumption steadily increased. During that time, production from the huge oil fields discovered in the 1930s in east Texas and in Wilmington, California, as well as from later discoveries, generally kept pace with demand. Since then, however, production has not kept up with demand. The last major discovery of oil in the United States was the giant field found in 1968 on the North Slope in Alaska. In 1982, a large but considerably smaller field was discovered off the coast of California, but because of environmental concerns and the high costs of offshore drilling, exploitation of this field has been limited. Worldwide, the most important discovery of the late 20th century was the enormous Tengis oil field in the Caspian region of Kazakhstan, part of the former Soviet Union.

The North Slope in Alaska includes the Arctic National Wildlife Refuge. Energy producers want to drill for oil in the refuge's coastal plain; conservationists are opposed to the idea because an oilfield would disrupt the area's abundant and diverse wildlife.

In the future, advances in technology and rising oil prices may make it feasible to obtain oil from wells that at present cannot be worked profitably. Also, as oil supplies dwindle, extraction of oil from oil shale and tar sands (Chapter 12) will probably become economically attractive. These measures, however, will serve only to postpone for a short time the inevitable exhaustion of available oil.

OPEC was founded in 1960. The 11 members of the organization today are Algeria, Indonesia, Iran, Iraq, Kuwait, Libya, Nigeria, Qatar, Saudi Arabia, United Arab Emirates, and Venezuela.

The Energy Crisis

By the early 1970s, North America, Western Europe, and Japan were heavily dependent on oil imported from the Middle East and thus were vulnerable to economic and political upheavals there. This vulnerability was demonstrated in 1973 and 1974, after the Arab-Israeli War. At that time, the five members of the Organization of Petroleum Exporting Countries (OPEC) (Saudi Arabia, Iraq, Iran, Kuwait, and Venezuela), which were supplying more than 50% of the world's oil, placed an embargo on shipments of oil to all nations that had supported Israel in the war. OPEC quadrupled the price of oil, causing shortages and economic disruption in the industrialized and developing nations of the world. In the United States, gasoline was in short supply and long lines of automobiles at gas stations became common. Oil prices, which were determined unilaterally by OPEC, continued to rise, reaching an all-time high in 1979.

In 1979, a second oil crisis occurred: After the fall of the Shah, Iran cut back production and suspended oil exports to the United States. The two energy shocks of the 1970s, coupled with rising oil prices, were a spur to the development of more efficient ways to use energy. Smaller cars became popular, thermostats in homes and offices were lowered in winter and raised in summer, and U.S. energy consumption fell in 1973 and again in 1979 (see Fig. 12.2). In the 1980s, to reduce

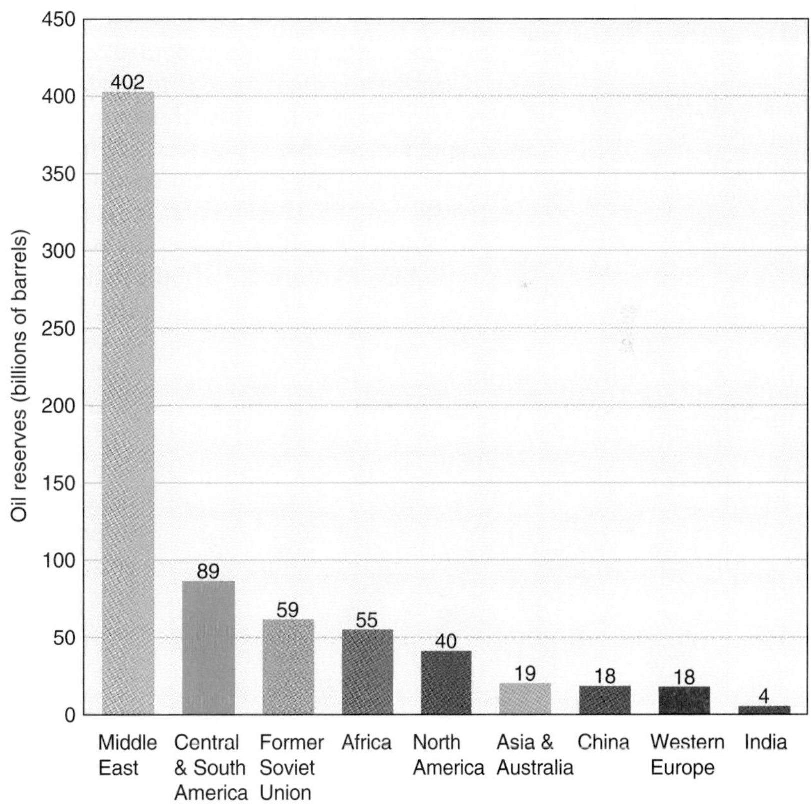

FIGURE 13.2

The world's oil reserves are very unevenly distributed. More than 50% of the known reserves are in the Middle East.

The Department of Energy was established by President Jimmy Carter in 1977 to develop long-range energy policy and conduct energy research.

its reliance on oil produced in the Middle East, the United States began importing more of its oil from other OPEC countries and from non-OPEC countries, primarily Mexico. In the 1980s, the members of OPEC (which by then numbered 13) could no longer agree on oil production goals; as a result, overproduction occurred, causing oil prices to fall dramatically. Concern about shortages of oil subsided; energy consumption began to rise again and has continued to do so until the present time. The drop in oil imports that occurred in the early 1980s has been reversed, and the United States is again becoming increasingly dependent on foreign oil.

The instability of the Middle East was again demonstrated in 1990 when Iraq invaded Kuwait. Although intervention by forces from the United States and other countries soon freed Kuwait, the firing of the oil wells by the retreating Iraqis jeopardized oil supplies (Chapter 12).

The energy crises in the 1970s and in 1990 were caused by the disruption of oil supplies. An energy crisis in the future is likely to be caused by an irreversible shortage of oil.

Natural Gas Reserves

The major import market for LNG in the future will be in the Pacific regions of the world where there are few gas pipelines.

Approximately 95% of the natural gas used in the United States is obtained from domestic sources. The remaining 5% is imported from Canada.

Natural gas, which supplies about 25% of the world's energy, is, like oil, unevenly distributed around the globe **(Fig. 13.3)**. The largest reserves are found in the republics of the former Soviet Union, which produce about 50% more natural gas than the United States does. Worldwide, natural gas consumption has been increasing rapidly.

The transportation and storage of natural gas remains a problem. In the United States, natural gas is distributed through pipelines; in many parts of the world, however, the great distances between the source of the gas and potential customers mean that pipelines are not a practical solution. Consequently, in the Middle East, Mexico, Venezuela, and Nigeria, much of the natural gas found with crude oil is burned as a waste product. However, if cooled to −162°C (−259°F) natural gas liquefies, and the volume it occupies decreases by a factor of 600, making it suitable for transportation by tanker. **Liquefied natural gas (LNG)** is expensive to produce, and there is always the risk of an explosion if leakage occurs. But as oil becomes scarce and demand increases, more LNG will undoubtedly be produced.

In the United States, natural gas production peaked in 1973. If consumption continues at its present rate, remaining U.S. reserves are expected to last until about the end of the 21st century. Supplies can be extended by importing gas by pipeline from Canada and, if the risk is deemed acceptable, by importing LNG from overseas. Synthesis gas can also be produced from coal and oil shale (Chapter 12).

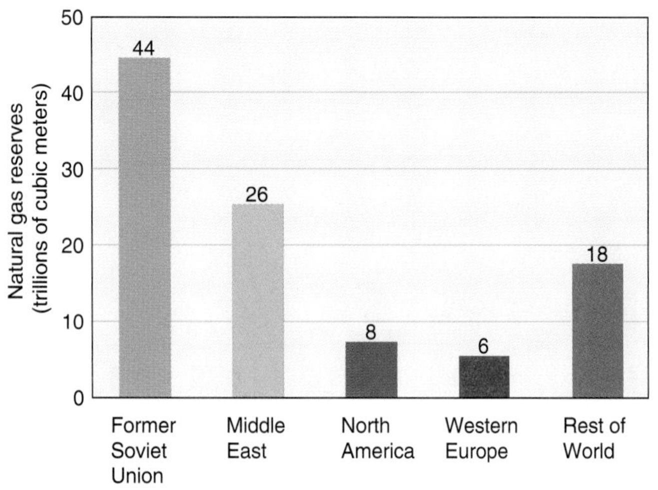

FIGURE 13.3

The world's reserves of natural gas are unevenly distributed. Most are located in the former Soviet Union.

Coal Reserves

Coal is the most abundant fossil fuel in the world. The largest proven recoverable reserves, estimated to be about 30% of the world's total, are found in the United States **(Fig. 13.4)**. Geologists believe that all the world's coal has been discovered. At the present rate of consumption, U.S. reserves are expected to last for approximately 250 years. Currently, coal supplies about 30% of the world's energy.

Could coal be a solution to the immediate energy problem of dwindling oil supplies? Coal is useful almost exclusively as a fuel for generating electricity, and most power plants are already using it for this purpose. Plants that once used oil changed to coal after the oil shortages in the 1970s. It is probable that coal will gradually replace petroleum as a source of organic chemicals needed by the petrochemical industry, and as oil becomes scarce and prices rise, coal liquefication (Chapter 12) is likely to become increasingly important. But coal will not be able to fill the energy gap. Apart from practical problems, the environmental costs of coal combustion are very high. Although emissions of air pollutants can be controlled with new technologies (Chapter 11), the release of carbon dioxide to the atmosphere cannot easily be controlled. Increasing coal combustion would inevitably worsen the problems associated with global warming (Chapter 11).

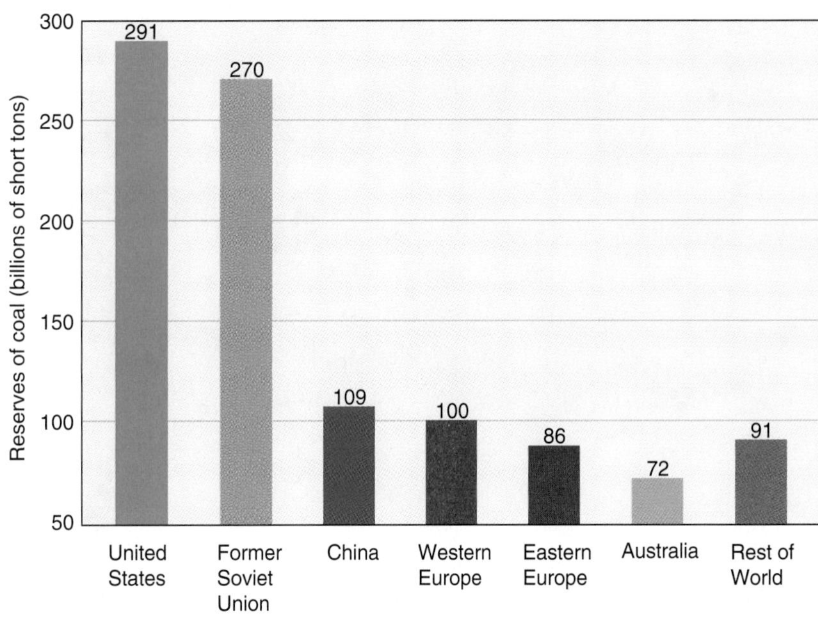

FIGURE 13.4

The largest reserves of coal in the world are in the United States.

13.2 Energy Conservation

Energy conservation is an important option in planning an energy policy for the future. Measures such as turning off lights, using less hot water, and turning down thermostats are important, but for the long term, conserving energy by improving the efficiency of energy use is far more effective. Improving efficiency means finding a way to decrease the amount of energy that is needed to perform a particular task. Already, great strides have been made by industry: Home appliances such as refrigerators, washers, and dryers are much more efficient, insulation in new homes and buildings has improved, and there has been an increase in the use of fluorescent light bulbs. In the 1970s, new passenger cars averaged 13 miles per gallon (mpg); by 1985, they were required by law to average 27.5 mpg. If the automobile industry makes use of available technology, new cars could average 60 mpg or more by 2100. Already, European manufacturers have produced cars that achieve over 80 mpg. In the United States, however, because of the increasing use of energy-inefficient sport utility vehicles (SUVs) and minivans, which are not yet required to meet the same standards as other passenger cars, fuel efficiency of new cars is declining. Without new laws mandating increased fuel efficiency, as long as gasoline prices remain low, few improvements in fuel economy are expected from United States automakers in the near future.

> Many utility companies encourage energy conservation. For example, they offer their customers free advice on how to save energy by improving insulation because it is more profitable to maintain current levels of energy production than to increase capacity.

Greater use of mass transportation, less intensive use of energy for agriculture, the manufacture of goods that are more durable, greater use of energy-efficient appliances, and more recycling are all measures that can be taken to save energy. The initial costs are often high, but savings will be considerable in the long run.

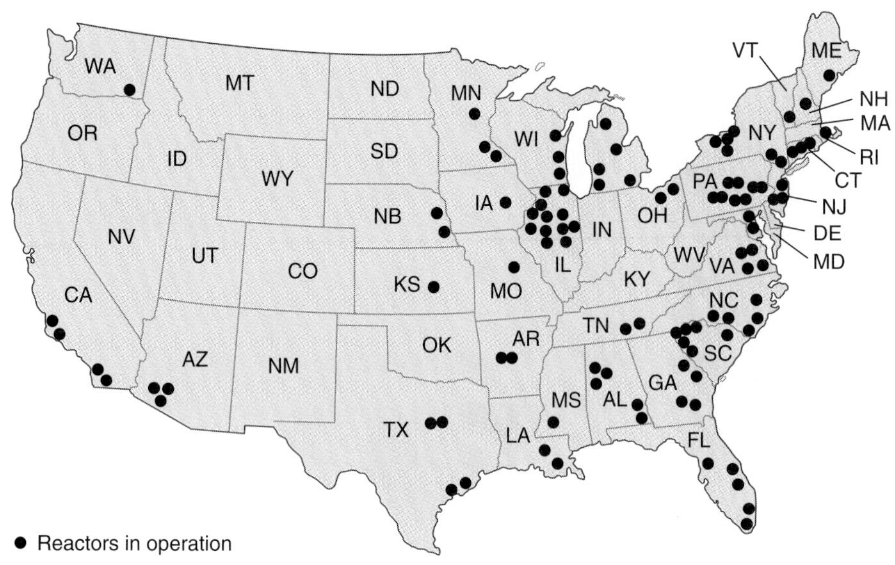

FIGURE 13.5

Nuclear power plants in operation in the United States in 2001.

Many industries have improved efficiency by using waste heat that previously was dissipated to supply energy for a second process. At some power plants, waste heat is being used to heat nearby buildings.

Energy conservation through increased efficiency is not an answer to the energy problem, but it is an essential step if we are to avoid severe shortages while alternative sources of energy are being developed to replace dwindling fossil fuel supplies.

13.3 Nuclear Energy

Since the mid-1970s, nearly all orders for nuclear plants have been cancelled.

The Shoreham nuclear power plant on Long Island, New York, which cost $5.5 billion to build, generated electricity for just 30 hours before it was shut down.

Few issues have generated as much controversy as the future of nuclear energy. When controlled nuclear fission reactions became a reality in the early 1940s (Chapter 5), nuclear energy appeared to hold great promise as a cheap, clean, and safe source of energy. In the 1950s, it was predicted that by the year 2000, nuclear power plants would supply up to 20% of the world's energy. However, because of fears about safety and the relatively low cost of other sources of energy, the actual figure was about 5%. In 2001, the United States had 103 operating nuclear power plants **(Fig. 13.5)** that were supplying approximately 8% of total energy (20% of electrical energy) (see Fig. 12.15). Other industrialized countries with significant numbers of plants are France and Japan, which are not well-endowed with coal and are highly dependent on nuclear power for their energy needs. Worldwide, 431 reactors were in operation in 1999.

Nuclear Fission Reactors

Recall from Chapter 5 that when uranium-235 is bombarded with neutrons ($^{1}_{0}n$), it splits into a variety of lighter atoms. At the same time, additional neutrons are formed, and a tremendous amount of heat energy is released. A typical fission reaction is as follows:

$$^{1}_{0}n + ^{235}_{92}U \longrightarrow ^{141}_{56}Ba + ^{92}_{36}Kr + 3\,^{1}_{0}n + \text{energy}$$

The neutrons formed can initiate further fission reactions, and the resulting chain reaction can lead to a tremendous explosion. In a nuclear power plant, the chain reaction is controlled so that energy is produced at a safe, steady rate.

Nuclear power is used primarily for the production of electricity **(Fig. 13.6a)**. A diagram of a *pressurized water reactor (PWR)*, the most common type of reactor in use, is presented in **Figure 13.6c**. The basic design is essentially the same as that for a coal-fired power plant (refer to Fig. 12.17a) except that nuclear fuel is used instead of coal as the source of heat for converting water to steam.

At the core of a nuclear reactor are several hundred steel **fuel rods (Fig. 13.6b)** containing the fissionable uranium fuel. Interspersed between the fuel rods are **control rods**

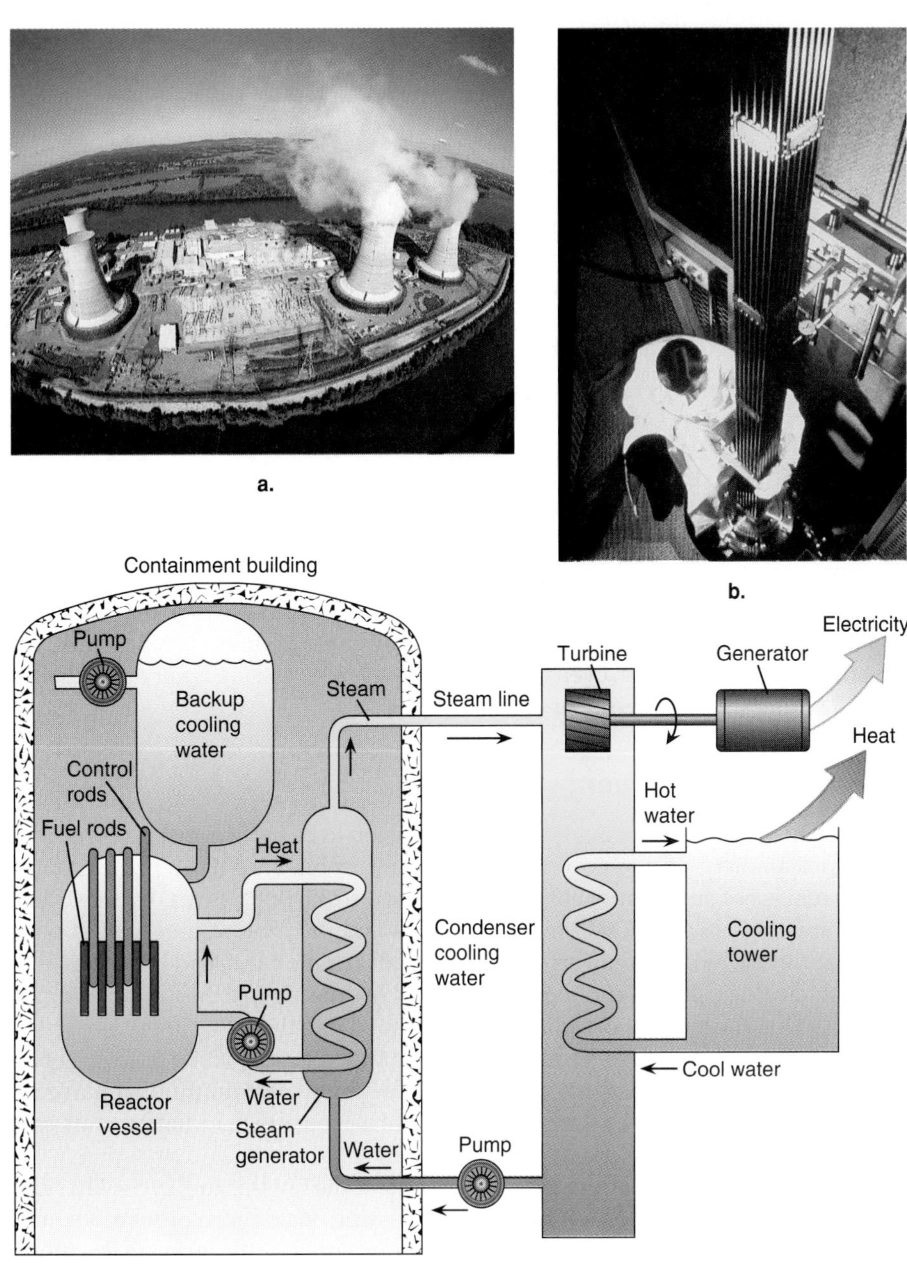

FIGURE 13.6

(a) Nuclear power plant at Three Mile Island in Pennsylvania. (b) Technician inspecting fuel rods. (c) Schematic diagram of pressurized water reactor.

made of a material—usually cadmium or boron—that absorbs neutrons. When the rods are withdrawn from the core, the rate of fission increases; when the rods are inserted, more neutrons are absorbed, and the rate slows. If there is an accident or a need to make repairs or remove spent fuel, operators can stop the reaction by inserting the rods to their limit. Water circulating around the fuel rods and control rods acts as a **moderator,** slowing the neutrons to speeds that are optimal for splitting uranium-235 atoms.

The water, which is circulated through a heat exchanger, also serves to keep the fuel rods cool and to prevent the reactor from overheating. The entire reactor is housed in a thick-walled containment building.

The Nuclear Fuel Cycle

Each pellet in a fuel rod weighs less than 1 gram but has the energy equivalent of 270 kilograms (595 pounds) of coal.

Because most uranium ore that is mined contains no more than 0.2% uranium, producing the fuel suitable for a nuclear power plant is expensive and energy-intensive. In the first steps, milling and chemical treatment convert the ore to a product that is approximately 80% uranium oxide (U_3O_8). The uranium in this oxide is 99% nonfissionable uranium-238 and only 0.7% fissionable uranium-235. The uranium content must be enriched to about 3% uranium-235 by means of a complex and technically difficult process; then the enriched material is fabricated into small pellets, which are packed into the fuel rods.

If the cooling system in a nuclear reactor fails, the core and fuel rods can heat to temperatures in excess of 3000°C (twice the temperature required to melt steel) and melt through the floor of the reactor into the ground.

As fission occurs in the reactor, the uranium-235 concentration decreases, and after about 3 years, fuel rods must be removed and replaced. Spent rods remain radioactive. They can either be permanently stored as wastes or be reprocessed to recover unreacted uranium-235 and the fissionable plutonium-239 produced during the nuclear reactions. Because of the danger that the recovered plutonium could be stolen from a reprocessing plant and used to make nuclear weapons, the United States currently does not reprocess nuclear fuel. However, even though the cost of reprocessing is very high, Japan, Britain, and France all reprocess spent fuel rods.

Problems with Nuclear Energy

Meltdown has been called "the China syndrome" because if it occurred in the United States, melting would be directed through the earth toward China. In fact, in a meltdown, molten material would penetrate only a few meters into the ground.

Most people consider the risk of a catastrophic explosion to be the greatest problem with the use of nuclear energy, but a power plant could never blow up like a nuclear bomb because the fuel is not sufficiently enriched in uranium-235. But, as was demonstrated in 1979 at Three Mile Island (TMI) in Pennsylvania and in 1986 at Chernobyl in what was then the Soviet Union, human error and mechanical failure can lead to serious accidents. In both incidents, loss of cooling water caused overheating of the reactor and core meltdown. At TMI, the amount of radiation that escaped into the atmosphere was small; in the Chernobyl accident, however, enormous quantities of radioactive gases and particles were released. Huge tracts of contaminated land will remain uninhabitable for hundreds of years. The health hazards associated with exposure to radiation were considered in Chapter 5.

EXPLORATIONS reveals the story behind the nuclear power plant disaster at Chernobyl (see pp. 406–407).

Some risk of radiation exposure exists at every step of the nuclear fuel cycle. The huge quantities of crushed waste rock left from the processing of uranium ores are a source of low-level radioactivity, which may leach into groundwater or be dispersed to the atmosphere in windblown dust. A major problem, which will be studied in Chapter 19, is disposal of the high-level radioactive wastes produced during the operation of a nuclear power plant. These wastes, which include fission products and spent fuel rods, must be stored so that they are completely isolated from the atmosphere.

Nuclear power plants have a limited life expectancy. After about 30 years of operation, continual bombardment of plant components with neutrons makes the metals brittle. The chances of cracking and of leakage of radiation are increased, and for safety reasons, the plant must be shut down (decommissioned). Even after spent fuel rods and circulating water have been removed, a plant is still radioactive; the usual procedure is to seal the plant permanently in reinforced concrete. By 2012, nineteen of the nuclear reactors operating in the United States will have come due for decommissioning. By 2030, all will be due for retirement.

Nuclear Breeder Reactors

The world's supply of uranium ores is not abundant; in the 1960s, when rapid expansion of the use of nuclear energy was expected, there were fears that shortages of uranium-235 would develop. The solution appeared to be the development of **nuclear breeder reactors,** which not only produce heat from fission but also yield a new supply of fissionable fuel.

If uranium-238 is bombarded with fast-moving neutrons, the following series of reactions occurs, and fissionable plutonium-239 is formed:

$$\underset{\text{nonfissionable}}{{}^{238}_{92}\text{U}} + {}^{1}_{0}\text{n} \longrightarrow \underset{\text{unstable}}{{}^{239}_{92}\text{U}}$$

$${}^{239}_{92}\text{U} \longrightarrow \underset{\text{unstable}}{{}^{239}_{93}\text{Np}} + {}^{0}_{-1}\beta$$

$${}^{239}_{93}\text{Np} \longrightarrow \underset{\text{fissionable}}{{}^{239}_{94}\text{Pu}} + {}^{0}_{-1}\beta$$

Fission of uranium-235 provides the neutrons needed to start the reaction sequence. Since two or, in some cases, three neutrons are produced in every uranium-235 fission (see the equation on p. 390), two (or three) plutonium-239 atoms may be formed from each uranium-235 atom **(Fig. 13.7)**. The amount of fuel produced thus exceeds the amount consumed. Water cannot be used as the moderator in a breeder reactor because it slows the neutrons needed to produce the plutonium-239. Instead, liquid sodium is used.

The future of breeder reactors is uncertain. In addition to the problems associated with conventional nuclear reactors, there are several other drawbacks. Plutonium-239 has an extremely long half-life (Chapter 5)—24,000 years—and is one of the most toxic substances known. Inhalation of even a minute quantity can cause lung cancer. Another problem is that plutonium-239 can be used more easily than uranium-235 to make nuclear weapons, thus increasing the need for security. The conversion of uranium-238 to fissionable plutonium-239 is difficult to control, and the sodium used as the moderator reacts explosively if it comes in contact with water. At the present time, no breeder reactors are operating commercially in the United States.

The Department of Energy's Clinch River Reactor in Tennessee was never completed and was closed down in 1983 largely because of cost overruns.

Nuclear Fusion

When two very light atomic nuclei are combined, or *fused,* a heavier nucleus is formed. There is a loss of mass, and an enormous amount of energy is released. Fusion of hydrogen atoms to form helium is the primary source of the energy emitted by the sun. Theoretically, for each gram of fuel, **nuclear fusion** releases four times as much energy as the fission of uranium-235 does and about a million times as much as the combustion of fossil fuels does. Many scientists believe that if controlled fusion could be achieved on earth, it would solve the world's energy problems. But enormous technical difficulties must be overcome before this source of energy can be exploited.

A temperature approaching that in the sun is required for hydrogen atoms to combine; even if this temperature could be attained, the problem of finding a container that could withstand the heat without vaporizing remains. A fission (or atomic) bomb was used to produce the high temperature needed for the hydrogen bomb, but this is hardly an option for controlled energy production.

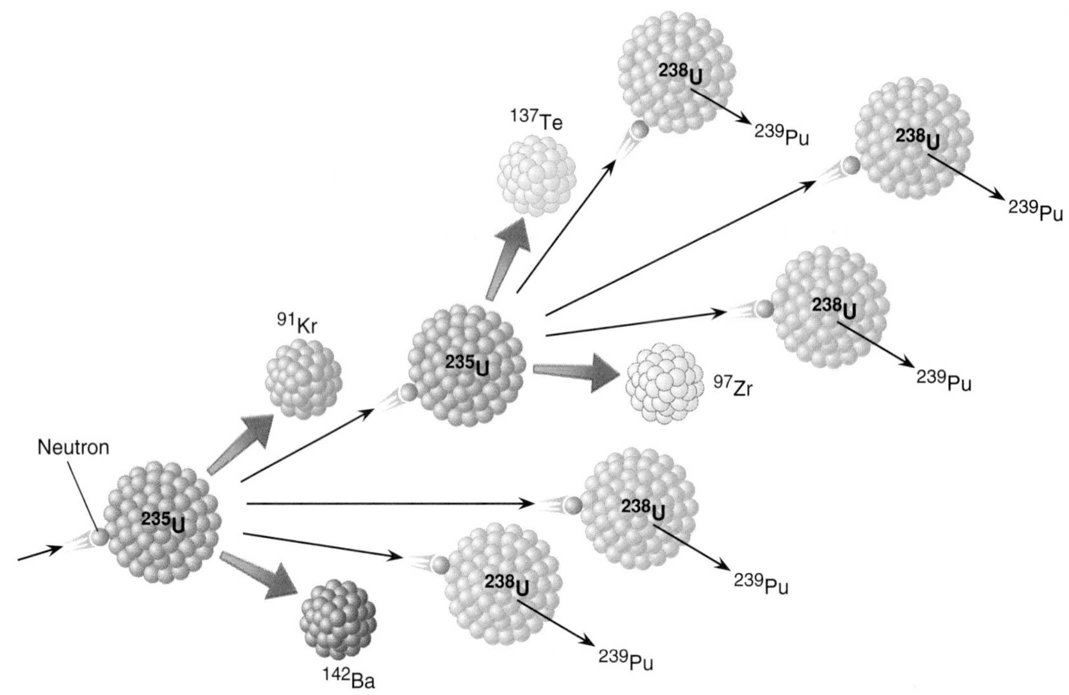

FIGURE 13.7

Typical reaction in a breeder reactor: When uranium-235 atoms are bombarded with neutrons, fission occurs. In each fission reaction, a uranium-235 atom produces two atoms of lighter elements and two or three neutrons. The neutrons bombard other uranium-235 atoms or uranium-238 atoms, converting the latter to atoms of fissionable plutonium-239. Fission of one uranium-235 atom can result in the formation of more than one fissionable plutonium-239 atom. Thus, the amount of fuel produced (plutonium-239) exceeds the amount of fuel consumed (uranium-235).

Fusion of the two isotopes of hydrogen, deuterium (^2_1H) and tritium (^3_1H) (Chapter 3), is receiving most attention because fusion of these two atoms requires a lower temperature (approximately 100,000,000°C) than other fusion reactions.

$$\underset{\text{deuterium}}{^2_1H} + \underset{\text{tritium}}{^3_1H} \longrightarrow \underset{\text{helium}}{^4_2He} + {^1_0n} + \text{energy}$$

Deuterium, a naturally occurring isotope, can be obtained in unlimited quantity from seawater. Tritium, however, is an unstable radioactive isotope and must be produced by bombardment of lithium with neutrons.

$$^1_0n + {^6_3Li} \longrightarrow {^3_1H} + {^4_2He}$$

Although nuclear fusion produces little radioactive waste, its drawbacks are the danger of leakage of tritium (half-life of 12.3 years) and the possibility of thermal pollution. Also, known reserves of lithium ores are very limited.

The word *tokamak* is an abbreviation of the Russian for "torroidal magnetic chamber."

Two approaches to achieving nuclear fusion are being tested. One is the *tokamak reactor* pioneered by Soviet physicists. In this device, very high temperatures strip electrons from deuterium and tritium atoms, creating a gaslike plasma of energetic nuclei and

a.

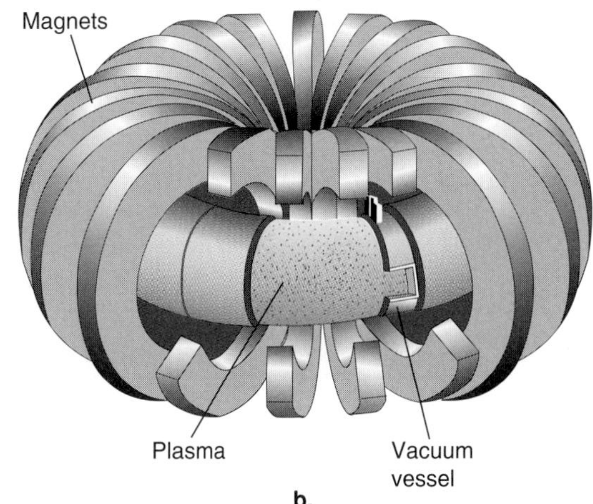

b.

FIGURE 13.8

(a) The Tokamak Fusion Test Reactor at Princeton University. (b) The plasma is contained in the tunnel inside the doughnut-shaped magnet.

free electrons. The greater the energy, the more likely the nuclei are to fuse. The hot plasma is contained within a powerful magnetic field **(Fig. 13.8)**. In the second approach, laser beams are focused narrowly on a minute pellet containing frozen deuterium and tritium **(Fig. 13.9)**. A rapid increase in pressure and temperature causes the nuclei to fuse. So far, neither method has achieved sustained controlled fusion, and nuclear fusion is unlikely to become a practical source of energy in the near future.

13.4 Solar Energy

Probably the most attractive source of energy is the sun. During daylight hours in sunny locations, huge quantities of **solar energy** reach the earth's surface. This energy comes to us free and is nonpolluting and, for all practical purposes, infinitely renewable. However, it is widely dispersed, and concentrating it and converting it to a usable form are both difficult and costly.

Solar Heating for Homes and Other Buildings

Solar energy can be used very simply to heat buildings and water. A building made of appropriate materials and suitably constructed and oriented can capture the sun's heat. In a typical **passive solar heating system,** solar energy enters through windows facing the sun, and convection currents passively distribute the heat around the building. Some of the heat is stored in rock below the building for release when the sun is not shining.

In an **active solar heating system,** heat gathered in solar collectors located on the roof of a building is circulated by means of pumps **(Fig. 13.10)**. A typical flat-plate collector

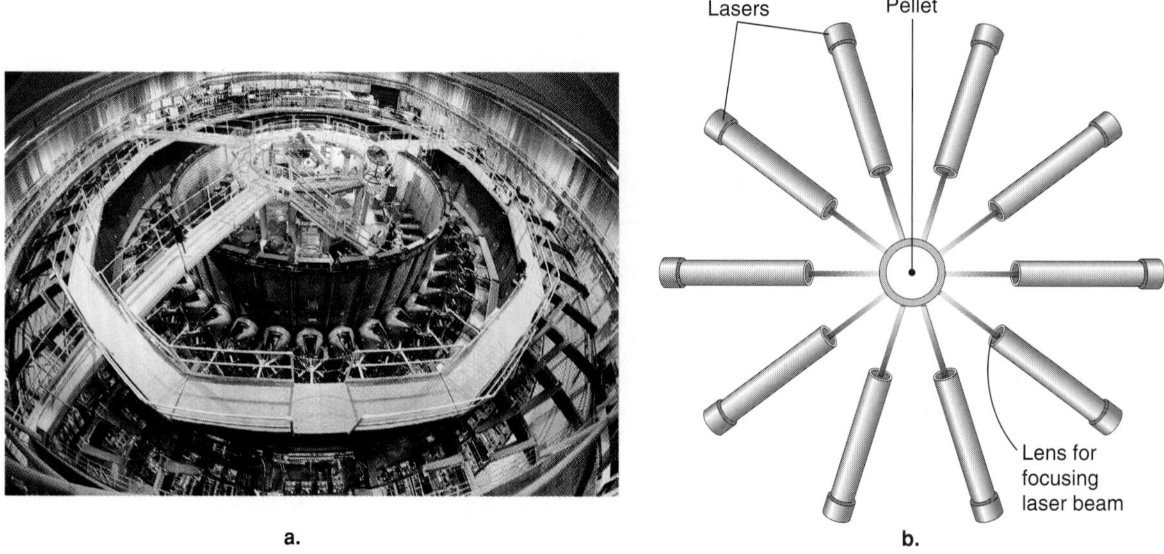

FIGURE 13.9

(a) The experimental laser fusion reactor at Lawrence Livermore Laboratory. (b) Powerful laser beams are focused on a tiny glass pellet containing a mixture of deuterium and tritium; the high pressure and temperature produced cause the hydrogen isotopes to fuse.

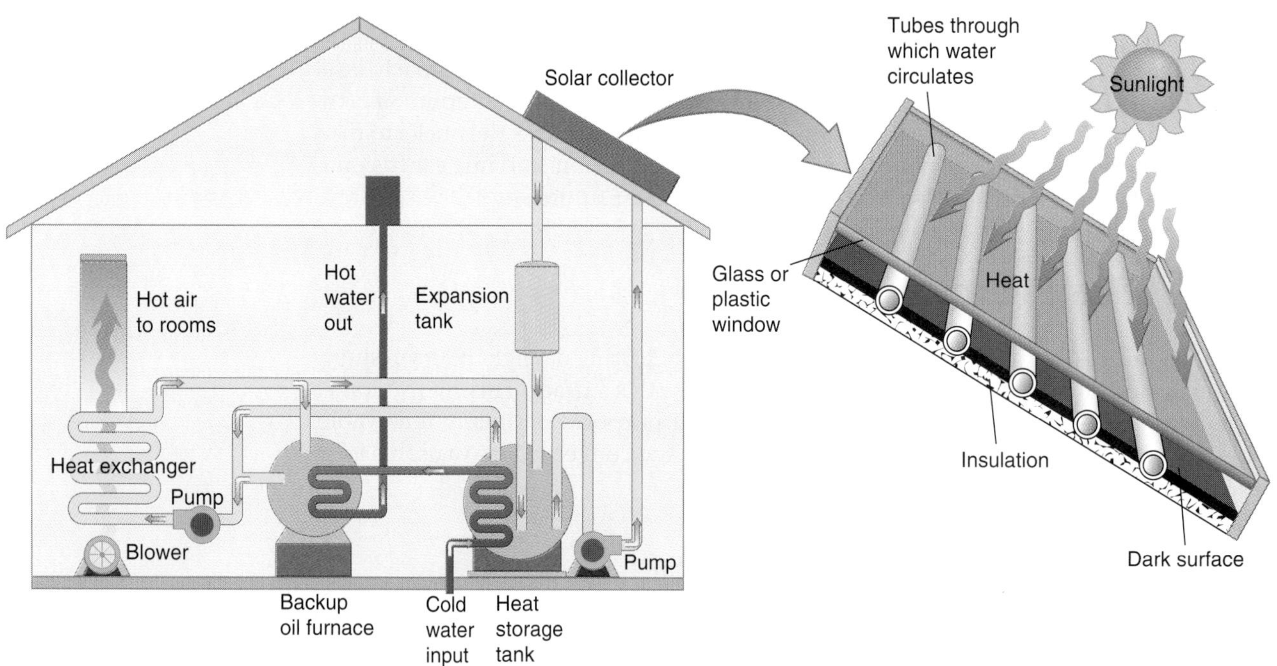

FIGURE 13.10

In an active solar heating system, water circulating through flat-plate collectors on the roof is heated by the sun and conveyed to a storage tank; the heat from the stored hot water is used to provide hot water, as well as hot air for space heating. The system is termed "active" because energy other than solar energy is required to pump the water through the house.

consists of a black surface covered with a glass or plastic plate. As anyone who has walked barefoot on asphalt in summer knows, sunlight is absorbed by a black surface and converted to heat. In a solar collector, the glass plate allows sunlight to enter but traps the heat that is produced. Circulating water is heated as it passes between the glass and the black surface.

Depending on climate and the availability of sunshine, passive and active solar systems can provide from 50% to 100% of home heating requirements. Although initial construction costs are high and a backup heating system may be needed, savings on energy bills are substantial. As oil becomes scarce and solar technology advances, solar heating should become increasingly attractive and affordable.

FIGURE 13.11

In a parabolic trough collector system for generating electricity, sunlight hitting the curved reflector is focused on the pipe running down its center, and oil flowing through the pipe is heated. The heated oil is used to boil water and produce steam to power a generator.

Electricity from Solar Energy

Heat from solar collectors can be used to generate electricity. One of the most successful developments is the parabolic trough collector **(Fig. 13.11)**. The trough is a reflector that focuses sunlight onto a pipe running down its center. Oil or other fluid in the pipe is heated to a temperature up to 400°C (750°F). The heat is used to boil water and produce steam that can be used to turn a turbine and generate electricity. The most up-to-date of these facilities convert about 20% of the sunlight reaching the troughs into electricity. There are several solar trough facilities in the Mojave desert in California, which are expected to increase production in the next few years.

The inventor and scientist Charles Abbott introduced the idea of a solar trough collector for generating electricity in 1930.

Electricity from Photovoltaic Cells

Another approach to converting solar energy directly into electricity is the **photovoltaic cell** (or *solar cell*). Most solar cells consist of two layers of almost pure silicon. The top, very thin layer contains a trace of arsenic; the lower, thicker layer contains a trace of gallium. (The addition of a trace of another element to silicon is known as *doping*.) Recall that silicon has four valence electrons, arsenic has five, and gallium has three (Chapter 4). In a pure silicon crystal, each silicon atom is covalently bonded to four other

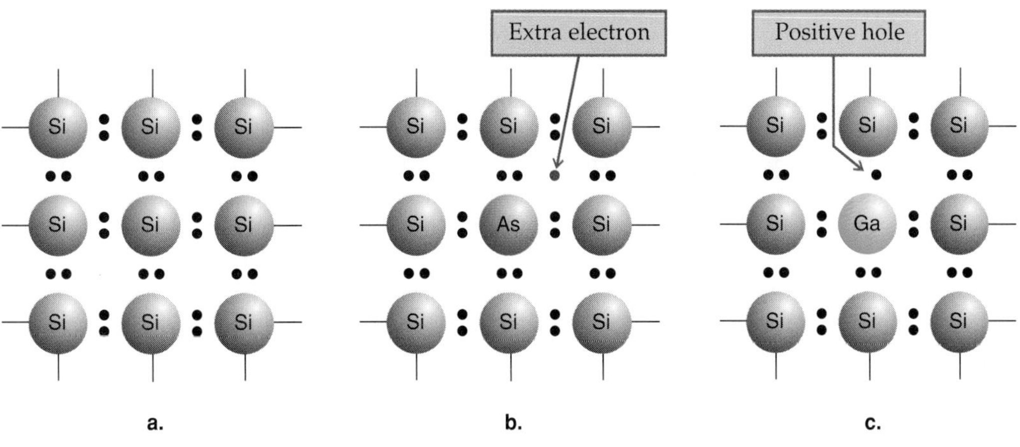

FIGURE 13.12

(a) Pure silicon crystal. (b) Silicon doped with arsenic. (c) Silicon doped with gallium.

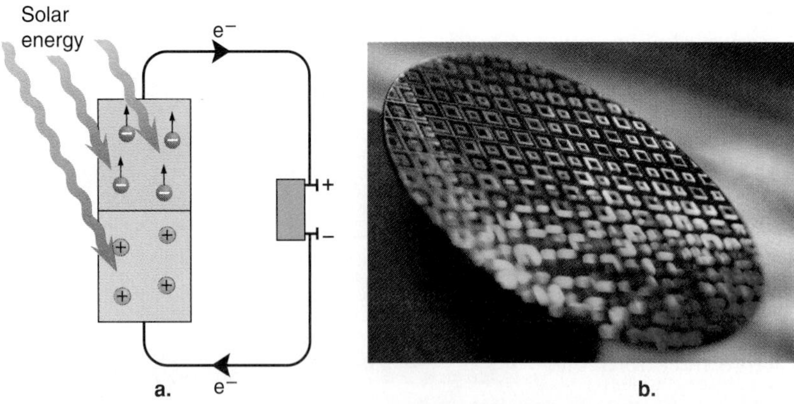

FIGURE 13.13

(a) In a photovoltaic cell, electrons are energized by sunlight and flow through an external circuit. (b) A solar panel. (c) The Sunraycer, an experimental car built by General Motors, is powered by solar cells. Silver-zinc batteries provide energy on cloudy days or if needed on steep hills.

silicon atoms **(Fig. 13.12a)**. When arsenic atoms are included in the silicon structure, four of each arsenic atom's electrons form bonds with silicon atoms. The fifth electron is relatively free to move about **(Fig. 13.12b)**. When gallium atoms are included, there is a shortage of one bonding electron around each gallium atom, and a positive hole is created **(Fig. 13.12c)**. As a result, there is a tendency for free electrons in the arsenic-doped layer to migrate to the gallium-doped layer to fill the holes. Once the holes near the interface between the two layers are filled, the flow of electrons ceases. If the mobile electrons are sufficiently energized by exposure to sunlight and the layers are connected by an external circuit, however, electrons will flow through the external circuit, producing an electric current that can do useful work **(Fig. 13.13a)**.

Solar cells were first used in the 1950s to provide power to space satellites. Today, they are widely used in calculators, watches, and other small devices. In the last 25 years, the cost of producing electricity from solar cells has fallen dramatically. At the same time the efficiency of solar cells has increased from about 10% to more than 20%. As new silicon materials and possible substitutes for silicon such as germanium (Ge), gallium arsenide (GaAs), and cadmium telluride (CdTe) are developed, efficiency is expected to increase further.

Since the 1950s, U.S. and Russian satellites and space vehicles have been powered by solar cells.

If technology continues to advance and prices continue to fall as projected, rooftop arrays of solar cells are expected to provide electricity to many new homes **(Fig. 13.13b)**. Already, experimental automobiles that run on solar energy have been built **(Fig. 13.13c)**. An even more significant breakthrough is the development in California of an electric generating plant powered by banks of solar cells that can provide enough electricity to meet the needs of approximately 10,000 people. Although solar-powered plants are not yet competitive with conventional fossil-fueled plants, they have many advantages: They are relatively inexpensive to construct, produce no pollution, require little maintenance, and have no fuel costs.

Solar cells hold enormous promise for the future, particularly in developing countries such as India, where installing photovoltaic systems to supply electricity to remote, rural communities is much cheaper than extending existing power lines. Lead-acid batteries can be used to store electricity at night.

As numerous aging nuclear-, petroleum-, and coal-fueled power plants in the United States near the end of their useful lives, many are hoping that serious consideration will be given to replacing them with solar-powered plants. Already, Japan and several European countries are investing heavily in solar technology.

13.5 Energy from Biomass

Biomass, which is defined as any accumulation of biological materials, can be used as a source of energy. Examples of biomass are wood and crop residues. Like fossil fuels, biomass can be burned directly to provide energy in the form of heat. It can also be converted to methane (natural gas) and the liquid fuels methanol and ethanol. Burning biomass or products made from it has the disadvantage that it adds to the concentration of carbon dioxide in the atmosphere.

Burning Biomass

Wood, a form of biomass, is the major source of energy for cooking and heating for about 80% of people in the less developed nations of the world. In many areas of these countries, the constant search for firewood has led to excessive deforestation, with resulting erosion and degradation of soil. If trees continue to be cut down faster than they are replaced, severe shortages of firewood are anticipated in the very near future in many places in the world.

Many sawmills obtain most of their power from burning wood waste, and wood-burning stoves are popular in homes. However, it is unlikely that burning wood or wood wastes will ever be more than a very minor source of energy in the United States or other industrialized countries. Cutting timber on a scale large enough to provide a significant amount of energy would have an adverse effect on the environment. Furthermore, wood stoves, unless they include control devices, are a source of both indoor and outdoor air pollution and are often dangerous. For this reason, London and cities in South Korea, among others, have banned wood burning.

A number of cities in the United States burn municipal trash, which is generally about 40% wastepaper, as a source of energy. The heat evolved by a typical modern waste incinerator can generate enough electricity to supply the needs of 50,000 homes (Chapter 19).

Crop residues are another source of energy that can be exploited. In Hawaii, the fibrous residue from sugar cane is burned to produce electricity. Another possibility is raising crops of fast-growing, high-energy-yield plants specifically to be used as fuel. However, these "energy plantations" have several disadvantages. Large areas of land are needed, and because of the energy that must be expended to raise, harvest, and dry the crop, net energy yield is very low.

Production of Biogas

Plants, organic wastes, manure, and other forms of biomass can be used as sources of methane. The process of converting biomass to methane (or other fuels) is called **bioconversion.** In the absence of air, digestion of the organic materials in biomass by anaerobic bacteria produces **biogas,** which is about 60–70% methane. The most suitable starting materials are sewage sludge or manure. In India, as many as 2 million digesters are producing biogas from cow dung. The biogas is used for cooking, lighting, and heating, or it is used to generate electricity. The nitrogen-rich slurry that is produced as a by-product provides an excellent fertilizer. Dung from two cows can supply sufficient fuel to meet the needs for cooking and lighting for the average rural family. China has over 5 million digesters, which use human waste as well as animal dung. Production of biogas is economically feasible only in locations where there is a large concentration of the starting materials.

Methane produced in the anaerobic digestion of organic wastes in landfills is collected at many sites and burned as fuel.

Alcohols from Biomass

The methane in biogas can be converted to methanol as shown in the following equations:

$$\underset{}{CH_4} + \underset{\text{steam}}{H_2O} \longrightarrow CO + 3\ H_2$$

$$CO + 2\ H_2 \longrightarrow \underset{\text{methanol}}{CH_3OH}$$

Natural gas or methane produced by coal gasification (Chapter 12) can be converted to methanol in the same way. Methanol can also be produced by destructive distillation of wood (Chapter 8).

Fermentation of sugars and starches in plants produces ethanol (Chapter 8).

$$C_6H_{12}O_6 \xrightarrow{\text{yeast}} 2\ CO_2 + \underset{\text{ethanol}}{2\ C_2H_5OH}$$

Suitable plant materials for bioconversion to alcohol are sugar cane, sugar beets, cassava, and sorghum, all of which have a high content of sugars and starches. The alcohol can be concentrated by distillation. A major problem with producing alcohol in this way is that large-scale use of land to grow fermentable crops could infringe on land needed to produce food crops.

Both methanol and ethanol have high octane ratings (see Table 12.6) and can be used directly as automobile fuels. Methanol requires major modifications in conventional engine design, but ethanol can be used in today's automobiles with only minor changes. Using sugar cane as the starting material, Brazil has pioneered the development of alcohol as fuel; most of its cars run on ethanol or gasohol, a mixture of gasoline and 10–20% (by volume) ethanol.

In the United States, as a result of the Clean Air Act of 1992, most major metropolitan areas are required to sell oxygenated gasoline during the winter months to reduce carbon monoxide emissions (see Chapter 11). The oxygen content of gasoline can be increased by adding ethyl alcohol, but most U.S. petroleum companies increased it by adding methyl *t*-butyl ether (MTBE), which has a higher octane rating than ethanol. Since the EPA recommended a ban on the use of MTBE because of health considerations, farmers in the Midwest who have grain surpluses are lobbying for legislation that would require that some of the reformulated gasoline contain alcohol produced from agricultural products.

Although it is unlikely that bioconversion will ever become a major source of energy, it is a useful way to supplement other sources. It is particularly valuable as a means of converting plant and animal agricultural wastes and wastepaper in municipal garbage into usable energy.

13.6 Wind Power

Wind power was one of the earliest forms of energy humans harnessed to do useful work. Until the 1930s, rural America relied heavily on windmills for pumping water, grinding corn, and generating electricity. Then, as rural electrification schemes brought cheap power to farming communities, most windmills fell into disuse. Now, windmills, which when used to generate electricity are called *wind turbines*, are making a dramatic comeback.

A modern wind turbine consists of three fiberglass blades mounted on a steel tower. The blades are much stronger and lighter than older, conventional ones, and the whole system is automated and very reliable. The turbines are usually arranged in *wind farms* consisting of several hundred turbines **(Fig. 13.14)**, each with blades measuring about 15 meters (45 feet), and capable of producing 10–15 kilowatts of electricity (1 kW = 0.001 MW). Single units can be used for individual homes and farms, but arrays of units are more efficient. The land between the wind turbines can still be cultivated or used for grazing. California already has three large wind farms, which together can generate sufficient electricity to meet the needs of San Francisco, and wind power is expanding into several other states.

FIGURE 13.14

A wind farm.

For wind power to be cost-effective, winds must blow fairly steadily at about 10 miles per hour (18 kilometers per hour). Many areas in the world meet this criterion, and wind power holds great promise for developing nations, especially for those that lack large reserves of fossil fuels. Today, over 50,000 wind turbines are in operation worldwide. Most of the wind turbines in the United States are located in California, but many other states are planning to invest in them.

In Mongolia, where winds blow steadily, nomadic people carry small wind-driven electric generators to provide electricity for lights and TV sets.

Because of dramatic decreases in the cost of the technology, in many areas electricity can now be generated more cheaply from wind than from coal. It is projected that wind power will supply over 10% of the world's electrical energy by 2050. Unlike the U.S. government, many European governments, including those of Spain and Germany, are investing heavily in research and development of wind technology.

The first electric power generating plant based on steam began operating in Larderello, Italy, in 1904 and has been in continuous use ever since.

Wind is free and abundant, and wind power produces no pollutants and no carbon dioxide. However, wind does have a number of disadvantages: It is intermittent, making it necessary to have a storage system or an alternative source of energy when the wind is not blowing; wind turbines are noisy, and they pose a danger to birds if they are located along migratory routes.

13.7 Geothermal Energy

Geothermal energy is heat energy that is generated deep within the earth's interior by the decay of radioactive elements (Chapter 2). Geothermal energy is abundant, but like solar energy, it is widely dispersed. It becomes accessible only in certain unstable regions of the world, where, as a result of geological activity, magma rises from great depths to near the earth's surface. The hot magma, at temperatures between 900 and 1000°C (1600 and 1800°F), heats rock and groundwater that comes in contact with the rock. Heated groundwater may emerge as a geyser **(Fig. 13.15)** or hot spring, or it may remain as a reservoir below the surface, sealed by a layer of impermeable cap rock.

FIGURE 13.15

Old Faithful geyser in Yellowstone National Park has erupted every 40–80 minutes for the past 100 years.

Reservoirs may contain hot water or steam. Geothermal wells, basically similar to oil wells, can be drilled into the reservoirs to release the hot water or steam. Hot water deposits are the most common and are used primarily to heat buildings. Steam is used to generate electricity. The largest known geothermal field is The Geysers, located 145 kilometers (90 miles) north of San Francisco. Between 1960 and 1988, this field was producing electricity more cheaply than fossil-fueled or nuclear-powered plants (Fig. 13.16). Since then, however, power output has fallen because the underground reservoir is drying up. The largest electric-generating plant based on steam is in New Zealand.

Both water and steam deposits are abundant in Iceland. In Reykjavik, the capital, nearly all the buildings and the numerous greenhouses that produce fruits and vegetables are heated with hot water obtained from geothermal wells. Hot water reservoirs also heat homes in Boise, Idaho, and Klamath Falls, Oregon.

Hot rock can also be exploited to obtain steam. Cold water is injected into wells drilled into the hot rock. The water returns to the surface as steam, which can be used to generate electricity.

The future of geothermal energy is uncertain. Many geothermal fields occur in rugged, inaccessible terrain; many are located in scenic areas, such as Yellowstone National Park, and cannot be developed. Most steam deposits contain hydrogen sulfide, small amounts of which are released during plant operation and pollute the atmosphere. Salts dissolved in the water released with steam from reservoirs can corrode pipes and other fixtures and, if not prevented from reaching streams or rivers, can cause severe ecological damage. Despite these problems, the use of geothermal energy is likely to increase in areas where it can be exploited as oil becomes scarce and more expensive.

FIGURE 13.16

At The Geysers in California, a geothermal system is used to generate electricity.

In the 1980s, The Geysers was generating enough electricity to supply a city of 2 million people.

13.8 Water Power

To produce large quantities of electricity from water, huge dams are built across rivers (Fig. 13.17). As water in the reservoir behind a dam is released, the flow is used to drive turbines that produce electricity. **Hydropower** is a very efficient (80% efficiency) means of producing electricity, and it is essentially renewable and nonpolluting. However, the dams that must be built to produce hydroelectric power create many environmental problems. Water impounded behind a dam may flood scenic stretches of river, valuable cropland, places of historical or geological interest, or people's homes. Downstream from a dam, as flow is adjusted to meet electrical demand, the constantly changing water level alters the natural ecosystem. Silt becomes trapped behind the dam, and the amounts of sediments and nutrients downstream are reduced. Dams disrupt the migration and spawning of fish. For example, although fish ladders have been installed around dams on the Columbia River in Washington state, the ladders have not been effective in preventing drastic reductions in the salmon population.

Worldwide, many new dams are planned, and there is great concern about the environmental impact of two of the largest projects: the Three Gorges Dam on the Yangtze River in China and the James Bay project in Quebec in Canada. In the United States, because most of the best dam sites have already been used and because of environmental concerns, few, if any, new large dams are expected to be constructed. About 8% of the electricity in the United States and about 20% worldwide is supplied by hydropower.

Tides are caused by gravitational attractions among the moon, the sun, and the earth. Their magnitude varies in cycles that depend on the relative positions of the three bodies.

13.9 Tidal Power

The twice daily rise and fall of ocean tides represents an enormous potential source of energy. This tidal power can be exploited by building a dam across the mouth of a bay or inlet. The incoming tide generates electricity as it flows through turbines constructed in the dam. The turbine blades are then reversed so that the outgoing tide also produces electricity. To be practical, the difference in level between low and high tide must be 6 meters (20 feet) or more; this difference ranges from 1 to 10 meters (3 to 30 feet) around the world. Three tidal power plants have been built—one in Canada, one in France, and the other in the former Soviet Union (**Fig. 13.18**).

Tidal power plants alter the normal flow of water and disrupt and disfigure the natural environment. Because of the adverse environmental effects and lack of suitable sites, tidal power is never likely to be more than a minor source of energy in the United States.

FIGURE 13.17

Hoover Dam.

13.10 Energy from Hydrogen Gas

A fuel that could possibly replace oil and natural gas is hydrogen gas. Hydrogen could be transported through pipes like natural gas. It could be used for heating homes and for heating water to produce steam for electric power generation. With minor changes to the carburetor, today's cars could run on hydrogen. Hydrogen burns cleanly, combining with oxygen in the air to produce water vapor, and it releases more energy gram for gram than coal, gasoline, or natural gas (see Table 12.3). Transporting hydrogen through pipelines is cheaper than transmitting electricity over power lines.

One problem is that hydrogen is even more flammable than gasoline, but one advantage is that, because hydrogen is much lighter than air, it dissipates rapidly. Being heavier than air, gasoline vapor tends to accumulate at the site of a leak. Another problem is how to store hydrogen in the vehicle. Tanks that can hold hydrogen gas under very high pressure are available, but they have limited capacity and some danger of rupture. A promising technique that is being investigated is absorption of hydrogen by certain metals to form hydrides, which then readily release the hydrogen as it is needed.

A major problem is that practically no free hydrogen exists on earth. Hydrogen (along with oxygen) can be produced by the electrolysis of water (Chapter 1), but this process requires the input of energy. In fact, because of the second law of thermodynamics (Chapter 2), it requires more energy to break the bonds between hydrogen and oxygen atoms in water molecules than is released when hydrogen is burned. However, it is projected that the decreasing cost of solar cells will make it economical to use solar-generated electricity to produce hydrogen by electrolysis. Seawater could provide an almost inexhaustible source of water.

If solar energy could be used *directly* to decompose water into hydrogen and oxygen, hydrogen could be produced much more economically. In the initial stages of photosynthesis, water is decomposed to hydrogen and oxygen. This photo dissociation has been duplicated with some success in the laboratory by exposing blue-green algae to sunlight. The hydrogen and oxygen produced were separated and collected. The process is still in the early stages of development but has promise for the future.

13.11 Fuel Cells

Batteries and fuel cells depend on oxidation-reduction reactions. Batteries store energy; fuel cells convert one form of energy (chemical) to another form (electrical).

Fuel cells are a more promising way of using hydrogen as an energy source for automobiles. Instead of fueling the vehicle by burning hydrogen gas, electricity generated by the oxidation of hydrogen in a fuel cell is used as a source of power. Fuel cells are lightweight and are well-suited for producing electricity in spacecraft.

Fuel cells depend on an oxidation-reduction reaction (Chapter 7) that converts chemical energy directly into electrical energy. The oxidation of hydrogen by oxygen is the basis of the fuel cells used in space:

$$2\ H_2 \quad + \quad O_2 \quad \longrightarrow \quad 2\ H_2O \quad + \quad \text{energy}$$

The cells consist of an electrolyte solution, usually potassium hydroxide (KOH), and two porous carbon electrodes containing platinum or some other suitable metal. Hydrogen and oxygen are fed continuously to the anode and cathode compartments **(Fig. 13.19)** where the following reactions occur:

At the anode: $2\ H_2 + 4\ OH^- \longrightarrow 4\ H_2O + 4\ e^-$

At the cathode: $O_2 + 2\ H_2O + 4\ e^- \longrightarrow 4\ OH^-$

Overall reaction: $2\ H_2 + O_2 \longrightarrow 2\ H_2O$

The electrodes are connected, and electrons flow from anode to cathode through the external circuit. The water produced may be contaminated with electrolyte but can be purified to make it suitable for drinking.

Fuel cells that convert chemical energy into electrical energy have an efficiency of about 60%. Calculation of the overall efficiency, however, must also take into consideration the energy used in producing the hydrogen and the oxygen. In the future, it may become possible to obtain both gases by the photodissociation of water, as described in Section 13.10.

As new technologies become available, it is expected that electricity from fuel cells will be used to power cars and to provide heat for buildings. In 1999, Daimler-Chrysler and Ford announced plans to introduce hydrogen-fuel-cell cars. Daimler-Chrysler's Necar and Ford's P2000 prototypes run on electricity generated by the reaction between hydrogen and oxygen shown above. Because the only product of the reaction is water vapor, the vehicles emit no pollution. Toyota and General Motors also have plans to market fuel-cell vehicles in the near future.

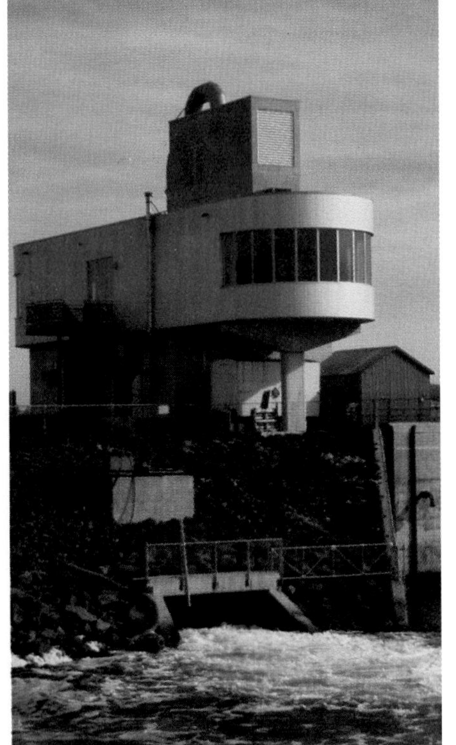

FIGURE 13.18

Tidal power plant in the Bay of Fundy in eastern Canada.

13.12 Clean Cars for the Future

Two factors have been responsible for the development of automobiles that are not powered by a gasoline engine: the eventual exhaustion of oil supplies, and the need to reduce tailpipe emissions. Legislation in California and several other states requires that 10% of all cars sold in the state meet zero emission standards by 2003. In addition to the promising hydrogen-fuel-cell cars just discussed, several other options are being developed. These include electric cars and hybrid cars that run on both gasoline and electricity (see Explorations in Chapter 9).

At the global warming conference in Kyoto, Japan in 1997, Toyota introduced a hybrid car, the Prius, which can achieve 50 mpg and cuts tailpipe emissions by almost 90%. The Prius uses a small gasoline engine and a nickel-metal battery in tandem. Unlike a completely electric car, it needs no recharging because it has its own generator. The Prius and a similar Honda hybrid, the Insight (**Fig. 13.20**), are now on sale in the United States. General Motors and Daimler-Chrysler are also planning to produce hybrids.

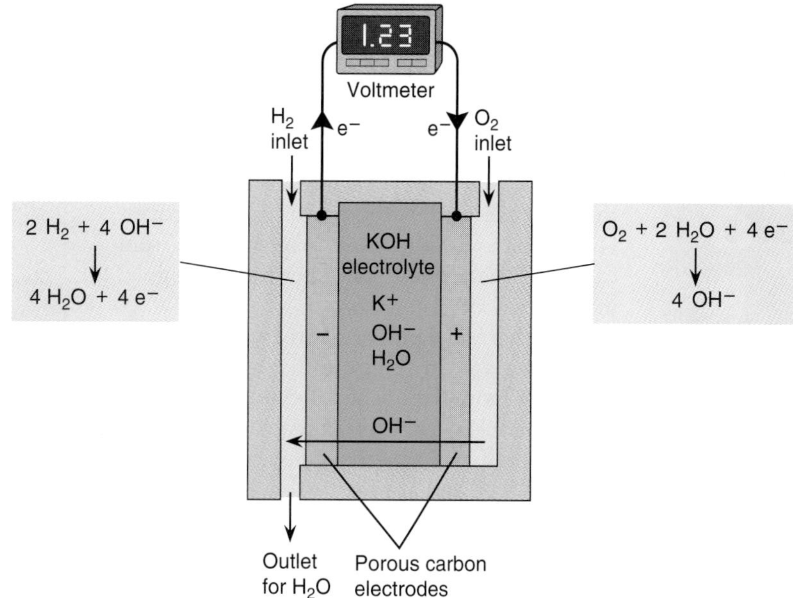

FIGURE 13.19

Cross section of a hydrogen-oxygen fuel cell, which is the primary source of electrical energy on spacecraft.

13.13 Energy Sources for the 21st Century

As oil supplies in the United States dwindle, the nation will become almost entirely dependent on imported oil. By the end of the 21st century, as recoverable oil and natural gas reserves worldwide are used up, the United States and other industrialized nations will be forced to shift to other sources of energy.

Coal is plentiful in the United States, and in the short term, more will be used, primarily to generate electricity. Some oil is expected to be obtained from oil shale, tar sands, and coal liquefication. But all these fossil fuel sources have a high environmental cost.

The best immediate option is energy conservation. Since there is not yet any suitable substitute for liquid fuels, developing cars that are more fuel-efficient to reduce consumption should be a top priority. Many believe that if another energy crisis is to be avoided, new laws must be passed, mandating substantially higher gas mileages than the present requirement of 27.5 mpg.

The most promising source of renewable energy for the future is solar energy. Already solar cells show great promise, and as they become cheaper and more efficient, solar-powered electric generating plants, automobiles, and numerous other products are likely to become common. Fuel cells and solar production of hydrogen are other promising long-term options. In suitable areas, wind power, geothermal energy, and hydroelectric power from small generating plants are expected to make an increasingly large local contribution. The future of nuclear energy is uncertain, but because of fear of accidents and problems with disposal of hazardous wastes, its use will probably decrease.

Without a strong commitment by the federal government to provide incentives for conservation and to fund the research and development of economically and environmentally acceptable new technologies, the transition from an economy based on fossil fuels to one based on alternative energy sources will not be easy. As the following quotation shows, the difficulties inherent in changing the status quo have been understood for centuries:

FIGURE 13.20

Honda's hybrid car, the Insight, is powered by both a gasoline engine and a battery.

> There is nothing more difficult to carry out, nor more doubtful of success, nor more dangerous to handle, than to initiate a new order of things. For the reformer has enemies in all who profit by the old order, and only lukewarm defenders in all those

EXPLORATIONS

The Chernobyl Nuclear Power Plant Disaster

On April 26, 1986, Unit 4 at the Chernobyl nuclear power station in the Ukraine, then part of the Soviet Union, exploded, sending 100 times more radioactive material into the atmosphere than was released by the bombs that destroyed Hiroshima and Nagasaki in World War II. Ironically, it all started with an experiment designed to increase safety.

The Chernobyl power station was gigantic. Four units were in service at the time of the accident; two more were scheduled to begin operating in 1988. The core of each reactor was made up of more than 1500 fuel rods filled with enriched uranium fuel, and approximately 200 boron-carbide control rods. The fuel rods and control rods were packed into a stack of graphite blocks that served as the moderator to slow down fast-moving neutrons to an optimal speed for effective atom splitting.

Causes of the Accident

Nuclear power involves maintaining a delicate balance between keeping the fission reaction going and preventing it from getting out of control. In Unit 4, excess heat in the core was removed by water circulated around the fuel rods; there was an additional water-cooling system to automatically shut down the reactor in case of an emergency. The rate of fission was regulated by the control rods. When the rods were withdrawn from the core, the reaction rate increased; when the rods were fully inserted, the reactor shut down.

Each unit had backup generators to keep water-cooling pumps running in case of a break in the normal flow of electricity from the turbine generators. However, about 40 seconds were required for the backup system to start up. On April 25, when Unit 4 was scheduled to be shut down for routine maintenance, it was decided to run an experiment to determine whether, when power was cut off, the gradually slowing turbine rotors could be made to supply enough energy to keep the pumps running during this critical time gap.

At the start of the experiment, as the reactor's power was slowly lowered, plant operators turned off the automatic emergency water-cooling system. They did this—in violation of safety rules—to avoid a complete shutdown, which would have prevented a repeat of the experiment if one had become necessary. Then, over the next few hours, because of a failure to set an automatic power control, they had difficulty regulating power, until finally a sudden surge of power triggered an uncontrolled chain reaction. The operators had broken a total of six safety regulations; consequently, none of the emergency systems that should have shut down the reactor was operable. The supervisor desperately slammed all the rods into the core, but it was too late. At exactly 1:23 A.M. on April 26, Unit 4 exploded.

The fuel rods, red hot from lack of coolant water, had broken through into the water-steam system causing a colossal explosion. The containment walls around the reactor were blasted apart, and a deadly plume of radioactive material shot into the night sky. A second fiery explosion occurred a few seconds later when hydrogen, produced by steam mixing with hot graphite, reacted violently with oxygen in the air. Red-hot graphite and chunks of burning material started numerous fires at adjacent buildings.

Unit 4 after the accident

The Immediate Aftermath

It was not until 2 days later, after high levels of cesium-134 (a fission product found only in nuclear reactors) had been detected in Sweden during routine monitoring for radioactivity at one of their own plants, that the Soviets reported the accident. In a brief announcement on April 28, they admitted that a reactor at Chernobyl had been "damaged" and stated that "measures were being taken to liquidate the consequences."

The fire in the reactor core proved very difficult to extinguish. Even after it had been smothered by 5000 tons of material dropped onto it from helicopters, heat continued to build up. Meltdown (complete melting of the rods and core) leading to melting of the concrete containment floor and rock beneath it became a possibility. To prevent this ultimate nuclear nightmare, operators drained the basin of cooling water that lay below the reactor—a very dangerous operation because the radioactivity beneath the core was intense—and then 400 miners tunneled under the reactor to install a heat exchanger mounted on a massive concrete base. By the end of June, the entire reactor was encased in 300,000 tons of concrete.

Immediately after the accident, few people in the rural community around the power station or in the nearby town of Pripyat understood the gravity of the situation. They were told to stay indoors with their windows shut, but schools and businesses stayed open, and marriages continued to be performed. The inhabitants were taken completely by surprise when, 36 hours after the accident—by which time radiation levels had soared to 400,000 times normal—an immediate evacuation of everyone within 6 miles of the power station was ordered. Many were reluctant to leave, but all were loaded into a fleet of buses that drove out over roads specially treated to prevent radioactive dust from being kicked up. Military trucks carrying livestock followed. Six days later, the evacuation zone was widened to 18 miles.

The accident at Chernobyl was by far the worst that has ever occurred at a nuclear power plant.

Initially, the radioactive cloud was carried in a northwesterly direction, and fallout was recorded over most of Europe, with the highest levels occurring where there was rainfall. Then, 9 days after the explosion, the wind veered to the south, and radiation levels almost 100 times the normal level were recorded in Kiev, a city of 2.4 million people 60 miles south of Chernobyl. Buildings and streets were hosed down daily; residents were advised to wash their hair every day, clean the soles of their shoes, and eat only food that had passed inspection. Not surprisingly, thousands of people fled the city despite efforts by the authorities to prevent them from leaving.

The Cleanup

In the evacuation zones, the task of cleanup was overwhelming. Tens of thousands of workers—who had to be rotated every few minutes because of high radiation levels—removed topsoil, sprayed trees with decontaminants, demolished contaminated houses, and erected barriers to stop radioactive water from reaching the Dnieper River. Even so, an area greater than 1000 square miles remains uninhabitable for the foreseeable future.

Thirty-one people died in the immediate aftermath of the accident. One man was killed by the explosion, and five died within hours from a combination of severe radiation burns and massive doses of radiation. The other twenty-five victims died in the next few months from radiation exposure. Most of those who died were heroic firefighters, who, well aware of the dangers, gave their lives to prevent fires from spreading to the nearby Unit 3. In 2000, it was estimated that more than 4,000 Ukrainians who participated in the cleanup had died and 70,000 had been disabled by radiation. More than 3 million Ukrainians are believed to have been affected by the accident.

After the investigation into the accident, which revealed reckless operating procedures and repeated violations of safety regulations, the director of the station and two engineers were each sentenced to 10 years hard labor. Six others received sentences of between 2 and 5 years.

The accident at Chernobyl was by far the worst that has ever occurred at a nuclear power plant. The radioactivity released—which represented only about 10% of what might have been released—was several million times greater than that released at Three Mile Island (TMI) in Pennsylvania in 1979. At Chernobyl, long-lived plutonium radioisotopes (which were not released at TMI) will continue to emit radiation for thousands of years. Of particular concern are strontium-90 and cesium-137, both of which have half-lives of approximately 30 years and readily move up through food chains.

Could another Chernobyl occur? The type of reactor at Unit 4 had several design faults including its fatal instability at low operating power. Unlike reactors in the United States and Western European countries, it used combustible graphite instead of water as the moderator; it lacked a strong concrete dome above the reactor; and the automatic computer-controlled emergency systems could be overridden manually. The last of the four units at Chernobyl nuclear power plant was closed down in 2000.

Even with the most up-to-date technology and every conceivable safety device, machinery can fail and human beings can make mistakes. As Chernobyl has shown, a nuclear accident, no matter where it occurs, can have disastrous and widespread consequences.

References: V. Haynes and M. Bojcun, *The Chernobyl Disaster: The True Story—An Unanswerable Indictment of Nuclear Power* (London: Hogarth Press, 1988); R. P. Gale and T. Hauser, *Final Warning: The Legacy of Chernobyl* (New York: Warner Books, 1988).

who would profit by the new order. This lukewarmness arises partly from fear of their adversaries, who have the law in their favor, and partly from the incredulity of mankind, who do not truly believe in anything new until they have had actual experience of it.

—Machiavelli, *The Prince,* 1517

Chapter Summary

1. Fossil fuels, which supply 80% of the world's energy, constitute a finite, nonrenewable resource that is being depleted rapidly.
2. The United States, which has 5% of the world's population, consumes 30% of the world's energy.
3. The uranium fuel in a nuclear reactor is used to heat water and convert it to steam. The steam turns the blades of a turbine, which generates electricity.
4. The core of a nuclear reactor consists of fuel rods, which contain fissionable uranium-235, and control rods, which absorb neutrons and control the rate of fission.
5. Nuclear breeder reactors produce fissionable plutonium-239 from nonfissionable uranium-238. No breeder reactors are in operation in the United States.
6. Nuclear fusion—the combining of two light nuclei to produce a heavier nucleus—releases an enormous amount of energy. Nuclear fusion is the primary source of the energy emitted by the sun but is not yet available for commercial use.
7. Solar energy can be used to heat homes, using either passive or active solar heating systems.
8. Photovoltaic cells (solar cells) consist of pairs of silicon wafers, one doped with arsenic and the other doped with gallium. These cells convert solar energy directly into electricity and are used in watches and other small devices and, on a larger scale, in communication satellites.
9. Biomass, such as crop residues and plants grown especially for use as fuel, can be burned to provide energy in the form of heat. Biomass can also be converted to methane (biogas) and the liquid fuels methanol and ethanol.
10. Where winds blow steadily, wind farms consisting of several hundred wind turbines are used to produce electricity.
11. Geothermal energy is heat generated deep within the earth by the decay of radioactive elements. It is used to heat buildings and generate electricity.
12. Hydropower is an efficient way to generate electricity, but the dams required to provide it create many environmental problems.
13. Tidal power can be used to generate electricity in regions where the difference in level between high and low tide is large.
14. Hydrogen gas has potential as a fuel to replace oil and natural gas, but obtaining an inexpensive source of hydrogen remains a problem.
15. Fuel cells use an oxidation-reduction reaction to convert chemical energy directly into electrical energy.

16. The most promising source of energy for the future is solar energy, which is free, nonpolluting, and infinitely renewable.
17. In suitable areas, wind power, geothermal energy, and electric power from small generating plants are expected to make an increasing contribution to energy supplies.
18. Stringent emission standards and diminishing oil supplies are forcing automakers to develop hybrid cars (which use both a gasoline engine and a battery), all-electric cars, and cars powered by hydrogen gas and by fuel cells.

Key Terms

active solar heating system (p. 395)
bioconversion (p. 399)
biogas (p. 399)
biomass (p. 399)
control rods (p. 390)
fuel cell (p. 403)
fuel rods (p. 390)
geothermal energy (p. 401)
hydropower (p. 402)
liquefied natural gas (LNG) (p. 388)
moderator (p. 391)
nuclear breeder reactor (p. 393)
nuclear fusion (p. 393)
passive solar heating system (p. 395)
photovoltaic cell (p. 397)
solar energy (p. 395)

Questions and Problems

1. What percentage of the total energy needs of the United States is supplied by petroleum? In approximately what year are U.S. petroleum reserves expected to be used up?
2. What events caused the U.S. oil shortage in 1974? In 1990?
3. What will cause oil shortages in the United States in the future?
4. What steps could be taken quickly to conserve a substantial amount of petroleum?
5. What is the most abundant fossil fuel in the world?
6. Give two reasons why coal is a poor choice to replace petroleum as the most commonly used source of energy.
7. It is estimated that air pollutants from coal-fired electric power plants cause more than 10,000 deaths a year. Carbon dioxide emissions from these plants are measured in millions of tons and may be contributing to global warming. In your opinion, should U.S. energy strategy favor increased use of coal-burning power plants? What are the alternatives?
8. How long are the natural gas reserves in the United States expected to last?
9. What are the disadvantages of liquefied natural gas (LNG)?
10. Which countries place a high reliance on nuclear power?
11. Draw a schematic diagram of a pressurized water reactor and label all parts.
12. In your opinion, should the United States commission more nuclear fission reactors to meet electrical demand in the future?
13. What percentage of uranium-235 must the fuel pellets used in a nuclear power plant contain?
14. When 3-year-old nuclear fuel rods are ready to be replaced, what fissionable material do they contain that they did not contain originally?
15. The licensing time for new nuclear power plants in the United States should be cut in half so that costs would be reduced. Do you agree with this statement?
16. Can nuclear bombs be made from uranium used in nuclear reactors? Explain.

17. Describe the nuclear fuel cycle for the production of uranium fuel for nuclear reactors.

18. How often do individual fuel rods in a nuclear reactor need to be replaced? What happens to spent fuel rods?

19. For how many years can a nuclear power plant be expected to generate electricity before it reaches the end of its useful life?

20. More nuclear power plants should be built in the United States to reduce dependence on imported petroleum and to slow global warming. Do you agree with this statement?

21. How does a nuclear breeder reactor produce more fuel than it uses?

22. List two reasons why a nuclear breeder reactor may be more dangerous than a nuclear fission reactor.

23. The U.S. government should develop the nuclear breeder reactor to conserve uranium resources and avoid becoming dependent on other countries for uranium. Do you agree? Why?

24. Describe the function of the following in a nuclear power plant:
    a. control rods b. fuel rods
    c. moderator d. containment building

25. What roles do deuterium and tritium play in nuclear fusion?

26. In a fusion reaction, a plasma (called "the fourth state of matter") is formed. Describe the composition of a plasma.

27. How much more energy can be obtained from a pound of fuel using nuclear fusion than using nuclear fission?

28. How are the very high temperatures necessary for a fusion reaction going to be contained?

29. What are some of the problems associated with the use of solar energy?

30. What are the differences between active and passive solar heating systems?

31. Draw a schematic diagram of a photovoltaic cell.

32. What chemical elements are used to dope silicon for use in photovoltaic cells? Which element creates an excess of electrons? Which element creates an excess of positive charges?

33. Should the U.S. government increase funding for the rapid development of photovoltaic cells? Can the U.S. generate most of the electricity it needs through direct solar energy?

34. In a photovoltaic cell, electrons from the portion of the cell containing __________ move to the portion of the cell containing ________.

35. Describe what is meant by *biomass*.

36. Biomass is the major source of energy for cooking and heating in many less developed countries. Describe just what type of "biomass" is being described.

37. Write chemical equations to show how each of the following is produced from biomass:
    a. methyl alcohol
    b. ethyl alcohol
    c. methane

38. List the advantages and disadvantages of using biomass as a source of energy.

39. List two disadvantages of wind power.

40. Will geothermal energy ever be a major source of energy worldwide? Explain.

41. Is the following statement true? "Electricity provides clean heat." Explain.

42. Rank the following energy sources in order according to what you think will be their importance in meeting U.S. energy needs in the year 2025. Briefly explain your ordering.
    a. nuclear fusion b. solar energy
    c. coal d. petroleum

43. List two advantages that fuel cells have over conventional batteries.

44. Will tidal power ever be a major source of energy in the United States? Explain.

45. A mandatory nationwide energy conservation program should be adopted by the U.S. government. Explain briefly why you agree, or disagree, with this statement.

Chapter 14

Biochemistry

The Molecules of Life

Chapter Objectives

In this chapter, you should gain an understanding of:

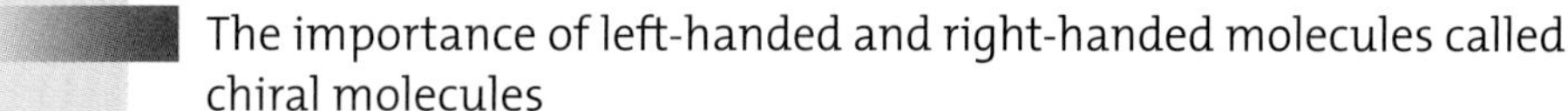

- The importance of left-handed and right-handed molecules called chiral molecules
- The classification and molecular structure of important carbohydrates
- The properties of lipids, saturated and unsaturated fatty acids, triglycerides, and steroids
- The vital importance of proteins and the amino acids from which they are made
- The structures of deoxyribonucleic acid (DNA) and ribonucleic acid (RNA), their role in the storage and transmission of genetic information, and the applications of genetic engineering

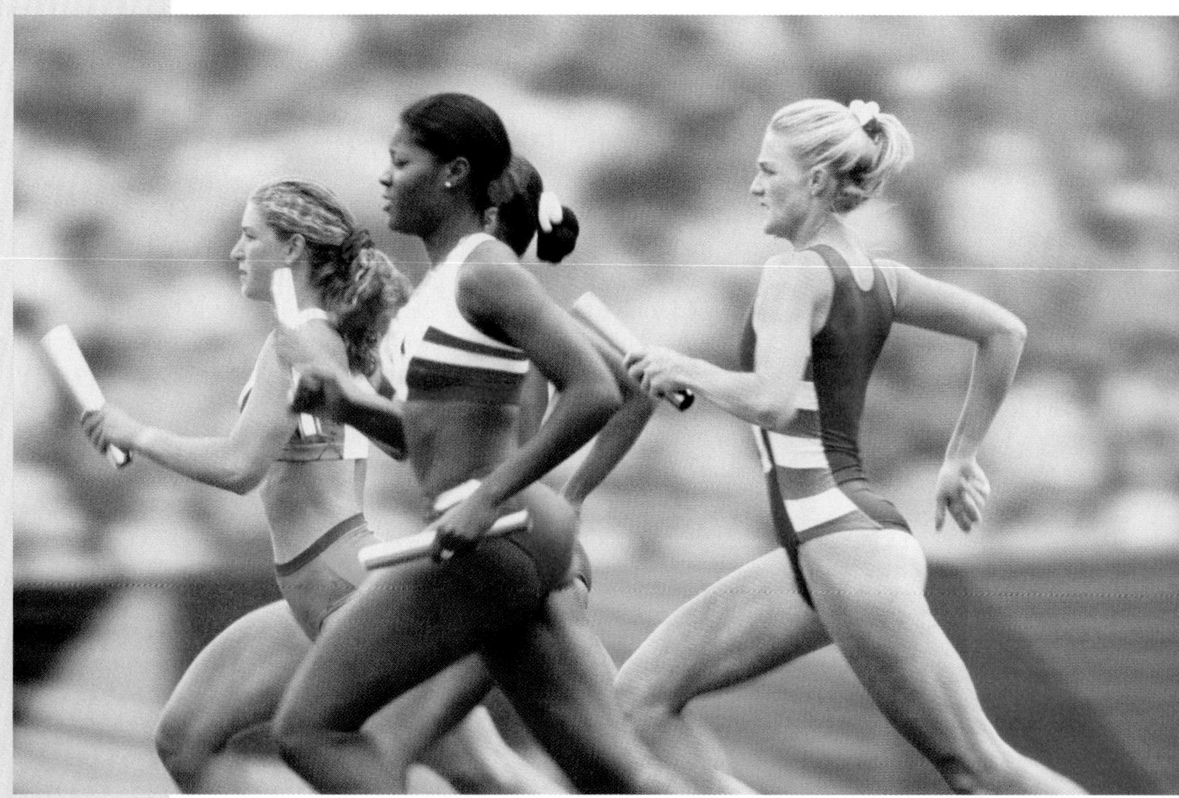

Biochemicals—carbohydrates, fats, proteins, and nucleic acids—are essential ingredients of life. They make up the structural materials of the human body and provide the energy being used by these women athletes.

ONLINE FEATURES

www.jbpub.com/chemistry

- Chemistry and the Environment
- eLearning
- Chemistry in the News
- Special Topics
- Research & Reference Links
- Ask the Authors

BIOCHEMISTRY DEALS WITH THE SUBSTANCES found in living organisms and the changes these substances undergo during life processes. Living things come in an amazing variety of forms—from bacteria and plants to fish and mammals—but all of these different organisms are composed of the same basic organic components: carbohydrates, fats, proteins, and nucleic acids. In addition to these four biochemicals, living organisms contain water, various minerals and salts, and a number of miscellaneous organic substances, including vitamins.

The four major groups of biochemicals provide a living organism with energy and with the components for the complex molecules needed for life processes. They also provide the biological catalysts that control the chemical reactions occurring within the organism and the molecules responsible for the transmission of genetic information from one generation to the next.

Until the early part of the 19th century, it was generally believed that living organisms contained a "vital force" that enabled them to synthesize the biochemicals found only in living matter and that it would be impossible to synthesize these biochemicals in the laboratory in the absence of this force. This belief was shaken in 1828 when the German chemist Friedrich Wohler synthesized urea—an organic compound (Chapter 8) found in the urine of most mammals—from inorganic starting materials. Since then, scientists have synthesized numerous biological compounds, including carbohydrates, lipids, proteins, hormones, and vitamins. We still may not understand the secret of life, but we do know that a biochemical synthesized in the laboratory has exactly the same structure as the naturally occurring biochemical and functions in the same way when introduced into an organism.

In this chapter, we study the chemical structures and functions of the four important classes of biological compounds: carbohydrates, lipids, proteins, and nucleic acids. First, however, we consider a structural feature that is common to many biological molecules—handedness.

14.1 Right-Handed and Left-Handed Molecules

The part of chemistry that is concerned with the structure of molecules in three dimensions is called *stereochemistry* (from the Greek word *stereos*, meaning "solid").

Many organic molecules, including many important biochemicals, exist in two forms that are related to each other in the same way that a right-hand glove is related to a left-hand glove. Two gloves in a pair are alike in every respect, but they cannot be superimposed, one upon the other. You cannot fit a right-hand glove on your left hand, or vice versa. The two gloves—as well as your two hands—are mirror images of each other. The mirror reflection of a left hand looks like a right hand **(Fig. 14.1)**.

Molecules that exist in these two related forms have the same chemical formula, and their atoms are joined in the same order; *they differ only in the three-dimensional spatial arrangement of their atoms.*

FIGURE 14.1

The mirror image of a left hand is a right hand.

Chiral Molecules

A molecule consisting of a tetrahedral carbon atom attached to four *different* atoms or groups of atoms can exist in two forms that are mirror images of each other. Two such molecules are represented by the ball-and-stick models in **Figure 14.2a**. The four different-colored balls represent four different atoms, or groups of atoms, attached to each carbon atom. The only difference between the model on the left and the one on the right is that the red and green balls have been switched. As you can see in the figure, the two models are mirror images of each other.

The two molecular models are not identical because no matter how they are turned, they cannot be superimposed, one on the other **(Fig. 14.2b)**. The two models represent two distinct molecules. They differ from each other in much the same way as a left hand does from a right hand. Because of this *handedness*, these molecules are called **chiral molecules** (from the Greek word *cheir*, meaning "hand"). A carbon atom to which four different atoms or groups are attached is called a chiral carbon atom.

Thousands of chiral molecules exist. Two simple examples, lactic acid and 2-butanol, are shown below.

mirror

$$\begin{array}{c} COOH \\ | \\ H-C-OH \\ | \\ CH_3 \end{array} \quad \Bigg| \quad \begin{array}{c} HOOC \\ | \\ HO-C-H \\ | \\ CH_3 \end{array}$$

lactic acid

mirror

$$\begin{array}{c} H \\ | \\ CH_3-C-OH \\ | \\ CH_2 \\ | \\ CH_3 \end{array} \quad \Bigg| \quad \begin{array}{c} H \\ | \\ HO-C-CH_3 \\ | \\ CH_2 \\ | \\ CH_3 \end{array}$$

2-butanol

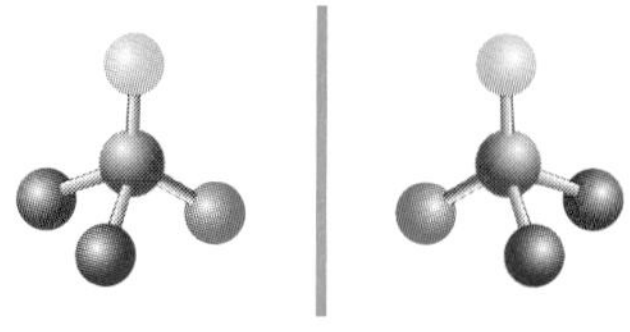

Mirror

a.

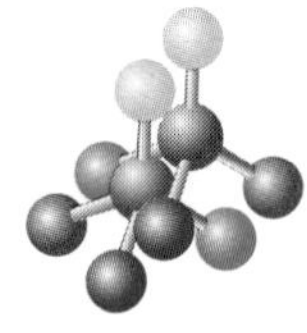

b.

FIGURE 14.2

The two ball-and-stick models in (a) represent simple molecules, each containing a chiral carbon atom. The models are identical, except that the groups represented by the red and green balls have switched positions. The models are related in the same way as an object and its mirror image. (b) The models cannot be superimposed on one another.

A chiral carbon has four different atoms or groups attached to it.

EXAMPLE 14.1 Drawing mirror-image forms of chiral molecules

The molecule 2-chlorobutane ($CH_3CHClCH_2CH_3$) is chiral. Draw the two mirror-image forms of this molecule.

Solution

1. Draw the condensed formula of butane, and then add the chlorine atom at carbon 2:

$$CH_3-CH_2-CH_2-CH_3$$

butane

$$\begin{array}{c} \overset{1}{CH_3}-\overset{2}{CH}-\overset{3}{CH_2}-\overset{4}{CH_3} \\ \quad\;\; | \qquad\qquad\quad \\ \quad\;\; Cl \qquad\qquad\quad \end{array}$$

2. Identify the chiral carbon: Carbon 2 is chiral because four different atoms or groups are attached to it (methyl group, ethyl group, chlorine, and hydrogen).

3. Draw a structural formula of the first molecule with carbon 2 in the center:

$$\begin{array}{c} CH_3 \\ | \\ H-C-Cl \\ | \\ C_2H_5 \end{array}$$

4. Draw the second molecule as a mirror image of the first:

$$\begin{array}{c} CH_3 \\ | \\ H-C-Cl \\ | \\ C_2H_5 \end{array} \quad \Bigg| \quad \begin{array}{c} CH_3 \\ | \\ Cl-C-H \\ | \\ C_2H_5 \end{array}$$

Practice Exercise 14.1

The molecule 2-chloro-2-bromobutane is chiral. Draw the two forms in which it can exist. (The formula of butane is given in Example 14.1.)

Chiral Molecules and Life Processes

Nearly all organic molecules in living organisms and in most of the foods we eat are chiral molecules, with the property of handedness. In nature, chiral molecules exist almost exclusively in one form or the other, not both. For example, naturally occurring carbohydrates and amino acids exist in just one of their two possible chiral forms. How and why this happened as life developed on earth is one of the mysteries of nature. As a result of this natural distinction, living organisms are capable of metabolizing, and making use of, only those molecules that are in the naturally occurring form. If a person were fed a synthetic diet composed of carbohydrates with the wrong handedness and proteins made from amino acids with the wrong handedness, that person would eventually die of starvation.

14.2 Carbohydrates

Carbohydrates are biological compounds of the elements carbon, hydrogen, and oxygen. Examples include glucose, sucrose (table sugar), starch, and cellulose. Carbohydrates are vitally important to both plants and animals. Plants are composed almost entirely of carbohydrates and water. Typically, two-thirds of the human diet is made up of carbohydrates. Rice, flour, and corn meal, which are the basic foods of most people in the world, are composed of carbohydrates.

Almost all carbohydrates are produced by photosynthesis in green plants. As we saw earlier (Chapters 2 and 6), green plants use energy from the sun to convert carbon dioxide from the air and water from the soil into simple carbohydrates, mainly glucose. Although photosynthesis is actually a series of complex steps, the overall reaction can be written as follows:

$$6\ CO_2 \quad + \quad 6\ H_2O \xrightarrow{\text{sunlight}} \underset{\text{glucose}}{C_6H_{12}O_6} \quad + \quad 6\ O_2$$

In plants, glucose is converted to cellulose and starch. In animals and humans, the oxidation of plant carbohydrates eaten as food provides both energy and the carbon, hydrogen, and oxygen atoms needed to synthesize fats, proteins, and other biological compounds.

As their name implies, carbohydrates were once regarded as hydrates of carbon—that is, compounds in which carbon is bonded to water molecules. The formula for glucose, $C_6H_{12}O_6$, for example, can be written as $C_6(H_2O)_6$, and that for sucrose, $C_{12}H_{22}O_{11}$, can be written as $C_{12}(H_2O)_{11}$. We now know that the three elements that make up all carbohydrates are arranged primarily as hydroxyl (—OH), and aldehyde $\left(-\overset{O}{\overset{\|}{C}}-H\right)$ or ketone $\left(-\overset{O}{\overset{\|}{C}}-\right)$ groups (Chapter 8).

Naturally occurring carbohydrates can be divided into three main classes: monosaccharides, disaccharides, and polysaccharides. **Monosaccharides,** or simple sugars, cannot be broken down into smaller carbohydrate units. **Disaccharides** are larger carbohydrates made up of two monosaccharide units bonded together. **Polysaccharides** are giant molecules, made up of hundreds or even thousands of monosaccharide units linked together somewhat like identical railroad cars coupled to form long freight trains. In biologically important polysaccharides, the monosaccharide units are glucose.

The term *saccharide* is derived from the Latin word *saccharum*, meaning "sugar."

14.3 Monosaccharides

Important monosaccharides include glucose, galactose, fructose, ribose, and deoxyribose. Glucose, galactose, and fructose all have the same molecular formula: $C_6H_{12}O_6$. Because they have six carbon atoms, they are known as *hexoses*. Their straight-chain structural formulas are shown in **Figure 14.3a**. Ribose ($C_5H_{10}O_5$) and deoxyribose ($C_5H_{10}O_4$) are five-carbon carbohydrates, known as *pentoses*. Ribose is a component of ribonucleic acid (RNA), and deoxyribose is a component of deoxyribonucleic acid (DNA); these key biochemicals will be considered later in this chapter in the section on nucleic acids.

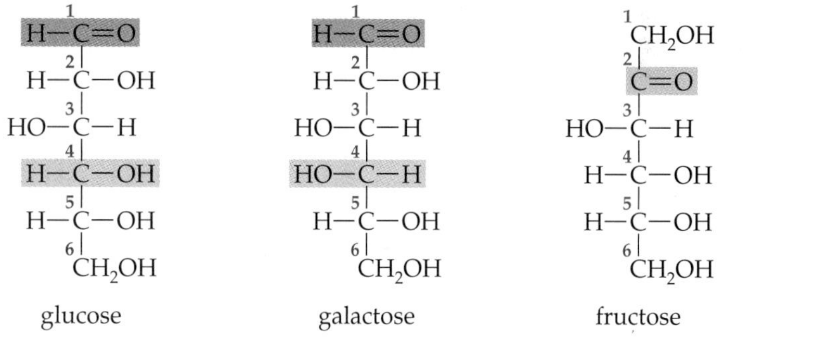

Molecular Structures of Monosaccharides

The formulas of fructose and glucose are identical from carbons 3 through 6, but fructose differs from both glucose and galactose in containing a ketone group $\left(-\overset{O}{\overset{\|}{C}}-\right)$ instead of an aldehyde group $\left(-\overset{O}{\overset{\|}{C}}-H\right)$.

The straight-chain structural formulas for glucose, galactose, and fructose in Figure 14.3a do not adequately represent their

glucose galactose fructose

a. Straight-chain forms

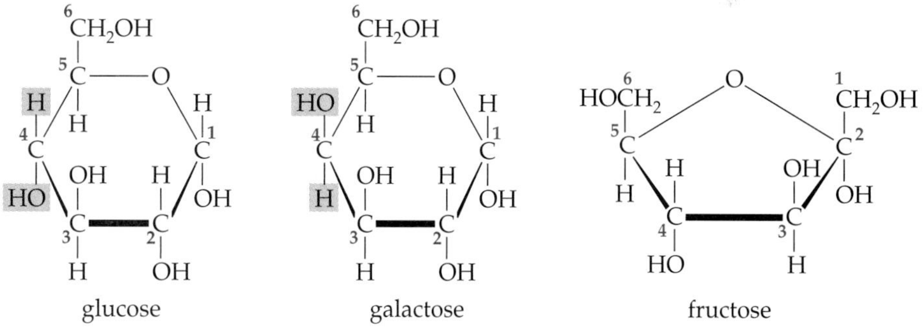

glucose galactose fructose

b. Cyclic forms

FIGURE 14.3

(a) The straight-chain structures of glucose, galactose, and fructose. Glucose and galactose have an aldehyde group at carbon 1; fructose has a ketone group at carbon 2. (b) The cyclic structures of glucose and galactose (six-membered rings) and fructose (five-membered ring). Note that the structures of glucose and galactose—both straight-chain and cyclic—differ only in the orientation of the —H and —OH on carbon 4.

actual molecular structures. In fact, the sugars exist mainly as cyclic structures **(Fig. 14.3b)**. Reactions between the aldehyde group at carbon 1 and the —OH group at carbon 5 in the straight-chain form of glucose or galactose result in a bond forming through the oxygen atom of the —OH group on carbon 5, and the formation of a six-membered ring. In fructose, a five-membered ring is formed by a linkage between carbons 2 and 5.

Some Important Monosaccharides

Glucose (also known as *dextrose, grape sugar,* or *blood sugar*) is by far the most biologically important monosaccharide. It is a moderately sweet sugar that occurs naturally in honey, grapes, dates, and other fruits. Because glucose contains numerous —OH groups that can form hydrogen bonds with water molecules, it is very soluble in water.

In the human body, glucose is formed when the disaccharides and polysaccharides in foods are broken down during the digestive process (Chapter 15). Glucose enters the bloodstream and is carried to all the tissues of the body. By means of a complex series of steps, glucose in body cells is oxidized to carbon dioxide and water, and energy is released:

$$C_6H_{12}O_6 \quad + \quad 6\ O_2 \quad \longrightarrow \quad 6\ CO_2 \quad + \quad 6\ H_2O \quad + \quad \text{energy}$$

This reaction is the reverse of the overall reaction that occurs in photosynthesis.

Glucose is the only carbohydrate that is used directly by the body as a source of quick energy. For this reason, patients in hospitals who are unable to take food by mouth are given a glucose (dextrose) solution intravenously.

Galactose is a constituent of many plant gums and pectins. It is also a component of the disaccharide lactose, which is sometimes called *milk sugar.* Although galactose differs from glucose only in the orientation of the —OH group on carbon 4 (see Fig. 14.3), galactose cannot be used by the body unless that group's orientation is changed so that the molecule becomes glucose.

Fructose is the sweetest of all the naturally occurring sugars. It is found in honey and many fruits and is a component of the disaccharide sucrose (common table sugar).

14.4 Disaccharides

Disaccharides consist of two monosaccharides joined together by the elimination of a molecule of water. The molecular formula of disaccharides is

$$C_{12}H_{22}O_{11} \qquad (C_6H_{12}O_6 + C_6H_{12}O_6 - H_2O)$$

The three most common disaccharides are

1. sucrose (one glucose unit and one fructose unit)
2. lactose (one glucose unit and one galactose unit)
3. maltose (two glucose units)

Sucrose occurs naturally in sugar cane and sugar beets. Over 80 million tons of pure sucrose are produced annually from these sources. On average, in the United States, sucrose provides more than one-fourth of the calories consumed daily. Because it contains numerous hydroxyl groups, which form hydrogen bonds with water molecules, sucrose dissolves readily in tea, coffee, and other aqueous solutions. During digestion (Chapter 15), enzyme action converts sucrose to glucose and fructose:

sucrose ⟶ glucose + fructose

Lactose is the major sugar in milk. In the mammary glands, glucose obtained from the blood supply is converted by enzyme action into lactose. When milk is drunk, enzyme action in the digestive system converts the lactose to glucose and galactose:

lactose ⟶ glucose + galactose

Further enzyme action changes the galactose to glucose.

Maltose, which is present in germinating grains, is found in beer, malted milk, and corn syrup. Maltose is important primarily as the main product of the digestion of starch. During digestion, maltose breaks down to yield two molecules of glucose:

maltose ⟶ glucose + glucose

EXPLORATIONS discusses the history and chemistry of beer making (see pp. 446–447).

14.5 Polysaccharides

The three most important polysaccharides found in nature are starch, glycogen, and cellulose. When completely broken down into monosaccharide units, all three yield glucose as the only product. Unlike monosaccharides, and most disaccharides, polysaccharides are not sweet.

Starch and Glycogen

Starch is a polysaccharide stored in plants as a food reserve and source of energy. Starch in potatoes, rice, wheat, and other cereal grains accounts for about two-thirds of the average human diet. During digestion in the body, starch is gradually broken down by enzyme action to yield glucose. When the starch you eat provides more glucose than your body requires for its immediate needs, some of the excess starch is converted to **glycogen** and stored in the liver and muscles. Large excesses of starch are converted to fats and stored in special fatty tissue called *adipose tissue.* Glycogen is the storage polysaccharide for animals in the same way as starch is the storage polysaccharide for plants. When needed by the body, glycogen is converted back to glucose.

Starch and glycogen are polymers (Chapter 9) of glucose. Starch is actually a mixture of two types of polysaccharides: amylose (20–30%) and amylopectin (70–80%). Amylose consists of about 200 glucose units joined together to form a straight chain **(Fig. 14.4a)**. A typical molecule of amylopectin consists of about 1000 glucose units, joined to form branched chains. Branching occurs at approximately every 25th glucose unit **(Fig. 14.4b)**. Structurally, glycogen is similar to amylopectin except that its chains are more highly branched, with a branch occurring at approximately every 10th glucose unit.

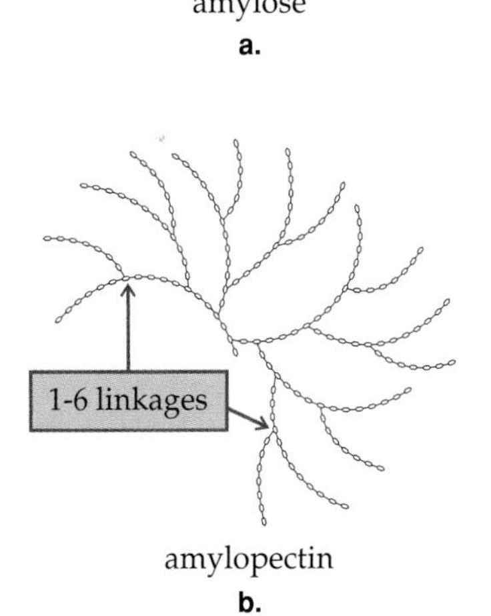

FIGURE 14.4

Starch is a mixture of amylose and amylopectin molecules. (a) In amylose, glucose units are joined in long straight chains through 1–4 linkages. (b) In amylopectin, the chains are branched; branching occurs by linkage through the —OH on carbon 1 of one glucose molecule and the —OH on carbon 6 of another (a 1–6 linkage).

Cellulose

Cellulose forms the structural component of the cell walls of plants. It is by far the most abundant polysaccharide in the natural world. Approximately 50% of the mass of a tree is cellulose; cotton is almost 100% cellulose. Cellulose provides us with many manufactured products, including paper and paper products, rayon, linen, and cellulose nitrate (a constituent of nail polish).

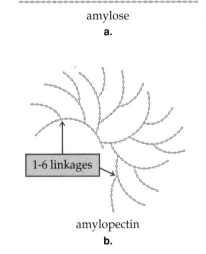

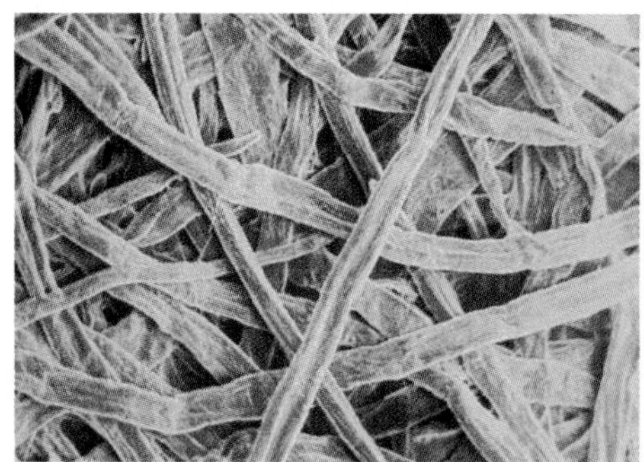

FIGURE 14.5

This electromicrograph of cellulose shows the cross-hatched pattern of fibers that gives cellulose strength in all directions.

Cellulose is composed of long unbranched chains containing from 100 to 10,000 glucose molecules. Groups of these chains, held together by hydrogen bonding between —OH groups on adjacent chains, are twisted into ropelike structures that make cellulose tough and fibrous **(Fig. 14.5)**. The absorbent properties of cotton and paper towels are due to capillary action and the formation of hydrogen bonds between water molecules and —OH groups on cellulose chains.

Cellulose is the main food source for horses, cows, sheep, other herbivores, and termites. In the intestinal tracts of these animals, microorganisms produce enzymes that break down cellulose into glucose units.

The main difference between starch and cellulose, as shown in **Figure 14.6**, is the way the glucose units are joined together. In starch, the linkages joining adjacent glucose units through an oxygen atom are angled; in cellulose, these linkages are not angled. The linkages in starch are termed α-linkages, and those in cellulose are called β-linkages. The result of this apparently minor difference in structure is that humans and carnivores cannot use cellulose as a food source because they lack the enzymes needed to break the β-linkages between glucose units. Although humans cannot digest the cellulose in celery, lettuce, and other green vegetables, these products provide fiber that is needed to facilitate the excretion of solid wastes (Chapter 15).

Capillary action refers to the way a liquid rises in a narrow-bore tube or spreads through the narrow spaces between the fibers of a paper towel.

14.6 Lipids

The term *lipid* is derived from the Greek word *lipos*, meaning "fat" or "lard."

Lipids, like carbohydrates, are biological compounds of carbon, hydrogen, and oxygen. Unlike carbohydrates (and proteins), however, lipids cannot be defined on the basis of their chemical structure. They are generally defined as the greasy substances in plants and animals that are insoluble in water but soluble in nonpolar solvents such as benzene, chloroform, and ether. Lipids include a large group of compounds of widely differing chemical composition, including fatty acids, triglycerides (or neutral fats), waxes, terpenes, steroids, and fat-soluble vitamins.

This chapter's discussion of lipids is limited to fatty acids, triglycerides, and steroids. Vitamins will be considered in Chapter 15.

14.7 Fatty Acids and Triglycerides

Fatty Acids

Fatty acids are long-chain carboxylic acids (compounds we encountered in Chapter 8) with an even number of carbon atoms. Common fatty acids have between 12 and 26 carbon atoms. An example is palmitic acid, which contains 16 carbon atoms.

$$CH_3-CH_2-CH_2-CH_2-CH_2-CH_2-CH_2-CH_2-CH_2-CH_2-CH_2-CH_2-CH_2-CH_2-CH_2-\overset{\overset{\displaystyle O}{\|}}{C}-OH$$

palmitic acid

The formula of palmitic acid, or any other fatty acid, is usually written in condensed form:

$$CH_3-(CH_2)_{14}-COOH$$

palmitic acid

FIGURE 14.6

The polysaccharides starch and cellulose differ in the way the glucose units are joined. The units are joined through α-linkages in starch and through β-linkages in cellulose.

As is the case for straight-chain alkanes (Chapter 8), the carbon atoms in a fatty acid are not in a straight line but are staggered:

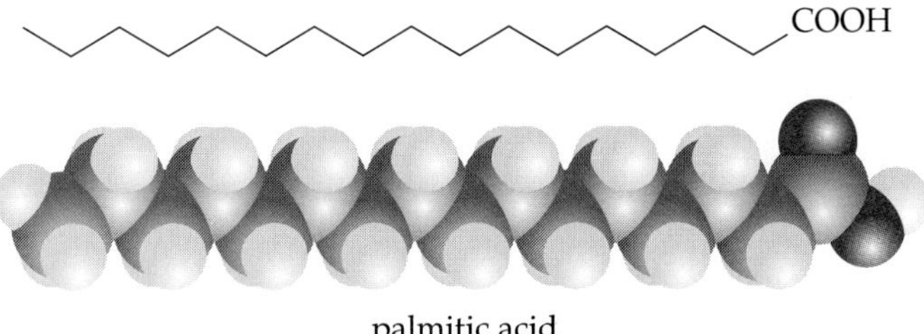

palmitic acid

The long chains of methylene groups ($-CH_2-$) make fatty acids, and most of their derivatives, nonpolar and thus insoluble in water.

Palmitic acid is a **saturated fatty acid:** one in which all the carbon-carbon bonds are single bonds. Other fatty acids, such as oleic acid, are **unsaturated** and contain one or more carbon-carbon double bonds.

$$CH_3-(CH_2)_7-CH=CH-(CH_2)_7-COOH$$

oleic acid

In most naturally occurring unsaturated fatty acids, the configuration of the double bonds is *cis* (for more information on *cis-trans* isomerism, see Chapter 8).

$$\begin{array}{ccc} -CH_2 & & CH_2- \\ & C=C & \\ H & & H \end{array} \qquad \begin{array}{ccc} -CH_2 & & H \\ & C=C & \\ H & & CH_2- \end{array}$$

cis   *trans*

The *cis* configuration causes a bend in the chain, as illustrated in skeleton form for oleic acid:

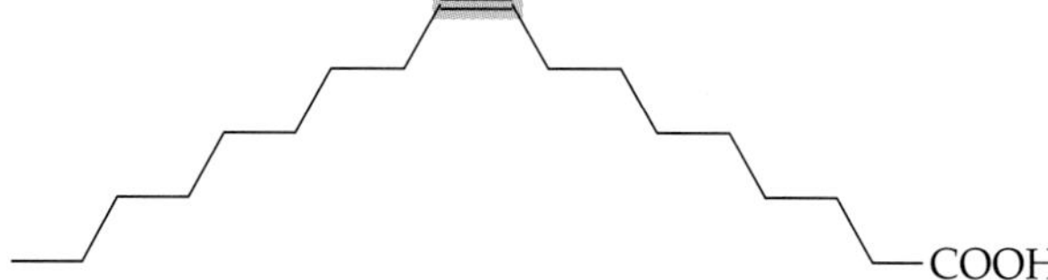

Fatty acids containing more than one double bond are often called *polyunsaturated fatty acids*. Biologically important polyunsaturated fatty acids contain up to four double bonds.

Important saturated and unsaturated fatty acids are shown in **Table 14.1**, along with their melting points and natural sources. The uniform straight-chain structure of saturated fatty acids allows their molecules to pack closely together in an orderly fashion.

Table 14.1 Structures, Melting Points, and Sources of Some Common Fatty Acids

Name	Number of Carbon Atoms	Formula	Melting Point (°C)	Sources
Saturated fatty acids				
Lauric	12	$CH_3(CH_2)_{10}COOH$	44	Coconut oil
Myristic	14	$CH_3(CH_2)_{12}COOH$	54	Butterfat, coconut oil
Palmitic	16	$CH_3(CH_2)_{14}COOH$	63	Beef fat, butterfat, cottonseed oil, lard
Stearic	18	$CH_3(CH_2)_{16}COOH$	70	Beef fat, butterfat, cottonseed oil, lard
Arachidic	20	$CH_3(CH_2)_{18}COOH$	76	Peanut oil
Unsaturated fatty acids				
Oleic	18	$CH_3(CH_2)_7CH=CH(CH_2)_7COOH$	13	Beef fat, lard, olive oil, peanut oil
Linoleic	18	$CH_3(CH_2)_4(CH=CHCH_2)_2(CH_2)_6COOH$	−5	Corn oil, cottonseed oil, linseed oil, soybean oil
Linolenic	18	$CH_3CH_2(CH=CHCH_2)_3(CH_2)_6COOH$	−11	Corn oil, linseed oil
Arachidonic	20	$CH_3(CH_2)_4(CH=CHCH_2)_4(CH_2)_2COOH$	−50	Animal tissues, corn oil, linseed oil

However, unsaturated fatty acid molecules, because of their bent structure, cannot pack as closely together. As a result, unsaturated fatty acids are less dense than saturated fatty acids, and attractions between adjacent molecules are weaker; melting points decrease as the number of double bonds per molecule increases.

Free fatty acids make up a small fraction of natural lipids. Fatty acids are important primarily as the building blocks from which the triglycerides in fat and oils are formed.

Triglycerides

The most abundant lipids in nature are fats and oils, which belong to the class called **triglycerides.** Triglycerides that are solids at room temperature, such as lard, butter, suet, and tallow, are generally known as **fats.** Those that are liquids are usually called **oils;** examples are corn oil, olive oil, soybean oil, and linseed oil.

Fats and oils are less dense than water and thus float on its surface.

We studied esters and the polyhydroxy alcohol glycerol in Chapter 8. Trigylcerides are triesters formed by the reaction of three fatty acid molecules with a glycerol molecule (**Fig. 14.7**). During the reaction, three molecules of water are eliminated. The three fatty acid molecules may be identical, or they may be different. Natural fats and oils are complex mixtures of many different triglycerides; their exact compositions vary depending on the source.

An organic ester, which has the general formula

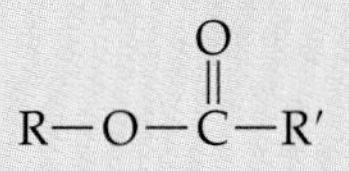

$$\mathrm{R{-}O{-}\overset{\overset{\displaystyle O}{\|}}{C}{-}R'}$$

is formed from the reaction of an alcohol with an organic acid.

Animal fats, such as lard, suet, and butter, are composed of triglycerides rich in saturated fatty acids and, in general, are solids at room temperature. Vegetable oils are liquids because they have a high content of triglycerides containing unsaturated fatty acids.

Studies have shown that a diet high in saturated fats increases the risk of developing *atherosclerosis,* the condition in which fatty deposits called *plaque* accumulate in the arteries. Thus, many people now avoid eating animal fats. To encourage consumers to cook with vegetable oils instead of butter or lard, advertisers of oils emphasize their high content of polyunsaturated fats.

Tristearin and other triglycerides are used to make soaps.

Hydrogenation of Oils

An oil can be converted to a semisolid fat by being partially hydrogenated—that is, by the addition of hydrogen atoms to some of its fatty acid double bonds, thus decreasing the degree of unsaturation. Soft-spread margarines are prepared by the partial

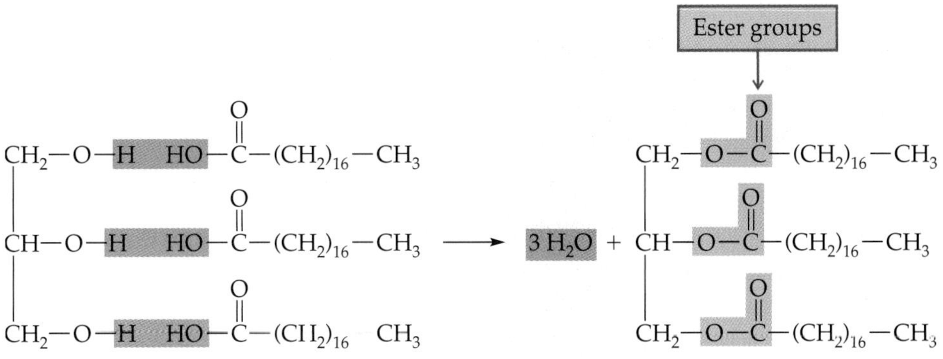

Ester groups

$$\begin{array}{lcl}
\mathrm{CH_2{-}O{-}H\quad HO{-}\overset{\overset{\displaystyle O}{\|}}{C}{-}(CH_2)_{16}{-}CH_3} & & \mathrm{CH_2{-}O{-}\overset{\overset{\displaystyle O}{\|}}{C}{-}(CH_2)_{16}{-}CH_3} \\
\mathrm{CH{-}O{-}H\quad HO{-}\overset{\overset{\displaystyle O}{\|}}{C}{-}(CH_2)_{16}{-}CH_3} & \longrightarrow\ 3\,H_2O\ + & \mathrm{CH{-}O{-}\overset{\overset{\displaystyle O}{\|}}{C}{-}(CH_2)_{16}{-}CH_3} \\
\mathrm{CH_2{-}O{-}H\quad HO{-}\overset{\overset{\displaystyle O}{\|}}{C}{-}(CH_2)_{16}{-}CH_3} & & \mathrm{CH_2{-}O{-}\overset{\overset{\displaystyle O}{\|}}{C}{-}(CH_2)_{16}{-}CH_3}
\end{array}$$

glycerol three stearic acid molecules tristearin

FIGURE 14.7

The formation of the triglyceride tristearin from glycerol and three molecules of stearic acid. During the reaction, three molecules of water are eliminated.

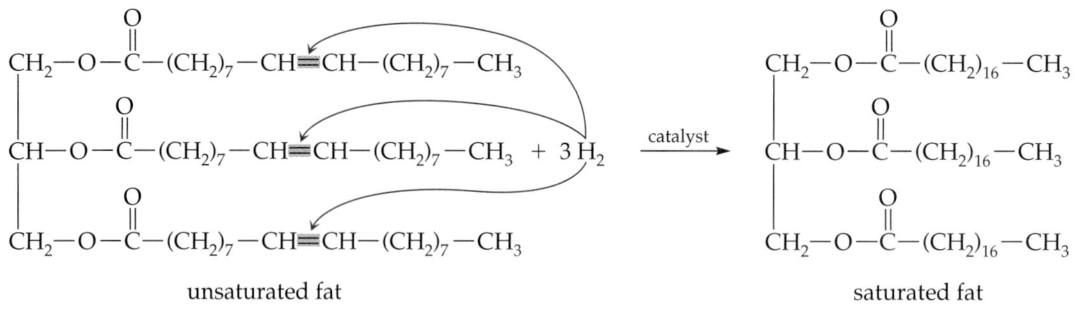

$$\begin{array}{l} CH_2-O-\overset{O}{\overset{\|}{C}}-(CH_2)_7-CH{=}CH-(CH_2)_7-CH_3 \\ | \\ CH-O-\overset{O}{\overset{\|}{C}}-(CH_2)_7-CH{=}CH-(CH_2)_7-CH_3 \\ | \\ CH_2-O-\overset{O}{\overset{\|}{C}}-(CH_2)_7-CH{=}CH-(CH_2)_7-CH_3 \end{array} + 3\,H_2 \xrightarrow{\text{catalyst}} \begin{array}{l} CH_2-O-\overset{O}{\overset{\|}{C}}-(CH_2)_{16}-CH_3 \\ | \\ CH-O-\overset{O}{\overset{\|}{C}}-(CH_2)_{16}-CH_3 \\ | \\ CH_2-O-\overset{O}{\overset{\|}{C}}-(CH_2)_{16}-CH_3 \end{array}$$

unsaturated fat saturated fat

FIGURE 14.8

Margarine is made from vegetable oils by hydrogenation of double bonds in the oil molecules. Hydrogenation converts liquid oils (polyunsaturated fats) into semisolid fats (partially saturated fats).

hydrogenation of vegetable oils, such as corn oil, soybean oil, and cottonseed oil **(Fig. 14.8)**. The reaction is carefully controlled to achieve the desired consistency for the product. Vegetable oils are relatively inexpensive, and margarines can be produced more cheaply than butter.

Margarines and other products containing partially hydrogenated vegetable fats are often labeled "made with 100% vegetable oil," suggesting to consumers that they are much healthier than butter. However, during partial hydrogenation, some double bonds of the fatty acids in oils are converted from the natural *cis* configuration to the *trans* configuration. Molecules of *trans* fatty acids do not have a bend in their structure and, like saturated fatty acids, can pack closely together. Also, like saturated fatty acids, they can increase the risk of heart disease. Consumption of fats containing *trans* fatty acids has been shown to increase blood cholesterol levels.

The Role Fats Play in Animal Metabolism

Adipose tissue, which is found primarily under the skin, around deep blood vessels, and in the abdominal cavity, stores carbon compounds consumed in excess of the body's need for energy and for the replenishment of glycogen reserves.

In humans and other animals, any excess glucose formed in the body from carbohydrates in foods is converted to fats and stored in adipose tissue. In this process, glucose is broken down into two-carbon units, which are then built up into fatty acids and finally triglycerides. Because they are built from two-carbon units, fatty acids always have an even number of carbon atoms.

Fats insulate an animal's body against changes in temperature and also protect vital internal organs from mechanical damage. Fat deposits represent the most efficient form in which energy is stored in the body. Triglycerides not only pack together more compactly in muscles and the liver than glycogen, they also yield more energy per gram. A typical triglyceride breaks down as follows:

$$2\ C_{55}H_{100}O_6 \quad + \quad 154\ O_2 \quad \longrightarrow \quad 110\ CO_2 \quad + \quad 100\ H_2O \quad + \quad \text{energy}$$

The gradual breakdown of stored fat supplies hibernating animals with the water and energy they need to survive the months they spend without food. Similarly, camels in the desert survive on fats stored in their humps.

Absorption of Toxic Substances by Fats

Because they are nonpolar, fats in animal tissues absorb and dissolve other nonpolar substances. The damaging effects of polychlorinated hydrocarbons, such as PCBs and the insecticide DDT, occur when these substances enter natural waters and then become concentrated in the fatty tissues of fish and fish-eating birds (Chapter 10).

14.8 Steroids

The important group of lipids called **steroids** includes cholesterol, bile acids, vitamin D, and many hormones. Steroids, which are present in both plants and animals, are completely different in chemical structure from the other lipids we have discussed so far. The basic structure shared by all steroids consists of three six-membered rings and one five-membered ring fused together. Steroids are distinguished from one another by the location of double bonds within this skeleton and the nature and location of substituents attached to the rings. The steroid skeleton and the formulas of cholesterol and some important hormones are shown in **Figure 14.9**.

Cholesterol

Cholesterol, the most abundant steroid in the human body, is the starting material for the synthesis of all the other steroids found in the body. Cholesterol is the main component of gallstones and also occurs in brain and nerve tissues, in blood, and in all cell membranes.

Cholesterol is present in animal fats; therefore, milk, butter, eggs, and meat are all sources of dietary cholesterol. Plants, however, do not contain cholesterol, and foods such as vegetable oils, peanut butter, and margarine are cholesterol-free. Although some of the body's cholesterol is obtained directly from cholesterol-containing foods, most is synthesized in the liver from carbohydrates, fats, and proteins. Reducing the intake of cholesterol-containing foods will not, therefore, automatically lower blood cholesterol levels. To achieve this result, the entire diet must be modified.

To reduce very high cholesterol levels, doctors prescribe drugs such as Zocor.

Cholesterol is a major factor in the development of atherosclerosis **(Fig. 14.10)**. The fatty deposits of plaque that build up in arteries are composed largely of cholesterol, along with some triglycerides, and their formation is associated with high levels of these substances in the blood. As plaque accumulates, the arteries become narrower; the heart must work harder to pump blood through the narrower passages. If blood vessels to the brain or heart become blocked, the flow of blood is reduced, and the resulting lack of oxygen can cause a stroke or a heart attack.

Cholesterol is the starting material for the synthesis of related steroids called *bile salts*, which are stored in the gallbladder. Bile salts are released into the intestine after a meal to aid in the digestion of fats and oils (Chapter 15). In some conditions, cholesterol in the gallbladder forms into stones, some as large as marbles, which can cause nausea and severe abdominal pain.

Bile acids, which are present in the fluid released into the small intestines during digestion, are steroidal acids derived from cholesterol.

Transport of Cholesterol in the Bloodstream

Like all lipids, cholesterol is insoluble in water. So that it can be transported through the bloodstream, it is bonded to water-soluble *lipoproteins*, complex mixtures of lipids and proteins that are classified according to their densities. Very-low-density lipoproteins (VLDLs) transport triglycerides. Low-density lipoproteins (LDLs) transport cholesterol from the liver to the tissues. High-density lipoproteins (HDLs) remove excess cholesterol from the tissues and transport it to the liver, where it is converted to bile salts and excreted.

A better indication of a person's risk of a heart attack than his or her cholesterol level is the relative amounts of HDLs and LDLs in the blood. LDLs, which carry most of the blood cholesterol, are called "bad cholesterol" because they deposit cholesterol on artery walls. HDLs, which carry about 25% of blood cholesterol, are referred to as

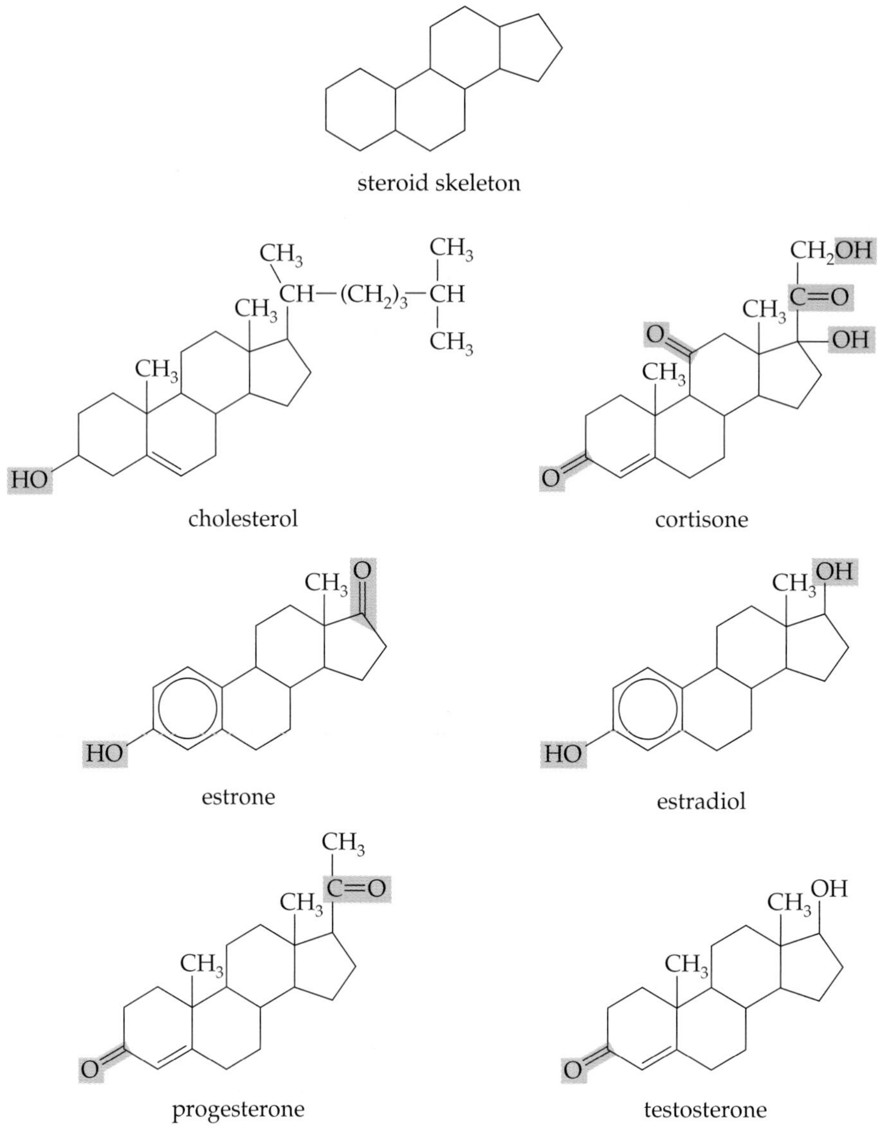

FIGURE 14.9

Steroids all share the same basic four-ring structure (three six-membered rings and one five-membered ring form the steroid skeleton). Cholesterol is the starting material for the synthesis of all other steroids in the body. Cortisone is used as an anti-inflammatory agent. Estrone, estradiol, and progesterone are female sex hormones; testosterone is the most important male sex hormone. (Functional groups are shaded blue.)

"good cholesterol" because they remove excess cholesterol from arteries and thus have a protective effect. Exercise has been shown to increase blood levels of HDLs. A diet high in saturated fats increases VLDLs and LDLs.

Steroid Hormones

Hormones are substances produced in specialized glands in the body: the pituitary, thymus, hypothalamus, thyroid, parathyroid, adrenals, pancreas, and gonads (testes and ovaries). These glands make up the endocrine system. Hormones are secreted directly into the bloodstream and carried to different locations in the body, where they direct and control reproduction, prenatal development, and growth and regulate a variety of other physiological processes. When a hormone reaches its correct location in the body, it binds with a specific receptor molecule. Once bound, the hormone initiates its action. Many, but not all, hormones are steroids and are formed from cholesterol.

An important steroid hormone secreted by the adrenal gland is *cortisone.* When applied to the skin or injected, synthetic cortisone, which is identical to the natural hormone, acts to reduce inflammation. It is also used to treat arthritis.

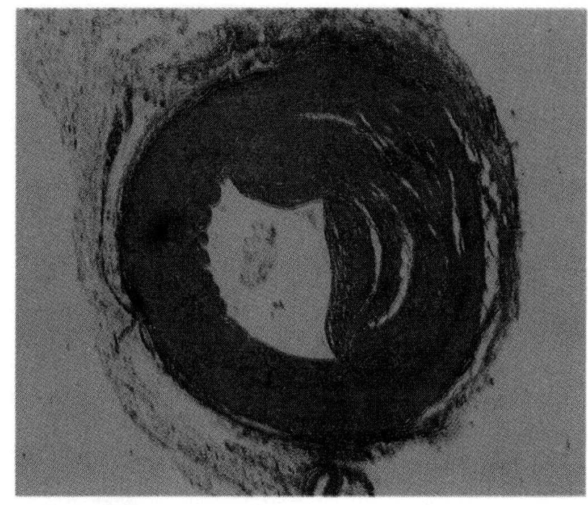

FIGURE 14.10

In atherosclerosis, plaque buildup can severely narrow coronary arteries and cause a heart attack.

The sex hormones produced by the ovaries in females and the testes in males are structurally related to cholesterol and cortisone. Two important female hormones are estradiol and estrone, which together are called **estrogens.** Unlike cholesterol and the other steroids shown in Figure 14.10, estrogens include an aromatic hydrocarbon ring (Chapter 8) in their structure. Estrogens control the menstrual cycle, breast development, and other female sex characteristics. Another female hormone is **progesterone,** which causes changes in the uterine wall that prepare the uterus for pregnancy and also prevents the further release of eggs from the ovary during pregnancy. Oral contraceptives contain synthetic steroids with molecular structures very similar to the structure of progesterone. Like progesterone, these steroids prevent ovulation.

Male sex hormones are called **androgens.** They are responsible for the development of male sex organs and masculine sex characteristics, such as facial hair and a deep voice. The most important male sex hormone is **testosterone,** which has both masculinizing (androgenic) and muscle-building (anabolic) effects. Although androgens predominate in males and estrogens predominate in females, both types of hormones are present in both sexes.

Anabolic Steroids

Synthetic anabolic steroids with molecular structures very similar to that of testosterone have been used for many years by some athletes to build muscle mass and improve performance. However, these products can have serious side effects. In men, they can cause testicular atrophy, impotence, breast enlargement, liver dysfunction, and acne and can increase the risk of heart disease. In women, they can cause irregularities in the menstrual cycle and the development of male characteristics such as body hair and a deep voice. Anabolic steroids are particularly dangerous during puberty when a person's body is first exposed to the effects of the sex hormones.

Synthetic anabolic steroids are banned by the world's major amateur athletic organizations. These steroids can cause sterility in men and masculinization in women, as well as having other adverse effects, but they are useful in preventing muscle wasting in patients recovering from surgery.

The anabolic steroid androstenedione, which is sold over-the-counter, has received a great deal of publicity mainly because St. Louis Cardinals baseball legend Mark McGwire took it as a dietary supplement for a time. Androstenedione is produced naturally in the testicles, ovaries, and adrenal glands and chemically is very closely related to testosterone. Little is known about its effects, but it is assumed that it is converted to testosterone in the body. Although Major League Baseball does not ban the use of androstenedione, most other athletic organizations, including the National Football League, do and also ban other anabolic steroids.

Endocrine Disruptors

There is growing concern that many of the persistent organic pollutants (POPs) (see Chapter 10), which have been widely disseminated through the environment, may be inter-

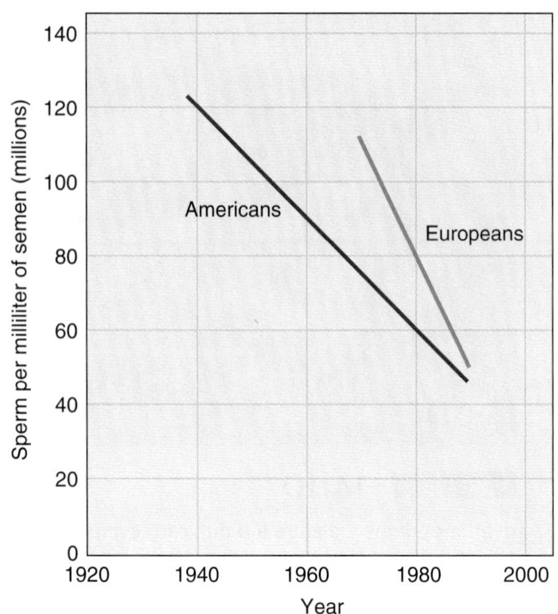

FIGURE 14.11

The average sperm count of American and European men dropped between 1938 and 1990.

fering with the endocrine system in both animals and humans and disrupting reproduction and fetal development. These pollutants, called **endocrine disruptors,** include PCBs, dioxins, DDT (see Fig. 10.19), methoxyclor and other pesticides, and phthalates, which are components in many plastics (see Chapter 9). They have been shown to cause infertility, reduced sperm counts, cancer, and neurological disorders.

Observations of wildlife exposed to POPs, along with studies on laboratory animals, have shown that, even at extremely low levels, many of these chemicals mimic natural estrogens and cause feminizing in males and accentuation of female characteristics in females. For example, male fish exposed to POPs produce ovaries and eggs instead of sperm. There is also growing evidence that POPs are responsible for some of the increasing number of abnormalities found in wildlife, such as alligators with underdeveloped reproductive organs.

Because hormones are common to many different species, it would be expected that chemicals that disrupt biochemical processes in animals would also disrupt them in humans. It is now thought that the dramatic drop in sperm counts among large groups of men in the United States and Europe since 1938 **(Fig. 14.11)** and the increase in the number of cases of undescended testicles may be due to estrogen-mimicking POPs.

Although the chemical structures of estrogen mimics are unlike those of natural estrogens (see Fig. 14.9), research has shown that they are able to bind to estrogen receptor sites. Once bound to these sites, they initiate the action of the natural hormone. Other hormone disruptors block receptor sites and depress normal hormone action. It is known that some combinations of hormone-disrupting chemicals act *synergistically;* that is, the effect of exposure to certain combinations of these chemicals far exceeds the sum of the effects of exposure to any two of them separately.

There are approximately 80,000 synthetic chemicals on the market today, and most of them have never been tested for toxicity. Some have been screened to determine if they are carcinogens, but only a minute fraction have been tested for their effects on the endocrine system. In 1998, the EPA was required by law to begin testing 15,000 chemicals for their possible role as endocrine disruptors.

14.9 Proteins

Proteins are essential components of all living cells. Their vital importance is indicated in the name *protein,* which is derived from the Greek word *proteios,* meaning "first" or "foremost." Proteins are present in the smallest cellular organisms as well as in more complex life forms. They are involved in practically every process that occurs in living tissues.

Proteins form the structural material of animals in much the same way that cellulose forms the structural material of plants. Skin, hair, wool, nails, tendons, and muscles are all made of proteins. As antibodies, proteins protect the body from disease; as enzymes, they catalyze numerous metabolic processes in the body. The blood protein hemoglobin transports oxygen throughout the body; the protein insulin is a hormone that regulates glucose metabolism. Proteins are also an important source of energy for animals. Foods that are high in proteins include meat, poultry, fish, cheese, eggs, beans, and nuts.

All proteins contain the elements carbon, oxygen, hydrogen, nitrogen, and sulfur. Some also contain phosphorus, iodine, iron, magnesium, or any of a number of other elements. The human body may contain as many as 100,000 different proteins, and each type of cell in the body makes its own particular kinds of proteins. The building blocks of proteins are the amino acids.

Amino Acids

Proteins are synthesized in body cells by enzyme-catalyzed reactions between amino acids. **Amino acids,** as their name implies, contain an amino group ($-NH_2$) and a carboxyl group ($-COOH$) (Chapter 8). In nearly all the amino acids that join together to form proteins, the amino group and the carboxyl group are bonded to the same carbon atom, as shown below for a representative amino acid.

Proteins are very large molecules with formula masses ranging from 5000 amu to several million amu.

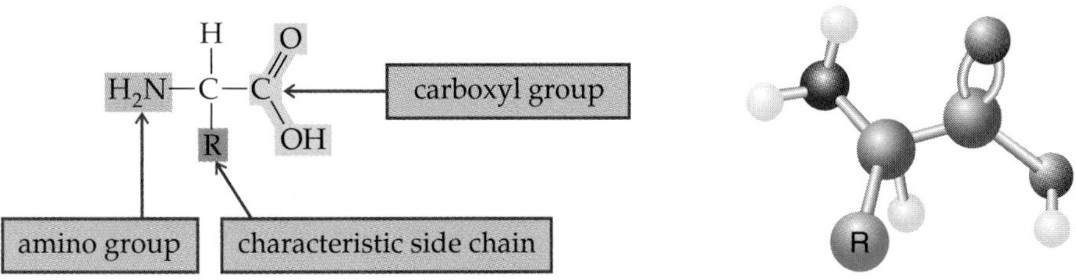

The R group of the side chain varies with each amino acid and takes many forms. In the simplest amino acid, glycine, R is a hydrogen atom. In the next simplest, alanine, R is a methyl group.

$$H_2N-\underset{H}{\overset{H}{\underset{|}{\overset{|}{C}}}}-COOH \qquad H_2N-\underset{CH_3}{\overset{H}{\underset{|}{\overset{|}{C}}}}-COOH$$

glycine — alanine (chiral carbon atom)

With the exception of glycine, all amino acids have one or more chiral carbon atoms in their structure, and, therefore, they can exist as chiral molecules. In humans and most other animals, all proteins are made from amino acids that have the same handedness.

Each protein is composed of a specific number of different amino acids that are linked together in a definite sequence. Virtually all proteins in the human body are made from the 20 amino acids listed in **Table 14.2**. All these amino acids are necessary constituents of human proteins. The ten whose names are printed in blue are called *essential amino acids,* because they cannot be synthesized in the body from other chemicals and must be obtained from food. The others—the nonessential amino acids—can be synthesized in the body. The names of amino acids are abbreviated as three letters, usually the first three letters, as indicated in Table 14.2.

Peptide Bonds

Amino acids become linked together in proteins by the formation of amide bonds (Chapter 8), also called **peptide bonds.** A peptide bond results when the amino group of one amino acid reacts with the carboxyl group of a second amino acid. A molecule of water is eliminated, and a **dipeptide** is formed. The formation of a dipeptide from alanine and glycine is shown below.

Peptide chains with formula masses up to 10,000 are called *polypeptides;* those with higher molecular weights are called *proteins.*

Table 14.2 Amino Acids Found in Human Proteins*

$H-CH(NH_2)-COOH$	Glycine (Gly)
$H_3C-CH(NH_2)-COOH$	Alanine (Ala)
$H_2C(CH_2-)(CH_2-N(H)-)$ ring: $CH_2-C(H)(-C(=O)OH)-N(H)-CH_2-H_2C$	Proline (Pro)
$H_2N-C(=NH)-NH-(CH_2)_3-CH(NH_2)-COOH$	Arginine (Arg)
ring $N-CH=C(-)-NH-HC=N$: $C-CH_2-CH(NH_2)-COOH$	Histidine (His)
$HO-CH_2-CH(NH_2)-COOH$	Serine (Ser)
$H_2N-C(=O)-CH_2-CH(NH_2)-COOH$	Asparagine (Asn)
$H_3C-CH_2-CH(CH_3)-CH(NH_2)-COOH$	Isoleucine (Ile)
$H_3C-CH(OH)-CH(NH_2)-COOH$	Threonine (Thr)
$HO-C(=O)-CH_2-CH(NH_2)-COOH$	Aspartic acid (Asp)
$H_3C-C(CH_3)(H)-CH_2-CH(NH_2)-COOH$	Leucine (Leu)
benzene ring fused to ring $C=CH-N(H)$: $C-CH_2-CH(NH_2)-COOH$	Tryptophan (Trp)
$HS-CH_2-CH(NH_2)-COOH$	Cysteine (Cys)
$H_2N-(CH_2)_4-CH(NH_2)-COOH$	Lysine (Lys)
$OH-C_6H_4-CH_2-CH(NH_2)-COOH$	Tyrosine (Tyr)
$HO-C(=O)-(CH_2)_2-CH(NH_2)-COOH$	Glutamic acid (Glu)
$H_3C-S-(CH_2)_2-CH(NH_2)-COOH$	Methionine (Met)
$H_3C-CH(CH_3)-CH(NH_2)-COOH$	Valine (Val)
$H_2N-C(=O)-(CH_2)_2-CH(NH_2)-COOH$	Glutamine (Gln)
$C_6H_5-CH_2-CH(NH_2)-COOH$	Phenylalanine (Phe)

*R groups are shaded purple; names of essential amino acids are printed in blue (histidine and arginine are essential for infants only).

$$H_2N{-}CH(CH_3){-}C(=O){-}OH + H{-}NH{-}CH_2{-}C(=O)OH \longrightarrow H_2N{-}CH(CH_3){-}C(=O){-}NH{-}CH_2{-}C(=O)OH + H_2O$$

alanine (Ala) glycine (Gly) dipeptide Ala-Gly

peptide bond

Peptide bond

Two different amino acids can bond to form two distinct dipeptides. For example, depending on which amino group bonds with which carboxyl group, glycine and alanine can join to form either Ala-Gly or Gly-Ala.

$$H_2N{-}CH(CH_3){-}C(=O){-}NH{-}CH_2{-}COOH \qquad H_2N{-}CH_2{-}C(=O){-}NH{-}CH(CH_3){-}COOH$$

Ala-Gly Gly-Ala

These two dipeptides, which are isomers, have different physical and chemical properties.

The number of possible isomers that can be formed increases very rapidly as the length of the peptide chain increases. For example, three different amino acid molecules such as glycine, alanine, and valine can bond to form six distinct tripeptides:

Gly-Ala-Val	Gly-Val-Ala	Ala-Gly-Val
Ala-Val-Gly	Val-Gly-Ala	Val-Ala-Gly

Four different amino acids can form 24 tetrapeptides. Eight can be sequenced in over 40,000 different ways. Sequences of up to 50 amino acid units are generally called *polypeptides*. Longer sequences are called *proteins*.

Since proteins typically contain from 50 to well over 1000 amino acids in sequence, the number of possible combinations that can be made from the 20 amino acids listed in Table 14.2 is almost endless. However, nature is very selective, and only those proteins required for specific functions are actually made. As we'll see in the next section, if just one amino acid in the sequence is altered, the character of a protein is changed, and it may then be unable to perform its normal function.

EXAMPLE 14.2 Drawing Structures of Dipeptides

Draw the structures of the dipeptides that can be formed from the amino acids phenylalanine and alanine. See Table 14.2 for the structures of the amino acids.

Solution

1. Two dipeptides can be formed:

   The amino group of alanine can bond with the carboxyl group of phenylalanine, *or*
   the amino group of phenylalanine can bond with the carboxyl group of alanine.

2. Show how the amino group of alanine combines with the carboxyl group of phenylalanine with the elimination of a molecule of water:

$$C_6H_5{-}CH_2{-}CH(NH_2){-}CO\boxed{OH + H}{-}N(H){-}CH(CH_3){-}COOH$$

Phenylalanine   Alanine

$$\downarrow$$

$$C_6H_5{-}CH_2{-}CH(NH_2){-}C(=O){-}N(H){-}CH(CH_3){-}COOH + \boxed{H_2O}$$

Dipeptide 1

3. Show how the amino group of phenylalanine combines with the carboxyl group of alanine:

$$C_6H_5{-}CH_2{-}CH(HOOC){-}N(H)\boxed{{-}H + HO}OC{-}CH(CH_3){-}NH_2$$

Phenylalanine   Alanine

$$\downarrow$$

$$C_6H_5{-}CH_2{-}CH(HOOC){-}N(H){-}C(=O){-}C(CH_3){-}NH_2 + \boxed{H_2O}$$

Dipeptide 2

Practice Exercise 14.2

Write the structures of the dipeptides that can be formed from the amino acids valine and alanine. (See Table 14.2 for the structures of the amino acids.)

The Structure of Proteins

The sequence of amino acids in a protein, its *primary structure,* determines the role the protein plays in the body. Also important is the *secondary structure,* the well-defined shape that a protein adopts as a result of hydrogen bonding between the amino acid units in its chain. For many proteins, including those in hair, wool, and nails, hydrogen bonding causes the amino acid chain to become twisted into a tightly coiled *helix* **(Fig. 14.12a)**, a structure that was first proposed by Linus Pauling (Chapter 4). In the helix, hydro-

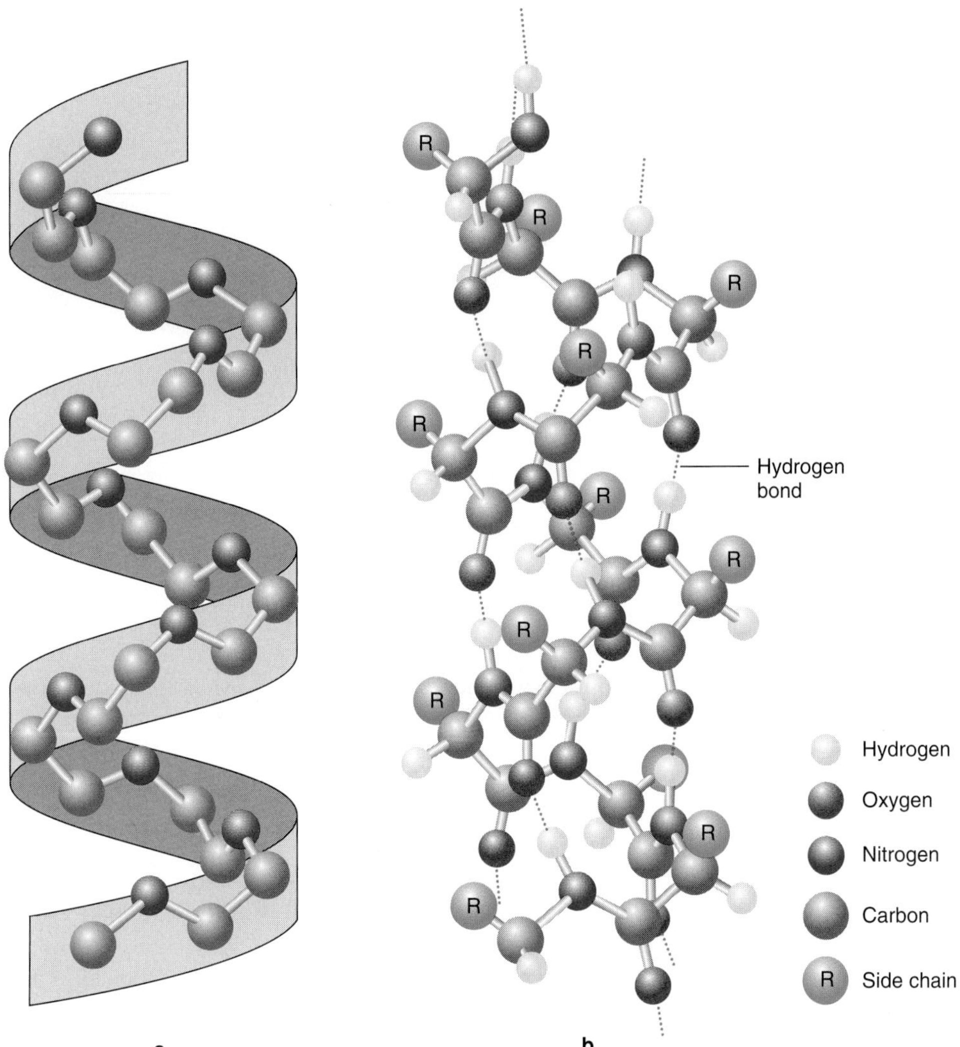

FIGURE 14.12

A helical structure for proteins was first proposed by Linus Pauling. (a) Ball-and-stick model of the peptide backbone with side chains omitted to emphasize the shape of the helix. (b) Ball-and-stick model showing how hydrogen bonding between N—H and C=O groups holds the helix together. (R represents any of the side chains shown in Table 14.2.)

gen bonding occurs between the oxygen atom in the carbonyl group (C=O) of one amino acid unit and the hydrogen atom in the —NH group in the amino acid unit that is four units farther along the chain **(Fig. 14.12b)**.

Another, less common secondary structure is a *pleated sheet*, in which amino acid chains are held together, side by side, by hydrogen bonds **(Fig. 14.13)**. This type of structure is found in silk and in myosin (a protein found in muscle). A third secondary structure, the *triple helix*, characterizes collagen, the fibrous protein of connective tissues such as cartilage, tendon, and skin. In this protein, three helical amino acid chains are braided together to form a strong, ropelike secondary structure.

Proteins also have a *tertiary structure*. This is the three-dimensional shape that results from further folding, crosslinking, and coiling of the amino acid chains caused by

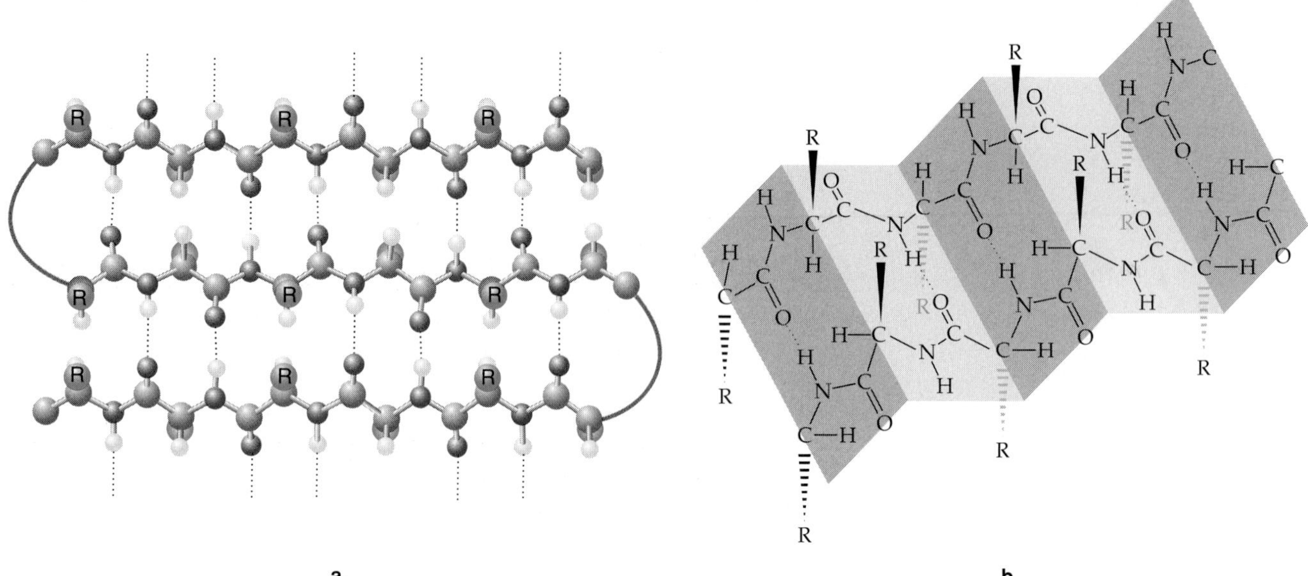

FIGURE 14.13

The pleated sheet structure characterizes some proteins. (a) Ball-and-stick model emphasizing the hydrogen bonding forming the sheet. (b) Structural formula representation emphasizing the pleating of the sheet.

interactions between the R groups of the amino acid units **(Fig. 14.14)**. Four major types of interactions that occur are

1. the formation of disulfide linkages (—S—S— covalent bonds) between pairs of cysteine side chains (see Table 14.2)
2. electrostatic attractions between amino acids with polar side chains, such as aspartic acid and lysine
3. hydrogen bonding between —OH, $—NH_2$, —COOH, and $—CONH_2$ groups on amino acid side chains
4. weak London forces (Chapter 4) between nonpolar side chains

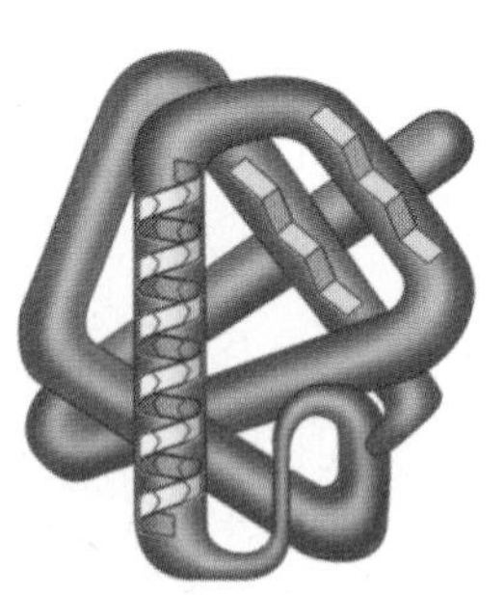

FIGURE 14.14

The tertiary structure of proteins is determined by the folding and coiling that result from interactions between amino acids that are far apart in the molecule.

As a result of these interactions, most proteins adopt a globular shape; others have a more linear, or fibrous, configuration.

We can use the cord connecting a personal computer to its keyboard **(Fig. 14.15)** as an analogy for the three levels of protein structures: The straight section of the cord represents the primary structure; the section where the cord is coiled into a helical arrangement represents the secondary structure; and the section where the coiled cord is folded on itself represents the tertiary structure.

Some proteins, including the oxygen-carrying blood protein hemoglobin, are made up of more than one polypeptide chain. The way in which these polypeptide chains are arranged to form the complete three-dimensional protein is referred to as the *quarternary structure*. The quarternary structure of hemoglobin, which consists of four polypeptide chains, is shown in **Figure 14.16**.

The vital importance of correct protein structure is demonstrated by the hereditary disorder sickle cell anemia. In this disorder, the nonpolar amino acid valine is substituted for one of the polar glutamic acid units in one of the polypeptide chains of hemoglobin. This substitution of one specific amino acid in the sequence of 146 in the chain alters the tertiary structure of hemoglobin. The protein assumes an abnormal three-dimensional shape, the red blood cells become distorted (**Fig. 14.17**), and the blood's ability to transport oxygen is reduced. The abnormal crescent shaped cells cannot flow through capillaries and block them, causing severe pain.

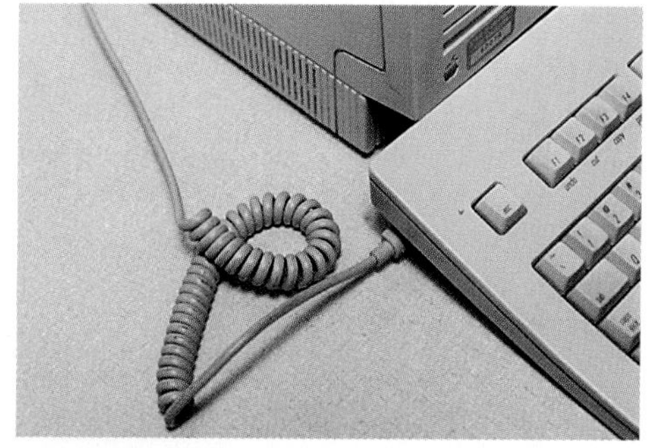

FIGURE 14.15

The coiling and folding back on itself of a cord that connects a personal computer to its keyboard are analogous to the secondary (helical) and tertiary structures of a protein.

14.10 Enzymes

Enzymes are biological catalysts that are synthesized by all living organisms. With very few exceptions, all are globular proteins. Enzymes are essential for nearly all the chemical reactions that occur in living systems: They take part in the digestion of food, the synthesis of complex carbohydrates, fats, and proteins, the process of photosynthesis, and the release of energy to tissues. Thousands of enzymes are present in the human body.

Like all catalysts, enzymes increase the rate of a reaction by lowering the activation energy (Chapter 6). Enzymes are extraordinarily efficient catalysts. In the human body, enzyme-catalyzed reactions occur extremely rapidly under very mild conditions—that is, at neutral pH and a temperature of 37°C (98.6°F). In the laboratory, in the absence of the enzymes and under the same mild conditions, these reactions occur only very slowly.

One reason vitamins and minerals are so important in the human diet is that they often act as coenzymes in vital metabolic processes.

Some enzymes are simple proteins. Others consist of a protein plus a nonprotein called a **coenzyme.** The coenzyme may be a metal ion (for example, Fe^{3+}, Mg^{2+}, or Cu^{2+}) or one of the water-soluble vitamins (Chapter 15). Without its coenzyme, an enzyme has no activity.

Many enzymes are very specific in their action and catalyze just one reaction or a few related reactions. The group of intestinal enzymes called *amylases* catalyze the breaking of the α-linkages that join glucose units in starches, but they cannot catalyze the breaking of the β-linkages that join glucose units in cellulose. The enzyme sucrase catalyzes a single reaction—the breakdown of sucrose to yield glucose and fructose.

The *lock-and-key model* is used to explain the action of specific enzymes such as sucrase, which catalyze just one reaction (**Fig. 14.18**). The substance with which the enzyme interacts (sucrose in this case) is called the **substrate.** The interaction between the substrate and the enzyme occurs at a small area on the enzyme called the **active site.** The active site, which has a specific shape, acts as the lock into which only one key, the substrate, can fit. A substrate-enzyme complex is first formed and then breaks apart as the products (glucose and fructose in this case) are formed, leaving the enzyme free to catalyze further reactions.

The enzyme trypsin is less specific than sucrase. It aids in the digestion of all proteins by breaking peptide bonds, and its action is explained by the *induced-fit model.* According to this model, as the substrate and active site come together, the active site is able to adjust its shape to fit the substrate (**Fig. 14.19**). The better fit is called an *induced fit.*

The lack of a particular enzyme can have serious health consequences and is the cause of a variety of hereditary conditions, including lactose intolerance. This condition affects many Asians and Africans, who traditionally drink little milk after they are weaned. When they become adults, their bod-

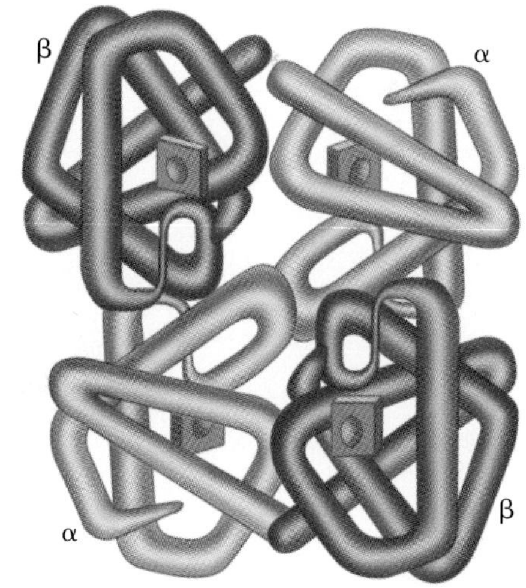

FIGURE 14.16

The quaternary structure of hemoglobin, which consists of four polypeptide chains (purple) and four heme units (red squares), the nonprotein parts of hemoglobin that carry oxygen.

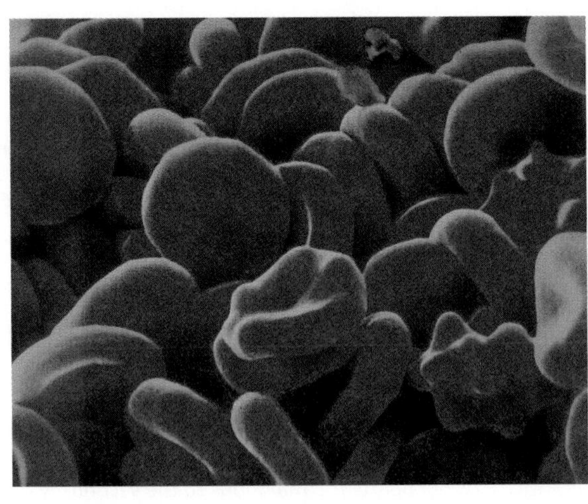

a.

b.

FIGURE 14.17

Electron micrographs showing (a) normal red blood cells and (b) sickled red blood cells.

ies stop producing lactase, the enzyme needed to digest the lactose in milk. As a result, if they consume milk or milk products, they suffer stomach pain and diarrhea.

We noted at the beginning of the chapter that a person fed a diet composed of sugars and amino acids with the wrong "handedness" would starve. The main reason is that the active sites on enzymes, including those that are associated with digestion, have a particular "handedness." Only molecules with the correct "handedness" can fit into the active sites on the enzymes and form products.

FIGURE 14.18

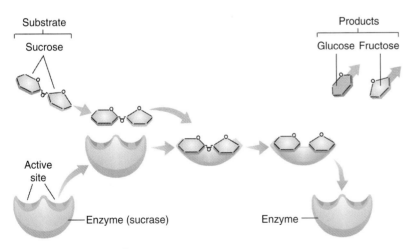

The lock-and-key model of enzyme action: The substrate (sucrose) fits the active site on the enzyme (sucrase) in the same way that a key fits a lock. The separation of the products (glucose and fructose) from the substrate-enzyme complex is analogous to the opening of the lock.

14.11 Nucleic Acids

The complex compounds called **nucleic acids** are found in all living cells except mammalian red blood cells. There are two kinds of nucleic acids: **deoxyribonucleic acid (DNA)** and **ribonucleic acid (RNA).** DNA is found primarily in the nucleus of a cell, while RNA is found primarily in the *cytoplasm*, the part of the cell surrounding the nucleus.

Functions of Nucleic Acids

DNA and RNA are responsible for the storage and transmission of genetic information in all living organisms. They allow genetic information to be transferred from one cell to another and genetic traits to be transmitted, via sperm and eggs, from parents to offspring. The main function of DNA, one in which RNA is also involved, is the control and direction of protein syn-

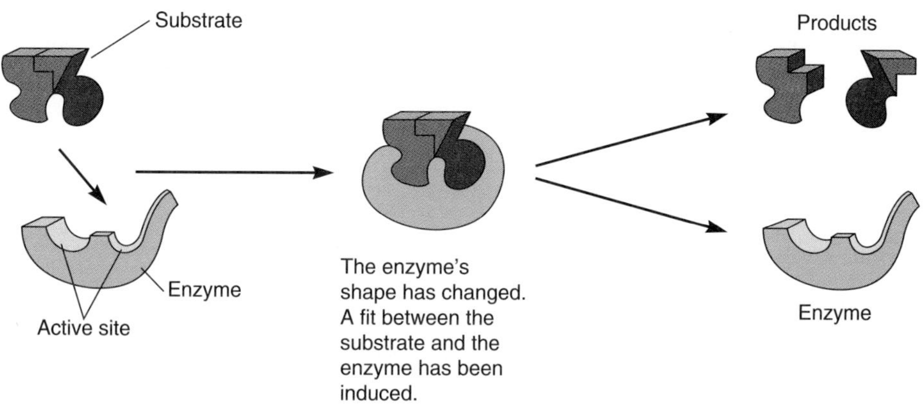

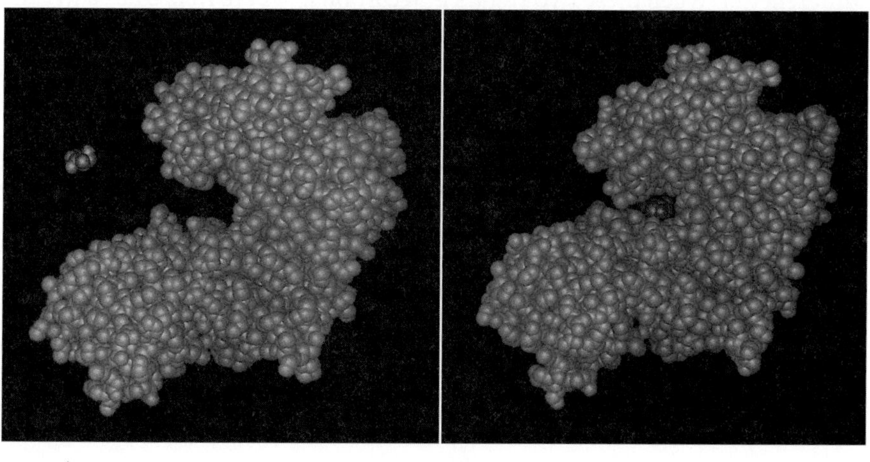

FIGURE 14.19

The induced fit model of enzyme action: (a) As the substrate approaches, the active site on the enzyme changes its shape to fit the substrate. After the product is released, the enzyme's active site resumes its original shape. (b) A computer simulation of the induced fit between the enzyme hexokinase (blue) and its substrate glucose (red). Note the difference in the shape of the enzyme before (left) and after (right) the substrate is bound.

thesis in body cells. Chemical information stored in the DNA of genes specifies the exact nature of the protein to be made and thus dictates the character of the organism.

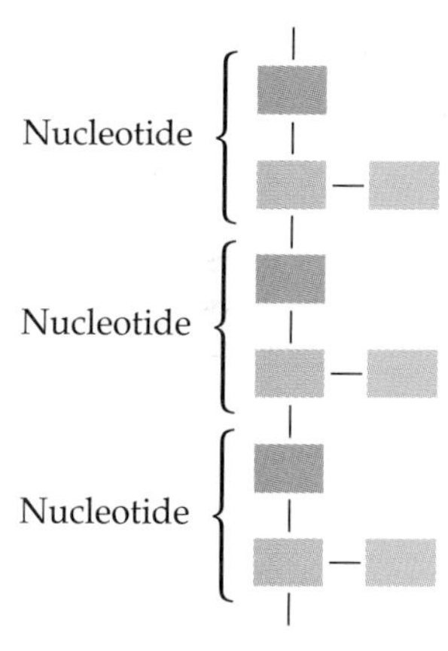

FIGURE 14.20

Nucleic acids consist of long chains of nucleotides. Each nucleotide in a chain is composed of a sugar, a base, and a phosphoric acid unit.

The Primary Structure of Nucleic Acids

Just as polysaccharides are composed of monosaccharides and proteins are composed of amino acids, nucleic acids are composed of repeating units called **nucleotides.** DNA molecules are the largest of the naturally occurring organic molecules.

Each nucleotide has three components: (1) a sugar, (2) a nitrogen-containing heterocyclic base, and (3) a phosphoric acid unit (Fig. 14.20).

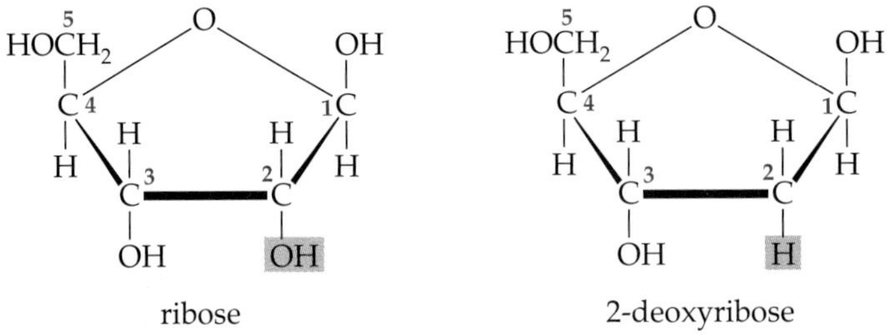

ribose 2-deoxyribose

FIGURE 14.21

The pentoses ribose and deoxyribose differ at carbon 2, where ribose has an H atom and an OH group and deoxyribose has two H atoms. Ribose is present in RNA; deoxyribose is present in DNA.

The sugar part of a nucleotide is a pentose, either ribose or deoxyribose **(Fig. 14.21)**. The only difference between these two sugars is at carbon 2, where ribose has a hydrogen atom and an —OH group but deoxyribose has two hydrogen atoms. As the names indicate, the sugar in DNA is deoxyribose, while that in RNA is ribose.

The five different bases found in nucleic acids are shown in **Figure 14.22**. Two—**adenine** (A) and **guanine** (G)—are double-ring bases and are classified as *purines*. The other

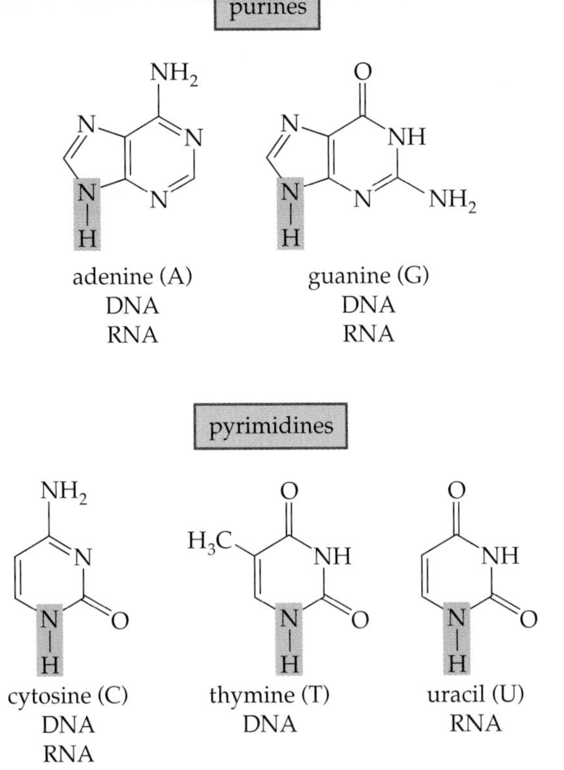

FIGURE 14.22

Of the five bases found in nucleic acids, adenine and guanine are purines, while cytosine, thymine, and uracil are pyrimidines.

three—**cytosine** (C), **thymine** (T), and **uracil** (U)—are single-ring bases and belong to the class of compounds called *pyrimidines.* Purines and pyrimidines are bases because their nitrogen atoms can accept protons (Chapter 7). Adenine, guanine, and cytosine are found in both DNA and RNA. Thymine is found in DNA, while uracil is found in RNA.

The formation of a representative nucleotide from its three components is shown in **Figure 14.23**. The pentose forms an ester bond with a phosphoric acid unit through the —OH on carbon 5 and forms another bond with the base through the —OH on carbon 1. Water is eliminated as these bonds are formed. When successive nucleotides bond to form long-chain nucleic acids, the phosphate unit of one nucleotide forms a second ester bond through the —OH on carbon 3 of the pentose unit of another nucleotide **(Fig. 14.24)**. The alternating phosphate-sugar units form the backbone of the nucleic acids.

The Double Helix

The sequence of the four bases in a nucleic acid, like the sequence of amino acids in a protein, is the primary structure. Like proteins, nucleic acids also have a secondary structure. The secondary structure of DNA was determined in 1953, at Cambridge University, by the American James D. Watson and the Englishman Francis H. C. Crick **(Fig. 14.25)**. Their discovery, with all its biological implications, was one of the most significant scientific achievements of the 20th century.

In the early 1950s, many facts were known about DNA. One particularly important observation was that when DNA from many different organisms was analyzed, the amount of adenine (A) always equaled the amount of thymine (T), and the amount of cytosine (C) always equaled the amount of guanine (G). X-ray studies by Maurice H. F. Wilkins and Rosalind Franklin, who were working at King's College (London), revealed the distances and angles between atoms in the DNA molecule, but the precise three-dimensional configuration of the nucleotide units in the molecule eluded scientists until 1953.

On the basis of the available information, Watson and Crick concluded that a DNA molecule must consist of two polynucleotide chains wound around one another to form a **double helix,** a structure that can be compared to a spiral staircase. The phosphate-sugar backbone represents the handrails, and the pairs of bases linked together by hydro-

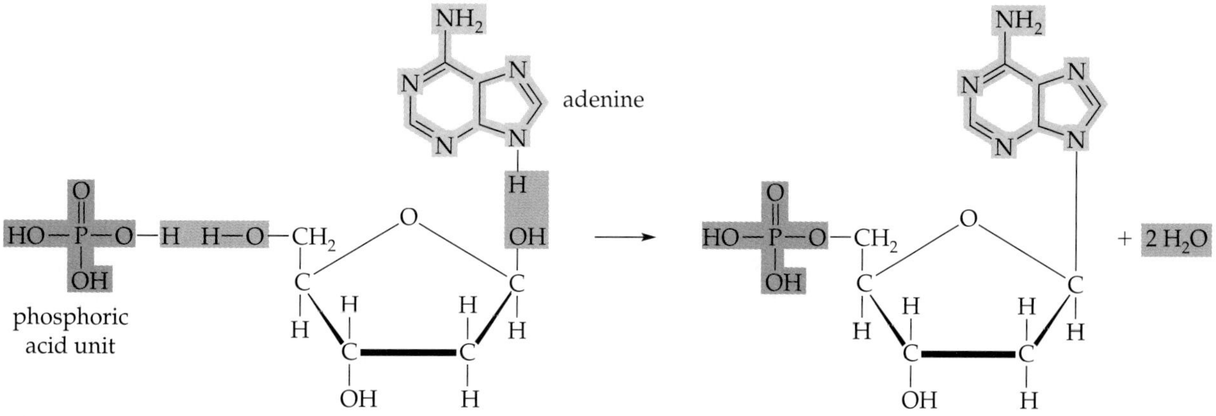

FIGURE 14.23

The formation of a nucleotide from its three components.

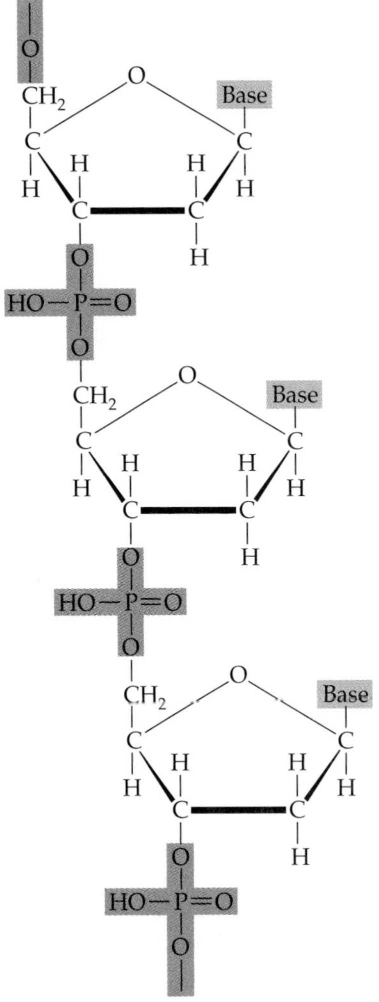

FIGURE 14.24

A three-nucleotide segment of a DNA strand, showing the alternating phosphate-sugar units that form the backbone of the molecule.

FIGURE 14.25

James D. Watson (left) and Francis H. C. Crick at the Cavendish Laboratory, Cambridge University, with their model of DNA.

gen bonds represent the steps **(Fig. 14.26a)**. Working with molecular scale models, Watson and Crick realized that, because of the dimensions of the space available between the two strands of the helix and the relative sizes of the bases, a guanine (G) on one chain must always be bonded to a cytosine (C) on the other chain, and an adenine (A) must always be bonded to a thymine (T). This arrangement gives "steps" of almost equal lengths and explains the equal amounts of C and G and of A and T found in DNA molecules **(Fig. 14.26b)**. Although hydrogen bonds are relatively weak, there are so many of them in a DNA molecule that, under normal physiological conditions, they hold the two chains together. Note that the C-G (or G-C) base pairs are held together by three hydrogen bonds, the A-T (or T-A) base pairs by two. **Figure 14.26c** shows a space-filling model of DNA.

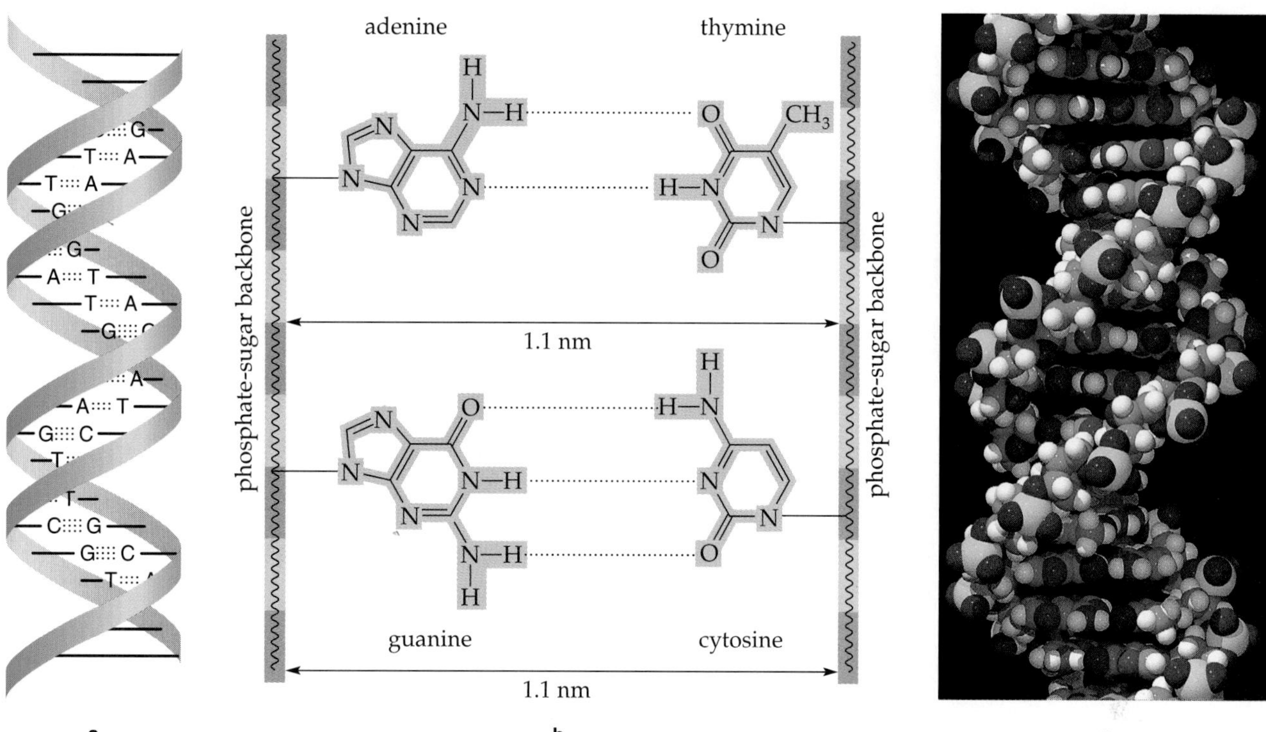

FIGURE 14.26

(a) Schematic representation of the double helix model of DNA. Hydrogen bonding between adenine (A) and thymine (T) and between cytosine (C) and guanine (G) is represented as dotted lines. (b) The steps in the spiral staircase structure of the DNA molecule are formed by two kinds of base pairs: Adenine (a purine) pairs with thymine (a pyrimidine) through two hydrogen bonds; guanine (a purine) pairs with cytosine (a pyrimidine) through three hydrogen bonds. These pairings give steps of equal width (1.1 nm) separating the two strands. (c) A space-filling model of DNA in which carbon atoms are blue, hydrogen atoms are white, oxygen atoms are red, nitrogen atoms are dark blue, and phosphorus atoms are yellow.

For their work in solving the DNA puzzle, Watson, Crick, and Wilkins were awarded a Nobel Prize in 1962.

Rosalind Franklin did not share the Nobel Prize with Watson, Crick, and Wilkins. She died of cancer at the age of 37, four years before it was awarded.

DNA, Genes, and Chromosomes

The DNA in the nuclei of cells is coiled around protein molecules called *histones* to form structures known as **chromosomes** (Fig. 14.27). The number of chromosomes varies with species. Humans have 46; each parent contributes 23.

Before the structure of DNA was understood, genes were known to be sections of chromosomes that determined inherited characteristics such as blue eyes and dark hair in humans. **Genes** are now defined as the segments of DNA molecules that control the production of all the different proteins in an organism. Proteins, in turn, control all the chemical reactions that occur in a living organism—including those that lead to blue eyes and dark hair. Each human being has approximately 30,000 genes, most of which are associated with a specific protein.

DNA molecules vary in the number and sequence of the base pairs they contain. The precise sequence of base pairs in the DNA molecule is the key to the genetic informa-

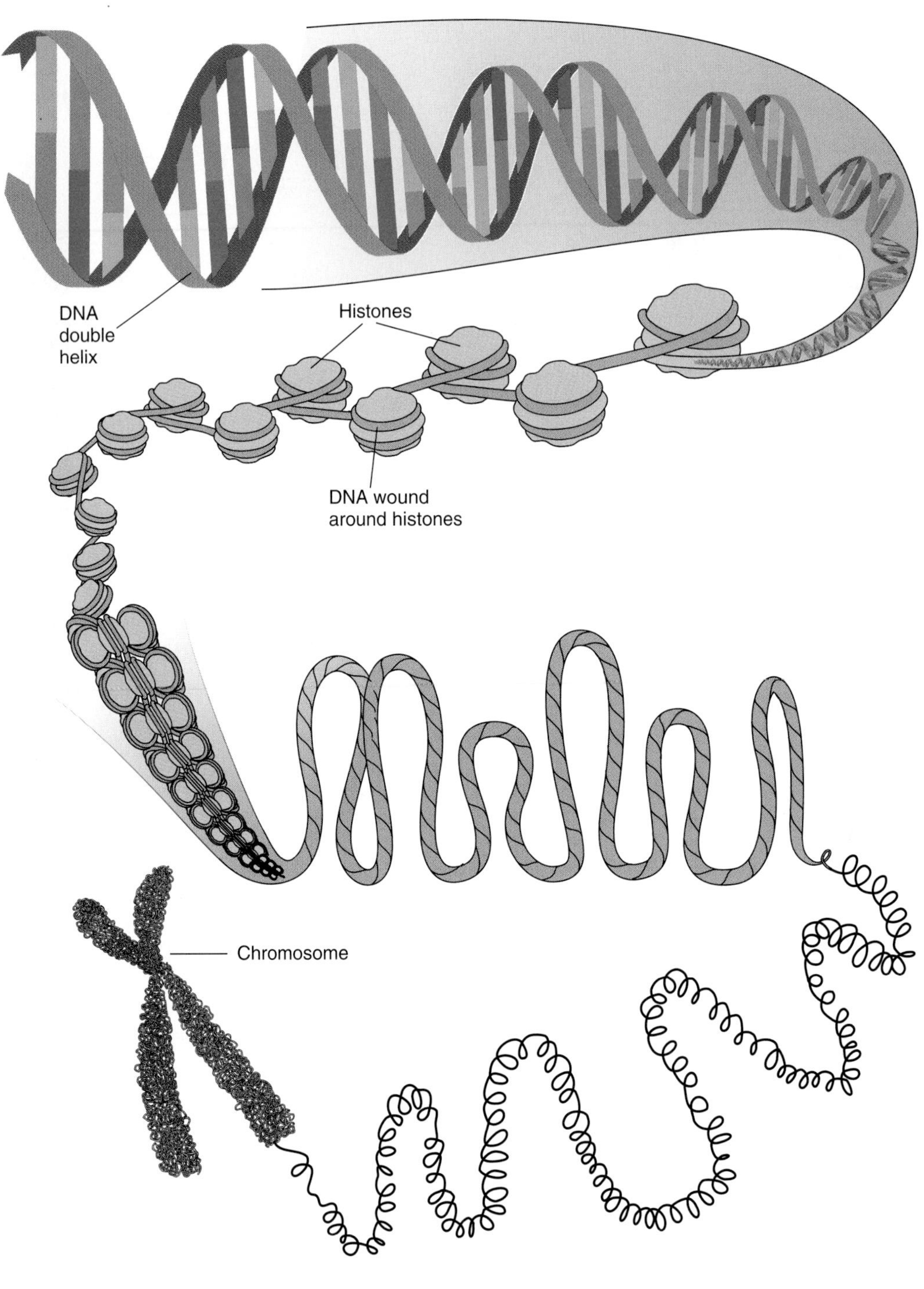

FIGURE 14.27

Chromosomes, which are found in cell nuclei, are tightly coiled strands of DNA.

tion that is passed on from one generation to the next; it is this sequence that directs and controls protein synthesis in all living cells. Each organism begins life as a single cell. The unique DNA in the nucleus of that cell determines whether the cell as it multiplies develops into a human, a bird, a rose, or a bacterium. The DNA carries all the information needed for making and maintaining the different parts of an organism—whether those parts are hearts, legs, wings, or petals.

Cell Replication

The double helix structure of DNA proposed by Crick and Watson explains very simply and elegantly how cells in the body divide to form exact copies of themselves. The two intertwined chains of the double helix are complementary. An A on one always pairs with a T on the other, and a C pairs with a G. Just before a cell divides, the double strand begins to unwind (**Fig. 14.28**). Each unwinding strand serves as a pattern, or *template*, for the formation of a new complementary strand. Nucleotides, which are always present in the cell fluid surrounding DNA, are attracted to the exposed bases and become hydrogen-bonded to them: A to T, T to A, C to G, and G to C. In this way, two identical DNA molecules are formed, one for each of the two daughter cells.

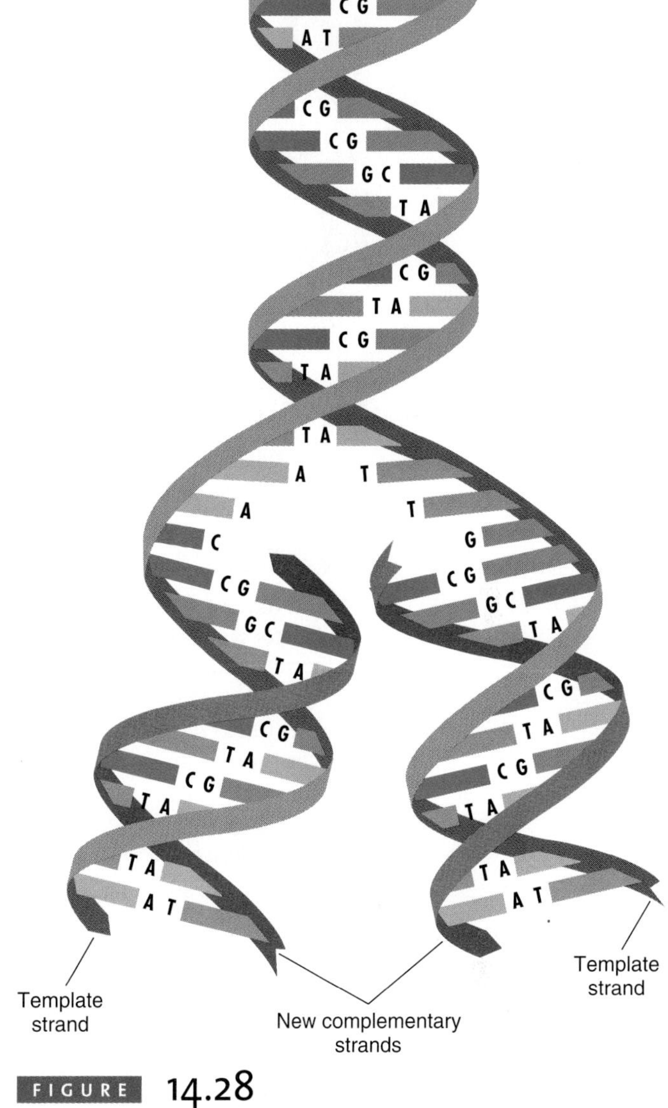

FIGURE 14.28

In DNA replication, the double helix gradually unwinds, and each strand acts as a template. Complementary nucleotides are attached to the single strands, forming two new DNA molecules identical to the original one.

14.12 Protein Synthesis

Protein synthesis is carried out in a series of complex steps involving RNA. As we noted earlier, the sugar in RNA is ribose, and the bases are uracil (U), adenine (A), guanine (G), and cytosine (C) (see Fig. 14.22). The primary structure of RNA is similar to that of DNA (see Fig. 14.20): phosphate-sugar units form the backbone, and each sugar (ribose) unit is bonded to one of the four bases. RNA molecules are much smaller than DNA molecules and contain from 75 to a few thousand base pairs. RNA molecules exist primarily as single strands rather than as double helixes.

Protein synthesis proceeds in two main steps:

$$\text{DNA} \xrightarrow[\text{transcription}]{} \text{mRNA} \xrightarrow[\text{translation}]{\text{tRNA}} \text{protein}$$

In the first step, **transcription,** a single strand of RNA is synthesized inside the cell nucleus, as follows. A segment of the DNA double helix separates into single strands (**Fig. 14.29a**), and the exposed bases on one strand act as the template for the synthesis of a molecule of RNA. The base sequence in this RNA, which is known as *messenger*

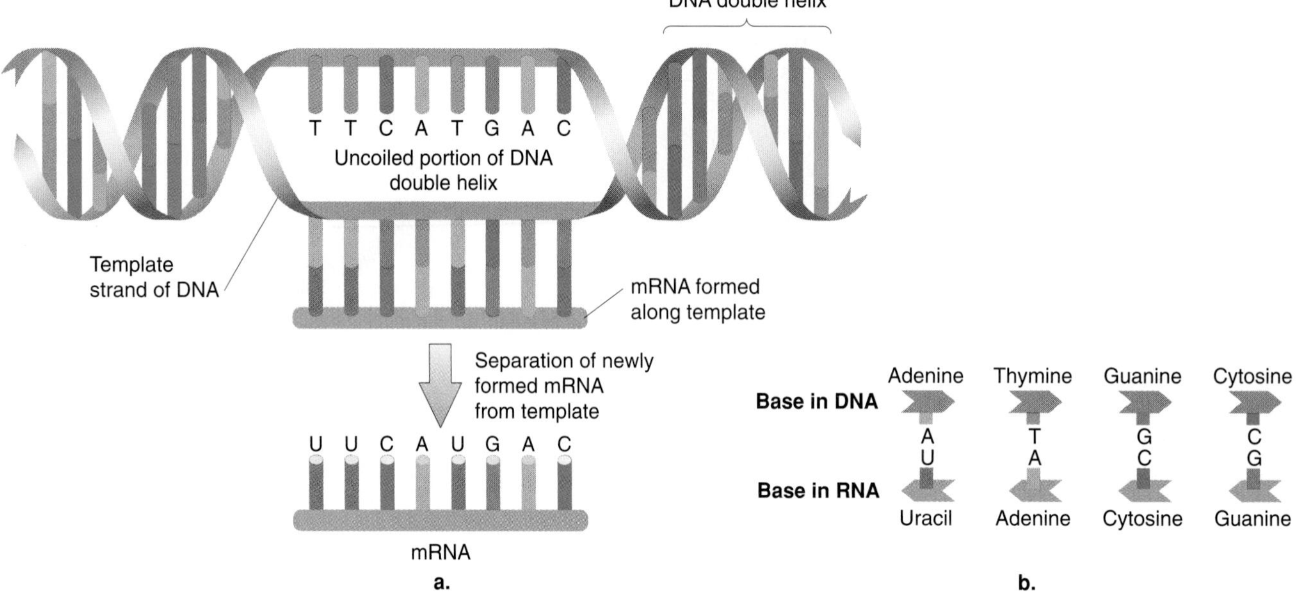

FIGURE 14.29

(a) In transcription, a segment of the DNA double helix unwinds, and one strand acts as the template for synthesis of messenger RNA (mRNA). Note that where there is an adenine (A) in the DNA, the mRNA transcribes uracil (U). (b) The base pairings between DNA and RNA.

RNA (mRNA), complements the base sequence on the DNA strand except that wherever there is an A in the DNA, the RNA will transcribe U instead of T (**Fig. 14.29b**).

EXAMPLE 14.3 Determining the sequence of bases in RNA

Show the sequence of bases in the mRNA that would be synthesized from the template strand of a DNA segment with the following sequence of bases:

DNA
—T—A—C—G—G—T—T—C—A—C—
—A—T—G—C—C—A—A—G—T—G— ← template strand

Solution The bases in the template strand would pair as follows: A with U, T with A, G with C, and C with G. Therefore, the mRNA would have the following sequence of bases:

—U—A—C—G—G—U—U—C—A—C—

(Notice how the sequence in the upper strand of DNA corresponds to the sequence in the mRNA, with U's instead of T's.)

Practice Exercise 14.3

Show the sequence of bases in the DNA strand that acted as the template for the synthesis of the following section of an mRNA molecule:

—C—G—G—U—A—U—C—U—A—

The next step is the **translation** of the code embodied in the new mRNA molecule to synthesize a particular protein. To direct this synthesis, the mRNA leaves the nucleus and takes its chemical message to the cytoplasm of the cell, where it binds with cellular structures called **ribosomes.** Taking part in translation are molecules of another kind of RNA called *transfer RNA (tRNA)*, which are responsible for delivering amino acids one by one to the mRNA. Each tRNA molecule carries a three-base sequence called an **anticodon** that determines the specific amino acid it will deliver. Three-base sequences along the mRNA strand, called **codons,** determine the order in which tRNA molecules bring the amino acids to the mRNA.

Guided by the first codon on the mRNA strand, a tRNA molecule with an anticodon that is complementary to this codon transports a specific amino acid to the mRNA codon. For example, if the bases are lined up as shown in **Figure 14.30a**, where the first codon is A-U-G, a tRNA with a U-A-C anticodon will transport the amino acid methionine (Met) to the mRNA, where the anticodon will pair up through hydrogen bonding with its complementary codon. Similarly, a second tRNA molecule will bring the amino acid phenylalanine (Phe) to the next mRNA codon, U-U-U. A peptide bond then forms between the two amino acids, and the methionine separates from the tRNA **(Fig. 14.30b)**. When a third amino acid, glutamic acid (Glu), has been added, the first tRNA molecule is released from the mRNA **(Fig. 14.30c)**. The actual protein synthesis occurs in the ribosomes, which move along the mRNA one codon at a time as the amino acid chain grows. In this way, the mRNA is read codon by codon, and the protein is built up one amino acid at a time in the correct sequence. When the synthesis is completed, the protein separates from the mRNA **(Fig. 14.30d)**.

14.13 The Genetic Code

The specific amino acid that each of the three-base sequences in mRNA codes for has been established. This information, which is shown in **Table 14.3**, is the **genetic code.** Sixty-four possible three-base codons can be formed from the four bases found in RNA (4^3, or $4 \times 4 \times 4 = 64$). Since all proteins are made from only 20 amino acids, there is some redundancy in the system; that is, several amino acids are designated by more than one codon. Some codons act as "stop" signals to terminate protein synthesis. The codon A-U-G not only encodes for methionine, it is also a "start" signal for protein synthesis.

In 1990, an international effort to map the location and sequence of all the genes on the chromosomes in the nucleus of human cells was launched. This project, which is called the *Human Genome Project*, was virtually completed in 2000. Final detailed maps will show the complete sequence of all the base pairs in human DNA and will identify the genes and the protein that each one makes. This information will give scientists a clearer understanding of the link between DNA sequence and disease, and it will speed up the search for ways to replace defective genes with normal ones.

The *genome* of an organism is its total complement of genes.

14.14 DNA Profiling

DNA profiling, also known as *DNA typing* or DNA *fingerprinting*, is becoming increasingly important as a means of identification in criminal investigations and paternity cases. The technique is based on the fact that, with the exception of identical twins, each person's DNA is unique. Blood, semen, saliva, skin, or hair follicles can be used for DNA profiling.

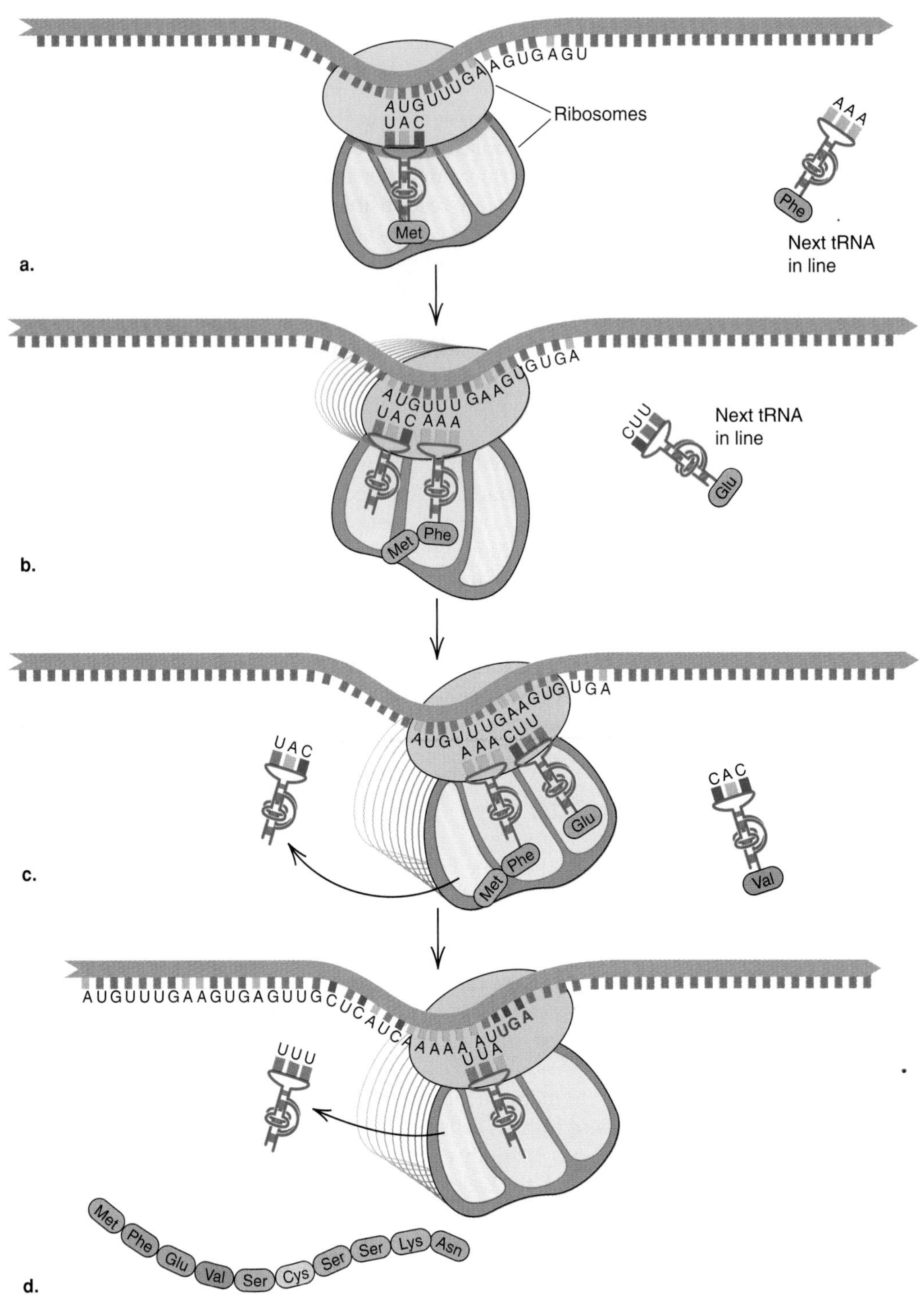

FIGURE 14.30

Translation and protein synthesis: (a) A transfer RNA (tRNA) molecule carrying the anticodon U-A-C delivers the amino acid methionine (Met) to the mRNA strand, where the anticodon bonds with the codon A-U-G. Another tRNA molecule carrying the anticodon A-A-A brings the amino acid phenyl alanine (Phe) to the next mRNA codon, U-U-U. (b) A peptide bond forms between Met and Phe, and Met separates from the tRNA. A third amino acid, glutamic acid (Glu), is brought to the next mRNA codon, G-A-A. (c) The tRNA molecule carrying the U-A-C anticodon is released from the mRNA. (d) When synthesis of the protein is complete, the protein molecule separates from the mRNA.

Table 14.3 The Genetic Code

First Base	Second Base: U		C		A		G		Third Base
U	U-U-U	Phe	U-C-U	Ser	U-A-U	Tyr	U-G-U	Cys	U
	U-U-C	Phe	U-C-C	Ser	U-A-C	Tyr	U-G-C	Cys	C
	U-U-A	Leu	U-C-A	Ser	U-A-A	Stop	U-G-A	Stop	A
	U-U-G	Leu	U-C-G	Ser	U-A-G	Stop	U-G-G	Trp	G
C	C-U-U	Leu	C-C-U	Pro	C-A-U	His	C-G-U	Arg	U
	C-U-C	Leu	C-C-C	Pro	C-A-C	His	C-G-C	Arg	C
	C-U-A	Leu	C-C-A	Pro	C-A-A	Gln	C-G-A	Arg	A
	C-U-G	Leu	C-C-G	Pro	C-A-G	Gln	C-G-G	Arg	G
A	A-U-U	Ile	A-C-U	Thr	A-A-U	Asn	A-G-U	Ser	U
	A-U-C	Ile	A-C-C	Thr	A-A-C	Asn	A-G-C	Ser	C
	A-U-A	Ile	A-C-A	Thr	A-A-A	Lys	A-G-A	Arg	A
	A-U-G	Met	A-C-G	Thr	A-A-G	Lys	A-G-G	Arg	G
G	G-U-U	Val	G-C-U	Ala	G-A-U	Asp	G-G-U	Gly	U
	G-U-C	Val	G-C-C	Ala	G-A-C	Asp	G-G-C	Gly	C
	G-U-A	Val	G-C-A	Ala	G-A-A	Glu	G-G-A	Gly	A
	G-U-G	Val	G-C-G	Ala	G-A-G	Glu	G-G-G	Gly	G

Many human genes are common to all of us. For example, we all have genes that code for the proteins hemoglobin, insulin, and myosin. We also have long segments of DNA that do not code for any protein but act primarily as spacers between the coding areas of DNA. This noncoding area of DNA varies considerably from person to person, and it is this area that is analyzed in DNA profiling.

In the first step in the analysis of a sample, DNA is extracted from a particular chromosome. Then enzymes called *restriction enzymes* are used to cut the DNA strands into fragments at specific base sequences. For example, there is one restriction enzyme that will cleave DNA wherever the sequence G-A-A-T-T-C occurs **(Fig. 14.31)**. The number and lengths of the fragments that are produced by the restriction enzymes vary from person to person. In the next step, the fragments are applied to a strip of gel and separated using *gel electrophoresis*, a technique based on the different rates at which charged particles move in an electric field. When the electric field is applied across the gel strip, the fragments, which contain negatively charged phosphate groups, migrate to the positive electrode at rates that vary according to their size. Shorter fragments move faster than longer ones. The gel strip is then treated to reveal the locations of the fragments.

In criminal cases, the pattern obtained from gel electrophoresis of the suspect's DNA is compared with the pattern obtained from a sample of hair, skin, other tissue, semen, or blood collected at the crime scene **(Fig. 14.32a)**. A DNA match is evidence that the suspect at least was present when the crime was committed. To increase the accuracy of the identification, DNA from more than one chromosome is analyzed.

EXPLORATIONS

Beer—America's Favorite Alcoholic Beverage

Alcoholic beverages have been consumed since before the dawn of recorded history; long before there was Budweiser or Miller, there was beer. Ancient records reveal that by 3000 B.C., beer was already an important commercial product in Babylon and a common dietary item in Egypt. Greek and Roman writers describe exuberant consumption of wine and beer in classical times by the gods and by mortals. In the first few centuries A.D., beer drinking became particularly popular among Germanic and Nordic tribes and among the Saxons and Celts of Northern Europe where the climate was unsuitable for cultivating grapes for making wine. By the Middle Ages, most larger households made beer for their own consumption. It was in the monasteries, however, that the real advances in the art of brewing occurred. Most monasteries maintained at least one brewhouse and monks received a daily ration of about a gallon of beer. Gradually, as towns grew, local commercial breweries were established, each one producing its own unique brew.

Alcoholic beverages have been consumed since before the dawn of recorded history; long before there was Budweiser or Miller, there was beer.

A vat of wort

Beer came to North America with the early English settlers, and in the years leading up to the American Revolution, breweries became an important part of the economy. Home brewing remained popular. George Washington had his own recipe for making beer on his Mount Vernon estate. The pale, light lagers or Pilsner-type beers, so popular today, came to the United States when German immigrants began arriving early in the 19th century.

Americans love beer, drinking more of it than either wine or hard liquor. Wine, the second most popular alcoholic beverage, is made primarily from grapes, while beer is made from sprouted grain, called *malt*. The principal ingredient in all these alcoholic beverages, and the cause of the intoxicating effects they produce, is the simple but mind-altering chemical ethanol.

The term *beer* usually implies a beverage of Western culture made principally from fermented barley and flavored with hops. But nearly all other cultures have a long history of beer making as well. Beer made from rice has been produced in Asia for thousands of years, and long before the discovery of the New World by Europeans, the Aztecs and Incas were making beer from fermented corn. In Africa, excellent beers are made from sorghum seeds.

Hops have been used as bittering agents to counteract the sweet taste of sugar in fermented grain since about 3000 B.C., but were probably not cultivated for this purpose until the 9th century A.D. The hop (*Humulus lupus*) is a perennial vine common in Europe, Asia, and North America. Male and female flowers grow on separate plants, and it is the cone-shaped flower clusters of the female plant that contain the oils and resin that give beer its characteristic bitter taste. Male plants are excluded from hop fields as far as possible to prevent fertilization, which results in the formation of unpleasant-tasting seeds. Hops require a mild climate and an abundant supply of water, and in the United States, they are cultivated primarily along the Pacific coast.

The Chemistry of Brewing

Today, the complex chemical processes of brewing are fully automated, but breweries still follow the four basic steps established in the Middle Ages: malting, mashing, wort boiling, and fermentation. To make hard liquor, such as vodka, gin, whiskey, rum, or brandy, an extra distillation step is required.

In malting, the grain is steeped in water, which softens it and initiates germination. As the grain sprouts, enzymes are formed and begin the conversion of starch to dextrins, malt sugar (maltose), and other fermentable sugars. At the appropriate stage in germination, the grain is kiln-dried to stop further growth. The kilning temperature is critical for the development of the desired amount of color in the final product. The higher the temperature, the darker the beer.

In the next step, mashing, the dried sprouted grain, or malt, is ground and mixed with hot water. Enzymes continue to work, and residual starch is broken down into simpler carbohydrates. If a Pilsner-type beer is being brewed, corn or rice, which are cheaper than malt, is added at this stage. Enzymes in the malt act on starch in the added grains, thus increasing the production of fermentable sugars. Soluble carbohydrates, together with various other organic compounds and minerals, dissolve in the water, which is then separated from the spent grain. (Rich in protein, the spent grain is sold as animal feed.) The liquid, called *wort,* is boiled with hop cones or an extract of hops to acquire the desired bitter flavor. After the wort has been cooled and strained, yeast is added. Soon fermentation, the most critical step in the brewing process, begins.

During fermentation, enzymes produced by yeast convert sugars to ethanol (ethyl alcohol) and carbon dioxide.

$$C_6H_{12}O_6 \xrightarrow{\text{enzymes}} 2\ C_2H_5OH + 2\ CO_2$$

Since the yeast greatly affects the flavor of the final product, each brewery has its own special strain. To make a typical Pilsner-type beer, or lager, the brewer uses a bottom-fermenting yeast, which sinks to the bottom when spent. When fermentation is complete, the beer is stored for several months in a cold cellar (the German word *lager* means "to store") where it becomes clear and acquires its mature flavor. To produce the slightly stronger, more highly flavored ales popular in the United Kingdom, a higher temperature and a top-fermenting yeast, which rises to the surface when spent, are used. Cask beers acquire their unique flavor as a result of residual fermentation that is allowed to occur in the cask. Before beer is bottled or canned, it is filtered, carbonated, and usually pasteurized.

The main ingredients in beer are ethanol and carbon dioxide, but trace amounts of many other compounds, including other alcohols, aldehydes, ketones, acids, esters, vitamins, and minerals, are also present. These ingredients, derived in varying amounts from hops, grain, yeast, and water, are responsible for subtle nuances of flavor in the final product. Although many beer drinkers express a preference for a particular brand of beer—Budweiser, Michelob, Miller, or Schlitz, for example—very few are able to recognize their proclaimed favorite in blind tests.

The Effects of Drinking Beer

When a person drinks a can of beer, most of the alcohol in it (4.0–5.0% by volume in the average American beer) is absorbed directly from the intestines into the bloodstream and carried to the liver. In the liver, enzymes begin the oxidation of the alcohol to acetaldehyde, then to acetic acid, and finally to carbon dioxide and water.

$$\underset{\text{ethanol}}{CH_3CH_2OH} \longrightarrow \underset{\text{acetaldehyde}}{CH_3CHO} \longrightarrow \underset{\text{acetic acid}}{CH_3COOH} \longrightarrow CO_2 + H_2O$$

If alcohol is consumed faster than it can be detoxified in the liver (more than about 1 ounce of alcohol per hour), blood going to the brain contains alcohol, which has an immediate effect on mental processes. At low blood concentration (slow consumption of a 12-ounce can of beer), alcohol relaxes tension and acts as a stimulant but, as the alcohol concentration in blood increases, the brain begins to process information in abnormal ways. Learning skills deteriorate, and the effects are increasingly depressant. If three or four beers are consumed rapidly, judgment is measurably impaired. After about six beers, a person is visibly drunk with slurred speech, unsteady gait, and confused thinking. After nine or ten beers in quick succession, most people lose consciousness.

The effects of alcohol vary considerably depending on a person's sex, size, genetic make-up, previous drinking habits, and many other factors. Women, because they lack a stomach enzyme that in men destroys some alcohol before it reaches the intestines, are approximately twice as vulnerable to alcohol as men. And they face another danger if they are pregnant. Any alcohol in their blood will reach the fetus, where it can cause stunted growth, disfigurement, and mental retardation, a condition known as *fetal alcohol syndrome.*

Reference: *Encyclopedia Brittanica* (1974 ed.); see "beer" and "brewing."

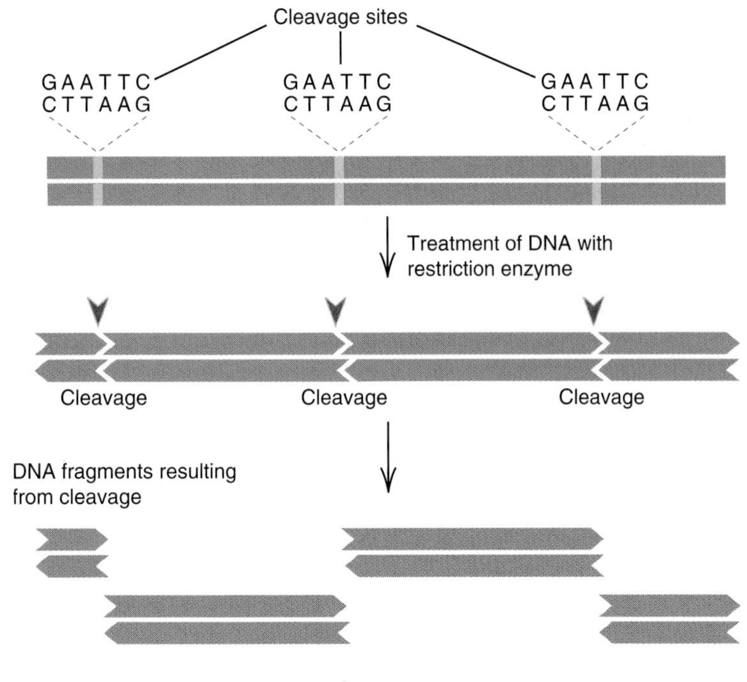

FIGURE 14.31

When DNA is treated with a restriction enzyme, the two strands are cut into fragments at specific sites. One restriction enzyme cleaves the DNA strands wherever the base sequence G-A-A-T-T-C occurs, producing a number of fragments of different lengths.

In paternity cases, blood from the child, the alleged father, and the mother are analyzed **(Fig. 14.32b)**. Half the child's DNA pattern will match that of the mother. If the other half matches that of the alleged father, it is proof that he is the child's biological father.

14.15 Genetic Engineering

Over 3000 hereditary diseases are known to be caused by an error in the nucleotide sequence in a single gene. Researchers have identified the defective genes responsible for several of these diseases, including sickle cell anemia, cystic fibrosis, and amyotrophic lateral sclerosis (ALS), also known as Lou Gehrig's disease. The development of diabetes and some cancers has also been traced to defective genes. Researchers hope that it will soon be possible to cure many hereditary diseases by replacing the defective genes with normal ones.

Another important area of research is the creation of plants that have been genetically altered to incorporate some desirable trait. For example, DNA from a bacterium that produces a substance toxic to some caterpillars has been inserted into the DNA of corn. When corn borers, which cause enormous crop damage, attempt to feed on this altered corn, they are killed. Genetic en-

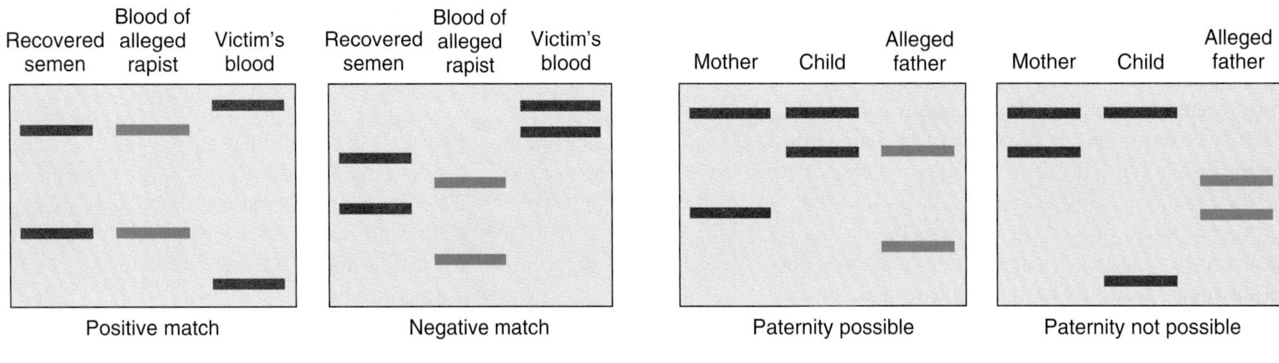

FIGURE 14.32

Representation of patterns produced in DNA profiling: (a) A match between the patterns obtained from an alleged rapist's blood and from semen recovered at the crime scene is evidence that the suspect could be the rapist. (b) A match between the patterns obtained from blood taken from an alleged father of a child and from blood taken from the child is evidence that the man could be the child's biological father.

gineering has also been used to produce crop plants with resistance to certain viruses and certain herbicides used as weed killers and to create bacteria that manufacture specific proteins, including human insulin.

Molecular engineering's potential for curing disease and improving crop yields is obviously enormous, but creating plants and animals with new characteristics also involves risk. There is concern that if not used wisely, genetic engineering could upset the balance of nature.

Chapter Summary

1. A molecule that contains a carbon atom attached to four different atoms or groups of atoms can exist in two forms that are mirror images of each other. Such a molecule is said to be chiral.
2. Carbohydrates can be divided into several classes:
   a. monosaccharides ($C_6H_{12}O_6$), such as glucose, galactose, and fructose
   b. disaccharides ($C_{12}H_{22}O_{11}$), which are made up of two monosaccahride units and include sucrose, maltose, and lactose
   c. polysaccharides, which are polymers of glucose, such as starch and cellulose
3. Glucose units are joined together through α-linkages in starch; through β-linkages in cellulose.
4. Fats and oils are composed of triglycerides, which are esters of fatty acids with glycerol. Fats are high in saturated fatty acids; oils are high in unsaturated fatty acids.
5. Unsaturated oils can be converted to semisolid fats, such as margarine, by hydrogenation—the addition of hydrogen atoms to some of the fatty acid double bonds.
6. Important steroids include cholesterol, the female hormones estrogen and progesterone, and the male hormone testosterone.
7. Persistent organic pollutants (POPs) are suspected of causing disruptions in reproduction and fetal development in humans and other animals.
8. Proteins are polymers of amino acids—compounds containing a carboxyl group ($-COOH$) and an amino group ($-NH_2$). Amino acids join together through peptide bonds to form proteins.
9. In humans, proteins are assembled from 20 different amino acids.
10. The primary structure of a protein is its sequence of amino acids.
11. Proteins also have secondary, tertiary, and quarternary structures, which together determine the proteins' shapes. The structures are the result of hydrogen bonding between amino acids and folding, cross-linking, and coiling of the chains.
12. Enzymes, which are biological catalysts, are involved in nearly all chemical reactions that occur in living systems.
13. Deoxyribonucleic acid (DNA), which is present in cell nuclei, and ribonucleic acid (RNA), which is present in cell cytoplasm, are responsible for storage and transmission of genetic information.

14. Nucleic acids are polymers of nucleotides, which are made up of three parts:
    a. a pentose sugar
    b. a nitrogen-containing heterocyclic base
    c. a phosphoric acid unit
15. The four bases in nucleic acids are adenine (A) and guanine (G), which are purines, and cytosine (C), thymine (T), and uracil (U), which are pyrimidines.
16. In DNA, the sugar is deoxyribose and the bases are A, G, C, and T. In RNA, the sugar is ribose and the bases are A, G, C, and U (uracil replaces thymine).
17. DNA consists of two phosphate-sugar chains wound around one another to form a double helix. A guanine attached to one chain always bonds with a cytosine attached to the other, and an adenine bonds with a thymine.
18. Genes are the segments of DNA molecules that control the production of proteins. The sequence of base pairs determines which protein is synthesized.
19. During cell replication, the two strands of the DNA double helix unwind, and each strand serves as the template for the formation of a new complementary strand.
20. In the first step of protein synthesis (transcription), a single strand of DNA acts as the template for the synthesis of messenger RNA (mRNA). The base A pairs with U, T with A, G with C, and C with G.
21. In the second step (translation), mRNA binds with ribosomes. Then, transfer RNA (tRNA) molecules carrying three-base sequences called anticodons bring specific amino acids, in turn, to the complementary three-base sequences along the mRNA, called codons. When all the amino acids have been added, the protein separates from the RNA.
22. The genetic code consists of the 64 possible codons in mRNA that specify individual amino acids. Some codons act as start and stop signals.
23. Genetic engineering involves inserting a gene taken from one organism into the DNA of another organism. Researchers hope to be able to cure hereditary diseases by replacing defective genes with normal ones.

Key Terms

active site (p. 433)
adenine (p. 436)
amino acid (p. 427)
androgens (p. 425)
anticodon (p. 443)
carbohydrate (p. 414)
cellulose (p. 417)
chiral molecule (p. 413)
cholesterol (p. 423)
chromosome (p. 439)
codon (p. 443)
coenzyme (p. 433)
cytosine (p. 437)
deoxyribonucleic acid (DNA) (p. 434)
dipeptide (p. 427)
disaccharide (p. 415)
DNA profiling (p. 443)
double helix (p. 437)
endocrine disruptor (p. 426)
enzyme (p. 433)
estrogens (p. 425)
fat (p. 421)
fatty acid (p. 418)
gene (p. 439)
genetic code (p. 443)
glucose (p. 416)
glycogen (p. 417)
guanine (p. 436)
hormone (p. 424)
lactose (p. 417)
lipid (p. 418)
monosaccharide (p. 415)
nucleic acid (p. 434)
nucleotide (p. 435)
oil (p. 421)
peptide bond (p. 427)
polysaccharide (p. 415)
progesterone (p. 425)
protein (p. 426)
ribonucleic acid (RNA) (p. 434)
ribosome (p. 443)
saturated fatty acid (p. 419)
starch (p. 417)
steroid (p. 423)
substrate (p. 433)
sucrose (p. 416)
testosterone (p. 425)
thymine (p. 437)
transcription (p. 441)
translation (p. 443)
triglyceride (p. 421)
unsaturated fatty acid (p. 418)
uracil (p. 437)

Questions and Problems

1. What is a chiral molecule?

2. Which of the following is a chiral molecule? Explain and draw the mirror-image molecule.

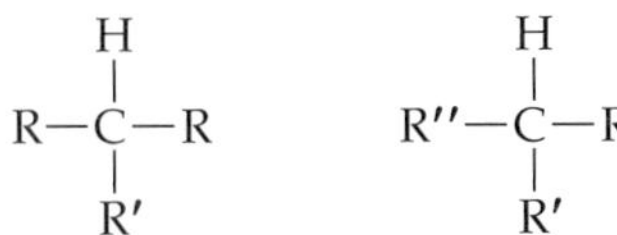

3. What are the three main classes of carbohydrates?

4. Both glucose and fructose have the same molecular formula, $C_6H_{12}O_6$. Draw their structural formulas. How do they differ?

5. Cellulose is made from glucose, but humans cannot digest cellulose. Explain.

6. How does your body store carbohydrates?

7. What similarities are there between starch and cellulose?

8. Glucose molecules exist mostly as rings, rather than straight chains. Draw the straight-chain and cyclic structures of glucose. Number the carbon atoms in each structure.

9. Give chemical names for each of the following:
   a. cane sugar **b.** grape sugar
   c. milk sugar **d.** sugar found in cola drinks

10. Indicate whether each of the following is a monosaccharide, disaccharide, or polysaccharide:
    a. sucrose **b.** glucose
    c. cellulose **d.** lactose

11. Name the two monosaccharides in each of the following:
    a. sucrose **b.** maltose **c.** lactose

12. What is glycogen? How is it formed, and what is its role in the human body?

13. Starch is a mixture of two types of polysaccharides. Name them.

14. Name the enzyme your body uses to break down table sugar. What are the products of this reaction?

15. What distinguishes starch, amylopectin, and glycogen?

16. Why are carbohydrates considered to be a source of energy?

17. How are lipids classified? List two physical properties of lipids.

18. What is the difference between unsaturated and saturated fats? Give an example of each.

19. What is a polyunsaturated fat?

20. What happens when a polyunsaturated fat is hydrogenated?

21. Name several foods that contain unsaturated fats.

22. What molecules react to form triglycerides? Write an equation showing how a triglyceride is formed.

23. Name an anti-inflammatory steroid produced naturally in your body. Draw the structure of a steroid.

24. In what organ is cholesterol converted to bile salts?

25. Name two important female hormones.

26. Estrogen mimics are POPs that are the focus of numerous animal studies. What characteristics are observed in animals that have ingested estrogen mimics?

27. The structures of estrogen mimics are not similar to the structure of estrogen. How do the mimics have effects similar to those of the real hormone?

28. What two functional groups are present in an amino acid molecule? Draw the structure of a representative amino acid.

29. Name the simplest amino acid and draw its structure. Is it a chiral molecule? Explain.

30. Approximately how many amino acids are needed to make the proteins found in the human body? How many are essential amino acids, and what does this term mean?

31. What element is always present in amino acids but is not present in sugars or fats?

32. What percentage of all the amino acids in the human body can be synthesized there?

33. Show how two amino acids join to form a dipeptide. Highlight the peptide bond.

34. If the three amino acids leucine, glycine, and histidine formed all possible different dipeptides, how many would there be? List them using the appropriate three-letter abbreviation to represent each amino acid.

35. a. What is meant by the primary, secondary, and tertiary structures of proteins? b. Draw a diagram showing the outline of the secondary structure of the protein in hair. Name this type of structure.
36. What type of bonds are responsible for the helix structure of some proteins?
37. Describe the lock-and-key model of enzyme action.
38. Describe what a substrate is and how it interacts with an enzyme.
39. Many adults cannot digest milk. What enzyme do they lack?
40. What factors determine the action of an enzyme?
41. What are the functions of nucleic acids?
42. What are the three components of every nucleotide? Draw a simple diagram showing how the components are bonded together.
43. Name two types of nucleic acids. How do the sugar molecules in these two types differ from each other?
44. Name the base that forms a pair with each of the following bases:
    a. uracil b. guanine
    c. adenine d. cytosine
45. The base sequence along one strand of DNA is A-C-T-G-T. What would be the sequence on the complementary strand of DNA?
46. Why does cytosine bond only with guanine?
47. Describe two ways in which RNA differs from DNA.
48. Which nucleotide bases have double rings?
49. Where in a living cell is DNA found?
50. Describe how replication occurs.
51. Describe the difference between coding and non-coding regions of the DNA strand.
52. How does gel electrophoresis separate DNA fragments for DNA profiling?
53. How would you analyze and compare the DNA contained in the following two blood samples: one drawn from the blood of a crime suspect and one from the crime scene?
54. How do scientists expect to cure genetic diseases using genetic engineering?

Answers to Practice Exercises

14.1

```
     CH3              CH3
     |                |
 Br—C—Cl          Cl—C—Br
     |                |
     C2H5             C2H5
```

14.2 Formation of dipeptide 1:

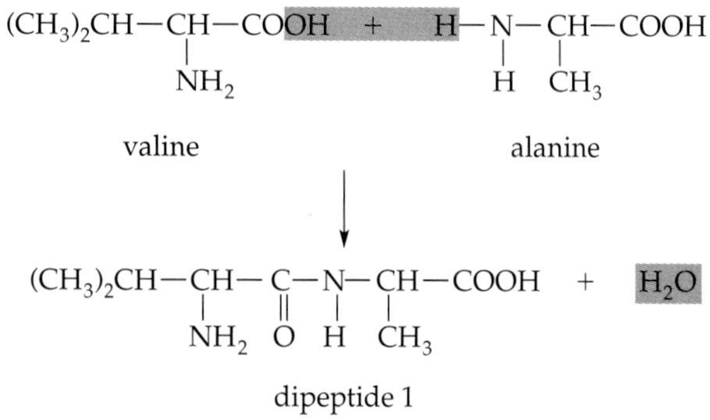

```
(CH3)2CH—CH—COOH  +  H—N—CH—COOH
         |             |  |
         NH2           H  CH3
      valine             alanine
                  ↓
(CH3)2CH—CH—C—N—CH—COOH  +  H2O
         |  ‖ |  |
        NH2 O H  CH3
         dipeptide 1
```

Formation of dipeptide 2:

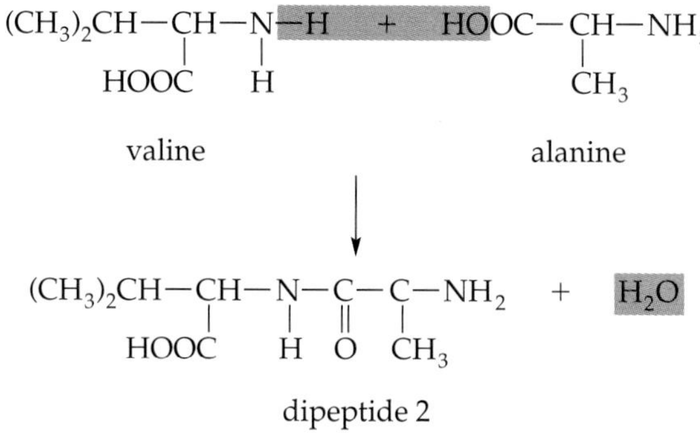

```
(CH3)2CH—CH—N—H  +  HOOC—CH—NH2
         |   |              |
      HOOC   H              CH3
      valine               alanine
                  ↓
(CH3)2CH—CH—N—C—C—NH2  +  H2O
         |   | ‖ |
      HOOC   H O CH3
          dipeptide 2
```

14.3 DNA sequence:

—G—C—C—A—T—A—G—A—T—

Chapter 15 Food and Nutrition

Chapter Objectives

In this chapter, you should gain an understanding of:

- The process by which food is digested
- How energy is obtained from food
- Dietary needs for carbohydrates, fats, and proteins
- Nutritional diseases and disorders
- How vitamins and minerals contribute to good nutrition
- The value and potential danger of food additives
- Dietary goals that promote good health

The bounty of nature—a varied harvest of fruits and vegetables.

ONLINE FEATURES

www.jbpub.com/chemistry

- Chemistry and the Environment
- eLearning
- Chemistry in the News
- Special Topics
- Research & Reference Links
- Ask the Authors

THE STUDY OF FOODS AND THE WAYS in which the body utilizes them is known as *nutrition*. It includes the identification of the components in foods that are essential for growth and for the maintenance of health. For humans, a nutritional diet must supply the proper balance of six basic ingredients: carbohydrates, fats, proteins, vitamins, minerals, and water.

These essential food components, or **nutrients,** serve three main functions: (1) provide energy for the performance of essential biological processes, (2) supply materials for building and replacing body tissues, and (3) provide substances needed for the regulation of life processes, such as the transport of oxygen and digestion. Fats and carbohydrates supply most of the body's energy needs, while proteins supply most of the building blocks for tissue construction. Small quantities of vitamins and minerals act as regulators of vital processes.

In this chapter, we examine the ways in which the body makes use of the different nutrients. We discuss nutritional disorders and the effects that deficiencies and overconsumption have on health. Lastly, we study food additives, chemicals that have little or no nutritional value but are added to foods to prevent spoilage or to flavor, color, or in some other way improve appearance or taste.

15.1 The Composition of the Body

The weight contributed by the different elements in the body is related to the atomic weights of their atoms (O = 16; C = 12; H = 1; N = 14).

We need food essentially to build and maintain our bodies. Food, therefore, must supply all the elements from which the body is composed. Broken down into its elements, the body is made up as shown in **Table 15.1**. Four elements account for 96% of the human body by weight: oxygen, 65%; carbon, 18%; hydrogen, 10%; and nitrogen, 3%. Two-thirds of the body's weight consists of water; therefore, most of the oxygen and hydrogen atoms

Table 15.1 Approximate Elemental Composition of the Human Body (including water)

Element	Percent by Weight
Oxygen (O)	65
Carbon (C)	18
Hydrogen (H)	10
Nitrogen (N)	3
Calcium (Ca)	1.5
Phosphorus (P)	1
Potassium (K)	0.35
Sulfur (S)	0.25
Chlorine (Cl)	0.15
Sodium (Na)	0.15
Magnesium (Mg)	0.05
Iron (Fe)	0.004
Trace elements	0.546

in the body are in water molecules. The remaining oxygen and hydrogen, together with the carbon and nitrogen, are combined to form organic compounds, primarily carbohydrates, fats, and proteins.

Other elements found in the body in significant amounts are calcium, phosphorus, potassium, sulfur, chlorine, sodium, and magnesium. Trace amounts of about 36 other elements are also present. Of these, about 20 are known to be essential for various life processes, but it is not yet known whether the remaining 16 also play essential roles.

15.2 The Fate of Food in the Body

The carbohydrates, fats, and proteins present in food are not in a form in which they can be immediately utilized by the body. They must first be broken down into simpler compounds and then recombined into specific compounds the body needs to build tissues and perform many other functions.

Starch is a polymer made up of many glucose molecules linked together.

Digesting Food

Maltose is a disaccharide composed of two molecules of glucose.

Breakdown of food occurs in a series of enzyme-catalyzed reactions as the food passes through the digestive tract (**Fig. 15.1**). Digestion of carbohydrates begins in the mouth when food is chewed and becomes mixed with saliva. Saliva contains the enzyme ptyalin, or salivary amylase, which catalyzes the partial breakdown of starch to yield the disaccharide maltose (Chapter 14). When food is swallowed, it enters the stomach and encounters the very acidic gastric juice. In this acid medium (pH from 1 to 3), the enzyme **pepsin** begins the digestion of proteins by catalyzing the breakdown of approximately 10% of the peptide bonds in the protein molecules. Polypeptides with molecular weights ranging from 600 to about 3000 are produced. Even though the lining of the stomach is protected from enzyme breakdown by a layer of mucus, it is slowly digested and must constantly be renewed. Digestion of carbohydrates is halted in the stomach because the acidic conditions there inactivate ptyalin.

Food passes from the stomach to the small intestine, where secretions from the pancreas neutralize the acid and provide the enzymes needed to complete the digestion of carbohydrates and proteins. Carbohydrates are broken down to yield the monosaccharides glucose, fructose, and galactose, and polypeptides yield amino acids. The relatively small monosaccharide and amino acid molecules are then absorbed through the intestinal wall into the bloodstream.

Digestion of fats occurs primarily in the small intestine, where triglycerides, which account for more than 90% of the lipids in food, are broken down to glycerol and fatty acids. The breakdown is brought about by the combined action of bile salts (Chapter 14) and the enzyme lipase. Bile salts,

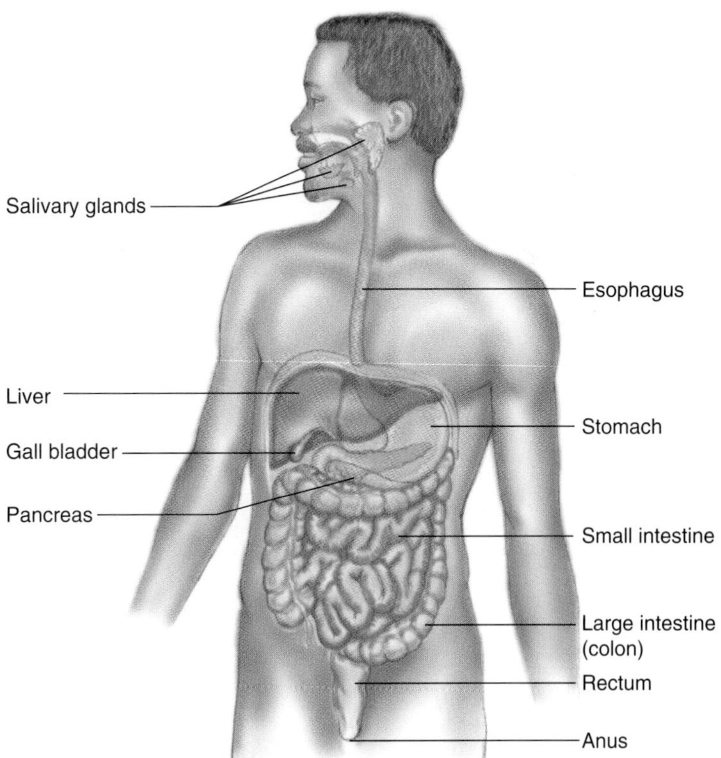

FIGURE 15.1
The digestive tract.

Bile, which is manufactured in the liver and stored in the gall bladder, contains bile salts, cholesterol, and bile pigments, which give bile its yellow to green color.

Cellulose, like starch, is a polymer of glucose but the linkages between glucose units in the two polymers are different. Humans do not have enzymes capable of breaking the linkages in cellulose.

which are released into the small intestine from the gall bladder, act rather like soaps (Chapter 17). Like soaps, bile salts have both polar and nonpolar areas in their molecular structures, which makes it possible for water-soluble lipase to react with water-insoluble triglycerides.

The end products of the digestion of food—monosaccharides, amino acids, glycerol, and fatty acids—are absorbed through the wall of the small intestine into the bloodstream and carried to the liver, where they undergo many further changes. Some glucose is released to general circulation, but most of it (along with fructose and galactose) is converted to glycogen and stored. Amino acids are converted to enzymes and other proteins, and fatty acids and glycerol are reformed into fats **(Fig. 15.2)**.

Digestion is very efficient; only a small amount of digestible food escapes conversion to small absorbable molecules and is excreted in the feces. Indigestible plant foods composed mainly of cellulose—usually referred to as **fiber** or *roughage*—account for most of the food material in feces.

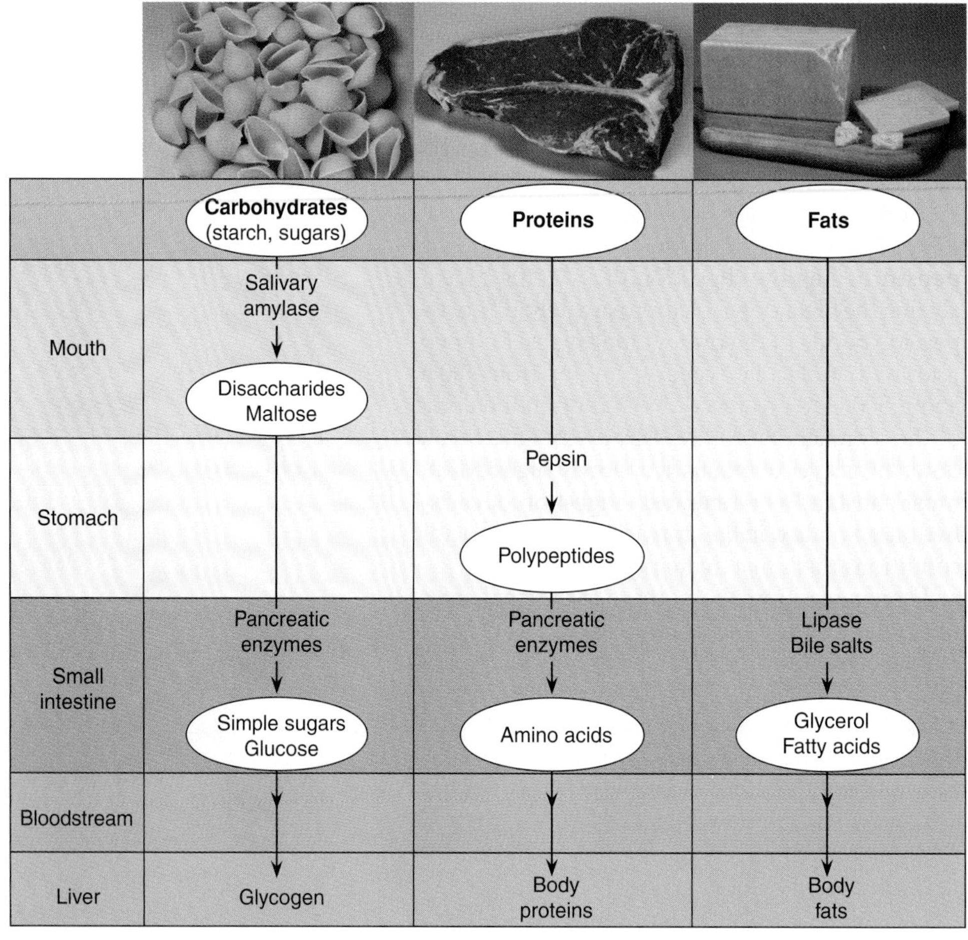

FIGURE 15.2

Summary of digestion of food, showing breakdown of carbohydrates, proteins, and fats into simple molecules that are absorbed into the bloodstream.

Building and Maintaining Body Tissues

The end products of digestion, primarily amino acids, are used to build and maintain the tissues and organs of the body. Body tissues are continually being broken down and replaced. Renewal occurs at different rates, depending on the type of tissue. Red blood cells, for example, have a life span of 120 days, while the cells lining the intestines are replaced as often as every 3–4 days. Some tissues last much longer. Collagen, a protein of tendons, has a life span of about 10 years. Many of the amino acids and other compounds formed when tissues break down are reused, but some new material must always be obtained from food.

15.3 Energy from Food

Food is to the body as fuel is to an engine. Food provides the body with the energy it needs to do work. As we have seen in earlier chapters (Chapters 2, 6, and 14), photosynthesis in plants is the first step in the production of the complex organic molecules—carbohydrates, fats, and proteins—on which all living things depend for energy. Most humans obtain energy by eating a combination of plants and animals. This energy is needed for muscular activity, transmission of messages from the brain to muscles, maintenance of body temperature, and other vital functions such as respiration, circulation, digestion, protein synthesis in the liver, and even thinking. When the body is at rest and not performing any external work, the energy liberated from foods is ultimately converted to heat.

The minimum amount of energy required to keep the body functioning normally when completely at rest is termed the *basal metabolic rate (BMR)*.

Energy derived from foods is generally measured in **Calories** (note the capital C) rather than in calories (see Table 12.1).

$$1 \text{ Cal} = 1 \text{ kcal} = 1000 \text{ cal} = 4180 \text{ joules}$$

The energy in a sample of food can be determined directly by measuring the heat released when the food is completely oxidized in a *bomb calorimeter* (**Fig. 15.3**). However, not all the energy in the food is available to the body because some food material is excreted in the feces. Also, some nitrogen-containing compounds derived from proteins in foods are never completely oxidized in the body and are excreted in the urine, mostly as urea (NH_2—CO—NH_2). **Table 15.2** shows the approximate amounts of energy provided by carbohydrates, fats, and proteins in foods, after allowance has been made for the material lost through excretion.

As shown in the following example, if the nutrient content of a food is known, the values given in Table 15.2 can be used to calculate the caloric value of the food.

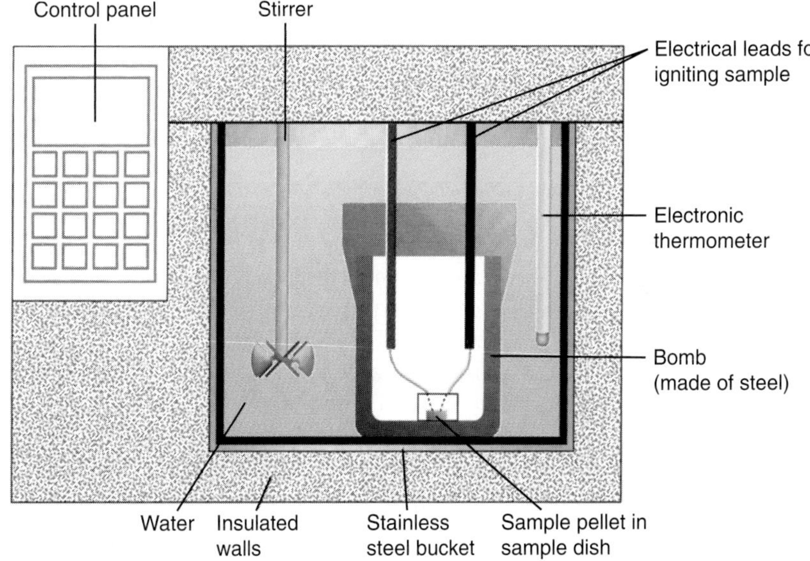

FIGURE 15.3

Cross section of a bomb calorimeter, which can be used to determine the caloric value of foods. A weighed sample of food is placed in the bomb, a heavy, sealed vessel made of steel. The bomb is filled with oxygen under pressure and placed in a calorimeter, an insulated container holding a measured quantity of water. The food is ignited, and the heat produced warms the water. From the increase in temperature and the known heat capacity of the calorimeter, the heat emitted per gram of food can be calculated.

Table 15.2 Calories Provided by Carbohydrates, Fats, and Proteins

Type of food	Cal/g
Carbohydrate	4
Fat	9
Protein	4

EXAMPLE 15.1 Calculating the caloric value of food

Whole milk is 3.5% protein, 3.5% fat, and 4.9% carbohydrate. Most of the rest is water. Calculate the number of Calories available from 100 grams of milk.

Solution

1. 100 g of milk contains 3.5 g protein, 3.5 g fat, and 4.9 g carbohydrate.
2. Calories from protein $= 3.5\text{ g} \times 4\text{ Cal/g} = 14.0\text{ Cal}$
   Calories from fat $= 3.5\text{ g} \times 9\text{ Cal/g} = 31.5\text{ Cal}$
   Calories from carbohydrate $= 4.9\text{ g} \times 4\text{ Cal/g} = 19.6\text{ Cal}$
3. Therefore, 100 g of milk provides 65.1 Cal.

Practice Exercise 15.1

White bread is 50.4% carbohydrate, 8.7% protein, and 3.2% fat. Approximately how many Calories from carbohydrate, protein, and fat does 200 grams of bread furnish?

The average normal-weight adult has enough body fat to supply the body's energy needs for 30–40 days.

Fats produce more than twice as much energy per gram as proteins and carbohydrates when they are oxidized to carbon dioxide and water, because, per gram, they contain less oxygen. The following equations show the oxidation of glucose and a typical fatty acid:

$$\underset{\text{glucose}}{C_6H_{12}O_6} + 6\ O_2 \longrightarrow CO_2 + 6\ H_2O + 670\text{ kcal (3.7 kcal/g of glucose)}$$

$$\underset{\text{palmitic acid}}{C_{16}H_{32}O_2} + 23\ O_2 \longrightarrow 16\ CO_2 + 16\ H_2O + 2385\text{ kcal (9.3 kcal/g of fatty acid)}$$

Daily Calorie Requirements

For normal activity, a young adult male requires about 3000 Cal per day; a young adult female requires about 2100 Cal per day. Some of the foods that are commonly eaten to provide these Calories, together with their caloric values and nutrient content, are listed in **Table 15.3**.

In the United States and other developed countries, fats generally provide 40% of the energy requirement, with carbohydrates supplying 48% and proteins 12%. In less developed countries, carbohydrates often provide as much as 80% of energy needs, with the remaining 20% coming equally from fats and proteins.

The distribution of fats, carbohydrates, and proteins in the average American diet is not ideal. As we shall see in the following sections, it is healthier to eat fewer fats and more carbohydrates.

EXAMPLE 15.2 Determining fat content of foods

A hamburger at a fast-food restaurant supplies a total of 575 Cal. If fat supplies 275 of these Calories, what percentage of the total Calories is derived from fat?

Solution To obtain the percentage of fat, divide the number of Calories derived from fat by the total number of Calories, and then multiply by 100:

$$\text{percentage of Calories from fat} = \frac{275 \text{ Cal}}{575 \text{ Cal}} \times 100$$
$$= 47.5\%$$

Practice Exercise 15.2

For lunch, you eat a McDonald's Big Mac that provides a total of 540 Cal. If 280 Cal are derived from fat, do you obtain more or less than half your Calories from fat?

15.4 Dietary Needs for Carbohydrates

We need carbohydrates in our diet primarily as a source of energy. In 1976, the U.S. Senate Select Committee on Nutrition and Human Health recommended that for good health, 58% of dietary Calories should be obtained from carbohydrates: 38% from complex car-

Table 15.3 Calorie Content and Approximate Percentages of Proteins, Fats, and Carbohydrates in Selected Foods

Food	Cal/100 g	Protein (%)	Fat (%)	Carbohydrate (%)
Meat, fish				
Lean beef, broiled	175	31.7	5.3	0
Chicken, whole, boiled	130	23.8	3.8	0
Cod, raw	73	17.6	0.3	0
Dairy products, eggs				
Milk, whole	65	3.5	3.5	4.9
Cheddar cheese	398	25.0	32.2	2.1
Eggs	160	12.9	11.5	0.9
Grains, grain products				
Whole-wheat bread	260	10.5	3.0	47.7
Brown rice, cooked	118	2.5	0.6	25.5
Vegetables				
Carrots, raw	85	1.1	0.2	19.7
Potatoes, cooked	96	2.6	0.1	21.1
Tomatoes, raw	25	1.1	0.2	4.7
Fruits, nuts				
Apples	64	0.2	0.6	14.5
Bananas	95	1.1	0.2	22.2
Oranges	55	1.0	0.2	12.2
Pecans	735	9.2	71.2	14.6

Note: There is a difference between raw and cooked food, because during cooking the water content changes.

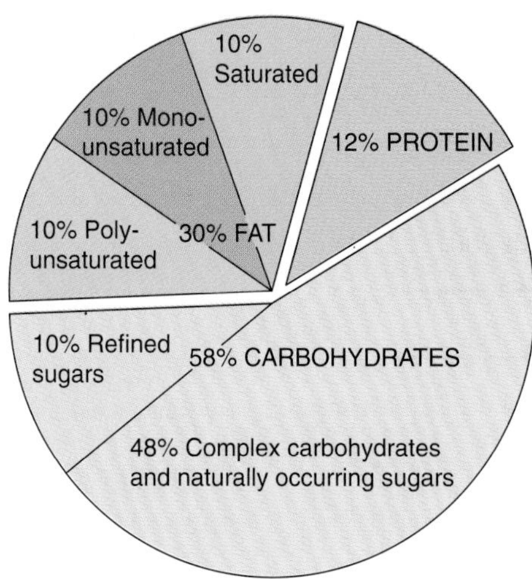

FIGURE 15.4

Recommended proportions of carbohydrates, fats, and proteins in the diet.

bohydrates, 10% from naturally occurring sugars, and no more than 10% from refined sugars (**Fig. 15.4**).

Sources of Carbohydrates

Complex carbohydrates in the form of starches are present in potatoes, yams, cereal grains (including rice, corn, wheat, barley, oats, and millet), and all the products such as pasta and bread that are made from grains. Fructose is present in honey, and both fructose and glucose are found naturally in many fruits and vegetables, with particularly high concentrations in grapes, dates, and figs. Lactose is found in milk and milk products. Refined sugar, or sucrose, is obtained from sugar cane and sugar beets. It is used in candy, cookies, preserves, soft drinks, many breakfast cereals, and numerous other processed foods. Many Americans have a "sweet tooth" and add large quantities of sucrose to coffee, tea, tart fruits, and other foods. In the American diet, sucrose often accounts for more than 20% of caloric intake, more than twice the recommended amount.

In addition to digestible carbohydrates, we need indigestible carbohydrates composed of cellulose to supply fiber.

EXPLORATIONS looks closely at one good source of complex carbohydrates—the potato (see pp. 476–477).

The aerobic sequence of reactions involves eight steps known as the *Krebs cycle* after Hans Krebs, who worked it out. Krebs shared the 1953 Nobel Prize for physiology and medicine with Fritz Lipmann, who discovered the roles played by ATP and ADP in energy storage.

Energy from Glucose

The energy available in carbohydrates is derived from the glucose produced in digestion and stored in the body as glycogen. When glucose is needed by the tissues, stored glycogen is converted to glucose and released. In this way, glucose in blood is maintained at a fairly constant level between 70 and 100 milligrams of glucose per 100 milliliters of blood. Glucose is carried in the bloodstream to all the body's tissues, where it supplies the energy cells need to perform their functions. In muscle tissue, some of the glucose received from the blood is converted back to glycogen and stored.

Body cells obtain energy from glucose by means of a complex series of chemical reactions that can be divided into two types: (1) anaerobic reactions (reactions that do not require oxygen), and (2) aerobic reactions (reactions that require oxygen). Each reaction in the overall series—whether anaerobic or aerobic—is catalyzed by a specific enzyme.

The anaerobic reactions involve the formation of a glucose-phosphate intermediate that subsequently breaks down to yield lactic acid. In these reactions, which require the participation of **adenosine diphosphate (ADP)** (**Fig. 15.5a**), some of the bond energy stored in glucose is transferred to ADP by changing it to **adenosine triphosphate (ATP)** (**Fig. 15.5b**). The overall reaction can be represented by the following equation:

$$\underset{\text{glucose}}{C_6H_{12}O_6} + 2\ ADP + 2\ H_3PO_4 \longrightarrow 2\ \underset{\text{lactic acid}}{CH_3-CH(OH)-C(=O)-OH} + 2\ ATP + 2\ H_2O$$

The energy of ADP and ATP molecules is stored in their phosphorus-oxygen bonds. As shown in Figure 15.5a, ADP contains one high-energy phosphorus-oxygen bond, while ATP contains two (Fig. 15.5b). In the overall reaction in the anaerobic pathway, a portion of the energy of one molecule of glucose is transferred to two molecules of ATP; the remainder of the energy is retained in the two molecules of lactic acid.

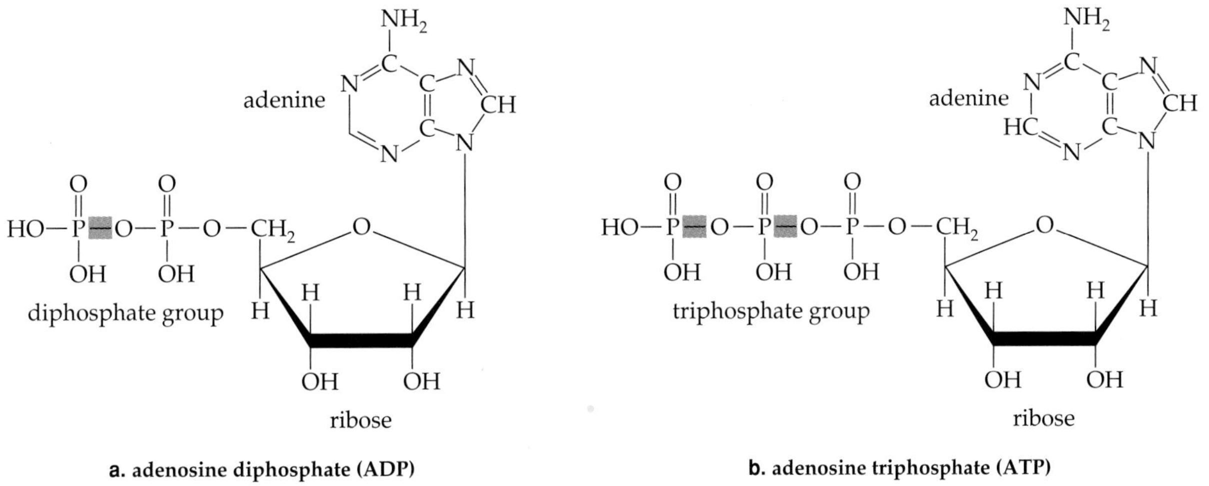

a. adenosine diphosphate (ADP)

b. adenosine triphosphate (ATP)

FIGURE 15.5

Structures of (a) adenosine diphosphate (ADP) and (b) adenosine triphosphate (ATP), with the high-energy bonds in the phosphate groups shaded red.

The aerobic reactions result in the conversion of lactic acid to carbon dioxide and water and the transfer of energy from lactic acid to additional molecules of ATP:

$$\underset{\text{lactic acid}}{C_3H_6O_3} + 18\ ADP + 18\ H_3PO_4 + 3\ O_2 \longrightarrow 3\ CO_2 + 21\ H_2O + 18\ ATP$$

During prolonged strenuous exercise, lactic acid is often formed more rapidly than it can be converted to carbon dioxide and water. As it accumulates in muscles, it causes fatigue and pain.

In body tissues, energy is obtained by the conversion of ATP back to ADP.

When any muscle is used—whether it is in the heart, the intestines, the leg, or any other part of the body—the energy powering the muscle contraction is obtained through the reactions just described.

The process by which energy is obtained from glucose is just one example of the extraordinarily complex processes involved in the functioning of the human body.

The Importance of Fiber in the Diet

In developed countries, where the diet is rich in processed foods but low in fiber, appendicitis, diverticulitis, and colon cancer are common. In less developed countries, however, where little processed food is consumed and the diet has a high fiber content, these conditions are uncommon. Good sources of dietary fiber are celery, the peel, seeds, and pulp of fruits, and wheat bran.

There is some evidence that potential carcinogens are absorbed by fiber in the small intestine and eliminated before they can cause any harm.

As indigestible carbohydrate material, or fiber, passes through the digestive tract, it absorbs water, swells, and becomes soft. This process promotes frequent bowel action, and usually the undigested material passes through the body within 2 days. If the diet lacks the bulk provided by fiber, however, bowel action is less frequent, and material may remain in the tract for as long as 4 days. This longer retention time gives bacteria—which make up about one-third of the weight of dry feces—more time to react with the material in the tract and produce toxic substances that can cause irritation and even colon cancer.

Problems with Processed Foods

Because many nutrients are lost during processing, most processed foods are less nutritious than unprocessed foods. This deficiency is often at least partially remedied by the manufacturer, by adding the removed nutrients to the final product.

In the United States, most bread products are made from milled white flour. The milling process removes the germ and husks of the grain, which are a good source of fiber and which contain proteins, vitamins, and minerals. Enriched flour is milled flour to which vitamins, minerals, and a variety of other additives have been added. Despite these additions, products made from whole grains are better food sources because they provide more fiber.

Brown sugar is more nutritious than white sugar because, in addition to sucrose, it contains molasses, a residue of sugar cane that is rich in minerals.

In a similar way, fiber and nutrients are removed during the production of refined white sugar from sugar cane and sugar beets. Compared with the sugars in fruits and other natural foods, refined sugar has the added disadvantage that it is digested much more rapidly, feeding glucose into the bloodstream quickly. The body utilizes glucose for energy less efficiently when it enters the bloodstream rapidly, and more of the glucose is converted to fat for storage.

A large egg contains about 270 milligrams of cholesterol.

15.5 Dietary Needs for Fats

Fats have two main functions in the body: to supply energy and to provide necessary components for the construction of cell membranes. For good health, it is recommended that fats make up no more than 30% of daily intake and that these fats be divided equally among saturated fats, monounsaturated fats, and polyunsaturated fats (refer to Fig. 15.4). In the United States, most diets have a higher fat content than this recommended proportion, and the fats are mainly saturated and monounsaturated. As we saw in Chapter 14, a diet high in saturated fats has been shown to increase the risk of atherosclerosis.

Synthetic prostaglandins and related compounds have many useful applications. They are used to induce labor, decrease gastric secretion, raise and lower blood pressure, and treat asthma. The name *prostaglandin* reflects the fact that they were originally extracted from prostate gland tissue.

The major sources of dietary fats are butter, suet and lard from meats, vegetable oils, and margarine prepared by hydrogenating vegetable oils. Although most vegetable oils are low in saturated fats, the tropical oils coconut oil and palm oil are exceptions (**Table 15.4**); these oils and products made from them should be avoided by anyone who wants to reduce intake of saturated fats. Cholesterol is found in animal tissues but not in vegetables. Products high in cholesterol include red meat, butter, and cheeses. Particularly high levels occur in egg yolks.

With the exception of linoleic acid (Chapter 14), the body can synthesize all the fatty acids that it needs. This one essential fatty acid is needed for the production of **prostaglandins,** a group of compounds that are produced in most tissues of the body and play a vital role in regulating hormonal activity. The average diet supplies sufficient linoleic acid to meet the body's needs.

Fats represent the body's most concentrated energy storehouse. As we have seen, glucose supplies the body's immediate energy needs, but fats must be mobilized from adipose tissue for sustained activity such as long-distance running. The metabolic pathways by which energy is obtained from fatty acids will not be considered; like those for glucose, they involve ADP and ATP.

Table 15.4 Percentage of Saturated Fatty Acids in Selected Vegetable Oils

Oil	Saturated Fatty Acids (%)
Coconut	93
Palm	57
Peanut	21
Olive	15
Corn	14
Soybean	14
Safflower	10

15.6 Dietary Needs for Proteins

Proteins have numerous functions in the body. Enzymes and antibodies are proteins; the materials from which most tissues and organs in the body are constructed are proteins. Hemoglobin, transferrin, and lipoproteins, which facilitate the transport of oxygen, iron, and lipids, respectively, are all proteins. Children need proteins for tissue building, while adults need them primarily for tissue repair. The daily requirement of protein for an adult is about 0.8 gram per kilogram of body weight. Pregnant women and growing children need approximately 1.5 and 2.0 grams per kilogram of body weight, respectively.

A lipoprotein is a combination of a lipid, such as a triglyceride or cholesterol, and a protein.

To meet the body's needs, the diet must include proteins that, when digested, yield all eight (ten for children) of the essential amino acids (identified in Table 14.2). Proteins from animal sources—meat, poultry, fish, eggs, milk, and cheese—are *complete proteins*. They provide adequate amounts of all the essential amino acids. In contrast, plant proteins are *incomplete proteins*. Cereal grains and legumes (beans, peas, and soybeans) are good sources of proteins, but plant proteins are deficient in one or more of the essential amino acids. Wheat, rice, and corn, for example, all lack lysine. Rice also lacks threonine; corn lacks tryptophan; and beans are deficient in methionine. A vegetarian diet, if it is to supply all the essential amino acids, must contain a complementary mix of several plant foods.

Many traditional diets that are mainly vegetarian provide all the essential amino acids by combining a cereal with a legume. For example, corn tortillas and refried beans are staples of the Mexican diet; rice and soybean curd (or tofu) form a major part of the diet in Japan. In these diets, the amino acids missing in the grain are present in the legume, and vice versa. Peanut butter sandwiches (with or without jelly), favorites among American children, also represent a combination of grain and legume.

15.7 Nutritional Diseases and Disorders

Nutritional diseases and disorders can be caused by (1) inadequate or excessive consumption of food or of some particular type of food or (2) an inability to assimilate a nutrient.

Kwashiorkor was the name used in Ghana for the evil spirit that was believed to cause the disease suffered by the firstborn child after the second child was born.

Deficiency Diseases

Except for vitamin-deficiency diseases, which are considered later in the chapter, most deficiency diseases are caused by a lack of protein in the diet. One of the most widespread and serious is *kwashiorkor* (pronounced kwash-ee-OR-core), which was first recognized in Ghana. This disease occurs in children between the ages of 1 and 4 who, for some reason—usually the birth of a sibling—are weaned from protein-rich breast milk to a protein-poor diet. The diet is usually high in starch and may supply adequate Calories but because it is deficient in essential amino acids, synthesis of proteins in the body is disrupted. Fluid collects in the tissues and the limbs, and the face and belly become swollen **(Fig. 15.6)**. The children look plump in the early stages and may not appear to be undernourished. If the condition is untreated, the skin becomes dry and scaly with sores, and growth is stunted. Because of a lack of antibodies, the children are very susceptible to infections. If the diet is deficient in both proteins and Calories, a more serious condition called *marasmus* develops. The belly becomes very bloated, the children have an old appearance, are apathetic and often mentally retarded, and are extremely underweight. However, if these children are treated with a balanced diet in the early stages, their chances for recovery are good.

FIGURE 15.6

Children with kwashiorkor, a disease caused by a lack of protein in the diet. Symptoms include swollen belly, retarded growth, and discolored, scaly skin with sores.

BMI is weight in kilograms divided by height in meters squared. A person with a BMI of 30 is about 30 pounds overweight.

Overconsumption of Food

In the less developed countries of the world, millions of people have inadequate diets and suffer from nutritional deficiency diseases. In the developed countries, many people have a different nutritional problem—obesity. **Obesity** is defined as a body mass index (BMI) of 30 or higher; *overweight* is defined as a BMI of 25 to 29.9. A healthy BMI ranges from 19 to 24. Obesity is caused primarily by long-term consumption of more Calories than are required by the body to meet its energy needs. The excess Calories are stored as fat that accumulates in the body. In 2000, the Centers for Disease Control (CDC) reported that in the United States obesity among adults had risen almost 60% since 1991. Almost 25% of Americans are obese, and 60% are either overweight or obese. Most overweight people get little exercise, and their diets are usually high in saturated fats, processed foods, and sucrose and low in fruits, vegetables, and fiber. They have a shorter life expectancy than slim people have and suffer disproportionately from diabetes, gall bladder disease (particularly women), high blood pressure, and heart disease.

Although there is no doubt that overconsumption of food is the main cause of obesity, other factors, including heredity and psychological problems, also play a part. In the United States, obesity has spawned an entire industry devoted to ways to reduce weight. Despite the millions of dollars spent yearly on drugs, surgery, exercise programs, and numerous fad diets, there is no escaping the fact that the loss of 1 pound of fat requires the expenditure of 3500 Calories. There is still only one sure way to lose weight: Reduce caloric intake and increase the expenditure of energy through exercise, while continuing to eat a nutritional diet.

Metabolic Malfunctions

Diabetes is a condition in which the body is unable to utilize glucose in the normal way; as a result, blood glucose levels become abnormally high. Normally, after a meal, the blood glucose level begins to rise as carbohydrates are digested and glucose enters the circulation. In response to this stimulus, the pancreas (see Fig. 15.1) releases the hormone *insulin* into the bloodstream. Insulin causes glucose to be transferred from the blood into tissue cells; within about an hour, the blood glucose level falls back to the fasting level (the level before eating the meal). In diabetes, this transfer process does not operate normally; in severe cases, the blood glucose level may rise to three or four times its normal level. If untreated, a diabetic may lapse into a coma.

Diabetes affects over 15% of Americans over the age of 40, and it is increasingly affecting younger people. According to the CDC, there was a 70% increase in the incidence of diabetes among Americans aged 30–39 during the 1990s, particularly in those who are overweight and who consume large quantities of sucrose and processed foods. This type of diabetes develops gradually and is usually caused by both a decrease in insulin production and a decrease in the number of insulin receptors (the molecules to which insulin binds) in cells. In most cases, the condition can be controlled by diet and exercise. The type of diabetes that develops suddenly in people under the age of 20—*called juvenile diabetes*—is much more serious. It is caused by an almost complete lack of insulin and must be treated with insulin injections. If given orally, insulin, which is a protein, would be broken down in the digestive tract.

In the United States, by law, all infants are tested for PKU at birth.

There are a number of inherited metabolic disorders in which the person's body lacks a specific enzyme that affects the utilization of food. One is *phenylketonuria (PKU)*, which occurs in 1 out of every 20,000 newborns and is caused by the lack of the enzyme that catalyzes the conversion of the amino acid phenylalanine to the amino acid tyrosine (see Table 14.2). When not converted, phenylalanine and related compounds, which are toxic to brain tissue, accumulate and cause severe mental retardation. If an affected infant is put on a low phenylalanine diet soon after birth, brain damage is minimized.

Another inheritable metabolic disorder is *galactosemia,* in which there is a deficiency of the enzyme that catalyzes the conversion of galactose to glucose. Infants with this condition are unable to utilize the galactose formed by the breakdown of lactose in milk and fail to thrive. They can be treated successfully with a special lactose-free diet.

The disacharide lactose is composed of a molecule of glucose linked to a molecule of galactose.

15.8 Vitamins

Vitamins are organic compounds that are required in small amounts for normal metabolism and the maintenance of good health. They are not a source of energy, but they have many vital bodily functions. Most are needed as components of enzyme systems. Vitamins are synthesized by plants but cannot be synthesized in the body and must, therefore, be obtained from food or food supplements.

Vitamins were originally called *vitamines* because they were recognized as essential for life (*vita* means "life" in Latin) and were thought to be amines (Chapter 8). The *e* was later dropped when it was realized that most of these substances are not amines.

Vitamins and Vitamin-Deficiency Diseases

For several hundred years, it was known that the absence of certain foods in the diet causes specific diseases, but it was not until the early part of the 20th century that the chemical identity of the missing substances was established. Five major diseases have been associated with the lack of a specific vitamin: (1) scurvy (vitamin C), (2) beri beri (vitamin B_1), (3) pellagra (vitamin B_3), (4) blindness (vitamin A), and (5) rickets (vitamin D).

One of the first vitamin-deficiency diseases to be recognized was **scurvy,** the scourge of seafarers who went without fresh fruits and vegetables for extended periods of time while on long ocean voyages. The sailors suffered from bleeding gums, loss of teeth, vomiting, weight loss, anemia, and slow-healing wounds; many died. In 1747, the Scottish naval surgeon James Lind showed that scurvy could be cured or prevented if citrus fruits were included in the diet (see Chapter 1). Shortly thereafter, the British Admiralty required all ships to carry stores of limes on long voyages—hence, the name "limeys" for British sailors. The factor in citrus fruits, and many other fruits and vegetables, that prevents scurvy is **vitamin C,** or **ascorbic acid.** Because vitamin C is destroyed during long storage and by cooking, fruits and vegetables should be eaten fresh.

Humans, other primates, and guinea pigs are the only mammals that are unable to synthesize ascorbic acid in their bodies.

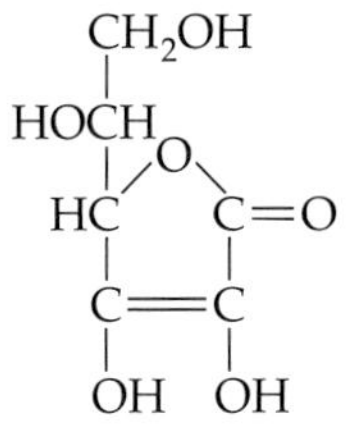

```
   CH2OH
    |
  HOCH
    | /O\
   HC     C=O
    |     |
    C ═══ C
    |     |
   OH    OH
```

vitamin C (ascorbic acid)

The disease *beri beri* became common in parts of Asia at the end of the 19th century as milled (or polished) rice began to replace unprocessed rice in people's diets. Milling improves the keeping qualities of rice, but it removes **vitamin B_1,** or **thiamine,** which is present in the husk and germ. People with beri beri suffer from stiffness of the limbs, heart disease, loss of appetite, and, in later stages, paralysis of the limbs and mental disorders.

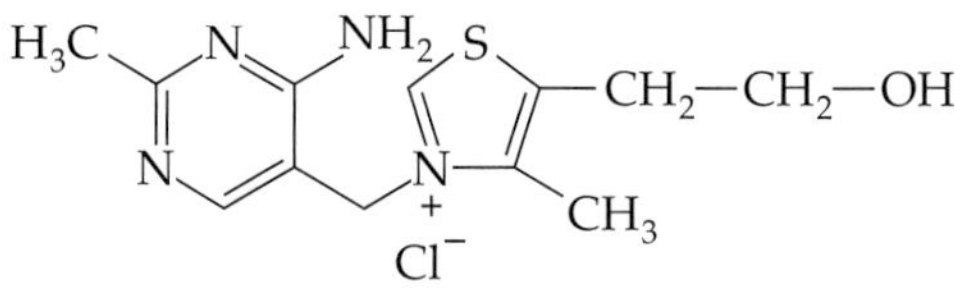

vitamin B_1 (thiamine)

Another disease associated with overdependence on one type of grain is *pellagra,* which at one time was common in the southern United States among people who subsisted mainly on corn. Corn, as was mentioned earlier, is deficient in the amino acid tryptophan, which is the precursor of **niacin** (sometimes called *vitamin* B_3). A lack of niacin causes reddening and drying of the skin and, in severe cases, gastrointestinal and nervous system disorders. Pellagra can be prevented by including whole wheat in the diet; it can be cured by administering niacin.

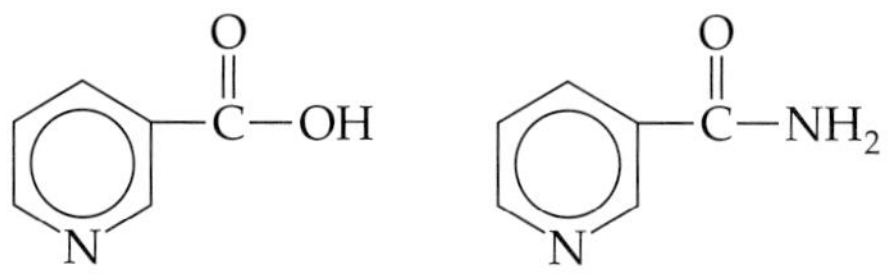

nicotinic acid nicotinamide

niacin

A major cause of blindness in young children in less developed countries is a deficiency in **vitamin A,** or **retinol.** An early sign of this deficiency is night blindness. Retinol, is required for the regeneration of *rhodopsin,* the photosensitive material that makes it possible for the eye to adapt to dim light. Retinol is also essential for the maintenance of the body's epithelial cells, those that cover and line the body and its organs. In severe retinol deficiency, the skin becomes scaly and hard, and the conjunctiva (membranes lining the eye socket and covering the front of the eyeball) and the cornea of the eye degenerate. Recently, it has been shown that vitamin A plays a role in the immune system. For those with a deficiency, treatment with vitamin A helps fight infection and also cuts the death rate from measles.

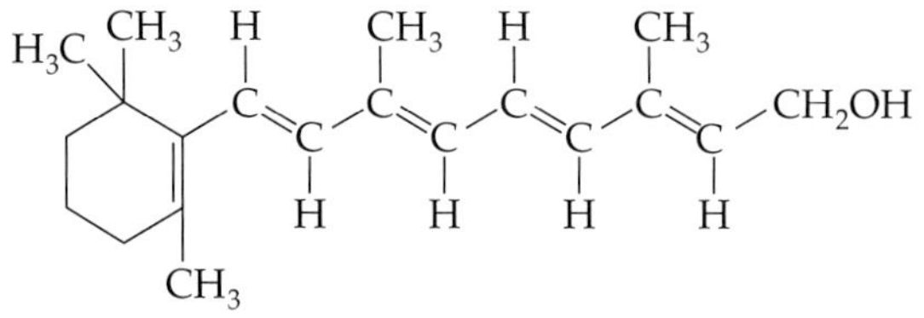

vitamin A (retinol)

Good sources of vitamin A are milk, cheese, eggs, liver, and some fatty fish. The body can also obtain vitamin A from green, leafy vegetables such as spinach and yellow and orange vegetables such as carrots and sweet potatoes. These vegetables contain β-carotene and other plant pigments that the body can convert to retinol.

A deficiency of **vitamin D (calciferol)** causes rickets in children and osteomalacia (soft bones) in adults. The diet need not supply all of the body's needs for vitamin D, because certain substances in the skin structurally related to cholesterol are converted to vitamin D on exposure to sunlight. Vitamin D is required for absorption of calcium and, to some extent, phosphorus from the intestinal tract, and it regulates the deposition of calcium and phosphorus in bones and teeth. If there is a deficiency, bones do not grow normally, and varous skeletal deformities, including knock knees and a protruding forehead, develop. Rickets was largely eliminated in developed countries when vitamin D was added to infant formula and milk. Foods that are naturally high in vitamin D are dairy products, fatty fish, and fish liver oils.

vitamin D (calciferol)

Rickets is practically unknown in the tropics, where there is plenty of sunlight, but it was common among poor children in northern industrial cities until the middle of the 20th century. Smog and the higher latitude combined to limit the children's exposure to sunlight, and poverty caused their diet to be poor.

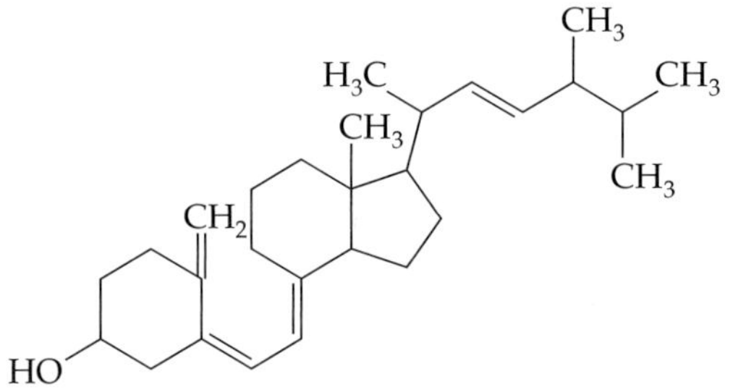

Polar bear liver contains such an exceptionally high concentration of vitamin D that it is poisonous to humans.

The B-Complex Vitamins

Since the discovery of the causes of beri beri and pellagra, several other B vitamins, which have the same food sources as vitamin B_1 and B_3, have been recognized and chemically identified. Collectively, the B vitamins are called the *B-complex vitamins.* They function primarily as *coenzymes* (molecules that are essential for enzyme activity) in metabolic reactions that lead to the release of energy in body cells, including the reactions described earlier that release energy from glucose. Vitamins B_{12} (cobalamin) and B_9 (folic acid) are required for the normal development of red blood cells. A deficiency of either leads to anemia.

Vitamins E and K

Vitamin E (tocopherol) is widely distributed throughout foods. An antioxidant, it is essential for the maintenance of cell membranes and is involved in the normal functioning of most organs of the body.

Vitamin K is needed for the formation of prothrombin, a protein necessary for blood clotting; a deficiency of vitamin K may result in delayed clotting and blood loss.

Many people consume large quantities of vitamin E because it has been suggested that it may postpone the aging process.

Classes of Vitamins

As we have seen, vitamins vary greatly in their functions and chemical structures. They are usually divided into two groups according to solubility. *Water-soluble vitamins* include the B-complex vitamins and vitamin C. They dissolve in water because their structures include polar hydroxyl groups (—OH). Water-soluble vitamins are not stored in the body but pass quite rapidly through it; thus, they must be replenished daily.

Fat-soluble vitamins, which include vitamins A, D, E, and K, have nonpolar hydrocarbon chains in their structures and are soluble in fats and oils. They do not need to be replenished as frequently as water-soluble vitamins because the body stores them in fatty tissues, particularly in the liver. If too much vitamin A or D is ingested, it builds up to toxic levels and causes serious problems. Too much vitamin D causes bonelike material to be deposited in the kidneys and other parts of the body.

The Recommended Daily Allowance (RDA), sources, and effects of deficiencies of the various vitamins are listed in **Table 15.5**.

Because water-soluble vitamin C is rapidly excreted, many scientists claim that large doses of it, which have been recommended as a preventative for the common cold, do little more than add to the amount excreted in the urine.

Table 15.5 Recommended Daily Allowance (RDA), Sources, and Effects of Deficiencies of Vitamins

Vitamin	RDA*	Sources	Effects of Deficiency
Water-soluble			
B_1 (thiamine)	1.5 mg	Whole-grain bread, milk, nuts, legumes	Beri beri: heart disease, mental disorders
B_2 (riboflavin)	1.7 mg	Milk, meat, eggs, whole-grain bread	Dermatitis
B_3 (niacin)	19.0 mg	Red meat, whole-grain bread, leafy vegetables	Pellagra: dry skin, intestinal and mental disorders
B_6 (pyridoxine)	2.0 mg	Eggs, liver, legumes, milk	Dermatitis; susceptibility to infection, anemia
B_9 (folic acid)	0.2 mg	Liver, leafy vegetables, whole-grain bread	Anemia, retarded growth
B_{12} (cobalamin)	2.0 μg	Meat, eggs, milk, fish	Pernicious anemia
Biotin	0.1 mg	Meat, peanuts, eggs	Dermatitis, fatigue
C (ascorbic acid)	60.0 mg	Citrus fruits, tomatoes, green peppers	Scurvy: bleeding gums, slow-healing wounds
Pantothenic acid	7.0 mg	Meat, eggs, milk	Retarded growth, mental disorders
Fat-soluble			
A (retinol)	1.0 mg	Milk, cheese, eggs, fatty fish, carrots, leafy vegetables	Night blindness
D (calciferol)	10.0 μg	Fish liver oils, fatty fish, milk	Rickets, skeletal disorders
E (tocopherol)	10.0 mg	Eggs, milk, green vegetables	Anemia, sterility
K	70.0 μg	Green leafy vegetables	Bleeding disorders

*RDA values are for adults.

15.9 Minerals

For nutritional purposes, **minerals** are defined as those elements, other than carbon, oxygen, hydrogen, and nitrogen, that are needed for normal growth and the maintenance of good health. Most minerals are present in the body as ions. Some, including calcium, phosphorus, and magnesium, are required in amounts of 1 gram or more per day, while others such as cobalt, copper, and zinc are needed only in trace quantities. Since many minerals are excreted in urine, feces, and sweat, they must be supplied daily in food. The average American diet is most likely to be deficient in calcium, iron, or iodine.

In many less developed countries, where meat and eggs are in short supply in the diet, iron-deficiency diseases are common.

Calcium and phosphorus are required for the proper development of bone and teeth. Growing children and women who are pregnant or are breastfeeding need about 1.5 grams of each per day, compared with approximately 1.0 gram needed by other people. The best source of calcium and phosphorus is milk. In addition to being a component of bone and teeth, calcium is required for the transport of ions in and out of body cells, the coagulation of blood, and the maintenance of the heart's normal rhythm.

Bone, like other parts of the body, is continually being broken down and rebuilt. As people grow older, bone breakdown tends to exceed bone formation, and bones may become thin and brittle and fracture easily. This condition, called *osteoporosis,* is common in women after menopause, when there is a drop in the secretion of the hormone estrogen (Chapter 14), which plays a role in preventing excessive bone loss.

Phosphorus, in addition to being present in bone, is a component of nucleic acids (Chapter 14) and ADP and ATP. Magnesium, like calcium, is stored in bone. Magnesium ions (Mg^{2+}) are present in all body cells and are an essential part of many enzyme systems.

Iron is an essential component of hemoglobin and myoglobin, the predominant protein in muscle. A deficiency of iron in the diet can cause anemia and is most likely to occur in women, who lose blood during menstruation, and in people with bleeding ulcers. Good sources of iron are meat and eggs.

Iodine is required for the synthesis of the hormones secreted in the thyroid gland in the neck; these hormones regulate the consumption of oxygen in the cells of the body. Fish and other seafood are excellent sources of iodine. A serious deficiency of iodine in children's diets results in mental retardation and stunted growth, a condition known as *cretinism.* An iodine deficiency in adults leads to *goiter,* a disorder characterized by enlargement of the thyroid gland **(Fig. 15.7)**. Iodine deficiencies occur mainly in inland regions—particularly mountainous areas and eroded river valleys—where iodine is lacking in both water and locally grown produce because it has been leached from the soil. Since the introduction of iodized salt (usually containing 0.1% potassium iodide, KI), iodine deficiency has become rare in developed countries. However, it remains a major cause of mental retardation in many developing countries.

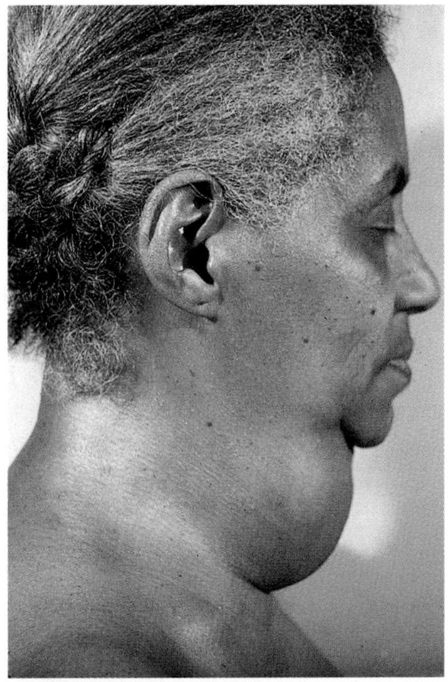

FIGURE 15.7

A lack of iodine in the diet causes goiter, or enlargement of the thyroid gland in the neck.

Goiter is quite common in mountainous areas such as the Alps and the Himalayas, where iodine has been leached from the soil.

Sodium, potassium, chloride, and bicarbonate ions are present in all body fluids and are essential for proper electrolyte balance. Too much salt (NaCl) in the diet, however, can cause water retention in the tissues, resulting in *edema* (swelling of the legs and ankles) and high blood pressure. Most American diets include too much salt but may be low in potassium because salt is added to processed foods to replace both the sodium and potassium lost in the manufacturing process.

The trace minerals copper, manganese, molybdenum, and zinc are essential as coenzymes in various enzyme systems. Cobalt is a component of vitamin B_{12}.

15.10 Food Supplements

Americans are so convinced that they need more nutrients than their diets provide that they spend almost $6 billion a year on vitamins, minerals, herbal remedies, and other food supplements. Although many people take them, vitamin supplements are unnecessary in most cases. Adults who regularly eat a well-balanced diet will obtain all the vitamins they need in their food, much of which in the United States is already fortified with vitamins. On the other hand, the elderly, dieters, alcoholics, and pregnant and lactating women may benefit from supplemental vitamins and minerals. Labels on multivitamin bottles list the amount of each individual vitamin and mineral in a tablet as a percentage of the U.S. RDA value for adults **(Fig. 15.8)**.

Directions: Adults: One tablet daily with food.

Supplement Facts
Serving Size: One tablet

	Amount Per Serving	% Daily Value		Amount Per Serving	% Daily Value		Amount Per Serving	% Daily Value
Vitamin A	5000 IU	100%	Biotin	30 mcg	10%	Chromium	65 mcg	54%
Vitamin C	60 mg	100%	Pantothenic Acid	10 mg	100%	Molybdenum	160 mcg	213%
Vitamin D	400 IU	100%	Calcium (elemental)	162 mg	16%	Chloride	72 mg	2%
Vitamin E	30 IU	100%	Iron	18 mg	100%	Potassium	80 mg	2%
Vitamin K	25 mcg	31%	Phosphorus	109 mg	11%			
Thiamin (B_1)	1.5 mg	100%	Iodine	150 mcg	100%	Nickel	5 mcg	*
Riboflavin (B_2)	1.7 mg	100%	Magnesium	100 mg	25%	Tin	10 mcg	*
Niacin	20 mg	100%	Zinc	15 mg	100%	Silicon	2 mg	*
Vitamin B_6	2 mg	100%	Selenium	20 mcg	28%	Vanadium	10 mcg	*
Folic Acid	400 mcg	100%	Copper	2 mg	100%	Boron	150 mcg	*
Vitamin B_{12}	6 mcg	100%	Manganese	3.5 mg	175%			

* Daily Value not established

FIGURE 15.8

The label on a bottle of multivitamins gives the percentages of RDA values.

15.11 Food Additives

Many people are concerned about the possible health hazards associated with food additives, the chemicals that are added during food processing and packaging. Of the several thousand additives used in the United States, very few have any nutritional value. They are added to prevent spoilage, sweeten, enhance flavor, color, emulsify, or in some other way make the product more marketable and attractive to consumers. Food additives generate approximately $4 billion in sales each year. The beverage industry is the largest user **(Fig. 15.9)**.

Food Preservation

Preservatives are the most useful of all food additives. Without some form of treatment, most foods soon spoil and become inedible. The main causes of food spoilage are (1) growth of bacteria and fungi (molds) and (2) air oxidation. Any method for improving the keeping qualities of foods must prevent or retard one or both of these processes.

Drying is an effective technique for food preservation because in the absence of water, microorganisms do not multiply, and air oxidation is slowed.

Since prehistoric times, drying has been used to prevent spoilage of meat, fish, grains, and other plant products. Salting to preserve meat and fish, and preservation of fruits in concentrated solutions of sugar are other ancient techniques. Refrigeration, bottling, and canning have been used for many years. More recent techniques include freezing, freeze drying, irradiation (Chapter 5), and the addition of various chemicals. The chemical preservatives that are used routinely today are of two kinds: (1) antimicrobial and (2) antioxidant.

ANTIMICROBIAL AGENTS The antimicrobial agents sodium nitrate ($NaNO_3$) and sodium nitrite ($NaNO_2$) have been added to cured meats such as bacon, hot dogs, bologna, and ham and smoked fish for at least 30 years. They are responsible for the pink color of the products. Without them, these foods would soon turn an unattractive gray.

Nitrites are particularly valuable in preventing the growth of *Clostridium botulinum*, the bacterium that produces deadly botulism poisoning. Recently, however, the use of both nitrates and nitrites has been questioned because of concerns that they may cause stomach cancer (Chapter 18). In the stomach, nitrates are reduced to nitrites, which react with hydrochloric acid in the stomach to form nitrous acid:

$$\underset{\text{sodium nitrite}}{NaNO_2} + \underset{\text{hydrochloric acid}}{HCl} \longrightarrow \underset{\text{nitrous acid}}{HNO_2} + \underset{\text{sodium chloride}}{NaCl}$$

Nitrous acid may then react with amines (Chapter 8) in the stomach to form *nitrosamines*, compounds that have been identified as cancer-causing agents.

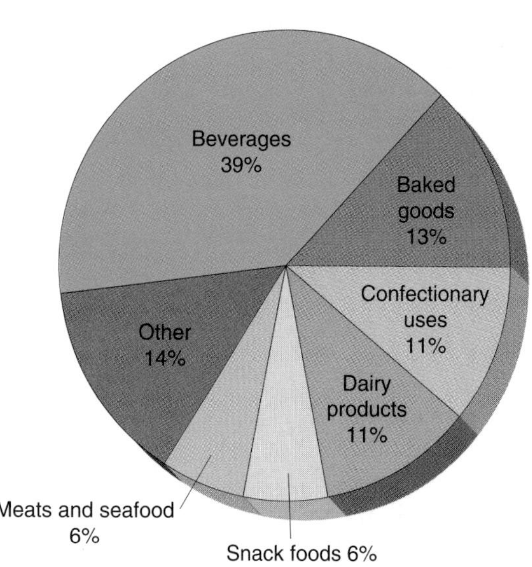

FIGURE 15.9

The beverage industry is the largest user of food additives.

$$\underset{\text{nitrous acid}}{HNO_2} + \underset{\text{amine}}{R{-}\overset{R'}{\overset{|}{N}}{-}H} \longrightarrow \underset{\text{nitrosamine}}{R{-}\overset{R'}{\overset{|}{N}}{-}N{=}O} + H_2O$$

The use of nitrates as a food additive has been banned by the FDA because nitrates are converted to nitrites in the body and thus have no value of their own.

In 1986, the FDA banned the spraying of fruits and vegetables with sulfites, and it issued regulations requiring that all food containing more than 10 ppm of sulfites be labeled to indicate the concentration.

Other controversial food preservatives are sulfites ($NaSO_3$ and KSO_3), which have been added to wine and dried fruits for centuries and have more recently been included in jellies and jams. In foods and wine, sulfites appear to be safe, but when sprayed on vegetables and fruits to produce a fresh appearance, sulfites have caused allergic reactions in asthmatics.

In the United States, the most widely used antimicrobial agents are the sodium, potassium, and calcium salts of benzoic acid, propionic acid, and sorbic acid **(Fig. 15.10a)**. Benzoic acid and benzoates are added to carbonated beverages, fruit juices, margarine, pickles, relishes, preserves and a variety of other products. Propionic acid, sorbic acid, and their salts are added to bread, cakes, chocolate, and cheese. They are very effective in retarding the growth of molds **(Fig. 15.10b)**.

ANTIOXIDANTS **Antioxidants** are chemicals added to foods to slow the oxidative process. Foods that deteriorate particularly rapidly because of air oxidation are fats and oils. The products of their oxidation include carboxylic acids, aldehydes, and ketones (Chapter 8), which give spoiled fats their characteristic unpleasant rancid taste and odor. The two antioxidants used most frequently in foods are the phenols (Chapter 8) **butylated hydroxyanisole (BHA)** and **butylated hydroxytoluene (BHT).**

[Structures: two isomers of BHA — OH, $C(CH_3)_3$, OCH_3 on benzene rings]

butylated hydroxyanisole (BHA)

[Structure: BHT — OH, two $C(CH_3)_3$, CH_3 on benzene ring]

butylated hydroxytoluene (BHT)

These compounds are added to vegetable oils, shortenings, and products that contain oils and fats, including breakfast cereals, potato chips, bread, and sausage. The oxidation of fats involves the formation of very reactive free radicals (Chapter 11),

$C_6H_5{-}\overset{O}{\overset{\|}{C}}{-}O^-Na^+$ sodium benzoate

$CH_3CH_2\overset{O}{\overset{\|}{C}}{-}O^-Na^+$ sodium propionate

$CH_3CH{=}CHCH{=}CH\overset{O}{\overset{\|}{C}}{-}O^-K^+$ potassium sorbate

a.

b.

FIGURE 15.10

(a) Antimicrobial agents that are widely used as food preservatives, with their carboxylate groups shaded red. (b) Bread that does not contain sorbates gets moldy quickly.

A study has shown that BHT may be beneficial. When rats were fed a diet high in BHT, their average life span increased by an amount equivalent to 20 years for humans.

which set up chain reactions. BHT and BHA interrupt the oxidation process by combining with the free radicals.

The safety of adding BHT and BHA to foods was first questioned in the 1950s, when a study suggested that large doses caused birth defects in rats. Although later studies were unable to confirm this finding, some people remain unconvinced that these additives are safe.

An effective natural antioxidant is vitamin E, which, like BHT and BHA, is a phenol. Vitamin E, however, is more expensive than the synthetic antioxidants and is used less frequently.

Flavorings

Many foods are very dependent on additives for their flavor. Flavorings, which may be natural or synthetic, represent the largest class of food additives. Of the more than 2000 flavors that have been approved by the FDA, approximately 1600 are synthetic.

A thousand years ago, peppercorns were so valuable and costly that they were often used in trading as a substitute for money.

For centuries, cloves, ginger, cinnamon, nutmeg, pepper, and many other plant products, have been used to flavor foods. Flavors such as vanilla, peppermint, and wintergreen are extracted from the crushed seeds or leaves with appropriate solvents. Natural flavors are complex mixtures of numerous compounds, many of which can now be synthesized in the laboratory. Synthetic and natural flavors are usually slightly different because the synthetic product lacks many of the components present in trace amounts in the natural product. The chemical structures of a number of commonly used flavorings are shown in **Figure 15.11**. Many flavorings are esters or aldehydes (Chapter 8).

Unlike the majority of other additives, most natural and synthetic flavors have not been extensively tested to determine if they might cause health problems. They are assumed to be safe in the small quantities added to foods and do not have to be identified on food labels.

Flavor Enhancers

A number of substances that have little or no flavor of their own enhance the flavors of other substances. One of the most widely used flavor enhancers is monosodium glutamate (MSG), the sodium salt of the amino acid glutamic acid (see Table 14.2). MSG has been approved as a food additive by the FDA, and it is added to thousands of processed foods. Although MSG occurs naturally in many foods, including tomatoes and mushrooms, studies have indicated that very large doses may cause brain damage in mice.

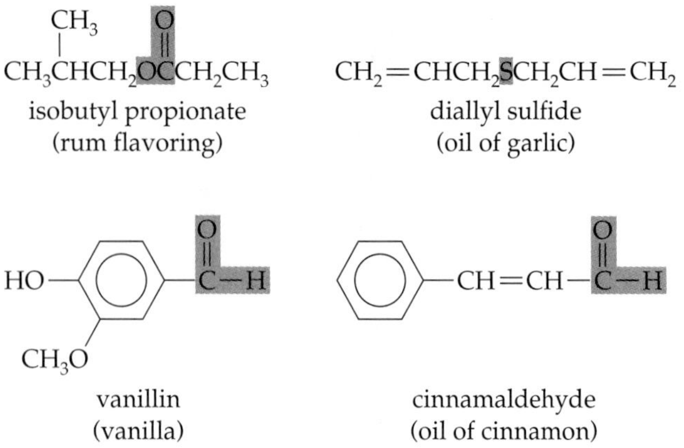

$CH_3CH(CH_3)CH_2OC(=O)CH_2CH_3$
isobutyl propionate
(rum flavoring)

$CH_2{=}CHCH_2SCH_2CH{=}CH_2$
diallyl sulfide
(oil of garlic)

vanillin
(vanilla)

cinnamaldehyde
(oil of cinnamon)

FIGURE 15.11

Examples of flavorings added to foods, with functional groups shaded red.

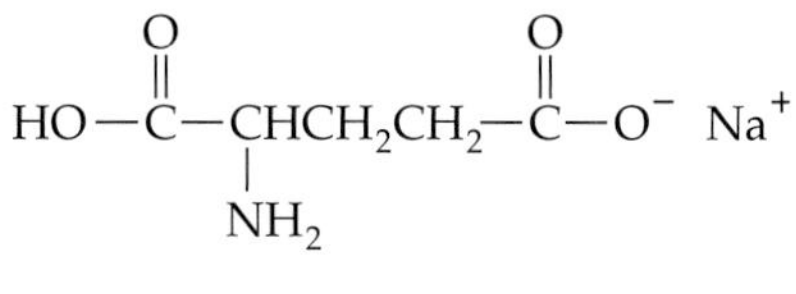

$HO{-}C(=O){-}CH(NH_2)CH_2CH_2{-}C(=O){-}O^-\ Na^+$

monosodium glutamate
(MSG)

Food Colorings

Food colorings increase the aesthetic appeal of foods but serve no other function. They are used to make products more attractive to consumers, who may reject foods if they are not colored as they expect.

About half the food colorings used in the United States are extracted from natural materials such as carrots, beets, grape skins, and the stamens of crocuses (saffron). The remainder are synthesized from coal tar (Chapter 12). Most synthetic food dyes consist of one or more aromatic rings joined by an azo group (—N=N—). FD&C Yellow No. 3 is a typical example.

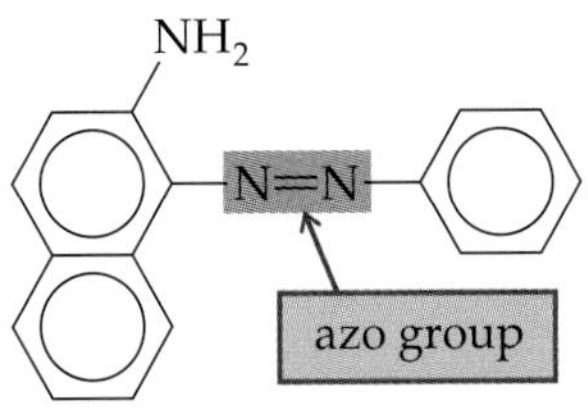

FD&C Yellow No. 3

Florida oranges often have green skins and, although it makes no difference to their flavor, are sometimes colored orange so that they can better compete with the naturally orange California fruit. Similarly, yellow dyes are added to chicken feed to give chicken skin and fat a yellow color.

Since 1906, food colorings have been regulated by the FDA, which sets limits on the amounts that can be added and requires that labels indicate that a color has been added. Several dyes once listed as safe have now been banned. Red Dye No. 2 was removed from the approved list in 1976, after large doses were shown to cause cancer in laboratory animals. A number of compounds in coal tar that are chemically related to certain food dyes have been shown to cause cancer. As a result, some people contend that all dyes obtained from this source should be banned.

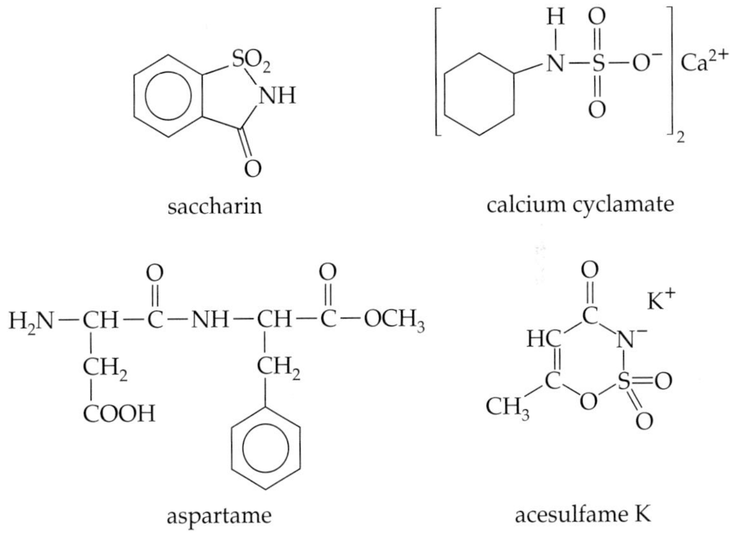

Artificial Sweeteners

In the United States, where so many people are overweight or weight-conscious, there is a thriving market for foods and beverages—particularly diet soft drinks—that are sweetened with artificial sweeteners instead of sucrose. These products are particularly useful for diabetics and others who must control their sugar intake. Most artificial sweeteners have no nutritional value and contribute few, if any, Calories.

Saccharin, which for many years was the only artificial sweetener available, was discovered accidentally in 1878. A chemist studying derivatives of toluene (Chapter 8), synthesized a new compound and noticed that it had a sweet taste. After it was tested for toxicity, the compound was marketed as saccharin (Fig 15.12). Saccharin is approximately 400 times sweeter than sucrose (Table 15.6), but it has a slightly bitter aftertaste.

In 1937, *cyclamates* (see Fig. 15.12) were discovered, again accidentally. These compounds are only about 30 times sweeter than sucrose, but they have no bitter aftertaste. Cyclamates were used extensively to sweeten beverages and foods until 1969, when they were banned because a study showed that very large doses could cause bladder cancer in rats. Although later work revealed that the study was flawed and that cyclamates are harmless to laboratory animals, attempts to have the ban lifted have so far failed.

FIGURE 15.12
Artificial sweeteners.

Table 15.6 Sweetness of Natural Sugars and Artificial Sweeteners Relative to Sucrose

Compound	Relative Sweetness*
Lactose	16
Glucose	74
Sucrose	100
Honey	145 (average)
Sodium cyclamate	3,000
Aspartame	18,000
Acesulfame K	20,000
Saccharin	40,000

*Sucrose is given an arbitrary value of 100.

The synthetic compound P-4000 is 400,000 times sweeter than sucrose but, because of its toxicity, cannot be used as a sweetener in foods.

The search for artificial sweeteners is still based on trial and error.

In 1978, the FDA recommended that saccharin, the only artificial sweetener still available at that time, should also be banned because studies indicated that very high doses could cause cancer in mice. However, in response to concerns that a ban would leave diabetics without a sugar substitute, the U.S. Congress decided that saccharin could remain on the market but would be labeled to indicate that it might be a health hazard.

Newer artificial sweeteners include aspartame and acesulfame K (Fig. 15.12). *Aspartame,* which is marketed as NutraSweet, is a methyl ester of the dipeptide formed between the amino acids aspartic acid and phenylalanine (see Table 14.2). It is metabolized as a protein. People with phenylketonuria (PKU), who lack the enzyme needed to metabolize phenylalanine, should avoid products containing aspartame. *Acesulfame K* is slightly sweeter than aspartame and has the added advantage that it is not broken down at the temperatures used in cooking.

Chemists still do not know what chemical structures impart a sweet taste. Very minor changes in the chemical structures of the known artificial sweeteners result in a loss of sweetness or the development of a bitter taste.

Other Food Additives

Other types of food additives besides those that have been described include sequestrants, acids, alkalis, emulsifiers, thickeners, stabilizers, and, in a different category, antibiotics.

Sequestrants are substances that are added to tie up trace metals that may get into foods accidentally during harvesting and processing. Sequestrants include citric acid and ethylenediaminetetraacetic acid (EDTA), which form complexes with metal ions and thus prevent them from catalyzing the oxidation reactions that cause the decomposition of foods.

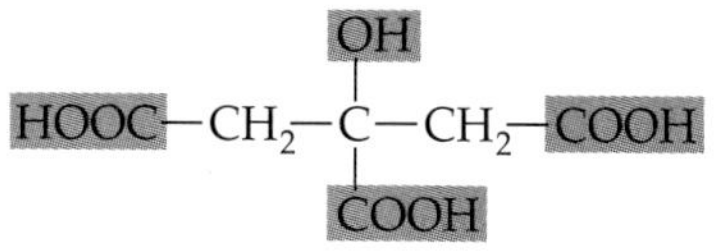

$$HOOC-CH_2-C(OH)(COOH)-CH_2-COOH$$

citric acid

$$(HOOC-CH_2)_2N-CH_2CH_2-N(CH_2-COOH)_2$$

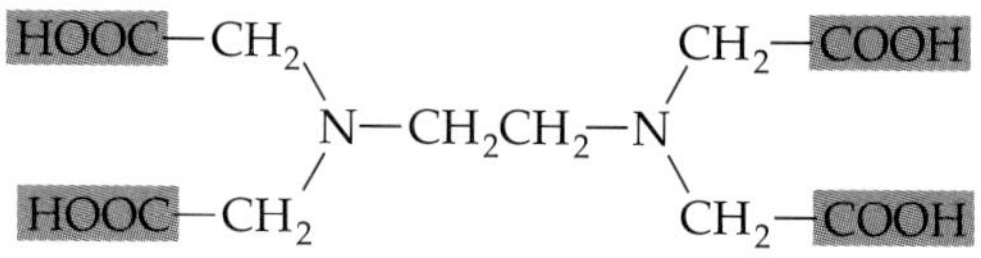

EDTA

Weak organic acids are added to soft drinks and other products to impart a tart taste and, in some cases, to mask undesirable aftertastes. These acids include citric, lactic, phosphoric, tartaric, and malic acids. They also help in preventing oxidation and the growth of microorganisms.

Emulsifiers are substances that keep very small droplets of a nonpolar liquid (usually an oil) evenly dispersed in a polar liquid (usually water). They are widely used in dairy products (including ice cream), mayonnaise, margarine, and peanut butter. Emulsifiers include mono- and diglycerides of fatty acids (Chapter 14), propylene glycol, polysorbates, and lecithin. The diglyceride, glycerol distearate is used in dairy products.

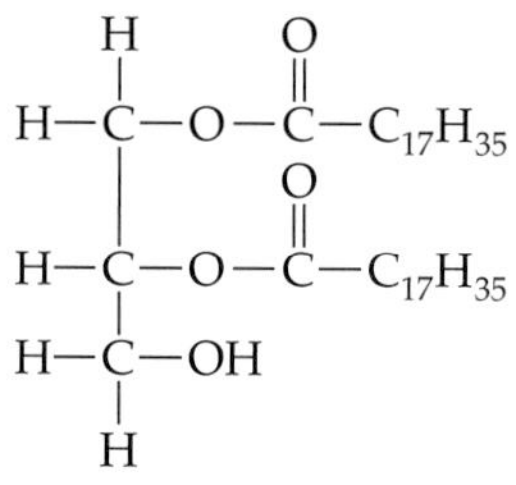

$$\begin{array}{l} H_2C-O-C(=O)-C_{17}H_{35} \\ \;\;| \\ HC-O-C(=O)-C_{17}H_{35} \\ \;\;| \\ H_2C-OH \end{array}$$

glycerol distearate

Emulsifiers function on the same principle as soaps and detergents: The nonpolar end of the emulsifier molecule dissolves in the oil, while the polar end dissolves in water. In this way, molecules of the two immiscible liquids are kept together and do not separate.

Antibiotics are present in meat and poultry as a result of the widespread practice of adding antibiotics to animal feed, primarily to promote growth. Many consumers are opposed to this practice because of the risk of increasing the opportunity for bacteria to become drug-resistant. The Centers for Disease Control and the World Health Organization have called for an end to the use of antibiotics to enhance the growth of livestock if the antibiotics are also used to treat human diseases.

How Safe Are Food Additives?

In the United States, food additives are regulated by the FDA. Before 1958, an additive could be used unless the FDA could prove that it was unsafe. But since then, federal law has required that the safety of any new additive must first be tested by the manufacturer and then approved by the FDA before it can be used in foods. Further, the Delaney Amendment, also passed in 1958, requires the FDA to ban any substance that has been shown to cause cancer in animals or humans.

The law did not apply to the several hundred additives already in use before 1958. Because it was impractical to attempt to test all those substances at that time, the FDA drew up a list of some 600 additives that were *generally regarded as safe* if used in specified foods in specified amounts. The list, which became known as the **GRAS list,** was based on the opinions of experts in toxicology and related fields. Since the list was published in 1959, many substances on it have been reevaluated; as noted earlier, some, including cyclamates and several food colorings, have been removed. Today, several thousand additives are approved by the FDA.

Many people are concerned that even if individual additives are harmless, several in combination may prove harmful. A single food product on the grocery shelf may contain as many as 100 additives, and it is obviously impossible to determine how, or whether, some of them might interact. However, even though some people would prefer all-natural foods, without additives of any kind, such an approach is unrealistic. Without preservatives, food would spoil, food supplies would be greatly reduced, costs would rise, and our already limited ability to feed the world's ever-growing population would be further jeopardized.

15.12 Dietary Goals for Good Nutrition

Good nutrition depends on eating a varied, properly balanced diet. The average American diet, unfortunately, does not meet this requirement. In general, it supplies more Calories than are needed, and, because of its reliance on convenience and processed foods, it is too rich in fats, cholesterol, sugar, and salt and too low in fiber. This diet has been linked to many of the leading causes of death in the United States, including heart disease, colon cancer, and diabetes.

For the sake of our health, we would all be well advised to follow the recommendations listed in the 1988 Surgeon General's Report on Nutrition and Health:

1. Maintain ideal weight. If overweight, reduce energy intake and increase energy expenditure.

EXPLORATIONS

The Lowly Potato—A Treasure from the New World

Potatoes are a favorite food. Baked, boiled, or as French fries and chips, we eat them almost every day. Confined to South America until the 16th century, the potato has since traveled the world, improving health, influencing population growth, and changing economies. Today, it joins rice, corn, and wheat as one of the world's four major food crops.

Origins and History

The potato was first cultivated high in the Andean mountains at least 8000 years ago. The plant belongs to the botanical family Solanacea, and its relatives include the tomato, pepper, eggplant, petunia, tobacco, and deadly nightshade plants. The potato (Solanum tuberosum) is a tuber, a swollen underground stem that stores food for the use of the plant above ground. New plants sprout from the "eyes," or buds, on the tuber. The name potato is derived from batata, the Spanish form of the Arawak Indian word for the sweet potato, or yam. (The potato and yam are not related, but the two tubers were often confused by the first Europeans who encountered them, and the name was misapplied.)

In the 16th century, when the Spanish conquistadors arrived in the New World in search of gold and souls to convert to Christianity, the potato was the main staple of the great Inca empire that stretched for almost 3000 miles along the western side of South America. The Spanish took potatoes home with them, little realizing that they carried an edible treasure of greater lasting value than all the gold they plundered from the Indians. From Spain, potatoes spread to the rest of Europe and, by the 17th century, reached Africa, India, China, Japan, and most of Southeast Asia. Potatoes traveled back across the Atlantic Ocean to North America with the early English settlers and were well established in Virginia and Pennsylvania by the end of the 17th century.

Europeans were slow to accept the strange tubers; for many years, they were viewed with deep suspicion. Scots refused to touch them because they were not mentioned in the Bible; when outbreaks of leprosy and tuberculosis occurred in regions where potatoes had been introduced, potatoes were blamed. The innocent potato was even accused of causing aphrodisiac effects deleterious to moral health.

Gradually, the potato was recognized as a very valuable food, and its cultivation during the 18th and 19th centuries had a profound effect on the history of Europe. Potatoes were easy to grow; they matured more quickly and provided more food per acre than either rye or wheat. With potatoes, a family could, for the first time, grow enough food on its own small plot to meet its basic needs. And potatoes were safer than grains. Although it was not understood at the time, the molds that often grew on stored grains produced potent cancer-causing substances called *aflatoxins*. By improving nutrition and health, potatoes were undoubtedly a significant contributing factor in the population explosion that provided the workforce needed to power the Industrial Revolution.

In no country has the potato had a greater impact than in Ireland. By the early 19th century, the poor were almost entirely dependent on it for their food, and when a deadly blight

Variety of potatoes grown in Peru

struck the potato crop in 1845, the results were devastating. The blight spread with terrifying speed through the land. Both harvested potatoes and those still in the ground rotted. Deprived of their only source of food, at least a million Irish died of starvation or sickness, and over a million emigrated to the United States. In time, the potato recovered, but it remained vulnerable to the blight, which we now know was caused by a fungus accidentally introduced from Mexico. Today, the disease is controlled with fungicides.

Varieties and Uses

Andean farmers have always grown thousands of different varieties of potatoes in an astonishing range of shapes, sizes, flavors, and colors, but in most countries only a few varieties have ever been cultivated. In the United States, just six varieties of potatoes, all scientifically bred, make up 80% of the crop, and one—the Russet Burbank—accounts for 40%. Lack of diversity can be a prescription for disaster, as was the case in Ireland in 1845. If a crop is genetically uniform, disease, pests, or an unexpected fluctuation in climate can completely wipe it out.

Potatoes are not just an excellent source of carbohydrate, they also contain minerals and vitamins and provide twice as much high-quality protein per acre as wheat.

Potatoes are not just an excellent source of carbohydrate, they also contain minerals and vitamins and provide twice as much high-quality protein per acre as wheat. Potatoes are 99.9% fat-free and an 8-ounce baked or boiled potato (if eaten without adding butter or sour cream) supplies approximately 100 Calories. The major potato-growing states in America are Idaho, Washington, and Maine.

More than half the American potato crop is processed into chips, frozen French fries, dehydrated products, and potato starch for use as a low-fat additive or binding agent in soups, ice cream, and bakery goods. Potato processing is an old technique that was first developed hundreds of years ago in the Andes. Potatoes were spread on the ground, and alternate freezing and drying during cold nights and warm days reduced their water content (potatoes are about 80 percent water). Villagers then stomped the potatoes with their bare feet to press out remaining water. Once dry, the dehydrated product, called *chuno,* could be stored for several years.

Potatoes are also used in the production of paper, textiles, adhesives, cosmetics, and pills and capsules. In the future, they are likely to find many more applications. Water-soluble, and thus readily disposable, potato starch materials may some day replace Styrofoam for packaging, and in the years to come, as petroleum becomes scarce, automobiles may run on potato gasohol.

The most exciting area of potato research is in genetic engineering. By extracting a desirable gene from one potato variety and introducing it into another, scientists are working to produce disease-resistant, high-yielding potatoes that can be grown in many different climates around the world. But genes cannot be invented; they can only be altered. Genetic reserves are kept at the International Potato Center, established in Lima, Peru, in 1971. The center has a collection of over 13,000 varieties of wild potatoes, which serve as a world bank of potato genes.

The humble potato, at times maligned and often scorned as poor people's food, is now universally respected. Nutritious, adaptable, hardy, versatile, and delicious, few foods can compare with it. This amazing vegetable, which can be grown almost anywhere in the world, may be our best hope for feeding the millions of hungry people that inhabit our planet.

References: M. Sayles Hughes, "Potayto, Potahto—either way you say it, they 'a' peel," *Smithsonian* (October 1991), p. 138; R. E. Rhoades, "The Incredible Potato," *National Geographic* (May 1982), p. 668; and "The World's Food Supply at Risk," *National Geographic* (April 1991), p. 74.

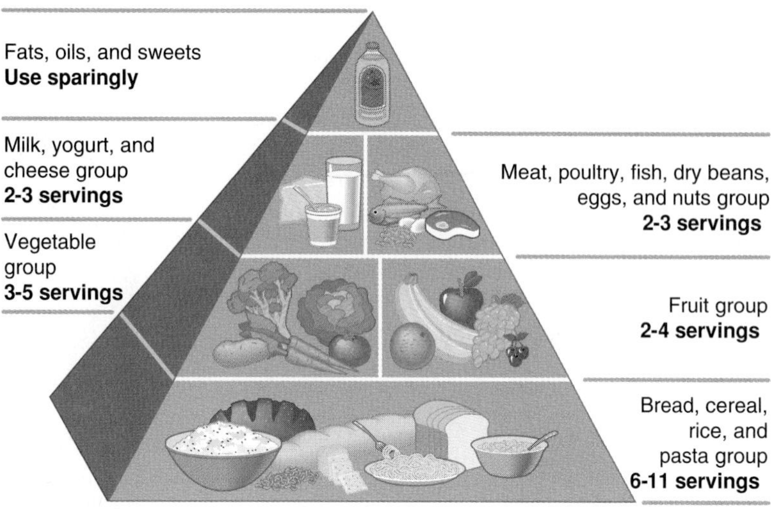

FIGURE 15.13

The food pyramid proposed by the U.S. Department of Agriculture, which shows how to plan daily meals for good nutrition.

2. Eat a varied diet, with energy intake made up as follows:
   a. Carbohydrates = 58%
      - complex carbohydrates = 38%
      - naturally occurring sugars = 10%
      - refined sugar, no more than = 10%
   b. Fats = no more than 30%
      - polyunsaturated fats = 10%
      - monounsaturated fats = 10%
      - saturated fats, no more than = 10%
   c. Protein = 12%
3. Avoid excess consumption of foods high in cholesterol.
4. Avoid excess sodium by keeping salt intake at about 5 mg/day.
5. Eat foods that provide adequate fiber.
6. Drink alcohol in moderation.
7. Increase calcium intake for adolescent girls and adult women.
8. Increase iron intake for children, adolescents, and women of child-bearing age, particularly in low-income families.

To obtain the ideal diet, a person should plan daily meals according to the diagram in **Figure 15.13**. The pyramid emphasizes the importance of increasing consumption of complex carbohydrates and fruits and vegetables, and the need to restrict consumption of fats and sweets.

Chapter Summary

1. For good nutrition, the diet should contain carbohydrates, fats, proteins, minerals, vitamins, fiber, and water.
2. During digestion of food, enzyme-catalyzed reactions break down carbohydrates to monosaccharides, proteins to amino acids, and fats to fatty acids.
3. Energy derived from food is measured in Calories (1 Cal = 1 kcal = 1000 cal).
4. A nutritional diet should be made up as follows: carbohydrates, 58% (primarily as complex carbohydrates); fats, 30% (equally divided between saturated, monounsaturated, and polyunsaturated fats); and proteins, 12%.
5. Body cells obtain energy from glucose in a series of complex reactions involving ATP and ADP.
6. Fats are the body's storehouse of energy.
7. Complete proteins provide all the essential amino acids the body needs. Incomplete proteins are deficient in one or more essential amino acids.

8. Lack of an adequate supply of protein leads to severe nutritional diseases in poor countries.
9. More than 60% of Americans are overweight, primarily because of overconsumption of food.
10. In diabetes, the pancreas fails to produce insulin, which is essential for the proper metabolism of glucose.
11. Most vitamins are coenzymes. They are required in small amounts for the maintenance of good health and cannot be synthesized by the body.
12. Diseases associated with the lack of a specific vitamin are scurvy (vitamin C), beri beri (vitamin B_1), pellagra (vitamin B_3), blindness (vitamin A), and rickets (vitamin D).
13. Water-soluble vitamins include the B-complex vitamins and vitamin C. Fat-soluble vitamins include vitamins A, D, E, and K.
14. Minerals that are needed for good health include calcium, phosphorus, magnesium, iron, iodine and small amounts of many other elements.
15. Conditions associated with a deficiency of particular minerals include osteoporosis (calcium and phosphorus), anemia (iron), and goiter and cretinism (iodine).
16. Food additives prevent spoilage, sweeten, or add color and flavor to foods.
17. Food preservatives are of two kinds: (a) antimicrobial agents, primarily benzoates, propionates, and sorbates, and (b) the antioxidants BHA and BHT.
18. Artificial sweeteners, which have little or no nutritional value and contribute few calories, include saccharin, cyclamates (now banned), aspartame and acesulfame K.
19. The GRAS list names food additives that have not been tested by the FDA but are generally regarded as safe.
20. For good nutrition, daily meals should be planned according to the food pyramid recommended by the U.S. Department of Agriculture.

Key Terms

adenosine diphosphate (ADP) (p. 460)
adenosine triphosphate (ATP) (p. 460)
antioxidants (p. 471)
ascorbic acid (vitamin C) (p. 465)
butylated hydroxyanisole (BHA) (p. 471)
butylated hydroxytoluene (BHT) (p. 471)
calciferol (vitamin D) (p. 466)
Calorie (p. 457)
diabetes (p. 464)
fiber (p. 456)
GRAS list (p. 475)
minerals (p. 468)
niacin (vitamin B_3) (p. 466)
nutrients (p. 454)
obesity (p. 464)
pepsin (p. 455)
prostaglandins (p. 462)
retinol (vitamin A) (p. 466)
saccharin (p. 473)
scurvy (p. 465)
thiamine (vitamin B_1) (p. 465)
tocopherol (vitamin E) (p. 467)
vitamin (p. 465)
vitamin K (p. 467)

Questions and Problems

1. List three main functions of nutrients in humans.
2. In addition to the eleven elements present in significant amounts in the body, how many other elements can be found in trace amounts?
3. Outline the steps in the digestion of carbohydrates. What are the end products?
4. What is the role of enzymes in digestion? Give examples.
5. How does the body store energy?
6. Briefly describe the role of ATP in the release of energy from glucose.
7. How many Calories of energy are released when 180 grams of glucose ($C_6H_{12}O_6$) are converted to carbon dioxide and water?
8. Does any food literally contain Calories? What does the manufacturer mean when the label on a candy bar says "contains 250 Calories"?
9. What compound causes soreness in your muscles after serious exercise? How is it formed?
10. Why are carbohydrates considered energy-rich?
11. Why do fats produce more than twice as much energy per gram as proteins and carbohydrates do when they are oxidized to carbon dioxide and water?
12. What foods supply fiber? Why do we need fiber in our diet?
13. There is one fatty acid the body needs but cannot synthesize. Name it.
14. What happens to fatty acids when they reach the liver?
15. What percentage of energy is derived from the consumption of fats by people living in **a.** the United States; **b.** developing countries?
16. Triglycerides are composed of a fatty acid and __________.
17. What is meant by a complete protein? Give examples of complete and incomplete proteins.
18. How much protein does an adult who weighs 80 kilograms need each day?
19. Are vitamins organic or inorganic compounds? Explain.
20. Identify each vitamin:
    a. tocopherol **b.** calciferol **c.** ascorbic acid
    d. cobalamin **e.** retinol
21. Identify the vitamin deficiency associated with each disease:
    a. pellagra **b.** night blindness **c.** rickets
    d. beri beri
22. Identify the deficiency disease associated with a diet lacking in each of the following vitamins:
    a. thiamine **b.** vitamin B_{12} **c.** niacin
23. Indicate whether each of the following is a water-soluble or fat-soluble vitamin:
    a. vitamin C **b.** vitamin K
    c. vitamin B **d.** vitamin A
    e. vitamin B_{12} **f.** riboflavin
    g. niacin **h.** tocopherol
    i. calciferol
24. Which vitamin is associated with blood clotting?
25. Which vitamin is considered to be a natural antioxidant?
26. Is excessive consumption of water-soluble vitamins more dangerous than excessive consumption of fat-soluble vitamins? Explain.
27. What is kwashiorkor? How is it caused?
28. What is marasmus? How is it caused?
29. What causes obesity?
30. Explain why blood glucose rises to abnormally high levels in untreated diabetics after they have eaten a meal.
31. What causes PKU? How is it treated?
32. What causes galactosemia? How is it treated?
33. Describe the problems associated with low-carbohydrate, high-fat diets.
34. Are there any nutritional problems associated with a strict vegetarian diet?
35. How do palm oil and coconut oil differ from other vegetable oils?
36. Indicate a biological function for each of the following:
    a. phosphorus **b.** iodine **c.** calcium
    d. iron

37. Which of the following elements would you expect to find in a healthy human? Justify each choice.
    a. Hg b. Na c. Zn
    d. Ca e. I f. P

38. What proportions of carbohydrates, fats, and proteins make up a healthy diet?

39. Why is it preferable to eat a whole apple rather than one that is peeled?

40. How do sport drinks, such as Gatorade, differ from soft drinks, such as Coke? (Check the label on a bottle of each drink.)

41. Sulfites, which are used to give fruits and vegetables a fresher appearance, have been shown to cause which of the following?
    a. cancer b. brain damage
    c. allergic reactions d. liver damage

42. From what raw ingredient are synthetic food colorings made?

43. What benefit do the food additives BHA and BHT provide in processed foods?

44. Three fast-food restaurants offer hamburgers and French fries. Each chain's offerings are tested, and the following data are collected:

Restaurant 1	**Hamburgers**	**Fries**
protein (g)	12	2
carbohydrates (g)	30	22
fat (g)	10	20
Restaurant 2		
protein (g)	11	3
carbohydrates (g)	30	26
fat (g)	12	12
Restaurant 3		
protein (g)	14	2
carbohydrates (g)	28	24
fat (g)	11	10

    a. Which restaurant's hamburger with fries offers the lowest Calorie intake?
    b. How many Calories would you consume if you ate the highest-Calorie offering?

45. Saltine crackers contain 0.2 g of protein and 10 g of carbohydrate per cracker. If your recommended daily allowance (RDA) is 70 g of protein, how many crackers would you need to eat to satisfy your RDA for protein? How many grams of carbohydrates would you also consume? Would you consider crackers to be a good source of protein?

46. A glass of milk contains 250 g of milk. Milk is about 3.5% protein. How much protein is contained in two glasses of milk? Is milk a reasonable source of dietary protein?

47. A cup of corn flakes (28.4 g of flakes) contains 2 g of protein, 24 g of carbohydrate, and 0 g of fat. How many Calories are contained in 1 cup of corn flakes? If 100 g of whole milk is added to a cup of corn flakes, how many Calories does this combination contain? Refer to Example 15.1 for the composition of milk.

48. What two types of agents are used as food preservatives? Give an example of each.

49. What preservative is added to bologna to give it a pink color? Are there any dangers associated with this additive?

50. What do the letters MSG stand for, and why is this substance added to food?

51. Compare the sweetness of the following substances: sucrose, honey, saccharin, aspartame.

52. What is a sequestrant? Give an example, and explain why it is added to foods.

53. What is the purpose of an emulsifier? Give an example.

54. How was the GRAS list established?

Answers to Practice Exercises

15.1 403 Cal from carbohydrate, 70 Cal from protein, 58 Cal from fat

15.2 More than half (51.8%)

Chapter 16

Agricultural Chemicals

Feeding The Earth's People

Chapter Objectives

In this chapter, you should gain an understanding of:

- How the human population is growing and food production is leveling off
- The composition and properties of soil
- The types and functions of plant nutrients
- The development of modern agriculture
- The production and use of synthetic inorganic fertilizers and the value of organic fertilizers
- The main classes of synthetic insecticides and herbicides and the benefits and problems associated with their use
- Insect control methods that do not rely on synthetic insecticides
- The development of new varieties of crop plants
- Approaches to farming that are less dependent on agricultural chemicals

Yields of farm products have been increased by the use of synthetic fertilizers and pesticides and the introduction of new varieties of crop plants.

ONLINE FEATURES

www.jbpub.com/chemistry

- Chemistry and the Environment
- eLearning
- Chemistry in the News
- Special Topics
- Research & Reference Links
- Ask the Authors

HOW MANY PEOPLE can the earth support? Is the population about to outrun food production? In the developed nations, large surpluses of food have been produced, and in many of the developing nations, food production has managed to keep pace with population growth. Food and resources are not evenly distributed, however, and despite the advances, millions of people in the poorer nations of the world suffer from hunger and malnutrition. Drought, war, and other disasters that disrupt food production and distribution continue to cause widespread starvation.

Significant increases in agricultural production began during the 19th century as farmers gained a better understanding of the needs of plants. Crop rotation, the use of natural fertilizers, improved irrigation, and the introduction of more efficient farm machinery—all these helped to increase crop yields. However, the dramatic increases in food production that have been achieved in recent years have been due primarily to (1) the application of synthetic fertilizers to restore soil fertility, (2) the use of pesticides to reduce the amount of crops lost to disease and pests, and (3) the development of improved crop plants.

In this chapter, we focus on the agricultural chemicals—fertilizers and pesticides—that help meet the world's increasing demand for food. We study the soil conditions and nutrients that plants need to grow and thrive. We also examine the risks to human health and the environment posed by the use of synthetic fertilizers and pesticides, and we consider some alternative farming techniques.

Today, despite decades of growth in agricultural productivity and steady expansion of cropland at the expense of forest and grassland, there are more hungry people on the earth than ever before.

16.1 Population Growth

In 1798, Thomas Robert Malthus, an English clergyman and economist, published an essay in which he argued that populations inevitably tend to grow faster than their food supply. He predicted that, unless birth rate was controlled, population would expand until it reached the limit of subsistence. After that, it would be kept in check by starvation, disease, pestilence, and war. Since Malthus's time, the world's population has increased from less than 1 billion to over 6 billion. However, because agricultural productivity has increased faster than Malthus thought possible, so far his pessimistic predictions have not been realized.

During most of the 3 million years that humans have lived on the earth, the population has grown very slowly, with numerous setbacks caused by famine and epidemic diseases (**Fig. 16.1**). One of the most devastating killers was the Black Death, or plague, which swept through Europe during the Middle Ages, causing the deaths of one-quarter of the population—approximately 25 million people. In the 18th and early 19th centuries, childhood diseases took a fearful toll, and many people died from waterborne diseases (Chapter 10). The introduction of vaccinations, the discovery of drugs that could

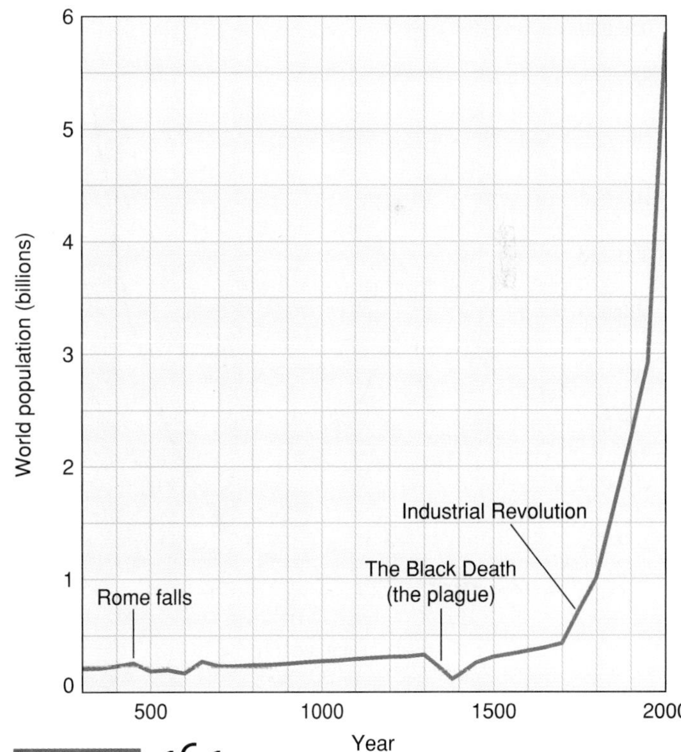

FIGURE 16.1

World population grew very slowly until the middle of the 17th century, when it began to grow very rapidly.

Table 16.1 Growth of World Population

Population	Year
1 billion	1850
2 billion	1930
4 billion	1975
6 billion	2001
9 billion (projected)	2050

cure many diseases, and the increased availability of clean water were the major factors responsible for the decrease in infant mortality and the increase in life expectancy that led to the rapid rise in population throughout the 20th century.

World population reached 1 billion in 1830. By 1930, just 100 years later, it had reached 2 billion. After another 45 years, it had doubled again to 4 billion (**Table 16.1**) and by 2001 it reached 6.1 billion. In 1996, a population of 9.4 billion was forecast for 2050. Now, mainly because of soaring deaths from the global AIDS epidemic particularly in Africa, this number has been revised down to 8.9 billion. According to U.N. demographers, one-fifth of the adult population in sub-Saharan Africa is infected with the human immunodeficiency virus (HIV).

Population Arithmetic

Malthus's dire prediction was based on the argument that population can grow geometrically, but food supplies can grow only arithmetically. **Geometric growth** occurs when something increases by a fixed percentage every year. For example, if $1000 is invested at 10% interest (compounded daily) at a child's birth, by the time the child is 7 years old, the account will be worth $2000 (**Fig. 16.2a**). Seven years later, it will have doubled to $4000. After 56 years, it will be worth over a quarter of a million dollars; at 70 years, it will be worth over $1 million.

The most striking characteristic of geometric growth is that it begins slowly, but once it has "rounded the bend," it proceeds at an ever increasing rate. This rapidly increasing rate occurs whether the growth is affecting money in a bank account or a country's population.

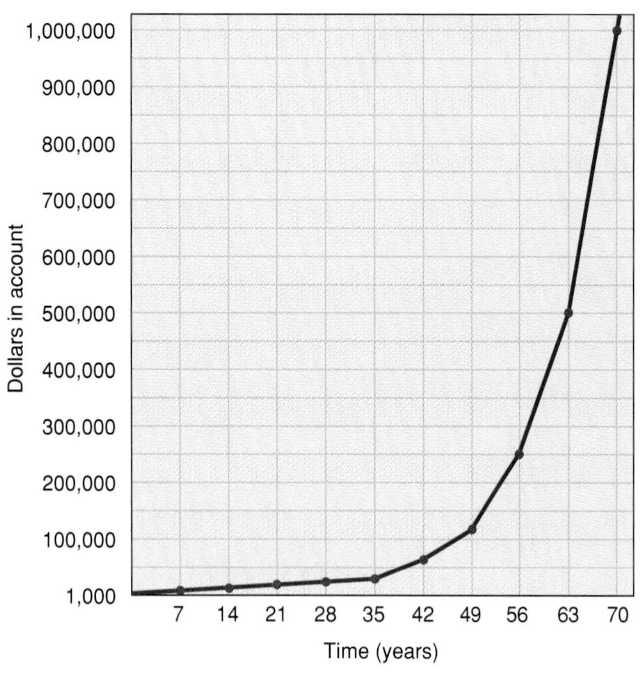

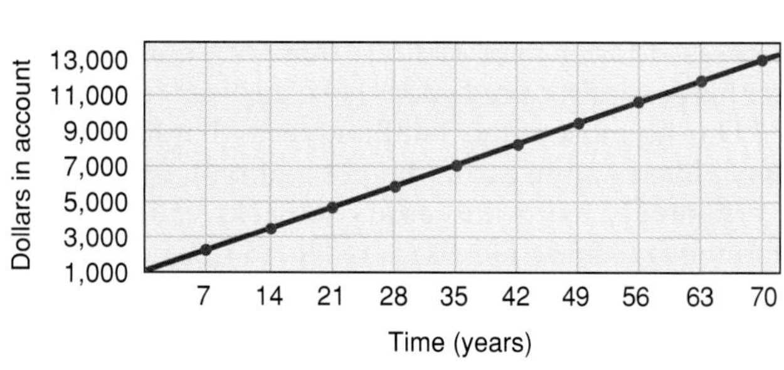

FIGURE 16.2

(a) Geometric growth of a savings account in which $1000 was invested at 10% interest compounded daily. A period of slow growth is followed by increasingly rapid growth. (b) Arithmetic growth of a savings account to which $1000 was added at regular intervals of 7 years.

Arithmetic growth occurs when a constant amount is added at regular intervals. For example, if a parent distrustful of banks puts aside $1000 at a child's birth and then adds further installments of $1000 at 7 year intervals, the sum will grow slowly and steadily over successive periods **(Fig. 16.2b)**. By the time the child reaches 70 years, the cache, if still intact, will amount to $11,000—far less than the $1 million produced over the same period if $1000 is invested at 10% compounded interest.

EXAMPLE 16.1 Analyzing the growth of savings

You set up a savings account as follows: You deposit $0.10 the first week and then double the amount you deposit each week thereafter ($0.20 the second week, $0.40 the third week, and so on).

a. How much money will be in the savings account at the end of 10 weeks?
b. Is this an example of geometric or arithmetic growth?

Solution

Week	Amount deposited ($)	Total in account ($)	Week	Amount deposited ($)	Total in account ($)
1	0.10	0.10	6	3.20	6.30
2	0.20	0.30	7	6.40	12.70
3	0.40	0.70	8	12.80	25.50
4	0.80	1.50	9	25.60	51.10
5	1.60	3.10	10	51.20	102.30

a. There will be $102.30 in the account at the end of 10 weeks.
b. The growth is geometric.

Practice Exercise 16.1

A child saves 10¢ every week and puts the money in her piggy bank.
a. How much will she have at the end of 10 weeks?
b. Are her savings growing arithmetically or geometrically?

An instructive perspective on population growth is provided by the **doubling time**—the number of years it takes for a population to double if its growth continues at the present rate. This number can be calculated by using the **Rule of 70**,—that is, by dividing the current annual growth rate (as a percentage) into 70. For example, world population is growing at the rate of 1.25% per year. If this seemingly slow growth rate continues, world population will double approximately every 57 years (70/1.25 = 56.8). The highest population growth rates are found in the poorest countries. The Indian subcontinent (India, Pakistan, and Bangladesh) is expected to add nearly 800 million people by 2050.

The rule of 70 is used in banking and finance to estimate the growth of investments.

EXAMPLE 16.2 Analyzing population growth

The population of a city is 16 million and is growing at an annual rate of 3.5%. In how many years will the population double if growth continues at the same rate?

Solution Use the rule of 70. Divide the annual growth rate into 70:

$$\frac{70}{3.5} = 20$$

The population will double in 20 years.

Practice Exercise 16.2

In 1999, it was estimated that the population of a country was 23.6 million and was growing at a rate of 4.1% a year. If the growth rate remains constant, in what year will the population reach 47.2 million?

16.2 The Development of Modern Agriculture

One of the earliest techniques for growing food was *slash-and-burn cultivation* **(Fig. 16.3)**, in which trees and other vegetation were cut down and burned. The ashes returned nutrients to the soil, making it sufficiently fertile to grow crops at the site for 2–5 years. When fertility waned, the site was abandoned, and a new plot was cleared. Slash-and-burn cultivation continues today in some less developed areas, including the tropical rain forests in the Amazon Basin of South America.

Legumes enrich the soil with nitrogen compounds because their root nodules contain bacteria that are able to fix atmospheric nitrogen.

For centuries, farmers in settled communities relied on the application of manure and the rotation of crops, which included nitrogen-fixing legumes (Chapter 2), to return nutrients to the soil, but the scientific basis for these practices was not understood until the 19th century. The German chemist Justus von Liebig (1803–1873), a pioneer in the field of agricultural chemistry, was the first to identify many of the chemical substances that plants require for healthy growth. As a result of his work and that of others, farmers began using increased amounts of natural fertilizers to supply plants with the needed nutrients. Crop yields increased dramatically, and new farm machinery introduced with the Industrial Revolution allowed food production to keep pace with the upsurge in population. The need for more and more fertilizer led directly to the introduction of synthetic inorganic fertilizers, on which agriculture now depends so heavily.

All our food comes ultimately from green plants. We eat either plants or animals that have fed on plants (Chapter 2). To grow and flourish, plants need sunlight, carbon dioxide, suitable temperatures, and soil that has an appropriate pH (Chapter 7) and can provide an adequate supply of water, oxygen, and essential nutrients. Before we consider the nutrients plants need, let's consider the properties that make soil productive.

FIGURE 16.3

With slash-and-burn cultivation, nutrients in the ashes make it possible to grow crops for 2–5 years before the land becomes exhausted and is abandoned.

16.3 Soil

Soil is a complex mixture of inorganic and organic materials. The inorganic part is made up of rock fragments formed over thousands of years by the weathering of bedrock. The organic part is derived from the decayed remains of plants and animals.

In the United States and other developed countries, soil is kept fertile largely by the application of commercial synthetic inorganic fertilizers. The productivity of soil is also greatly influenced by its texture and by its content of organic matter, both living organisms and dead material.

Soil Texture: The Inorganic Component

The mineral (inorganic) particles of soil are composed primarily of silicates (Chapter 2), but, depending on location, soil may also contain phosphates and limestone ($CaCO_3$). The size of the particles in a soil determines its texture. Relatively coarse particles (diameters of 0.10–2.0 mm) form sand; slightly finer particles form silt; and the finest particles (0.002 mm or less) form clays. A typical soil contains all three types of particles. **Loam,** a productive soil that crumbles easily, contains approximately equal proportions of sand and silt and organic matter.

Plant roots absorb water, nutrients, and oxygen from the soil and release carbon dioxide to it. Thus, soil should allow good circulation of air and provide continuous access to water and nutrients. In sandy soils, the relatively large soil particles have large spaces between them. The large spaces promote good air circulation but allow water to percolate rapidly through the soil, taking dissolved nutrients beyond the reach of the roots. Given a specific volume of soil, the surface area of a large number of small particles is greater than the surface area of a smaller number of large particles. Therefore, as the size of soil particles decreases, more water adheres to a given volume of soil. At the same time, the size of the spaces between particles becomes smaller. Smaller spaces do not allow water to flow through as rapidly and tend to hold it by capillary action **(Fig. 16.4)**. The disadvantage of small particles is that air circulation is less efficient. In clay soils, the very fine particles are packed so closely together that the soil becomes impermeable to water. Rainfall collects at the surface and in the upper layers, which soon become so waterlogged that gases cannot circulate.

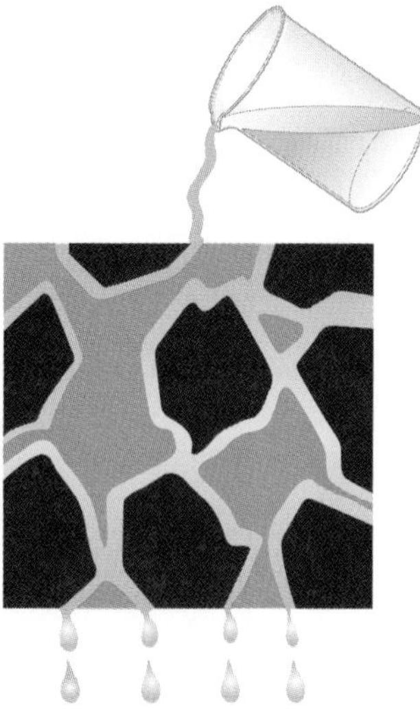

Water runs through.

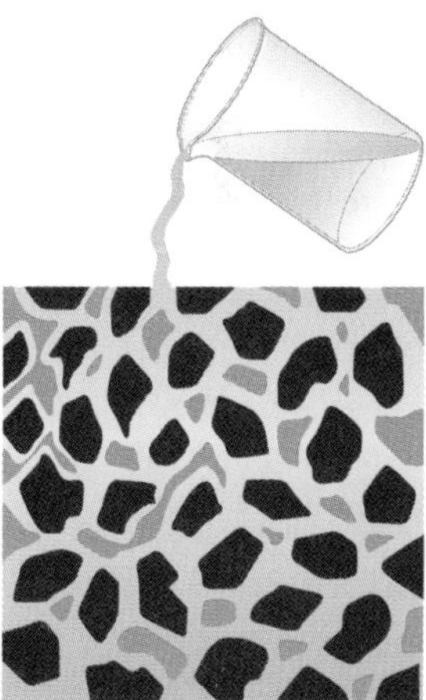

Water is held.

FIGURE 16.4

The water-holding capacity of soil increases as the size of the soil particles decreases. As particle size decreases, relative surface area increases, and more water can cling to a given quantity of soil.

Humus: The Organic Component

Soil is greatly improved by the addition of decayed organic matter called **humus.** As leaves, twigs, dead plants and animals, and other organic materials accumulate at the soil surface, they are attacked by bacteria and other detritus feeders and are partially decomposed. The partially decomposed organic product is humus, a dark-colored mixture that is resistant to further decomposition.

As humus forms, it becomes mixed with soil particles, and the soil acquires a light spongy texture that retains water well. Humus acts as a reservoir of water and nutrients, which are slowly released to plants in a form they can absorb. In an undisturbed natural ecosystem, nutrients are constantly recycled (Chapter 2), and humus is replenished. If crops are repeatedly grown on the same plot of land, however, nutrients are permanently removed as each crop is harvested, and the soil soon becomes infertile.

A typical soil is made up of three main layers, or *horizons* **(Fig. 16.5)**. The uppermost, or A, horizon, which includes the **topsoil,** is dark in color and contains most of the humus found in the soil. The A horizon is made up of decayed plant and animal materials, and it is rich in microorganisms and other creatures including earthworms, insects, and small burrowing animals that keep the soil aerated. Below the topsoil is the lighter-colored, humus-poor, and more compacted B horizon, or **subsoil.** This layer, which may be several feet thick, is made from inorganic particles of the parent rock and contains minerals and organic materials leached from the A horizon. The C horizon consists of a layer of fragmented bedrock mixed with clay that rests on solid bedrock.

Compost, which gardeners produce from vegetable wastes, leaves, grass, and other plant materials by allowing them to be digested by soil bacteria and worms, is a form of humus.

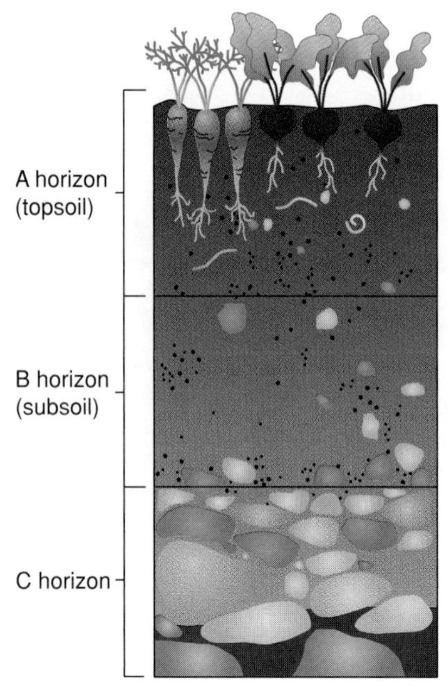

FIGURE 16.5

There are three main soil layers, or horizons, whose thickness varies considerably depending on the type of soil.

Soil pH

Plants vary considerably with respect to the pH range they can tolerate. Most crop plants prefer a soil with a pH close to neutral. However, some, such as corn, wheat, and tomatoes, grow best in slightly acidic soil, and potatoes and most berries prefer even more acidic conditions.

Soils tend to become acidic if the carbon dioxide released by roots or formed by the oxidation of organic matter accumulates below the soil surface, where it can react with water:

$$CO_2 + H_2O \longrightarrow H^+ + HCO_3^-$$

The acid can be neutralized by the application of lime (a process called *liming*) in the form of quicklime (CaO) or slaked lime [$Ca(OH)_2$], either of which removes hydrogen ions:

$$2\,H^+ + CaO \longrightarrow Ca^{2+} + H_2O$$

Some soils are alkaline and are neutralized by adding sulfur. Soil bacteria gradually convert the sulfur to sulfuric acid.

Acidic soils are often called *sour,* and slightly basic soils are described as *sweet.*

Soil Erosion

In the process of natural weathering, soil particles, primarily from topsoil, are dislodged and carried away by wind and flowing water. This process is termed **erosion,** and the extent to which it occurs depends primarily on wind force, the amount of precipitation, and the amount of ground cover.

Topsoil is a renewable resource if it is not removed more rapidly than it can be replaced. The roots of plants generally protect the topsoil from undue erosion, but whenever ground cover is removed, erosion by both wind and rain occurs at an accelerated rate. Regeneration of topsoil is extremely slow—just a few centimeters are created every thousand years. Thus, in many parts of the world, topsoil has become a nonrenewable resource. Soil scientists have estimated that about one-third of the topsoil that was originally found on cropland in the United States has been blown away or washed into rivers and lakes **(Fig. 16.6)**. Human activities that lead to erosion and the loss of valuable and irreplaceable topsoil include construction, mining, clear-cutting of timber, agricultural development, and overgrazing.

16.4 Plant Nutrients

Carbon, hydrogen, and oxygen are often referred to as the *nonmineral nutrients.*

Plants obtain carbon, hydrogen, and oxygen from air and water and, with energy from the sun, convert them to simple carbohydrates and oxygen through the process of photosynthesis:

$$6\,CO_2 + 6\,H_2O \longrightarrow C_6H_{12}O_6 + 6\,O_2$$

The simple carbohydrates are then converted into the complex carbohydrates that form the basic structural material of all plants (Chapter 14).

In addition to carbon, hydrogen, and oxygen, plants require at least 13 other elements, called *mineral nutrients* because they are obtained from the soil. The mineral nutrients are divided into three groups according to the relative amounts of each that plants need to thrive: **primary nutrients** (nitrogen, phosphorus, and potassium), **secondary nutrients** (calcium, magnesium, and sulfur) and **micronutrients (Table 16.2)**. Dissolved in water, mineral nutrients are absorbed from the soil through plants' roots.

FIGURE 16.6 Soil erosion in Arkansas.

Nitrogen

Plants need nitrogen to manufacture amino acids and proteins necessary for the formation of leaves and stems. After carbon, hydrogen, and oxygen, nitrogen is the element needed in greatest quantity. As a bushel (equal to 8 gallons) of corn ripens, it removes approximately 0.5 kilogram (1 pound) of nitrogen from the soil. If this nitrogen is not replaced, the soil's supply of nitrogen is gradually exhausted. Soil is more likely to be deficient in nitrogen than in any other nutrient.

Although 78% of the atmosphere is nitrogen gas (N_2), very few plants can exploit this source directly. For most plants, atmospheric nitrogen must first be changed by **nitrogen fixation** into a water-soluble form that plants can absorb through their roots. As we saw in Chapter 2, nitrogen fixation is accomplished primarily by: (1) bacteria in the soil or in nodules on the roots of leguminous plants, (2) blue-green algae (cyanobacteria) in soil and water, and (3) lightning. Nitrogen-fixing bacteria convert atmospheric nitrogen to ammonia (NH_3), which in the slightly acidic pH of most soils forms ammonium ions (NH_4^+). By the process of **nitrification,** other soil bacteria bring about the oxidation of ammonium ions to nitrate ions (NO_3^-), the preferred form for nitrogen for most plants.

During thunderstorms, electrical discharges in the atmosphere cause nitrogen to react with oxygen to produce nitric oxide (NO) and nitrogen dioxide (NO_2):

$$N_2 + O_2 \longrightarrow 2\,NO$$

$$2\,NO + O_2 \longrightarrow 2\,NO_2$$

Nitrogen dioxide dissolves readily in water, forming nitrous acid (HNO_2) and nitric acid (HNO_3):

$$2\,NO_2 + H_2O \longrightarrow HNO_2 + HNO_3$$

Nitric acid in rainfall replenishes the supply of nitrates in soil. The amount of nitric acid produced by thunderstorms is small and does not contribute significantly to the problems created by acid rain.

By the 19th century, in addition to fertilizing with manure, many farmers were practicing crop rotation, alternating a nitrogen-consuming crop, such as corn or wheat, with

Table 16.2 Chemical Nutrients Required by Plants

Name	Chemical Form Absorbed
Nonmineral	
Carbon (C)	CO_2
Hydrogen (H)	H_2O
Oxygen (O)	CO_2, H_2O, O_2
Primary nutrients	
Nitrogen (N)	NH_4^+, NO_3^-
Phosphorus (P)	$H_2PO_4^-$, HPO_4^{2-}
Potassium (K)	K^+
Secondary nutrients	
Calcium (Ca)	Ca^{2+}
Magnesium (Mg)	Mg^{2+}
Sulfur (S)	SO_4^{2-}
Micronutrients	
Boron (B)	$H_2BO_3^-$, $B(OH)_4^-$
Chlorine (Cl)	Cl^-
Copper (Cu)	Cu^{2+}
Iron (Fe)	Fe^{2+}, Fe^{3+}
Manganese (Mn)	Mn^{2+}
Molybdenum (Mo)	MoO_4^{2-}
Zinc (Zn)	Zn^{2+}

In aquatic ecosystems, cyanobacteria are the most important nitrogen fixers.

It is estimated that about 90% of nitrogen fixation is due to biological processes and 10% is due to lightning.

a nitrogen-fixing leguminous crop, such as beans, peas, alfalfa, or soybeans. The discovery of huge deposits of sodium nitrate (called *Chile saltpeter*) in the deserts of Chile provided a valuable additional source of nitrogen, but it was not until 1913 that a seemingly inexhaustible supply of nitrogen became available. The breakthrough came with the development by the German chemist Fritz Haber of a nitrogen-fixing process, which led directly to the production of synthetic nitrogen-containing fertilizers. In the Haber process (Chapter 6), nitrogen obtained from the atmosphere reacts directly with hydrogen to produce ammonia:

$$N_2 + 3\,H_2 \longrightarrow 2\,NH_3$$

In the United States, ammonia was first produced commercially by the Haber process in 1921.

During World War I (1914–1918), the Germans used ammonia to make ammonium nitrate for the manufacture of explosives. It was not until the war was over that the process was used as the first step in the manufacture of fertilizers.

Today, ammonium salts, urea, and liquid ammonia are the major compounds used to restore nitrogen to the soil, and all have the Haber process as the first step in their production. The pathways by which nitrogen compounds reach the soil are shown in Figure 2.17.

Scientists are working to understand how the unique enzymes in bacteria in legumes are able to fix atmospheric nitrogen. They hope to use genetic engineering to create nitrogen-fixing bacteria that can live on the roots of cereal plants as they do on those of legumes. Other approaches include improving the nitrogen-fixing abilities of legumes now being grown and treating soil with nitrogen-fixing blue-green algae (cyanobacteria).

Anhydrous Ammonia At normal temperatures and pressures, ammonia is a gas. At higher pressure, it is easily compressed into a liquid called *anhydrous ammonia* that can be stored and transported in tanks. The liquid ammonia is injected directly into the soil, where it is converted into ammonium ions.

Ammonium Salts The first step in the production of the fertilizer ammonium nitrate is oxidation of ammonia to nitric acid:

$$NH_3 + 2\,O_2 \longrightarrow HNO_3 + H_2O$$

The nitric acid then reacts with more ammonia to produce ammonium nitrate, the ammonium salt of nitric acid:

$$HNO_3 + NH_3 \longrightarrow \underset{\text{ammonium nitrate}}{NH_4NO_3}$$

Ammonium sulfate and ammonium phosphate are prepared by treating ammonia with sulfuric acid and phosphoric acid, respectively.

Urea Under high pressure, ammonia and carbon dioxide react to form urea:

$$2\,NH_3 + CO_2 \longrightarrow \underset{\text{urea}}{NH_2{-}\overset{\overset{\displaystyle O}{\|}}{C}{-}NH_2} + H_2O$$

Urea is a major constituent of urine.

On contact with water in the soil, urea gradually decomposes, releasing ammonia as it does so. This slow release gives urea an advantage over the inorganic ammonium salts (Fig. 16.7).

The use of synthetic fertilizers has been the main factor that has allowed food production to keep pace with population growth. But the cost of manufacturing them is high. Although there is an unlimited supply of nitrogen in the atmosphere, the hydrogen required to synthesize ammonia is obtained from petroleum (Chapter 12). Small hydrocarbon molecules, such as propane (C_3H_8), are reacted with steam in the presence of suitable catalysts to yield carbon dioxide and hydrogen:

$$C_3H_8 \quad + \quad 6\,H_2O \quad \longrightarrow \quad 3\,CO_2 \quad + \quad 10\,H_2$$

Thus, for the present, the cost of synthetic fertilizers is tied to the cost and availability of petroleum. In the future, it may be possible to use solar energy to obtain hydrogen from the electrolysis of water (Chapter 13).

FIGURE 16.7

Application of urea-based fertilizer. As the urea decomposes, it releases ammonia into the soil.

Phosphorus

Plants need phosphorus for the synthesis of DNA and RNA (Chapter 14) and the synthesis of ATP and ADP (Chapter 15), the compounds involved in energy transfer in many metabolic processes, including photosynthesis. Phosphorus is particularly important for plants such as tomatoes that are cultivated for their fruit.

Plants absorb phosphorus from the soil in the form of phosphate ions. In natural soils at near neutral pH, phosphates exist mainly as HPO_4^{2-} and $H_2PO_4^-$ ions, derived from the weathering of phosphate rock and from ancient deposits of skeletal remains of sea creatures and other animals. Phosphates are common in nearly all soils but are often present in low concentrations.

Although it was not until the 19th century that a plant's need for phosphorus was established by scientific studies, records show that the practice of improving soil by adding ground bones was used in China as early as 2000 B.C.

Because the phosphorus in bones and in phosphate rock is tightly bound to calcium and is only slightly soluble in water, it is not readily available to plants. But the phosphate can be converted to a more soluble form, called *superphosphate*, by treatment with sulfuric acid:

$$\underset{\text{phosphate rock or bone}}{Ca_3(PO_4)_2} \quad + \quad 2\,H_2SO_4 \quad \longrightarrow \quad \underset{\text{superphosphate}}{Ca(H_2PO_4)_2} \quad + \quad 2\,CaSO_4$$

Today, to replenish the phosphorus in soil, farmers mainly use superphosphate obtained by treating the phosphate mineral fluoroapatite [$Ca_5(PO_4)_3F$], with sulfuric acid. An alternative fertilizer is ammonium phosphate [$(NH_4)_3PO_4$], which supplies both phosphorus and nitrogen.

Fish meal and guano, the concentrated phosphate-rich droppings of sea birds and bats, have also been used to add phosphorus to soil.

Although deposits of phosphate rock are widespread throughout the world, with major deposits in Morocco, China, and the United States, world reserves of high-grade phosphate may be depleted within the next 40 years unless new deposits are discovered. As existing supplies dwindle, more low-grade phosphate rock containing significant amounts of cadmium must be used. Cadmium, which is toxic to many life forms (Chapters 10 and 18) and is subject to bioaccumulation in food chains (Chapter 2), must be removed before the phosphate can be used; this step adds considerably to the cost of the fertilizer.

Potassium

Potassium, the third primary nutrient, is taken up by plant roots as the ion K^+. Potassium takes part in the formation of starches and cellulose from glucose and is essential for the normal functioning of many plant enzymes. Potassium may also be involved in the movement of carbohydrates within the plant and in the synthesis of proteins.

Potassium is distributed widely in the earth's crust and is abundant in most soils. But if crops are grown repeatedly on the same land without renewing the potassium, the soil's supply of the nutrient will eventually be depleted. Commercial fertilizers usually contain potassium in the form of potash (K_2CO_3) or potassium chloride (KCl). Large deposits of these salts are found in the United States, Canada, and Germany. The potassium chloride deposits under the Canadian prairies are enormous, but because they are 1.5 kilometers (nearly 1 mile) underground, they are difficult and expensive to mine. Although there is no immediate shortage of potassium salts, it should be remembered that the deposits are nonrenewable.

Large deposits of potassium salts are located at Searles Lake, California, and Carlsbad, New Mexico.

Secondary Nutrients: Calcium, Magnesium, and Sulfur

Calcium is abundant in nearly all soils. Not only is it a component of many common minerals (limestone, dolomite, and many silicates), it is frequently spread on soil in the form of lime to neutralize acidity. Also, the application of superphosphate adds calcium ions as well as phosphate ions. Plants absorb calcium as the Ca^{2+} ion.

Magnesium, like calcium, is abundant in most soils, but in some soils (particularly acidic ones), it becomes tightly bound, and there may be too few magnesium ions available in solution. The deficiency can be corrected by liming with crushed dolomite ($CaCO_3 \cdot MgCO_3$) or by adding magnesium sulfate ($MgSO_4$). Magnesium, which is absorbed through plant roots as Mg^{2+}, is a constituent of chlorophyll. A magnesium deficiency causes *chlorosis,* (yellowing of leaves).

Sulfur deficiencies occur most frequently in sandy, well-drained soils. The problem is treated by adding ammonium sulfate, potassium sulfate, or some other sulfate. Plants absorb sulfur as SO_4^{2-} ions. Sulfur is a constituent of amino acids and thus is essential for the synthesis of proteins.

In the United States, soils most likely to be deficient in sulfur are located in the Northwest and the Southeast.

Micronutrients

Although required in only trace amounts, micronutrients are just as essential as macronutrients for healthy plant growth; the effects of deficiencies can be severe. Onions grown in soil lacking zinc, for example, are stunted and develop abnormally.

It is often difficult to determine which, if any, micronutrient is in short supply in a particular soil and how much of a micronutrient should be applied to remedy a deficiency. Another problem is that the applied nutrient may bond to chemicals already in the soil and become unavailable to the plant. For example, agriculturists have found that it is preferable to add iron in the form of its ethylenediamine tetraacetate (EDTA) complex rather than as a simple inorganic salt. In the complexed form, the iron is protected from the soil chemicals that otherwise would bind it but is held loosely enough that it can be released gradually for uptake by plant roots.

Because plants vary considerably in their requirements, micronutrients must be applied with care. For example, the amount of boron that is ideal for sugar beets is toxic to soybeans and many other plants. Micronutrients are not always applied to the soil. Molybdenum compounds, for example, are often sprinkled on seeds before planting; other micronutrients are applied to leaves.

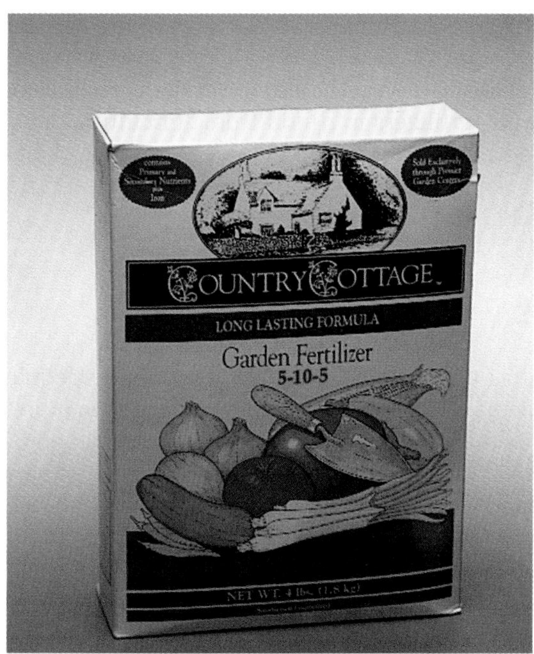

FIGURE 16.8

The three numbers on a bag of mixed fertilizer refer to the percentages by weight of nitrogen (as N), phosphorus (as P_2O_5), and potassium (as K_2O).

16.5 Mixed Fertilizers

Farmers and home gardeners usually correct soil deficiencies by applying fertilizers that contain a mixture of the three primary nutrients. Bags of such *mixed fertilizer* are labeled with a set of three numbers that indicate, in the following order, the percentages of (1) nitrogen (N), (2) phosphorus as P_2O_5, and (3) potassium as K_2O **(Fig. 16.8)**. For example, if the numbers are 5-10-5, the fertilizer is composed of 5% N, 10% P_2O_5, and 5% K_2O.

The choice of fertilizer depends on the crop being grown. For lawns, the most important nutrient is nitrogen, and a 20-10-10 fertilizer is suitable. Fruits and vegetables require a fertilizer rich in phosphorus, and a 10-30-10 mix is a better choice.

Elemental potassium and phosphorus are not actually present in mixed fertilizers as K_2O and P_2O_5. Potassium, for example, is present as potassium chloride (KCl). These designations date back to the late 19th century when plant requirements were first studied. Plants were burned, and the concentrations of the oxides K_2O and P_2O_5 in the ashes were determined. These concentrations are the basis for calculating the percentages of the elements in mixed fertilizers.

16.6 The Use of Synthetic Fertilizers

Synthetic inorganic fertilizers were first manufactured on a large scale in the late 1930s; since then, farmers' dependence on them to return nutrients to the soil has grown dramatically. Between 1955 and 1975, fertilizer use in the United States more than tripled. The application of massive amounts of synthetic fertilizers has undoubtedly been a major factor in the huge increases in food production that were achieved worldwide in the second half of the 20th century. Unfortunately, however, the use of synthetic fertilizers can cause severe environmental problems.

If synthetic inorganic fertilizers are applied repeatedly without the simultaneous addition of sufficient organic materials to sustain the formation of humus, soil becomes compacted and loses its nutrient- and water-holding properties and its ability to fix nitrogen; thus, increasing amounts of the fertilizers must be added to maintain crop yields. Eventually, the soil becomes mineralized and increasingly susceptible to erosion.

Although India's population increased by 130 million people during the 1970s, application of synthetic fertilizers increased grain production to such an extent that the country was able to export grain by 1980.

Another problem is that most commonly used fertilizers do not supply the needed micronutrients.

Water pollution is another problem associated with the use of synthetic fertilizers. As you learned in Chapter 10, if inorganic fertilizers in runoff from agricultural land enter rivers and lakes, they can cause eutrophication (the excessive growth of plants and algae). Also, very soluble ions such as nitrate ions, which are toxic above certain levels, can percolate down through the soil and contaminate groundwater.

In many Asian countries, human wastes, often called *night soil*, are collected and used as fertilizer.

16.7 Organic Fertilizers

Because of the problems associated with synthetic inorganic fertilizers, many people advocate a greater use of organic fertilizers, including animal manure, green manure, and biosolids. Animal manure can include dung and urine from cattle, horses, and poultry. *Green manure* consists of a crop of green plants, preferably legumes, that is not harvested but is plowed into the soil—a practice that has the disadvantage of keeping the land unproductive for a year. Biosolids, or composted sewage sludge (Chapter 10), if free from toxic heavy metals, are a valuable source of nutrients.

Manures return nitrogen and many other nutrients to the soil; at the same time, they improve soil structure by adding valuable organic materials. They also stimulate the growth of soil bacteria. However, weight for weight, organic fertilizers cannot produce the crop yields achieved with synthetic inorganic fertilizers. A typical manure is a 1-2-1 fertilizer. Another problem with using manures is cost. In modern agricultural practice, livestock and crops are seldom raised on the same farm, and packaging and transportation costs generally make organic fertilizers more expensive than synthetic inorganic fertilizers.

16.8 Pesticides

Ever since humans first appeared on the earth, they have had to contend with pests. The biblical account (found in *Exodus*) of the liberation of the Hebrews from bondage in Egypt in the 13th century B.C. includes descriptions of plagues of lice, flies, and locusts in the Nile Valley. The Black Death was transmitted to humans by fleas on rats carrying bubonic plague. The failure of the potato crop in Ireland in 1845, which caused widespread starvation there and mass emigration to the United States, was caused by the potato blight (Chapter 15). Malaria, spread by mosquitoes, still incapacitates or kills millions of people annually in the less developed countries, and swarms of locusts continue to devastate crops in many areas of the world **(Fig. 16.9)**. Plants on which we depend for food are under attack from insects, fungi, bacteria, viruses, and other microorganisms. Mice, rats, rabbits, and other animals also take a share of many crops.

EXPLORATIONS describes a tree that provides a natural pesticide and other benefits (see pp. 504–505).

Before 1940, only a few pesticides were available. Among them were several naturally occurring insect poisons extracted from plants; these include *pyrethrins*, compounds obtained from the pyrethrum flower (a member of the chrysanthemum family). Other early insecticides were nicotine sulfate obtained from tobacco, rotenone from the tropical derris plant, and garlic oil. A number of inorganic chemicals, primarily compounds of arsenic and lead, were also used. These are rarely used today because they persist in the environment and are toxic to humans and other animals besides insects.

Synthetic compounds related to natural pyrethrins are being used in increasing quantities as pesticides. These compounds are biodegradable and nontoxic to birds and mammals.

The massive use of pesticides began at the end of World War II with the introduction of DDT and escalated as other synthetic organic pesticides were developed.

Today, in the United States, approximately 500,000 metric tons of pesticides (insecticides and herbicides) are applied annually to control pests.

FIGURE 16.9

Swarms of locusts can completely destroy crops.

16.9 Synthetic Insecticides

The three main classes of synthetic insecticides in use today are all organic compounds: (1) chlorinated hydrocarbons, (2) organophosphates, and (3) carbamates.

Chlorinated hydrocarbons and most organophosphates kill a wide range of insect types—including many that are beneficial—and are termed **broad-spectrum insecticides.** Most carbamates are **narrow-spectrum insecticides,** which means they are toxic to only a few types of insects. Unfortunately, one of the insects they kill is the honeybee.

Chlorinated Hydrocarbons

The chlorinated hydrocarbon dichlorodiphenyltrichloroethane, better known as DDT **(Fig. 16.10)**, was first prepared in 1873, but it was not until shortly before the start of World War II that the Swiss chemist Paul Müller discovered that it was an effective insecticide.

Combined with talcum powder, DDT was used by Allied forces in World War II to kill body lice and fleas, and it successfully stopped a wartime outbreak of typhus (a disease carried by lice) in Naples, Italy.

After the war, DDT came into widespread use for agricultural as well as public health purposes. For almost 20 years, DDT and related chlorinated hydrocarbons, such as aldrin, dieldrin, and chlordan (Fig. 16.10), were major factors in increasing food production and in suppressing insect-borne diseases. In 1948, Müller was awarded the Nobel Prize in medicine and physiology for his discovery.

When first introduced, DDT appeared to be an ideal insecticide. As early as the 1950s, however, it was evident that this was not the case. Insects soon became resistant to it, and because it did not decompose readily, it persisted in the environment. Microorganisms living in water contaminated with DDT ingested the pesticide. Being soluble in fats but not in water, DDT dissolved in organisms' fatty tissues and bioaccumulated upward in food chains. Concentrations of DDT in fatty tissues in fish-eating birds (who are at the top of a food chain) were found to be several million times greater than the concentration in the water (Fig. 10.17). The DDT disrupted the birds' calcium metabolism, causing egg shells to be so thin and fragile that most of them broke, and only a few young birds hatched. As a result, ospreys, peregrine falcons, and other fish-eating birds nearly became extinct in some parts of the United States.

Because of these adverse effects, DDT was banned in the United States in 1972; since then, the threatened birds have made a dramatic recovery. Later in the 1970s, most of the other chlorinated hydrocarbons were also banned for most uses. But the United States continues to export DDT to developing countries, where its value in controlling malaria is seen to outweigh its disadvantages.

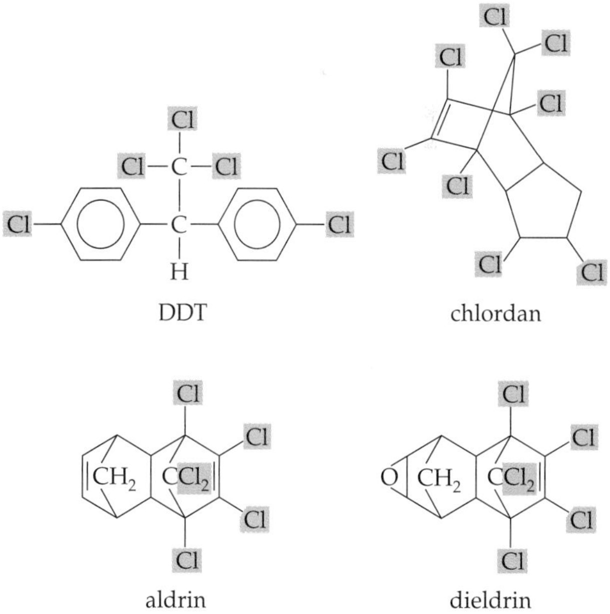

DDT chlordan

aldrin dieldrin

FIGURE 16.10

Examples of chlorinated hydrocarbons used as insecticides.

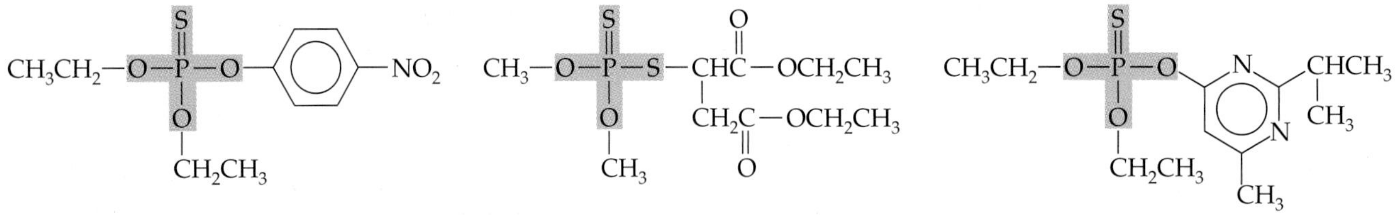

parathion malathion diazinon

FIGURE 16.11

Examples of organophosphates used as insecticides. Typical functional groups that characterize these compounds are shaded blue.

Organophosphates

DDT is credited with having saved more lives than any other chemical ever synthesized.

Tomatoes and other plants defend against insect attacks by producing compounds in their leaves that interfere with an insect's ability to digest the plant tissue.

Organophosphates used as insecticides include parathion, malathion, diazinon, and other related compounds (**Fig. 16.11**). These are cheap to produce and are very effective against many different insects. They are, however, toxic to humans and mammals; parathion, for example, is more than 20 times as toxic to rats as DDT is (Chapter 18). These insecticides are *neurotoxins,* or nerve poisons, that inactivate *cholinesterase,* the enzyme that plays a vital role in the transmission of impulses between nerve fibers (Chapter 18). Ingestion of organophosphates by humans can result in irregular heartbeat, convulsions, and even death. Because of this, farmers must wear respirators and protective clothing when applying them. Another disadvantage of organophosphates is that, because they break down rapidly in the environment, they must be applied frequently to be effective. Because of unacceptable risks to children, the EPA has called for a phase-out by 2003 of the use of diazinon, the pesticide most widely applied on lawns and gardens in the United States.

Carbamates

Carbamates, which are derivatives of carbamic acid (**Fig. 16.12**), include the following functional group in their structures:

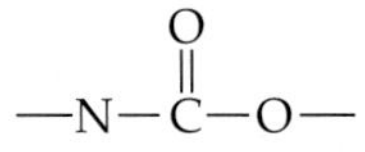

$$-N-\overset{\overset{\displaystyle O}{\|}}{C}-O-$$

Examples are carbaryl (Sevin) and Aldicarb (Temik) (Fig. 16.12).

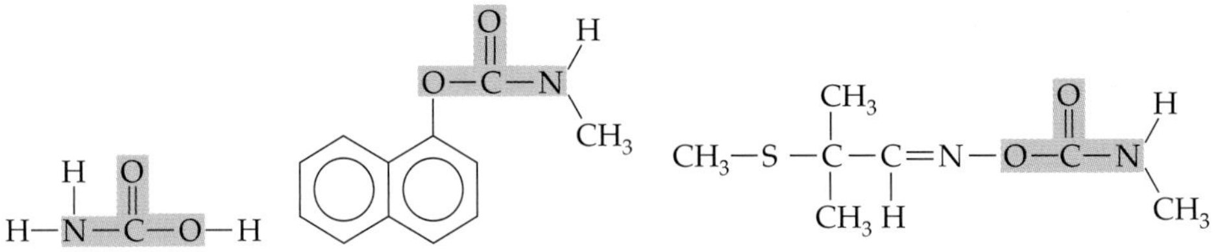

carbamic acid carbaryl (Sevin) aldicarb (Temik)

FIGURE 16.12

Carbamate insecticides such as Carbaryl and Aldicarb are derived from carbamic acid and contain the same functional group (shaded blue).

Carbamates, like organophosphates, inactivate cholinesterase. They are rapidly broken down in animal tissues and in the environment, and most are much less toxic to humans and other mammals than organophosphates are.

16.10 Herbicides

In addition to insect and animal pests, farmers must contend with weeds, which compete with crops for nutrients and water. The traditional method for controlling weeds is *tillage*. Typically, plowing turns over the top 10–15 inches of soil and buries and smothers the weeds. The soil is then harrowed to break up large clods before the crop is planted. During the growing season, row crops must be cultivated to destroy weeds that grow up between the rows.

With the introduction of selective herbicides in the 1950s, the need for many of these steps was reduced, and *no-till agriculture* is now practiced extensively in most of the developed countries. Herbicides are applied to a field to kill weeds, and then, in a single operation, a planting machine cuts a furrow, adds fertilizer and seeds, and covers the seeds. The weeds killed by the herbicide remain as mulch and protect the soil from erosion and water loss. The amount of energy and manual labor saved by using herbicides in no-till farming is enormous.

No-till agriculture reduces soil disturbance to a minimum, which helps prevent erosion.

The first chemicals used for weed control were nonselective and included solutions of sodium chlorate ($NaClO_3$), sulfuric acid, copper salts, and various other inorganic chemicals. These herbicides were of limited value in agriculture since they kill both weeds and crop plants with which they came in contact. The massive use of herbicides in farming began in the 1950s with the introduction of the selective organic herbicides, 2,4-D and 2,4,5-T (**Fig. 16.13a**) and related chlorophenoxy compounds. Today, approximately 50% of all pesticides used in the world are herbicides.

Selective herbicides kill only a particular group of plants, such as grasses or broad-leaved plants. Most selective herbicides are *systemic*—that is, they are absorbed by

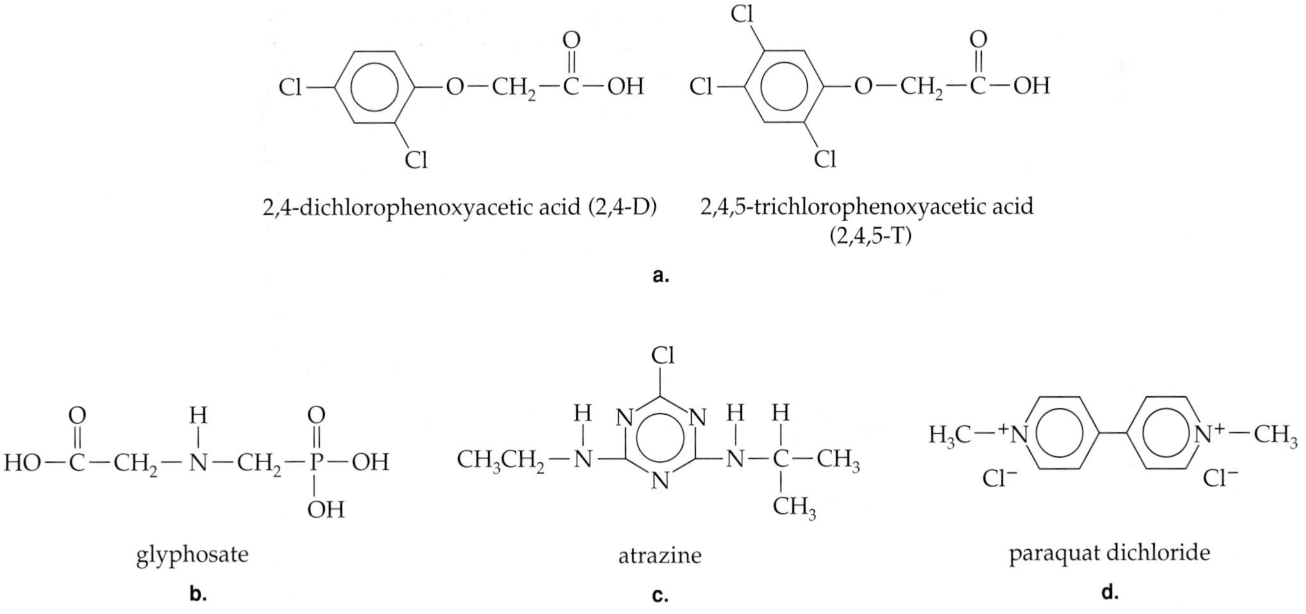

2,4-dichlorophenoxyacetic acid (2,4-D)

2,4,5-trichlorophenoxyacetic acid (2,4,5-T)

a.

glyphosate

b.

atrazine

c.

paraquat dichloride

d.

FIGURE 16.13

Some common herbicides.

leaves or roots and then are transferred throughout the plant. They act like growth hormones, causing a plant to grow so fast that it dies. 2,4-D is very effective against emerging broad-leaved weeds; 2,4,5-T is used to control woody plants. A mixture of these two herbicides, known as Agent Orange, was used extensively as a defoliant during the Vietnam War (Chapter 10). Because of health problems thought to be associated with its use, the EPA banned it in 1985.

As part of the war on drugs, marijuana has been sprayed with paraquat to destroy it. There have been instances in which sprayed marijuana has been harvested before the herbicide could kill it. Such contaminated marijuana, if smoked, can cause severe lung damage.

A popular systemic herbicide used to control perennial grasses is glyphosate (Roundup) **(Fig. 16.13b)**, which is a phosphate derivative of the amino acid glycine. It works by inhibiting the enzymes necessary for the synthesis of the essential amino acids tyrosine and phenylalanine.

Another important herbicide is atrazine **(Fig. 16.13c)**, which is used extensively in the no-till production of corn. Atrazine acts by interfering with photosynthesis, but it has no effect on corn and a number of other crop plants because these plants have the ability to change the herbicide into a harmless form. Weeds, in contrast, do not have this ability and quickly die.

Paraquat **(Fig. 16.13d)** and several related compounds are also widely used as herbicides. These herbicides are toxic to most plants, but because they break down rapidly in the soil, they can be used to kill weeds that emerge before the crop plant does. Paraquat is toxic to humans and must be used with care.

16.11 New Varieties of Crop Plants

The dramatic increase in crop yields that occurred during the 1960s was known as the *Green Revolution*.

World grain production increased dramatically during the second half of the 20th century **(Fig. 16.14)**. Next to the use of synthetic fertilizers and pesticides, the third most important factor responsible for the increase has been the development of new varieties of crop plants. In the 1960s, new high-yielding grains developed by agricultural scientists through selective cross-breeding were introduced into India, South America, and other developing countries. The result was that yields per acre were tripled. Equally important has been the development of crop plants resistant to attack by fungi and insect pests.

Breeding plants with a new trait is a slow process that can take at least 10 years of painstaking work. Genetic engineering techniques can reduce the time required to develop a new plant to as little as 1 year. Recently, scientists genetically engineered a tomato that can be grown in salty water and is said to taste like a conventional tomato. If approved for human consumption, this tomato could be grown in naturally brackish water or on land made salty through irrigation (see Chapter 10).

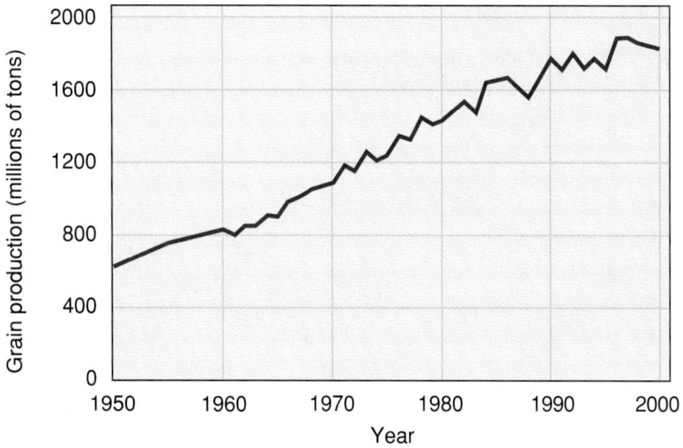

Grain production (millions of tons)
2000
1600
1200
800
400
0
1950 1960 1970 1980 1990 2000
Year

FIGURE 16.14

World grain production from 1950 to 2000.

16.12 Problems with Synthetic Pesticides

Synthetic pesticides have been used successfully to control insect-borne diseases and have been responsible for large increases in crop yields, but their use has also created many problems. The major problems are the resistance insects develop to the pesticides and the threats pesticides pose to the environment and to human health.

The Pesticide Treadmill

Because of genetic variations, some insects in a population have more resistance to pesticides than others do. When an area is sprayed with a pesticide for the first time, most insects are killed, but the resistant ones survive to breed. With repeated spraying of the same pesticide, each succeeding generation contains a higher percentage of resistant insects. Control can be maintained only if larger and larger doses of pesticide are applied. Eventually, a more potent pesticide may be required. Another problem is that, after a pest population has been brought under control, it often undergoes a resurgence, returning in even greater numbers than before.

Since the widespread use of synthetic insecticides began in the early 1950s, more than 400 kinds of insect pests have become resistant to one or more of them.

If beneficial insects are destroyed together with the target pests, the natural ecosystem is upset, leading to other problems. Honeybees that normally pollinate crops are killed, and minor pests that previously were controlled by natural predators multiply, causing serious secondary pest outbreaks. Farmers find themselves on a pesticide treadmill (**Fig. 16.15**), continually spending more money for larger quantities and more varieties of pesticides, which become increasingly ineffective.

Health Problems Due to Pesticide Use

The people at greatest risk for health problems due to pesticides are workers on farms and in pesticide manufacturing plants. Numerous pesticide-related illnesses and deaths have occurred—particularly in less developed countries—among farm workers who have failed to take adequate precautions when applying pesticides or who have entered sprayed areas before a pesticide has broken down to a harmless form. Organophosphates are especially dangerous in this respect.

According to the EPA, between 10,000 and 20,000 cases of pesticide poisoning are reported annually in the United States. The agency has issued regulations requiring protective clothing for workers applying pesticides and restricting access to newly sprayed areas.

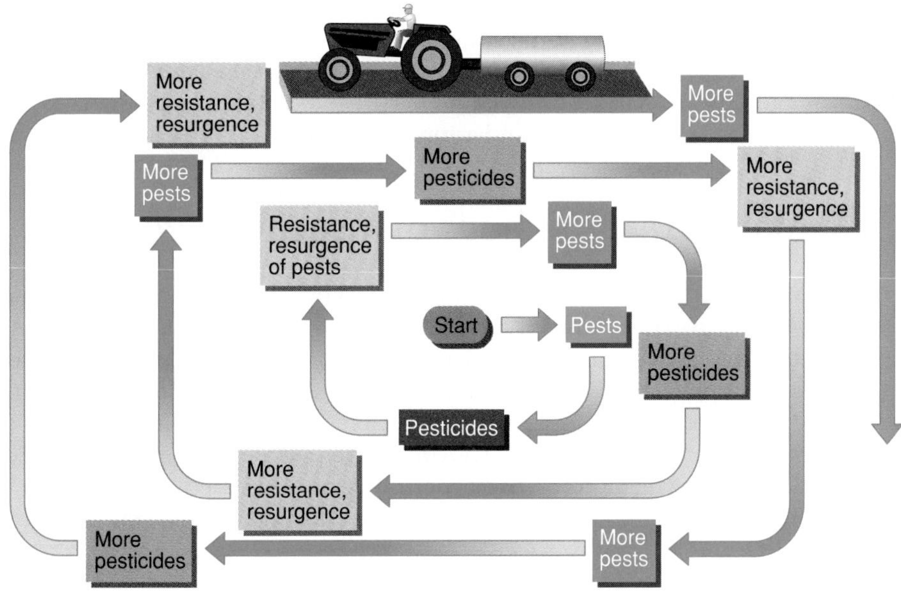

FIGURE 16.15

The pesticide treadmill: When a pesticide is first used, it kills most of the target insects. With repeated use, the number of insects resistant to the pesticide increases, and a more potent pesticide must be applied. Once an insect population is controlled, resurgence often occurs, and more and more pesticide of ever increasing potency must be applied in an unending cycle.

The worst pesticide accident on record occurred in 1984 when highly toxic methyl isocyanate (CH_3NCO), a gas used in the manufacture of carbaryl (Sevin) and other pesticides, leaked from a Union Carbide plant in Bhopal, India. More than 3,000 people were killed; 14,000 suffered severe injuries, including blindness, brain damage, sterility, and liver and kidney damage; many thousands more suffered less severe injuries.

EDB pumped into the soil during the 1970s and early 1980s has contaminated wells in California, Florida, and Massachusetts.

The possibility of water contamination by persistent pesticides that leach through soil into groundwater is a continuing concern for the general public. A disturbing example was the contamination of well water by ethylene dibromide, (EDB; $BrCH_2CH_2Br$), which occurred in several areas where the pesticide had been used as a soil fumigant to control root nematodes (microscopic soil-dwelling worms). EDB was also used to protect stored grain, and unacceptable levels were detected in flour and other foods. Although it had been shown in 1970 that EDB caused cancer in laboratory animals, opposition from users and manufacturers delayed an EPA ban on its use until 1984.

There is great concern that pesticide residues in foods may cause adverse health effects; in 1996, the EPA began reassessing the legal limits it sets for pesticides in food. The agency is expected to call for a ban on most uses of methyl parathion on fruits such as peaches which, in processed form, are eaten frequently by infants and young children. Children are of particular concern because, weight for weight, they usually eat more fruits and vegetables than adults do, and because they tend to be more sensitive than adults to neurotoxins and carcinogens.

Biochemist Bruce Ames, who developed a screening test for carcinogens (Chapter 18), believes that the risk from pesticide residues is exaggerated. He has pointed out that many common foods and beverages, including potatoes, tomatoes, peanut butter, mushrooms, celery, wine, and beer, contain toxic chemicals, including carcinogens. Thus, most people's exposure to natural toxins far exceeds any exposure to pesticide residues.

A report linking a form of cancer in dogs to regular treatment of lawns with 2,4-D has led to concern that children who play on treated lawns may also be at risk.

In addition to possible exposure via their water and food, people are often exposed to pesticide mists that drift beyond the target areas. Other problems are caused by the enormous quantities of herbicides and insecticides that are applied annually to lawns, golf courses, and other nonagricultural land. In the United States, legislation requires that all pesticides be registered with, and approved by, the EPA.

16.13 Alternative Methods of Insect Control

One-third of all the food grown in the world is destroyed by pests each year. In the United States alone, crop losses due to insects amount to over $4 billion annually. Obviously, without some form of pest control, there would be a drastic drop in food production, and world food supplies would be jeopardized. Fortunately, some alternatives to the massive use of synthetic pesticides are now available. There are biological and nonchemical methods to control insects that do not threaten either the environment or human health. These alternative methods of insect control are not as immediately effective, or as easy to apply, as the commonly used synthetic organic insecticides, but they have many advantages.

Pheromones

Pheromones are chemical substances that are released by insects and other animals as a means of communication. They are used to mark trails and territories and to attract mates. Approximately 30 species-specific pheromones that are secreted by female insects to attract males have been extracted from the insects or synthesized in the lab-

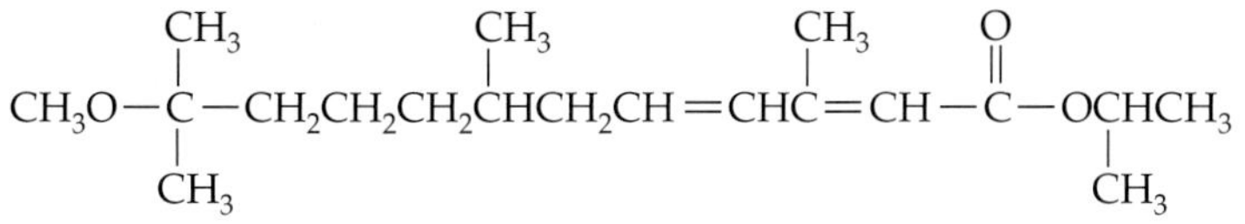

$$CH_3O{-}\overset{\displaystyle CH_3}{\underset{\displaystyle CH_3}{C}}{-}CH_2CH_2CH_2\overset{\displaystyle CH_3}{C}HCH_2CH{=}CH\overset{\displaystyle CH_3}{C}{=}CH{-}\overset{\displaystyle O}{\overset{\|}{C}}{-}O\underset{\displaystyle CH_3}{C}HCH_3$$

methoprene

FIGURE 16.16

Methoprene, the first commercially available juvenile hormone.

oratory and used to control insect pests. The pheromones, which are effective in amounts as small as a picogram (1×10^{-12} gram), can be used to lure male insects into baited traps, or they can be sprayed over an infested area. Males in a sprayed area smell a female in all directions and become so confused that they fail to mate.

Pheromones represent an environmentally attractive approach to pest control. They are species-specific and thus affect only the target pest; they are biodegradable, non-toxic, and effective at very low concentrations. But they are costly and difficult to produce. Many thousands of insects must be collected to obtain sufficient material for analysis, and identification and synthesis of the desired sex attractant present many problems. Also, sex attractants are effective only against adult insects. They do not control juvenile forms, such as larvae and caterpillars, which often do the most harm.

Juvenile Hormones

Another biological approach to insect control is the use of **insect growth regulators (IGRs)**, compounds that disrupt development at various stages. An important class of IGRs is **juvenile hormones.** Hormones (Chapter 14) are chemicals produced in special plant and animal cells. They are transported in the bloodstream to specific target tissues, where they control and mediate many physiological processes. Juvenile hormones secreted by immature insects regulate the early stages of development. As the insect matures, production of juvenile hormones ceases, and the insect develops into an adult. If immature insects are sprayed with juvenile hormone, they remain at an immature stage and are unable to mate.

The first commercially available juvenile hormone was methoprene (Fig. 16.16), approved by the EPA in 1975. In various formulations, methoprene and related compounds have been used to control mosquitoes, fleas, cockroaches, and a number of other insects that are pests in the adult stage. Juvenile hormones are not as useful against insects that are agricultural pests in their immature stage. When used against caterpillars, for example, the immediate effect is to cause the caterpillars to grow larger and eat more of the crop for a longer period of time.

Other IGRs are also available, and new ones are constantly being developed. Chitin synthesis inhibitors, which prevent the growth of chitin, the substance that forms an insect's shell, are a promising recent development. Without a protective shell, the insect dehydrates and dies.

The cottony-cushion scale, an insect pest that destroyed citrus crops when it was accidentally brought to California from Australia in 1873, was brought under control by importing two of the insect's natural predators from Australia.

Biological Control

Biological control entails keeping insect pests in check with their natural enemies, including predators, parasites, and disease-causing bacteria or viruses. To be effective, predators must be carefully chosen. Praying mantises and ladybugs that are sold commercially to home gardeners are generally of little value in controlling pests. Praying mantises are rapidly reduced in number because they devour each other, and they are as likely to eat useful insects as they are to eat pest insects. Ladybugs, if not at the feeding stage when released, simply fly away.

The naturally occurring soil bacterium *Bacillus thuringiensis* (Bt) is toxic to many leaf-eating caterpillars. Marketed as a powder under the name Dipel, it has been used by home gardeners for many years. It is also used as a spray to control gypsy moths. In 1972, a virus was approved for use against the *Heliothus zea* species of bollworms, the moth larvae that attack the tips of corn kernels (corn earworms) and burrow into cotton bolls (cotton bollworms).

Although a great deal of research is needed to identify suitable biological agents and production is difficult and costly, biological controls have many advantages. In general, they destroy only the target species and are nontoxic to other species. Once established, they are self-perpetuating, and pests do not become resistant to their natural enemies.

Sterilization

Sterilization is another option for controlling pest insects. In this technique, large numbers of the pest insects are bred and then sterilized by exposure to radiation. The irradiated insects are released in the target area where they mate with normal insects but produce no offspring. This method is only effective if the number of sterile insects released is sufficient to overwhelm the natural population of nonsterile insects. The procedure is expensive and of limited applicability, but it has been used with great success against the screwworm fly, a deadly pest that lays its eggs in the open wounds of cattle. The developing larvae feed on the wound, and resulting infections often can kill the animal. Sterilization has also been used successfully to control certain tropical fruitflies.

16.14 Future Farming: Less Reliance on Agricultural Chemicals

Pressure from environmental groups and consumers and fear that supermarkets may reject products that contain pesticide residues have encouraged farmers to adopt farming practices that rely less heavily on the massive use of synthetic fertilizers and pesticides. In California, where approximately 50% of the vegetables and 40% of the fruits and nuts produced in the United States are grown, the number of farmers using no agricultural chemicals has greatly increased during the last 10 years. To control pests, these farmers rely on **integrated pest management (IPM),** a technique that was first introduced in the 1960s.

In IPM, several pest control methods are integrated into a carefully timed program. Success depends on knowledge of the metabolism of the crop plant, the life cycles of the pests to which the crop is susceptible, and the interactions of the pests with their environment. IPM makes use of sex attractants, male sterilization, natural pesticides such as pyrethrins, and the release of natural predators. Since many pests feed on only one crop, crop rotation and crop diversity are employed to prevent insects, fungi, and microbes from building up in the soil and in crop residues from year to year. Planting and harvesting times are adjusted to maximize the number of insects that die of starvation between successive crops.

The aim of IPM is not to eliminate all pests but to keep them at a level low enough to prevent economic loss. The technique does not completely exclude synthetic pesticides but reserves them for selective use at the lowest possible dose and only where absolutely necessary. IPM programs, which usually must be individualized to suit soil and climate conditions for each crop, are expensive to set up and more labor-intensive than conventional methods, but are less expensive on a long-term basis.

In 1989, IPM received support from the Board of Agriculture of the National Research Council, which recommended greater emphasis on biologically based methods of farming, including IPM, the use of manure and crop rotations to return fertility to the soil, and mulching and other nonchemical means of controlling weeds. In 1996, IPM was endorsed by the National Research Council which issued a report recommending a reduction in the use of synthetic pesticides on farmland.

Individuals could play an important part in reducing pesticide use if they would tolerate weeds in their lawns (or switch to a ground cover other than grass) and if they would accept fruits and vegetables with some blemishes. At present, many pesticides are applied, not to increase yield, but solely to achieve the perfect appearance that many consumers demand.

16.15 Can We Feed Tomorrow's World?

Demand for food will continue to grow as population increases. Meeting this demand will be very difficult because the technologies that enabled farmers to increase grain production so dramatically in the 20th century—fertilizers, new crop varieties, pesticides, and irrigation—have already been fully exploited in many countries. World grain production has begun to level off (see Fig. 16.14) and world grain production per person reached a maximum in 1984 **(Fig. 16.17)**. Plowing up more land might seem to be an obvious option, but soil degradation and conversion of land to other uses have severely limited the amount of new cropland available. And because water tables are falling on all continents and many rivers are running dry before reaching the sea, irrigating arid land to grow more crops is no longer possible in most countries.

Additional use of fertilizer can be expected to increase crop yields in some developing nations, but the developed nations are already using the maximum amounts of fertilizer that are effective with existing crop varieties.

Genetic engineering is not expected to increase yields much more because in today's crop plants, the share of the plant's *photosynthate* (the product of photosynthesis) going to the seed is already between 50% and 55%. Plant physiologists do not believe this share can be increased above about 60%. One way scientists hope to increase grain production is by developing crop plants that are resistant to drought and can grow in salty soils.

Recently, scientists developed plants, including corn, potatoes, rice, and cotton, that are genetically engineered to produce *Bacillus thuringiensis* (Bt). The plants produce high levels of the insect-killing toxin throughout the growing season. Bt corn, which was approved by the EPA in 1995, has been grown extensively in the United States. Although the EPA has concluded that Bt corn poses no threat to the environment or to human health, many people oppose planting these crops because they fear that insects will develop

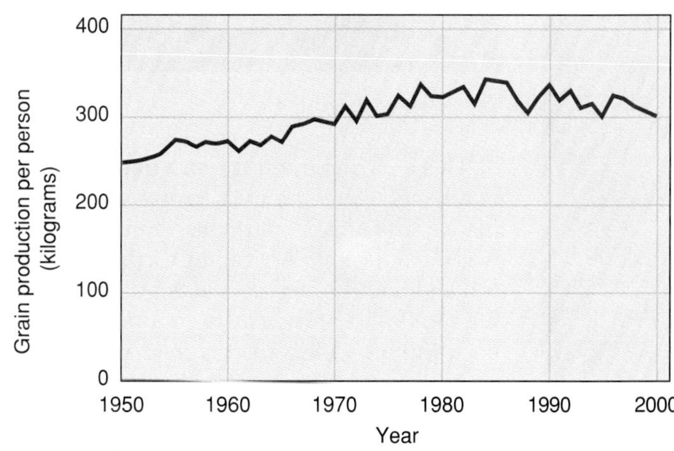

FIGURE 16.17

World grain production per person from 1950 to 2000. Production per person reached a maximum in 1984.

EXPLORATIONS

The Neem Tree—The World's Pharmacy?

Scientists are calling the neem a wonder tree. According to a report issued by the National Academy of Sciences (NAS) in February 1992, this remarkable tree has the potential to improve pest control, promote health, check population growth, and reduce deforestation.

Western scientists have only recently taken an interest in the neem tree, but it has been revered and utilized for generations in India. The earliest Sanskrit writings cite the value of neem seeds and leaves for healing and for pest control. As Noel Vietmeyer, the director of the NAS project, reports: "For centuries, millions have cleaned their teeth with neem twigs, smeared skin disorders with neem leaf juice, taken neem tea as a tonic, and placed neem leaves in their beds, books, grain bins, cupboards, and closets to keep away troublesome bugs. The tree has relieved so many different pains, fevers, infections, and other complaints that it has been called 'the village pharmacy.'"

Neem trees.

This amazing multipurpose tree is a member of the mahogany family. It grows widely in many parts of Asia, particularly in India, and was introduced to Africa, the Caribbean, Central and South America, and Saudi Arabia during the 19th century. The neem tree is a beautiful broad-leaved evergreen that can grow up to 90 feet tall with spreading leafy branches that provide dense, year-round shade. The neem bears masses of sweet-scented white flowers that are attractive to bees. Its fruit resembles olives, and the sweet pulp that surrounds the extremely bitter seeds is food for birds and many other creatures. Neem trees grow in the hottest climates but cannot withstand freezing or extended periods of cold. They thrive in both semiarid and wet tropical regions. Helped by their extensive and unusually deep roots, neem trees are able to extract nutrients even from poor soils.

Extensive Research Efforts

Scientists in India have studied neem products since the 1920s, but research did not begin elsewhere until the German entomologist Heinrich Schmutterer published a report describing a plague of locusts that he witnessed in the Sudan in 1959. Schmutterer observed that the insects devoured every green thing in their path with the exception of the neem trees. Swarms of locusts landed on the trees but flew off without feeding.

By 1990, research in many parts of the world, including the United States, had shown that extracts of neem seeds and leaves are effective against more than 200 species of insect pests as well as some species of mites and nematodes. The extracts appear to be nontoxic to humans, other mammals, birds, and insects such as bees that do not feed on plants. Furthermore, they are biodegradable, and insects do not seem to develop genetic resistance to them.

Active Ingredients

Neem extract contains four major and many minor active ingredients that act on insects in a variety of ways—although none kill outright. The major active ingredients are limonoids, compounds that are distantly related to steroids. The most potent ingredient is azadirachtin, which is found in greatest concentration in seed kernels and is also present in leaves. It is effective at concentrations of less than 1 ppm. Azadirachtin blocks production of hormones that control normal insect development; as a result, larvae do not develop into pupae. Azadirachtin is also so distasteful to many insects that they will starve to death rather than eat plants that bear traces of it. Two other neem ingredients (melantriol and salanin) also inhibit feeding. The latter appears to be a more effective insect repellent than DEET (N,N-diethyl-*m*-toluamide), the substance present in most commercial mosquito repellents. Still other neem compounds disrupt maturation at various stages and thus prevent breeding. Neem pesticides are promising candidates for inclusion in the increasingly popular IPM programs.

Potential Health Benefits

Asians have used neem medicinally for thousands of years and are convinced of its efficacy in treating numerous ailments. However, although neem's value as a general antiseptic is well established and certain extracts have demonstrated antifungal, antiviral, and some antibacterial activity, many health claims for neem have not yet been substantiated. Preliminary evidence suggests that neem has analgesic, antipyretic, antiinflammatory, hypotensive, and possibly antiulcerative effects, but these claims have not been tested under controlled conditions.

The potential for neem seems almost limitless, particularly in less developed countries where populations are expected to explode and where neem trees generally thrive.

There is little doubt that neem has value in dentistry. Even the poorest Indians generally have healthy teeth, although their only dental hygiene practice is to scrub their teeth with the frayed end of a neem twig. Research in Germany has shown that compounds in neem bark are effective in preventing tooth decay and healing inflammation of gums, and neem toothpaste is now marketed in both Germany and India.

Neem may also have value for birth control. Oil pressed from neem seeds is a powerful spermicide. Introduced into the vagina before intercourse, it has prevented pregnancy in laboratory animals and a small number of women volunteers. Moreover, neem leaf extracts show promise as a means of male birth control. Reportedly, the extracts reduce fertility in monkeys without inhibiting sperm production, and the effects are temporary.

Other Uses

That is not the end of neem's benefits: Its termite-resistant wood is valued for construction; its flowers are a source of high-quality honey; neem oil is used as a lubricant, a fuel for oil-burning lamps, and an ingredient in soap; neem cake, the material left after oil is pressed from seeds, improves soil fertility and protects against nematodes; and fast-growing neem trees can be planted to alleviate deforestation and desertification.

The potential for neem seems almost limitless, particularly in less developed countries where populations are expected to explode and where neem trees generally thrive. Neem oil could provide an acceptable, inexpensive, and easily available contraceptive, and neem preparations could protect crops from destructive pests. Effective pesticides can be made very simply by steeping crushed seeds or leaves in water. The resulting suspension can be used directly in the field or can be filtered and sprayed.

Before neem products can achieve wide acceptance, however, more research is needed. One problem is that neem extracts vary from region to region because of genetic differences in the trees, and no standard of potency has yet been established. It is difficult to evaluate claims made in different studies because harvesting and extracting methods have varied. Also, although neem products appear to be environmentally safe and harmless to humans and other mammals, rigorous toxicity tests have not been conducted, and the persistence of neem residues on food crops is not known. The only neem product currently registered in the United States (called "Margosan-O" after the Spanish name for neem) is authorized for use only against certain insects that do not feed on food crops.

Despite the uncertainties, many scientists are convinced that the neem tree is an extremely valuable resource that could benefit the world. The NAS report stresses the urgent need to provide more funds for research and to make the industrialized nations more aware of neem's extraordinary potential. In the future, the amazing neem tree could play a significant role in helping to solve some of our planet's seemingly intractable problems.

Reference: National Research Council, *Neem: A Tree for Solving Global Problems*, National Academy Press, Washington, D.C., 1992.

resistance to Bt, which is one of very few effective natural pesticides. There is also concern that these crops could cause unforeseen health problems.

As gains in grain productivity continue to slow, world demand for food continues to expand. Unless there is a reduction in the rate of population growth, food shortages appear to be inevitable in the 21st century.

Chapter Summary

1. Thomas Malthus concluded that because population increases geometrically and food production increases arithmetically, food production would be unable to keep up with population growth.
2. The time it takes for a population to double can be calculated using the rule of 70 (divide 70 by the annual growth rate as a percentage).
3. The productivity of soil depends on its texture and its content of organic matter (humus).
4. Erosion and loss of topsoil are caused by construction, mining, clearcutting of timber, agricultural development, and overgrazing.
5. To thrive, plants need three classes of nutrients:
   a. primary nutrients (nitrogen, phosphorus, and potassium)
   b. secondary nutrients (calcium, magnesium, and sulphur)
   c. micronutrients
6. Nitrogen fixation (the conversion of atmospheric nitrogen to a form plants can use) is accomplished by bacteria in soil and in nodules on the roots of legumes, cyanobacteria, and lightning.
7. Ammonia, the starting material for the manufacture of nitrogen fertilizers, is made by the Haber process: $N_2 + 3\,H_2 = 2\,NH_3$.
8. Bags of mixed fertilizer are labeled with three numbers that indicate the fertilizer's content of nitrogen, phophorus, and potassium.
9. Overuse of synthetic inorganic fertilizers causes environmental problems including soil mineralization and water pollution as a result of runoff from the land.
10. Organic fertilizers have advantages over synthetic fertilizers, but a given weight of organic fertilizer is less effective than the same weight of inorganic fertilizer.
11. The three main classes of insecticides used in agriculture are
    a. chlorinated hydrocarbons
    b. organophosphates
    c. carbamates
12. DDT and most other chlorinated hydrocarbons used as insecticides are banned in the United States because they bioaccumulate upward through food chains and harm wildlife.
13. Organophosphates and carbamates are neurotoxins and must be applied with caution.

14. Selective herbicides kill only specific types of plants, such as grasses or broad-leaved plants.
15. The three factors most responsible for the enormous increase in grain production during the second half of the 20th century were fertilizers, pesticides, and the development of new varieties of crop plants.
16. A major problem with pesticides is that insects develop resistance to them.
17. Alternatives to synthetic pesticides include pheromones, insect growth regulators (IGRs), biological controls, and insect sterilization.
18. There is growing concern that pesticide residues in foods can cause adverse health effects, particularly in children.
19. Integrated pest management (IPM) is a farming technique that relies less heavily on the use of synthetic fertilizers and pesticides.
20. Unless there is a reduction in the present rate of population growth, food shortages appear to be in inevitable in the 21st century.

Key Terms

arithmetic growth (p. 485)
biological control (p. 501)
broad-spectrum insecticide (p. 495)
carbamate (p. 496)
doubling time (p. 485)
erosion (p. 488)
geometric growth (p. 484)
humus (p. 487)
insect growth regulator (IGR) (p. 501)
integrated pest management (IPM) (p. 502)
juvenile hormone (p. 501)
loam (p. 487)
micronutrient (p. 489)
narrow-spectrum insecticide (p. 495)
nitrification (p. 489)
nitrogen fixation (p. 489)
pheromone (p. 500)
primary nutrient (p. 489)
rule of 70 (p. 485)
secondary nutrient (p. 489)
subsoil (p. 487)
topsoil (p. 487)

Questions and Problems

1. A coin collector adds two new coins to his collection every week.
   a. How many weeks will it take for him to acquire 100 coins?
   b. Is the collection growing arithmetically or geometrically?
2. You raise carrier pigeons. Starting with two, you have four at the end of the first month and eight at the end of the second month. Assuming that the pigeon population doubles every month:
   a. How many pigeons will you have at the end of the first year?
   b. Is the pigeon population growing arithmetically or geometrically?
3. The populations of some Latin American countries are growing at a rate of 2% per year. At this rate, how many years will it take for the population to double?
4. If the population of India, now at 1 billion, grows at an annual rate of 2%, how many years will it take for the population to double?
5. What prediction did Thomas Malthus make?
6. If a population doubles every 25 years, what is the annual rate of growth?
7. From the dawn of human history until about 200 years ago, the world population was almost constant. Why was this, and why is world population increasing now?

8. What causes a soil to be "sweet" or "sour"?
9. Draw a diagram to show the horizons in a typical soil.
10. If you spread lime on your vegetable garden, will it raise or lower the pH of the soil?
11. What soil pH range do most crop plants prefer?
12. Why are calcium compounds often added to soil even if the soil is not deficient in calcium?
13. Which has more air per cubic foot of soil: a soil that has a high clay content or a soil with a high sand content? Explain.
14. What is the name given to the decayed organic matter that is a key component of soil?
15. Nitrogen is a primary nutrient. In what form is nitrogen used by plants?
16. What three elements are obtained from the air and water as nonmineral plant nutrients?
17. Describe how ammonia is made commercially. Write balanced equations for its preparation.
18. List three means by which nitrogen can be fixed.
19. Describe how ammonium nitrate fertilizer is made. Start with ammonia and oxygen.
20. Urea is a component of urine and a good source of nitrogen for plants. Why is urea a better fertilizer than inorganic ammonium salts?
21. What is anhydrous ammonia, and for what is it used?
22. What is the source of phosphorus used in most commercial fertilizers?
23. What is superphosphate, and how is it made?
24. Why must cadmium be removed from phosphate ore deposits before they can be used for fertilizer?
25. What element (metal) is an important part of chlorophyll?
26. There are three major classes of synthetic organic insecticides. In which of these classes does parathion belong?
27. What is meant by *persistent pesticide?*
28. Are carbamates broad-spectrum insecticides?
29. Describe the trade-off that must be accepted in using DDT in underdeveloped countries.
30. Explain the following numbers printed on a fertilizer bag: 30-30-10.
31. Describe the two long-term problems that occur when farmers use large amounts of synthetic fertilizers.
32. Which is a bigger problem for a farmer? A soil with a shortage of nitrogen, phosphorus, and potassium or a soil with a shortage of calcium, magnesium, and sulfur? Explain your answer.
33. If solar energy were used to electrolyze water into hydrogen (H_2) and oxygen (O_2), the hydrogen produced could be used to make ammonia. Describe two advantages this process would have over current practice.
34. In what form is phosphorus readily absorbed by plants?
35. List three plant secondary nutrients. How are they supplied to the soil?
36. List three plant micronutrients. Are they supplied directly to the soil?
37. Why are farmers inclined to use synthetic fertilizers rather than organic fertilizers such as animal manure?
38. What is the content of nitrogen, phosphorus and potassium in organic fertilizers, and how does this compare with the content in commercial fertilizers?
39. Describe how chlorinated hydrocarbons become concentrated in a food chain.
40. How much of the world's food supply is destroyed by pests?
41. What is a narrow-spectrum insecticide? Give an example.
42. What is a pheromone? How is a pheromone used to control insects?
43. Are trail-marking substances used by ants pheromones? How could such a substance be used as a method of controlling ant infestations?
44. What is an insect growth regulator (IGR)?

45. The powder Dipel is a biological control agent. Describe how it is used to control leaf-eating caterpillars.

46. What are juvenile hormones, and how are they used to control insect pests?

47. Describe how sterilization can be used as a technique to reduce an insect population.

48. How do alternative methods of insect control differ from those used by insect exterminators?

49. How does the objective of integrated pest management (IPM) differ from that of applying synthetic pesticides?

50. **a.** Describe the natural pesticides that are produced by the neem tree. Are these materials poisonous to humans?
    b. What other useful agents have been extracted from the neem tree?

Answers to Practice Exercises

16.1 **a.** \$1.00 **b.** arithmetically

16.2 2016

Household Chemicals

Chapter Objectives

In this chapter, you should gain an understanding of:

- How soap is made and how it works
- The composition of laundry detergents and their advantages
- The different types of bleaching agents used in laundering clothes
- How shampoos and other hair care products work to enhance the appearance of hair
- The composition of deodorants, antiperspirants, and toothpastes and how they work

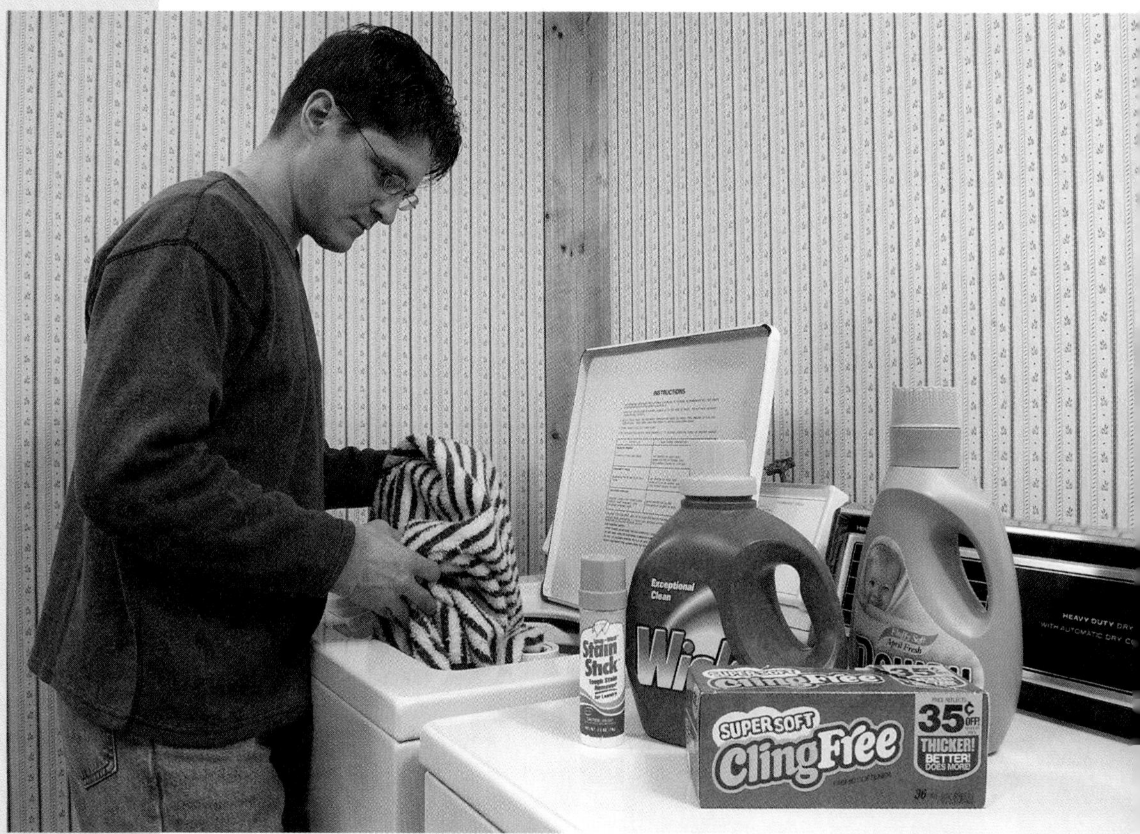

The products used in a typical laundry room in an American home include detergents, bleach, spot cleaners, and fabric softeners.

ONLINE FEATURES

www.jbpub.com/chemistry

- Chemistry and the Environment
- eLearning
- Chemistry in the News
- Special Topics
- Research & Reference Links
- Ask the Authors

MOST AMERICAN HOMES are full of chemicals. These are under kitchen sinks and in laundry rooms, bathroom cabinets, basements, and garages. Detergents, bleaches, cosmetics and other personal care products, pesticides, fertilizers, solvents, and paint are all commonly found in the average American home.

Many of these products are hazardous, and accidents often occur when consumers fail to follow the directions for their proper use. Also, although most communities make special collections of hazardous household chemicals such as pesticides, solvents, and batteries, many people often discard them with their regular trash or pour them down a drain—practices that can harm the environment.

In earlier chapters, we studied many of these chemicals, including pesticides, fertilizers, and paints. In this chapter, we concentrate on personal care products: soaps, detergents, bleaches, shampoos and other hair care products, deodorants, and toothpastes. We examine the chemicals from which they are made and consider the properties that make them effective.

17.1 Soap

How Soap Is Made

The earliest method used to clean clothes was to beat them with stones and then rinse them in a stream. Primitive societies also used certain plants for cleaning, including soapworts and soapberries, which contain carbohydrates called *saponins*. When mixed with water, saponins produce a soapy lather. Another early cleaning agent was made from the ashes of wood fires, which contain potassium carbonate (K_2CO_3) and sodium carbonate (Na_2CO_3). Mixing either of these substances with water produces a basic solution that acts like a detergent.

Soapberries are native to tropical and subtropical regions in Asia and America. Soapworts, which are found in temperate regions, are members of the pink family.

$$K_2CO_3 + H_2O \longrightarrow KHCO_3 + KOH$$

Soap is thought to have been used by the Phoenicians and the ancient Egyptians, but the Romans are credited with being the first to record how it was prepared. They first made soap by heating goat tallow with an extract of wood ash. Later, they improved the method by replacing the ash extract with the stronger base sodium hydroxide (NaOH), also known as *lye.*

Fats contain triglycerides (Chapter 14), which react with sodium hydroxide to form glycerol and soap as shown in the following equation. This reaction is called **saponification.**

$$\begin{array}{l} CH_2{-}O{-}\overset{\displaystyle O}{\overset{\|}{C}}{-}(CH_2)_{16}CH_3 \\ | \\ CH{-}O{-}\overset{\displaystyle O}{\overset{\|}{C}}{-}(CH_2)_{16}CH_3 \\ | \\ CH_2{-}O{-}\overset{\displaystyle O}{\overset{\|}{C}}{-}(CH_2)_{16}CH_3 \end{array} + 3\,NaOH \longrightarrow \begin{array}{l} CH_2{-}OH \\ | \\ CH{-}OH \\ | \\ CH_2{-}OH \end{array} + 3\,CH_3(CH_2)_{16}COO^-\,Na^+$$

tristearin (glycerol tristearate) — glycerol — sodium stearate (a soap)

FIGURE 17.1

Many different soaps are sold, but they all contain the sodium or potassium salts of fatty acids.

A **soap** is the sodium salt of long-chain fatty acids that were part of a triglyceride.

The early American pioneers made soap by boiling beef and mutton tallow in a basic solution of wood ash in water. They then added salt to the mixture, which lowered the solubility of the soap so that it separated out as a solid. Until it became a commercial operation, soap making was a household art. It was an unpleasant job because the housewife had to put her hands in the basic solution, and the lye attacked the skin.

Although the way soap is made has been known for hundreds of years, soap did not come into common use for bathing and washing until the early 19th century. Until then, people were unaware of the existence of microorganisms and how infections are spread, and, in most countries, very little attention was paid to personal cleanliness.

The fats that are used to make today's soaps include tallow from beef and mutton, coconut oil, palm oil, cottonseed oil, and olive oil. Bar soaps contain the sodium salts of one or more fatty acids, including lauric, myristic, palmitic, and stearic acids (see Table 14.1). Potassium hydroxide is used to make the softer soaps from which shaving cream and liquid soaps are made. Toilet soaps contain additives such as perfumes, dyes, oils, and germicides. Some soaps have air pumped into them to make them float; others are made transparent by the addition of alcohol (Fig. 17.1).

During the Renaissance in Europe, very few people ever bathed.

Soaps for scouring contain finely divided pumice or silica.

How Soap Works

Dirt, whether on our bodies or our clothes, is greasy because it picks up oils and fats from the skin. Many of the food stains that get on clothes are also greasy. Greasy dirt cannot be removed by simply washing in water; its removal requires the addition of soap. The cleansing action of soap is explained by the structure of a typical soap molecule:

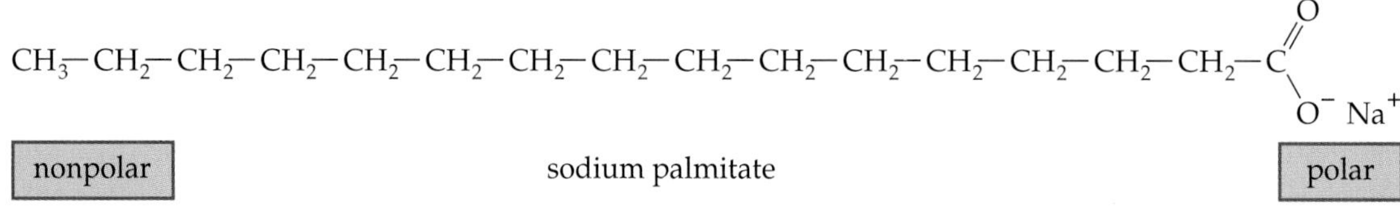

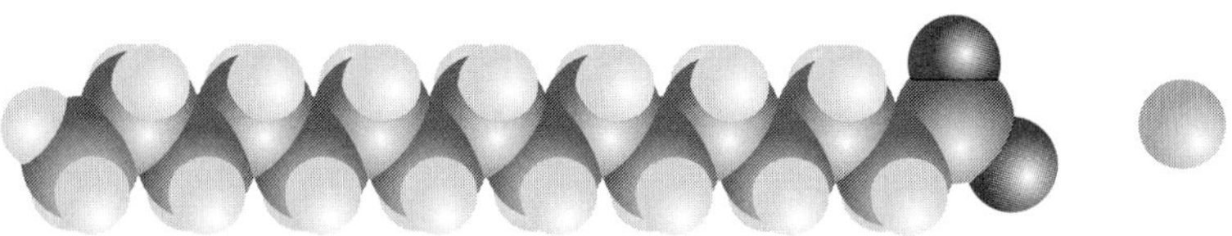

The hydrocarbon end of a long soap molecule, such as sodium palmitate, is nonpolar, while the ionized carboxylate end is highly polar. The nonpolar end dissolves in oily materials; the polar end dissolves readily in water. When greasy, oily dirt is vigorously mixed with soapy water, oily particles become surrounded by soap molecules. A grease-soap droplet called a **micelle** (Fig. 17.2) is formed as the nonpolar end of the soap enters the oily material and the polar end remains dissolved in water. The micelles cannot coalesce into

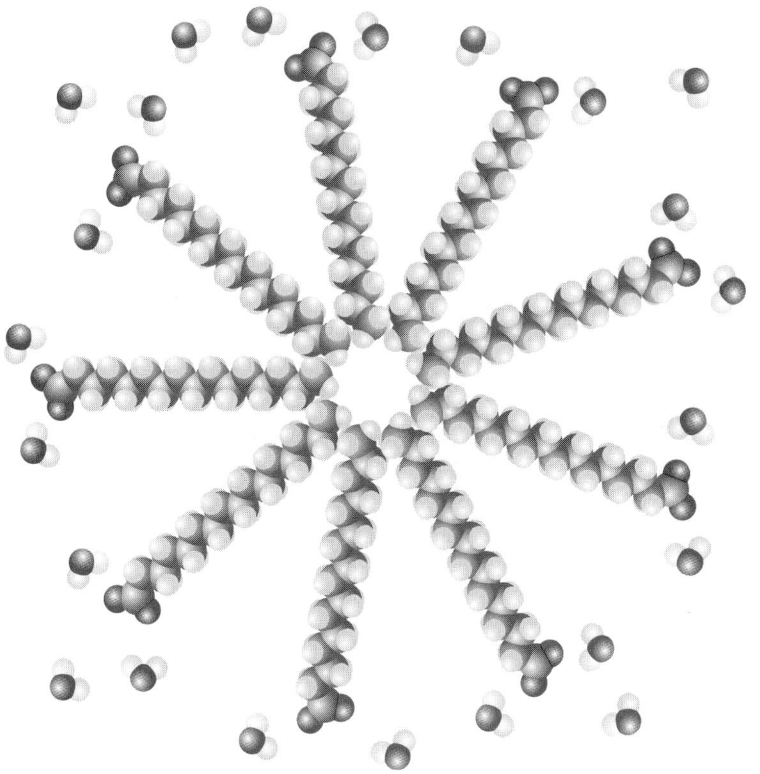

FIGURE 17.2

Cross section of a micelle. When soapy water is mixed with greasy dirt, the nonpolar hydrocarbon ends of soap molecules dissolve the grease; the polar carboxylate ends of soap molecules dissolve in water. The micelles, which repel each other because of their like charges, are washed away during rinsing.

larger droplets because the negative charges on their outer surfaces repel one another. They are washed away during rinsing, leaving behind a clean, grease-free surface.

Problems with Soap

One problem with soap is that it does not work in water that is acidic. Acid reacts with soap to form free fatty acids, which, because they do not have an ionic end, are insoluble in water and precipitate out as a greasy scum.

$$\underset{\text{a soap}}{CH_3(CH_2)_{20}COO^- \ Na^+} + \underset{\text{acid}}{H^+} \longrightarrow \underset{\text{fatty acid}}{CH_3(CH_2)_{20}COOH} + Na^+$$

Another more serious problem is that soaps do not work efficiently in hard water. **Hard water** is water that contains a relatively high concentration of Ca^{2+}, Mg^{2+}, or Fe^{2+} ions. Water that has a low concentration of these hard-water ions is called **soft water.** Soap forms abundant suds in soft water (Fig. 17.3a), but little sudsing occurs in hard water because the metal cations in the water react with the anions in the soap to form an insoluble precipitate (Fig. 17.3b). The "scum" that forms as a ring on bathtubs in areas with hard water is produced in this way.

Ferric ions (Fe^{3+}) dissolved in water can give the water a metallic taste and can cause rust stains.

a.

b.

FIGURE 17.3

(a) In soft water, soap suds are plentiful. (b) In hard water, there is little sudsing, and an insoluble scum forms.

Water hardness is expressed as the concentration of calcium carbonate (ppm by mass) that could be formed from the Ca^{2+} ions in solution if sufficient CO_3^- ions were also present in the water.

Hardness due to calcium or magnesium bicarbonate [$Ca(HCO_3)_2$] or [$Mg(HCO_3)_2$], is produced when water containing dissolved carbon dioxide comes in contact with limestone ($CaCO_3$) or dolomite ($CaCO_3 \cdot MgCO_3$).

$$CaCO_3 + CO_2 + H_2O \longrightarrow Ca^{2+} + \underset{\text{bicarbonate ion}}{2\ HCO_3^-}$$

$$CaCO_3 \cdot MgCO_3 + 2\ CO_2 + H_2O \longrightarrow Ca^{2+} + Mg^{2+} + 4\ HCO_3^-$$

Hardness due to these ions can be removed by boiling the water. On boiling, carbon dioxide and carbonate ions (CO_3^{2-}) are produced.

$$2\ HCO_3^- \longrightarrow CO_2 + \underset{\text{carbonate ion}}{CO_3^{2-}} + H_2O$$

The CO_3^{2-} ions then react with Ca^{2+} and Mg^{2+} ions to form insoluble carbonates ($CaCO_3$ and $MgCO_3$) which precipitate out of solution and can be removed. Here's the overall reaction, with M standing for Ca or Mg:

$$M^{2+} + 2\ HCO_3^- \longrightarrow MCO_3 + CO_2 + H_2O$$

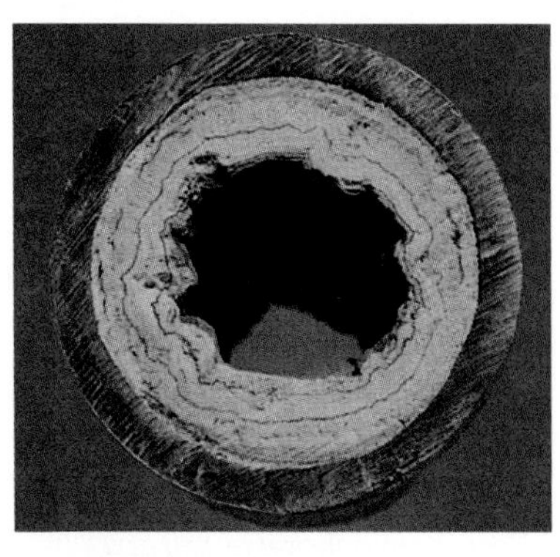

FIGURE 17.4

Boiler scale deposits can build up in hot water pipes and clog them.

The scaly deposits that form in teakettles are created in this way. Similar deposits in boilers and water pipes can be a serious problem when they build up and clog pipes (Fig. 17.4). Several methods are available to soften hard water.

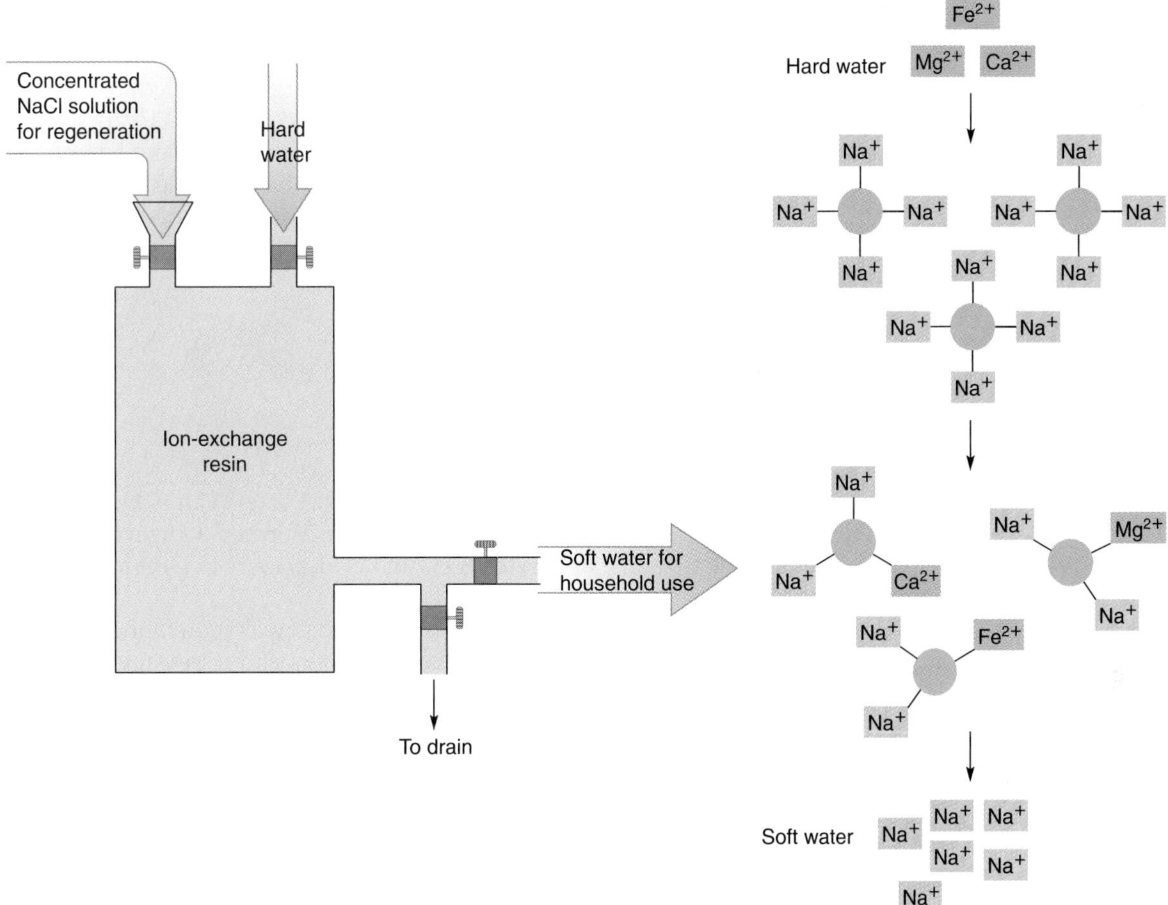

FIGURE 17.5

As hard water flows through a tank containing an ion-exchange resin, hard-water ions (Ca^{2+}, Mg^{2+}, and Fe^{2+}) replace sodium ions (Na^+) on the resin beads and are retained there. The water leaving the tank contains Na^+ ions, which do not form a scum with soap.

Water Softening

The most frequently used method for softening water for home use is **ion exchange,** a process in which hard-water ions (Ca^{2+}, Mg^{2+}, and Fe^{2+}) are exchanged for sodium ions (Na^+). In this method, hard water is passed through a tank containing specially prepared polystyrene beads (ion-exchange resin) covered with Na^+ ions (**Fig. 17.5**). As the water flows through the tank, Na^+ ions on the beads exchange with the hard-water ions. The hard-water ions are thus removed, and the water leaving the tank contains Na^+ ions, which do not form a scum with soap or produce scaly deposits.

When all the sites on the beads are occupied by hard-water ions, the material can be regenerated by flushing with a concentrated solution of sodium chloride (NaCl). Sodium ions replace the hard-water ions, which are flushed down the drain.

Water softened in this way contains an increased concentration of sodium ions. People who must restrict their sodium intake for health reasons may find it advisable to drink hard water and use the softened water for washing and heating purposes only.

17.2 Laundry Detergents

The word *detergent* is derived form the Latin word *detergere* meaning to "wipe off" or "cleanse."

The term *alkyl* refers to groups such as methyl ($—CH_3$), ethyl ($—C_2H_5$), and propyl ($—C_3H_8$), which result from removing a hydrogen atom from the corresponding alkane (Chapter 8).

Surfactants (surface-active agents) reduce the surface tension of water and make it a better wetting agent.

In the late 1940s, the introduction of synthetic laundry **detergents** helped to solve the problem of washing clothes in hard water. The first inexpensive detergents were sodium salts of alkylbenzenesulfonic acid, which, like soaps, have a polar end and a nonpolar end. In an **alkylbenzene sulfonate (ABS),** the ionic end of the molecule is a sulfonate group ($-SO_3^-$). These compounds, like soaps, are **surfactants** (or *surface-active agents*). They make it possible for water—a polar substance—to form a stable suspension with nonpolar greasy dirt.

$$CH_3-\underset{CH_3}{\underset{|}{CH}}-CH_2-\underset{CH_3}{\underset{|}{CH}}-CH_2-\underset{CH_3}{\underset{|}{CH}}-CH_2-\underset{CH_3}{\underset{|}{CH}}-C_6H_4-SO_3^-\ Na^+$$

an alkylbenzene sulfonate (ABS)

In hard water, Ca^{2+}, Mg^{2+}, or Fe^{2+} ions replace the Na^+ ions in an ABS detergent, and soluble products that do not precipitate out as scum are formed. However, because the hard-water ions tie up some of the detergent molecules, more detergent has to be used than would be required in soft water.

ABS detergents were very popular with consumers, but it was soon found that they caused serious problems. Microorganisms in natural waters and in sewage treatment plants were slow to degrade the branched chains, and the detergents passed through sewage treatment plants almost unchanged. Masses of sudsy foam began appearing in the streams and rivers into which the plants discharged treated water (**Fig. 17.6**). In some instances, suds were getting back into the water supply and coming out of faucets in homes.

Since 1965, ABS detergents have been replaced with detergents that contain a biodegradable **linear alkylbenzene sulfonate (LAS).** The unbranched chains of LAS detergents are much more easily broken down by microorganisms than the branched chains of ABS detergents are.

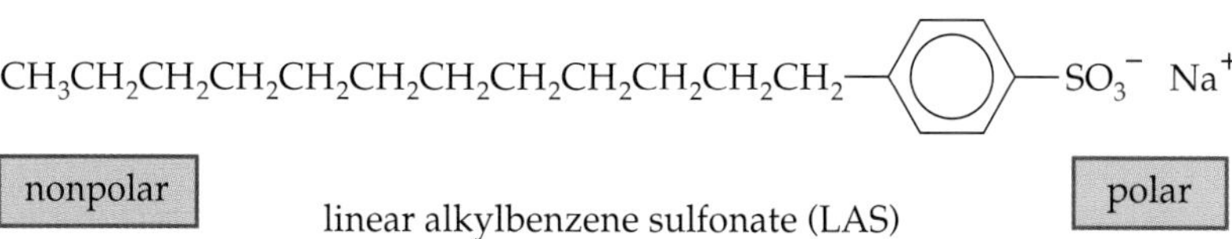

linear alkylbenzene sulfonate (LAS)

FIGURE 17.6
Sudsy foam was produced when water containing alkylbenzene sulfonate (ABS) detergents was discharged into waterways.

Linear alkylbenzene sulfonates are now the most commonly used surfactants. Like soaps and alkylbenzene sulfonates, they are called **anionic surfactants** because their polar ends are negatively charged. There is some concern that LAS break down to form phenol, a toxic substance that is harmful to fish.

In addition to surfactants, laundry detergents contain other ingredients, including builders and brighteners.

Builders

The function of a detergent's **builder** is to increase the efficiency of the surfactant. The first builders were polyphosphates, which soften water by binding with hard-water ions to form large water-soluble ions. For example, large water-soluble magnesium tripolyphosphate ions are formed when the builder sodium tripolyphosphate ($Na_5P_3O_{10}$) is added to water containing hard-water magnesium ions:

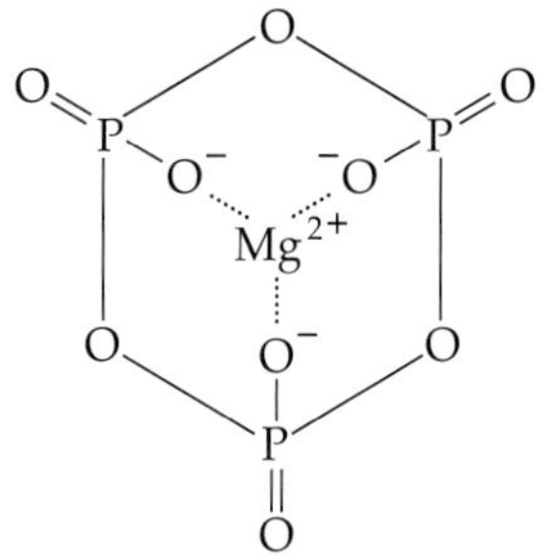

magnesium tripolyphosphate

In addition, polyphosphates increase the efficiency of the cleaning process by making the water slightly basic. In early detergents, the surfactant made up 15–20% of the product and the builder constituted 20–65%.

When first introduced, polyphosphates appeared to have no drawbacks. They were safe for all kinds of fabrics (white and colored) and noncorrosive to washing machines, and they biodegraded to simple phosphates, compounds that are harmless to humans and aquatic life. However, it soon became evident that the addition of phosphates to natural waterways caused eutrophication (Chapter 10). As a result, 20 states banned phosphate-containing detergents, and several others limited the amount of phosphate detergents can contain. Because of the difficulty of manufacturing different formulations to sell in various states, manufacturers turned to alternative builders. Most detergents now contain sodium carbonate or zeolites, or both.

Soaps have some advantages over detergents: They are made from animal and vegetable fats, which are natural ingredients, and they are biodegradable.

Sodium carbonate (Na_2CO_3), or washing soda, removes calcium and magnesium ions by precipitating them as their carbonates. It has the disadvantage that the precipitates tend to deposit on clothes. Also, as it dissolves in water, it forms a strongly basic solution, which can irritate the skin and be a hazard to young children.

$$2\,Na^+ + CO_3^{2-} + Ca^{2+} \longrightarrow CaCO_3 + 2\,Na^+$$

$$Na^+ + CO_3^{2-} + H_2O \longrightarrow Na^+ + OH^- + HCO_3^-$$

The most promising new builders are *zeolites,* complex sodium aluminum silicates that function as ion-exchange agents. Calcium ions become trapped when they penetrate the silicate's three-dimensional structure (Chapter 2) and exchange with sodium ions.

Zeolites are natural minerals that occur in sedimentary and volcanic rock deposits.

$$Ca^{2+} + Na_2Al_2Si_2O_8 \longrightarrow CaAl_2Si_2O_8 + 2\,Na^+$$

Although less efficient than phosphates, zeolites are not toxic or harmful to the environment and do not form strongly basic solutions.

Optical Brighteners

Optical brighteners are added to laundry detergents to overcome the tendency of fibers to become slightly yellow when washed. During washing, the brightener, which is a fluorescent compound, is incorporated into the fibers. It acts by absorbing invisible ultraviolet (UV) radiation from sunlight (refer to Fig. 1.13) and then emitting longer-wavelength light in the blue region of the visible spectrum (**Fig. 17.7**). The fibers acquire a bluish tint, which makes white fabric appear whiter and colored fabric appear brighter. Skin contact with brighteners can cause a rash on some people.

A fluorescent compound absorbs light of a certain wavelength and then emits light of a longer wavelength.

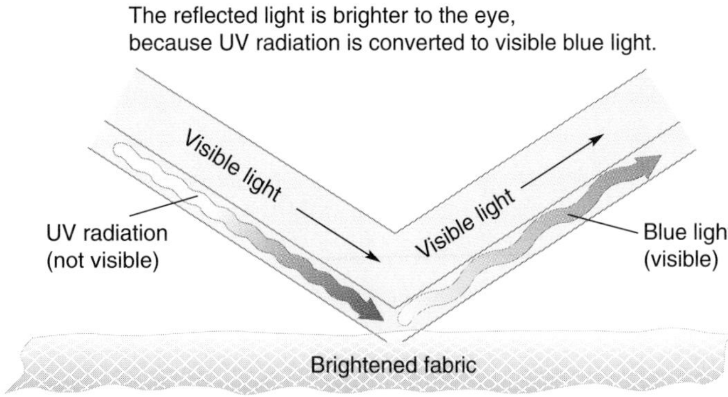

FIGURE 17.7

Optical brighteners absorb invisible ultraviolet (UV) radiation and then emit visible light of longer wavelengths. This makes fabric appear brighter.

Fabric Softeners

Many of today's washers are programmed to add a fabric softener during the final rinse. Fabric softeners are quaternary ammonium salts that contain two long hydrocarbon chains. A **quaternary ammonium salt,** or quat, is an ionic compound in which the cation consists of a nitrogen atom joined to four carbon atoms. An example is dioctadecyldimethyl-ammonium chloride:

$$\begin{array}{c} CH_3 \quad Cl^- \\ | \\ CH_3(CH_2)_{16}CH_2-N^+-CH_3 \\ | \\ CH_3(CH_2)_{16}CH_2 \end{array}$$

dioctadecyldimethylammonium chloride

Because the polar end of the molecule has a positive charge, quats are called **cationic surfactants.** However, they are not efficient surfactants and are not used as detergents. Clothes tend to carry a negative charge, and this attracts the positively charged quats, which become attached to the fabric fibers, making the clothes feel soft. For use in washing machines that do not automatically dispense fabric softener, manufacturers have developed sheets that are impregnated with a quaternary ammonium salt. A sheet is put in the dryer with each load of wet laundry and as the temperature rises, the fabric softener vaporizes and becomes attached to the laundry.

Other Detergent Ingredients

Laundry detergents may also contain a number of other ingredients, including fillers, enzymes, bleaches, and perfumes **(Fig. 17.8)**. A **filler** is an inert substance such as sodium sulfate (Na_2SO_4), that adds bulk to the powder and acts as a drying agent to keep it flowing freely. The concentrated detergents sold today as compacts, ultradetergents, or superconcentrates contain much less filler than earlier formulations, which gives them the advantage of using less packaging material.

Enzymes are added to detergents to help remove certain stains. Amylases break down starch-based stains, and proteases break down protein-based stains.

The bleach most often added to detergents is sodium hyperborate, an oxygen-releasing compound that does not harm fabrics or dull colors. Borax ($Na_2B_4O_7$) is also added to create the alkaline conditions that oxygen-releasing bleaches require. These bleaches are discussed in more detail in Section 17.3.

FIGURE 17.8

Laundry detergent ingredients include fillers, enzymes, bleaches, and perfumes.

Liquid Laundry Detergents

Some consumers prefer liquid laundry detergents, which, like powders, usually contain linear alkylbenzene sulfonates. In addition, they often contain other detergents, including nonionic surfactants such as alcohol ethoxylates. Molecules of these compounds do not have an ionic end, but they are very soluble in water because of hydrogen bonding.

$$CH_3(CH_2)_{10}CH_2(OCH_2CH_2)_8OH$$

a typical alcohol ethoxylate

Some liquid detergents contain no builders; others contain sodium carbonate and zeolites. Because liquid detergents are concentrated and relatively small amounts are needed for each load of laundry, savings are made on packaging.

17.3 Bleaches

Chlorine Bleaches

To make white clothes whiter or to remove stains from them, many people use a **bleach** containing chlorine, such as Clorox or Purex, which are 5.25% solutions of sodium hypochlorite (NaOCl). When dissolved in water, NaOCl produces hypochlorite ions (OCl^-), which are the bleaching agents. Hypochlorite ions also kill bacteria.

After many washings, white clothes look dull and slightly grayed or yellowed, and bleaches, which are oxidizing agents (Chapter 7), remove the unwanted color by oxidizing the molecules responsible for it to colorless products. During bleaching, hypochlorite ions gain electrons from these molecules and are reduced to chloride and hydroxide ions:

$$OCl^- + H_2O + 2\,e^- \longrightarrow Cl^- + 2\,OH^-$$

Chlorine bleaches work well on cotton and linen but can cause yellowing instead of whitening when used on some synthetic fibers. Also, the chlorine released by bleaches containing sodium hypochlorite can damage some fabrics. Thus, other chlorine-containing bleaches that produce chlorine more slowly and are less damaging to fabrics are sometimes preferred.

Chlorine bleaches can be very dangerous if mixed with other household products. For cleaning windows, most people use products that contain ammonia, and for cleaning toilet bowls, many use cleaners that contain hydrochloric acid. If a hypochlorite bleach is mixed with either of these products, chlorine-containing gases and chlorine, which are very toxic, are produced. The following equation shows how chlorine gas is formed if hypochlorite bleach is mixed with hydrochloric acid:

$$ClO^- + 2\,HCl \longrightarrow Cl^- + H_2O + Cl_2$$

Fabrics look colored to us because molecules in dyes reflect some wavelengths in the visible part of the spectrum and absorb others. A green dye, for example, reflects maximally in the green part of the spectrum and absorbs light at other wavelengths. If chlorine bleach is accidentally spilled on a colored fabric, it changes the dye molecules in such a way that they lose their ability to absorb visible light. As a result, all the visible wavelengths are reflected, and we see a white spot on the fabric.

For bleaching paper and fabric on an industrial scale, bleaching powder that contains calcium hypochlorite [$Ca(OCl)_2$] is used. This compound functions in the same way as sodium hypochlorite.

Before the use of chlorine for bleaching, textiles were bleached by exposing them to sunlight and air (which contains oxygen).

Oxygen-Releasing Bleaches

Oxygen-releasing bleaches have the advantage of being less damaging to clothes and to the environment than chlorine bleaches. They usually contain sodium hyperborate, a complex of sodium metaborate ($NaBO_2$) and hydrogen peroxide (H_2O_2). In water, sodium hyperborate releases hydrogen peroxide, which acts as the bleaching agent and decomposes to form oxygen:

$$2\ H_2O_2 \longrightarrow 2\ H_2O + O_2$$

Like chlorine bleaches, oxygen-releasing bleaches kill bacteria.

Although sodium hyperborate requires a more alkaline solution and must be used at higher concentration and temperature than chlorine bleaches, it works well for synthetic fabrics. Sodium percarbonate, which in Europe has mostly replaced sodium hyperborate, is receiving increasing attention. Although less stable than hyperborates, percarbonates have the advantage that, unlike borates, carbonates help in the washing process by acting as builders and by raising pH.

17.4 Hair Care Products

Shampoos and Conditioners

Hair is formed of long strands of protein. Each hair grows from a root in the scalp, which is contained in a follicle **(Fig. 17.9a)**. The root is alive, but the hair is lifeless. The hair shaft is lubricated with **sebum,** an oily secretion produced by the sebaceous gland, which lies near the follicle. The coating of sebum prevents hair from drying out and gives it a gloss. If there is too much sebum, the hair feels greasy and dirty; if there is too little, the hair feels dry, looks dull, and is unmanageable. Washing hair removes sebum and any dirt adhering to it. Although we want to remove the dirt and excess oil, we do not want to use a detergent that is strong enough to remove all the sebum.

When applied to the scalp, Minoxidol, which is sold under the trade name Rogaine, produces a growth of fine hair wherever hair follicles are present.

Most modern shampoos for adults contain sodium dodecyl (lauryl) sulfate, a synthetic detergent that, like a linear alkylbenzene sulfonate, is an anionic surfactant and has both a polar and a nonpolar end.

$$CH_3(CH_2)_{10}CH_2{-}O{-}SO_3^-\ Na^+$$

sodium lauryl sulfate

A sudsy lather is a disadvantage in washing machines, but consumers expect plenty of lather from their shampoo.

Sodium lauryl sulfate is milder than a laundry detergent but acts in exactly the same way to remove greasy dirt. Shampoos for oily hair have a higher concentration of this kind of detergent than those formulated for normal hair.

"Tear-free" shampoos for children contain an amphoteric detergent that is less irritating to the eyes. Amphoteric detergents have both a positive and negative charge in the same molecule, and thus can neutralize both acids and bases.

$$CH_3(CH_2)_{10}{-}\overset{\displaystyle H}{\underset{\displaystyle H}{N^+}}{-}CH_2CH_2{-}O^-$$

amphoteric detergent

Hair is composed primarily of the protein **keratin,** which contains both acidic and basic functional groups. Hair is slightly acidic. If the pH of a shampoo is too high or too low, the structure of the hair can be damaged. Most shampoos have a pH between 5 and 8. The pH also affects the ability of the shampoo to make clean hair look shiny. The glossiness of hair depends on a thin layer called the *cuticle,* which covers the surface of each hair **(Fig. 17.9b)**. The cuticle, which consists of overlapping transparent scales, can be held tightly to the hair shaft or ruffled out. If the shampoo is too basic, the cuticle scales fluff out and reflect light in a way that makes hair look dull. On the other hand, if the pH is slightly acidic, the scales are held tightly against the surface, the light is reflected differently, and the hair has an attractive luster.

Shampoos contain many ingredients in addition to the detergent. These include an ingredient that adjusts the pH; EDTA (Chapter 18), which softens water by binding with hard-water ions; and fragrances, which are added to make the products appealing to the consumer. Some shampoos also contain a conditioning agent such as oleyl alcohol or mineral oil, which is added to replace some of the oils lost in shampooing.

Prior to World War II, people used ordinary hand soap to wash their hair, and it often left a dull soap film on the hair. To remove the film, the hair was rinsed with lemon juice or vinegar. Today, most people use a conditioner after shampooing to add lubricants and luster to hair. Modern conditioners rely on quaternary ammonium salts such as tricetylammonium chloride to remove remaining detergent residues. The quats in a conditioner are large organic cations that electrostatically attract the lauryl sulfate anions

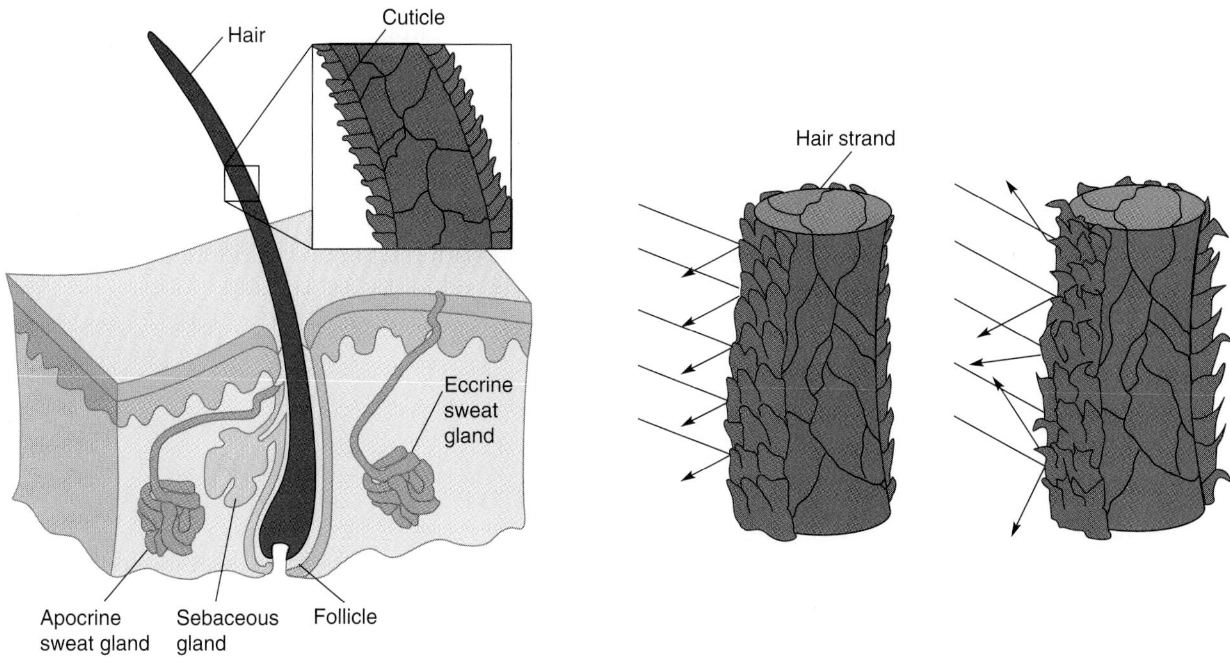

FIGURE 17.9

(a) Cross section showing a strand of hair and the glands that lie below the skin surface. (b) The cuticle scales that cover a hair are held tightly to the shaft when the pH is neutral or acidic, making the hair look shiny. At higher pH values, the cuticle scales are ruffled, and light is reflected differently, which makes hair look dull.

in shampoo. The positively charged quaternary ammonium ion and the negatively charged lauryl sulfate ion form a water-soluble ion pair, which is removed from the surface of the hair as it is rinsed.

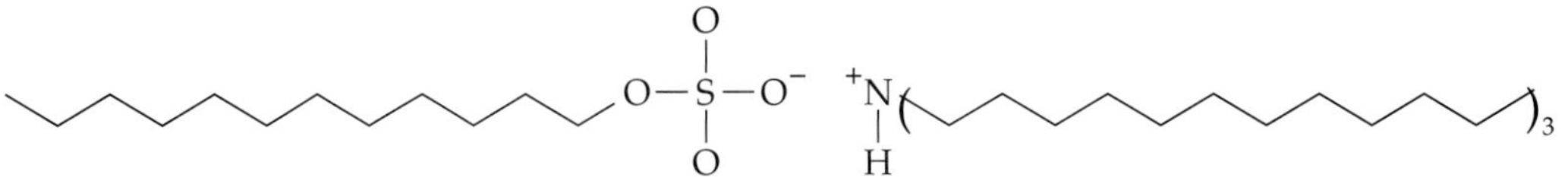

lauryl sulfate ion
(anionic shampoo)

tricetylammonium ion
(quaternary ammonium ion conditioner)

Just as quats in fabric softeners attach to fibers, these quats coat hair, making the hair soft and shiny.

Hair Curling and Permanents

In hair, adjacent strands of protein are held together by disulfide bonds (—S—S—) (see Chapter 14) and by weaker, but very numerous, hydrogen bonds (Chapter 4). Hair can be curled to produce temporary or permanent waves by breaking and rearranging some of the bonds. The simplest, and least permanent procedure is to wet the hair, which disrupts the hydrogen bonds holding the proteins in shape but does not break the stronger, covalent disulfide bonds. If the wet hair is allowed to dry while wound on rollers, hydrogen bonds reform in new positions, and the hair is set in temporary waves. Because heat, like water, can disrupt hydrogen bonds, curling irons can also be used to induce a temporary wave. The change in shape achieved in these two procedures is quickly reversed if there is an increase in humidity. Waves that have taken considerable time to create can disappear in minutes if there is a sudden change in weather.

Longer-lasting waves can be obtained by spraying the hair after it has been set with a resin-type polymer dispersed in a solvent. The polymer can also be applied to the hair as a mousse. After the polymer is sprayed or rubbed on the hair, the solvent evaporates, leaving the hair coated with the polymer. Commonly used products contain poly(vinyl pyrolidine) (PVP) or poly(vinyl acetate).

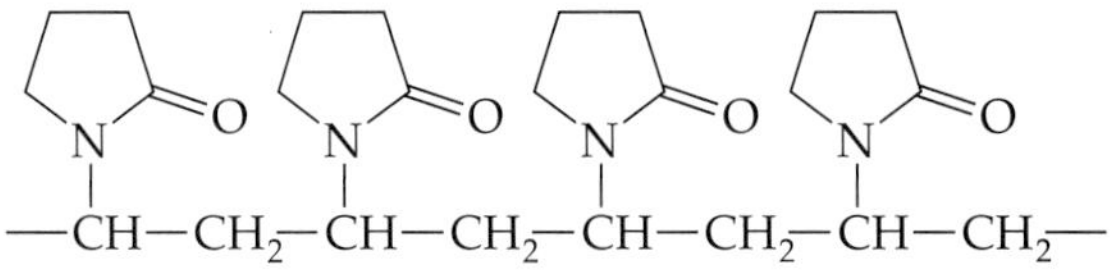

PVP

The polymer coating holds the hair in place longer than does a heat-induced set, but the effect lasts for only a matter of hours.

A permanent wave can be achieved by initiating a chemical reaction that breaks the disulfide linkages that hold the protein strands in hair together. This is done by applying a lotion containing the reducing agent thioglycolic acid ($HSCH_2COOH$) to the hair. The thioglycolic acid ruptures disulfide linkages in the hair protein and reduces them to sulfhydryl (—SH) groups **(Fig. 17.10)**, a step that is accompanied by the release of hydrogen sulfide (H_2S), which smells like rotten eggs. Next, the hair is set on rollers and treated with a mild oxidizing agent, such as hydrogen peroxide (H_2O_2). This agent causes disulfide linkages to reform in new positions. The strands of protein lock together in the shape of the roller and leave the hair with a permanent curl.

A permanent wave lasts as long as the hair. As new hair grows in, it assumes its natural straightness, and the permanent wave treatment must be applied to the new hair.

The same process can be used to straighten moderately curly hair. Instead of being curled on rollers after treatment with thioglycolic acid, the hair is rinsed and combed straight. For very curly hair, a *relaxer* such as sodium or lithium hydroxide must be used. Chemical relaxers are strong bases that break disulfide bonds. Very curly hair has so many disulfide bonds that a strong base is needed to rupture them in a reasonable time. Care must be taken not to burn the scalp with the strong base.

Hair Dyes

The color of both skin and hair depends on the relative amounts of two natural pigments: *melanin,* which is dark brown, and *phaeomelanin,* an iron-based pigment that is red-brown or yellow-brown. People with jet-black hair have large quantities of melanin, brunettes have a moderate amount, and blondes have very little of either pigment. The hair of redheads is low in melanin but high in phaeomelanin. (Because of the iron in this pigment, "Rusty" is an especially appropriate nickname for a redhead.) The loss of pigment is responsible for gray and white hair.

Brunettes can become blondes very easily by applying the oxidizing agent hydrogen peroxide (H_2O_2), which oxidizes hair pigments to colorless products. Dark hair can be converted to a lighter shade of brown by first bleaching the hair with hydrogen peroxide and then dying it with a light pigment. The addition of brown dyes to unbleached hair darkens its color.

Hair can be colored with either a temporary or a permanent dye **(Fig. 17.11)**. Temporary hair dyes are mixtures of several dyes, including FD&C Blue, Yellow, and Red, which are combined to give a pleasing color when coated on hair. Because the molecules in these water-soluble dyes are large, they do not penetrate the hair shaft but remain on the surface of the cuticle and are easily removed by repeated washing.

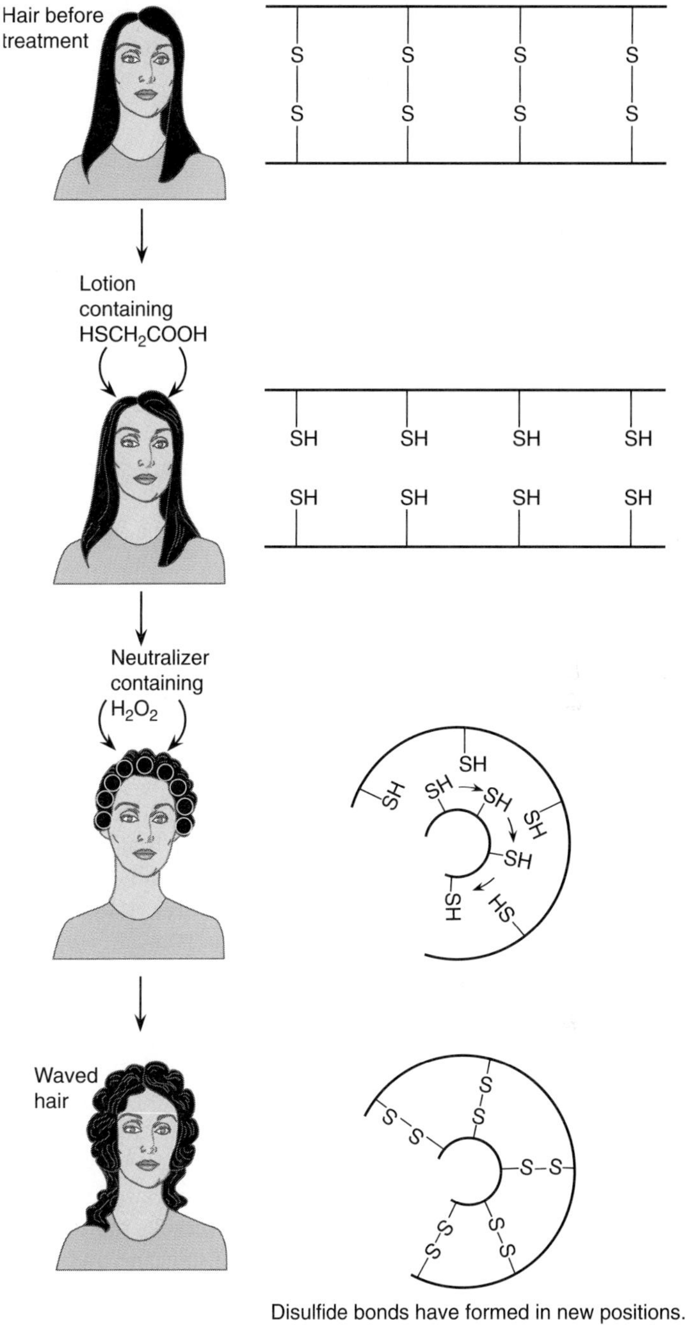

Permanent dyes penetrate the hair shaft and last as long as the hair does. They consist of three reagents: a primary intermediate; a secondary intermediate, or *coupler;* and an oxidizing agent, usually hydrogen peroxide. The primary and secondary intermediates are small, colorless organic molecules, which can migrate into the hair shaft between the protein (keratin) strands. Commonly used primary intermediates are the colorless compounds,

Hair before treatment

Lotion containing $HSCH_2COOH$

Neutralizer containing H_2O_2

Waved hair

Disulfide bonds have formed in new positions.

FIGURE 17.10

A reducing agent breaks the disulfide bonds that hold protein chains in hair together. When the hair is set on rollers and treated with an oxidizing agent, the disulfide bonds reform in new positions, and the hair acquires a permanent wave.

FIGURE 17.11
Hair dye in use.

p-phenylenediamine and *p*-aminophenol, which have functional groups on both ends of their molecules.

$NH_2-C_6H_4-NH_2$ *p*-phenylenediamine

$NH_2-C_6H_4-OH$ *p*-aminophenol

Once inside the hair, one end of the primary intermediate reacts with the keratin and is oxidized by the hydrogen peroxide. The other end then attaches to the coupler. This multistep process results in the formation of a dye that becomes permanently attached to the hair. *p*-Phenylenediamine produces a black color, but lighter colors can be achieved by using various derivatives of this compound. Permanent dyes color only the exposed part of hair. As new hair grows from the scalp, it has its natural color, and "roots" begin to show.

There are some risks associated with using permanent dyes. One derivative of *p*-phenylenediamine, *p*-methoxy-*m*-phenylenediamine (MMPD), has been shown to be carcinogenic when fed to laboratory mice.

NH_2, NH_2, OCH_3 on benzene ring

p-methoxy-*m*-phenylenediamine (MMPD)

The terms *ortho-*, *meta-*, and *para* (abbreviated to *o-*, *m-*, *p-*) indicate the position of a second constituent on a benzene ring (see Chapter 8).

Because of the cancer risk, this compound has been removed from most commercial hair dyes. Those that continue to include it must carry a warning on the label. Because MMPD is structurally similar to many of the other chemicals used for permanently dying hair, there is concern that these chemicals may also pose a risk. None of them are completely harmless.

Grecian Formula does not cause lead poisoning because the lead ion is so big that it is not readily absorbed through the skin.

Inorganic hair dyes, such as Grecian Formula, darken the hair slowly and are often used daily to gradually deepen the color. Grecian Formula is formulated as a shampoo or a cream that contains colorless lead acetate $Pb(CH_3COO)_2$. When applied to the hair, lead ions in the solution penetrate the hair shaft and react with sulfur in the keratin to form lead sulfide (PbS), which is black. Repeated application deepens the color as more lead and sulfur react. The percentage of sulfur-containing amino acids in hair protein varies from person to person. If the percentage is high, the hair will darken rapidly.

The name Grecian Formula is a reminder of some early history about darkening hair. The ancient Greeks made their aqueducts watertight by lining them with a lead sheet. The soft, low-melting metal was easy to fabricate and formed an effective seal, but flowing water slowly dissolved the lead, and it was noticed that people who bathed often in this water had very dark hair. People, of course, also drank the water, and lead poisoning (Chapters 10 and 18) is believed to have contributed to the decline of both the Greek and Roman empires.

17.5 Deodorants and Antiperspirants

EXPLORATIONS reveals some interesting facts about perfumes (see pp. 528–529).

Americans are very concerned about body odor and spend a lot of money on deodorants, antiperspirants, and perfumes (**Fig. 17.12**), even though, for most people, a daily shower or bath is just as effective in controlling body odor as these products.

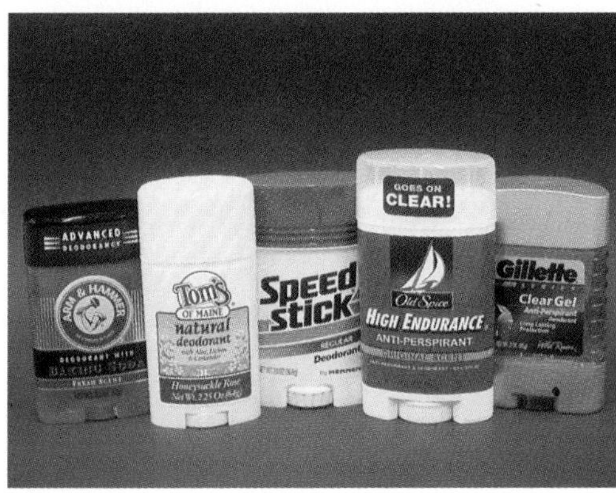

FIGURE 17.12

Examples of deodorants and antiperspirants.

Sweating and perspiring are natural body functions that are essential for the regulation of body temperature. When the surrounding temperature rises or the body heats up during exercise, water in sweat evaporates and the body is cooled (Chapter 10). One type of sweat glands, the **eccrine glands,** cover almost all surfaces of the body. Their odorless secretions are 99% water with small amounts of sodium chloride, urea, glucose, and a few other organic componds. The other type of sweat glands, the **apocrine glands** (see Fig. 17.9a), occur almost exclusively in the armpits and groin. These glands, like the sebaceous glands, have a duct that opens into the upper part of the hair follicle and empties the gland's secretion in response to emotional stimuli such as fear, pain, or anxiety. The milky apocrine secretion is mainly water but contains sodium chloride and small amounts of proteins, carbohydrates, and ferric and ammonium salts.

Like eccrine secretions, apocrine secretions are initially odorless. But in the warm, moist environment of the armpit and groin, bacteria proliferate, and the bacteria break down apocrine secretions and oils secreted by the sebaceous glands into unpleasant-smelling fatty acids and amines.

Because the apocrine glands do not develop until puberty, young children do not have a strong body odor.

Deodorants and antiperspirants are two distinct types of products. **Deodorants** act by controlling the growth of the bacteria that produce the odors. **Antiperspirants** act by reducing the amount of sweat that is produced.

Most deodorants contain triclosan (2,4,4′-trichloro-2′-hydroxydiphenyl ether), a broad-spectrum *bacteriostat* (a reagent that kills a wide range of bacteria) that works at low concentrations. Deodorants also contain cosmetic oils, alcohol, and fragrances that help mask body odor:

Cl HO

Cl—⟨ring⟩—O—⟨ring⟩—Cl

triclosan

One of the first antiperspirants was Everdry, introduced in 1902. It consisted of a solution of aluminum chloride ($AlCl_3$), and if it were available today, it would probably be the most effective antiperspirant on the market. But not many consumers would want to use it since it was a messy liquid that irritated the underarm and damaged clothing. The aluminum ion (Al^{3+}) made the product effective, and it was later found that if some of the chloride ions (Cl^-) in aluminum chloride were replaced with hydroxide ions (OH^-), the product was less irritating. Today, aluminum chlorohydrates such as aluminum dichlorohydrate [$Al_2(OH)_4Cl_2$] and aluminum chlorohydrate [$Al_2(OH)_5Cl$] are used as the drying agent in antiperspirants. These compounds, which are alkaline, also help to kill odor-causing bacteria.

Aluminum ions act as astringents; they reduce the amount of sweat produced by closing the ducts from the eccrine and aponine glands. They also coagulate proteins. Before electric razors were introduced, aluminum ions in alum or septic pencils were used to coagulate proteins in blood and thus stop bleeding from small cuts caused by razor blades.

17.6 Toothpastes

Tooth decay has plagued humans for centuries, even though the *enamel* that makes up the outer layer of teeth is the hardest substance in the body. The 2-millimeter thick layer of enamel is composed primarily of hydroxyapatite [$Ca_{10}(PO_4)_6(OH)_2$]. In the mouth, the enamel is continually and gradually being demineralized, as calcium, phosphate, and hydroxide ions dissolve in saliva. At the same time, the reverse process, mineralization, occurs, and the ions recombine to form hydroxyapatite.

$$\underset{\substack{\text{hydroxyapatite}\\\text{(solid)}}}{Ca_{10}(PO_4)_6(OH)_2} \underset{\text{mineralization}}{\overset{\text{demineralization}}{\rightleftharpoons}} \underbrace{10\ Ca^{2+} + 6\ PO_4^{3-} + 2\ OH^-}_{\text{(dissolved ions)}}$$

In children, the rate of formation of enamel exceeds the rate of demineralization. Tooth decay occurs when the equilibrium is upset by various factors causing the rate of demineralization to exceed that of mineralization.

Bacteria in the mouth (**Fig. 17.13**) convert sugars in food to polysaccharides. Soon after eating, teeth become coated with a thin adhesive film called **plaque,** which consists of bacteria and polysaccharides. Other bacteria convert carbohydrates in the plaque to carboxylic acids such as lactic acid, which dissolve the enamel. The more sugar that is present, the more the bacteria multiply, and the more acid they produce. If the enamel becomes severely eroded, bacteria penetrate into the interior of the tooth and cause decay.

Plaque does not form evenly on all teeth. Little accumulates on front teeth, which are regularly in contact with the tongue. Much more accumulates in the crevices in teeth in the back of the mouth, where the tongue does not reach as often. Plaque combines with calcium and phosphate ions in saliva to form a hard, yellowish-white material called **tartar.** The best way to prevent cavities is to remove accumulated plaque and food particles by brushing and flossing the teeth regularly, particularly after meals. Tartar can only be removed by professional cleaning.

FIGURE 17.13

Electromicrograph of bacteria that cause tooth decay.

Toothpastes (Fig. 17.14) contain many ingredients, including an abrasive, a detergent, sweeteners, flavors, fluoride, and preservatives (Table 17.1). The most important ingredient is the abrasive, a gritty substance that has to be hard enough to grind away plaque but not hard enough to damage the tooth enamel. Commonly used abrasives are calcium carbonate ($CaCO_3$), hydrated alumina ($Al_2O_3 \cdot nH_2O$), hydrated silica ($SiO_2 \cdot nH_2O$), titanium oxide (TiO_2), and various calcium phosphates.

The detergent in most toothpastes is sodium dodecyl (lauryl) sulfate, the same detergent that is used in shampoos. During brushing, it forms suds that suspend the particles removed by the abrasive so that they can be rinsed away.

Stannous fluoride (SnF_2) or sodium fluoride (NaF) is present in most toothpastes. Studies have shown that children develop markedly fewer cavities in regions where the water contains at least 0.1 ppm of fluoride ions. Fluoride ions in solution replace some of the hydroxide ions in the hydroxyapatite in enamel to give fluorapatite:

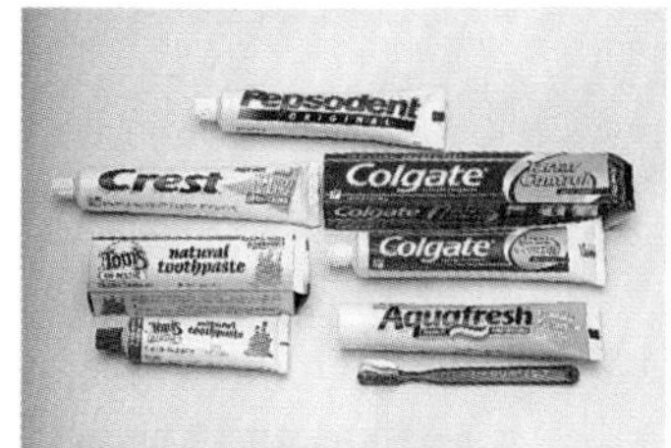

FIGURE 17.14

Some of the many available brands of toothpaste. The most important ingredients in toothpaste are the abrasive and the detergent.

$$\underset{\text{hydroxyapatite}}{Ca_{10}(PO_4)_6(OH)_2} + 2\,F^- \longrightarrow \underset{\text{fluorapatite}}{Ca_{10}(PO_4)_6F_2} + 2\,OH^-$$

Fluorapatite is harder and denser than hydroxyapatite, and it is 100 times less soluble in acid. Fluoride ions also suppress the ability of bacteria to generate acid.

The natural fluoride content of drinking water in the United States ranges from 0.05 ppm to 8 ppm. In regions where the fluoride concentration is low, fluoride is typically added to drinking water to bring the level up to 1 ppm. Although this level has been shown to be safe, at much higher concentrations fluoride ions are toxic, and they can cause *dental fluorosis*, a condition in which white or brownish patches develop on teeth and the enamel becomes brittle. Because of concerns about these problems, fluoridation remains controversial, and some municipalities do not add fluoride to their water. Where water is not fluoridated, young children can obtain protection by taking fluoride tablets, or a dentist can apply a gel containing fluoride to their teeth. Fluoridated toothpaste and fluoride tablets are not as effective as fluoridated water in reducing tooth decay.

Table 17.1 Comparison of Ingredients in a Typical Toothpaste and in a Natural Toothpaste

	Typical Toothpaste	Natural Toothpaste
Abrasive	Calcium carbonate	Calcium carbonate from limestone
Detergent	Sodium lauryl sulfate	Lauryl alcohol from coconut
Fluoride	Sodium fluoride	Sodium monofluorophosphate from the ore fluorspar
Sweetener	Saccharin	Glycerin (vegetable oil by-product)
Anticaries agent	None	Xylitol from birch trees
Flavoring	Peppermint	Spearmint or peppermint oils from natural sources
Thickener	Cellulose gum	Carrageenan from seaweed
Preservative	Sodium benzoate	None
Coloring	Artificial colors	None
	Water	Water

EXPLORATIONS

The Allure of Perfume

The fascination with perfume goes back many thousands of years. Perfumes were made in ancient China, India, and Egypt, and they were an integral part of life in classical Greece and Rome. Assyrians perfumed their beards, and ancient Egyptians anointed their bodies with fragrant oils. Cleopatra, intent on seducing the Roman general Mark Antony, greeted him on a barge with sails soaked in perfume.

The earliest perfume was probably incense. The resins from cedar of Lebanon, sandalwood, and salai trees (frankincense) that were burned in ancient religious ceremonies imparted aromatic scents to the smoke that wafted up to the gods. The word *perfume* is believed to be derived from the Latin *perfum*, which means "through smoke."

Perfumes can be alluring, evocative, mysterious, and seductive, and they can create different moods for different occasions. We detect scents through our noses. The olfactory receptors, which are located in the roof of the nasal cavity behind the bridge of the nose, can distinguish several thousand odors. Airborne molecules of a particular scent, on reaching the receptors, bind to a protein that is specific for that scent. This causes a nerve impulse to travel to the area of the brain that is responsible for many emotions, including pain, hunger, thirst, and pleasure, and it is here that the impulse is interpreted. Some odors, like the smells of lavender and baking bread, give us pleasure; others, like the smells of dead fish and manure, are distasteful.

Essences of Flowers

The traditional sources of fragrant essences for perfumes are plants, mainly flowers such as rose, jasmine, lavender, lily of the valley, violet, and gardenia. For centuries, France was the center of perfumery and Provence was the main region where the raw materials—roses, jasmine, and lavender—were grown. Flowers used to make perfumes must be picked by hand, and labor accounts for as much as 60% of the cost of a fragrant oil. Petals from as many as 800 roses are required to make just 1 pound of concentrate, and the roses have to be picked at exactly the right moment after the flowers have opened and before they begin to fade. A pound of oil from the exquisite-smelling *rose de mai* (rose of May), grown in France, costs $3,650. Today, most flowers for perfumes are grown in developing countries, where land and labor are very much cheaper than they are in France. A pound of oil from roses grown in Morocco, Turkey, or Bulgaria costs about $600. Jasmine, which is now grown primarily in India, Egypt, and Morocco, must be picked at dawn before the heat of the day can evaporate any of the fragrant oils, and about 2.5 million flowers are required to produce 1 pound of jasmine concentrate. Although the subtler fragrance of French jasmine is preferred by many perfumers, only a few houses can afford to purchase it.

Animal products that have been used as a base for perfumes include ambergris from sperm whales and musk from musk deer and civet cats. Some farmers in Ethiopia and other developing countries raise civet cats. The yellowish musk called *civet*, which they scrape from the cats'

A field of lavender being harvested

anal glands, has the consistency of butter and makes an excellent base for perfume.

Synthetic Ingredients

Perfumes can be alluring, evocative, mysterious, and seductive, and they can create different moods for different occasions.

At the end of the 19th century, perfumes contained only natural fragrances. Today, because of the extremely high cost of natural fragrances, the vagaries of weather, and concerns for conservation and animal welfare, synthetic chemicals are being used increasingly. Synthetics can now replicate the scent of sandalwood and many floral fragrances, and they have replaced the oils that were formerly obtained from animals. In 1930, 15% of the ingredients in perfumes were synthetic. Today, synthetics make up as much as 85% of fragrances.

A fragrance is a mixture of oils in 75–95% alcohol. A perfume is more than 22% oil; an eau de parfum contains between 15% and 22%, and a eau de toilette between 8% and 15%. Some colognes are less than 5% oil, and after-shave lotions and splash colognes contain as little as 0.5–2%. Perfumes are included in practically all cosmetics and toiletries, and products such as paints and cleaning materials contain perfumes designed to cover up their unpleasant odors. The scent of leather is often added to plastic furniture coverings to make them more attractive to consumers.

The Blending of Notes

Creating a perfume is an art that is often compared to composing a piece of music. A fine perfume may contain as many as 500 ingredients, which are known as "notes." A perfumer has about 2,000 notes from which to choose, but most of these fall into just a few fragrance families. There are floral fragrances such as rose, violet, and lavender; spicy blends that include nutmeg, clove, and cinnamon; sweet Oriental fragrances like balsam and vanilla, and a woody group that includes sandalwood and cedar. All perfumes contain three basic notes: a top note, a middle note, and a base note. The top note is the volatile scent first released from the skin. It lingers briefly in the air and is usually light, refreshing, and floral. The middle note has a solid character. It lasts for several hours and generally contains spicy, woody, or herbal essences. The base (or dry-down) note, which often contains musk or sandalwood, adds depth to the perfume and lasts for up to one or two days. Perfumes are usually applied to the skin over the pulse points on the wrists, behind the ears, or on the inside of the elbows where the skin is warm and the scent evaporates quickly.

Perfumes are designed for different occasions and to create different moods. They are given enticing names like Pleasures, White Diamond, and Dolce Vita, and their appeal is enhanced by the exotically shaped bottles in which they are presented. Fragrances can be blended to give perfumes that are fresh and light, subtle and elegant, or sensual and romantic. And they are evocative. For example, if your grandmother wore a distinctive perfume, the scent of this perfume will immediately transport you back to her home and evoke memories of childhood.

Perfume is an indulgence, an extravagance, that serves no useful purpose. But a beautiful fragrance is an ephemeral delight that adds a little mystery and magic to our lives.

Reference: Cathy Newman, "Perfume: The Essence of Illusion," *National Geographic* (October 1998), p. 94.

The other ingredients in toothpaste are not essential, but without some of them toothpaste would not be palatable. Nearly all toothpastes contain saccharin as a sweetener and flavorings such as mint and peppermint, which taste pleasant and make the mouth feel fresh. Most toothpastes also contain colorings and thickeners, and some contain antibacterials that kill bacteria other than those in plaque and help control breath odor. Tartar-control toothpastes contain sodium pyrophosphate ($Na_2P_2O_7$), which helps control the formation of tartar.

"Natural" toothpastes contain no perservatives, saccharin, or food colorings (see Table 17.1). They use chalk (powdered limestone, $CaCO_3$) as the abrasive. Other ingredients such as flavorings, sweetener (glycerin), fluoride, and thickener (carrageenan from seaweed) are all obtained from natural sources.

Today, the major cause of tooth loss is not tooth decay but gum disease. Some toothpastes now include peroxide, which releases bubbles of oxygen during brushing. The purpose of the oxygen is to kill bacteria that contribute to gum disease.

Chapter Summary

1. Soap, which is the sodium salt of a long-chain fatty acid, is made by reacting a fat with sodium hydroxide (or lye).
2. A soap is able to remove greasy dirt because its molecules have a nonpolar end, which dissolves in greasy, oily materials, and a polar end, which dissolves in water.
3. In hard water containing Ca^{2+}, Mg^{2+}, or Fe^{2+} ions, soap forms a scaly deposit or scum.
4. Water can be softened by exchanging hard-water ions for Na^+ ions.
5. Synthetic laundry detergents are long-chain alkylbenzene sulfonates. Those with linear hydrocarbon chains are biodegradable and have replaced those with branched chains, which are not.
6. Builders, brighteners, fabric softeners, bleaches, fillers, and various other ingredients are added to laundry detergents.
7. Builders remove hard-water ions. Modern builders include sodium carbonate and zeolites.
8. Optical brighteners make dull-looking white and colored fabrics look brighter by giving them a blue tint.
9. In addition to linear alkylbenzene sulfonates, liquid laundry detergents contain nonionic surfactants, which are very soluble in water.
10. Fabric softeners, which are quaternary ammonium salts (or quats), are cationic surfactants. They become attached to negative charges on fabrics, which makes clothes feel soft.
11. Bleaches are oxidizing agents. Hypochlorite (OCl^-) bleaches release chlorine; hyperborates and percarbonates release oxygen.
12. Hair, which is composed of the protein keratin, is lubricated with sebum stored in the sebaceous glands in the scalp.
13. Shampoos contain the anionic surfactant sodium lauryl sulfate, which acts in the same way as a laundry detergent. They also contain fragrances and ingredients that adjust pH and soften water.

14. Conditioners contain lubricants and quats, which remove dull soap from hair.
15. Hair can be permanently curled by first treating it with a reducing agent, which breaks disulfide bonds in the hair's protein, and then treating it with an oxidizing agent to reform the disulfide bonds in different positions.
16. The color of hair depends on the relative amounts of the pigments melanin (dark brown) and phaeomelanin (red-brown or yellow-brown). Blondes have little of either pigment.
17. To change its color, hair is first bleached with the oxidizing agent hydrogen peroxide and then colored with either a temporary or a permanent dye.
18. Sweat is produced in the eccrine and apocrine glands.
19. Deodorants contain triclosan, a bacteriostat that controls the growth of odor-causing bacteria. Antiperspirants contain aluminum compounds, which reduce sweat production and also help kill bacteria.
20. Bacteria in the mouth cause the formation of plaque and organic acids that dissolve tooth enamel. Plaque combines with calcium and phosphate ions to form tartar.
21. Fluoride ions in drinking water help prevent dental caries.
22. A typical toothpaste contains an abrasive, a detergent, fluoride, a sweetener, coloring, and flavoring.

Key Terms

alkylbenzene sulfonates (ABS) (p. 516)
anionic surfactant (p. 516)
antiperspirant (p. 525)
apocrine gland (p. 525)
bleach (p. 519)
builder (p. 516)
cationic surfactant (p. 518)
deodorant (p. 525)
detergent (p. 516)
eccrine gland (p. 525)
filler (p. 518)
hard water (p. 513)
ion exchange (p. 515)
keratin (p. 521)
linear alkylbenzene sulfonate (LAS) (p. 516)
micelle (p. 512)
optical brightener (p. 517)
plaque (p. 526)
quaternary ammonium salt (p. 518)
saponification (p. 511)
sebum (p. 520)
soap (p. 512)
soft water (p. 513)
surfactant (p. 516)
tartar (p. 526)

Questions and Problems

1. What chemicals present in wood ash were used to make a cleaning agent in ancient times?
2. What organic functional groups are present in each of the following?
   a. triglycerides
   b. glycerol
   c. stearic acid
3. Describe the saponification reaction. What are the reactants and products?
4. What are the sources of the fats used to make soaps today?
5. At what date did soap come into common use for bathing?
6. What base is used to make softer soaps that are used for shaving products?
7. What is a micelle? Draw a picture of one.
8. Does soap work in acidic water? Explain.

9. How does hard water affect soap? What is bathtub scum?
10. Describe how ion exchange is used to soften hard water.
11. How is the ion-exchange resin in a water softener regenerated once all the sites on the resin are occupied by hard-water ions?
12. Compare the acid end of a soap molecule (sodium stearate) with the acid end of a detergent molecule (sodium alkylbenzene sulfonate). Now compare the organic end of each molecule.
13. What environmental problems did alkylbenzene sulfonate (ABS) detergents cause when they were discharged into estuaries and streams?
14. How does a linear alkylbenzene sulfonate (LAS) differ from an alkylbenzene sulfonate (ABS)?
15. What is meant by *anionic surfactant?*
16. What environmental problem do phosphate builders, which are used in laundry detergents, cause when they are released into estuaries and streams?
17. How does the builder sodium carbonate (Na_2CO_3) help make a laundry detergent clean more efficiently?
18. What are zeolites? Why are they replacing phosphates as builders in laundry detergents?
19. What color does a white fabric acquire from an optical brightener?
20. Describe the fluoresence that takes place in an optical brightener that has been absorbed on a white shirt.
21. What is a cationic surfactant?
22. How do fabric softeners work?
23. Why is a filler added to a laundry detergent? Do superconcentrated detergents contain as much filler as conventional detergents do?
24. Why are enzymes added to some laundry detergents?
25. What organic functional groups are present in the nonionic surfactants alcohol ethoxylates?
26. What ion is responsible for the bleaching action in chlorine bleaches such as Chlorox?
27. In the bleaching reaction of chlorine bleaches, what is oxidized? What is reduced?
28. What is the active chemical in oxygen-releasing bleaches?
29. Why is it very dangerous to mix chlorine-containing bleach with other household cleaning products such as drain cleaners that contain hydrochloric acid?
30. What is sebum?
31. What is the most common detergent used in shampoo?
32. What is hair composed of? Is hair acidic or basic?
33. What is the pH range of most shampoos?
34. How does the pH of shampoo affect the glossiness of hair?
35. How do hair conditioners remove soap film from the hair?
36. Why do hot curling irons cause hair to become wavy?
37. How does a mousse or gel produce longer lasting waves in your hair?
38. What chemical bond is responsible for a permanent wave in hair?
39. When a home permanent is applied, several sequential chemical reactions are carried out.
    a. What happens when thioglycolic acid is applied?
    b. What happens when hydrogen peroxide is applied?
40. What is the active chemical that is contained in the relaxers used to straighten very curly hair?
41. What precautions need to be taken when chemical relaxers are used?
42. What natural pigments are responsible for dark hair?
43. What is the composition of temporary hair dyes?
44. Why does hydrogen peroxide cause brown hair to turn blonde?
45. Permanent hair dye penetrates the hair shaft and lasts as long as the hair does. Describe the function of each of the following components of such dyes:

a. primary intermediate
b. secondary intermediate
c. oxidizing agent

46. What is the active ingredient in Grecian Formula hair dye? How does Grecian Formula color hair?

47. What is the origin of the name Grecian Formula?

48. Name two different types of sweat glands. Why does the sweat from armpits have an unpleasant odor?

49. How do deodorants differ from antiperspirants?

50. What is the most common bacteriostat contained in deodorants?

51. What chemicals are used as the drying agents in antiperspirants?

52. What is an astringent?

53. What are the chemical formulas of the abrasives used in toothpastes?

54. Why is fluoride added to toothpaste? How does it work?

55. Compare the ingredients in a typical toothpaste to those of a natural toothpaste.

56. Why is peroxide-containing toothpaste becoming so popular?

Chapter 18

Toxicology

Chapter Objectives

In this chapter, you should gain an understanding of:

- How the LD_{50} test is used to determine toxicity
- Routes by which toxic substances can enter the body
- Corrosive poisons and their effects
- How metabolic poisons cause harm
- Poisoning with toxic metals and with neurotoxins
- How chemicals are detoxified in the liver
- The similarities and differences among teratogens, mutagens, and carcinogens

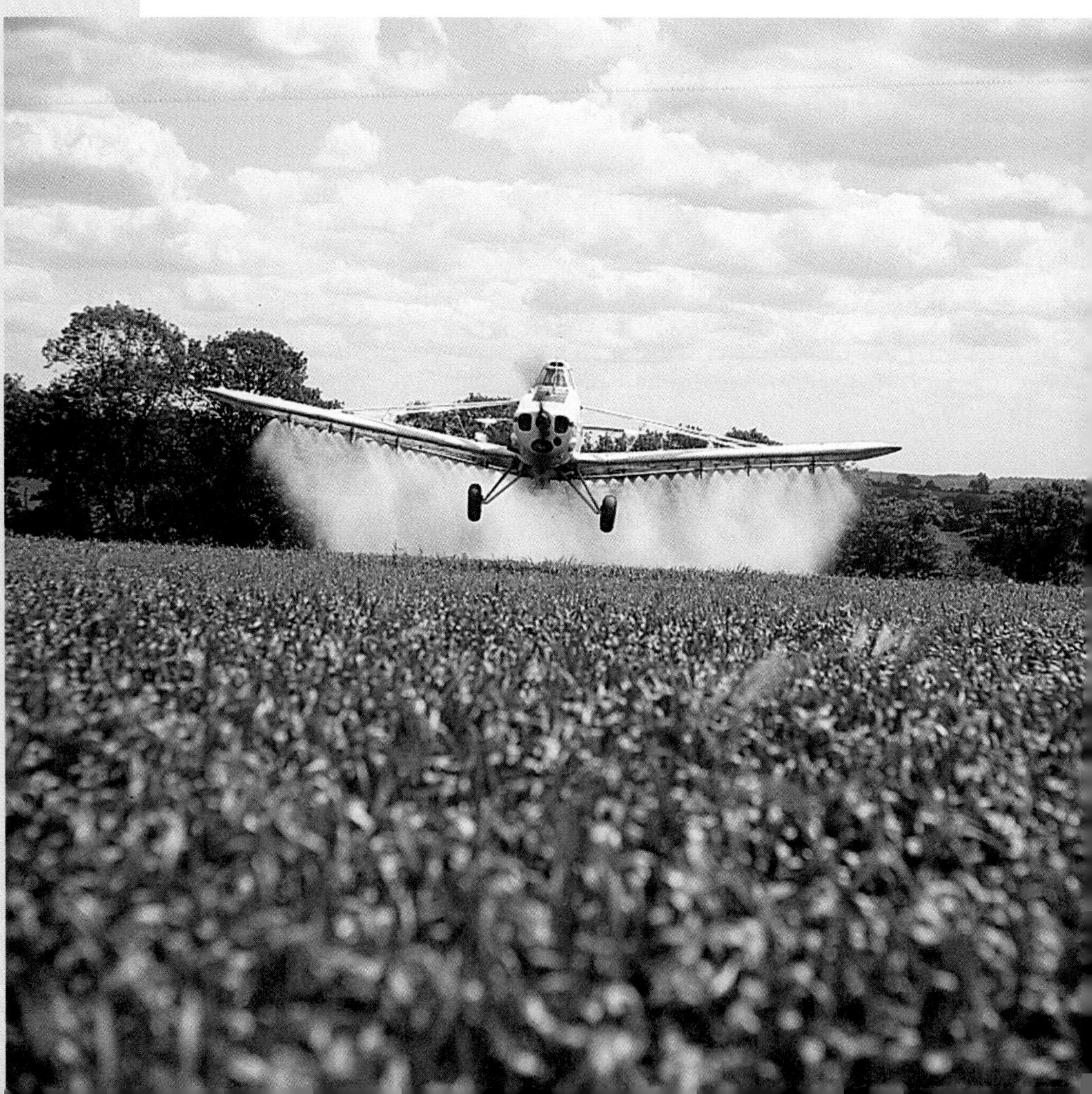

Insecticides sprayed on crops contain toxic chemicals that are particularly poisonous to insects.

ONLINE FEATURES

www.jbpub.com/chemistry

- Chemistry and the Environment
- eLearning
- Chemistry in the News
- Special Topics
- Research & Reference Links
- Ask the Authors

HUMANS HAVE BEEN FAMILIAR with the harmful effects of animal venoms and poisonous plants for thousands of years. This knowledge was used in hunting and in warfare. The earliest known document to give information about poisons is the Ebers papyrus (circa 1500 B.C.), which describes early medical practice in Egypt. This document includes nearly 1000 recipes, many containing recognizable poisons. For example, it mentions hemlock, the poison used in the execution of the Greek philosopher Socrates (470–399 B.C.); aconite, an arrow poison used by the Chinese; opium, used both as a poison and antidote; and plants containing belladonna alkaloids, which can cause cardiac arrest. During the Roman Empire and continuing through the Middle Ages and into the Renaissance, poisoning was a common practice. The Borgia family in Italy removed many of its enemies in this way.

It was not until the beginning of the 19th century that any systematic attempt was made to identify the agents responsible for the toxicity of venoms and poisonous plants. One of the first to identify the chemical makeup of a poison was the French physiologist François Magendie. In 1809, while studying arrow poisons, he isolated the alkaloid strychnine from the plants used to make the poison and showed that it was responsible for the convulsions that victims suffered before they died.

Today, we are concerned with the risk of exposure to toxic substances that may be present in water supplies, the atmosphere, and the workplace. There is particular concern about the possible harmful effects of pesticide residues in foods. In this chapter, we examine the sources and effects of exposure to toxic substances. We focus on what makes certain substances toxic and how the risk associated with each is measured.

18.1 What Is Meant by Toxic?

A substance is said to be *toxic* if it causes harm to a biological system. Only substances that cause harm when present at very low concentration (parts per million or less) are generally described as toxic. Many substances, including common table salt, are harmful at high concentration but are not labeled as toxic. Toxic substances act in many different ways: Some upset metabolism, and others the nervous system; still others cause genetic changes, cancers, or birth defects.

Although the terms *poison* and *toxin* have different meanings, they are often used interchangeably. **Poison** is a general term for any toxic substance, whether synthetic or of natural origin. The term **toxin** refers to poisons of biological origin. Toxins can be present in spoiled foods. Aflatoxins, for example, are an extremely potent group of toxins present in molds that can grow on peanuts and grains while they are in storage. Insect stings and snake bites can also introduce toxins into humans and other animals.

The word *poison* is derived from an Old English word, *poysoun*, meaning a poisonous drink.

Aspirin, the most widely used drug in the world, is an invaluable medicine, but is toxic in large doses. For many years before the introduction of child-proof containers, aspirin caused more cases of accidental poisoning in young children than any other substance. Because the attractively flavored tablets looked like candy, unsupervised children often ingested a toxic quantity of them.

Testing for Toxicity: The LD_{50} Test

The *dose* is the amount of a chemical, per unit of body mass, to which an animal is exposed. The *response* is the effect this chemical exposure has on the animal.

Because any chemical can be harmful if consumed in a large enough quantity, we need a measure of the level at which a particular substance becomes toxic. The toxicity of a chemical is usually tested by determining the **dose** (the amount per unit of body mass) that kills a laboratory animal, usually a rat or mouse.

FIGURE 18.1

Specially bred laboratory rats are used to determine the LD_{50} values of test chemicals.

Since 1920, the standard method for determining the toxicity of a chemical has been the **LD_{50} test** (LD stands for "lethal dose"). In this procedure, different doses of the chemical being tested are fed to large groups of laboratory animals **(Fig. 18.1)**. The result is expressed as the LD_{50} value—the dose that kills 50% of the animals in a group. The LD_{50} values of some common chemicals are listed in **Table 18.1**. Comparing the LD_{50} values allows us to assess the relative toxicity of the chemicals.

The smaller the LD_{50} value, the greater the toxicity of the substance. For example, the LD_{50} for aflatoxin-B fed to rats is 0.009 mg/kg, while that for caffeine is 130 mg/kg. This means that aflatoxin-B is about 10,000 times as toxic to rats as caffeine is.

The LD_{50} values in Table 18.1 are a measure of the toxicity of the listed chemicals to rats or mice. Other animal species may vary in their response to these substances. For example, guinea pigs are 10,000 times as sensitive to dioxin (TCDD) as are dogs. Because of these species differences, LD_{50} values for rats, or other animals, cannot be extrapolated to determine the amount of a chemical that would be lethal to 50% of a human population. Nevertheless, the values for rats are useful in predicting the probable toxicity in humans. A substance that is extremely toxic to animals is likely to be toxic to humans, too.

The method used to introduce a chemical into an animal's body affects the LD_{50} test and must be taken into account. For example, in mice, the LD_{50} for procaine, a local anesthetic, is 800 mg/kg when the drug is injected under the animal's skin, 500 mg/kg when given by mouth, and only 45 mg/kg when injected directly into a vein.

EXAMPLE 18.1 Determining LD_{50} values

Assuming that humans and rats are equally sensitive to the harmful effects of strychnine, how many grams of this chemical would be lethal to 50% of a group of: **a.** 220-lb adults? **b.** 22-lb children?

Table 18.1 LD_{50} Values of Selected Chemicals

Chemical	LD_{50}(mg/kg)*
Sugar	29,700
Ethyl alcohol	14,000
Vinegar	3,310
Sodium chloride	3,000
Malathion (insecticide)	1,200
Aspirin	1,000
Caffeine	130
DDT (insecticide)	100
Arsenic	48
Strychnine	2
Nicotine	1
Aflatoxin-B	0.009
Dioxin (TCDD)	0.001
Botulinum toxin	0.00001

*For rats or mice.

Solution

a. Convert the weight in pounds to kilograms. The conversion factor is 2.2 lb = 1 kg. Therefore, each adult's weight in kilograms is

$$220 \text{ lb} \times \frac{1 \text{ kg}}{2.2 \text{ lb}} = 100 \text{ kg}$$

Find the LD_{50} of strychnine in Table 18.1. It is 2 mg/kg. Therefore, the lethal dose for 50% of 220-lb adults is

$$100 \text{ kg} \times \frac{2 \text{ mg}}{1 \text{ kg}} = 200 \text{ mg} = 0.200 \text{ g}$$

b. Similarly, a 22-lb child weighs 10 kg. The lethal dose for 50% of such children is

$$10 \text{ kg} \times \frac{2 \text{ mg}}{1 \text{ kg}} = 20 \text{ mg} = 0.020 \text{ g}$$

Practice Exercise 18.1

How many grams of arsenic would be lethal to 50% of a group of: **a.** 121-lb adults? **b.** 44-lb children?

EXAMPLE 18.2

The average cup of regular coffee contains 10 mg of caffeine, and the LD_{50} for caffeine (for rats or mice) is 130 mg/kg. Assuming that the LD_{50} for humans is the same as for rats and mice, determine the number of cups of regular coffee that each person in a group of 150-lb adults would have to drink all at one time for the caffeine to be fatal to 50% of the group.

Solution

1. Convert 150 lb to kilograms:

$$150 \text{ lb} \times \frac{1 \text{ kg}}{2.2 \text{ lb}} = 68.2 \text{ kg}$$

2. Calculate the dose that would be lethal for 50% of the adults in the group:

$$68.2 \text{ kg} \times \frac{130 \text{ mg}}{1 \text{ kg}} = 8860 \text{ mg}$$

3. Calculate the number of cups of coffee that would have to be drunk to provide 8860 mg of caffeine:

$$\frac{1 \text{ cup}}{10 \text{ mg}} \times 8860 \text{ mg} = 886 \text{ cups}$$

Practice Exercise 18.2

The LD_{50} for the anti-inflammatory drug ibuprofen is 495 mg/kg. How many 350-mg ibuprofen tablets taken all at one time would cause death for 50% of a group of 174-lb adults?

18.2 Human Exposure to Toxic Substances

Toxic substances in the environment can enter the human body in three main ways: *inhalation* (through the lungs), *ingestion* (through the mouth), or *dermal contact* (through the skin). When a toxic substance enters the body, it may cause harm locally at the site of entry, or it may not exert its effect until it is absorbed into the bloodstream. Once in the bloodstream, a toxic substance is transported to all parts of the body and, depending on its nature, may then disrupt one or more of the body's normal functions.

Toxic gases in the atmosphere, such as carbon monoxide, enter the body primarily by inhalation. Fine particulates that are often present in industrial emissions also reach the lungs via this route. Particulates and gases such as sulfur dioxide and ozone damage lung tissue and cause respiratory problems. Carbon monoxide is absorbed into the bloodstream, where it disrupts the oxygen-carrying capacity of the blood.

Acceptable levels for inhalation of chemicals in the working environment were determined in response to the Occupational Safety and Health Act (OSHA) of 1970.

Toxic substances ingested in food and water pass from the mouth into the digestive tract. Most cause harm after they have been absorbed into the bloodstream through the walls of the intestines.

Toxic chemicals that cause harm through dermal contact are absorbed through the skin into the tissues below and enter the blood that supplies those tissues. Penetration through the skin is time-dependent. The longer the skin is in contact with the chemical, the greater the amount transferred through the skin and into the circulatory system. An individual splashed by a chemical should always try to wash it off immediately with copious amounts of water.

Acute exposure to a chemical is defined as exposure for less than 24 hours. *Chronic exposure* is defined as exposure for more than 3 months.

18.3 Corrosive Poisons

Corrosive poisons are substances that destroy tissues on contact. These include strong acids such as hydrochloric acid (HCl) and sulfuric acid (H_2SO_4), strong bases such as sodium hydroxide (NaOH) and potassium hydroxide (KOH) (Chapter 7), and strong oxidizing agents such as ozone (O_3). A list of some common corrosive poisons is provided in Table 18.2.

Acids and Bases

Many strong acids and bases are found in the home. Hydrochloric acid (also called *muriatic acid*) is often used to clean calcium carbonate deposits from toilet bowls; sulfuric acid is present in automobile batteries; both sulfuric acid and sodium hydroxide are used in drain cleaners because they destroy hair; sodium hydroxide and potassium hydroxide are ingredients in oven cleaners. There have been many tragic results from children swallowing these products. The strong acids and bases in them destroy the tissues in the mouth and throat and cause severe pain. Swallowing as little as 0.5 ounce (15 milliliters) of concentrated (98%) sulfuric acid can be fatal.

Acids and bases destroy tissues by catalyzing the breakdown of peptide bonds in protein molecules in the tissues (Chapter 14). In acids, the catalyst is the hydrogen ion; in bases, it is the hydroxide ion. Acids add to the destruction by dehydrating the tissues.

$$\text{R—C(=O)—N(H)—R}' + H_2O \xrightarrow{H^+ \text{ or } OH^-} \text{R—C(=O)—OH} + \text{H—N(H)—R}'$$

peptide bond

Sulfur dioxide (SO_2), which is released into the atmosphere when sulfur-containing coal is burned (Chapter 11), is another corrosive poison. In the atmosphere, sulfur dioxide is converted to sulfur trioxide (SO_3), which then combines with moisture in the air to form sulfuric acid. If the contaminated air is inhaled, the acid damages lung tissue.

Table 18.2 Corrosive Substances and Their Properties

Name	Effects
Nitric acid (HNO_3)	Reacts with protein in exposed skin to form lesions
Sulfuric acid (H_2SO_4)	Dehydrates protein in skin to form lesions
Hydrochloric acid (HCl)	Splash to the eye causes severe damage, inhalation of HCl gas damages lungs
Hydrofluoric acid (HF)	Produces severe skin burns and causes painful ulcers
Metal hydroxides (NaOH, KOH)	Dissolve tissue and cause severe burns
Halogen gases (Cl_2, Br_2, F_2)	Produce corrosive acids when in contact with moist tissue; inhalation damages lungs
Hydrogen peroxide (H_2O_2)	Oxidizes tissue and causes severe burns

Oxidizing Agents

Other corrosive poisons destroy tissues by oxidation. For example, ozone, a component of photochemical smog (Chapter 11), causes damage to tissues by deactivating enzymes and oxidizing the lipid layers in cell membranes.

18.4 Metabolic Poisons

Metabolic poisons are more specific in their action than corrosive poisons. They cause harm, and frequently death, by interfering with some essential metabolic process in the body. Examples of metabolic poisons are cyanide and carbon monoxide (which was discussed in Chapter 11).

Cyanide

Many of us have watched a movie in which a captured spy being interrogated suddenly pops something into his or her mouth and dies instantly. The spy probably bit down on a cyanide capsule. A dose as small as 90 milligrams (0.003 ounce) can be fatal.

Gaseous hydrogen cyanide (HCN) and cyanide salts such as sodium cyanide (NaCN) have many valuable uses. Hydrogen cyanide is used as a fumigant to destroy rodents and insects in warehouses and grain storage bins. Solutions of sodium and potassium cyanide are used in electroplating baths; metal that is dissolved in the bath is plated onto the surface of an object placed into the bath. This process is used to make gold-plated jewelry. The polymer polyacrylonitrile, which is used to make fibers for carpeting and fabrics (Chapter 9), contains cyanide, and deaths in fires are often caused by hydrogen cyanide being released from these products as they burn.

If cyanide enters the body, it inhibits an enzyme called *cytochrome oxidase*, which is essential for a key step in the process that allows oxygen to be used by the cells. Cyanide binds to the iron(III) in the enzyme and prevents its reduction to iron(II). If this reduction does not occur, oxygen cannot be utilized. Cellular respiration ceases, and death results within minutes.

Iron can form two ions: Fe^{3+}, iron(III) ion, and Fe^{2+}, iron(II) ion. When Fe^{3+} gains an electron, it is reduced to Fe^{2+}.

An antidote for cyanide poisoning is sodium thiosulfate ($Na_2S_2O_3$), which reacts with cyanide as shown in the following equation:

$$\underset{\text{cyanide}}{CN^-} + \underset{\text{thiosulfate}}{S_2O_3^{\ 2-}} \longrightarrow \underset{\text{thiocyanate}}{SCN^-} + \underset{\text{sulfite}}{SO_3^{\ 2-}}$$

The transfer of a sulfur atom from the thiosulfate ion to the cyanide ion results in the formation of the relatively harmless thiocyanate ion. Because cyanide acts so rapidly, however, thiosulfate must be administered almost immediately after the poison is ingested to be effective.

18.5 Toxic Metals

Metals are widely used in our society, and thus there are many opportunities for exposure to those that are toxic. People who live near incinerators and industrial plants are concerned that they may be exposed to fine metal particles released into the atmosphere when materials containing metals are burned. There is also concern that metal-containing wastes produced by industry and released into rivers and lakes may pollute water.

The presence of toxic metals in incinerator ash makes safe disposal of the ash very difficult.

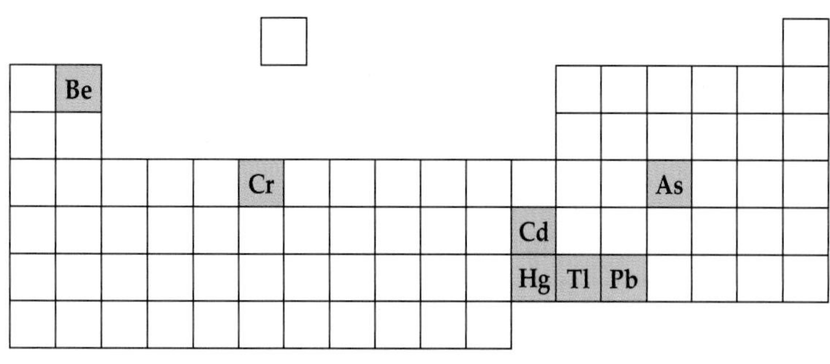

FIGURE 18.2

Periodic table showing the positions of toxic metals, which are dangerous in very small amounts.

Most metals, including those such as iron that are essential for life, are toxic if consumed in large enough quantities. Some metals are toxic in very small amounts. Metals that are of particular concern include beryllium, cadmium, chromium, lead, mercury, thallium, and arsenic, which is not actually a metal but has many metallic properties (Fig. 18.2).

Heavy metals—those found near the bottom of the periodic table—derive their toxicity from their ability to react with sulfhydryl groups (—SH). These groups are present in the active sites of enzyme systems involved in oxygen transport and the cellular production of energy. An atom of a heavy metal such as mercury, for example, reacts by replacing the hydrogen atoms in two sulfhydryl groups (Fig. 18.3): The free sulfhydryl groups are essential for the enzyme's activity; once the bond with the metal is formed, the enzyme becomes inactive.

Arsenic

There are many recorded instances of murderers who used arsenic-containing pesticides to dispose of their victims.

Arsenic is a by-product in the manufacture of certain chemicals and of some mining operations. Many subsoils naturally contain arsenic compounds, and arsenic can seep into groundwater from this source. In parts of India and Bangladesh, where many wells for irrigation were dug during the last 25 years, there have been epidemics of acute arsenic poisoning because of severe contamination of groundwater with naturally occurring arsenic. Most cases of arsenic poisoning occur slowly over many years and are difficult to diagnose. Arsenic, a known carcinogen, greatly increases the risk of bladder cancer. Until recently, in the United States, the limit for arsenic in drinking water was 50 ppb, but, as a result of a 1999 National Academy of Science study, it was lowered to 20 ppb in 2001.

In the past, arsenic compounds were used extensively as pesticides. Because of their toxicity, inorganic arsenic salts containing arsenate ions (AsO_4^{3-}) or arsenite ions (AsO_3^{3-}) were used to destroy rodents, insects, and fungi. In recent years, there has been

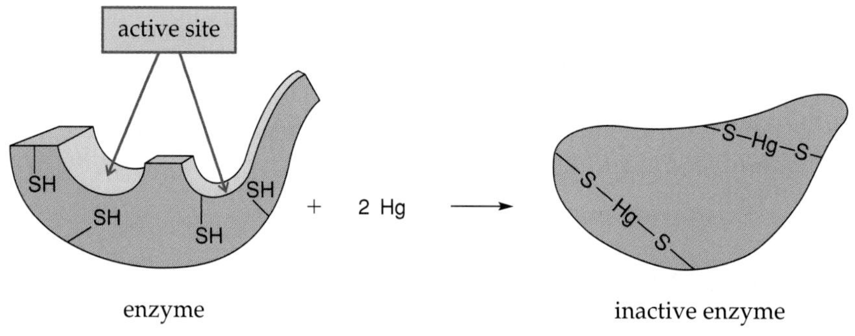

FIGURE 18.3

Heavy metals such as mercury combine with sulfhydryl groups (— SH) at the active sites of certain enzymes. This alters the shape of the active site, which inactivates the enzyme.

a decline in the use of these arsenic compounds as pesticides. The less toxic organic compound arsphenamine was the first drug used successfully to treat syphilis.

Like heavy metals, arsenic compounds inactivate enzymes by reacting with sulfhydryl groups in enzyme systems:

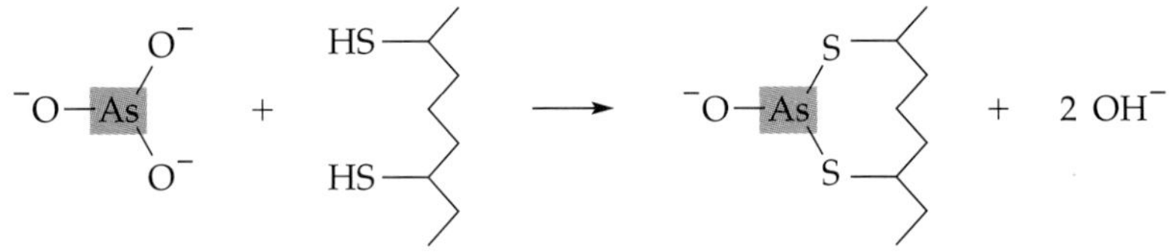

$$^{-}O-As(O^{-})_2 + (HS)_2\text{-enzyme} \longrightarrow {}^{-}O-As(S)_2\text{-enzyme} + 2\ OH^{-}$$

arsenite ion — enzyme active site — inactivated enzyme

During World War I, the U.S. Army was caught offguard when the German army first used "mustard gas" **(Fig. 18.4)**. The word *mustard* refers to the yellow-green color of this gas, which was chlorine (Cl_2), the first chemical agent used in modern warfare. Chlorine is a corrosive poison that forms hydrochloric acid in the wet mucous membranes of the nose and lung. Extensive research to find a "more effective" chemical warfare agent led to the production of an arsenic-containing gas, Lewisite, named for W. Lee Lewis, who discovered it.

Lewisite was never used in warfare.

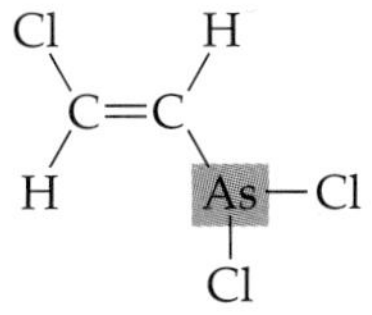

$$ClCH=CH-AsCl_2$$

Lewisite

Anticipating the introduction of Lewisite into the battlefield, British scientists began a search for an antidote to it. Knowing that arsenic reacts with sulfhydryl groups, they synthesized a compound containing these groups. It could compete with enzyme molecules for arsenic and thus prevent the arsenic from disrupting the enzyme system. The compound, British anti-Lewisite (BAL), also proved effective as an antidote for heavy metal poisoning. BAL is a **chelating agent** (from the Greek *chela,* meaning "claw"). It surrounds and binds a heavy metal atom in the same way that a claw can surround and grip an object.

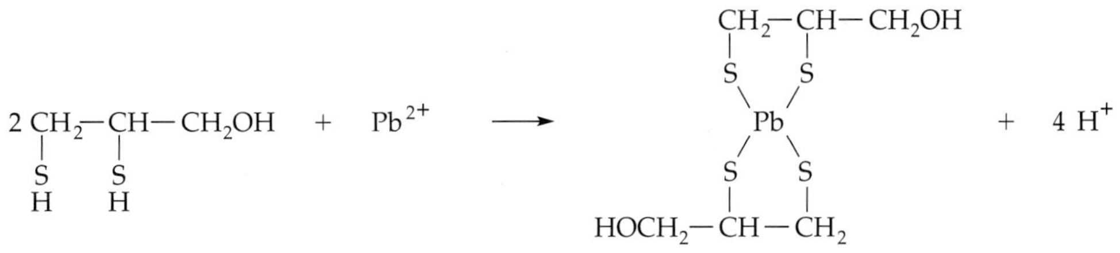

$$2\ CH_2(SH)-CH(SH)-CH_2OH + Pb^{2+} \longrightarrow Pb(S_2C_3H_5OH)_2 + 4\ H^{+}$$

BAL — lead chelated by BAL

Once the metal ions are removed from the enzyme sites, the sulfhydryl groups reform and the enzyme resumes its normal functions.

FIGURE 18.4

During World War I (1914–1918), military personnel were issued gas masks to protect them from toxic gases.

BAL, in combination with another chelating agent, ethylenediaminetetraacetic acid (EDTA), is a standard treatment for heavy metal poisoning.

$$(HOOC{-}CH_2)_2N{-}CH_2CH_2{-}N(CH_2{-}COOH)_2$$

EDTA

The four carboxylic acid groups ($-$COOH) and the two nitrogen atoms in EDTA form bonds with the metal ion. EDTA is administered as the calcium complex because in its uncomplexed form, it could cause problems by combining with calcium ions in the blood. In the body, the calcium in the complex is displaced by the heavy metal, which binds more strongly to EDTA than calcium does. The heavy metal–EDTA complex is water-soluble and is excreted in the urine. To be effective, treatment must begin as soon as possible after ingestion of the metal.

Mercury

Magic properties have been attributed to mercury. For thousands of years, the strange, silvery liquid was used to treat every imaginable ailment.

Mercury is the only metal that is a liquid at room temperature. The metal is used in electric discharge tubes (mercury lamps), pressure gauges in laboratories, and in dental amalgams. Mercury salts are used as fungicides; in the past, they were used in the felt hat industry (Chapter 10).

Inorganic mercury salts are not very toxic. If swallowed, most of the salts are excreted unchanged from the body. Mercury vapor, however, is very toxic. If inhaled, it passes into the bloodstream and is transported to the brain, where it causes serious damage to the nervous system. Symptoms of mercury poisoning include depression, insomnia, irritability, shaking of the hands, and psychotic behavior.

Organic mercury compounds are much more toxic than inorganic salts. The most notorious case of environmental poisoning with mercury occurred in the late 1950s in Japan, when a chemical plant discharged mercury-containing wastes into Minimata Bay (Chapter 10). Aquatic organisms became contaminated with mercury, and a toxic organic mercury compound was formed, which passed up the food chain to small fish and then to larger fish. Through bioaccumulation, the mercury became more concentrated at each level, and more than 100 people who ate fish taken from the bay were poisoned. Forty-four people died, and many children were born with birth defects.

Lead

Acidic foods can leach lead compounds from lead-containing glazes on ceramic dishes.

Of all the toxic metals, lead is the most widely distributed in the environment, and the one most likely to be encountered by the average person. In the United States, lead is used primarily to manufacture batteries for automobiles. It is also used to make plumbing fixtures, pigments, solder, and coverings for cables, and it is present in glazes on some imported ceramic articles. At one time, lead poisoning was common among workers involved in the production of lead or the manufacture of lead products, but now precautions are taken to limit workers' exposure. Today, those at greatest risk are members of the general public, particularly young children, who live in older buildings or in an area close to a battery-recycling plant, a lead smelter, or a factory that is likely to release lead-containing particulates.

Traces of lead are often present in drinking water (Chapter 10). In older buildings that have lead pipes, small amounts of lead are dissolved from the pipes as the water passes through them, particularly if the water is slightly acidic. Lead may also be dissolved from lead solder on copper pipes. As a precaution, hot water from the faucet should not be used for cooking, and cold water should be allowed to run for several minutes first thing in the morning.

Another source of lead is old paint. Although the use of lead-based paints was banned in 1978, children continue to be poisoned because many older buildings contain lead-based paint under layers of newer paint. If the painted surfaces are sanded, lead-containing dust contaminates the air in the building. In deteriorating buildings in poor neighborhoods, where paint is often peeling from the walls or woodwork (Fig. 18.5), lead poisoning is common. Undernourished, often hungry, children frequently develop an appetite for strange things. This condition, which is termed *pica,* leads them to eat paint.

FIGURE 18.5

Paint peeling from a windowsill in an old building.

Lead has been detected in foods in concentrations up to 300 ppb and in municipal water supplies in concentrations up to 100 ppb. These levels do not pose much of a threat because the average adult can excrete 2 mg of lead per day, and most of us do not generally take in this much from food, water, and air. If intake exceeds excretion, however, the excess lead is rapidly transported to bone marrow and then stored in the bones. Some lead remains in the liver and kidneys, but 90% eventually ends up in the bones. Lead disrupts hemoglobin metabolism, causing anemia, and, like mercury, it inhibits enzymes containing sulfhydryl groups, which causes damage to the central nervous system. The concentration of lead in blood is measured to detect lead poisoning. Concentrations as low as 10 mg/dL are considered to be dangerous.

In the 1970s, a concentration of 50 mg/dL of lead in a child's blood was considered acceptable.

A deciliter (dL) is equal to 100 milliliters (mL).

Children are at greater risk than adults because they cannot excrete as much lead. Consequently, they retain a higher percentage of ingested lead than adults do. Also, their growing bones do not absorb lead as rapidly as full-grown bones; the lead remains in the bloodstream longer and has more opportunity to damage developing organs. Exposure to lead can stunt a child's intellectual, behavioral, and physical development. Studies have shown that infants exposed to lead have IQ scores that are 5% lower by age 7 than the scores of unexposed children; they are six times more likely to have reading disabilities and seven times more likely to drop out of school.

18.6 Neurotoxins

Neurotoxins are metabolic poisons that act by disrupting the normal transmission of nerve impulses in the central nervous system. These poisons include the alkaloid coniine in hemlock, chemical warfare agents (nerve gas), atropine, strychnine, nicotine, organophosphate insecticides, and botulinum toxin.

The Transmission of Nerve Impulses

Nerve impulses travel along nerve fibers as electrical signals. To pass from the end of one nerve fiber to receptors on the next nerve fiber, the impulse must cross a small gap, called the **synapse** (Fig. 18.6). When an impulse reaches the end of a nerve fiber, chemicals called **neurotransmitters** are released that allow the impulse to cross the gap and

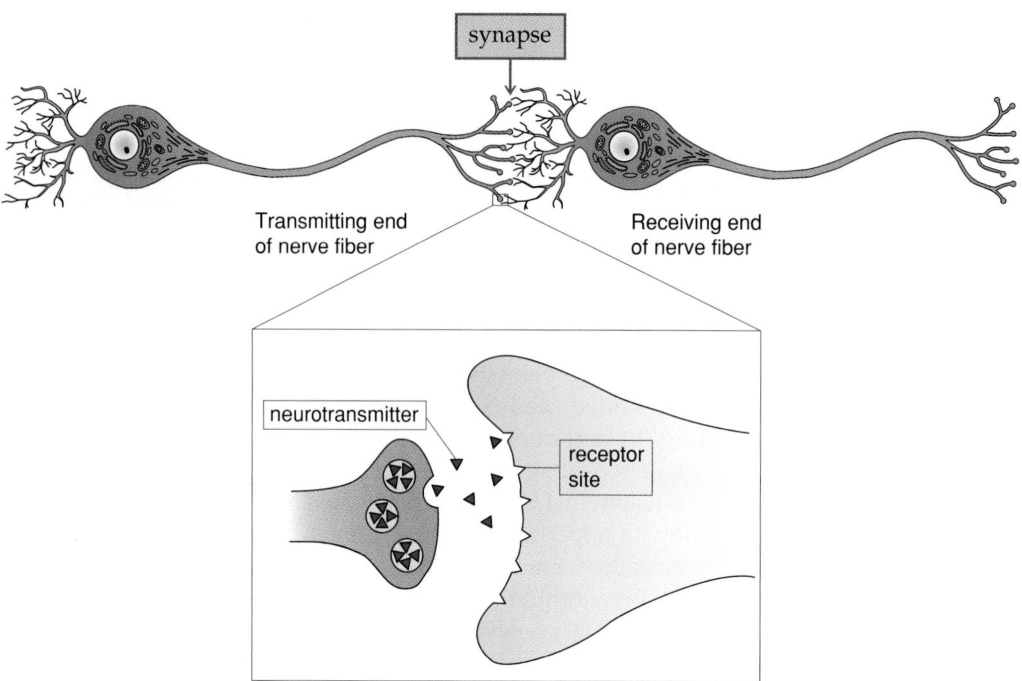

FIGURE 18.6

Transmission of a nerve impulse from one nerve fiber to another occurs when an electrical impulse stimulates the release of neurotransmitter molecules from the transmitter end of one nerve fiber. The neurotransmitter molecules cross the synapse and fit into receptor sites on the receiving end of another nerve fiber.

travel to receptor cells on the receiving nerve fiber. Each neurotransmitter must fit into a specific receptor to bring about the transfer of the message. Once the nerve impulse has been received, the neurotransmitter is destroyed; the synapse is cleared and is ready to receive the next electric signal.

An important neurotransmitter is **acetylcholine.** Once it has mediated the passage of an impulse across the synapse, it is broken down to acetic acid and choline. The reaction is catalyzed by the enzyme cholinesterase.

$$\underset{\text{acetylcholine}}{CH_3\overset{\overset{O}{\|}}{C}OCH_2CH_2-\overset{\overset{CH_3}{|}}{\underset{\underset{CH_3}{|}}{N^+}}-CH_3} + H_2O \xrightarrow{\text{cholinesterase}} \underset{\text{acetic acid}}{CH_3\overset{\overset{O}{\|}}{C}-OH} + \underset{\text{choline}}{HOCH_2CH_2-\overset{\overset{CH_3}{|}}{\underset{\underset{CH_3}{|}}{N^+}}-CH_3}$$

Other enzymes convert acetic acid and choline back to acetylcholine, which can then transmit another impulse across a synapse.

Different neurotoxins disrupt the acetylcholine cycle at different points. They may block the receptor sites, block the synthesis of acetylcholine, or inhibit the enzyme cholinesterase.

Neurotoxins That Block Acetylcholine Receptors

Neurotoxins such as atropine act by occupying the receptor sites on nerve endings that are normally available to the neurotransmitter acetylcholine. By occupying those sites, the neurotoxin stops the transmission of impulses across the synapse. The message is sent but cannot be received. Small amounts of atropine can be used as a local anesthetic. Applied to the skin, the chemical relieves pain by deactivating the nerve endings on the skin.

Another compound that blocks receptor sites is nicotine. Nicotine first stimulates and then depresses the central nervous system by blocking receptor sites. Nicotine is a good example for illustrating how the toxicity of a chemical can depend on the route by which it enters the body. Experiments with mice have shown that nicotine is 1000 times more toxic if it is absorbed into the bloodstream through the skin than if it is administered orally. If a human were to absorb nicotine through the skin, as little as 50 milligrams would be fatal in a few minutes. During smoking, the high temperature of the cigarette oxidizes this potent poison to less toxic products, thus rendering cigarette smoke addictive, but not lethal.

In 1988, the U.S. Surgeon General announced that nicotine was an addictive substance.

Neurotoxins That Block the Synthesis of Acetylcholine

Botulinum toxin is produced in improperly canned foods by the bacterium *Clostridium botulinum*, which grows naturally in soil (Fig. 18.7). There is a very small chance of finding the toxin in commercially canned food; most of the known cases of botulinum poisoning are from home-canned, nonacidic vegetables that were improperly heated. Botulinum toxin, one of the most toxic materials known, binds irreversibly to the nerve endings, preventing the synthesis of acetylcholine. As a result, impulses are not sent to the muscles involved in respiration, and the victim dies from respiratory failure. Fortunately, botulinum poisoning can be avoided by using proper canning techniques including cooking foods at 100°C (212°F) for an adequate time. This treatment kills the bacterium.

Minute quantities of botulinum toxin are used to relieve many muscle disorders, including strabismus (squint eyes).

The toxicity of some mushrooms is due to their ability to disrupt the cholinesterase cycle.

Neurotoxins That Inhibit Cholinesterase

Anticholinesterase poisons prevent the breakdown of acetylcholine by bonding to and inactivating the enzyme cholinesterase. As the acetylcholine builds up, nerve impulses are transmitted in quick succession, and nerves, muscles, and other organs are overstimulated. The heart begins to beat erratically, causing convulsions and death. Anticholinesterase poisons include organophosphate nerve gases and insecticides.

Sarin is one of the most toxic chemicals ever synthesized.

Chemical Warfare Agents At the end of World War I, there was general agreement that poisonous gases should never again be used in warfare. Despite this agreement, many nations continued to develop new and increasingly deadly military poisons.

During World War II, German chemists developed the organophosphate nerve gases Tabun and Sarin. Tabun has a fruity odor, but Sarin, which is four times as toxic as Tabun, is odorless, and thus more difficult to detect. After the war, the U.S. army began manufacturing these poisons (and named them agent GA and agent GB), as well as several related poisons.

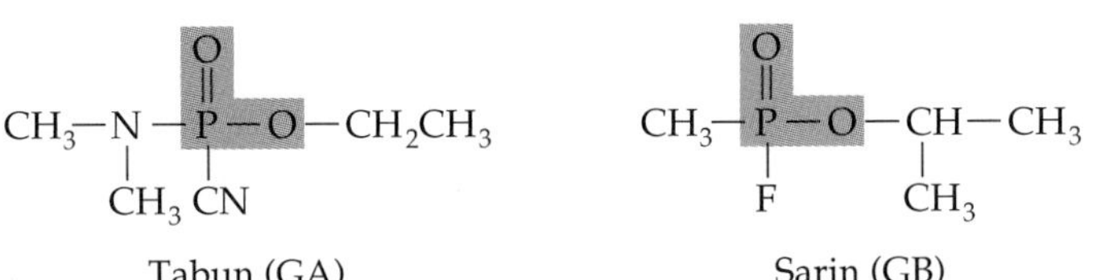

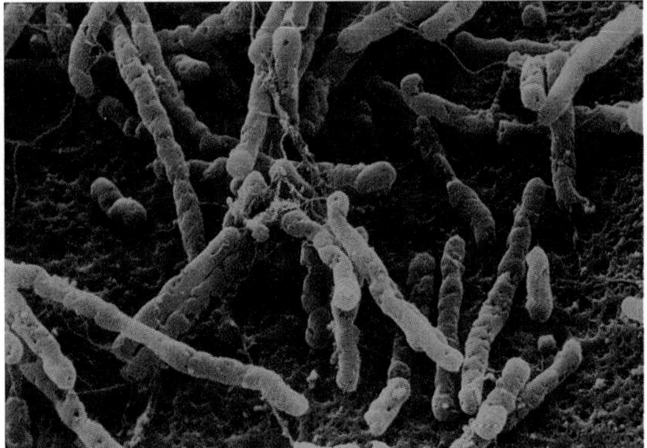

FIGURE 18.7

An electron micrograph of *Clostridium botulinum*, the microorganism that produces botulin, one of the most toxic substances known.

Tabun and Sarin are cholinesterase inhibitors. Both contain an organic group bonded to a $\overset{\overset{\displaystyle O}{\|}}{P}-O$ group. If troops were exposed to the gases, the gases would be absorbed through the skin as well as through the lungs. Victims would lose muscle control and die in a few minutes from suffocation.

ORGANOPHOSPHATE INSECTICIDES The organophosphate insecticides malathion and parathion (Chapter 16) are structurally similar to the nerve gases. If absorbed into the body, these compounds are converted to cholinesterase inhibitors.

$$\underset{\text{parathion}}{CH_3CH_2-O-\overset{\overset{\displaystyle S}{\|}}{\underset{\underset{\displaystyle CH_3CH_2-O}{|}}{P}}-O-C_6H_4-NO_2} \qquad \underset{\text{malathion}}{CH_3-O-\overset{\overset{\displaystyle S}{\|}}{\underset{\underset{\displaystyle CH_3-O}{|}}{P}}-S-\underset{\underset{\displaystyle CH_2-\underset{\underset{\displaystyle O}{\|}}{C}OCH_2CH_3}{|}}{CH}-\overset{\overset{\displaystyle O}{\|}}{C}OCH_2CH_3}$$

FIGURE 18.8

Agricultural workers who apply anticholinesterase insecticides, such as parathion, must wear protective clothing.

Although parathion and malathion are considerably less toxic than Tabun, they must be used with extreme caution. Acute poisoning by these phosphate-based insecticides can cause tremors, convulsions, and cardiac or respiratory failure. Parathion is available for application only by licensed pesticide applicators **(Fig. 18.8)**.

18.7 Detoxification of Chemicals by the Liver

Because the liver has the ability to detoxify (render harmless) many chemicals, humans can tolerate moderate amounts of some poisons. The liver can detoxify a chemical by oxidizing or reducing it or by coupling it to a chemical such as a sugar or an amino acid that is naturally present in the liver.

Oxidation-Reduction Reactions

Ethyl alcohol (ethanol) is detoxified by an oxidation reaction. After an alcoholic beverage is consumed, ethyl alcohol is absorbed into the bloodstream from the intestines and transported to the liver, where enzymes catalyze the oxidation of the alcohol, first into acetaldehyde, which is further oxidized into acetic acid, and then into carbon dioxide and water:

$$\underset{\text{ethyl alcohol}}{CH_3CH_2OH} \longrightarrow \underset{\text{acetaldehyde}}{CH_3\overset{\overset{\displaystyle O}{\|}}{C}-H} \longrightarrow \underset{\text{acetic acid}}{CH_3\overset{\overset{\displaystyle O}{\|}}{C}-OH} \longrightarrow CO_2 + H_2O$$

Alcohol is the most abused drug in the United States; it is calculated that 10 million Americans are alcoholics.

Chronic alcohol consumption causes liver enzymes to build up. The same enzymes that oxidize alcohol also oxidize the male sex hormone testosterone. Alcoholic impotence, a well-known symptom of the disease of alcoholism, is a direct result of the oxidation of testosterone by the high concentration of liver enzymes.

The end product of liver oxidation is not always less toxic than the chemical being oxidized. For instance, methyl alcohol (methanol, CH_3OH) is oxidized to the more toxic chemical formaldehyde (HCHO). Methyl alcohol poisoning is known to cause blindness, respiratory failure, and death. It is not the methyl alcohol itself that causes the problems but its oxidation product, formaldehyde. Ethyl alcohol is often administered intravenously as an antidote for methyl alcohol poisoning and for ethylene glycol (CH_2OHCH_2OH, antifreeze) poisoning. Because ethyl alcohol competes with these other alcohols in the oxidation reactions in the liver, it can be used to slow their conversion to more harmful oxidation products, thus giving the body a chance to excrete them before damage is done.

Coupling Reactions

The cytochrome P-450 enzymes in the liver play an important role in detoxifying poisons. These enzymes act on fat-soluble substances to make them water-soluble and thus more easily excreted. For example, benzene, a component of gasoline, is a toxic aromatic compound that is insoluble in water. If benzene is ingested, it tends to be deposited in fatty tissues in the liver. The cytochrome P-450 enzymes oxidize it to phenol, which is more soluble in water. The phenol then couples to glucuronic acid, a sugar that is naturally present in the liver, to form phenyl glucuronide, which is even more water-soluble than phenol, and is readily excreted in the urine.

OH — O—glucuronide

benzene —oxidation→ phenol —glucuronic acid→ phenyl glucuronide

18.8 Teratogens, Mutagens, and Carcinogens

The toxic chemicals we have discussed so far act quickly and cause harm almost immediately. The harmful effects of other toxic substances are often not apparent until as much as 10 years or more after exposure to them. Toxic substances of this kind include **carcinogens** (chemicals that cause cancer), **teratogens** (chemicals that cause birth defects by damaging embryonic cells), and **mutagens** (chemicals that are capable of altering genes and chromosomes sufficiently to cause inheritable abnormalities in offspring). Some chemicals are both mutagens and carcinogens, while others are only carcinogens.

EXPLORATIONS describes how the EPA analyzes toxic chemicals to determine their cancer-causing potential (see pp. 552–553).

Teratogens

From the 3rd to the 8th week of pregnancy, the embryo is particularly sensitive to the harmful effects of teratogens. During this critical period, the different parts of the body are differentiated, and the limbs, eyes, ears, and the internal organs are developed. If the embryo is exposed to teratogens at this time, abnormalities in development can result. Examples of teratogens are listed in Table 18.3.

The word *teratogen* is derived from the Greek word *teras*, meaning "monster."

Probably the most notorious teratogen is thalidomide, a drug that was prescribed extensively as a tranquilizer and sleeping pill in Europe and Japan in 1960 and 1961. On the basis of animal studies, thalidomide was considered so safe at the time that it was sold in what was then West Germany without a prescription. In 1961, it became

Table 18.3 Teratogenic and Mutagenic Substances

Teratogens	Mutagens
Arsenic (As)	Aflatoxin
Cadmium (Cd)	Benzo(a)pyrene
Cobalt (Co)	Lysergic acid diethylamide (LSD)
Mercury (Hg)	Nitrous acid (HNO_2)
Diethylstilbestrol (DES)	Ozone (O_3)
Polychlorinated biphenyls (PCBs)	Tris(2,3-dibromopropyl) phosphate
Retinoic acid	
Thalidomide	

shockingly apparent that the drug could cause deformities. If it was taken between days 34 and 50 of pregnancy, children were born with no arms or legs or with abnormally short limbs, and with other deformities. Approximately 10,000 children were affected.

thalidomide

Thalidomide is now being used in the United States as an experimental drug for the treatment of AIDS. The drug has also shown promise in the treatment of multiple myeloma.

Thalidomide was never approved for distribution in the United States, thanks to the action of Frances Kelsey, an investigator with the Food and Drug Administration (FDA). Dr. Kelsey was not convinced that the animal testing done by the company that was promoting the drug was complete, and, despite pressure from the company, refused to issue approval. As a result, only 20 "thalidomide babies" were born in the United States.

Another drug that has been shown to be a teratogen is Accutane (retinoic acid), which is used to treat acne. In 1988, the FDA concluded that Accutane could cause heart defects, facial abnormalities, and mental retardation if a developing embryo was exposed to the drug for just a few days. Accutane may have caused birth defects in approximately 1000 children born to women who took the drug between 1982 and 1986.

One of the most dangerous, and most commonly used, teratogens is alcohol. Consumption of even small quantities of alcohol can lead to the development of fetal alcohol syndrome (FAS). Infants with this syndrome are abnormally small, are often mentally retarded, and have facial deformities (**Fig. 18.9**). Cigarette smoking during pregnancy also affects the fetus. Smoking raises a woman's blood levels of carbon monoxide, nicotine, and benzo(a)pyrene (a known carcinogen), and these chemicals can then pass into the developing infant's blood. Since any chemical that can pass across the pla-

FIGURE 18.9

A child whose mother drinks alcohol during pregnancy is usually underweight and may be born with fetal alcohol syndrome, which involves mental retardation and facial abnormalities.

centa from mother to child is a potential teratogen, pregnant women should not drink alcohol or smoke, and they should take only medications prescribed by their doctors.

Mutagens

Mutagens are chemicals that alter the sequence of bases in the nucleic acids that make up DNA, the material that carries genetic traits from one generation to the next (Chapter 14). Changes in DNA may also be brought about by exposure to radiation, and some occur spontaneously. Changes in DNA are called **mutations.** Examples of mutagens are listed in Table 18.3.

All the cells in the body contain chromosomes, threadlike structures composed of DNA and proteins. Genes are sections of DNA. Humans have 46 chromosomes (23 pairs), which contain approximately 30,000 genes **(Fig. 18.10)**. Each gene contains the information required to produce one specific protein. These proteins determine hereditary traits such as eye and hair color and height. They also determine whether an individual has a hereditary disease.

If a chemical substance, or some other agent, causes a mutation, a protein with an incorrect sequence of amino acids may be formed. If the mutation occurs in a body (somatic) cell, very little damage may result, or there may be uncontrolled cell growth, a cancer. If the mutation occurs in a germ cell (egg or sperm), the alteration in the amino acid sequence is passed on to a person's offspring and may result in a child's being born with a hereditary disease such as sickle cell anemia, hemophilia, or cystic fibrosis. The abnormal genes responsible for cystic fibrosis and several other hereditary diseases have been identified, but how they cause the symptoms associated with the diseases is still not understood.

Many chemicals that alter chromosomes and produce mutations in plants, viruses, bacteria, insects, mice, and other animals have been identified. Although there is no conclusive proof that any chemical has caused mutation in human germ cells, there is concern that chemicals that cause mutations in microorganisms and animals may also cause them in humans.

One chemical that is a focus of concern is sodium nitrite ($NaNO_2$), which is used to retard spoilage and to give a pink color to processed meats such as hot dogs, bologna,

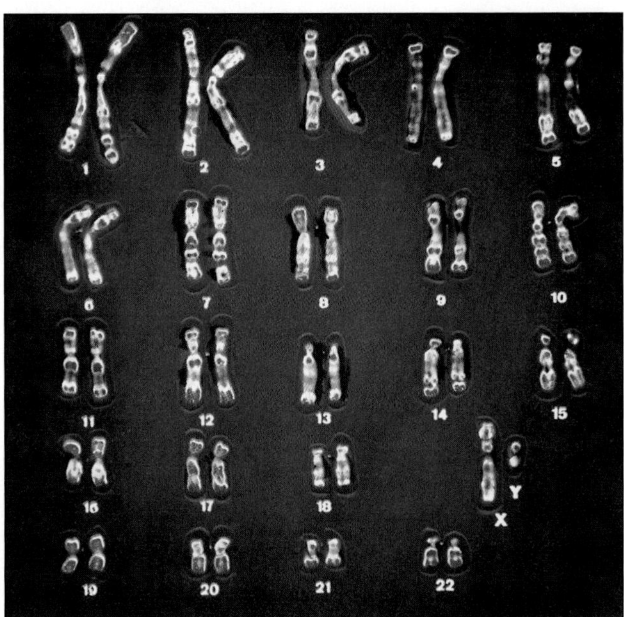

FIGURE 18.10

The 23 pairs of chromosomes of a normal human male. A normal human female would have two X chromosomes instead of an X and a Y chromosome, as shown here.

FIGURE 18.11

Processed meats such as bologna, bacon, and hot dogs are colored pink by the addition of sodium nitrite. Without this additive, the meats would have a gray appearance.

During curing, nitrites are converted to nitric oxide (NO), which reacts with brown pigments in muscle and blood in meats, converting them to pink compounds.

bacon, and smoked ham **(Fig. 18.11)**. When these foods are eaten, sodium nitrite is converted to nitrous acid (HNO_2) by stomach acid:

$$\underset{\text{sodium nitrite}}{NaNO_2} + HCl \longrightarrow \underset{\text{nitrous acid}}{HNO_2} + NaCl$$

Nitrous acid has been shown to be a bacterial mutagen, and as a result, residual levels of nitrites allowed in foods have been decreased fivefold over the last 20 years. In 1996, the American Cancer Society stated that nitrites in food were not a significant cause of cancer.

Another mutagen that has received attention is tris(2,3-dibromopropyl) phosphate, commonly called "tris." It was once used as a flame retardant in children's sleepwear made from synthetic polymer fibers. Tris was an excellent flame retardant, but unfortunately laboratory studies showed that it caused mutations (and also cancer) in animals, and its metabolites were found in the blood of children wearing treated pajamas. The presence of the metabolites indicated that tris is absorbed through the skin, and the U.S. government ordered it removed from the market.

Carcinogens

Chemical carcinogens are compounds that cause cells in an organism to reproduce and grow abnormally to produce a **malignant tumor.** Cells in a malignant tumor may divide more rapidly or more slowly than normal cells do. The growth of the tumor is usually uncontrolled, and tumor cells migrate through the body to other tissues, which they destroy. Tumors are characterized as *benign* if they grow slowly and do not spread to other tissues. Benign tumors, unlike malignant tumors, often regress spontaneously.

CHEMICALS THAT CAUSE CANCER Chemicals that are known to cause cancer vary widely in their structures, and scientists still do not know which chemical structures are likely to make a compound carcinogenic. In 2000, nearly 50 chemicals had been shown conclusively to cause cancer in humans **(Table 18.4)**. Another 166 had been listed as reasonably anticipated to be human carcinogens. As with mutagens, there is concern that any animal carcinogens may also be human carcinogens.

As early as 1775, soot was suspected as the cause of the high incidence of testicular cancer in chimney sweeps in England. Later, it was shown that 3,4-benzo(a)pyrene, a component of soot and coal tar, is a carcinogen.

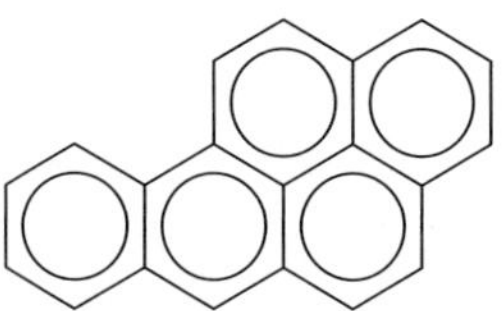

3,4-benzo(a)pyrene

Multiringed compounds such as 3,4-benzo(a)pyrene are called *polycyclic aromatic hydrocarbons* (PAHs; see Chapter 8). PAHs have been studied extensively, and many of them have been shown to be carcinogenic. (There appears to be some correlation between the size and shape of the PAH molecule and carcinogenicity.) Often formed when organic materials are burned, they are present in automobile exhaust, cigarette smoke, charcoal-grilled meat, and stack gases from industrial incinerators.

Table 18.4 Substances Known to be Human Carcinogens

Substance	Source or Use
Aflatoxins	Produced by fungi on peanuts
4-Aminobiphenyl	Used to retard oxidation in rubber
Analgesic mixtures containing phenacetin	Used in over-the-counter pain relievers
Arsenic compounds	Wood preservative, insecticide
Asbestos	Insulation
Azathioprine	Drug developed to prevent organ rejection in kidney transplants
Benzene	Solvent in paint, gasoline additive
Benzidine	Intermediate in the production of dyes
Bis(chloromethyl) ether	Used in the synthesis of certain plastics
1,3-Butadiene	Used in the manufacture of synthetic rubber, and other synthetic compounds
1,4-Butanediol dimethylsulfonate (Myleran, Busulfan)	Chemotherapeutic agent for leukemia
Cadmium and cadmium compounds	Used in making alloys, solders, batteries, photoelectric cells, and in dyeing and electroplating
Chlorambucil	Chemotherapeutic agent for lymphoma
1-(2-chloroethyl)-3-(4-methylcyclohexyl)-1-nitrosourea (MeCCNU)	Chemotherapeutic agent for malignant melanoma and and cancer of the brain, lungs, and digestive tract
Chloromethyl methyl ether	Synthesis of chloromethylated compounds
Chromium hexavalent compounds	Steel manufacture, electroplating, and numerous other industrial uses
Coal tar	By-product in the destructive distillation of coal
Conjugated estrogens	Drug for uterine bleeding
Creosote	Obtained from coal and wood
Cyclophosphamide	Chemotherapeutic drug for leukemia
Cyclosporin A	Immunosuppressive agent
Diethylstilbestrol	Synthetic hormone, used for menopause and menstrual disorders
Dyes that metabolize to benzidine	Direct Black 38 and Direct Blue 6
Erionite	Catalyst used for cracking crude oil
Ethyl alcohol	Beverages
Ethylene oxide	Used as fumigant, fungicide, and sterilizing agent
Melphalan	Chemotherapeutic agent for ovarian cancer
Methoxsalen with UV-A therapy	Used in the treatment of severe psoriasis
Mustard gas	Chemical warfare agent
2-Naphthylamine	Intermediate in the production of dyes
Radon	Radioactive gas emitted from soils containing uranium ores
Silica, crystalline (respirable size)	Used to make abrasives, glass, ceramics, and petroleum products
Solar radiation	Exposure to sun, sunlamps, and sunbeds
Strong sulfuric acid mists	Industrial processes
Tamoxifen	Chemotherapeutic agent for breast cancer
Thorium dioxide	Medium for X-ray imaging
Tobacco, smoke	Smoking and breathing tobacco smoke
Tobacco, smokeless	Chewing tobacco
Tris(1-aziridinyl)phosphine sulfide (Thiotepa)	Insect sterilant
Vinyl chloride	Monomer in the production of vinyl plastic

Source: National Institute of Environmental Health Sciences, "Ninth Annual Report on Carcinogens," 2000.

EXPLORATIONS

Assessing the Risk of Hazardous Chemicals

In the mid-1970s, the Environmental Protection Agency (EPA) began risk analysis to determine the cancer risks associated with toxic chemicals and pesticides. The EPA has a four-step procedure for risk analysis: (1) hazard assessment, (2) dose-response assessment, (3) exposure assessment, and (4) risk characterization. Because the information available to the EPA is often incomplete, the analysis is often imprecise. By constantly reassessing the risks as new information becomes available, the EPA can review and rewrite regulations to protect public health.

Hazard Assessment

The process of *hazard assessment* involves examining evidence and determining whether a potential hazard has caused harm. Often historical data are available and have been used to calculate risks. For example, it has been determined that 3.6 of every 1000 people who smoke a pack of cigarettes a day will die from smoking-related diseases each year. Although there are historical data describing exposure of certain individuals to chemicals, it is very difficult to correlate that information with cancer deaths. Because the onset of cancer usually occurs long after exposure, many other factors affecting the person's health for the last 20 years may be involved.

In cases where the cause-and-effect relationship is obvious, epidemiological studies are used. To establish a link between a particular chemical and a certain cancer, an epidemiological study compares a large number of people who were exposed to the chemical and their cancer rates with a large number of people who were not exposed and their cancer rates. The statistical correlation of this type of information has been used by the National Cancer Institute to label particular chemicals as "known carcinogens."

If no historical data exist, a second way to assess a cancer risk is to test the chemical of interest on animals. For example, before a new artificial sweetener can be marketed, it must be tested on several thousand mice or rats. Rodents have much shorter lifetimes (2–4 years) than humans, so even though they may develop cancers late in their lives, the test takes only 3–5 years. If a significant number of rodents develop tumors after being fed the sweetener, the test indicates that the sweetener may be a carcinogen. The three main criticisms of these studies are (1) large numbers of animals are needed, (2) rodents and humans may respond differently to a particular chemical, and (3) the doses fed to the animals are often very high. Proponents of animal testing point out that all chemicals shown to be human carcinogens are also carcinogenic to animals.

A third way to assess the risk of cancer is to conduct a bacterial screening test. The Ames test uses *Salmonella typhimurium* bacteria to test for mutations. Although the test is simple to perform and inexpensive, it is the least accurate of the three methods for assessing cancer risk. The Ames test assumes that the first step in the development of cancer is a mutation, and, therefore, a chemical that causes a mutation is likely to be a carcinogen as well.

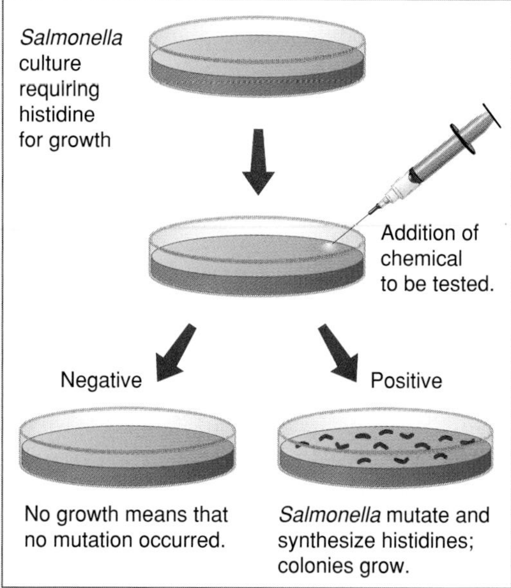
Salmonella culture requiring histidine for growth
Addition of chemical to be tested.
Negative
Positive
No growth means that no mutation occurred.
Salmonella mutate and synthesize histidines; colonies grow.

The Ames test for mutations.

The bacteria that are used in the Ames test are modified biologically so that they are unable to synthesize the amino acid histidine. In the absence of histidine, the bacteria cannot grow. In the test, the modified bacteria are placed in a medium that contains all the ingredients they need to grow, except histidine. The suspected chemical carcinogen is then added to the medium. If the test chemical is not a mutagen, no bacterial growth will occur. But if it is a mutagen, a mutation will occur, and the bacteria will revert to a form that can synthesize histidine, and grow. The growth of bacteria in the test dish identifies the chemical as a mutagen.

Because of the strong correlation between chemicals that cause cancer and those that cause

A Ranking of Hazards According to the Degree of Risk

Hazard	Annual Risk*
Cigarette smoking (1 pk/day)	3.6 per 1000
All cancers	2.8 per 1000
Motor vehicle accident	2.4 per 10,000
Police killed in action	2.2 per 10,000
Air pollution, eastern United States	2.0 per 10,000
Home accidents	1.1 per 10,000
Alcohol, light drinker	2 per 100,000
Radiation, sea level	2 per 100,000
Electrocution	5.3 per 1,000,000
Drinking water containing EPA limit of chloroform	6 per 10,000,000

*Probability of dying.

mutations (about 90% appear to do both), the Ames test is useful for screening for potential chemical carcinogens. If a chemical is identified as a mutagen in the Ames test, it is then tested on animals to confirm its carcinogenicity.

Dose-Response Assessment

The process of *dose-response assessment*, the second step in analyzing risk, begins with establishing a correlation between exposure of animals to a chemical and a harmful effect. Next, the relationship between the dose (concentration of chemical) and the effect is studied. Both the frequency of incidence (the number of times the chemical produces an adverse effect) and the severity of the effect are noted. Once this information is available, statisticians use it to predict the effect of the chemical on human health.

By constantly reassessing the risks as new information becomes available, the EPA can review and rewrite regulations to protect public health.

Exposure Assessment

If the chemical being studied is shown to be harmful, an *exposure assessment* is undertaken. Information on groups of people who have already been exposed, where and how they came to be exposed, the dose to which they were exposed, and the duration of exposure must be collected. Only when this process is completed can the risks be characterized.

Risk Characterization

The information gathered in the first three steps of the risk analysis is compiled, and the risk posed by the chemical in question is calculated. This step is called *risk characterization,* and the risk is expressed as a probability. The 1990 Clean Air Act regulates chemicals that pose a cancer risk greater than 1 in 1 million people who are exposed to high doses. The same standard is used by the FDA for regulating chemicals in food and drugs. Because it is difficult to express this information, risks are often presented as a reduction in life expectancy or the increased probability of dying, as shown in the table.

Reference: R. Wilson and E. A. Crouch, "Risk Assessment and Comparisons: An Introduction," *Science,* 236 (1987): 267.

Other compounds that were studied because they were suspected of causing cancer in humans were the aromatic amines. In 1895, a German physician noted a number of cases of bladder cancer among workers in a dye-making factory. In 1937, it was shown that 2-naphthylamine, one of the amines to which the workers were exposed, causes bladder cancer in dogs. Since then, related compounds, including benzidine, which is an intermediate in the manufacture of magenta-colored dyes, have been shown to cause bladder cancers in animals.

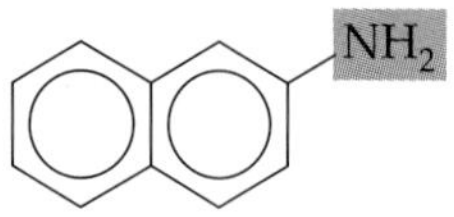

NH_2 NH_2 NH_2

2-naphthylamine 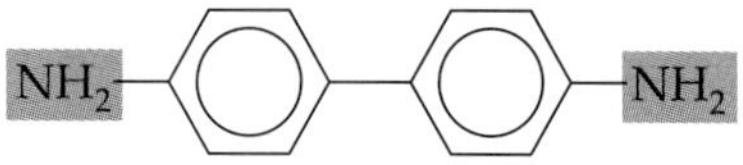 benzidine

FD&C stands for "Food, Drug and Cosmetic." A chemical identified by an FD&C number has been approved by the FDA for use in food, drugs, or cosmetics.

Other dyes that have received attention are the azo dyes, several of which have been found to be carcinogens. The structures of these compounds include two benzene rings joined by a $-N{=}N-$ group. Azo dyes have been used for many years to color foods (Chapter 15). Recently, a number of them, including FD&C Yellow No. 3, were banned because they can react with stomach acid to produce known carcinogens.

Not all carcinogens are aromatic compounds. The nonaromatic organic compound vinyl chloride ($CH_2{=}CHCl$), which is used in huge quantities to make the polymer poly(vinyl chloride) (Chapter 9), is a well known-carcinogen. Its carcinogenicity was confirmed after it was suspected of causing a rare form of liver cancer in workers who were exposed to large quantities of it in the atmosphere in their workplace.

Inhalation of very fine asbestos fibers causes lung cancer and various respiratory diseases (Chapter 2). Certain metals and metal compounds are also carcinogenic. Metals that can cause cancer in animals include beryllium, chromium, lead, nickel, and titanium.

Carcinogens are present in the natural environment in many plants and microorganisms. Aflatoxin-B, which causes liver cancer in rats, is present in molds that grow on grains. Food items that contain carcinogens include certain mushrooms, pepper, mustard, celery, and citrus oils. It is generally accepted that the concentration of carcinogens in these foods is too low to be of concern (Chapter 15).

The Development of Cancer Metabolic poisons and neurotoxins produce effects almost immediately on entering the body, but carcinogens behave differently. Many years may elapse before a person exposed to a carcinogen shows signs of cancer. For example, poly(vinyl chloride) was first manufactured in the 1940s, but it was not until about 25 years later that liver cancer began to appear in workers who had been involved in its production. Similarly, workers exposed to asbestos and to benzidine did not develop cancers until 20 years after exposure.

Some chemicals cause cancer at the point where they make contact with the body; for example, inhaled carcinogens in cigarette smoke and inhaled asbestos fibers cause lung cancers. Other chemicals cause cancer in an organ that is distant from their entry point into the body. The liver, which receives most of the toxic chemicals that enter the body, and the colon, where solid wastes collect, are particularly susceptible to carcinogens.

How cancers develop is still not understood. The pathways leading to the growth of a tumor are complex, and there is evidence that different carcinogens produce cancers in different ways. It is generally agreed that *carcinogenesis* is a two-step process. In the first step, which is called *initiation*, a reaction occurs between the carcinogen and the DNA in a cell, and, as a result, an abnormal cell is formed. The abnormal cell has the potential to grow into a malignant tumor. It may start to divide and grow into a tumor immediately after it is formed, but more often it continues its normal functions and does not proliferate until a second stage, called *promotion*, occurs months or years later. Promotion is thought to occur in response to exposure to a toxic agent (the promoter). The abnormal cell becomes a cancerous cell and develops into a tumor.

$$\textbf{Step 1} \quad \text{normal cell} \xrightarrow{\text{initiation}} \text{abnormal cell}$$

$$\textbf{Step 2} \quad \text{abnormal cell} \xrightarrow{\text{promotion}} \text{cancer}$$

The World Health Organization has estimated that lifestyle and environmental factors are responsible for the development of up to 90% of all cancers. About 10% are thought to be caused by inherited genetic traits. Cigarette smoking is blamed for 40% of cancers, diet for 25–30%, occupational exposure for 10–15%, and environmental pollutants for 5–10%.

Cancer strikes almost a million Americans each year, and one-third of those now living will eventually develop some type of cancer. As the life expectancy of the average person continues to rise above 70 years, the number of deaths due to cancer will increase.

Fortunately, in the United States, about half the people who develop cancer are cured. Many people could reduce the risk of developing cancer by making lifestyle changes.

It is estimated that 150,000 deaths per year in the United States are related to cigarette smoking.

Chapter Summary

1. The LD_{50} test is used to determine the toxicity of a chemical.
2. The toxicity of a substance varies, depending on the route by which it enters the body.
3. Corrosive poisons destroy tissues on contact. They include strong acids and bases and strong oxidizing agents.
4. Cyanide and carbon monoxide are metabolic poisons. Cyanide inhibits vital enzymes.
5. Arsenic and heavy metals such as mercury and lead inactivate enzymes by tying up sulfhydryl groups (—SH) in enzymes.
6. Organic mercury compounds bioaccumulate toward the top of food chains and are much more toxic than inorganic mercury compounds.
7. Very low levels of lead in blood in children cause mental retardation.
8. Neurotoxins act by disrupting the transmission of nerve impulses in the central nervous system.
9. Atropine and nicotine block receptor sites on nerve endings that are normally occupied by the neurotransmitter acetylcholine.
10. Botulinum toxin prevents the synthesis of acetylcholine.
11. Chemical warfare agents and organophosphate insecticides inhibit the enzyme cholinesterase.
12. The liver detoxifies a chemical by oxidizing or reducing it or by coupling it to another chemical to form a less toxic product.
13. The toxic effects of teratogens, mutagens, and carcinogens take many years to become apparent.

14. Teratogens cause birth defects. Examples are thalidomide (prescribed as a tranquilizer) and alcohol.
15. Mutagens alter the sequence of bases in DNA and can cause hereditary diseases and cancer.
16. Carcinogens cause cells in an organism to grow abnormally, resulting in a malignant growth. Nearly 50 chemicals have been identified as definite human carcinogens.
17. Cancer is believed to develop in two stages: initiation (the formation of an abnormal cell) and promotion (proliferation of the abnormal cell to form a tumor).
18. Lifestyle and environmental factors are responsible for the development of most cancers

Key Terms

acetylcholine (p. 544)	dose (p. 535)	mutation (p. 549)	synapse (p. 543)
carcinogen (p. 547)	LD_{50} test (p. 536)	neurotoxin (p. 543)	teratogen (p. 547)
chelating agent (p. 541)	malignant tumor (p. 550)	neurotransmitter (p. 543)	toxin (p. 535)
corrosive poison (p. 538)	mutagen (p. 547)	poison (p. 535)	

Questions and Problems

1. Define:
   a. toxin **b.** poison **c.** LD_{50} **d.** dose
2. Is ethanol (grain alcohol) poisonous? Is methyl alcohol poisonous? Explain.
3. Does the toxicity of a chemical depend on the amount administered to the individual?
4. What are the three main ways a chemical can enter the body? Can you think of another way?
5. A pain relief medication contains 500 mg of acetaminophen per tablet. Assume that the LD_{50} of 338 mg/kg for mice applies to humans as well. How many tablets, taken all at once, would produce a 50% chance of a lethal dose of acetaminophen in a 154-lb (70-kg) person?
6. An anti-inflammatory preparation contains 350 mg of ibuprofen per tablet. Assume that the LD_{50} of 1050 mg/kg for rats applies to humans as well. How many tablets, taken all at once, would produce a 50% chance of a lethal dose in a 44-lb (20-kg) child?
7. The herbicide Paraquat was once sprayed on marijuana plants by the U.S. government to kill the plants so that they could not be sold. Paraquat has an LD_{50} of 100 mg/kg for rats. Another herbicide, Silvex, has an LD_{50} of 650 mg/kg in rats. Is Paraquat more toxic to humans than Silvex?
8. Dioxin (TCDD) was a toxic by-product formed in the production of Agent Orange during the Vietnam War. Dioxin has an LD_{50} of 0.022 mg/kg in rats. How much dioxin would be lethal to a 154-lb (70-kg) person?
9. Describe how nitric acid can act as a corrosive poison.
10. Write a reaction that shows the effect of sodium hydroxide on protein. Is sodium hydroxide a corrosive poison?
11. How does cyanide act as a poison?
12. Why was the gas warfare agent used during World War I called "mustard gas"?
13. Give two causes of mercury poisoning.
14. In the past, arsenic compounds were used for many purposes. List three of these uses.
15. Which toxic metal is the most widely distributed in the environment?
16. Mercury is a silvery liquid that has been used for many applications. List three uses for mercury.
17. What is the mechanism of mercury poisoning?
18. How does BAL act as an antidote for mercury poisoning?
19. How does EDTA act as an antidote for lead poisoning?

20. What is pica?

21. Why are children at greater risk of lead poisoning than adults are?

22. List three different ways children can accidentally ingest lead.

23. What are the symptoms of lead poisoning?

24. In which of the following is botulinum toxin most likely to be found: commercially canned food, home-canned nonacidic vegetables, home-canned acidic vegetables?

25. What is a neurotoxin?

26. Neurotoxins can interfere with the action of the neurotransmitter acetylcholine. List three different ways in which this can occur.

27. Nerve gases are anti-cholinesterase poisons. Name two nerve gases.

28. How do nerve gases act on cholinesterase?

29. What is a synapse?

30. Why is atropine an antidote for nerve gas poisoning?

31. What do the molecular structures of the nerve gases Sarin and Tabun have in common with the molecular structure of the insecticide malathion?

32. The liver removes toxins from our bodies. How is ethanol removed from the body?

33. Why is ethyl alcohol used as an antidote for methyl alcohol poisoning?

34. Where are P-450 cytochrome enzymes found? What function do they serve?

35. List three problems that occur in infants with fetal alcohol syndrome (FAS).

36. How do the P-450 enzymes participate in detoxifying water-insoluble poisons?

37. Why is there concern about using sodium nitrite as a food additive?

38. How many chromosomes do humans have?

39. List the following in order of size, starting with the largest: DNA, gene, chromosome.

40. Explain the difference between a teratogen and a mutagen.

41. Explain the difference between a mutagen and a carcinogen.

42. What are the differences between benign and malignant tumors?

43. What are some sources of polycyclic aromatic hydrocarbons (PAHs)?

44. What are the two general steps in the process of cancer development?

45. Describe what is involved in an epidemiological study.

46. What two organs are particularly susceptible to the development of cancer?

47. Describe how a dose-response assessment is carried out.

48. Describe how the EPA performs a hazard assessment.

49. Describe how the Ames test determines if a chemical is a mutagen.

50. Are all carcinogens also mutagens?

51. Many government agencies that regulate chemicals calculate a "risk characterization" for each chemical. List the steps the agency takes to do a risk characterization.

52. The city of New Orleans takes its drinking water from the Mississippi River. It has been shown that the water in the Mississippi contains many carcinogens at low levels (parts per billion). Why isn't the cancer rate in New Orleans much higher than that for the rest of the country?

Answers to Practice Exercises

18.1 **a.** 2.64 g **b.** 0.96 g

18.2 112 tablets

The Disposal of Dangerous Wastes

Chapter Objectives

In this chapter, you should gain an understanding of:

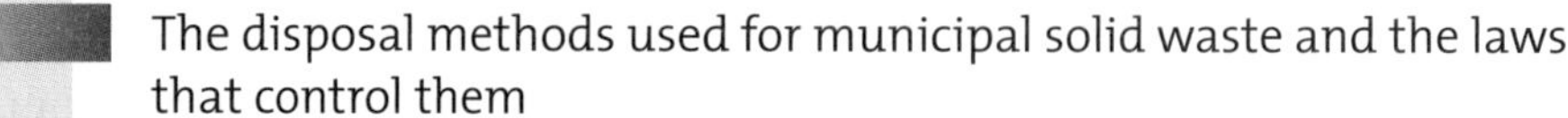

- The disposal methods used for municipal solid waste and the laws that control them
- The categorization of hazardous wastes and the policies that control their disposal
- The sources, classification, and regulation of radioactive wastes

Some radioactive wastes take thousands of years to decay to safe levels. This deep-underground repository at Yucca Mountain, Nevada, is intended for permanent storage of these wastes.

INDUSTRIAL SOCIETIES GENERATE enormous quantities of wastes, and the United States produces more waste per person than any other country in the world. Americans discard products that are worn out and items they no longer want; they throw out food, paper, empty containers, and yard waste. Industrial processes produce chemical wastes; many of these contain toxic substances. The nation's nuclear weapons program and the nuclear power industry generate wastes that are radioactive. Safe disposal of these wastes is one of the most serious environmental problems facing the United States today.

ONLINE FEATURES

www.jbpub.com/chemistry

- Chemistry and the Environment
- eLearning
- Chemistry in the News
- Special Topics
- Research & Reference Links
- Ask the Authors

Until legislation was enacted in the 1960s and 1970s to regulate waste disposal and protect the air and water from contamination, wastes of all kinds were discarded with little concern for human health or the environment. Hazardous wastes were incinerated, contaminating the atmosphere with toxic emissions. They were discarded in sewers, rivers, and streams, in leaking landfills, in abandoned buildings and mines, or they were simply dumped on vacant lots and along roadways. Many people were exposed to polluted air and contaminated water; today, thousands of sites in the United States remain severely contaminated with dangerous chemicals.

For purposes of management and disposal, wastes are divided into three categories: (1) solid waste, (2) hazardous waste, and (3) radioactive waste. In this chapter, we study these kinds of wastes and discuss the advantages and disadvantages of the disposal methods that have been recommended for each category. We examine the legislation that has been enacted to regulate the management and disposal of wastes being generated today and to clean up wastes produced in the past.

19.1 Careless Waste Disposal in the Past

Numerous examples of irresponsible, often criminal, instances of waste disposal can be cited. For example, in 1969, oil and other flammable, greasy wastes floating on the surface of the Cuyahoga River, which flows through Cleveland, Ohio, actually caught on fire and destroyed seven bridges. In Toone and Teague, Tennessee, groundwater cannot be used for drinking because it is contaminated with pesticide wastes, which are leaking from 350,000 drums that were dumped in a landfill years ago. Between 1952 and 1972, about 130 million liters (34 million gallons) of chemical wastes, including by-products from DDT manufacture, were dumped at the Stringfellow Acid Pits, a 22-acre site near Riverside, California. Contaminated liquid leaking from this site continues to migrate and threaten groundwater supplies. Many of 17,000 waste drums located at a site in Kentucky known as the "Valley of the Drums" (**Fig. 19.1**) have leaked, contaminating soil and nearby water. The huge 560-square-mile Hanford weapons complex in Washington has over 1000 separate sites where radioactive and hazardous wastes have been stored or dumped. Contamination at the complex is so extensive that complete cleanup may never be possible.

EXPLORATIONS tells the story of the chemical waste disaster at Love Canal (see pp. 578–579).

FIGURE 19.1

At a site in Kentucky known as the "Valley of the Drums," dangerous chemicals are leaking from thousands of drums and contaminating the soil and nearby water supplies.

The most notorious case of hazardous waste dumping in the United States occurred at Love Canal in Niagara Falls, New York. In the winter of 1978, toxic chemical wastes buried years before began oozing from the ground, forcing the evacuation of many of the town's residents.

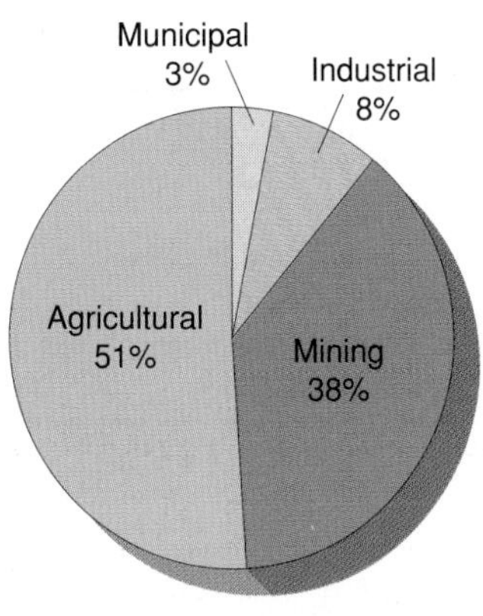

FIGURE 19.2

In the United States, agriculture is the major source of solid waste (51% of the total). Mining accounts for 38%, and industry for 8%. Municipal waste, which is produced by commercial operations, homes, schools, and other community activities, accounts for 3%.

The EPA estimates that, in 1996, the United States produced more than 200 million tons of municipal waste.

19.2 Definition of Solid Waste

According to the Environmental Protection Agency (EPA), **solid waste** includes garbage, refuse, sludge from waste-treatment plants, wastes from air pollution control facilities, and any other discarded material not excluded by regulation. Although called "solid," these wastes may actually contain solid, liquid, semisolid, and even gaseous materials. Agricultural, mining, industrial, and commercial operations and community activities all produce solid waste that must be disposed of (Fig. 19.2). Most agricultural wastes, which consist primarily of manure and crop residues, are plowed back into the land. Mining wastes that are produced when ores are processed are generally left at the mine site (Fig. 19.3). Nonhazardous industrial wastes, municipal refuse generated by community activities, and sludge from sewage-treatment plants that is not composted into fertilizer, are usually disposed of in landfills or incinerated. Increasingly, some industrial and municipal waste is being recycled.

19.3 Disposal of Municipal Solid Waste (MSW)

Municipal solid waste (MSW)—more commonly referred to as "trash" or "garbage"—is the solid waste collected locally from homes, institutions, and commercial establishments. It includes food wastes, paper, glass, metal cans, plastic containers, yard clippings, and various other discarded household items. The composition varies depending on whether it is commercial or residential waste and whether it comes from an affluent or a poor neighborhood. Typically, MSW collected in the United States has the composition shown in Figure 19.4. It also includes discarded items such as furniture, household appliances, automobiles, and tires. Americans generate approximately 1 ton (2000 pounds) of municipal waste per person per year, far more than is generated by most other industrialized nations (Fig. 19.5).

Regulation of MSW Disposal

Until the 1960s, most MSW was trucked to open dumps, where it was discarded and often burned (Fig. 19.6). The dumps became breeding grounds for rats, flies, and other pests, and the burning trash polluted the air with smoke and unpleasant odors. Since the passage of the Solid Waste Disposal Act in 1965, disposal of MSW has been subject to increasingly strict controls. Particularly important was the enactment in 1976 of the Resource Conservation and Recovery Act (RCRA, pronounced "rickra"). The section of the law dealing with solid waste required that all existing open dumps in the United States be closed or upgraded; it banned the creation of new open dumps and set tough standards for landfills. In an effort to reduce the volume of waste generated, the law stressed the need for reuse and regeneration. Federal procurement agencies were directed to make use of recycled materials wherever possible, and the Department of Energy (DOE) was instructed to conduct research to develop ways of converting solid wastes into energy. Since 1983, the EPA has required that new landfills meet even stricter standards. Today, about 65% of all MSW is discarded in regulated landfills. The remainder is recycled, incinerated, or composted.

FIGURE 19.3

Waste rock produced in mining operations is usually left in large piles at the mine site, as shown here for coppermine tailings.

Landfills

The **sanitary landfills** in use prior to 1983 consisted of a trench excavated in the ground in which trash was dumped and then compacted by bulldozers. Each day's load of refuse was covered with a layer of soil. Covering the waste minimizes access by vermin and

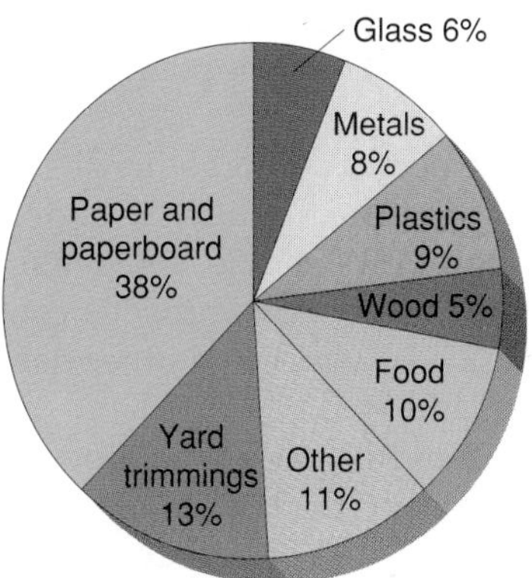

FIGURE 19.4

The composition of typical municipal solid waste by weight. Percentages vary, depending on the affluence of the neighborhood and whether it is primarily residential or commercial. At certain times of the year, yard wastes (leaves, branches, grass clippings) account for more than 25% of the total waste.

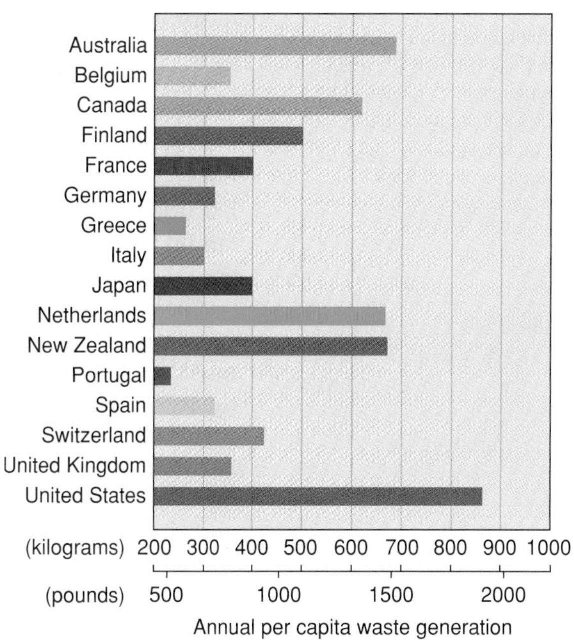

FIGURE 19.5

The United States produces more waste per capita than any other industrialized nation.

FIGURE 19.6

In the past, municipal waste was often burned in open dumpsites. The burning trash produced unpleasant odors and polluted the atmosphere near the dumpsite.

Little biodegradation occurs at the center of a landfill. Whole hot dogs and readable newspapers have been found in 40-year-old landfills.

Because most landfills do not collect the methane (CH_4) they emit, landfills are responsible for one-third of the methane emissions in the United States.

In 1992, despite incentives such as jobs, tax breaks, and economic development, several rural communities in New York State rejected a waste management company's plan to build a regional landfill.

the problem of odors. When completely filled in, such landfills are not suitable for building because the ground continues to settle for many years, but they can be planted with grass and made into attractive recreation areas.

In sanitary landfills, there is always the danger that any metal salts and other chemicals in the liquid that leaches from rotting garbage might percolate through the underlying soil and contaminate nearby groundwater. Although MSW should not contain hazardous materials, many householders often increase the chance of groundwater contamination by discarding unused pesticides, paint, cleaning solvents, and other dangerous wastes together with their nonhazardous refuse (**Table 19.1**).

The EPA regulations introduced in 1983 require that all landfills meet the following standards: They must be located well above the water table, and the bottom of the fill must be covered with a layer of impermeable clay or a plastic liner or both to contain the leachate (the fluid that seeps through the trash). The landfill must be fitted with a system for collecting this leachate and treating it, and groundwater in the vicinity must be monitored regularly. The methane produced by the anaerobic decomposition of the organic waste must be vented or, preferably, collected and used as fuel for generating electricity. After each day's delivery of trash, the trash must be covered with a layer of dirt to minimize odors and discourage rodents. When filled, the site must be covered with a layer of clay and then topsoil (**Fig. 19.7**).

Municipal waste disposal has become a major problem for many big cities in the United States. Not only will numerous existing landfills be filled to capacity within the next few years, but because of high costs, many municipalities have chosen to close their landfills rather than upgrade them to meet the new standards. Some municipal authorities are turning to incineration and recycling as alternatives. Others are trucking their MSW across state borders to state-of-the-art mega-fills that, since the beginning of the 1990s, have opened in many rural areas. Although communities receiving the waste have benefited economically, many citizens resent having other people's trash in their backyard.

Table 19.1 Hazardous Materials Often Found in Household Trash

Products	Hazardous Materials
Pesticides	
Insect sprays	Insecticides, organic solvents
Other pesticides	Insecticides, herbicides
Paint	
Oil-based	Organic solvents
Latex	Organic polymers
Automotive products	
Gasoline	Organic solvents
Motor oil	Organic compounds, metals
Antifreeze	Organic solvents, metals
Batteries	Lead, sulfuric acid
Miscellaneous	
Mercury batteries	Mercury
Nail-polish remover	Organic solvents
Moth balls	*p*-Dichlorobenzene

Incineration

At the present time, about 15% of the over 200 million tons of MSW produced annually in the United States is incinerated. Incineration of MSW has a number of advantages. The volume of the waste is reduced by approximately 85%, the high temperature kills disease-causing organisms, there is no risk of groundwater pollution, and the heat produced can be used as a source of energy. At some facilities, ferrous metals are removed from the trash before combustion, and other valuable metals, including aluminum, are recovered from the noncombustible ash.

The ash remaining after combustion must be tested. If it contains no toxic metals or other hazardous materials, it is disposed of in a landfill. If the ash fails this test, it must be managed as hazardous waste, which is much more expensive (see below). The metals most likely to be in the ash are lead and cadmium.

The main disadvantage of incinerating MSW is its potential for polluting the atmosphere with toxic chemicals, particularly metals. When discarded items such as cans, jar lids, and batteries are incinerated, many toxic metals, includ-

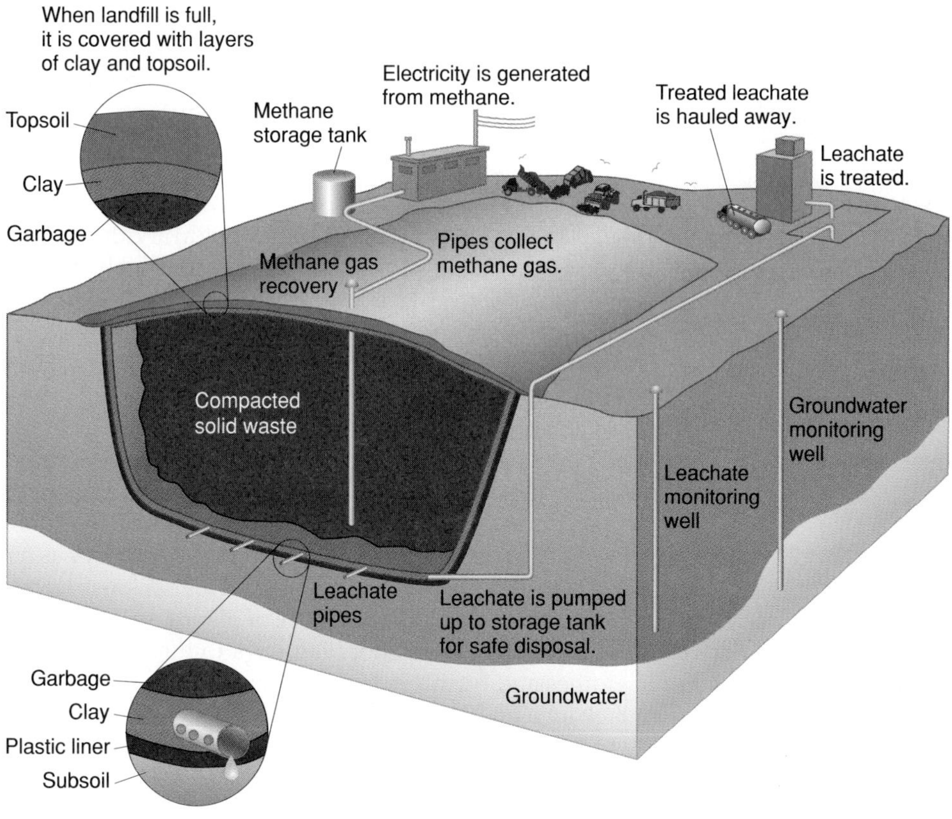
When landfill is full, it is covered with layers of clay and topsoil.
Topsoil
Clay
Garbage
Methane storage tank
Electricity is generated from methane.
Treated leachate is hauled away.
Leachate is treated.
Methane gas recovery
Pipes collect methane gas.
Compacted solid waste
Groundwater monitoring well
Leachate monitoring well
Leachate pipes
Leachate is pumped up to storage tank for safe disposal.
Groundwater
Garbage
Clay
Plastic liner
Subsoil

FIGURE 19.7

Modern sanitary landfills are designed so that any leakage is contained and cannot percolate into groundwater.

ing beryllium, cadmium, chromium, lead, and mercury, are released and carried up the incinerator stack in fly ash. In the past, municipal incinerators were a significant source of air pollution (Chapter 11), but today, in compliance with strict air-quality standards, they are equipped with efficient electrostatic precipitators and other control devices to prevent the escape of fly ash. Because these facilities often operate intermittently, however, there is concern that even if yearly average emissions are within allowed limits, emissions at peak periods may be excessive.

Modern incinerators, which are called *combustion facilities* or *waste-to-energy (WTE) plants* if they generate electricity (Fig. 19.8), burn cleanly and consume up to 3000 tons of trash a day. They produce enough electricity to power 70,000 homes. Although combustion of trash is never likely to contribute more than 1% of total U.S. energy needs, it can be a useful local source of energy.

Recycling and Resource Recovery

More and more solid waste is being recycled. Many communities require businesses and homeowners to collect discarded items such as newspapers, office paper, aluminum cans, glass containers, and certain plastic items in separate containers so that they can be collected and recycled. Some municipalities have found it more economical to build high-tech **resource recovery plants** that sort the trash after it has been collected.

The EPA estimates that 96% of discarded lead-acid automobile batteries are recovered and recycled.

FIGURE 19.8

A waste-to-energy plant may burn 3000 tons of solid waste per day and produce electrical power to supply 70,000 homes. Pollution-control devices prevent hazardous materials from escaping into the atmosphere from the smokestack.

Everyone is in favor of recycling, but to be successful, it must be cost-effective. Dealers and manufacturers will not be willing to accept items for recycling if they cannot count on steady markets for the recycled products.

Source Reduction

Disposing of the enormous quantity of MSW that Americans produce each year is becoming increasingly difficult and costly. Obviously, a key way to address this problem would be for everyone to produce less trash. This would lower disposal costs, save valuable resources, and reduce adverse effects on the environment.

There are many ways to reduce waste. For example, businesses can use e-mail in place of paper for interoffice memos, both sides of paper when printing and copying, and durable mugs instead of Styrofoam cups for coffee. Homeowners can let grass clippings remain on their lawns as fertilizer, use cloth napkins instead of paper ones, avoid wasting food, minimize their use of disposal products like plastic plates and flatware, and bring their own shopping bags to the grocery store. Unwanted items such as old clothes and furniture can be donated for reuse. Already several states, including Pennsylvania and New York, have found an effective way to reduce their trash loads: They have introduced user fees based on the volume of MSW discarded.

The EPA estimates that nearly 30% of municipal solid wastes is being recycled.

19.4 The Problem of Hazardous Waste

If the public is to be protected from the dangers associated with hazardous waste, two problems must be solved: (1) how to properly manage and dispose of waste generated now and in the future, and (2) how to clean up wastes produced in the past. The major pieces of legislation that address these two issues are RCRA (amended in 1984) and the Comprehensive Environmental Response, Compensation and Liability Act (CERCLA) of 1980.

19.5 RCRA: Regulation of Hazardous Wastes

Prior to the passage of RCRA, the two main pieces of legislation enacted to protect the environment were the 1970 Clean Air Act and the 1972 Clean Water Act. RCRA was passed to establish safe methods for the disposal of all wastes, especially **hazardous wastes**. Under RCRA, the EPA was given the responsibility for defining hazardous wastes and for issuing and enforcing regulations to protect public health and the environment from improper management and disposal of such wastes.

What Are Hazardous Wastes?

According to RCRA, wastes are described as hazardous if they "(a) cause or significantly contribute to an increase in mortality or an increase in serious irreversible, or incapacitating reversible, illness or (b) pose a substantial present or potential hazard to human health or the environment when improperly treated, stored, transported, or disposed of, or otherwise managed." Such wastes may be in solid, liquid, or gaseous form and may result from industrial, commercial, mining, or agricultural operations or from community activities.

The EPA defines hazardous wastes in two different ways. A waste is hazardous if, because of its inherent properties, the EPA has specifically *listed* it as hazardous. A waste is also hazardous if it exhibits any of the following *characteristics:* (1) ignitability, (2) corrosivity, (3) reactivity, and (4) toxicity.

If a "characteristic" waste is treated in such a way that it loses its hazardous characteristic, it ceases to be a hazardous waste and can be disposed of in a regulated landfill. However, in the original regulations, a "listed" waste, once generated, retained its hazardous classification regardless of the steps taken to dilute it or render it nonhazardous. A listed waste was not permitted in a landfill but had to be disposed of according to strict EPA regulations. In 1996, the EPA ruled that if the concentration of a listed chemical in a waste falls below an established acceptable risk level, it need not be managed as a hazardous material.

The responsibility for deciding if a waste is hazardous lies with the producer of the waste. If the waste has not been listed by the EPA, the producer must determine if the waste possesses any of the four characteristics that define it as hazardous. Once a waste has been identified as hazardous, the proper method for disposing of it must be selected. Any company that generates or transports a hazardous waste is required to keep detailed records of the handling and transferring of the waste from point of origin to point of ultimate disposal or destruction. Chemical plant managers and incinerator operators are required to keep records of all hazardous chemicals that are emitted into the air, land, or water. Landfill operators must maintain groundwater monitoring data. By law, "cradle-to-grave" responsibility for a hazardous waste is assigned to the generator of the waste.

The generator of a hazardous waste must obtain an identification number for the waste from the EPA. The number is used to track the waste from generation to disposal.

Listed Hazardous Wastes

The wastes that the EPA, by rule, has specifically listed as hazardous number nearly 400. The wastes are classified according to source and include (1) nonspecific chemicals—primarily solvents—produced by many industries (for example, spent halogenated solvents used in degreasing such as tetrachloroethylene and carbon tetrachloride); (2) sludges and by-products produced by specific industries [for example, waste water treatment sludge from the production of chrome yellow (a lead compound) and orange pigments]; (3) discarded chemicals (for example, barium cyanide); and (4) chemicals produced as intermediates in manufacturing processes (for example, acetaldehyde).

Characteristic Hazardous Wastes

The EPA has developed special tests that must be used to determine if a waste exhibits any of the four hazardous characteristics.

IGNITABLE WASTES Wastes are classified as **ignitable wastes** if they present a significant fire hazard. Ignitable wastes include (1) liquids such as gasoline, hexane, and other organic solvents that have a flash point (the lowest temperature at which their vapors ignite when tested with a flame) below 60°C (140°F); (2) ignitable compressed gases such as propane; (3) various solid materials that are liable to cause persistent, vigorously burning fires as a result of absorption of moisture (for example sodium and potassium metal), friction, or spontaneous chemical change; and (4) oxidizers. Oxidizers are included because they are likely to intensify an already burning fire. Perchloric acid ($HClO_4$), for example, is a strong oxidizing agent that poses an explosion hazard because it reacts very vigorously with organic matter.

CORROSIVE WASTES **Corrosive wastes** include acidic aqueous wastes with a pH equal to or less than 2.0 and basic aqueous wastes with a pH equal to or greater than 12.5. Wastes with these very low or very high pH values are likely to react dangerously with other wastes, and they may leach metals and contaminants from those wastes. If allowed to leak into waterways, corrosives can have a devastating effect on aquatic habitats (Chapter 10).

Also included as corrosive are wastes that can corrode steel at a rate equal to or greater than 0.635 centimeter (0.25 inch) per year. This criterion was included to prevent the kind of disasters that occurred in the past when acids and other hazardous materials leaked from corroded steel drums in which they had been stored.

REACTIVE WASTES **Reactive wastes** include materials that explode readily when subjected to shock or heat, that tend to undergo violent, spontaneous chemical change, or that react violently with water. Explosions at a dump site in Cheshire, England, were caused by discarded sodium metal, which reacted with water to produce hydrogen gas:

$$2\ Na + 2\ H_2O \longrightarrow 2\ NaOH + H_2$$

The hydrogen ignited and combined explosively with oxygen in the air:

$$2\ H_2 + O_2 \longrightarrow 2\ H_2O$$

Also classified as reactive are wastes that generate dangerous quantities of toxic gases when exposed to water, weak acids, or weak bases. For example, wastes containing cyanides or sulfides can produce toxic hydrogen cyanide (HCN) or hydrogen sulfide (H_2S).

TOXIC WASTES Most characteristic wastes are hazardous because of their toxicity. The terms *toxic* and *hazardous* are often used interchangeably, but they are not synonymous. Toxic substances are those that cause harm to living organisms by interfering with normal physiological processes (Chapter 18). They include substances that are carcinogenic, mutagenic, teratogenic (cause birth defects), or phytotoxic (poisonous to plants).

Toxic wastes are of particular concern because of their potential as groundwater contaminants. The test used to determine whether a waste is toxic was designed to model the leaching that would be likely to occur if the waste were disposed of on land. In the test, a nonliquid waste is treated with an acid solution (pH 5.0 ± 0.2) under specified conditions, and the amount of toxic material that dissolves in the acid is determined. If the concentration of toxic substances in the extract (or in the liquid in the case of a liquid waste) exceeds the regulatory level set by the EPA, the waste is considered to be toxic.

Originally, the EPA characterized just 14 chemicals as toxic: 8 metals (arsenic, barium, cadmium, chromium, lead, mercury, selenium, and silver) and 6 pesticides (endrin, lindane, methoxychlor, toxaphene, 2,4-D, and 2,4,5-TP [Silvex]). By the end of 1991, a total of 39 chemicals were regulated as toxic, including common chemicals such as benzene, carbon tetrachloride, chloroform, nitrobenzene, vinyl chloride, and several pesticides.

19.6 Sources of Hazardous Waste

Industry is responsible for most of the nearly 300 million tons of hazardous waste that is generated annually in the United States. More than 90% of this waste is produced by chemical, petroleum, and metal-related industries. Small business establishments such as dry cleaners, gas stations, and electroplating shops account for a small but significant part of the remaining 10%. Household and farm wastes are exempt from hazardous waste regulations, but any business with a monthly output of hazardous waste in excess of 100 kilograms (220 pounds) must dispose of it according to EPA regulations.

Thousands of different chemicals and processes are used by the petrochemical industry in refining crude oil and producing the synthetic organic chemicals needed for manufacturing polymers, pesticides, medicines, and other consumer products. All these processes inevitably produce wastes, and it has been estimated that 10–15% of them are hazardous.

Of greatest concern are (1) toxic metals and (2) synthetic organic chemicals, particularly chlorinated hydrocarbons. As you can see in **Table 19.2**, these hazardous substances are generated by a wide variety of industries. Metal wastes are produced primarily by metal-processing industries and by paint, textile, and other manufacturing industries that make or use pigments. Chlorinated hydrocarbons are used in the manufacture of polymers (Chapter 9), pesticides (chlordane, heptachlor, and others; see Chapter 16), electrical insulation (PCBs), and refrigerants (chlorofluorocarbons). They are also used as solvents (chloroform, carbon tetrachloride) in many industrial processes.

19.7 National Policy for Management and Disposal of Hazardous Wastes

In 1983, the National Academy of Sciences issued an influential report that has been a major factor in establishing hazardous waste management policy in the United States. The report described three basic ways for managing hazardous wastes, which, in order of desirability, are (1) minimize the amount produced, (2) convert to a less hazardous or nonhazardous form, and (3) isolate in a secure perpetual-storage site **(Fig. 19.9)**. Let's consider each of these options in turn.

19.8 Waste Minimization: Process Manipulation, Recycling, and Reuse

The high cost of disposing of hazardous waste in compliance with EPA regulations has made it economically attractive for industries to minimize waste production by modifying their manufacturing processes. For example, some electroplating plants use ion-exchange resins to selectively remove contaminant metals from electroplating baths. The plating solution, which previously would have been discarded as waste, is returned to the bath for reuse. Many industries are also purifying wastes to recover chemicals that can be reused, recycled, or sold to another industry for use as raw materials. For example, the pesticide industry has found it can recover hydrochloric acid from its hazardous chlorinated wastes and sell it to other industries. In Europe, several countries have established *waste exchanges*, where waste chemicals are successfully traded.

It may be cheaper for a plant to change its process to avoid generating a listed waste than to have to dispose of the waste according to EPA regulations.

Table 19.2 Hazardous Wastes Produced by Industry

Industry or Product	Hazardous Wastes
Electrical insulation	Polychlorinated biphenyls (PCBs)
Electroplating	Toxic metals, cyanide
Fertilizers	Sulfuric, nitric, and phosphoric acids; sodium hydroxide, ammonia
Insecticides and herbicides	Polychlorinated hydrocarbons, organophosphates, carbamates, ethylene dibromide
Iron and steel	Acids (including hydrofluoric acid), bases, phenols, benzene, toluene
Leather	Toxic metals, organic solvents
Medicines	Toxic metals, organic solvents
Nonferrous metals	Acids, metals (cadmium, zinc, lead)
Organic chemicals	Many ignitable, corrosive, reactive, and toxic compounds
Paints	Toxic metals, pigments (chrome yellow), organic solvents
Petroleum refining	Oil, phenols, other organic compounds, toxic metals, corrosives (acids, bases)
Plastics	Monomers used in polymer manufacture (ethylene, propylene, styrene, vinyl chloride, phenol, formaldehyde), phthalate plasticizers
Power industry; steam generation (fossil-fuel and nuclear-powered)	Fly ash and bottom ash (organics and heavy metals), sulfur dioxide, wet sludge
Soaps and detergents	Corrosives (sodium hydroxide, sulfuric acid)
Synthetic rubber	Ignitable monomers (styrene, butadiene, isoprene); antioxidants, fillers, etc. in tires
Textiles	Polymer monomers, dyes (azo, nitroso compounds), organic solvents, toxic metals
Wood products, paper	Phenol, formaldehyde, corrosives, sodium sulfide

19.9 Conversion of Hazardous Waste to a Less Hazardous or Nonhazardous Form

Since 1990, in compliance with the 1984 amendments to RCRA, nearly all hazardous wastes have had to be treated in some way to make them less hazardous before they can be buried on land. Several options are available (see Fig. 19.9); the choice depends on the type of waste to be treated.

Incineration and Other Thermal Treatment

If adequately controlled to prevent air pollution, incineration has a number of advantages as a means of detoxifying many hazardous wastes. At sufficiently high temperatures, incineration destroys 99.999% of toxic organic compounds by decomposing

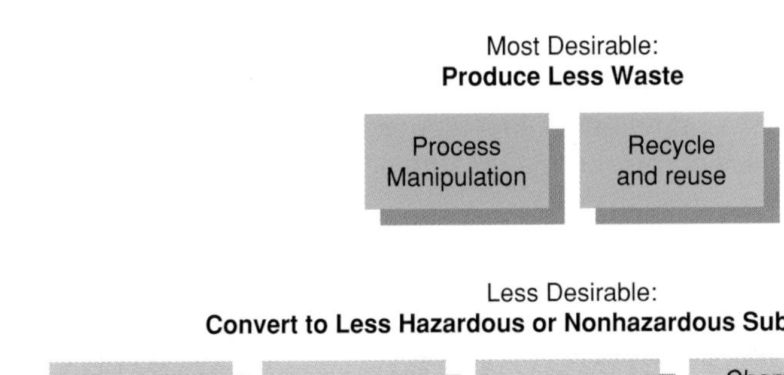

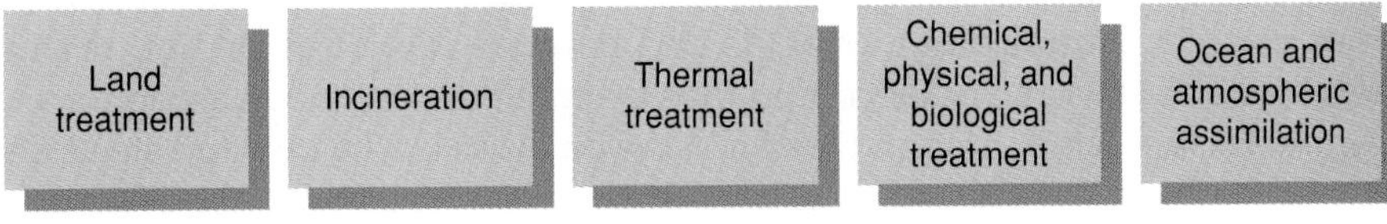

Least Desirable:
Perpetual Storage

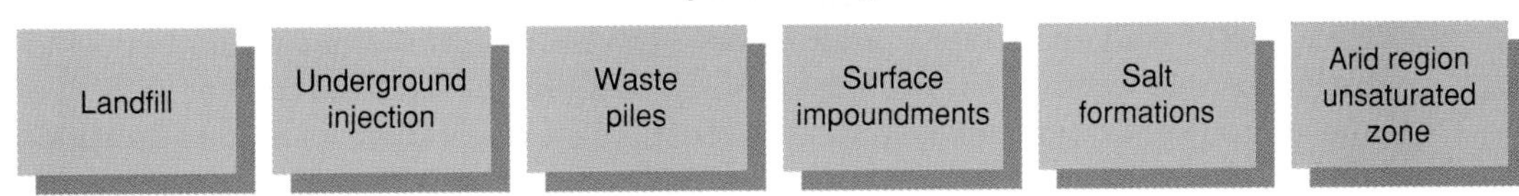

FIGURE **19.9**

Three levels of options for dealing with hazardous wastes. The top-level methods are the most desirable. The least desirable methods are often the least expensive.

During the incineration of Agent Orange at sea, the EPA monitored the air to ensure that dangerous quantities of hazardous products were not released to the atmosphere.

them to carbon dioxide, water, and harmless gases. At the same time, the volume of the waste is greatly reduced, and the energy released can be used either as a source of heat or to generate electricity. Combustible liquid and solid hazardous wastes such as solvents, pesticides, and many of the organic compounds present in petroleum refinery wastes can all be destroyed by incineration.

In the Netherlands, Germany, and Denmark, between 50% and 80% of all hazardous wastes are incinerated. In the United States, however, only about 5% of hazardous wastes are treated in this way. The scrubbers and other devices that must be installed to prevent the escape of toxic gases and particulates make incineration very expensive, but the major factor in preventing construction of more incinerators in the United States is intense community opposition. Many believe that, even with the protection of pollution controls, the potential for air pollution still exists.

During the 1970s and 1980s, hazardous wastes, including Agent Orange (the defoliant used in the Vietnam War; see Chapters 10 and 16), were destroyed at sea on ships equipped with incinerators. This practice was discontinued in 1988, pending adoption by the EPA of measures to regulate incineration at sea. Since 1985, the EPA has operated several mobile incinerators (**Fig. 19.10**) for the on-site destruction of small quantities of particularly hazardous materials, including dioxin wastes in soil and water.

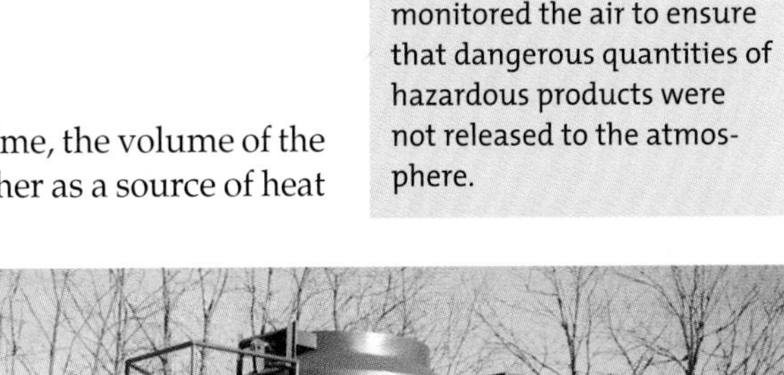

FIGURE **19.10**

A mobile incinerator for destruction of hazardous wastes on site. The incinerator can be disassembled and moved to another site on a flatbed truck. Use of this type of incinerator is becoming more common as a result of requirements that off-site transport of hazardous wastes from Superfund cleanup projects be minimized.

Certain hazardous organic wastes do not need to be incinerated to render them nonhazardous. They can be decomposed to nonhazardous biodegradable compounds by being heated under pressure at relatively low temperatures (450–600°C [840–1100°F]). In some cases, the products of this thermal treatment can be used to synthesize other compounds.

Chemical and Physical Treatment

Adsorption is the attachment of a gaseous substance, or a substance in solution, onto the surface of a solid, such as finely divided carbon.

Chemical methods that can be used to convert hazardous wastes to nonhazardous materials include neutralization of acids and bases, oxidation-reduction reactions, and removal of metals and other compounds by precipitation or by adsorption on carbon. Acidic wastes, which are produced in large quantity by the iron and steel, electroplating, and other metal-related industries, can be neutralized with lime. This process also serves to precipitate out heavy metals that can then be collected and recycled. Toxic chromium(VI) salts in wastes can be chemically reduced to far less dangerous chromium(III) salts. Cyanide, which is used in many industrial processes, including the extraction of gold and silver from their ores, can be oxidized with sodium hypochlorite ($NaOCl$) to far less toxic cyanates. The cyanates can be oxidized further with chlorine to produce carbon dioxide and nitrogen. In the textile industry, carbon adsorption can be used to remove toxic dyes from liquid wastes.

Certain wastes can be made nonhazardous or less hazardous by immobilizing them in a solid mass. For example, waste material can be mixed with cement or lime to form a concrete block, which can then be disposed of in a landfill. If the concrete is likely to be exposed to acidic conditions, leaching can be prevented by encapsulating the block in an impermeable plastic container. This technique is useful for certain liquid wastes and for the flue-gas-cleaning sludges that are produced in enormous quantity by coal-burning power stations. The sludges, which are formed as a result of treatment to control sulfur dioxide emissions (Chapter 11), contain calcium sulfate ($CaSO_4$) and fly ash that may include toxic metals.

Bioremediation

Increasingly, microorganisms are being used to clean up soil that is contaminated with toxic organic compounds. The microorganisms, often mixed with nutrients, decompose the wastes and thus detoxify them. Oxidation by sunlight further aids in the degradation. This technique, which is termed **bioremediation,** is used primarily to treat soil contaminated with oil. It is also used to treat petroleum-refinery and paper-mill wastes. Bioremediation is inexpensive, but precautions must be taken to ensure that the wastes do not contain toxic metals or other nonbiodegradable chemicals that could contaminate the soil or groundwater. Using genetic engineering, microbiologists are working to develop specialized microorganisms that can effectively detoxify polycyclic aromatic hydrocarbons (PAHs), PCBs, trichorethylene, and other compounds that are resistant to biodegradation.

19.10 Perpetual Storage of Hazardous Wastes

Hazardous wastes that cannot be recycled or converted to nonhazardous forms must be stored. Perpetual-storage methods include (1) burial in a secured landfill, (2) deep-well injection, and (3) surface impoundment. None of these methods, however, provides entirely secure storage, and the 1984 amendments to RCRA included restrictions on land disposal of hazardous waste. A 1990 deadline was set for the end of disposal of untreated hazardous materials on land except in specific cases where the EPA has determined that no other option is feasible. Since 1990, wastes not so designated have been automatically banned from land disposal facilities and must be disposed of according to EPA regulations. As a result of legislation, there has been a sharp reduction in improper disposal.

Secured Landfills

A **secured landfill** is a burial site for the long-term storage of hazardous wastes contained in drums **(Fig. 19.11)**. Strict EPA specifications for construction minimize the chance of contaminants migrating in the event of a drum leak. The landfill must be located in thick, impervious clay soil at least 165 meters (500 feet) from a water source in an area that is not subject to floods, earthquakes, or other disturbances. Ideally, the site should be in an arid region, far from any water supplies. The bottom and sides of the trench holding the drums must be lined with two layers of impervious, reinforced plastic, and the drums must rest on a layer of gravel. Below the gravel, there must be a network of pipes that can collect any leakage from drums and any rainwater seeping through the fill; leachate must be pumped to the surface and treated. When filled, the site must be covered with a layer of impervious plastic material topped, in turn, with layers of sand, gravel, and clay. The cap must be contoured to allow water to drain away from the site. The quality of groundwater in the area of a secured landfill must be monitored for at least 30 years.

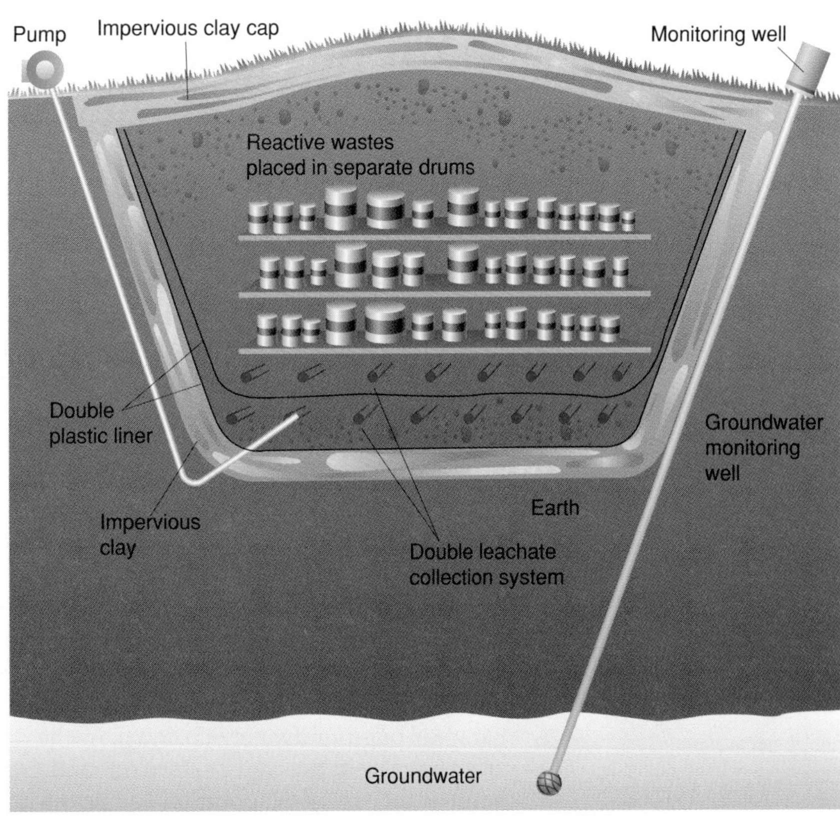

FIGURE 19.11

In a secured landfill for long-term storage of hazardous wastes contained in drums, the bottom and sides of the fill are lined with plastic. Below the gravel on which the drums rest are pipes that collect any leachate; groundwater below the fill is constantly monitored for contamination. The landfill is capped with impervious clay to prevent water from seeping into it.

Soils vary in their ability to retain water. Clays are impervious to water; gravel and sand are very porous.

Many critics contend that despite the EPA's tough regulations, all landfills will leak eventually, and therefore they should not be used to store materials that retain their hazardous properties almost indefinitely. The critics maintain that, in time, some drums will corrode and leak, clay layers will be invaded by burrowing animals, and plastic liners will be torn by freezing temperatures or settling or will be disintegrated by leachate.

Deep-Well Injection

In the past, liquid hazardous wastes were frequently disposed of by **deep-well injection,** a technique that was originally developed by the petroleum industry as a means of disposing of brines (salt solutions) that were brought to the surface with crude oil.

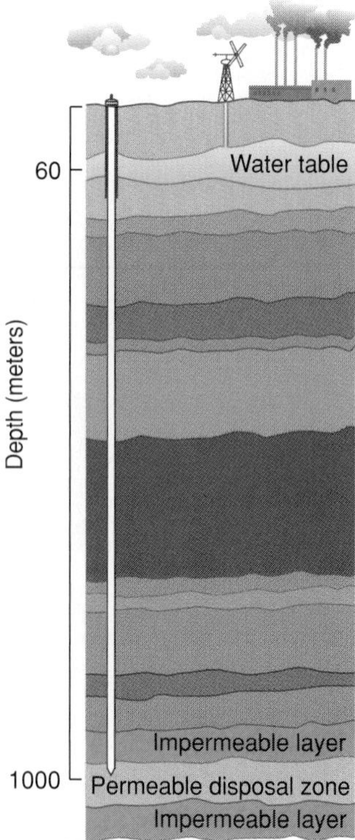

FIGURE 19.12

Deep-well injection is used for disposing of liquid hazardous wastes. Liquid waste is pumped into a deep well drilled into a layer of dry, porous rock lying well below any groundwater and below a layer of impervious rock. The well shaft is surrounded by a sealed casing to prevent groundwater contamination.

Liquid waste is pumped into a deep well drilled into a layer of dry porous rock located far below any usable groundwater, and below a layer of impervious rock (Fig. 19.12). The wastes gradually seep into the porous material and are trapped below the impervious layer of rock. The well shaft is surrounded by a sealed casing to prevent contamination as the waste is pumped through the groundwater region.

The RCRA restrictions on land disposal apply to deep-well injection, but there is a loophole in the law that permits injection of untreated hazardous waste if a company can prove that there will be no migration of the waste from the injection zone during the time the waste remains hazardous. Deep-well injection is one of the least expensive waste-disposal methods, and, with the implementation of the rules restricting disposal in landfills, companies have made use of the loophole to increase the amount of waste they dispose of in deep wells. Critics claim that, even if properly executed, deep-well injection has the potential for groundwater contamination. In time, the well casing is likely to corrode, and there is always a danger that earth tremors or earthquakes may fracture the impervious layers and lead to the migration of hazardous materials.

Surface Impoundment

Surface impoundment is an inexpensive disposal method used primarily to manage relatively small quantities of hazardous wastes contained in large volumes of wastewater. The wastewater is usually pumped directly from the plant that produced it to a pond lined with impervious clay, a plastic material, or both, to prevent leakage into underlying soil or groundwater (Fig. 19.13). As the water evaporates, the small quantity of hazardous waste in the water settles out and gradually accumulates on the bottom of the pond. If the rate of evaporation equals the rate of waste input, surface impoundments can be used almost indefinitely. Like all methods of disposal on land, surface impoundment is likely to cause environmental contamination sooner or later.

If there is exceptionally heavy rainfall, a surface impoundment pond may overflow, spreading hazardous substances into the surrounding area.

According to the EPA, toxic chemicals released by industry were reduced by 26% between 1987 and 1990.

19.11 The Unsolved Problem

Society wants the products that generate hazardous wastes—electricity, plastics, gasoline, and pesticides, for example—but no one wants a waste disposal site in his or her neighborhood. This so-called NIMBY (Not In My Back Yard) syndrome has hampered waste disposal but, along with economic considerations, has been an important factor in forcing manufacturers to find ways to minimize waste production. Since the 1980s, in-plant process modifications, recycling, and reuse have steadily reduced the production of hazardous waste. But a huge quantity continues to be generated, and there are still no completely acceptable ways of dealing with it.

In 1979, it was estimated that only 10% of hazardous waste was being disposed of in an environmentally safe manner. Today, the situation is very much better, but because of the high cost of proper disposal, large quantities of waste are still dumped illegally. The EPA does not have the resources to keep track of the thousands of operations that generate hazardous wastes, and consequently, it is often unable to enforce its "cradle-to-grave" provisions. The EPA must also contend with organized crime, which is known to be involved in waste disposal. Mob-controlled trash collectors in New Jersey and New York, for example, have been discovered mixing household refuse with hazardous wastes and illegally disposing of the combined wastes in municipal landfills.

19.12 Superfund: Cleaning Up Hazardous-Waste Dumpsites

Exposure of the Love Canal disaster in 1978, along with increasing evidence that numerous other dangerous abandoned dumpsites existed, made it clear that legislation was urgently needed to deal with the problems created in the past. CERCLA, passed in 1980, established the **Superfund program,** which was aimed primarily at (1) identifying existing hazardous waste dumpsites and (2) establishing a trust fund to finance the cleanup of the sites.

According to this legislation, the parties responsible for creating a hazardous dumpsite are responsible for cleaning it up. If they fail to do so, the EPA is authorized to undertake the cleanup and then sue the delinquent parties for three times the costs incurred. In cases where responsibility cannot be established, the EPA is authorized to undertake the cleanup using funds provided jointly by the federal government and state governments and by a tax on the chemical and petrochemical industries.

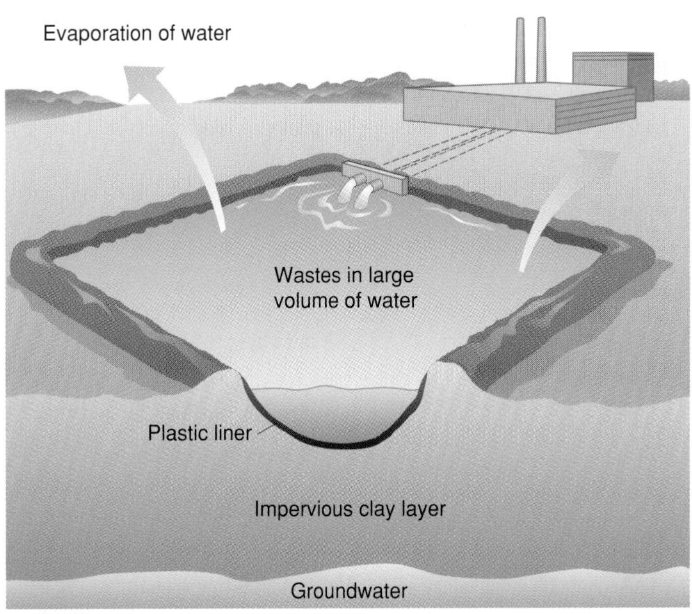

FIGURE 19.13
Surface impoundment is used for the disposal of small amounts of hazardous waste present in large volumes of water. The water gradually evaporates, and the hazardous materials accumulate at the bottom of the pond, which is lined with clay or plastic to prevent leakage into the ground.

The EPA drew up a **National Priorities List** identifying sites with the most urgent need of cleanup, but it soon became evident that the $1.6 billion trust fund provided for cleanup during the first 5 years of the Superfund program was totally inadequate. Many more hazardous sites than expected were discovered, and cleanup was found to be far more difficult than anticipated. By 1990, CERCLA had been amended twice and a trust fund of $13.6 billion had been authorized through 1994. The EPA originally listed 38,000 sites for cleanup but has since reduced this number to 14,000. By 2000, about half of the 1231 sites the EPA had placed on the National Priorities List had been cleaned up or methods for cleanup had been selected.

19.13 Radioactive Wastes

Although radioactive wastes are undoubtedly hazardous to health and the environment, very few are regulated under RCRA. Instead, they are covered by the Atomic Energy Act of 1954 (and later amendments), which is administered by the Department of Energy (DOE). Created in 1977, the DOE received as a legacy from its predecessor, the Atomic Energy Commission, responsibility for production of the nation's nuclear weapons. The DOE is also responsible for regulating the radioactive wastes generated by nuclear weapons facilities and by commercial nuclear power plants. (Radioactivity and nuclear power were discussed in Chapters 5 and 13, respectively.)

Like hazardous waste, radioactive waste presents two kinds of problems: (1) what to do about waste produced in the past and not disposed of properly and (2) how to dispose of waste currently being generated.

Table 19.3 Locations of U.S. Nuclear Weapons Facilities

State	Site
California	Lawrence Livermore National Laboratory
Colorado	Rocky Flats
Florida	Pinellas
Idaho	Idaho Falls
Kansas	Kansas City
Nevada	Nevada Test Site
New Mexico	Los Alamos National Laboratory, Sandia National Laboratories, Waste Isolation Pilot Plant (WIPP)
Ohio	Fernald, Mound Plant
South Carolina	Savannah River
Tennessee	Oak Ridge
Texas	Pantex
Washington	Hanford

Sources of Radioactive Wastes

Nuclear weapons manufacturing facilities and nuclear power plants are the principal sources of the nation's most dangerous radioactive wastes. The nuclear weapons facilities, which are located in 12 states (Table 19.3), are responsible for at least 10 times more waste than the nuclear power industry. Relatively small quantities of far less dangerous radioactive wastes are generated by hospitals, research laboratories, and some industries and by the mining and processing of the uranium used as nuclear fuel.

Nuclear reactors produce wastes that contain a mixture of radioisotopes with half-lives varying from a few days to many years. These include strontium-90 (half-life of 28 years), cesium-137 (half-life of 30 years), and plutonium-239 (half-life of 24,400 years). Since 10 half-lives must elapse before radioactivity is reduced to negligible levels, wastes containing plutonium-239, which emits an alpha particle as it disintegrates (Chapter 5), remain deadly for tens of thousands of years. Plutonium-239 is used as a nuclear fuel in weapons programs, and, for many years prior to 1977, it was generated during the reprocessing of uranium. Plutonium-239 is also produced in breeder reactors (Chapter 13), which are used in several European countries to generate electricity. In the United States, however, only a few experimental breeder reactors are in operation.

Classification of Radioactive Wastes

Depending primarily on how they are generated, radioactive wastes are classified as high-level or low-level. **High-level radioactive wastes,** which all contain plutonium-239, consist of spent fuel rod assemblies from nuclear reactors in both commercial power plants and weapons plants and certain other highly radioactive wastes generated by nuclear weapons facilities. All other radioactive wastes are classified as **low-level radioactive wastes.**

Low-level radioactive wastes are produced by hospitals, research laboratories, and certain industries. Most are dilute and contain radioisotopes with half-lives measured in no more than hundreds of years. Low-level wastes also include transuranic wastes (wastes containing elements whose atomic numbers are higher than uranium's), which are produced at nuclear weapons plants. These wastes are usually dilute but dangerous because they contain radioisotopes, including plutonium-239, with long half-lives.

19.14 The Legacy from the Past

Since World War II, radioactive waste, much of it liquid high-level waste, has been accumulating at nuclear weapons facilities (Table 19.3), which produced the plutonium needed to construct nuclear warheads. For over 40 years, weapons production took precedence over all other considerations, including contamination of the environment and dangers to public health. During this period, most wastes were either dumped directly into the ground, stored in ponds, or deposited in tanks that frequently developed leaks.

As a result of these practices, the environment surrounding all nuclear weapons facilities in the United States is extensively contaminated with radioactive material. The situation is even worse at nuclear weapons sites in the former Soviet Union, where radioactive wastes were discarded even more irresponsibly. At several plants there, billions of gallons of liquid radioactive wastes were pumped into the ground near major rivers, threatening surface waters and groundwater. One weapons facility in the Ural Mountains, known as Chelyabinsk-65, discharged so much nuclear waste into nearby Lake Tarachay that a person standing on its shores for an hour will receive a fatal dose of radiation.

The Savannah River Weapons Plant near Aiken, South Carolina, which produced the radioactive tritium gas used to enhance the explosive power of nuclear weapons, is known to have released tritium gas (half-life of 12.3 years) into the air and water around the plant starting in the 1950s. It is also suspected of having deliberately discharged liquid radioactive wastes into the environment for many years. Leaks from old waste storage tanks continue to threaten the Savannah River and nearby groundwater with radioactive contaminants. All four nuclear reactors at the site are now closed down. At another weapons plant near Fernald, Ohio, it is estimated that at least 3 million pounds of uranium dust and other radioactive materials were released into the atmosphere between 1951 and 1985.

Tritium (^3_1H) is an isotope of hydrogen. It decays at a rate of 5.5% a year.

The largest, most contaminated, and most dangerous weapons facility in the United States is the 560-square-mile Hanford complex near Richland, Washington, which until operations were halted in 1987, produced plutonium for nuclear weapons. In the past, this facility discharged billions of gallons of liquid radioactive wastes directly into the ground. Wastes that were stored in tanks have leaked, contaminating several cubic miles of soil not only with radioactive material but also with other dangerous by-products of weapons production, including toxic solvents and heavy metals. Some of these wastes continue to migrate through the soil into the Columbia River.

Another severely contaminated weapons facility is the Rocky Flats plant near Denver, Colorado, which, until it was shut down in 1989, produced the plutonium "triggers" that set off the nuclear chain reactions in warheads. Wastes at the plant have leaked from corroded storage tanks for many years.

In 1992, Rockwell International Corporation agreed to plead guilty to having illegally disposed of radioactive waste at the Rocky Flats nuclear weapons plant and to pay $18.5 million in fines.

Of particular concern are over 70 million gallons of liquid high-level wastes from past weapons production, which are buried underground in steel tanks, some at Hanford and the rest at Savannah River (**Fig. 19.14**). Many of the tanks are leaking, and the wastes in some tanks are potentially explosive. In one notorious tank at Hanford that contains a mixture of radioactive sludge, toxic solvents, and unidentified chemicals, hydrogen gas builds up at intervals and must be vented. Some scientists believe that an electric spark could easily ignite the hydrogen, setting off explosions at nearby tanks and causing the release of large amounts of radioactivity.

One of the biggest problems is finding out exactly what chemicals are in the tanks. Eventually, the DOE plans to remove the waste from the tanks, vitrify it (fuse it with sand to form a radioactive glass) and then store it in a permanent repository.

It has been estimated that the massive task of cleaning up the nation's nuclear weapons plants and other sites contaminated with radioactive wastes will ultimately cost at least $600 billion and take decades to accomplish. Cleanup is hampered by the lack of adequate disposal technology and the difficulty of finding an acceptable site for long-term storage.

In 1993, a tank containing high-level radioactive wastes exploded at a weapons plant in Tomsk, Siberia, contaminating a wide area. The explosion was caused by a rapid rise in temperature that resulted when nitric acid was added to the tank.

a.

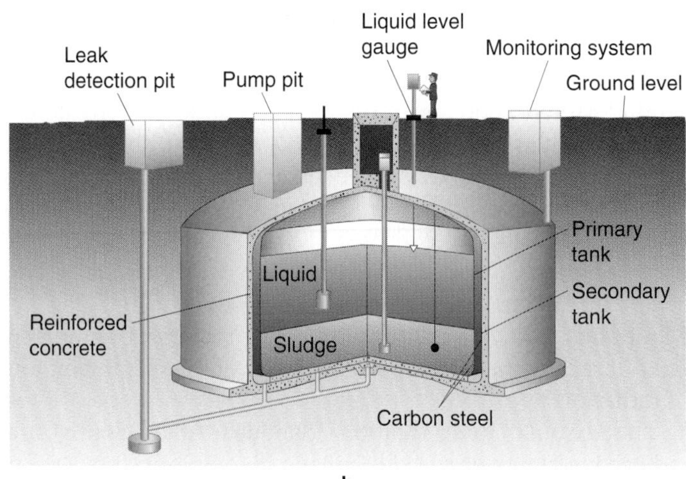

b.

FIGURE 19.14

High-level liquid radioactive wastes are stored in double-shell steel tanks, each of which can hold 1 million gallons. (a) Tanks at various stages of construction. When completed, the tanks will be covered with 7–10 feet of earth. (b) Cross-section of a typical double-shell steel tank.

According to a study by the National Academy of Science, completed in 2000, more than two-thirds of the sites involved in nuclear bomb production will never be completely cleared of contamination. These sites will require monitoring for the foreseeable future.

19.15 Disposal of Radioactive Wastes

Regulations governing disposal of radioactive wastes vary depending on whether the wastes are high-level or low-level and whether they were produced by commercial operations or by nuclear weapons facilities. In all cases, the wastes must be shipped long distances across the country to central disposal sites.

FIGURE 19.15

Low-level radioactive wastes are stored permanently in steel drums that are buried in lined trenches.

Low-Level Radioactive Wastes

Until 1970, most low-level radioactive wastes generated in the United States were disposed of at sea. When this practice was banned, low-level wastes were buried in shallow, often unlined trenches. The wastes, which in some cases included highly radioactive liquid military wastes, were supposed to be contained by impermeable rock, but leaks developed at many locations. Today, commercially generated low-level radioactive wastes must be buried in steel drums in lined, carefully sited trenches in accordance with rules set by the Nuclear Regulatory Commission (NRC) (**Fig. 19.15**). During the 1970s, states in which the few NRC-licensed facilities were located objected to receiving wastes generated outside their borders, and there was increasing concern about the dangers inherent in transporting radioactive wastes over long distances. In 1980, in response to these problems, Congress passed the Low-Level Radioactive Waste Policy Act, which set 1986 (later extended to 1992) as the deadline for each state to take responsibility for disposal of its own low-level wastes. If states preferred, they had the option of joining in multi-state compacts in which one state would provide a disposal site for the other states in the compact. The search for sites touched off furious environmental and political battles, and many states are still not in compliance with the law. A site in South Carolina has been receiving low-level radioactive wastes from 38 states since the 1970s. However, in 1999, the governor of the state began proceedings to limit access to the site to just his own state and Connecticut and New Jersey. If access is limited, the other 35 states will have to find another state to accept their low-level wastes.

To provide a permanent storage site for low-level radioactive wastes produced by nuclear weapons plants, the DOE in 1983 began building a repository in a salt bed half a mile under federal land near Carlsbad, New Mexico. The repository, known as the **Waste Isolation Pilot Plant (WIPP),** consists of a 1-square-mile network of tunnels and chambers. In 1999, after years of opposition from environmental groups and others concerned about long-term safety, the site, which cost almost $2 billion to build, finally received its first shipment of wastes. These wastes, which are contained in steel drums, consist mainly of protective clothing, tools, equipment, and soil that are contaminated with plutonim-239. Deliveries of wastes from the nation's nuclear weapons plants are expected to continue for at least the next 30 years.

If technetium-99 (half-life of 210,000 years), a radioisotope often present in nuclear weapons wastes, is bombarded with neutrons, it is transmuted to technetium-100. which decays to stable, nonradioactive ruthenium-100.

High-Level Radioactive Wastes

Several technologies for the disposal of high-level radioactive wastes have been considered. They include burial of the waste in deep ocean trenches, shooting the waste into the sun or into outer space, and transmutation of dangerous radioisotopes to harmless or less harmful isotopes by means of neutron bombardment. None of these options is realistic either because of the cost involved, because of safety considerations, or because the technology is not presently available.

Some utilities have already run out of space for storing used fuel rods and have begun storing them in above-ground facilities that cost $2 million to build and must be guarded and monitored.

Currently, the recommended procedure for disposing of high-level radioactive waste is to concentrate it, vitrify it—that is, incorporate it into a glass material by fusing it with sand at about 950°C (2000°F)—and then pour the molten mixture into stainless steel corrosion-resistant canisters (**Fig. 19.16**). The glass will be radioactive, but it will be very resistant to leaching. Until the canisters can be transferred to a permanent repository, they would be stored in concrete.

Pending establishment of a permanent repository for high-level radioactive wastes, spent fuel rods from nuclear reactors are stored under water in tanks at the reactor sites (**Fig. 19.17**). The water shields the radiation and dissipates the heat that is generated by the rods. At present, 40,000 metric tons of highly radioactive fuel rods are being stored at 72 nuclear power plants around the country, many of which are running out of storage space.

FIGURE 19.16

Stainless steel containers for glass-fused high-level radioactive waste.

To deal with the mounting problem, Congress in 1982 passed the Nuclear Waste Policy Act, which set up a timetable for building an underground repository, deep underground in a geologically stable area for permanent storage of high-level waste. The repository was originally intended for wastes from commercial nuclear reactors only, but under new plans, it will also accept defense weapons wastes, including plutonium from dismantled nuclear war heads, after they have been vitrified.

In 1987, the DOE chose Yucca Mountain, 90 miles (150 kilometers) northwest of Las Vegas in Nevada, as the site for the underground repository. The site met two important criteria: It was very dry, with an annual

EXPLORATIONS

Love Canal—The Nation's Most Infamous Chemical Waste Dumpsite

On August 2, 1978, at a public meeting in Albany, New York, Dr. Robert Whalen, the state commissioner of health made the following statement: "A review of all the available evidence respecting the Love Canal Chemical Waste Landfill site has convinced me of the existence of a great and imminent peril to the health of the general public residing at or near the site." These ominous words confirmed the worst fears of the people living in Love Canal and introduced the general public to the nation's most infamous chemical waste dumpsite.

Historical Roots of the Disaster

The disaster at Love Canal, a neighborhood of modest homes in the city of Niagara Falls, New York, began many years before Dr. Whalen's announcement. The stage was set in the late 1800s when William Love, attracted to the area by the swiftly flowing waters of the Niagara River and the prospect of cheap power, formed a company to build a model city on the shores of Lake Ontario. He planned to divert water from the river into a 7-mile-long navigation canal and build a hydroelectric generating plant at the site where the water returned to the river. But Mr. Love's ambitious scheme was never completed. After a few homes had been built and a canal—60 feet wide, 10 feet deep, and 3000 feet long—had been dug, his company failed. The canal bed, fed by streams originating in the Niagara River, became the center of a popular recreation area.

In 1905, after inexpensive hydroelectric power had become available, the Hooker Electrochemical Company opened a plant near the city of Niagara Falls for the manufacturing of chlorine and caustic soda. In 1942, with permission from the city, Hooker began disposing of chemical wastes in the old canal; a few years later, the company purchased the canal and some of the surrounding land. By the 1950s, Hooker was producing pesticides, plastics, and other chemicals and had become one of the area's biggest employers, with annual sales in excess of $75 million.

Chemical wastes, which included caustic substances, chlorinated hydrocarbons, and solvents, were poured directly into the canal or into metal and fiber drums that were then deposited in it. Twenty-two thousand tons of chemicals had been dumped into the canal by 1952, and it was almost full. The Hooker company, having no further use for the canal, covered it with dirt and clay, and deeded it, along with 16 acres of surrounding land, to the Niagara Falls School Board for the token sum of $1. The deed of sale included an interesting disclaimer: The purchaser assumed all risk and liability incident to the use of the land. This meant that no claim could ever be made against Hooker for injury or death caused by the buried wastes.

Despite the disclaimer and the concern of a contractor who encountered chemicals as he dug into the construction site, the school board built an elementary school on land bordering the canal. The school playground and adjoining fields where children played lay directly over the canal dumpsite. Housing developments grew up around the site until eventually nearly 1000 homes were built, some with backyards bordering the school board's land.

Disturbing Events

In the ensuing years, many unusual and disturbing occurrences were noted. Black liquid was occasionally observed seeping out of the ground; grass and backyard vegetables failed to grow; and noxious fumes were often a problem. Children running barefoot in the fields near the school complained of irritated sore feet, and many suffered from bronchitis and asthma. Complaints to local officials were disregarded, and the residents, unaware of the existence of the canal and its history, did not appreciate the seriousness of these problems. Then, in the winter of 1977–1978, as a result of several years of unusually heavy precipitation, the canal dumpsite became saturated with water. Thick, black, malodorous liquid began oozing to the surface in backyards and the school's playing fields and seeping through cinderblock walls into basements of nearby homes. Corroded drums were brought to the surface, and chemical wastes leaked into storm sewers, contaminating groundwater. Complaints from alarmed citizens could no longer be ignored.

Health Problems

Studies carried out by the EPA in 1978 identified 82 different chemicals, including harmful polychlorinated hydrocarbons, in the area near the dump. Eleven of the chemicals were known or suspected carcinogens. In families living close to the dumpsite, state epidemiologists discovered rates of miscarriages four times greater than average. They found an unusually large number of children with birth defects and a high incidence of cancer and of respiratory, nervous, and kidney diseases.

Chemical wastes, which included caustic substances, chlorinated hydrocarbons, and solvents, were poured directly into the canal or into metal and fiber drums that were then deposited in it.

Residents were confused and frightened. Test results were not properly explained to them, but they knew that their health—particularly that of their children—was threatened and that their property values had plummeted. A fence bearing a "No Trespassing" sign was erected around part of the canal area, and the advice given to those living close to the dumpsite—to stay out of their basements and refrain from eating vegetables grown in their backyards—only added to their anxiety.

Despite numerous promises of help from elected representatives and other officials, nothing further was done until August 1978 when public denunciations from outraged residents and adverse publicity in the media finally forced the state to take action. A health emergency was declared, the school was closed, and 239 families in the first two rings of houses around the canal were evacuated. By May 1980, an additional 710 families in neighboring blocks, who had angrily protested that they too were in danger, had also been relocated. Fifty other eligible families refused to leave their homes. President Carter declared the site a disaster area, and federal money was made available to pay resettlement costs and buy almost 800 contaminated homes. Houses nearest the dumpsite were demolished, and approximately 500 others were boarded up. The surface of the canal was sealed with clay and planted with grass; a drainage system was installed to carry leaking chemicals to a treatment plant. A federal study indicated that chemicals had not migrated away from the dumpsite through underground channels.

In 1990, after more than $275 million had been spent on buying homes from residents, preparing studies, and cleaning up, the New York State Department of Health declared that much of the area was again habitable. Despite the concerns of environmental organizations that health hazards still remained, the EPA supported the state's decision to encourage resettlement. Homes became available at 20% below their market value because of the Love Canal stigma, and many people with modest incomes were eager to buy them. Banks, however, were unwilling to hold mortgages on these homes, and in 1991, only a few people had moved back into the area, which had been renamed "Black Creek Village."

Area of 1978 evacuation at Love Canal. The Niagara River can be seen at the top; the canal ran from top to bottom between the two rows of houses.

The U.S. Government sued Occidental Chemical Corporation, the successor to the Hooker company, for $610 million to cover the cost of cleanup and for punitive damages. In 1983, former Love Canal residents settled out of court with Occidental, the city of Niagara Falls, and its school board, regarding claims of health problems arising from their exposure to the Love Canal chemicals. But for the rest of their lives, these people must worry about what the long-term effects of the years spent living next to the nation's most notorious hazardous waste dumpsite will be on their health and that of their children.

Reference: A. G. Levine, *Love Canal: Science, Politics, and People* (Lexington Books, Lexington, Massachusetts), 1982.

FIGURE 19.17

After being removed from a nuclear reactor, spent fuel rods are stored under water, where they glow as they continue to emit radiation.

rainfall of only 6 inches, and the rock formation was tuff, a compacted volcanic ash that is considered to be one of the best materials for absorbing heat and containing radioactivity. Construction was supposed to be completed by 1998 but there were delays because of strong opposition from the people of Nevada. Also, there is concern that during the 10,000 years that, according to the EPA, is required for spent fuel rods to decay to safe levels, earthquakes and volcanic eruptions may occur. Some geologists believe that water could seep into storage chambers, become contaminated with radioactive materials and move into groundwater. The repository, which has already cost $19 billion, is not expected to be open until 2010 at the earliest. The DOE, under terms of the Nuclear Waste Policy Act, has been collecting fees from the nuclear power industry for many years to build the permanent storage facility.

Disposal of radioactive wastes is one of the most intractable problems the DOE faces. For environmentalists and the general public alike, opposition to radioactive waste disposal sites is not just NIMBY— it is more like NOPE (Not On Planet Earth).

19.16 The Post–Cold War Challenge

In the late 1980s, with the end of the Cold War, the United States discontinued production of plutonium-239. Disposal of huge quantities of surplus plutonium from dismantled nuclear warheads presents a serious problem in both the United States and Russia.

The DOE has rejected secure storage of plutonium as a solution because of the danger of theft by terrorists or rogue nations seeking to make nuclear weapons. It has decided that two disposal strategies should be adopted. In one, the plutonium would be immobilized in glass, as described in the previous section, and ultimately stored underground. This strategy treats the plutonium as waste and does not make use of its energy value. In the second strategy, the plutonium would be reprocessed. It would be oxidized and mixed with conventional uranium fuel to form a *mixed oxide*, or *MOX*, which could be used as fuel to produce energy at commercial nuclear power plants. The MOX would be used only once, and would not be reprocessed to recover the plutonium that would remain in the spent fuel rods after burning. A glass immobilization facility and a plant for manufacturing MOX are expected to be open at the Savannah River site by 2013. Russia is expected to dispose of its weapons-grade plutonium by using the second option.

Many people are unhappy about the DOE's decision to make MOX because it appears to contradict long-standing U.S. policy against reprocessing plutonium. As you learned in Chapter 13, spent fuel rods at commercial nuclear power plants contain unreacted uranium and plutonium that can be reprocessed to make new nuclear fuel. Japan, Britain, and France all reprocess nuclear fuel. However, the United States does not do so because of the possibility that the plutonium could be stolen and used to make nuclear weapons.

Japan is committed to using plutonium as an energy source and has already begun importing plutonium from fuel reprocessing plants in Europe.

Chapter Summary

1. Municipal solid waste (MSW) is the solid waste collected locally from homes and commercial establishments.
2. Since 1983, landfills in the United States have been required to meet strict standards: They must be located well above the water table, have a liner to prevent leakage, have a system for venting methane, and cover each day's load with dirt.
3. In the United States, about 15% of MSW is incinerated. Incineration kills microorganisms, reduces volume, and produces heat that can be used as an energy source, but it has the potential to pollute the atmosphere.
4. In the United States, about 20% of MSW is recycled.
5. RCRA and CERCLA were passed to regulate the management and disposal of wastes generated now and in the future and the cleanup of wastes generated in the past.
6. The EPA defines a waste as hazardous if the EPA has *listed* it as hazardous because of its inherent properties *or* if it has any of the *characteristics* of ignitability, corrosivity, reactivity, and toxicity.
7. By law, cradle-to grave responsibility for a hazardous waste is assigned to the generator of the waste.
8. More than 90% of the hazardous waste generated in the United States is produced by the chemical, petroleum, and metal-related industries.
9. The hazardous wastes of greatest concern are toxic metals and polychlorinated hydrocarbons.
10. In order of desirability, the three basic ways to manage hazardous waste are
    a. minimize production
    b. convert to a less hazardous form
    c. isolate in a secure perpetual-storage facility

11. The Superfund program was established to identify existing hazardous dumpsites and finance their cleanup.
12. The National Priorities List identifies the sites in most urgent need of cleanup.
13. The DOE is responsible for regulating radioactive wastes produced by nuclear weapons facilities and by commercial nuclear power plants.
14. High-level radioactive wastes all contain plutonium-239. Low-level radioactive wastes are more dilute and contain radioisotopes with relatively short half-lives.
15. The DOE chose WIPP in Mew Mexico as the permanent storage site for low-level radioactive wastes produced by nuclear weapons facilities and Yucca Mountain in Nevada as the site for underground storage of all high-level radioactive wastes.
16. The recommended procedure for disposing of high-level radioactive waste is to concentrate it, vitrify it, and store it underground in steel canisters.
17. The DOE has recommended two procedures for disposal of surplus plutonium-239 from dismantled nuclear warheads: The first is to vitrify it and store it; the second is to reprocess it to form MOX, a mixed oxide of plutonium and uranium that can be reused as fuel in commercial nuclear power plants.

Key Terms

bioremediation (p. 570)
corrosive waste (p. 566)
deep-well injection (p. 571)
hazardous waste (p. 564)
high-level radioactive waste (p. 574)
ignitable waste (p. 565)
low-level radioactive waste (p. 574)
municipal solid waste (MSW) (p. 560)
National Priorities List (p. 573)
reactive waste (p. 566)
resource recovery plant (p. 563)
sanitary landfill (p. 560)
secured landfill (p. 571)
solid waste (p. 560)
Superfund program (p. 573)
surface impoundment (p. 572)
toxic waste (p. 566)
Waste Isolation Pilot Plant (WIPP) (p. 577)

Questions and Problems

1. Describe how a modern sanitary landfill is constructed. How does such a landfill differ from a traditional garbage dump?
2. Name the most significant federal legislation enacted to regulate solid waste.
3. List the advantages and disadvantages of trash incinerators. Would you recommend that the refuse generated by your family be disposed of in a landfill or incinerated? Explain your choice.
4. What is fly ash, and why is it dangerous?
5. What is the most dangerous pollutant released by incinerators?
6. Why isn't a larger percentage of MSW in the United States recycled? List three steps that your hometown could take to encourage recycling.
7. Does CERCLA regulate cleanup of abandoned waste dumps or operation of municipal dumps?
8. a. Define *hazardous waste.*
   b. Who has the responsibility for deciding that a waste material is hazardous?
   c. Who is assigned cradle-to-grave responsibility for hazardous waste?
9. Give definitions for each of the following terms:
   a. ignitable wastes
   b. corrosive wastes
   c. reactive wastes
   d. toxic wastes

10. Is sulfuric acid from discarded automobile batteries classified as ignitable waste or corrosive waste?
11. Should waste sulfuric acid be stored in steel drums prior to proper disposal? What type of container should be used?
12. Picric acid, which was once used by printers to etch copper, is known to explode when heated or subjected to shock. How would picric acid be classified as a waste?
13. Many printers use hydrocarbon solvents to clean ink from their hands and equipment. These solvents have a flash point below 60°C. How would they be classified as wastes?
14. How would lead from discarded automobile batteries be classified?
15. How would waste sodium hydroxide be classified?
16. You have to dispose of some nitric acid. What could you do to make it less corrosive?
17. Why is there more incineration of toxic waste in Europe than in the United States?
18. List three methods used for perpetual storage of hazardous wastes. If you had to make a choice among the three for storage of dioxin-contaminated waste, would you be able to recommend one?
19. Why is deep-well injection a problem for communities near the well?
20. What is surface impoundment?
21. What is the Superfund program? What does it regulate?
22. From where does the money for the Superfund trust fund come?
23. What is the National Priorities List?
24. What is meant by the acronym NIMBY? Apply NIMBY to hazardous wastes.
25. Identify two ways in which manufacturing companies can minimize the amount of hazardous wastes they generate.
26. Some people advocate ocean dumping as a method for disposing of toxic materials far away from human populations. On the basis of what you have learned so far, do you agree that this is a viable option?
27. Before 1970, where were most low-level radioactive wastes generated in the United States dumped?
28. What federal government agency regulates almost all radioactive wastes?
29. What are the two major sources of radioactive wastes?
30. Are radioactive materials regulated by RCRA?
31. Cesium-137, which is produced in nuclear reactors, has a half-life of 30 years. If a nuclear reactor stopped operating today, how long will it be until the cesium it contains would be considered no longer dangerous?
32. How many half-lives does it take for the radiation from radioactive waste to be considered at negligible levels? How many years for plutonium-239? For strontium-90?
33. How has radioactive waste been disposed of at the Hanford nuclear weapons facility in Richland, Washington?
34. Describe what is meant by vitrification of high-level radioactive wastes.
35. Nuclear disarmament should result in the retrieval of large quantities of plutonium from nuclear weapons that are being destroyed. Suggest a peaceful use for this plutonium.
36. What is the Low-Level Radioactive Waste Policy Act? Should a heavily populated state such as New York be held responsible for the disposal of radioactive waste from its nuclear power plants?
37. What is WIPP? Should the state of New Mexico have a say in what goes into WIPP?
38. What happens to spent fuel rods from nuclear power plants? Where will the radioactive material in these rods be permanently stored?
39. What type of radioactive waste does the Nuclear Waste Policy Act regulate?
40. In what state does the Department of Energy plan to permanently store waste from nuclear power plant reactors?
41. Why isn't radioactive material from spent fuel rods recycled?
42. Describe the recommended procedure for the disposal of high-level radioactive waste.

43. Some people have suggested that radioactive waste should be shot into the sun or outer space. List two reasons why this disposal method is not likely to be used.

44. Some have suggested that decommissioned nuclear power plants be entombed in concrete rather than being dismantled. List the advantages and disadvantages of entombment and dismantling. Consider the exposure of workers and the public to radiation and the cost.

45. What is MOX, and what is its source?

46. With the end of the Cold War, many old nuclear warheads were decommissioned, and their plutonium had to be disposed of. List three ways in which the disposal was carried out.

47. From what materials should a container designed to hold high-level radioactive waste be made?

48. It is difficult to find a location for the disposal of high-level radioactive waste. List three criteria that should be used to pick a location.

49. What is the most common way to dispose of medical waste?

50. What is meant by the acronym NOPE?

Appendix A Exponential Notation

Scientists frequently use numbers that are very large or very small, for example:

(a) 12 g of carbon contains **602,000,000,000,000,000,000,000** atoms of carbon
(b) a single molecule of water weighs **0.000 000 000 000 000 000 000 03** g

Such numbers are cumbersome to work with and difficult to write without losing track of the zeros. For convenience and to avoid errors, scientists express very large and very small numbers in a compact form called **exponential notation,** or **scientific notation.**

Exponential notation is a system in which numbers are expressed in this form:

$$N \times 10^x$$

N, which is called the **coefficient,** is a number between 1 and 10; x is the **exponent,** often referred to as the **power of ten;** and 10^x is the **exponential term.**

The exponent (or power of ten) may be positive or negative. Positive exponents indicate numbers equal to 10 or greater; negative exponents indicate numbers less than 1. (Numbers between 1 and 10 do not need an exponential term.)

The two numbers (a) and (b), written above in long form, are written as follows in exponential notation:

(a) 6.02×10^{23} (coefficient = 6.02, exponent = 23)
(b) 3.0×10^{-24} (coefficient = 3.0, exponent = −24)

Positive and Negative Exponents

A **positive exponent** tells how many times the coefficient must be *multiplied* by ten to get the long form of the number, for example:

$$5.43 \times 10^4 = 5.43 \times 10 \times 10 \times 10 \times 10 \quad \text{(4 times)}$$
$$= 54300$$
$$2.1 \times 10^3 = 2.1 \times 10 \text{ x } 10 \times 10 \quad \text{(3 times)}$$
$$= 2100$$

For (a), above, 6.02 must be multiplied by ten, 23 times to get the long form.

A **negative exponent** tells how many times the coefficient must be *divided* by ten to get the long form of the number, for example:

$$1 \times 10^{-1} = \frac{1}{10} \quad \text{(1 time)}$$
$$= 0.1$$
$$8.06 \times 10^{-5} = \frac{8.06}{10 \times 10 \times 10 \times 10 \times 10} \quad \text{(5 times)}$$
$$= 0.0000806$$

For (b) above, 3.0 must be divided by ten 24 times to get the long form.

Rules for Writing Numbers in Exponential Notation

1. *For numbers* **equal to or greater** *than 10:* Move the original decimal point to the *left* to obtain a number between 1 and 10. (The decimal point may not be shown, but its position is understood.)

   760 000 000 000. ← original position of decimal

   The original decimal point must be moved 11 places to the left to obtain the number 7.6. The number of decimal places moved (11) gives the exponent (or power of ten); the exponent is *positive.*

   $$760\ 000\ 000\ 000. = 7.6 \times 10^{11}$$

2. *For numbers* **less** *than 1:* Move the original decimal point to the *right* to obtain a number between 1 and 10.

   0.00631

   The original decimal point must be moved 3 places to the right to obtain the number 6.31. The number of decimal places moved (3) gives the exponent; the exponent is *negative.*

   $$0.00631 = 6.31 \times 10^{-3}$$

EXAMPLE A.1

Express the following numbers in scientific notation:

a. 8400 kg **b.** 6,000,000 m **c.** 0.792 mL **d.** 0.00036 g

Solution

a. 8400 kg $= 8.4 \times 10^3$ kg

To obtain a number between 1 and 10, the original decimal point must be moved 3 places to the *left*. Therefore, the exponent is +3.

b. 6,000,000 m $= 6.0 \times 10^6$ m

To obtain a number between 1 and 10, the original decimal point must be moved 6 places to the *left*. Therefore, the exponent is +6.

c. 0.792 mL $= 7.92 \times 10^{-1}$ mL

To obtain a number between 1 and 10, the original decimal point must be moved 1 place to the *right*. Therefore, the exponent is −1.

d. 0.00036 g $= 3.6 \times 10^{-4}$ g

To obtain a number between 1 and 10, the original decimal point must be moved 4 places to the *right*. Therefore, the exponent is −4.

To Convert from Exponential Notation to Decimal Notation

The value of the exponent indicates the number of places the decimal point must be moved.

If the exponent is *positive*, the decimal point must be moved to the *right*.
If the exponent is *negative*, the decimal point must be moved to the *left*.

Zeros may have to be added as the decimal point is moved.

EXAMPLE A.2

Write each number in decimal notation:

a. 2.69×10^5 **b.** $4.0 + 10^{-6}$

Solution

a. The exponent is +5. Therefore, the decimal point must be moved 5 places to the right, and 3 zeros must be added.

added zeros

$2.69000 = 269000$

b. The exponent is −6. Therefore, the decimal point must be moved 6 places to the left, and 5 zeros must be added (not counting the one before the decimal point).

added zeros

$000004.0 = 0.000004$

Addition and Subtraction

To add or subtract numbers expressed in exponential notation, proceed as follows:

EXAMPLE A.3

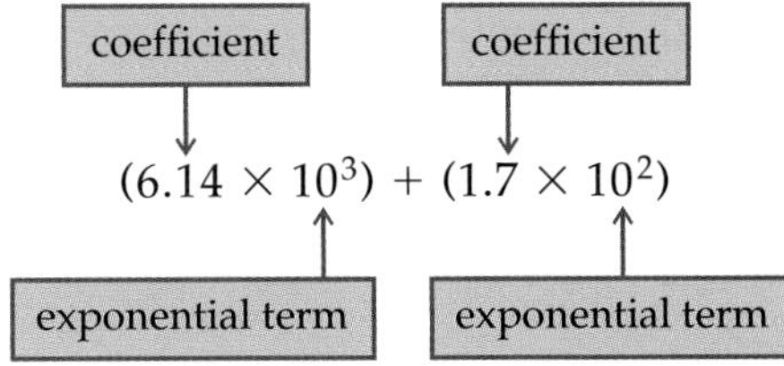

Solution

1. Adjust the coefficients so that both exponential terms are the same (in this case, you can use either 10^2 or 10^3):

$$(61.4 \times 10^2) + (1.7 \times 10^2)$$

2. Add the new coefficients:

$$(61.4 + 1.7) \times 10^2 = 63.1 \times 10^2$$

3. Adjust the answer to exponential notation if necessary.

$$6.31 \times 10^3$$

EXAMPLE A.4

Do the following subtraction:

$$(2.39 \times 10^{-4}) - (8.0 \times 10^{-6})$$

Solution

1. Adjust the coefficients so that both exponential terms are the same (either 10^{-4} or 10^{-6} in this example):

$$(2.39 \times 10^{-4}) - (0.08 \times 10^{-4})$$

2. Subtract the second new coefficient from the first:

$$(2.39 - 0.08) \times 10^{-4} = 2.31 \times 10^{-4}$$
$$= 2.31 \times 10^{-4}$$

Multiplication and Division

When numbers are expressed in exponential notation, multiplication and division are greatly simplified. In multiplication, exponents are *added;* in division, they are subtracted.

MULTIPLICATION To multiply two or more numbers, proceed as follows:

EXAMPLE A.5

$$(1.2 \times 10^3)(3.0 \times 10^2)$$

Solution

1. Multiply the coefficients in the usual manner:

$$1.2 \times 3.0 = 3.6$$

2. To obtain the new exponent, add the two exponents together:

$$10^3 \times 10^2 = 10^{3+2} = 10^5$$

3. Combine the new coefficient with the new exponential term to obtain the answer:

$$3.6 \times 10^5$$

EXAMPLE A.6

$$(5.2 \times 10^9)(7.4 \times 10^{-7})$$

Solution

1. $5.2 \times 7.4 = 38$
2. $10^9 \times 10^{-7} = 10^{9+(-7)} = 10^2$
3. Combine coefficient and exponential term:

$$38 \times 10^2$$

4. Express in correct exponential notation:

$$3.8 \times 10^3$$

DIVISION To divide two or more numbers, proceed as follows:

EXAMPLE A.7

$$\frac{4.8 \times 10^4}{2.4 \times 10^2}$$

Solution

1. Divide the coefficients in the usual manner:

$$\frac{4.8}{2.4} = 2.0$$

2. To obtain the new exponent, subtract the exponent of the denominator from the exponent of the numerator:

$$\frac{10^4}{10^2} = 10^{4-2} = 10^2$$

3. Combine coefficient and exponential term to obtain the answer:

$$2.0 \times 10^2$$

EXAMPLE A.8

$$\frac{7.64 \times 10^{10}}{8.91 \times 10^{-4}}$$

Solution

1. $\frac{7.64}{8.91} = 0.857$

2. $\frac{10^{10}}{10^{-4}} = 10^{10-(-4)} = 10^{10+4} = 10^{14}$

3. Combine coefficient and exponential term:

$$0.857 \times 10^{14}$$

4. Express in correct exponential notation:

$$8.57 \times 10^{13}$$

Common Logarithms

The *common, or base 10, logarithm* (abbreviated log) of a number N is the power to which 10 must be raised to give N. For example, the logarithm of 1000 is 3 because raising 10 to the third power gives 1000.

$$10^3 = 1000 \qquad \log 1000 = 3$$

Thus, a logarithm is an exponent.

When a number N is an integral power of 10, its log is a simple integer; if N is less than 1, the integer is negative. For example:

$$\begin{array}{llll} N = 1 & = 10^0 & \log 10^0 & = 0 \\ N = 10 & = 10^1 & \log 10^1 & = 1 \\ N = 100 & = 10^2 & \log 10^2 & = 2 \\ N = 0.01 & = 10^{-2} & \log 10^{-2} & = -2 \\ N = 0.0001 & = 10^{-4} & \log 10^{-4} & = -4 \end{array}$$

Logarithms of numbers that are integral powers of 10 are easily determined by inspection. However, logarithms of numbers that are not integral powers of 10, such as 6, 34.2, or 0.00851, must be obtained from a log table or by using a calculator.

Logarithms of nonintegral numbers are not required to solve any of the problems given in this text and will not be considered.

pH

The pH is a measure of the acidity, or hydrogen ion concentration, of a solution (Chapter 7); it is a logarithmic relationship that is frequently used in chemistry.

By definition, the pH of a solution is equal to the negative logarithm of the hydrogen ion concentration, $[H^+]$:

$$\text{pH} = -\log [H^+]$$

The pH of a solution is, therefore, equal to the negative power to which 10 must be raised to give the hydrogen ion concentration.

If $\qquad [H^+] = 10^{-x}$

then $\qquad \text{pH} = x$

The pH of a solution whose hydrogen ion concentration is an exact power of 10 is given by the negative of the exponent.

Example A.9

What is the pH of a solution for which $[H^+] = 0.0001$?

Solution

$$0.0001 = 10^{-4}$$
$$\log 10^{-4} = -4$$

Therefore,

$$-\log 10^{-4} = -(-4) = 4$$
$$\text{pH} = 4$$

pH values that are not whole numbers will not be considered in this text.

Questions and Problems

1. Express the following in scientific notation:

a. 423 million

b. 903,584

c. 0.000617

d. 0.00000006

e. 0.00000000000000749

f. 37,693,528

g. 0.04020

h. 0.000000001

i. 6.003

j. 2,300,000,000

2. What is the pH of a solution that has a hydrogen ion concentration of 0.00001?

3. The pH of a solution is 7. Give $[H^+]$ in exponential notation.

Appendix B

Measurement and the International System of Units (SI)

In science and many other fields, it is very important to make accurate measurements. For example, the establishment of the law of conservation of matter was dependent on accurate weight measurements; medical diagnosis relies on accurate measurements of factors such as temperature, blood pressure, and blood glucose concentration; a carpenter building a deck must make careful measurements before cutting the wood.

All measurements are made relative to some *reference standard*. For example, if you measure your height using a ruler marked in meters, you are comparing your height to an internationally recognized reference standard of length called the *meter*.

Most people in the United States use the English system of measurement and think in terms of English units: feet and inches, pounds and ounces, gallons and quarts, and so forth. If a man is described as being 6 foot 4 inches tall and weighing 300 pounds, we immediately visualize a large individual. Similarly, we know how much to expect if we buy a half-gallon of milk or a pound of hamburger at the grocery store. Most of us, however, have a far less clear idea of the meaning or size of meter, liter, and kilogram, common metric units of measurement that are used by the scientific community and by every other major nation in the world. Although the United States is committed to changing to the metric system, the pace of change, so far, has been extremely slow.

The International System of Units (SI)

The International System of Units, or SI (from the French, *Le Système International d'Unités*), was adopted by the International Bureau of Weights and Measures in 1960. The SI is an updated version of the metric system of units that was developed in France in the 1790s, following the French Revolution.

The *standard unit of length* in the SI is the **meter.** Originally, the meter was defined as one ten-millionth (0.0000001) of the distance from the North Pole to the equator measured along a meridian. This distance, however, was difficult to measure accurately, and for many years the meter was defined as the distance between two lines etched on a platinum-iridium bar kept at 0°C (32°F) in the International Bureau of Weights and Measures in Sevres, France. Today, the meter is defined even more precisely as equal to 1,650,763.73 times the wavelength of the orange-red spectrograph line of $^{86}_{36}Kr$.

The *standard unit of mass* is the **kilogram.** It is defined as the mass of a platinum-iridium alloy bar, which, like the original meter standard, is kept at the International Bureau of Weights and Measures. (Note: The difference between mass and weight is explained at the end of this appendix).

Table B.1 SI Base Units

Quantity Measured	Name of Unit	SI Symbol
Length	meter	m
Mass	kilogram	kg
Time	second	s
Electric current	ampere	A
Thermodynamic temperature	kelvin	K
Amount of a substance	mole	mol
Luminous intensity	candela	cd

BASE UNITS There are seven **base units** of measurement in the SI. They are shown in Table B.1.

PREFIXES The SI base units are often inconveniently large (or small) for many measurements. Smaller (or larger) units, defined by the use of *prefixes*, are used instead. Multiple and submultiple SI prefixes are given in Table B.2. Those most commonly used in general chemistry are shaded.

For example:

$$1\ \textit{kilo}\text{gram} = 1000 \text{ grams} \quad (\text{or } \frac{1}{1000} \text{ or } 0.001 \text{ kg} = 1 \text{ g})$$

$$1\ \textit{centi}\text{meter} = \frac{1}{100} \text{ or } 0.01 \text{ meter} \quad (\text{or } 100 \text{ cm} = 1 \text{ m})$$

DERIVED SI UNITS In addition to the seven SI base units, many other units are needed to represent physical quantities. All are derived from the seven base units. For example, *volume* is measured in cubic meters (m^3), *area* is measured in square meters (m^2),

Table B.2 SI Prefixes

Factor: Exponential Form	Factor: Decimal Form	Prefix	SI Symbol
10^6	1 000 000	mega	M
10^3	1 000	kilo	k
10^2	100	hecto	h
10	10	deka	da
10^{-1}	0.1	deci	d
10^{-2}	0.01	centi	c
10^{-3}	0.001	milla	m
10^{-6}	0.000 001	micro	μ
10^{-9}	0.000 000 001	nano	n
10^{-12}	0.000 000 000 001	pico	p

Table B.3 SI Derived Units

Physical Quantity	Unit	SI Symbol	Definition
Area	square meter	m^2	
Volume	cubic meter	m^3	
Density	kilogram per cubic meter	kg/m^3	
Force	newton	N	$kg{\cdot}m/s$
Pressure	pascal	Pa	$N/m^2 = kg/m{\cdot}s^2$
Energy, or quantity of heat	joule	J	$N{\cdot}m = kg{\cdot}m^2/s^2$
Quantity of electricity	coulomb	C	$A{\cdot}s$
Power	watt	W	$J/s = kg{\cdot}m^2/s^3$
Electric potential difference	volt	V	$J/A{\cdot}s = W/A$

Note: In 1964, the liter (L) was adopted as a special name for the cubic decimeter (dm^3).

and *density* is measured in mass per unit volume (kg/m^3). Derived SI units commonly used in general chemistry are listed in **Table B.3**.

Probably the least familiar of the derived units are the ones used to represent force (force = mass × acceleration), pressure (pressure = force/area), and energy, which is defined as the ability to do work (work = force × distance).

The SI unit of force, the **newton (N),** is defined as the force that, when applied for 1 second, will give a 1-kilogram mass a speed of 1 meter per second.

The SI unit of pressure, the **pascal (Pa),** is defined as the pressure exerted by a force of 1 newton acting on an area of 1 square meter. Often, it is more convenient to express a pressure in kilopascals (1 kPa = 1000 Pa). For example, atmospheric pressure at sea level is approximately equal to 100 kPa.

The SI unit of energy (or quantity of heat), the **joule (J),** is defined as the work done by a force of 1 newton acting through a distance of 1 meter. Often it is more convenient to express energy in kilojoules (1 kJ = 1000 J). (Note: The relationship between the joule and the more familiar calorie is discussed later in this appendix).

CONVERSIONS WITHIN THE SI Because the SI is based on the decimal system, conversions within it are much easier than conversions within the English system. Subunits and multiple units in the SI always differ by factors of 10, as shown in Table B.2. Thus, conversions from one unit to another are made by moving the decimal point the appropriate number of places. This procedure is best explained by some examples.

EXAMPLE B.1

Convert 0.0583 kilograms to grams.

Solution

1. Obtain the relationship between kilograms and grams from Table B.2:

$$1 \text{ kg} = 1000 \text{ g}$$

2. From the relationship, determine the conversion factor by which the given quantity (0.0583 kg) must be multiplied to obtain the answer in the required unit (g):

$$\text{conversion factor} = \frac{1000 \text{ g}}{1 \text{ kg}}$$

[The factor-unit method of problem solving (also called *dimensional analysis* or the *unit-conversion method*) is explained in more detail in Appendix C.]

3. Multiply 0.0583 kg by the conversion factor to obtain the answer:

$$0.0583 \cancel{\text{kg}} \times \frac{1000 \text{ g}}{1 \cancel{\text{kg}}} = 58.3 \text{ g}$$

EXAMPLE B.2

Convert 72 600 grams (g) to kilograms (kg).

Solution

1. 1 kg = 1000 g
2. The answer is required in kilograms. Therefore, the conversion factor is

$$\frac{1 \text{ kg}}{1000 \text{ g}}$$

(not 1000 g/1 kg as in Example B.1, where the answer was required in grams).

3.
$$72\ 600 \cancel{\text{g}} \times \frac{1 \text{ kg}}{1000 \cancel{\text{g}}} = 72.6 \text{ kg}$$

EXAMPLE B.3

Change 4236 millimeters (mm) to kilometers (km).

Solution

1. Table B.2 does not give a direct relationship between mm and km, but it does give relationships between mm and m, and between m and km:

$$1000 \text{ mm} = 1 \text{ m} \qquad 1000 \text{ m} = 1 \text{ km}$$

2. The given quantity (4236 mm) must be multiplied by *two* factors to obtain the answer in the required unit (km). In order for the proper terms to cancel out the two factors must be:

$$\frac{1 \text{ m}}{1000 \text{ mm}} \quad \text{and} \quad \frac{1 \text{ km}}{1000 \text{ m}}$$

3.
$$4236 \cancel{\text{mm}} \times \frac{1 \cancel{\text{m}}}{1000 \cancel{\text{mm}}} \times \frac{1 \text{ km}}{1000 \cancel{\text{m}}} = \frac{4236 \text{ km}}{1\ 000\ 000} = 0.004236 \text{ km}$$

CONVERSION FROM THE SI TO THE ENGLISH SYSTEM, AND VICE VERSA Units commonly used in the English system of measurement for length, mass, and volume are given in Table B.4.

Relationships that must be used to convert from SI units to English units, and vice versa, are given in Table B.5.

Table B.4 Units of Measurement in the English System

Length	Mass	Volume
12 inches (in.) = 1 foot (ft)	16 ounces (oz) = 1 pound (lb)	16 fluid ounces (fl oz) = 1 pint (pt)
3 feet = 1 yard (yd)	2000 pounds = 1 ton	2 pints = 1 quart (qt)
1760 yards = 1 mile (mi)		4 quarts = 1 gallon (gal)

EXAMPLE B.4

Convert 25 inches to centimeters.

Solution

1. Table B.5 gives the relationship between inches (in.) and centimeters (cm).

$$1 \text{ in.} = 2.54 \text{ cm}$$

2. The answer is required in centimeters. Therefore, the conversion factor is

$$\frac{2.54 \text{ cm}}{1 \text{ in.}}$$

3. $$25 \cancel{\text{in.}} \times \frac{2.524 \text{ cm}}{1 \cancel{\text{in.}}} = 63.5 \text{ cm}$$

EXAMPLE B.5

Convert 60 pounds to kilograms.

Solution

1. From Table B.5:

$$1 \text{ lb} = 0.454 \text{ kg}$$

2. The answer is required in kilograms. Therefore, the conversion factor is

$$\frac{0.454 \text{ kg}}{1 \text{ lb}}$$

3. $$60 \cancel{\text{lb}} \times \frac{0.454 \text{ kg}}{1 \cancel{\text{lb}}} = 27.24 \text{ kg}$$

Table B.5 Conversion Factors for Common SI and English Units

Length	Mass	Volume
1 inch (in.) = 2.54 centimeters (cm)	1 ounce (oz) = 28.4 grams (g)	1 fluid ounce (fl oz) = 29.6 milliliters (mL)
1 yard (yd) = 0.914 meter (m)	1 pound (lb) = 454 grams (g)	1 U.S. pint (pt) = 0.473 liter (L)
1 mile (mi) = 1.61 kilometers (km)	1 pound (lb) = 0.454 kilogram (kg)	1 U.S. quart (qt) = 0.946 liter (L)
		1 gallon (gal) = 3.78 liters (L)

EXAMPLE B.6

Convert 3.5 liters to quarts.

Solution

1. From Table B.5:

$$1 \text{ qt} = 0.946 \text{ L}$$

2. The answer is required in quarts. Therefore, the conversion factor is

$$\frac{1 \text{ qt}}{0.946 \text{ L}}$$

3.
$$3.5 \cancel{\text{L}} \times \frac{1 \text{ qt}}{0.946 \cancel{\text{L}}} = 3.7 \text{ qt}$$

EXAMPLE B.7

Convert 4.83 meters to feet.

Solution

1. From Table B.5:

$$1 \text{ yd} = 0.914 \text{ m} \qquad \text{and} \qquad 1 \text{ yd} = 3 \text{ ft}$$

2. The answer is required in feet. Therefore, the two conversion factors needed are

$$\frac{1 \text{ yd}}{0.914 \text{ m}} \qquad \text{and} \qquad \frac{3 \text{ ft}}{1 \text{ yd}}$$

3.
$$4.83 \text{ m} \times \frac{1 \cancel{\text{yd}}}{0.914 \text{ m}} \times \frac{3 \text{ ft}}{1 \cancel{\text{yd}}} = 15.8 \text{ ft}$$

USEFUL APPROXIMATIONS BETWEEN SI AND ENGLISH UNITS To make rough estimates of English units in terms of SI units and to get a feel for the meaning of SI units, you will find it useful to memorize the following approximations:

$$1 \text{ kg} \approx 2 \text{ lb} \qquad 1 \text{ m} \approx 1 \text{ yd} \qquad 1 \text{ L} \approx 1 \text{ qt} \qquad 1 \text{ km} \approx \frac{2}{3} \text{ mi}$$

Other Commonly Used Units of Measurement

For certain measurements, including those of temperature and energy, scientists continue to use units that are not SI units.

TEMPERATURE The SI unit for temperature is the kelvin (K) (Table B.1), but scientists use the **Celsius scale** for many temperature measurements. On this scale, the unit of temperature is the degree (°C); 0°C corresponds to the freezing point of water, and 100°C corresponds to its boiling point, both at atmospheric pressure. The 100 degrees between these two reference points are of equal size.

On the Kelvin scale, zero temperature corresponds to the lowest temperature it is possible to attain, or **absolute zero.** Absolute zero, as determined theoretically and confirmed experimentally, is equal to −273°C (more accurately −273.15°C). The unit of temperature on the Kelvin scale is the same size as a degree on the Celsius scale. Therefore, to convert from °C to K, it is only necessary to add 273:

$$K = °C + 273 \quad \text{or} \quad °C = K - 273$$

Thus, the boiling point of water is 373 K, and its freezing point is 273 K.

In the United States, most temperatures, including those given in weather reports and cooking recipes, are measured on the **Fahrenheit scale.** On this scale, the freezing point of water is 32°F, and its boiling point is 212°F. Thus, there are 180 (212 − 32) degrees between the two reference points, compared to 100 degrees on the Celsius scale.

CONVERSIONS BETWEEN °C AND °F

$$100 \text{ divisions in } °C = 180 \text{ divisions in } °F$$

or, dividing both sides of the equation by 20:

$$5 \text{ divisions } °C = 9 \text{ divisions } °F$$

To convert from °C to °F, the following equation is used:

$$°F = \frac{9}{5}(°C) + 32$$

Here, 32 is added to the °C because the freezing point of water is 32° on the Fahrenheit scale compared to 0° on the Celsius scale.

To convert from °F to °C, the following equation is used:

$$°C = \frac{5}{9}(°F - 32)$$

In this case, 32 must be subtracted from the °F before multiplying by $\frac{5}{9}$.

EXAMPLE B.8

Convert 77°F to °C.

Solution

$$\begin{aligned} °C &= \frac{5}{9}(77 - 32) \\ &= \frac{5}{9} \times 45 \\ &= 25 \end{aligned}$$

$$77°F = 25°C$$

EXAMPLE B.9

Convert 35°C to °F.

Solution

$$\begin{aligned} °F &= \frac{9}{5}(35) + 32 \\ &= 63 + 32 \\ &= 95 \end{aligned}$$

$$35°C = 95°F$$

ENERGY The SI unit for measuring heat, or any form of energy, is the joule (J) (Table B.3), but scientists often use the more familiar **calorie (cal).** For measuring the energy content of food, the kilocalorie (kcal) or **Calorie (Cal)** is used.

A calorie is the amount of heat required to raise the temperature of 1 gram of water 1°C.

$$1 \text{ cal} = 4.18 \text{ J}$$

The Difference Between Mass and Weight

Although the terms *mass* and *weight* are frequently used interchangeably, they have basically different meanings.

Mass is a measure of the quantity of matter in an object.
Weight is a measure of the force exerted on an object by the pull of gravity.

The difference between mass and weight was dramatically demonstrated when astronauts began to travel in space. An astronaut's mass does not change as he or she is rocketed into space, but once free of the gravitational pull of the earth, the astronaut becomes weightless. On the moon, an astronaut's weight is approximately one-sixth of what it is on earth because of the moon's much weaker gravitational pull.

Questions and Problems

1. Convert 825 mL to liters.
2. Convert 153,000 mg to kilograms.
3. Convert 0.00061 cm to nanometers.
4. Convert 56800 mg to ounces.
5. Convert 1.7 m to feet and inches.
6. Convert 3 lb 8 oz to kilograms.
7. A road sign informs you that you are 60 km from Calais. How many minutes will it take you to arrive there if you travel at 60 miles per hour?
8. A doctor orders a 1 fluid ounce dose of medicine for a patient. How much is this dose in milliliters?
9. You need to purchase approximately 2 lb of meat at an Italian grocery store. Should you ask for $\frac{1}{2}$, 1, or 2 kg?
10. Normal body temperature is 98.6°F. A patient has a temperature of 40°C.
    a. Is this temperature above or below normal?
    b. By how many °F?
    c. By how many °C?
11. The room temperature is 15°C. Should you turn on the air conditioning or the heat?

Appendix C

Problem Solving Using Dimensional Analysis

Solving mathematical problems, including many chemical problems, often requires one or more conversions from one unit of measurement to another—for example, from feet to inches, pounds to kilograms, or grams to moles. Such problems can often be solved with the aid of the method variously called **dimensional analysis,** the *unit-conversion method,* or the *factor-unit method.*

This method uses the fact that *units* in the denominator and numerator of a fraction cancel in the same way that numbers cancel. For example, if the units in both numerator and denominator are inches, they cancel:

$$\frac{9\ \cancel{\text{inches}}}{5\ \cancel{\text{inches}}} = \frac{9}{5}$$

To convert from one unit to another, it is necessary to derive a **conversion factor** based on the known relationship between the units involved. This is best explained by an example. Consider the conversion of 6 feet to inches. The relationship between feet and inches is

$$1\ \text{ft} = 12\ \text{in.}$$

From this relationship, we can derive *two* conversion factors, A and B. Since 1 ft equals 12 inches, it follows that

$$\frac{1\ \text{ft}}{12\ \text{in.}} = 1 \qquad \text{conversion factor A}$$

Similarly:

$$\frac{12\ \text{in.}}{1\ \text{ft}} = 1 \qquad \text{conversion factor B}$$

The key to solving problems using dimensional analysis is choosing the correct conversion factor.

To convert 6 feet to inches, we need to multiply by either factor A or factor B. We choose the factor so that feet cancel out and the answer is expressed in inches. We use factor B:

$$6\ \cancel{\text{ft}} \times \frac{12\ \text{in.}}{1\ \cancel{\text{ft}}}$$

feet (ft) cancel

$$6\ \text{ft} = 72\ \text{in.}$$

Factor A would have given the wrong numerical answer and the wrong unit:

$$6\ \text{ft} \times \frac{1\ \text{ft}}{12\ \text{in.}} = 0.5\ \text{feet}^2/\text{in.}$$

In general terms, the formula for converting from one unit to another is

(given quantity) × (conversion factor) = (answer in desired unit)

The conversion factor is chosen so that when it is multiplied by the given unit, the desired unit results:

$$\cancel{\text{given unit}} \times \frac{\text{desired unit}}{\cancel{\text{given unit}}} = \text{desired unit}$$

Example C.1

Convert 100 pounds to kilograms (1 lb = 0.454 kg).

Solution

1. Given quantity: 100 lb
2. Desired unit: kg
3. Conversion factor: $\frac{0.454 \text{ kg}}{1 \text{ lb}}$
4. $100 \cancel{\text{lb}} \times \frac{0.454 \text{ kg}}{1 \cancel{\text{lb}}} = 45.4 \text{ kg}$

Example C.2

The distance between two cities in Mexico is 80 kilometers. This is equal to how many miles (1 mi = 1.61 km)?

Solution

1. Given quantity: 80 km
2. Desired unit: mi
3. Conversion factor: $\frac{1 \text{ mi}}{1.61 \text{ km}}$
4. $80 \cancel{\text{km}} \times \frac{1 \text{ mi}}{1.61 \cancel{\text{km}}} = 50 \text{ mi}$

Example C.3

Convert 7.0 liters to fluid ounces (1 qt = 0.946 L).

Solution

1. Given quantity: 7.0 L
2. Desired unit: fl oz

3. Two conversion factors are required, one to convert liters to quarts and another to convert quarts to fluid ounces (1 qt = 32 fl oz).

$$\frac{1 \text{ qt}}{0.946 \text{ L}} \qquad \frac{32 \text{ fl oz}}{1 \text{ qt}}$$

4. $$7.0 \not{L} \times \frac{1 \not{qt}}{0.946 \not{L}} \times \frac{32 \text{ fl oz}}{1 \not{qt}} = 240 \text{ fl oz}$$

Conversion Factors Based on Equivalence

In the above examples, the numerator of the conversion factor was always *equal* to the denominator. For example, in the factor 12 in./1 ft, 12 inches equal 1 foot; in the factor 1 mi/1.61 km, 1 mile equals 1.61 kilometers.

Conversion factors based on units that are *equivalent* to each other, but not equal, can also be used to solve problems, as shown in Example C.4.

Example C.4

A worker is paid at the rate of $10.50 per hour. How many hours will he have to work to earn $84?

Solution

1. Given quantity: 84 dollars
2. Desired unit: hours
3. There is an equivalence between time and dollars:

$$1 \text{ hour} \sim \$10.50$$

The sign ~ is used to represent "is equivalent to." The conversion factor is

$$\frac{1 \text{ hour}}{10.50 \text{ dollars}}$$

4. $$\frac{84 \not{\text{dollars}} \times 1 \text{ hour}}{10.50 \not{\text{dollars}}} = 8 \text{ hours}$$

The worker will have to work 8 hours to earn $84.

Example C.5

A solution contains 0.040 g of salt per 100 mL. How many grams of salt are contained in 250 mL of the solution?

Solution

1. Given quantity: 250 mL of solution
2. Required quantity: g of salt
3. In any problem involving solution concentration (g/L, mg/mL, etc.), the conversion factor is based on the relationship between the mass and the vol-

ume. Depending on the quantity that has to be determined, it is either mass/volume or volume/mass. In this problem, the conversion factor is

$$\frac{0.040 \text{ g}}{100 \text{ mL}}$$

4. $$250 \cancel{\text{mL}} \times \frac{0.040 \text{ g}}{100 \cancel{\text{mL}}} = 0.10 \text{ g}$$

0.10 g of salt are contained in 250 mL.

EXAMPLE C-6

When heated, 434 g of mercuric oxide (HgO) yield 32 g of oxygen (O_2). How many grams of HgO are needed to produce 100 g of O_2?

Solution

1. Given quantity: 100 g O_2
2. Required quantity: g HgO
3. To solve this problem, we need a conversion factor based on the relationship between the quantities of two substances (HgO and O_2) measured in the same unit (g):

$$\frac{434 \text{ g HgO}}{32 \text{ g } O_2}$$

4. $$100 \cancel{\text{g } O_2} \times \frac{434 \text{ g HgO}}{32 \cancel{\text{g } O_2}} = 1360 \text{ g HgO}$$

1360 g HgO are needed to produce 100 g of O_2.

Questions and Problems

(Conversion tables and more problems are presented in Appendix B.)

1. Convert 754.0 centimeters to meters.
2. Convert 0.036 kilogram to milligrams.
3. Convert 6 feet 3 inches to meters.
4. Convert 3 pints to milliliters.
5. It takes a secretary a half-hour to type 4 pages. How long will he take to type 60 pages?
6. In 1996, the permitted level of arsenic in drinking water was 50 ppb (50 μg/L). A 12-oz glass of water contained 30 μg of arsenic.
   a. Was this above or below the permitted level?
   b. By how much, in μg/L?
7. A solution contains 4.0 g of sodium hydroxide (NaOH) per liter. How many grams are contained in 50 mL?
8. The longest and shortest waves of visible light have wavelengths of 0.000067 cm and 0.000037 cm. Convert these values to nanometers.
9. An athlete runs 100 yards in 10 seconds. What is her speed in miles per hour?
10. A solution contains 6.4 g of barium nitrate per liter. How many milliliters of this solution are needed in order to obtain 1.6 g of barium nitrate?
11. The reaction of 87.0 g of manganese(IV) oxide (MnO_2) with excess hydrochloric acid yields 71.0 g of chlorine (Cl_2). How many grams of MnO_2 are required to obtain 56.8 g of Cl_2?
12. When 114 g of octane (C_8H_{18}), an ingredient of gasoline, are burned, 44 g of carbon dioxide (CO_2) are produced. When 20 kg of octane are burned, calculate how much CO_2 is produced in:

   a. g b. kg c. tons

Appendix D

Significant Figures

A measurement is never exact. Depending on the type of instrument used, there is a limit to the accuracy with which any measurement can be made. For example, if the piece of wood shown in Figure D.1 is measured with the ruler shown, it is not possible to determine with certainty if its length is 3.23, 3.24 or 3.25 cm. There is no uncertainty about the first two digits (3 and 2), but there is some degree of uncertainty about the last digit. The measurements 3.23 cm, 3.24 cm, and 3.25 cm are said to have three significant figures: the first two, which are known with certainty, and the last one which is an estimate.

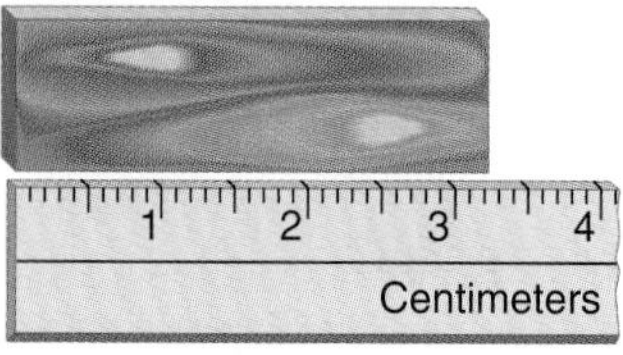

FIGURE D.1

The **significant figures** in a number can be defined as *the numbers that express reasonably reliable information*. They include all the numbers known with certainty plus one that is uncertain.

Rules for Determining Significant Figures

The following rules are used to determine how many significant figures there are in a number. It is assumed that the number represents a measurement that has been accurately made and recorded.

1. All nonzero digits (1, 2, 3, 4, 5, 6, 7, 8, and 9) are significant.

   478 has 3 significant figures.
   523.61 has 5 significant figures.

2. Zeros between nonzero digits are significant.

   1.063 has 4 significant figures.
   5002.08 has 6 significant figures.

3. Zeros to the *left* of nonzero digits are not significant. They serve only to indicate the position of the decimal point.

   0.00205 has 3 significant figures.
   0.09 has 1 significant figure.

4. Zeros *following* other digits in a number that contains a decimal point are significant.

   0.5300 has 4 significant figures.
   0.070 has 2 significant figures.
   58.00 has 4 significant figures.

5. Zeros at the end of a number with *no* decimal point may or may not be significant. A number such as 9,400 can usually be assumed to have 2 significant figures unless otherwise indicated. The uncertainty can be removed by expressing the number in exponential notation (see Appendix A):

9.4×10^3 denotes 2 significant figures.
9.40×10^3 denotes 3 significant figures.
9.400×10^3 denotes 4 significant figures.

Accuracy and Precision

In science, it is important to make measurements that are both accurate and precise. **Accuracy** refers to the extent to which a measurement agrees with the true value of the quantity being measured. **Precision** refers to the closeness of agreement between successive measurements of a quantity. The difference between these two terms is best explained by an example.

Three students each make three measurements of the concentration of sodium chloride in a solution that is known to contain 2.79 g/L, with the following results:

Measurement	Student 1	Student 2	Student 3
1	2.41 g/L	2.24 g/L	2.80 g/L
2	2.43 g/L	3.15 g/L	2.79 g/L
3	2.42 g/L	2.98 g/L	2.78 g/L
Average:	2.42 g/L	2.79 g/L	2.79 g/L

Student 1's measurements are very precise because they vary from each other by no more than 0.02 g, but they are not accurate because they differ from the true value (2.79 g/L) by as much as 0.38 g/L. Student 2's measurements are very imprecise, although their average gives an accurate value. Student 3's measurements are both accurate and precise.

Rounding Off Numbers

When measured quantities are added, subtracted, multiplied, or divided, the results must be reported with the correct number of significant figures. This means that any nonsignificant numbers must be *rounded off*. The rules for rounding off are as follows:

1. If the first nonsignificant figure is *5 or greater,* increase the last significant figure by 1 and drop the nonsignificant figures.

EXAMPLE D.1

Round the number 24.381 to 3 significant figures.

Solution

The first nonsignificant figure, 8, is greater than 5. The last significant figure, 3, is therefore increased by 1.

$$24.381 = 24.4 \quad \text{(to 3 significant figures)}$$

EXAMPLE D.2

Round the number 1273 to 2 significant figures.

Solution

$$1273 = 1300 \quad \text{(to 2 significant figures)}$$

Note that in this case the dropped nonsignificant figures must be replaced by zeros to maintain the correct dimensions.

---

2. If the first nonsignificant figure is *less than 5,* drop all the nonsignificant figures and do not change the last significant figure.

EXAMPLE D.3

Round the number 1.6392 to 2 significant figures.

Solution

The first nonsignificant figure, 3, is less than 5.

$$1.6392 = 1.6 \quad \text{(to 2 significant figures)}$$

EXAMPLE D.4

Round the number 855,216 to 3 significant figures.

Solution

$$855{,}216 = 855{,}000 \quad \text{(to 3 significant figures)}$$

ADDITION AND SUBTRACTION In addition and subtraction, the result should be rounded off so that it has only as many digits after the decimal point as the number being added or subtracted that has the *fewest* number of digits after the decimal point. In these operations, the number of significant figures in the numbers being added or subtracted is irrelevant.

EXAMPLE D.5

Add the following numbers:

49.8426, 6.07, 103.5

Solution

The number with the fewest number of digits after the decimal point is 103.5. Therefore, the sum should have only one digit after the decimal point.

$$\begin{array}{r} 49.8426 \\ 6.07 \\ \underline{103.5} \\ 159.4126 \end{array}$$

Answer: 159.4

Notice that the correct answer (159.4) has 4 significant figures even though one of the numbers (6.07) has only 3 significant figures.

EXAMPLE D.6

Add the following numbers:

$$737, 6.5, 15.316$$

Solution

The number 737 has no digits after the decimal point (which is understood to be at the right of the last digit). Therefore, the sum should be rounded off so that it has no digits after the decimal point.

$$\begin{array}{r} 737 \\ 6.5 \\ 15.316 \\ \hline 758.816 \end{array}$$

Answer: 759

EXAMPLE D.7

Subtract 1.526 from 5.2308.

Solution

$$\begin{array}{r} 5.2308 \\ -1.526 \\ \hline 3.7048 \end{array}$$

Answer: 3.705

MULTIPLICATION AND DIVISION In multiplication and division, the result should be rounded off to the same number of significant figures as the number being multiplied or divided that has the *fewest* number of significant figures. In these operations, the number of digits after the decimal point in any of the numbers is not taken into consideration.

EXAMPLE D.8

Multiply 3.9 by 1.483.

Solution

$$3.9 \times 1.483 = 5.7837$$

Answer: 5.8

EXAMPLE D.9

Divide 27.832 by 1.51

Solution

$$\frac{27.832}{1.51} = 18.431788$$

Answer: 18.4

Questions and Problems

Perform the indicated mathematical operations and give the answers with the proper number of significant figures.

1. 46.105 g + 27.3 g + 208.17 g
2. 0.143 + 123 + 1.51
3. 93.007 mL − 0.024 mL
4. 254.710 mg − 2.71 mg
5. 63.8 L + 5.42 L − 0.095 L
6. 32 m × 53.7 m × 0.41 m
7. 0.293 × 0.028
8. $\frac{2.35\text{ g}}{4600\text{ g}}$
9. $\frac{283.9\text{ g}}{51.0\text{ mol}}$
10. $1.236\text{ L} \times \frac{373\text{ K}}{263\text{ K}}$

Appendix E

Solving Problems Based on Chemical Equations

Many problems based on mass and mole relationships in chemical equations can be solved very easily using dimensional analysis (see Appendix C).

As you saw in Chapter 6, a properly balanced chemical equation provides a great deal of information. For example, from the equation for the reaction of hydrogen with oxygen to produce water, we can derive the following information:

$$2\ H_2 + O_2 \longrightarrow 2\ H_2O$$

$$2 \text{ molecules } H_2 + 1 \text{ molecule } O_2 \longrightarrow 2 \text{ molecules } H_2O$$

$$2 \text{ mol } H_2 + 1 \text{ mol } O_2 \longrightarrow 2 \text{ mol } H_2O$$

$$2 \text{ formula masses } H_2 + 1 \text{ formula mass } O_2 \longrightarrow 2 \text{ formula masses } H_2O$$

$$2 \times 2 \text{ amu } H_2 + 32 \text{ amu } O_2 \longrightarrow 2 \times 18 \text{ amu } H_2O$$

$$2 \text{ molar masses } H_2 + 1 \text{ molar mass } O_2 \longrightarrow 2 \text{ molar masses } H_2O$$

$$2 \times 2 \text{ g } H_2 + 32 \text{ g } O_2 \longrightarrow 2 \times 18 \text{ g } H_2O$$

$$4 \text{ g } H_2 + 32 \text{ g } O_2 \longrightarrow 36 \text{ g } H_2O$$

$$4 \text{ lb } H_2 + 32 \text{ lb } O_2 \longrightarrow 36 \text{ lb } H_2O$$

Thus, a chemical equation shows the relationships between the various reactants and products taking part in the reaction in terms of molecules, moles, and mass (amu, g, lb or any other unit of mass). These relationships provide the conversion factors needed for solving problems using dimensional analysis. They enable us, for example, to determine the number of moles or grams of a given reactant that are required to produce a certain number of moles or grams of product.

Mass-to-Mass Calculations

As we have just seen, a balanced chemical equation tells us the *mass relationships* between the substances taking part in a reaction. Knowing these relationships, we can calculate, for example, the number of kilograms of product that can be obtained from a given number of kilograms of a reactant or the number of pounds of reactant needed to produce a required number of pounds of product.

Example E.1

How many grams of water can be obtained from the complete reaction of 10 g of hydrogen with oxygen?

Solution

1. Write the balanced equation for the reaction.

$$2\ H_2 + O_2 \longrightarrow 2\ H_2O$$

2. From the equation, determine the mass relationship between the *given substance* and the *required substance.*

   a. Determine the molar masses of the given substance (H_2) and the required substance (H_2O). [Remember that the molar mass of a substance is the mass in grams that is numerically equal to the substance's formula mass. The atomic masses (atomic weights) that are needed to calculate formula masses can be found inside the cover of any chemistry textbook.]

   $$\text{Molar mass of } H_2 = 2 \times 1 = 2\text{ g}$$
   $$\text{Molar mass of } H_2O = (2 \times 1) + 16 = 18\text{ g}$$

   (The formula mass of O_2 is not required since the question does not ask for the mass of oxygen.)

   b. Substitute these values in the equation:

   $$\begin{array}{ccccc} 2 \times \text{molar mass } H_2 & + & 1 \times \text{molar mass } O_2 & \longrightarrow & 2 \times \text{molar mass } H_2O \\ 2 \times 2\text{ g } H_2 & & & & 2 \times 18\text{ g } H_2O \\ 4\text{ g } H_2 & & & & 36\text{ g } H_2O \end{array}$$

   Thus, 4 g of H_2 will react with O_2 to yield 36 g of H_2O.

   c. State the relationship between the given substance and the required substance:

   $$4\text{ g of } H_2 \text{ are } \textit{equivalent} \text{ to } 36\text{ g of } H_2O$$
   $$4\text{ g } H_2 \sim 36\text{ g } H_2O$$

   Express the relationship as a ratio:

   $$\frac{4\text{ g } H_2}{36\text{ g } H_2O} \quad \text{or} \quad \frac{36\text{ g } H_2O}{4\text{ g } H_2}$$

3. Solve the problem by dimensional analysis using the following formula:

   $$(\text{given quantity}) \times (\text{conversion factor}) = (\text{required quantity})$$

   Given quantity: 10 g H_2

   Required quantity: g H_2O

   Conversion factor: The correct conversion factor is the ratio shown above (in c) that when multiplied by the given quantity will give the required quantity:

   $$\frac{36\text{ g } H_2O}{4\text{ g } H_2}$$

   $$10\text{ }\cancel{\text{g } H_2} \times \frac{36\text{ g } H_2O}{4\text{ }\cancel{\text{g } H_2}} = 90\text{ g } H_2$$

   90 g of water can be obtained from the complete reaction of 10 g of hydrogen with oxygen.

EXAMPLE E.2

How many kilograms of iron are needed to prepare 116 kg of the magnetic oxide of iron, Fe_3O_4?

Solution

1. Balanced equation:

$$3\ Fe + 2\ O_2 \longrightarrow Fe_3O_4$$

2. From the equation, determine the mass relationship between Fe and Fe_3O_4.

   a. Molar mass of Fe $= 56$ g
   Molar mass of $Fe_3O_4 = (3 \times 56) + (4 \times 16) = 168 + 64$
   $= 232$ g

   b. Substitute these values in the equation:

$$\begin{array}{ccccc} 3 \times \text{molar mass Fe} & + & 2 \times \text{molar mass } O_2 & \longrightarrow & 1 \times \text{molar mass } Fe_3O_4 \\ 3 \times 56\text{ g} & & & & 232\text{ g} \\ 168\text{ g} & & & & 232\text{ g} \end{array}$$

   c. State the relationship between the given substance and the required substance:

$$168\text{ g Fe} \sim 232\text{ g } Fe_3O_4$$

   The answer is required in kilograms. Express the relationship using kilograms:

$$168\text{ kg Fe} \sim 232\text{ kg } Fe_3O_4$$

   Express the relationship as a ratio:

$$\frac{168\text{ kg Fe}}{232\text{ kg } Fe_3O_4} \quad \text{or} \quad \frac{232\text{ kg } Fe_3O_4}{168\text{ kg Fe}}$$

3. Solve by dimensional analysis.

   Given quantity: 116 kg Fe_3O_4

   Required quantity: kg Fe

   Conversion factor: $\dfrac{168\text{ kg Fe}}{232\text{ kg } Fe_3O_4}$

$$116\ \cancel{\text{kg } Fe_3O_4} \times \frac{168\text{ kg Fe}}{232\ \cancel{\text{kg } Fe_3O_4}} = 84\text{ kg Fe}$$

   84 kg of iron are needed to prepare 116 kg of Fe_3O_4.

Example E.3

How many grams of hydrochloric acid are required to convert 80 g of sodium hydroxide to salt and water?

Solution

1. Balanced equation:

$$HCl + NaOH \longrightarrow NaCl + H_2O$$

2. From the equation, determine the mass relationship between HCl and NaOH.

   Molar mass of HCl $= 36.5$ g
   Molar mass of NaOH $= 40$ g

Substitute these values in the equation:

$$1 \times \text{molar mass HCl} + 1 \times \text{molar mass NaOH} \longrightarrow 1 \text{ molar mass NaCl} + 1 \text{ molar mass } H_2O$$
$$36.5 \text{ g HCl} \qquad\qquad 40 \text{ g NaOH}$$

Express the relationship as a ratio:

$$\frac{36.5 \text{ g HCl}}{40 \text{ g NaOH}} \quad \text{or} \quad \frac{40 \text{ g NaOH}}{36.5 \text{ g HCl}}$$

3. Solve by dimensional analysis.

Given quantity: 80 g NaOH

Required quantity: g HCl

Conversion factor: $\frac{36.5 \text{ g HCl}}{40 \text{ g NaOH}}$

$$80 \cancel{\text{g NaOH}} \times \frac{36.5 \text{ g HCl}}{40 \cancel{\text{g NaOH}}} = 73 \text{ g HCl}$$

73 g of HCl are required to convert 80 g of NaOH to salt and water.

Mole-to-Mole Calculations

As we saw at the beginning of this appendix, when describing the reaction of hydrogen with oxygen to produce water, a balanced chemical equation shows the *mole relationships* between reactants and products. Knowing these mole relationships, we can calculate, for example, the number of moles of product that can be obtained from a given number of moles of a reactant or the number of moles of a reactant that are needed to produce a given number of moles of product.

We will go through the logical steps to be followed in solving this type of problem, although, as the following examples show, most mole-to-mole problems can be solved by inspection.

EXAMPLE E.4

How many moles of water are produced by the complete combustion of 360 moles of methane (CH_4)?

Solution

1. Balanced equation:

$$CH_4 + 2\,O_2 \longrightarrow CO_2 + 2\,H_2O$$

2. Determine the mole relationship between the given substance (CH_4) and the required substance (H_2O).

   a. Enter the numbers of moles of the substances taking part in the reaction in the equation:

$$1 \text{ mol } CH_4 + 2 \text{ mol } O_2 \longrightarrow 1 \text{ mol } CO_2 + 2 \text{ mol } H_2O$$

   b. State the mole relationship between the given substance and the required substance:

$$1 \text{ mol } CH_4 \sim 2 \text{ mol } H_2O$$

Express the relationship as a ratio:

$$\frac{1 \text{ mol } CH_4}{2 \text{ mol } H_2O} \quad \text{or} \quad \frac{2 \text{ mol } H_2O}{1 \text{ mol } CH_4}$$

3. Solve the problem using dimensional analysis.

Given quantity: $360 \text{ mol } CH_4$

Required quantity: $\text{mol } H_2O$

Conversion factor: $\frac{2 \text{ mol } H_2O}{1 \text{ mol } CH_4}$

$$360 \cancel{\text{mol } CH_4} \times \frac{2 \text{ mol } H_2O}{2 \cancel{\text{mol } CH_4}} = 720 \text{ mol } H_2O$$

720 moles of water are produced by the complete combustion of 360 moles of methane.

Example E.5

How many moles of hydrogen are needed to react completely with nitrogen to produce 6 moles of ammonia (NH_3)?

Solution

1. Balanced equation:

$$N_2 + 3\,H_2 \longrightarrow 2\,NH_3$$

2. Mole relationship between H_2 and NH_3:

$$1 \text{ mol } N_2 + 3 \text{ mol } H_2 \longrightarrow 2 \text{ mol } NH_3$$

$$3 \text{ mol } H_2 \sim 2 \text{ mol } NH_3$$

3. Solve the problem by dimensional analysis.

Given quantity: $6 \text{ mol } NH_3$

Required quantity: $\text{mol } H_2$

Conversion factor: $\frac{3 \text{ mol } H_2}{2 \text{ mol } NH_3}$

$$6 \cancel{\text{mol } NH_3} \times \frac{3 \text{ mol } H_2}{2 \cancel{\text{mol } NH_3}} = 9 \text{ mol } H_2$$

9 moles of hydrogen are needed to react completely with nitrogen to produce 6 moles of ammonia (NH_3).

Mole-to-Mass and Mass-to-Mole Calculations

A balanced chemical equation gives both mass and mole relationships for the substances taking part in the reaction. This information makes it possible to calculate, for example, the number of *grams* of a reactant that will produce a given number of *moles* of product, or the number of *grams* (or kg, or lb) of product that can be obtained from a given number of *moles* of reactant.

EXAMPLE E.6

How many moles of O_2 are produced by heating 49 grams of potassium chlorate ($KClO_3$)?

Solution

1. Balanced equation:

$$2\ KClO_3 \longrightarrow 2\ KCl + 3\ O_2$$

2. Determine the relationship between the required quantity, *moles* of O_2, and the given quantity, *grams* of $KClO_3$.

   a. Enter the number of moles of O_2 and $KClO_3$ taking part in the reaction in the equation.

   $$2\ \text{mol}\ KClO_3 \qquad 3\ \text{mol}\ O_2$$

   b. Express the formula mass of $KClO_3$ in grams:

   $$\text{formula mass of}\ KClO_3\ \text{in g} = 122.5\ \text{g} = \text{mass of 1 mol}\ KClO_3$$
   $$2\ \text{mol}\ KClO_3 = 2 \times 122.5\ \text{g}\ KIO_3$$
   $$= 245\ \text{g}\ KClO_3$$

   c. From a and b, we can see that the relationship between the required quantity (mol O_2) and the given quantity (g $KClO_3$) is

   $$3\ \text{mol}\ O_2 \sim 245\ \text{g}\ KClO_3$$

3. Solve the problem by dimensional analysis.

   Given quantity: $49\ \text{g}\ KClO_3$

   Required quantity: $\text{mol}\ O_2$

   Conversion factor: $\dfrac{3\ \text{mol}\ O_2}{245\ \text{g}\ KClO_3}$

   $$49\ \cancel{\text{g}\ KClO_3} \times \frac{3\ \text{mol}\ O_2}{245\ \cancel{\text{g}\ KClO_3}} = 0.60\ \text{mol}\ O_2$$

   0.60 moles of O_2 are produced by heating 49 grams of $KClO_3$.

EXAMPLE E.7

In photosynthesis, carbon dioxide combines with water to produce glucose and oxygen. How many grams of glucose ($C_6H_{12}O_6$) can be produced by the reaction of 10 moles of CO_2 with water?

Solution

1. Balanced equation:

$$6\ CO_2 + 6\ H_2O \longrightarrow C_6H_{12}O_6 + 6\ O_2$$

2. Determine the relationship between moles and grams.

   a. Mole relationships:

   $$6\ \text{mol}\ CO_2 \qquad 1\ \text{mol}\ C_6H_{12}O_6$$

b. Convert moles of glucose to grams of glucose:

$$\text{mass of 1 mol of } C_6H_{12}O_6 = \text{formula mass of } C_6H_{12}O_6 \text{ in g} = 180 \text{ g}$$

c. Relationship between mol CO_2 and g $C_6H_{12}O_6$:

$$6 \text{ mol } CO_2 \sim 180 \text{ g } C_6H_{12}O_6$$

3. Solve the problem by dimensional analysis.

Given quantity: 10 mol CO_2

Required quantity: g $C_6H_{12}O_6$

Conversion factor: $\dfrac{180 \text{ g } C_6H_{12}O_6}{6 \text{ mol } CO_2}$

$$10 \cancel{\text{mol } CO_2} \times \frac{180 \text{ g } C_6H_{12}O_6}{6 \cancel{\text{mol } CO_2}} = 300 \text{ g } C_6H_{12}O_6$$

300 grams of glucose can be produced by the reaction of 10 moles of carbon dioxide with water.

Questions and Problems

Problems 1–4 involve mass-to-mass calculations.

1. How many tons of potassium hydroxide (KOH) are needed to completely react with chlorine (Cl_2) to yield 49 tons of potassium chlorate ($KClO_3$)?

$$3\ Cl_2 + 6\ KOH \longrightarrow 5\ KCl + KClO_3 + 3\ H_2O$$

2. Calculate the number of kilograms of hydrogen required to reduce 89 pounds of lead oxide (PbO) to lead (454 g = 1 lb).

$$PbO + H_2 \longrightarrow Pb + H_2O$$

3. How many milligrams of $BaSO_4$ are produced by the reaction of an excess of Na_2SO_4 with 6.24 g of $Ba(NO_3)_2$?

$$Na_2SO_4 + Ba(NO_3)_2 \longrightarrow BaSO_4 + 2\ NaNO_3$$

4. $Mg(OH)_2$, the active ingredient in many common antacids, reacts with stomach acid (HCl) to produce $MgCl_2$ and water. How many grams of $Mg(OH)_2$ are necessary to react with 1.0 g HCl?

$$Mg(OH)_2 + 2\ HCl \longrightarrow MgCl_2 + 2\ H_2O$$

Problems 5 and 6 involve mole-to-mole calculations.

5. How many moles of oxygen (O_2) are needed to convert 12 moles of ethane (C_2H_6) to carbon dioxide and water?

$$2\ C_2H_6 + 7\ O_2 \longrightarrow 4\ CO_2 + 6\ H_2O$$

6. Calculate the number of moles of hydrogen produced when 15.9 moles of iron react with water as shown in the following equation.

$$4\ H_2O + 3\ Fe \longrightarrow 4\ H_2 + Fe_3O_4$$

Problems 7–10 involve mole-to-mass and mass-to-mole calculations.

7. The catalytic converter, which is mandatory equipment in all automobiles in the United States, converts CO to CO_2 according to the following reaction:

$$2\ CO + O_2 \longrightarrow 2\ CO_2$$

How many pounds of CO_2 are produced from 25 moles of CO?

8. How many moles of O_2 are needed to convert 320 grams of SO_2 to SO_3?

$$2\ SO_2 + O_2 \longrightarrow 2\ SO_3$$

9. How many kilograms of nitric acid can be obtained from the reaction of 40.0 moles of ammonia with oxygen?

$$NH_3 + 2\ O_2 \longrightarrow HNO_3 + H_2O$$

10. How many moles of water must be reacted with calcium to produce 296 g of slaked lime, $Ca(OH)_2$?

$$Ca + 2\ H_2O \longrightarrow Ca(OH)_2 + H_2$$

Glossary

abiotic Pertaining to factors or things that are separate and independent from living things; nonliving.

absolute zero The temperature at which all gases, if they did not condense, would have zero volume: 0 K = −273.15°C.

acetylcholine A neurotransmitter.

acid A substance that releases hydrogen ions (H^+) in solution (old definition); a proton donator (new definition).

acid deposition Any precipitation, including rain, snow, and fog, that is more acidic than pH 5.6; caused by sulfur dioxide (SO_2) and oxides of nitrogen (NO_x) in the atmosphere.

acid mine drainage Acid released during mining operations, particularly the mining of pyrite (FeS_2).

acid rain Rain with a pH less than 5.6.

activation energy The amount of energy reactants must have in order to overcome the energy barrier to their reaction.

active site The part of an enzyme to which the substrate binds, similar to the way in which a key fits a lock.

active solar heating system Heating system in which heat collected in solar collectors is circulated through a building by pumps.

addition polymer A polymer that is formed when many units of a small molecule (the monomer) bond together without the loss of any atoms from the monomer units.

adenine One of the five organic bases present in nucleic acids; a purine.

adenosine diphosphate (ADP) The diphosphate of adenosine; the energy molecule that, along with adenine triphosphate, is involved in metabolism.

adenosine triphosphate (ATP) The triphosphate of adenosine; the energy molecule that, along with adenine diphosphate, is involved in metabolism.

adsorption The process in which chemicals (ions or molecules) stick to the surface of other materials.

aerobic Requiring oxygen.

aerosols Small particles (diameters less than 1 μm) that are dispersed in air.

alcohol An organic compound that contains an —OH functional group covalently bonded to a carbon atom; general formula is ROH.

aldehyde An organic compound that has a carbonyl group (—C=O) bonded to one carbon atom and one hydrogen atom; general formula is RCHO.

alkali metals Elements in group IA of the periodic table.

alkaline earth metals Elements in group IIA of the periodic table.

alkane A hydrocarbon containing only carbon-carbon single bonds; general formula is C_nH_{2n+2}.

alkene A hydrocarbon containing one or more carbon-carbon double bonds; general formula is C_nH_{2n}.

alkyl group Organic group formed by removing a hydrogen atom from an alkane; for example, $—CH_3$ is the methyl group.

alkylbenzene sulfonate (ABS) A surfactant consisting of a nonpolar branched hydrocarbon chain attached to a polar sulfonate group.

alkyne A hydrocarbon containing one or more carbon-carbon triple bonds; general formula is C_nH_{2n-2}

alpha particle A helium nucleus; symbolized as α or ${}^4_2He$.

amide An organic compound that has the general formula $RCONR'_2$, where R′ may be a hydrogen atom or an alkyl group.

amine An organic derivative of ammonia (NH_3) in which one or more of the hydrogen atoms have been replaced by an alkyl or aromatic group; general formula is RNH_2, R_2NH, or R_3N.

amino acid Difunctional organic compound that contains both a carboxyl group and an amino group.

anaerobic Not requiring oxygen.

androgens Male sex hormones.

anion A negatively charged ion.

anionic surfactant A surfactant with a nonpolar hydrocarbon end and a polar, water-soluble negatively charged end.

anode A positively charged electrode at which oxidation occurs.

antacid Medication used to neutralize excess stomach acid; usually containing hydroxide or carbonate ions.

anthracite Hard coal, which produces a lot of heat and little smoke when burned.

anticodon A sequence of three ribonucleotides on transfer RNA (tRNA) that is complementary to a sequence on messenger RNA (mRNA).

antioxidants Chemicals that slow down or stop oxidation.

antiperspirant Personal care product that reduces or stops perspiration.

apocrine glands Sweat glands found in the armpits and groin.

aqueous solution A solution of a substance in water.

aquifer An underground layer of porous rock, sand, or other material that allows the movement of water between layers of nonporous rock or clay.

arithmetic growth The process of adding a constant amount at regular intervals.

aromatic compound A compound that contains a benzene ring.

artificial transmutation The changing of one element into another by bombarding it with high-energy particles.

asbestos A class of fibrous long-chain silicates.

ascorbic acid Vitamin C.

asthenosphere The layer of the upper mantle that underlies the lithosphere; its rock is very hot and therefore weak and easily deformed.

atom The fundamental unit of all elements.

atomic mass The average mass of the naturally occurring isotopes that make up an element.

atomic mass unit (amu) A unit based on the value of exactly 12 amu for an atom of the isotope of carbon that has six protons and six neutrons in its nucleus.

atomic number The number of protons in the nucleus of an atom of an element.

Avogadro's number The number of particles (atoms, formula units, ions, or molecules) in 1 mole of a substance; 6.02×10^{23}.

bag filtration A method for removing particulates from smokestacks by filtering smoke through fabric bags.

base A substance that releases hydroxide ions (OH^-) in aqueous solution (old definition); a proton acceptor (new definition).

battery A series of electrochemical cells.

beta particle Energetic electron emitted from the nucleus; symbolized as β or ${}_{-1}^{0}\beta$.

BHA Abbreviation for butylated hydroxyanisole; food additive that prevents spoilage.

BHT Abbreviation for butylated hydroxytoluene; food additive that prevents spoilage.

big bang The tremendous explosion that is believed to have released the matter from which the galaxies and stars were formed.

binding energy Energy obtained from the conversion of a very small amount of mass when protons and neutrons pack together to form atomic nuclei.

bioaccumulation The buildup of higher and higher concentrations of chemicals in organisms toward the top of a food chain.

bioconversion The use of biomass as fuel, either by burning materials such as wood, paper, or plant wastes directly to produce energy or by converting the materials into fuels such as alcohol or methane.

biogas A mixture of gases, about two-thirds methane and one-third carbon dioxide, that results from anaerobic (without air) digestion of organic matter; the methane content enables it to be used as a fuel.

biological control Control of a pest population by introduction of predatory, parasitic, or disease-causing organisms.

biomass The total mass of a particular category of biological material in an ecosystem; for example, the biomass of producers.

bioremediation The use of microorganisms to clean up soil contaminated with toxic organic substances.

biosolids The organic matter (sludge) that is removed in sewage treatment and used as fertilizer.

biotic Living or derived from living things.

bitumen Heavy black oil, high in sulfur, that is extracted from tar sands.

bituminous coal Soft coal, which produces less heat and more smoke than hard (anthracite) coal when burned.

bleach A substance used to remove unwanted color from fabric, hair, and other materials.

bond energy The energy needed to separate two covalently bonded atoms.

Boyle's law At constant temperature, the volume of a fixed amount of gas is inversely proportional to the gas pressure.

broad-spectrum insecticide An insecticide that kills a wide range of insect types.

buffer A substance that will maintain the pH of a solution by reacting with excess acid or base.

builder A substance in a detergent that softens water and thus increases the efficiency of the surfactant.

butylated hydroxyanisole (BHA) A food additive that prevents spoilage.

butylated hydroxytoluene (BHT) A food additive that prevents spoilage.

calciferol Vitamin D.

calorie Fundamental unit of energy into which all forms of energy can be converted; the amount of heat required to raise the temperature of 1 gram of water 1°C (Calories used in connection with food are kilocalories, equal to the amount of heat required to raise the temperature of 1 liter of water 1°C).

canal rays Positively charged rays made up of protons that were discovered in gas discharge experiments.

carbamate A group of narrow-spectrum insecticides that kill by inactivating cholinesterase.

carbohydrate An organic compound composed of the elements carbon, hydrogen, and oxygen; for example, sugars and starches.

carbon cycle Processes by which carbon is continually recycled through the environment.

carbon monoxide (CO) A colorless, odorless, poisonous gas which poisons by displacing oxygen from hemoglobin.

carbonyl group The —C=O functional group; present in aldehydes and ketones.

carboxyl group The —COOH functional group.

carboxylic acid An organic compound that has a carbonyl group bonded to one carbon and one —OH group; general formula is RCOOH.

carcinogen A substance that causes cancer, at least in animals and by implication in humans.

carnivore An animal that feeds more-or-less exclusively on other animals.

catalyst A substance that affects the rate at which a chemical reaction occurs.

catalytic converter The device used by automobile manufacturers to reduce the amount of carbon monoxide, nitrogen oxides, and hydrocarbons in car exhaust; it contains catalysts that change these compounds to carbon dioxide, nitrogen, and water as the exhaust passes through it.

catalytic reforming Process that increases the octane rating of gasoline by increasing its content of branched-chain, and aromatic, hydrocarbons.

cathode A negatively charged electrode, at which reduction occurs.

cathode ray tube Apparatus in which streams of electrons are produced when high voltage is applied to electrodes in an evacuated tube.

cation A positively charged ion.

cationic surfactant A surfactant with a nonpolar hydrocarbon end and a polar, water soluble, positively charged end.

cellulose A polysaccharide composed of glucose units joined by β-linkages; it is the prime constituent of plant cell walls and cannot be digested by humans but has dietary value as fiber.

chain reaction A self-sustaining nuclear reaction.

Charles's law At constant pressure, the volume of a fixed amount of gas is directly proportional to its temperature.

chelating agent A compound that surrounds and bonds to a metal atom.

chemical bond The force of attraction that holds two atoms together in a compound.

chemical change A transformation of one or more substances into one or more different substances.

chemical equation A written expression of a chemical reaction in which symbols and formulas are used to represent reactants and products, which are balanced in terms of the numbers of atoms of the various elements that are present.

chemical formula Representation with symbols and subscript numbers of the types and numbers of atoms in a chemical compound.

chemical property A characteristic of a substance that describes a chemical reaction the substance undergoes that results in a change in its composition.

chemistry The study of the nature, properties, and transformations of matter.

chiral carbon atom A carbon atom bonded to four different atoms or groups of atoms.

chlorinated hydrocarbons Synthetic organic compounds used as insecticides (for example, DDT) that are harmful to the environment because they bioaccumulate upward through food chains.

chlorofluorocarbons (CFCs) Synthetic organic molecules that contain one or more carbon atoms attached to chlorine and fluorine atoms.

chlorophyll A plant pigment that plays a major role in converting solar energy to chemical energy in photosynthesis.

cholesterol The steroid that is the precursor of all other steroids in the human body, including the male and female sex hormones.

chromosome Structure consisting of DNA coiled around protein molecules and present in the nuclei of all cells; humans have 46 chromosomes.

***cis-trans* isomers** A pair of organic compounds that differ only in having a pair of atoms or groups on the same side of the double bond (*cis* isomer) or on opposite sides of the double bond (*trans* isomer).

codon A sequence of three ribonucleotides on mRNA that codes for a specific amino acid.

coenzyme A small, organic molecule that acts as an enzyme's cofactor.

compound A substance formed by the chemical combination of two or more elements in fixed proportions.

condensation polymer A polymer formed when two or more monomers link together with the elimination of a small molecule, usually water.

consumer An organism in an ecosystem that derives its energy from feeding on other organisms or their products.

continuous spectrum The pattern where one color of light merges into the next that is produced when white light is passed through a glass prism.

control rods The rods in the core of a nuclear reactor that control the rate of fission by absorbing neutrons.

copolymer A polymer that is formed from two or more different monomers.

core Central portion of the earth.

corrosive poison A poison that acts by destroying tissue; examples are mineral acids, such as sulfuric acid, and strong bases, such as sodium hydroxide.

corrosive waste Acidic waste material with a pH less than 2.0 or basic waste material with a pH greater than 12.5.

cosmic rays Energetic particles, primarily protons, that bombard earth from space.

covalent bond A chemical bond formed by the sharing of electrons between two atoms.

cracking The process used in petroleum refining in which long-chain hydrocarbons are broken down into shorter-chain hydrocarbons.

critical mass The amount of fissionable (radioactive) material necessary to sustain a chain reaction; amount needed to make a nuclear bomb.

crust The outermost solid layer of the earth.

cyanobacteria Blue-green algae; photosynthetic plants that live and reproduce entirely immersed in water.

cycloalkane An organic compound in which carbon atoms are joined to form a ring.

cyclone separation A method for removing particulates from smokestack emissions by spinning the gases, forcing the particulates against the container walls, where they separate out.

Dalton's atomic theory The theory that all matter is composed of indivisible particles called atoms.

DDT A synthetic chlorinated hydrocarbon used as a pesticide in the United States until it was banned in 1972.

decomposer Any organism that feeds on dead or decaying plant and animal material; the primary decomposers are fungi and bacteria.

deep-well injection A method for disposing of hazardous wastes by injecting them into deep wells.

denitrification The process by which bacteria convert nitrates in soil to nitrogen gas.

density The mass of an object per unit volume.

deodorant A personal care product that prevents body odor by killing odor-causing bacteria.

deoxyribonucleic acid (DNA) The macromolecule found in the nucleus of every cell, which contains the genetic information needed for the development of each individual.

desalination The removal of salts from water.

detergent A synthetic surfactant consisting of an alkylbenzene sulfonate, in which the alkyl chain may be branched or unbranched.

detritus The dead organic matter, such as fallen leaves, twigs, and other plant and animal wastes, that is found in an ecosystem.

diabetes Condition in which the body is unable to utilize glucose because of a deficiency of insulin.

dipeptide A compound formed when the amino end of one amino acid reacts with the carboxylic acid end of a second amino acid with the elimination of a molecule of water.

dipole A molecule with a partial positive charge at one end and an equal but opposite negative charge at the other end.

dipole-dipole interactions Relatively weak attractions between polar molecules.

disaccharide A carbohydrate formed when two monosaccharides join together with the elimination of a molecule of water; table sugar ($C_{12}H_{22}O_{11}$) is a disaccharide.

disinfectant by-products Compounds formed when chlorine used in sewage treatment reacts with residual organic and inorganic substances present in the treated water.

distillation A process for purifying water, or other liquids, by boiling the liquid and collecting the condensed vapor, leaving the impurities behind in the boiler.

DNA profiling A method of identifying a person by determining unique patterns in his or her DNA; also called DNA typing or DNA finger printing.

dose A measure of the concentration times the length of exposure for any given material or form of radiation; effects correspond to the product of these two factors.

double bond A covalent bond between two atoms, formed by sharing two pairs of electrons; represented as $=$.

double helix A structure consisting of two strands coiled around one another, such as the two polynucleotide strands of DNA.

doubling time The time it takes for a population that is increasing at a known rate to double.

dynamic equilibrium The state in a chemical reaction when the rate at which product is being formed equals the rate at which it is being decomposed.

eccrine glands Sweat glands that cover nearly all body surfaces.

ecosystem A grouping of plants, animals, and other organisms interacting with each other and the environment in such a way as to perpetuate itself more-or-less indefinitely.

elastomer A polymer that has elastic properties; that is, when stretched, it tends to return to its original form.

electrochemical cell A device for producing electricity in which electrons released in a redox reaction are passed through an electric circuit.

electrode A metal surface on which oxidation and reduction reactions occur.

electromagnetic radiation A form of energy that has wave characteristics and that propagates through a vacuum at the speed of 3.0×10^{10} cm/s.

electron A subatomic particle with a relative negative charge of one (-1), and a negligible mass.

electron cloud The region surrounding the nucleus of an atom that is occupied by the atom's electrons.

electron configuration The specific arrangement of an atom's electrons in energy levels; represented by notation such as $1s^2 2s^2 2p^6$, for a neon (Ne) atom.

electron-dot symbol An element's symbol surrounded by dots representing the element's valence electrons.

electronegativity The ability of an atom in a molecule to attract electrons; a numerical value on the electronegativity scale.

electrostatic precipitation A method for removing particulates from smokestacks by passing emissions through a high-voltage chamber, where the particulates acquire a charge, are attracted to the oppositely charged chamber walls, and are collected.

element A fundamental building block of matter, which cannot be decomposed into simpler substances by ordinary chemical means.

endocrine disruptors Synthetic organic compounds that persist in the environment and are suspected of disrupting reproduction and fetal development.

endothermic reaction A chemical process that absorbs heat from the surroundings.

energy The ability to do work; energy has many forms, for example, heat, chemical, electrical, mechanical, and radiant energy.

energy level Any of several regions surrounding the nucleus of an atom in which the electrons move.

entropy A measure of the randomness of a system; the greater the disorder, the greater the entropy.

environment The combination of all things and factors external to a given individual or population of organisms.

enzyme A compound, usually a protein, that acts as a catalyst for biological reactions.

erosion The loss of soil due to particles being carried away by wind or water.

essential amino acids Amino acids that humans cannot biosynthesize and that must be present in the food they consume.

ester A derivative of a carboxylic acid (RCOOH) in which the hydrogen atom of the carboxyl group is replaced with an alkyl group (R′); general formula is RCOOR′.

estrogens A class of female sex hormones.

ether An organic compound containing an oxygen atom bonded to two alkyl groups; general formula is $R{-}O{-}R'$.

eutrophication The process whereby a body of water becomes nutrient-rich and supports abundant growth of algae and other aquatic life at its surface.

excited state State in which an electron has acquired sufficient energy to move to a higher energy level.

exothermic reaction A chemical process that releases heat to the surroundings.

fat A mixture of triglycerides that is solid because it contains a high proportion of saturated fatty acids.

fatty acid A long chain carboxylic acid with an even number of carbon atoms.

fiber Cellulose component of many foods.

filler An inert substance such as sodium sulfate (Na_2SO_4) that is added to a detergent to add bulk and to act as a drying agent.

first law of thermodynamics Matter and energy are neither created nor destroyed but can change from one form into another.

flue-gas desulfurization (FGD) A process used to remove sulfur dioxide (SO_2) from smokestack gases.

fluidized bed combustion (FBC) A process used to remove sulfur dioxide (SO_2) from smokestack gases.

fluorescence The property of a substance to absorb light of a certain wavelength and then emit light of a longer wavelength.

fly ash Solid particulates, including soot, that are released when coal is burned and are present in smokestack gases.

food chain Transfer of energy and material through a series of organisms as each one is fed upon by the next.

formula mass The sum of the individual atomic masses of all atoms represented in the formula of a chemical compound or polyatomic ion.

fossil fuels Fuels formed by the breakdown, over millions of years, of plant and animal material; coal, oil, and natural gas are the three fossil fuels.

fractional distillation The separation of components of petroleum by boiling followed by condensation into fractions with similar boiling ranges; low boiling, small molecules emerge first, followed by higher-boiling, larger molecules.

free radicals Highly reactive but uncharged chemical species that can be atoms or groups of atoms having an odd number of electrons.

frequency (υ) The number of crests or troughs of a wave that pass a fixed point in a given unit of time.

freshwater Water containing up to 0.1% (1000 ppm) of dissolved solids.

fuel A substance that burns readily and releases a relatively large amount of energy as heat on burning.

fuel cell A device that directly converts chemical energy into electrical energy through chemical reactions.

fuel rods The steel tubes containing pellets of uranium or other fissionable material, which, with the control rods, form the core of a nuclear reactor.

functional group The reactive part of an organic molecule.

galaxy A group or cluster of millions or even billions of stars.

gamma rays High-energy form of electromagnetic radiation given off by many radioactive substances; symbolized as γ.

gas One of the states of matter; gases have no definite shape or volume.

gas-discharge tube Apparatus in which streams of electrons are produced when high voltage is applied to electrodes in an evacuated tube.

gasohol A mixture of gasoline and 10–20% (by volume) ethanol (or methanol) that produces less carbon monoxide when burned than pure gasoline does.

Geiger counter A device used to detect and measure nuclear radiation.

gene A segment of DNA that directs the synthesis of a protein in a living organism.

genetic code The 64 possible three-base sequences in messenger RNA (mRNA) that code for specific amino acids or act as stop or start signals.

geometric growth Growth that is increasing at a fixed percentage during a set growth period.

geothermal energy Heat energy originating deep within the earth's interior.

global warming The possibility that the earth's atmosphere is gradually heating up as a result of the greenhouse effect produced by carbon dioxide and other gases.

glycogen A polysaccharide formed of branched chains of glucose.

GRAS list A list of food additives that are generally considered by the FDA to be safe.

greenhouse effect A rise in the temperature of the earth's atmosphere caused by the increasing amounts of carbon dioxide and certain other gases, which absorb and trap heat energy that normally escapes from the earth.

greenhouse gases Gases in the earth's atmosphere that absorb infrared radiation and contribute to the greenhouse effect; they include carbon dioxide, water vapor, methane, nitrous oxide, and chlorofluorocarbons.

ground state The state of an atom in which all its electrons are in their lowest posssible energy levels.

groundwater Water that has accumulated in the ground, completely filling and saturating all the pores and spaces in the rock and/or soil.

group One of the 18 vertical columns in the periodic table; elements in a particular group tend to have similar properties.

guanine One of the five organic bases present in nucleic acids; a purine.

half-life The time required for half of the atoms originally present in a sample of a radioactive isotope to decay.

halogens Elements in group VIIA of the periodic table.

hard water Water containing a relatively high concentration of the ions of calcium, magnesium, and iron.

hazardous waste Waste material that poses a substantial present or potential hazard to human health or the environment if improperly treated, stored, transported, or disposed of.

heat capacity The quantity of heat required to raise the temperature of a fixed quantity of matter by 1°C.

heat of fusion The amount of heat necessary to convert 1 gram (or 1 mole) of a solid to a liquid.

heat of vaporization The amount of heat necessary to convert 1 gram (or 1 mole) of a liquid into a gas.

herbicide Substance used to kill or slow down the growth of plants, especially weeds.

herbivore An animal that feeds on green plants or plant products such as nuts and berries.

heterogeneous mixture A mixture that is visually nonuniform.

high-density polyethylene (HDPE) A hard, tough polymer composed of long unbranched chains of ethylene monomer units.

high-level radioactive wastes Wastes containing plutonium-239, including spent fuel rods from nuclear reactors and certain wastes generated by nuclear weapons facilities.

homogeneous mixture A mixture that appears uniform to the naked eye.

homologous series A series of compounds in which each member differs from the previous one by a constant increment.

homopolymer A polymer formed from identical monomer units.

hormone A chemical substance that is formed in one organ of the body and carried to another organ, where it has a specific action.

humus A dark-brown or black, soft, spongy residue of organic matter that remains after the bulk of dead leaves, wood, and other organic material has decomposed.

hydrocarbon An organic compound composed of only carbon and hydrogen atoms.

hydrogen bonds Intermolecular forces due to attraction between an electronegative atom (oxygen, nitrogen, or fluorine) and a hydrogen atom bonded to an electronegative atom (usually oxygen or nitrogen) in another molecule.

hydrologic cycle The processes of evaporation, transpiration, condensation, and precipitation by which the earth's water is continually purified and recycled.

hydronium ion The ion formed by the reaction of a hydrogen ion with a water molecule; H_3O^+.

hydroxyl group The —OH group

hydroxyl radical Very reactive uncharged hydroxyl species containing an unpaired electron (·OH).

hypothesis An educated guess concerning the cause of an observed phenomenon, which is subjected to experimental tests to prove its accuracy or inaccuracy.

ideal gas law The volume of a fixed amount of gas is inversely proportional to the pressure of the gas and directly proportional to its temperature.

ignitable waste Waste material that poses a serious fire hazard.

industrial smog The grayish mixture of moisture, soot, and sulfurous compounds that occurs in localities where industries are concentrated and coal is the primary energy source.

inorganic Consisting of compounds of all elements other than those of carbon.

insect growth regulator (IGR) A compound that kills insects by disrupting their development at various stages.

insecticide Any chemical that kills insects.

integrated pest management (IPM) Two or more methods of pest control combined into a unified program with the objective of minimizing the use of synthetic pesticides.

intermolecular forces Forces of attraction and repulsion between partial charges in different molecules.

ion An atom that has lost or gained electrons and therefore has acquired a positive or negative charge.

ion exchange Process in which ions in solution are exchanged for corresponding ions held to the surface of an ion exchange material; for example, Ca^{2+} and Mg^{2+} ions in hard water can be exchanged for Na^+ ions.

ionic bond A bond formed by the transfer of electrons between metal and nonmetal atoms.

ionizing radiation Radiation capable of dislodging an electron from (ionizing) an atom or a molecule that it strikes.

isomers Compounds that have the same molecular formula but different structures and properties.

isotopes Atoms of an element that contain the same number of protons but different numbers of neutrons and therefore have different masses.

juvenile hormone A hormone that controls larval development in insects and thus can be used to prevent insects from growing into adults.

keratin A protein of which hair and other natural fibers are composed.

kerogen A hydrocarbon material contained in oil shale that vaporizes when heated and can be recondensed into a material similar to crude oil.

ketone An organic compound that has a carbonyl group bonded to two carbon atoms; general formula is R_2CO.

kinetic energy Energy of motion.

kinetic-molecular theory A model explaining the behavior of the three states of matter based on molecular motion.

lactose A disaccharide present in milk.

law of conservation of mass In an ordinary chemical reaction, matter is neither created nor destroyed.

law of definite proportions Different samples of any pure compound contain the same elements in the same proportions by mass.

law of multiple proportions The masses of one element that can combine chemically with a fixed mass of another element are in a ratio of small whole numbers.

LD_{50} test Test carried out to determine the lethal dose (LD) of a substance for 50% of the population being tested.

Lewis symbol The symbol of an element surrounded by dots representing its valence electrons.

line spectrum A spectrum produced by an element that appears as a series of bright lines at specific wavelengths, separated by dark bands.

linear alkylbenzene sulfonate (LAS) A surfactant consisting of a nonpolar linear hydrocarbon chain attached to a polar sulfonate group.

lipid Biochemical substance found in plant and animal tissues that is insoluble in water but soluble in nonpolar solvents; may be a fat or another type of compound that resembles a fat in physical properties.

liquefied natural gas (LNG) Natural gas that has been cooled and condensed into liquid form.

liquefied petroleum gas (LPG) Liquefied propane and butane gas used as heating fuel.

liquid One of the states of matter; a liquid has a definite volume and an indefinite shape.

lithosphere Outer shell of the earth; composed of the crust and the solid layer above the asthenosphere.

loam A soil consisting of a mixture of about 40% sand, 40% silt, and 20% clay.

London forces Intermolecular forces of attraction between short-lived, temporary dipoles.

low-density polyethylene (LDPE) A transparent, flexible polymer consisting of irregularly branched chains of ethylene monomer units.

low-level radioactive wastes Radioactive wastes other than those containing plutonium-239.

malignant tumor A cancerous tumor that invades and destroys neighboring tissue.

mantle The portion of the earth beneath the crust and surrounding the core.

mass number The sum of the number of protons and neutrons in the nucleus of an atom.

matter The physical material that makes up the universe; anything that has mass and volume.

mesosphere Atmospheric layer above the stratopause, where temperature decreases with altitude.

metals Malleable elements with a lustrous appearance that are good conductors of heat and electricity and that tend to lose electrons to form cations.

metalloids Elements that have properties intermediate between those of metals and nonmetals; also called semimetals.

micelle A group of molecules that have an ionic end and a nonpolar organic end and that cluster together in order to maximize the contact of their ionic ends with water molecules and minimize contact of their nonpolar organic ends with water molecules; soap and detergent molecules form micelles in water.

micronutrients A group of elements essential in trace amounts for healthy plant growth.

mineral Naturally occurring inorganic substance found in the earth's crust as solid material.

mixture A combination of two or more pure substances in which each substance retains its own identity; for example, air is a mixture containing primarily oxygen, nitrogen, and carbon dioxide.

moderator The material in a nuclear reactor that slows down neutrons produced in fission reactions so that they travel at the right speed to trigger another fission reaction.

molarity (M) A concentration unit for solutions; moles of solute per liter of solution.

molar mass The atomic or formula mass of a substance expressed in grams.

mole A quantity of a substance consisting of 6.023×10^{23} particles (atoms, formula units, ions, molecules).

molecule The smallest unit of a compound that retains the characteristics of that compound.

monomer A small molecule that can combine with others like it or different from it to form a polymer.

monosaccharide A simple carbohydrate that cannot be broken down into smaller carbohydrate units.

municipal solid waste (MSW) The solid waste collected locally from homes, institutions, and commercial establishments.

mutagen A substance that causes mutations.

mutation A random change in one or more genes of an organism, which may occur spontaneously in nature but is often induced by exposure to radiation and/or certain chemicals.

narrow-spectrum insecticide An insecticide that kills only a few insect types.

National Priorities List EPA list of hazardous waste sites in the United States most in need of clean-up.

neurotoxin A metabolic toxin that acts by disrupting the normal transmission of nerve impulses in the central nervous system.

neurotransmitter A chemical that carries nerve impulses across the synapse between nerve cells.

neutralization reaction The exchange reaction between an acid and a base to yield water and a salt.

neutron An electrically neutral subatomic particle found in the nuclei of atoms.

niacin Vitamin B_3.

nitrification The conversion of ammonia in the soil to nitrates by soil bacteria.

nitrogen cycle The cycle by which nitrogen is continually recycled through the environment.

nitrogen fixation The process of chemically converting nitrogen gas (N_2) from the air into compounds such as nitrates (NO_3^-) or ammonia (NH_3), which can be used by plants in building amino acids and other nitrogen-containing organic molecules.

noble gases The inert gases in group VIIIA of the periodic table.

nonmetals Elements that lack the properties of metals.

non-point source A wide area from which pollutants are discharged, for example, feedlots and farmlands.

nonrenewable resource A resource such as an ore or a fossil fuel that exists as a finite deposit in the earth's crust and is not replenished by natural processes as it is removed.

nuclear breeder reactor Fission reactor that produces more fuel than it consumes by converting uranium-238 to plutonium-239.

nuclear fission The splitting apart of the nucleus of an atom to give two lighter nuclei and several neutrons.

nuclear fusion The combination of two light nuclei to give a heavier nucleus.

nucleic acid A biological polymer composed of repeating units of linked nucleotides; examples are DNA and RNA.

nucleotide The repeating unit of a nucleic acid, consisting of a sugar, an organic nitrogen base, and phosphoric acid.

nucleus The very small, dense, positively charged central core of an atom, composed of protons and neutrons.

obesity The condition of being severely overweight; having a body mass index (BMI) of 30 or higher.

octane rating A measure of the amount of knocking produced by a gasoline when it is burned in an automobile engine.

octet rule In forming compounds, atoms gain, lose, or share electrons in such a way that they achieve the electron configuration of the nearest noble gas in the periodic table.

oil A naturally occurring combustible liquid composed of hundreds of hydrocarbons and other organic compounds; also known as petroleum.

oil shale Rock that contains oil.

omnivore An animal that feeds on both plant and animal material.

optical brightener A compound used in detergents that absorbs invisible ultraviolet radiation from sunlight and then emits longer-wavelength light in the blue region of the visible spectrum.

orbital A region in space in which an electron can be found 90% of the time.

ore A mineral rich in a particular metal such as iron, copper, or aluminum that can be economically mined and refined to produce the metal.

organic Consisting of compounds of the element carbon.

osmosis The movement of solvent molecules (usually water) through a semipermeable membrane from a more dilute solution into a more concentrated one.

oxidation Chemical process that involves gaining oxygen, losing hydrogen, or losing electrons.

oxygen-consuming wastes Organic waste materials present in a body of water that consume oxygen as they decompose.

oxygen cycle Processes by which oxygen is continually recycled through the environment.

ozone hole A region in the stratosphere over the Antarctic that has been depleted of its normal level of ozone (O_3) because of emissions of chlorofluorocarbons (CFCs) from anthropogenic sources.

ozone layer The layer of ozone (O_3) gas in the stratosphere that screens out harmful ultraviolet radiation coming from the sun.

particulates Small solid and liquid particles suspended in the air; those from anthropogenic sources are a major source of air pollution.

passive solar heating system Heating system in which heat gathered in solar collectors is circulated passively by convection currents.

pathogen An organism, usually a microbe, capable of causing disease.

pepsin An enzyme in the stomach that catalyzes the breakdown of proteins.

peptide bond The type of bond that holds amino acids together in proteins.

period One of the horizontal rows of elements in the periodic table.

periodic table The arrangement of elements in order of increasing atomic number, with elements having similar properties placed in vertical columns.

persistent organic pollutants (POPs) Widely disseminated synthetic organic compounds that are suspected of interfering with endocrine systems in humans and other animals.

pesticide A chemical that kills or inhibits the growth of pests (insects, rodents, weeds).

pH The negative logarithm of the hydrogen ion concentration; a measure of acidity.

pheromone A chemical substance used by insects to attract a mate; some insect pheromones have been synthesized and used for pest control.

photochemical smog A brown haze resulting from chemical reactions between air pollutants primarily from automobile exhausts; sunlight is a catalyst for these reactions.

photosynthesis The process by which green plants, using energy from the sun, produce glucose and oxygen from carbon dioxide and water.

photovoltaic cell A device that converts solar energy into an electric current; also called a solar cell.

physical change A change in a substance that occurs with no alteration in the chemical composition of the substance.

physical properties Properties that can be measured without changing the composition of a substance.

plaque Film consisting of bacteria and polysaccharides that forms on teeth after eating.

plastic A polymeric substance that can be molded into various shapes and forms.

plasticizer A substance that is added to plastics to make them more pliable.

point source A specific point from which pollutants are discharged, for example, a factory drain or an outlet from a sewage treatment plant.

poison A general term for any toxic substance, whether synthetic or of natural origin.

polar covalent bond A covalent chemical bond in which there is unequal sharing of electrons between atoms.

polar molecule A molecule that has partial positive and partial negative charges.

polyacrylonitrile An addition polymer in which a nitrile group (—C≡N) has replaced one of the hydrogen atoms in the starting ethylene monomer.

polyamide A condensation polymer made from a diamine and a dicarboxylic acid; an example is nylon.

polyatomic ion An electrically charged group of two or more atoms.

polychlorinated hydrocarbons Stable synthetic organic compounds that by bioaccumulation have contaminated most food chains on earth; many are known carcinogens.

polyester A condensation polymer made from a dialcohol and a dicarboxylic acid; an example is polyethylene terephthalate (PET).

polyethylene An addition polymer made from ethylene (C_2H_4) monomer units.

polymer A large organic molecule made up of repeating units of smaller molecules (monomers).

polymerization A reaction in which monomers combine to form a polymer.

polypropylene An addition polymer made from propylene monomer units.

polysaccharide A carbohydrate containing more than two simple sugar units.

polystyrene An addition polymer whose monomer is styrene, a derivative of ethylene in which one hydrogen has been replaced by a benzene ring.

polytetrafluoroethylene (Teflon) An addition polymer whose monomer is a derivative of ethylene in which all four hydrogen atoms have been replaced with fluorine atoms.

polyurethane A rearrangement polymer made from a dialcohol and a diisocyanate.

polyvinyl chloride (PVC) An addition polymer whose monomer is vinyl chloride, a derivative of ethylene in which one hydrogen atom is replaced by a chlorine atom.

positron A positively charged subatomic particle with the same mass as an electron.

potential energy Stored energy; energy of position.

power The rate at which energy is used.

primary air pollutants The substances responsible for 90% of all air pollution in the United States—carbon monoxide, sulfur dioxide, nitrogen oxides, volatile organic compounds, and suspended particulate matter.

primary nutrient Any of the three elements plants need to thrive—nitrogen (N), phosphorus (P), and potassium (K).

primary treatment Sewage treatment process that consists of passing wastewater very slowly through a large tank, which permits the solid organic material to settle out.

principal energy level One of the energy levels to which electrons are limited.

producer Any organism in an ecosystem, usually a green plant, that uses light energy to construct its organic constituents from simple inorganic compounds present in its environment.

products The substances formed in a chemical reaction.

prostaglandins Compounds that are derived from arachidonic acid and are involved in contractions of muscle, increases in blood pressure, and other physiological processes.

protein An important type of biological molecule composed of amino acids linked together through peptide bonds.

proton A subatomic particle with a relative positive charge of one (+1) unit and a mass of 1 amu; a hydrogen ion (H^+).

quantum The smallest increment of radiant energy that may be absorbed or emitted.

quaternary ammonium salt An ionic compound containing two hydrocarbon chains in which the positively charged ammonium nitrogen atom is joined to four carbon atoms; also called a "quat" and used as a fabric softener.

rad A unit for measuring the amount of radiation absorbed by living tissue, regardless of type of radiation.

radioactive decay Steps by which a radioisotope spontaneously emits radiation and is converted to a different element.

radioactivity The particles and energy (for example, alpha particles, beta particles, and gamma rays) released by the nucleus of an atom as it undergoes nuclear decay.

radioisotope An isotope of an element that is unstable and tends to gain stability by giving off radiation.

radon A radioactive gas that is produced by natural processes in the earth and often seeps into buildings; it can be a hazard within homes and is a known carcinogen.

reactants The starting materials in a chemical reaction.

reactive waste Waste material that explodes when heated or shocked or that reacts violently with water.

rearrangement polymer A polymer formed by rearrangement of atoms in the monomer units without the loss of any atoms.

reduction Chemical process that involves gaining hydrogen atoms, losing oxygen atoms, or gaining electrons.

rem (sievert) A unit of radiation that takes into account the potential damage to tissue caused by different types of ionizing radiation.

renewable resource A biological resource, such as trees, that may be replaced by reproduction and regrowth.

representative elements Elements in group IA through group VIIIA of the periodic table.

respiration The process by which living organisms take in oxygen and give off carbon dioxide.

retinol Vitamin A.

reverse osmosis Process for removing dissolved salts from water by applying pressure in excess of the osmotic pressure to force water molecules through a semipermeable membrane; salt molecules are too large to pass through the membrane.

ribonucleic acid (RNA) A nucleic acid involved in the transfer of genetic information needed for the synthesis of proteins.

ribosome Structure in a cell where protein synthesis occurs.

risk assessment The process of evaluating the risks associated with a particular hazard.

rule of 70 Method for calculating the doubling time for a process that is increasing exponentially; 70 divided by the growth rate (as a percentage) equals the doubling time in years.

saccharin An artificial sweetener.

salt The ionic compound produced by the neutralization of an acid with a base.

sanitary landfill A depression in the ground in which trash is dumped and then compacted; such a landfill must meet EPA standards to prevent leakage into the surroundings.

saturated fatty acid An organic carboxylic acid in which all the carbon-carbon bonds are single bonds.

saponification The reaction of a triglyceride with a strong base to produce glycerol and a soap.

saturated hydrocarbon An organic compound that contains only carbon and hydrogen atoms and in which all the carbon-carbon bonds are single bonds.

science The study of aspects of the world around us.

scientific method The method by which scientific information is gathered; involves observing, formulating specific questions and a hypothesis, then testing the hypothesis through experimentation.

scurvy Disease caused by a long-term lack of Vitamin C; once common on long sea voyages.

sebum The oily secretion that lubricates hair.

second law of thermodynamics It is never possible to get 100% usable energy from any energy source, because some energy is always lost as waste heat.

secondary air pollutants Substances formed as a result of chemical reactions between primary air pollutants and other constituents in the atmosphere.

secondary nutrients Three elements plants need to thrive (though not in as great quantities as the primary nurients)—calcium (Ca), magnesium (Mg), and sulfur (S).

secondary treatment Sewage treatment process by which organic material is removed from sewage by enabling organisms to feed on it and oxidize it through respiration; also called biological treatment.

secured landfill Burial site for long-term storage of hazardous wastes contained in stainless steel drums.

semiconductors The elements silicon, germanium, and arsenic, which conduct electricity better than nonmetals but not as well as metals.

sequestrant A chemical that is added to water or to products containing water to bind any metal ions that are in solution.

shale oil A brown liquid that is obtained from kerogen in oil shale and can be refined like petroleum.

silicates A category of minerals composed of silicon and oxygen combined with other elements.

silicones Polymers with a silicon backbone; formed from the reaction of silicon with organic molecules.

sludge The mudlike material that is formed of the solids removed from wastewater during sewage treatment.

soap The sodium or potassium salt of a long-chain fatty acid made by treating a triglyceride with sodium or potassium hydroxide.

solar energy Energy derived from the sun.

soft water Water with a low content of calcium, magnesium, and/or iron ions.

solid One of the states of matter; a solid has a definable volume and shape.

solid waste Refuse, "trash," and other discarded materials that are not excluded from landfills by EPA regulations; most liquids are excluded.

solute The substance dissolved in a solvent.

solution A mixture of substances that has uniform composition; a homogeneous mixture.

solvent The substance present in greatest quantity in a solution.

starch A polysaccharide that is composed of glucose units and stored in plants as a food reserve.

steroid A type of lipid that has a structure based on a tetracyclic (four-ring) carbon skeleton; examples are cholesterol and many hormones.

strategic metal A metal that is essential for industry or defense and is stockpiled.

stratosphere Atmospheric layer directly above the troposphere.

structural isomers Compounds that have the same molecular formula but different molecular structures.

suborbitals Subdivisions within energy levels of atoms that are occupied by electrons of different energies; designated as *s*, *p*, *d*, and *f*.

subsoil The soil lying beneath the topsoil, which is more compacted and contains little or no humus or other organic material, living or dead.

substrate The substance with which an enzyme reacts.

sucrose Disaccharide composed of glucose and fructose; common table sugar.

Superfund program The mechanism authorized and funded by the Comprehensive Environmental Response, Compensation and Liability Act (CERCLA) for cleaning up hazardous waste sites in the United States.

surface impoundment Method for disposing of small quantities of hazardous waste contained in a large volume of water by pumping the contaminated water into a pond lined with plastic and allowing the water to evaporate.

surfactant A substance whose molecules have a polar end and a nonpolar end and which allows water (a polar substance) to form a stable suspension with a nonpolar substance such as oil or grease.

synapse Narrow gap between nerve cells across which an electrical impulse is carried.

synthesis gas Gas, primarily methane, that is produced by treating crushed coal with superheated steam.

tar sands Sedimentary material containing bitumen, a tar-like oil, which can be extracted and then refined in the same way as crude oil.

tartar The hard material that forms on teeth as a result of the combination of plaque with calcium and phosphate ions present in saliva.

technology The application of scientific knowledge to solve practical problems or reach a desired goal.

temperature inversion A weather phenomenon in which a layer of warmer air overlies cooler air that is closer to the ground and prevents pollutants from rising and dispersing.

teratogen Any agent that causes birth defects in a fetus when ingested by the mother.

terpenes Aromatic, volatile hydrocarbons that evaporate from the leaves of pine, eucalyptus, and sandalwood trees.

terrestrial radiation Radiation from rocks and soil other than radon.

tertiary treatment Third stage of sewage treatment, designed to remove nitrogen and/or phosphorus (plant nutrients) from wastewater so as to reduce eutrophication.

testosterone A male hormone secreted by the testes; responsible for the development of secondary male sex characteristics.

theory A statement that provides a rational explanation for a series of observations.

thermal pollution The discharge of abnormal amounts of heat to air or water, primarily the release of waste heat from electric generating plants.

thermoplastic A polymer that can be repeatedly softened by heating and formed into different shapes that harden on cooling.

thermosetting plastic A polymer that decomposes when heated; once formed, it cannot be melted and remolded into a different shape.

thermosphere The highest thermal layer in the earth's atmosphere, above the mesopause.

thiamine Vitamin B_1.

thymine One of the four organic bases present in DNA; a pyrimidine.

topsoil The surface layer of soil, which is rich in humus and other organic material, both living and dead.

toxic substance A substance that causes harm to an organism that ingests it by interfering with the organism's metabolism.

toxic waste Waste material that causes harm to living organisms by interfering with normal physiological processes.

toxin A natural substance, from a biological source, that can cause harm to an organism.

transcription The first step in protein synthesis in which genetic information in DNA is read and used to synthesize mRNA.

transition elements The B groups of elements in the periodic table, which lie between the two blocks of representative elements.

translation The process in which mRNA directs the synthesis of proteins.

transmutation Conversion of one kind of atomic nucleus to another.

triglyceride A fat consisting of the triester formed by the reaction of three fatty acid molecules with a molecule of glycerol.

triple bond A covalent bond involving three electron pairs; represented as $\equiv$.

trophic level Feeding level with respect to the primary source of energy in a food chain.

troposphere The layer of the earth's atmosphere extending from the surface to about ten miles in altitude.

unsaturated fatty acid Organic acid containing one or more carbon-carbon double bonds.

unsaturated hydrocarbon An organic compound that contains only carbon and hydrogen atoms and has at least one carbon-carbon double bond or carbon-carbon triple bond.

uracil One of the four organic bases present in RNA; a pyrimidine.

valence electrons The outermost electrons of an atom, which are used for bonding.

vitamin An organic compound that is required in small amounts by living organisms for normal metabolism and the maintenance of good health.

volatile organic compounds (VOCs) Compounds, including many hydrocarbons, that enter the atmosphere in automobile exhausts and play a major role in the formation of photochemical smog.

vulcanization The process by which rubber is heated with sulfur in order to form a polymer with cross-linking, which gives the rubber better elastic properties.

Waste Isolation Pilot Plant (WIPP) Repository for the permanent storage of low-level radioactive wastes produced by nuclear weapons plants; located in New Mexico.

wavelength (λ) The distance between successive crests or troughs in a wave.

wave-mechanical model The model of the atom in which electrons are treated as both particles and waves and complex mathematical equations are used to describe their arrangements and motion.

weathering The gradual breakdown of rock into smaller and smaller particles, caused by natural chemical, physical, and biological factors.

work Any change in motion or state of matter; requires the expenditure of energy.

X-rays Electromagnetic radiation of extremely short wavelength and high energy.

Answers to Even-Numbered Questions and Problems

Chapter 1

2. Physical change

4. **a.** heterogeneous **b.** heterogeneous
   c. homogeneous **d.** homogeneous
   e. heterogeneous **f.** homogeneous

6. **a.** homogeneous mixture
   b. homogeneous mixture
   c. heterogeneous mixture

8. **a.** element **b.** compound
   c. element **d.** element
   e. compound **f.** element
   g. compound **h.** compound
   i. element **j.** compound
   k. compound **l.** element
   m. compound **n.** element

10. **a.** magnesium **b.** copper
    c. cobalt **d.** carbon
    e. boron **f.** nickel
    g. phosphorus **h.** silicon
    i. sulfur **j.** aluminum
    k. bromine **l.** iodine
    m. mercury **n.** oxygen

12. **a.** neutral **b.** anion
    c. cation **d.** anion
    e. neutral **f.** cation
    g. neutral **h.** cation
    i. neutral **j.** anion

14. Silicon

16. **a.** A carbon atom reacts with an oxygen molecule to give a carbon dioxide molecule.
    b. Four iron atoms react with three oxygen molecules to give two iron oxide molecules.
    c. Two sodium atoms react with a chlorine molecule to give two sodium chloride molecules.

18. **a.** two **b.** eight **c.** four
    d. two **e.** eight

20. Solids maintain shape and volume.
    Liquids maintain volume but not shape.
    Gases maintain neither shape or volume.

22. **a.** $\lambda = \text{cm}$ **b.** $\nu = \text{cycles/s} = \text{hertz}$
    c. $c = 3.0 \times 10^{10}\ \text{cm/s}$
    $c = \lambda\nu$

24. **a.** radiant energy the sun transmits through space
    b. capacity for doing work
    c. 10^{-9} meter
    d. distance between adjacent wave crests

26. **a.** X-rays are ionizing radiation.
    b. yes
    c. They are low energy and do not pose as great a risk.
    d. yes
    e. UV light is higher energy than visible light; yes, there is a risk.

28. Boiling water

30. Sight, hearing

Chapter 2

2. **a.** Mercury, Venus, Earth, Mars
   b. Jupiter, Saturn, Uranus, Neptune, Pluto
   c. The terrestrial planets have high density and are composed of metals and minerals. The giant planets have low density and large amounts of gases, such as hydrogen and helium.

4. Because fusion reactions are occurring in stars today, it is reasonable to assume that these atom-building reactions have been going on for millennia. The large amount of hydrogen in the universe is an excess of reacting material, not the product of other reactions.

6. Iron, oxygen, silicon and magnesium. The distribution of elements according to mass did not happen because some elements combined with others to form compounds. The melting points and densities of the compounds determined how the elements were distributed on earth.

8. Venus never formed oceans. When Venus's interior warmed, carbon dioxide and water vapor were released and collected near the surface. Because Venus is closer to the sun than the Earth is, its surface never cooled enough to allow water vapor to condense and form rain. Carbon dioxide stayed in the atmosphere (since there was no ocean in which to dissolve) and kept Venus too hot (300°C, 572°F) to support life.

10. Through the process of photosynthesis, blue-green algae, or cyanobacteria, converted carbon dioxide and water into carbohydrates and oxygen. As the cyanobacteria multiplied, more and more carbon dioxide was removed from the atmosphere and was replaced with oxygen.

12. The basic unit of all silicates is the SiO_4 tetrahedron in which the silicon atom at the center is bonded to four oxygens located at the corners. See Figure 2.6.

14. **a.** Strong, flexible, resistant to corrosion and an excellent-thermal insulator
    b. Insulate pipes, fireproof clothing, brake shoes, roofing materials, electrical insulation and firebrick
    c. Friable asbestos is a much greater inhalation hazard.

16. Bauxite, Al_2O_3; Jamaica and Australia

18. **b.** 50%

20. Beverage cans, doors, window frames, siding, pots and pans, household appliances

22. Strategic metals are those metals that are essential for industry and defense.

24. State your own opinion.
26. *Biotic* refers to the living factors in the environment, such as plants, animals, fungi, and bacteria. *Abiotic* refers to the nonliving factors in the environment, such as temperature, rainfall, nutrient supply, and sunlight.
28. **a.** Producers are living factors that are able to convert sunlight into chemical energy through the process of photosynthesis. Examples are green plants and blue-green algae.
 b. Consumers are unable to harness energy from the sun to manufacture their own food. They must consume plants or other creatures to obtain nutrients and energy
30. Decomposers feed on freshly dead or partly decomposed remains of plants and animals. They include bacteria, fungi, earthworms, and many plants. Decomposers break down complex organic compounds in the dead plants and animals into simpler chemicals that are returned to the soil.
32. First law of thermodynamics: Energy can be neither created nor destroyed; it can only be transformed from one form to another. Example: electricity is converted to chemical energy in charging a battery. When the battery is used, the stored chemical energy is converted back into electrical energy.
34. Heat energy
36. 100 units
38. **a.** fourth **b.** first **c.** blueberries
40. See Figure 2.14.
42. C, H, N, O; S, P, Fe, Mg, or Ca
44. Carbon dioxide from the atmosphere
46. Decomposers feed on the dead remains of plants and animals and through respiration release carbon dioxide and water into the atmosphere. Respiration:

$$C_6H_{12}O_6 + 6\ O_2 \longrightarrow 6\ CO_2 + 6\ H_2O$$

48. Sugar, $C_6H_{12}O_6$
50. $N_2 + O_2 \xrightarrow{\text{lightning}} NO_3^-$ [nitrate]
52. Nitrogen-containing fertilizers from farm fields are carried by rainwater runoff to rivers and streams. Untreated human sewage contains large amounts of nitrogen.
54. Plant clover, spread manure.
56. State your own opinion.

Chapter 3

2. Because of the law of conservation of mass
4. In a chemical reaction, matter is neither created nor destroyed.
6. **a.** cathode rays **b.** canal rays
8. Dalton's theory defined atoms as hard, indivisible particles. Rutherford concluded that there was a dense nucleus with a positive charge at the center of the atom and that electrons were distributed in the space surrounding the nucleus.
10. The law of definite proportions
12. The law of multiple proportions
14. 6 grams
16. C_3H_8
18. **a.** mass = 1, charge = +
 b. mass = 1, no charge
 c. no mass, charge = −
20. **a.** electrons **b.** protons, electrons **c.** 1
22. Although the two atoms have the same total mass, they are different elements. The atomic number determines the identity of the element.
24. **a.** Atomic number is the number of protons; atomic mass is the sum of the number of protons and neutrons.
 b. Mass number is the sum of the number of protons and neutrons for a particular isotope of an element. The atomic mass is the average mass of the individual isotopes that make up an element's natural distribution on earth.
 c. atomic mass units (amu)
26. **a.** 11 **b.** 13 **c.** 11
28.

	Protons	Neutrons	Electrons
a.	82	126	82
b.	82	122	82
c.	17	20	17
d.	13	14	13
e.	15	16	15

30. Each atom contains 15 protons and 16 neutrons.
32.

Atom	Protons	Neutrons	Electrons	Symbol
Neon	10	11	11	Ne
Barium	56	82	56	Ba
Scandium	21	24	21	Sc
Phosphorus	15	16	15	P

34.

	Protons	Neutrons
^{21}Ne	10	11
^{22}Ne	10	12
^{23}Ne	10	13

36. N, Cl, Zn, Xe, Hg
38. Density, melting point, boiling point
40. **a.** metal **b.** nonmetal
 c. nonmetal **d.** metal
 e. metal **f.** metal
 g. metal **h.** nonmetal
 i. metal **j.** metal
 k. nonmetal **l.** nonmetal
42. Calcium, Ca; strontium, Sr; beryllium, Be
44. He or Ar
46. Semimetals
48. a, d, f, h
50.

	Neutrons	Protrons
a.	8	7
b.	2	1
c.	125	82
d.	88	63
e.	60	47
f.	62	47

52. Titanium, a metal

Chapter 4

2. The single electron can move to different orbits, or energy levels.

4. **a.** As: $1s^2\,2s^2\,2p^6\,3s^2\,3p^6\,4s^2\,3d^{10}\,4p^3$
b. Na: $1s^2\,2s^2\,2p^6\,3s^1$
c. Br: $1s^2\,2s^2\,2p^6\,3s^2\,3p^6\,4s^2\,3d^{10}\,4p^5$

6. **a.** 18 **b.** *s, p, d, f*
c. 1, 4, 9, 6 **d.** 2

8. **a.** Mg **b.** K
c. P

10. **a.** $2p$ fills before $3s$
b. The second level does not have a *d* orbital.

12. **a.** Li **b.** F
c. O **d.** B

14. To gain a noble (inert) gas electron configuration
a. Al^{3+} **b.** S^{2-}
c. B^{3+} **d.** Cs^+

16. A filled octet, 8 electrons

18. **a.** K **b.** Ca

20. **a.** Ba: **b.** :C̈l: (7 dots)
c. •Al• (3 dots) **d.** :S̈: (6 dots)

22. **a.** CsBr **b.** $SrBr_2$

24. **a.** ionic **b.** ionic
c. ionic **d.** polar covalent

26. **a.** NaCl **b.** $CaCl_2$
c. $AlCl_3$

28. **a.** BeF_2, beryllium fluoride
b. AlP, aluminum phosphide
c. $MgBr_2$, magnesium bromide

30. **a.** Polar covalent bonds are formed by the unequal sharing of bonding electrons.
b. For molecules with three or more atoms, the orientation of individual bonds determines whether or not the molecule as a whole is polar.

32. Electrostatic forces

34. An intramolecular force is one that occurs between molecules. An example is London forces. An intermolecular force occurs within a molecule.

36.

```
   H
   ••
H:C:H
   ••
   H
```

38. CF_4

40. Single bond consists of one shared e^- pair.
Double bond consists of two shared e^- pair.
Triple bond consists of three shared e^- pair.

42. **a.** :F̤̈:N:::N:F̤̈:

b.

```
H.         .H
  :C::C::C:
H'         'H
```

44. **a.** CaF_2 **b.** CCl_4
c. $MgBr_2$ **d.** NCl_3

46. Nonpolar covalent bond

48. **a.** F **b.** N
c. O

50. **a.** H—O, H—Cl, H—F
b. N—O, P—O, Al—O
c. Br—Br, B—N, H—Cl
d. N—P, S—O, Be—F

52. Carbonate ion, $CO_3{}^{2-}$

54. $CaPO_4$

56. $LiBrO_4$, $NaIO_4$

Chapter 5

2. **a.** ${}^{0}_{-1}\beta$ or ${}^{0}_{-1}e$ **b.** ${}^{4}_{2}\alpha$ or ${}^{4}_{2}He$ **c.** ${}^{0}_{0}\gamma$

4. Beta particles have a charge of −1 and almost no mass, just like an electron. Unlike an electron, they come from the nucleus of an atom.

6.

	Protons	Neutrons
a. ^{37}Cl	17	20
b. ^{115}Ag	47	68
c. ^{17}O	8	9
d. ^{136}Cs	55	81
e. ^{99}Mo	42	57
f. ^{13}C	6	7

8. There is no difference in the chemical reactivity of the two isotopes.

10. A beta particle: ${}^{210}_{82}Pb \longrightarrow {}^{210}_{83}Bi + {}^{0}_{-1}\beta$

12. **a.** ${}^{206}_{81}Tl$ **b.** ${}^{1}_{1}p$ **c.** ${}^{226}_{88}Ra$

14.

	Mass number	Atomic number
a.	stays the same	increases by 1
b.	decreases by 4	decreases by 2
c.	no change	no change

16. The chemical reactivity of sulfur-35 is exactly the same as that of sulfur-32.

18. Cobalt-60

20. **a.** ${}^{42}_{19}K \longrightarrow {}^{42}_{20}Ca + {}^{0}_{-1}\beta$
b. 0.031
c. The potassium isotope has a short half-life, and potassium salts are very soluble and widely distributed in the body.

22. 3024 years, 7 half-lives

24. Radiation enters Geiger tube and ionizes argon gas, causing electric current to flow; see Figure 5.6.

26. Only if the tools were made from once-living material, such as wood. If the tools were made from stone or metal, carbon-14 couldn't be used.

28. **a.** Plot disintegrations vs. time; half-life = 6 h
b. 42 h

30. No, the rate of nuclear decay is not temperature-dependent.

32. Possibly. As an approximation, 4% of 5730 is 230 years: 1990 − 230 = 1770. Use rate equation to get exact date.

34. a. the splitting of a nucleus into two or more fragments
b. a self-sustaining reaction in which one of the products of the reaction causes a new reaction to occur
c. The mass of an isotope necessary to sustain a chain reaction

36. Uranium-235

38. 0.7% uranium-235

40. Radon

42. Gamma ray

44. Radiation can cause ionization of the atoms it encounters. It can disrupt the normal working of the cells in living tissue and can cause abnormalities in genetic material (DNA).

46. Alpha particles are large and, because of their size, are easily stopped. If a radioisotope that emits alpha particles is ingested, the large alpha particles will cause damage to internal delicate internal tissues.

48. The rem takes into account the potential damage to living tissues caused by the different types of ionizing radiation.

50. The fission chain reaction is slowed if the neutrons produced are not used to break apart other uranium atoms.

Chapter 6

2. a. $4\,Al + 3\,O_2 \longrightarrow 2\,Al_2O_3$
b. $4\,Fe + 3\,O_2 + 6\,H_2O \longrightarrow 4\,Fe(OH)_3$
c. $2\,CH_3OH + 3\,O_2 \longrightarrow 2\,CO_2 + 4\,H_2O$
d. $CH_4 + 2\,O_2 \longrightarrow CO_2 + 2\,H_2O$

4. a. $2\,Na + Cl_2 \longrightarrow 2\,NaCl$
b. $H_2 + I_2 \longrightarrow 2\,HI$
c. $Si + 2\,Br_2 \longrightarrow SiBr_4$

6. a. $2\,Mg + CO_2 \longrightarrow 2\,MgO + C$
b. $Br_2 + 2\,K \longrightarrow 2\,KBr$
c. $SiH_4 + 2\,O_2 \longrightarrow 2\,H_2O + SiO_2$

8. 66.6 kg of copper

10. a. 6.66 b. 4.16 c. 7.2
d. 10 e. 81

12. a. 17.0 g/mol b. 120.4 g/mol
c. 46.0 g/mol d. 159.6 g/mol
e. 98.1 g/mol

14. C_2H_4

16. a. 58.5 g/mol b. 342 g/mol
c. 241.2 g/mol d. 262.9 g/mol

18. CH_4

20. a. 9 mol b. 10 mol
c. 3 mol

22. $\dfrac{3.17\ g}{74.9\ g/mol} = 0.042\ mol\ As$

$\dfrac{7.51\ g}{71.0\ g/mol} = 0.105\ mol\ Cl_2$

$\dfrac{0.105\ mol\ Cl_2}{0.042\ mol\ As} = 2.5$

$As + 2.5\,Cl_2 \longrightarrow AsCl_5$

$2\,As + 5\,Cl_2 \longrightarrow 2\,AsCl_5$

24. 2 mol of methane

26. $S + O_2 \longrightarrow SO_2$

$1000\ kg \times 0.012 = 12\ kg = 12{,}000\ g\ S$

$\dfrac{12{,}000\ g}{32\ g/mol} = 375$ mol of S, which means 375 mol of SO_2

375 moles × 64 g/mol = 24,000 g = 24 kg of SO_2

28. iron reacting to form rust

30. a. The rate of a chemical reaction is the rate or speed at which reactants are converted into products.
b. The activation energy must be applied to a reaction to get it started.
c. Entropy describes the degree of disorder, or randomness.
d. A reaction in which heat is produced is an exothermic reaction.

32. a. decreases b. increases c. increases d. increases

34. The minimum amount of energy reactants must possess in order to react is called the *activation energy*.

36. Temperature, concentration, presence of a catalyst

38. The lower temperature of the refrigerator will slow the rate of the souring reaction.

40. a. The candle is the fuel for combustion in oxygen; the reaction is spontaneous.
b. Reactions slow down as the temperature is lowered because the reacting molecules or atoms move more slowly.
c. Increasing the concentration of reactants speeds a chemical reaction because the probability of two reactants hitting each other increases with concentration.

42. If the concentration of the reactants is increased, the probability of the two reactants colliding, in solution, is also greater. The reactants cannot form products without colliding.

44. As the temperature is increased, the speed of the reactant molecules increases. Their increased speed increases the probability of their colliding to form products.

46. As the temperature is decreased, the speed of the reactant molecules decreases. Their decreased speed decreases the probability of their colliding to form products. As the temperature drops, the viscosity (thickness) of the solution increases, and this makes it more difficult for reacting molecules to collide.

48. a. Flour that is dispersed as small particles has lots of surface area and is dispersed in the air, which is the other reactant.
b. Ground solids have a larger surface area and can be dissolved by the water more rapidly. Frequency of collisions between reactants also increases.

50. A reaction that has a higher rate goes to completion faster than a reaction with a lower rate. The fast reaction does not produce more product than the slow reaction; it just makes that quantity of product faster.

Chapter 7

2. a. K^+, Br^- b. Mg^{2+}, CO_3^{2-} c. Na^+, SO_4^{2-} d. NH_4^+, Cl^- e. Mg^{2+}, NO_3^-

4. 2.48 M; 0.12 M

6. 10 ppb; yes, this violates EPA standards.

8. 0.05 ppm

10. red (by acids), blue (by bases)

12. base

14.

	Brønsted-Lowry Acid	Brønsted-Lowry Base
a.	HF	H_2O
b.	H_2O	S^{2-}
c.	H_2CO_3	H_2O
d.	H_2O	CH_3NH_2

16. a. acetic acid

18. $H_2O \longrightarrow H^+ + OH^-$

20. a. acidic b. almost neutral c. slightly acidic

22. a. pH = 4 b. pH = 8 c. pH = 1

24. a. 1.0×10^{-10} M b. 1.0×10^{-7} M c. 1.0×10^{-4} M d. 1.0×10^{-2} M

26. a. Solution A is more acidic.
b. 1000 times more acidic

28. Acetic acid (vinegar), ascorbic acid (Vitamin C), citric acid (citrus fruits), calcium carbonate (stomach antacid), sodium hydroxide (liquid drain cleaner)

30. $NH_3 + H_2O \longrightarrow NH_4^+ + OH^-$
Since it forms a hydroxide ion, ammonia is a strong base. This is a hydrolysis reaction.

32. $HCl + NaOH \longrightarrow H_2O + NaCl$
$HNO_3 + NaOH \longrightarrow H_2O + NaNO_3$
Acids and bases react to form water and a salt in a neutralization reaction.

34. The pH scale was introduced to make the negative exponents of the hydrogen ion concentration easier to describe and use. Since

$$H_2O \rightleftharpoons H^+ + OH^-$$
$$[H^+] = [OH^-] = 10^{-7} \text{ in neutral water at } 20°C$$

36. Buffers are substances that resist changes in pH. The bicarbonate system is a good example; see Section 7.10.

38. a salt

40. Carbonic acid, H_2CO_3

42. 7.3–7.5

44. pH = 5.6; because carbon dioxide is dissolved in it

46. a. $CaCO_3 + H^+ \longrightarrow Ca^{2+} + HCO_3^-$
b. $2\ KAlSi_3O_8 + 2\ H^+ + H_2O \longrightarrow Al_2Si_2O_5(OH)_4 + 4\ SiO_2 + 2\ K^+$

48. Besides air pollution, the natural process of soil formation is also important in causing acidification. Changes in land use and consequent changes in vegetation on the land are known to acidify soils.

50. Acid mine drainage is neutralized with lime before it is released. Many mines have been sealed to exclude oxygen and water so that acids cannot form.

52. a. oxidation b. oxidation c. reduction

54. a. neither b. reduction c. oxidation e. oxidation

56. $Zn \longrightarrow Zn^{2+} + 2\ e^-$ [oxidation]

58. Zinc and mercury

60. a. $SnCl_4$ b. O_2

62. a. $Sn^{2+} \longrightarrow Sn$ [reduction]
b. $Zn \longrightarrow Zn^{2+}$ [oxidation]
c. $Fe^{2+} \longrightarrow Fe^{3+}$ [oxidation]
$Cr_2O_7^{2-} \longrightarrow Cr^{3+}$ [reduction]
d. $MnO_4^- \longrightarrow Mn^{2+}$ [reduction]

Chapter 8

2. a. 2,3-dimethylbutane b. 3-ethylpentane c. 2-methylbutane d. 4-methylheptane

4. ethane CH_3CH_3
propane $CH_3CH_2CH_3$
butane $CH_3CH_2CH_2CH_3$
pentane $CH_3CH_2CH_2CH_2CH_3$
hexane $CH_3CH_2CH_2CH_2CH_2CH_3$

6. $CH_3CH_2CH_2CH_2CH_3$
pentane

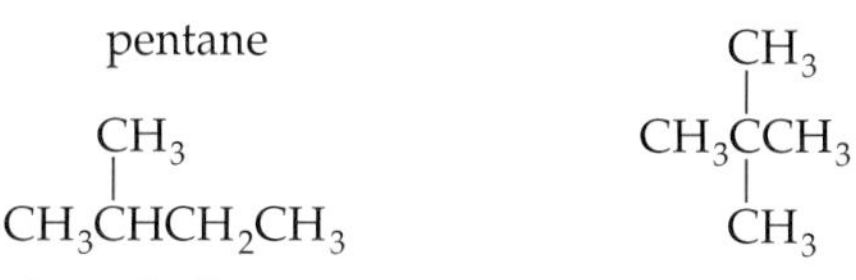

2-methylbutane 2,2-dimethylpropane

8. a. $CH_3(H)C{=}C(H)CH_3$
b. $H_2C{=}CH{-}CH_2{-}CH_2{-}CH_2{-}CH_3$
c. $H_2C{=}CH{-}CH_3$
d. $H_2C{=}CH{-}CH_2{-}CH_2{-}CH_3$

10. a. $CH_3{-}CH_2{-}CH_2{-}CH_2{-}CH_2{-}CH_3$
b. $CH_3{-}CH(CH_3){-}CH_2{-}CH_2{-}CH_3$
c. $CH_3{-}CH_2{-}CH(CH_3){-}CH_2{-}CH_3$
d. $CH_3{-}C(CH_3)_2{-}CH_2{-}CH_3$
e. $CH_3{-}CH(CH_3){-}CH(CH_3){-}CH_3$
f. $CH_3{-}CH(CH_2CH_3){-}CH_2{-}CH_3$

12. c. 2-butene
d. 1-chloropropene

14. a. $CH_3-CH(Cl)-CH(Cl)-CH_2-CH_3$

b. 1,4-dichlorocyclohexane (Cl at top and bottom of cyclohexane ring)

c. $CH_3-CH_2-CH(CH_2CH_3)-CH_2-CH_3$

d. $Br-CH_2-CH_2-CH_2-CH_3$

16. a. ketone b. carboxylic acid c. ester

18. a. none (hydrocarbon) b. ketone
c. alcohol d. ketone and alcohol
e. ester f. ether

20. a. 2,2-dimethyl-3-hexene
b. 3-nonene
c. 3-hexene

22. a. RCOOH b. RCOR′
c. RCH_2OH d. RCOOR′

24. a. isopropyl alcohol b. methyl alcohol
c. ethyl alcohol d. ethylene glycol

26. Acetaldehyde, CH_3CHO

28. Carboxylic acid

30. Ester (ethyl acetate)

32. Both have two carbon atoms, but ethylene glycol has two hydroxyl groups (—OH) and ethanol has only one.

34. Benzene ring bonded to C(=O)OH (benzoic acid)

36. Sugar and yeast

38. Acetic acid

40. An aldehyde has a hydrogen and an alkyl group bonded to the carbonyl group. A ketone has two alkyl groups bonded to the carbonyl group.

42. a. HCHO b. $CH_3COOCH_2CH_3$
c. CH_3CH_2OH d. CH_3COCH_3

44. Butyric acid has a sharp vomit-like smell; methyl butyrate, an ester of butyric acid, has a sweet smell.

46. a. amine
b. aldehyde (carbonyl)
c. carboxyl (carboxylic acid)
d. ester

48. a. $H-O-CH_2-CH=CH-CH_3$

b. $CH_3-C(=O)-CH_3$

c. $CH_3-CH_2-CH_2-CH_2-CH_2-CH=CH-CH_2-CH_2-CH_2-CH_2-C(=O)OH$

50. a. Benzene ring with CH_3 b. Benzene ring with CH_3 and CH_3, *meta* c. Benzene ring with CH_3 and CH_3 at opposite (para) positions

52. Only two

54. Toluene

56. *para* position

Chapter 9

2. a. Monomers are small molecules that can be linked together to form long polymer chains.
b. *Vulcanization* is the name given to the chemical reaction between rubber and sulfur. Sulfur cross-links the polymer chains of the rubber.
c. An addition polymer is formed when monomers link together without losing any of their atoms.

4. Orlon is a homopolymer and made by the polymerization of acrylonitrile. Polyethylene is a homopolymer made by the polymerization of ethylene. Nylon is a copolymer made by the polymerization of adipic acid and 1,6-diaminohexane.

6. Cellulose nitrate

8. $CH_2{=}CHCl \longrightarrow \left(-\underset{H}{\overset{H}{C}}-\underset{Cl}{\overset{H}{C}}- \right)_n$

10. $CF_2{=}CF_2$, addition polymer

12. Vulcanization is a chemical reaction of rubber and sulfur. The sulfur cross-links the long polymer chains of rubber and changes the rubber from being soft and elastic to hard, tough and more resistant to heat and cold.

14. Latex, the polymeric material in latex paints, is named after the natural, milky juices of the rubber plant. Modern latex paints contain poly(vinyl acetate) (PVA) or acrylic polymers based on methyl methacrylate.

16. Latex paints use water as the solvent. Oil-based paints use organic solvents.

18. a. poly(vinyl chloride) (PVC)
b. polyacrylonitrile
c. polyethylene terphthalate (PET)
d. poly(vinyl chloride) (PVC)

20. 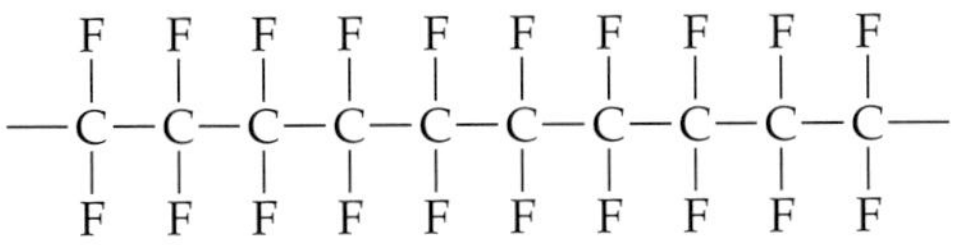 $-\underset{F}{\underset{|}{\overset{F}{\overset{|}{C}}}}-\underset{F}{\underset{|}{\overset{F}{\overset{|}{C}}}}-\underset{F}{\underset{|}{\overset{F}{\overset{|}{C}}}}-\underset{F}{\underset{|}{\overset{F}{\overset{|}{C}}}}-\underset{F}{\underset{|}{\overset{F}{\overset{|}{C}}}}-\underset{F}{\underset{|}{\overset{F}{\overset{|}{C}}}}-\underset{F}{\underset{|}{\overset{F}{\overset{|}{C}}}}-\underset{F}{\underset{|}{\overset{F}{\overset{|}{C}}}}-\underset{F}{\underset{|}{\overset{F}{\overset{|}{C}}}}-\underset{F}{\underset{|}{\overset{F}{\overset{|}{C}}}}-$

22. Teflon

24. **a.** PVC **b.** Teflon
c. polyacrylonitrile, Orlon **d.** polystyrene

26. The sulfur inserts itself between the long polymer chains of rubber and links them together. This cross-linking changes the rubber from being soft and elastic to being hard, tough, and more resistant to heat and cold.

28.

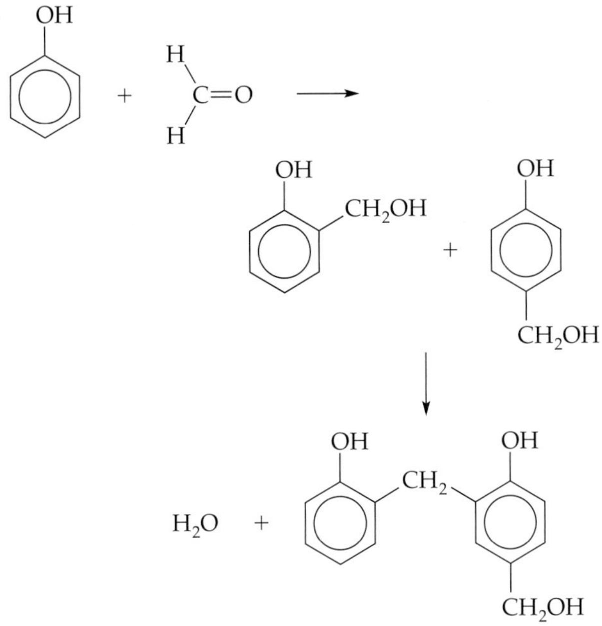

```
   OH
   |             H
  ( )     +       \C=O     ---->
                 /
                H

            OH                      OH
            |   CH2OH               |
           ( )/           +        ( )
                                    |
                                    CH2OH
                     |
                     v
              OH            OH
              |     CH2     |
  H2O  +     ( ) /     \   ( )
                            |
                            CH2OH
```

30. The solvent in the paint, which holds the pigment, binder and other ingredients in suspension, evaporates as the paint dries. The paint's binder polymerizes and hardens as the paint dries, and forms a continuous film which holds the paint to the surface.

32. The main ingredients of paint are 1) pigment, 2) solvent, and 3) binder.

34. 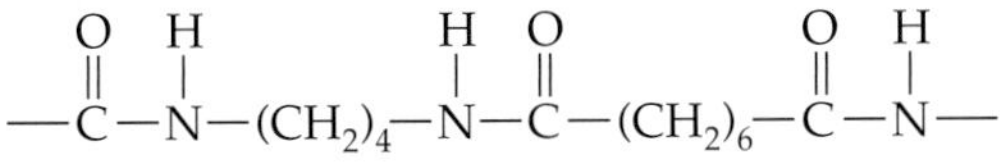 $-\overset{O}{\overset{\|}{C}}-\overset{H}{\overset{|}{N}}-(CH_2)_4-\overset{H}{\overset{|}{N}}-\overset{O}{\overset{\|}{C}}-(CH_2)_6-\overset{O}{\overset{\|}{C}}-\overset{H}{\overset{|}{N}}-$

36. Dacron is a polyester that is blended with cotton to make permanent-press fabrics. HDPE is tough, but not a good choice for making clothing fibers. It is not very flexible or soft and comfortable.

38. **a.** Thermosetting: Bakelite, Formica; structures are shown on pages 266 and 267.
b. Thermoplastic: polyethylene, polystyrene (Styrofoam), poly(vinyl chloride) (vinyl plastic), polyacrylonitrile (Orlon, Acrilan), polytetrafluoroethylene (Teflon); structures are shown in Table 9.1.
c. Silicone: Silly Putty, silicone lubricants; structures are shown on page 269.

40. Rearrangement polymerization

42. See page 269.

44. Recycling, reuse, biodegradability, reducing plastic use

46. Light-sensitive carbonyl groups can be incorporated into the polymer chains. Starch or cellulose can be incorporated into the polymer during production so that it can be digested by micro-organisms.

48. Recycling: Separation according to type before remelting. Only thermoplastics can be recycled.
Incineration: Carbon dioxide is released when all plastics are burned. Toxic gases are released when some plastics (e.g., PVC) are burned.
Degradable plastics: Not as strong as regular plastics. When buried, they degrade.
Reducing plastic use: Stop the "throw-away society"; encourage large corporations to curb plastic use.

50. State your own opinion.

52. Give your plan, use the uniform code for identification of plastic types.

Chapter 10

2. **a.** The heat of vaporization is the amount of heat required to convert 1 g of a liquid to a vapor at its boiling point.
b. The high heat capacity of water allows the oceans to absorb large amounts of heat without a corresponding rise in the water temperature.
c. Because of its high heat of vaporization, a large amount of heat is required to evaporate a small volume of water. Water is much more effective as a coolant than another liquid with a lower heat of vaporization.

4. Sea water is salty because of the constant evaporation of water from the surface of the ocean. This evaporation increases the concentration of the dissolved ions.

6. If the water molecule were linear, the open lattice arrangement of hydrogen and oxygen atoms could not form when water froze. The open-lattice arrangement places the molecules of water farther apart in solid water than they are in liquid water and makes ice less dense than liquid water. Solid water forming from linear water molecules would be more dense than the liquid.

8. **c.** evaporation

10. The hydrologic, or water, cycle: Evaporation from the surface of lakes and ponds, cloud formation by condensation, rainfall by precipitation, rain seeping into the soil and collecting in porous rock called an *aquifer*.

12. Increased temperature would cause evaporation from the surface of lakes, ponds, and the oceans to increase. More clouds and rainfall would be expected.

14. If the level of the Mississippi were raised, bigger boats would be able to navigate the river and commerce would benefit. If water from the Great Lakes were diverted, the level of water

in the lakes would fall, and plant and animal life would be affected.

16. When massive amounts of groundwater are removed from natural aquifers, their porous rock structure can collapse and cause a sinkhole. The land will not return to its original level if the water removal is stopped.

18. Large amounts of water evaporate as the water is exposed to sunlight. As pure water is removed by evaporation, the concentration of dissolved solids in the remaining water increases, and the water becomes more brackish.

20. Wash your car with the water from your bath. Water your vegetables with the rainwater. Cook your dinner with the well water.

22. Any of the ten types listed at the beginning of Section 10.8.

24. Cholera, typhoid, dysentery, polio: None of these is a serious problem in the United States today. As a result of sewage treatment and disinfection of water supplies, waterborne diseases have been virtually eliminated in developed countries.

26. **a.** An increase in water temperature decreases oxygen solubility.
b. A decrease in atmospheric pressure decreases oxygen solubility.

28. Organic detritus in aquatic ecosystems is decomposed by aerobic (oxygen-consuming) decomposers, primarily bacteria and fungi. If the water is overloaded with organic wastes, aerobic decomposers proliferate, and DO is consumed more rapidly than it can be replaced from the atmosphere.

30. Point sources discharge pollutants into a small area (usually from a pipe). Nonpoint sources discharge pollutants over a wide area (runoff).

32. When eutrophication occurs, dense mats of rooted and floating plants are formed. Blue-green algal blooms that appear on the water surface release unpleasant-smelling, bad-tasting substances.

34. Sediments fill irrigation ditches and clog harbors. Toxic substances can adsorb on the surface of sediments and become concentrated. Spawning grounds become buried. Photosynthesis in turbid water decreases.

36. Bioaccumulation. Because aquatic ecosystems have from four to six trophic levels, bioaccumulation is a more serious problem in aqueous habitats than in terrestrial ones, which usually have only two or three trophic levels.

38. Metals, because they are elements, cannot be broken down into simpler, less toxic forms. They persist in the environment for years and bioaccumulate through food chains.
Mercury: Used in the production of chemicals, paints, plastics, pharmaceutical, photographic and electrical equipment.
Lead: Leaded gasoline, batteries, paint, lead pipes, and solder
Cadmium: Used in the production of paints, plastics, and nickel-cadmium batteries and in electroplating.

40. Fish and shellfish become contaminated with mercury. The mercury in the water bioaccumulates in the fish and shellfish. Humans who eat the fish and shellfish ingest a much higher concentration of mercury than is present in the water of the bay.

42. The main concern for the United States is the mercury content of fish.

44. No, the consumption of a half pound of fish would not be fatal. A 55-kg woman would require a 27.5-g dose; 30.3 lb of fish would contain this amount.

46. Agent Orange was a herbicide that was sprayed to defoliate the jungles of South Vietnam. Agent orange contained a low concentration of dioxin, a very toxic chemical that is a by-product of its synthesis.

48. Heated water can be made to flow through a cooling pond, or it can be sprayed into a cooling tower and cooled by evaporation.

50. State your own opinion.

52. **a.** $Cl_2 + H_2O \longrightarrow HClO + H^+ + Cl^-$
b. Cl_2 reacts with organic material in water to form chlorinated hydrocarbons, which are suspected carcinogens.
c. O_3, ozone

54. One method adds aluminum sulfate ($AlSO_4$), which results in the precipitation of phosphate as insoluble aluminum or calcium phosphates):
$3\ PO_4^{3-} + Al^{3+} + 3\ Ca^{2+} \longrightarrow AlPO_4 + Ca_3(PO_4)_2$

Chapter 11

2. See Figure 11.6.

4. **a.** Diameter decreases.
b. Diameter decreases.

6. Because a gas, such as air, is compressible. When pressure is applied to the brake pedal, that pressure should be transmitted by the brake fluid to the brakes in the wheels. Air in the brake line compresses, the pressure is not transmitted, and the car doesn't stop.

8. 1520 mm

10. −4.8°C, 268.2 K

12. 3226 mL

14.
$$O_2 + \text{sunlight} \longrightarrow 2\,O\cdot$$
$$O\cdot + O_2 \longrightarrow O_3$$

16. The ozone layer forms the earth's main shield against the sun's dangerous UV radiation, and without it life on earth could not exist.

18. An *air pollutant* is a substance that is present in the atmosphere at a concentration sufficient to cause harm to humans, other animals, vegetation, or building materials. Primary air pollutants are emitted directly into the troposphere. Secondary air pollutants are produced by chemical reactions between primary pollutants and other constituents of the atmosphere.

20. **a.** diesel particulates, oxides of carbon, oxides of nitrogen
b. oxides of carbon, oxides of sulfur

22. **a.** See Figure 11.10.
b. As population continues to increase, more and more automobiles will be on the road. The more cars, the more pollutants will be emitted.

24. **a.** false **b.** false **c.** false

26. The major nitrogen oxide air pollutant is NO_2. It is removed from the atmosphere by one of the following pathways.

Combination with water vapor to form nitric acid:

$$4\,NO_2 + 2\,H_2O + O_2 \longrightarrow 4\,HNO_3$$

Reaction with hydroxyl radicals to form nitric acid:

$$NO_2 + OH\cdot \longrightarrow HNO_3$$

28. Exposure to sulfur dioxide causes irritation of the eyes and respiratory passages and aggravates symptoms of respiratory disease in humans.

30. State your own opinion.

32. Reactions that lead to the production of photochemical smog are initiated by sunlight and involve hydrocarbons and nitrogen oxides.

Nitrogen oxides:

$$NO_2 \longrightarrow NO + O$$
$$O + NO \longrightarrow NO_2$$
$$O + O_2 \longrightarrow O_3$$
$$O + H_2O \longrightarrow 2\,OH\cdot$$

Hydrocarbons (RH):

$$RH + OH\cdot + O_2 \longrightarrow RO_2\cdot + H_2O$$
$$RO_2\cdot + NO \longrightarrow \text{aldehydes and ketones} + NO_2 + HO_2\cdot$$

34. Ozone is very toxic to plants, and even brief exposure to a concentration of 0.1 ppm reduces photosynthesis and causes yellowing of green leaves. Ozone irritates the eyes and nasal passages. Exposure of humans to 0.3 ppm for 1 or 2 hours can cause fatigue. People with asthma or heart disease are particularly susceptible.

36. **a.** A *mist* is a cloud of minute droplets. A fog that does not reduce visibility is called a *mist* or a *haze.*

b. *Fog* is a collection of tiny water droplets that float in the air. Fog is similar to clouds, except that clouds do not touch the earth's surface.

c. *Soot* is finely divided impure form of carbon that is formed from the incomplete combustion of coal or diesel fuel.

d. *Fly ash* consists of solid particulates that are released when coal is burned.

e. *Smog* is a mixture of fly ash, soot, sulfur dioxide, and some VOCs.

38. The greater the surface area of a particle or aerosol, the greater the chance that the particle will react with the wall of the lung once it is inhaled.

40. The smaller the size of a particle, the greater the chance it will be inhaled deep into the lung. Smaller particles also have larger surface areas on which chemical reactions can take place.

42. (1) Particulates with diameters greater than 1 μm intercept and scatter incoming light away from the earth's surface and exert a cooling effect.

(2) Aerosols act as nuclei for the formation of droplets in clouds.

44. A free radical has an unpaired electron that is not used for bonding.

46. NASA found ClO in the region of the ozone hole in Antarctica.

48. The ozone hole is located over Antarctica.

50. The greenhouse gases are carbon dioxide (CO_2), methane (CH_4), nitrous oxide (N_2O) CFCs (e.g., $CFCl_3$), and ozone (O_3). The combined total concentration of methane, nitrous oxide, ozone and CFCs is considerably less than that of carbon dioxide, but these gases absorb and reemit more infrared radiation than does carbon dioxide. If their levels in the atmosphere continue to rise at the present rate, it is estimated that their warming effect will soon equal that of carbon dioxide.

52. Indoor air pollutants include cigarette smoke, asbestos fibers, radon, nitrogen oxides, carbon monoxide, VOCs, formaldehyde, fungi, bacteria, pet dander.

54. State your own opinion.

Chapter 12

2. Steam engines were fueled with wood or coal.

4. Petroleum originated from microscopic marine organisms that once lived in great numbers in shallow coastal waters.

6. Gas, oil, and water are found together. Water usually occupies one-half of the pore space in the reservoir rock.

8. Natural gas burns cleanly, leaves no residue, and, weight-for-weight, has a higher heat output when burned than any other common fuel.

10. Sulfur deactivates the catalysts used by refineries in the cracking step. Sulfur in refined gasoline produces sulfur dioxide when burned.

12. 1920 kcal

14. $2\,C_4H_{10} + 13\,O_2 \longrightarrow 8\,CO_2 + 10\,H_2O$

Bonds broken

20 C—H	99 kcal/mol		1980 kcal/mol
6 C—C	83		498
13 O—O	118		1534
		Total	4012 kcal/mol

Bonds made

16 C=O	192 kcal/mol		3072 kcal/mol
20 O—H	111		2220
		Total	5292 kcal/mol

$5292 - 4012 = 640$ kcal/mol

16. Gasoline is a lower-boiling fraction of petroleum than kerosene is. The hydrocarbon molecules in gasoline have shorter chains than do those in kerosene.

18. Catalytic reforming is a process in which hydrocarbon vapors in straight-run gasoline are heated in the presence of suitable catalysts, such as platinum or other metals. In this process, straight-chain hydrocarbon molecules are converted to branched hydrocarbon molecules.

20. Toluene

22. Tetraethyl lead was a cheap and effective octane booster. As little as 3 g of TEL per gallon of gas could increase the octane rating by 10 to 15 points.

24. 2,2,4,-trimethylpentane (isooctane) was assigned an octane rating of 100. The straight-chain hydrocarbon heptane was given a rating of 0. The octane rating of a particular gasoline is determined by burning the gasoline in a standard engine and comparing the resulting knocking with that from standard mixtures of isooctane and heptane. A gasoline that performs in the same way as a mixture containing 90% isooctane and 10% heptane is assigned an octane rating of 90.

26. Anti-knock agents are octane enhancers, such as tetraethyl lead (TEL), methanol, ethanol and MTBE.

28. Oil shale is a common sedimentary rock that contains a solid organic material called *kerogen*. When this rock is heated in the absence of air, a thick brown liquid called *shale oil* is produced. Shale oil can be refined like petroleum after its sulfur and nitrogen impurities are removed.

30. If petroleum prices are low, there is no economic incentive to pursue new sources of petroleum, such as shale oil.

32. Crude natural gas is composed of the hydrocarbons methane (60–90%), ethane (5–9%), propane (1–2%), and butane (1-2%). It also contains carbon dioxide, hydrogen sulfide, nitrogen, and noble gases. Before crude natural gas can be used as a commercial fuel, carbon dioxide, sulfur compounds, water vapor, and hydrocarbons larger than ethane must be removed.

34. Peat contains too much water (90%), and it has a low carbon content (5%).

36. Strip mining produces large quantities of waste rock, and acid leaching from this waste can contaminate streams and rivers.

38. Coal is dirty and bulky and expensive to transport. When burned, the large quantities of fly ash produced present a disposal problem. All coal contains some sulfur, and sulfur dioxide emitted during its combustion is the primary cause of acid rain.

40. Coal gasification is the process that changes solid coal into synthesis gas.

$$C + H_2O \longrightarrow H_2 + CO$$
$$CO + 3\,H_2 \longrightarrow CH_4 + H_2O$$
$$CO + H_2O \longrightarrow H_2 + CO_2$$

Synthesis gas is a useful fuel, but it produces only about one-third of the heat produced by the combustion of natural gas.

42. Crushed coal is first mixed with a catalyst, potassium hydroxide or potassium carbonate. The material is mixed with carbon monoxide and hydrogen and heated to 700° C.

$$2\,C + 2\,H_2O \longrightarrow CH_4 + CO$$

Conditions are adjusted to give the highest yield of methane. Unreacted carbon monoxide and hydrogen are recycled through the system.

44. (1) Burning fossil fuels produces carbon dioxide, a greenhouse gas.
(2) The extraction of fossil fuels has adverse environmental consequences.

46. All of the types of industry are increasing demand for fossil fuels.
   a. farming: petroleum products
   b. food processing: electricity from coal, nuclear, oil, natural gas
   c. manufacturing: electricity from coal, nuclear, oil, natural gas
   d. transportation: petroleum products

48. See Figures 12.16 and 12.17. Electricity is produced whenever a coil of wire is moved through a magnetic field. The movement induces the flow of electrons in the wire. Most electric generators operate by rotating a coil of wire within a circular arrangement of magnets.

50. The 30 railroad cars of ash contain the noncombustible material that is part of the coal. The contents of the other 70 railroad cars was combusted in the power plant and went up the smokestack as gases.

$$\underset{\text{coal}}{C} + \underset{\text{oxygen}}{O_2} \longrightarrow \underset{\text{carbon dioxide}}{CO_2}$$

Chapter 13

2. In 1974, oil-exporting countries greatly reduced their output in order to drive up the price of oil. In 1990, Iraq invaded Kuwait. The United States and other countries intervened, and retreating Iraqi soldiers started the Kuwaiti oil wells on fire.

4. Increase the efficiency of automobiles and of consumer devices that use electricity

6. The environmental damage mining causes. Coal is difficult to handle, and it contains other material that is not combustible.

8. 40 to 60 years

10. Japan and France

12. State your own opinion.

14. Pu-239

16. No. Reactor-grade uranium is enriched to only 3% U-235.

18. The fuel rods are replaced every three years. Presently, they are permanently stored as wastes. Other countries, France and Japan, reprocess the fuel rods and recover the fissionable material for new fuel rods.

20. State your own opinion.

22. (1) The breeder reactor is cooled with liquid sodium. If there is a leak, sodium will explosively react with water.
(2) Pu-239 is extremely toxic and has an extremely long half-life (24,000 years).

24. a. Control rods are interspaced between the fuel rods and are made of material that absorbs neutrons (carbon or boron). When they are inserted into the core of the reactor, they absorb neutrons, and the rate of the fission reaction slows.
   b. Fuel rods are steel rods containing fissionable uranium fuel (enriched to 3% U-235).
   c. The moderator is water circulating around the fuel rods, which slows the neutrons to speeds that are optimal for splitting U-235 atoms.
   d. The containment building is the reinforced concrete structure that surrounds the core of a reactor. The problems at Chernobyl were made worse because the reactor had no containment.

26. In the plasma, the very high temperatures strip deuterium and tritium atoms of their electrons, creating a gas-like plasma of very energetic nuclei and free electrons. The greater the energy, the more likely the nuclei are to fuse together to form a new, heavier element.

28. Magnetic fields will be used to contain the reaction.

30. In an active solar heating system, heat collected in solar collectors located on the roof of a building is circulated by means of pumps. In a passive solar heating system, the system relies on convection currents to distribute the heat passively around the building. See Figure 13.10.

32. Arsenic, when doped into silicon, produces a semiconductor that has an excess of electrons. Boron, when doped into silicon, produces a semiconductor that has a shortage of electrons and thus positive holes.
34. arsenic, boron
36. *Biomass* is any accumulation of biological material. Biomass, like fuel, can be burned directly to provide energy in the form of heat.
38. Advantages: Biomass is renewable fuel. It can be burned directly to provide energy in the form of heat. Biomass can also be converted to methane or alcohols, such as methanol and ethanol.
    Disadvantages: When biomass is burned, carbon dioxide, which was slowly removed from the atmosphere by photosynthesis, is rapidly released back into the atmosphere.
40. The future of geothermal energy is uncertain. Many geothermal fields occur in inaccessible areas. Most steam deposits contain hydrogen sulfide, which can pollute the atmosphere when released from the plant. Salts dissolved in the water that is released with the steam can corrode the pipes of the generating plant. Despite these problems, geothermal energy will be increasingly exploited as oil becomes scarce.
42. c, d, b, a
44. Tidal power is never likely to be more than a minor source of energy in the United States.

Chapter 14

2. The structure on the left does not have four different groups bonded to the central carbon atom, so it is not chiral. The structure on the right does have four different groups and is chiral.
4. See Figure 14.3. They differ at carbons 1 and 2; fructose contains a ketone group, and glucose contains an aldehyde.
6. Excess starch is converted to glycogen and stored in the liver and in muscles. Large excesses are converted to fats and stored in the adipose tissue. Glycogen is the storage polysaccharide in animals.
8. See Figure 14.3.
10. **a.** sucrose = disaccharide
    b. glucose = monosaccharide
    c. cellulose = polysaccharide
    d. lactose = disaccharide
12. Glycogen is a polysaccharide formed of branched chains of glucose. When the starch we eat provides more glucose than the body requires for its immediate needs, some of the excess starch is converted to glycogen and stored in the liver and in muscles.
14. Insulin; glucose and fructose
16. When carbohydrates are oxidized in the body with molecular oxygen, the reaction yields carbon dioxide, water, and energy.
18. In a saturated fatty acid, all the carbon bonds are single bonds. An example of a saturated fatty acid is stearic acid. An unsaturated fatty acid contains one or more carbon-carbon double bonds. An example of an unsaturated fatty acid is oleic acid. See Fig. 14.7.
20. When the polyunsaturated fatty acids are hydrogenated, the double bonds are converted to carbon-carbon single bonds, and the liquid polyunsaturated fat is converted into a semisolid saturated fat. See Fig. 14.8.
22. Triglycerides are formed by the reaction of three fatty acid molecules with a glycerol molecule. See Figure 14.7.
24. Gall bladder
26. Lower sperm counts
28. An amino group, $-NH_2$, and a carboxyl (carboxylic acid) group, $-COOH$
30. Virtually all the protein in our bodies is made from the 20 amino acids listed in Table 14.2. The 10 that are highlighted are the essential amino acids that cannot be synthesized in the body from other chemicals and must be obtained from food.
32. Ten of the twenty amino acids listed in Table 14.2 cannot be synthesized by the body.
34. The total possible number of dipeptides is equal to $n!$ For three amino acids, this would be $3 \times 2 \times 1 = 6$

Leu-Gly-His	Gly-Leu-His	Leu-His-Gly
His-Gly-Leu	His-Leu-Gly	Gly-His-Leu

36. Hydrogen bonds
38. A substrate is the substance with which an enzyme interacts.
40. The shape of the active site of an enzyme will determine what substrate molecules can fit there. Some enzymes require a coenzyme, which may be a metal ion or a water-soluble vitamin.
42. Phosphoric acid unit, sugar, base
44. **a.** uracil-adenine
    b. guanine-cytosine
    c. adenine-thymine
    d. guanine-cytosine
46. Cytosine and guanine are held together by three hydrogen bonds. Adenine and thymine use only two hydrogen bonds.
48. Adenine and guanine
50. The two intertwined chains of the double helix are complimentary. Just before cell division, the double helix begins to unwind. See Figure 14.28. Each unwinding strand serves as a pattern, or template, for the formation of a new complimentary strand.
52. The DNA fragments are attracted to the positive electrode. They move at different rates that are determined by their sizes.
54. See Section 14.15.

Chapter 15

2. K, S, Cl, Na, Mg, Fe; see Table 15.1.
4. Enzymes are catalysts that facilitate the breakdown of food. The enzyme ptyalin catalyzes the breakdown of carbohydrates, and the enzyme pepsin catalyzes the digestion of proteins. The enzyme lipase is involved in the digestion of fats.
6. Glucose reacts with ADP to form lactic acid. Lactic acid reacts with ADP to form CO_2, H_2O, and ATP. In body tissues, energy is formed in the conversion of ATP back to ADP.
8. The energy derived from foods is measured in calories; 250 calories of heat is released when the candy bar is completely oxidized in a bomb calorimeter.

10. Energy from carbohydrates comes from glucose produced in digestion and stored as glycogen. Glycogen is released, as needed by the tissues, and enters the circulatory system. The blood glucose level is maintained between 70 and 100 mg of glucose per 100 mL of blood. Glucose is carried to all the body's tissues, where it supplies the energy cells need to perform their functions.
12. Celery; peel, seed, and pulp of fruit; wheat bran. In countries where the diet is low in fiber, appendicitis, diverticulitis, and colon cancer are common.
14. Fatty acids and glycerol, which were produced in the small intestine, are resynthesized to fats.
16. Glycerol
18. 64 g
20. **a.** vitamin E **b.** vitamin D **c.** vitamin C **d.** vitamin B_{12} **e.** vitamin A
22 **a.** beri beri **b.** pernicious anemia **c.** pellagra
24. Vitamin K
26. No, since water-soluble vitamins are not stored, but pass rapidly through the body, they must be replenished more often.
28. A disease caused by protein deficiency
30. The blood sugar rises as carbohydrates are digested and glucose enters the bloodstream. When a diabetic person eats, the pancreas does not function normally to provide insulin to reduce the concentration of glucose.
32. Galactosemia is an inheritable disease in which there is a deficiency of the enzyme that catalyzes the conversion of galactose to glucose.
34. Vegetarians need to eat a varied diet. The combination of cereals with legumes provides all the amino acids.
36. **a.** bone and teeth **b.** thyroid hormones **c.** bone and teeth **d.** hemoglobin
38. Carbohydrates 58%, fats 30%, proteins 12%
40.

	Coke	Gatorade
	(per 240 mL or 8 fl oz serving)	
Calories	140	50
Fat	0 g	0 g
Sugar	39 g	14 g
Sodium	35 mg	110 mg
Potassium	0 mg	30 mg

42. Coal tar
44. Lowest is restaurant 3 (461 cal); highest is restaurant 1 (534 cal).
46. 17.5 g; yes
48. Antimicrobial additives (sodium nitrate and sodium nitrite) and antioxidants (BHA and BHT)
50. Monosodium glutamate. It is used as a meat tenderizer and to enhance the flavor of foods.
52. Sequestrants, such as citric acid and EDTA, are substances that are added to tie up trace metals that may accidentally get into food.
54. Established by the U.S. Congress in 1958 as a list of additives considered safe by the Food and Drug Administration (FDA).

Chapter 16

2. **a.** 8,192 **b.** geometric
4. 35 years
6. 2.8%
8. Acidic soils are sour, and basic soils are sweet.
10. Raise the pH
12. Since calcium compounds are basic, they neutralize acid found in the soil.
14. Humus
16. Carbon, hydrogen, oxygen.
18. (1) soil bacteria, (2) rhizobium bacteria that live in nodules on the roots of leguminous plants, and (3) blue-green algae
20. When wet, urea decomposes slowly, releasing ammonia.
22. Superphosphate from phosphate rock or bone
24. Cadmium is toxic to humans, and it bioaccumulates in plants.
26. Organophosphate insecticides
28. Yes
30. 30% nitrogen, 30% phosphorous as P_2O_5, 10% potassium as K_2O
32. N, P, and K. Ca, Mg, and S are abundant in most soils, but liming easily replaces these elements very cheaply.
34. Superphosphate
36. Boron, copper, iron, manganese, molybdenum, zinc. No, sometimes applied to seeds.
38. 1-2-1; much lower than synthetic fertilizer
40. One-third of all the food grown in the world is destroyed by pests each year.
42. A pheromone is a chemical communicating substance that is used between members of the same species. Sex attractants and trail- marking substances are two examples. A sex attractant can be used to attract an insect into a trap.
44. Insect growth regulators disrupt insect development at various stages. An agent that inhibits the formation of the insect's exoskeleton (chitin) is an example of an IGR.
46. Juvenile hormones regulate the development of an insect from juvenile to adult. If juvenile insects are treated with juvenile hormones, they will remain immature and unable to mate.
48. The goal of alternative methods is not to eliminate all pests but to keep them at a level low enough to minimize economic losses.
50. **a.** Four major components: liminoids, azadirachtin, melantriol, salanin. They are nontoxic to humans.
    b. Neem oil: lubricant, fuel for oil-burning lamps, making soap, neem cake

Chapter 17

2. **a.** ester **b.** alcohol **c.** carboxylic acid
4. Tallow from beef and mutton, coconut oil, palm oil, cottonseed oil, and olive oil
6. Potassium hydroxide is used to make the softer soaps for shaving products.

8. No. Acid reacts with soap to form free fatty acids, which, because they do not have an ionic end, are insoluble in water and precipitate out as a greasy scum.
10. Hard water is passed through an ion-exchange resin with Na^+ ions on the resin. The Na^+ ions exchange with the hard-water ions, removing them from the water.
12. Acid end: soap has carboxylic acid ($-COO^-$); detergent has sulfonic acid ($-SO_3^-$).
Organic end: soap has a nonpolar straight-chain hydrocarbon ($CH_3(CH_2)_{20}-$), detergent has a nonpolar alkyl benzyl chain ($CH_3(CH_2)_{11}C_6H_4-$).
14. The nonpolar hydrocarbon end of an ABS has many branches; the LAS has a straight-chain hydrocarbon.
16. Phosphate builders cause eutrophication in natural waterways.
18. Zeolites are complex sodium aluminumsilicates that function as ion-exchange agents.
20. Optical brighteners absorb UV radiation from sunlight and then emit longer wavelength light in the blue region of the spectrum.
22. Fabric softeners contain quaternary ammonium salts (quats). Clothes tend to carry a negative charge, and the quats, which are positively charged, attach to the clothes and make them feel soft.
24. Enzymes are added to detergents to help remove stains. Amylases break down starch-based stains, and proteases break down protein-based stains.
26. Sodium hypochlorite produces hypochlorite ion (OCl^-).
28. Hydrogen peroxide (H_2O_2)
30. Sebum is an oily secretion produced by the sebaceous gland, which lies at the base of each hair follicle.
32. Hair is a long strand of protein. Hair is slightly acidic.
34. If the pH of the shampoo is slightly acidic, it gives the hair an attractive luster. If the pH of the shampoo is basic, the hair will look dull.
36. The heat from the curling iron disrupts hydrogen bonds in the hair and can cause a curl to be set temporarily.
38. The disulfide bond ($-S-S-$).
40. Sodium hydroxide
42. Black and dark brown hair are colored by melanin. Red-brown hair is colored by phaeomelanin.
44. Hydrogen peroxide oxidizes pigments to colorless products.
46. Lead acetate, $Pb(CH_3COO)_2$
48. Eccrine glands, apocrine glands. Bacteria break down the secretions from the apocrine glands into unpleasant smelling fatty acids and amines.
50. Triclosan (2,4,4-trichloro-2-hydroxydiphenyl ether)
52. Aluminum ions
54. Children develop fewer dental cavities if they are given fluoride.
56. Peroxide toothpaste releases bubbles of oxygen while the person is brushing.

Chapter 18

2. Ethyl alcohol is considered to be nonpoisonous, but if administered in high amounts, it can cause death. Methyl alcohol is poisonous.
4. Dermal contact, inhalation, and oral ingestion
6. (1050 mg/kg)(20 kg) = 23,660 mg of ibuprofen, or 68 tablets
8. (0.022 mg/kg)(70 kg) = 1.54 mg
10. $$R-\overset{\overset{\displaystyle O}{\|}}{C}-NHR + H_2O + OH^- \longrightarrow R-\overset{\overset{\displaystyle O}{\|}}{C}-OH + RNH_2$$
12. Chlorine gas (Cl_2), which was used as a chemical warfare agent in World War I, has a mustard color.
14. Pesticide, rodent poison, drug for humans
16. Mercury lamps, pressure gauges, dental amalgams, fungicides.
18. BAL surrounds the metal as a claw would surround a ball.
20. Undernourished children develop an appetite for unusual things. This condition, called pica, leads them to eat paint.
22. Lead paint, drinking water, food
24. Home-canned nonacidic vegetables
26. Neurotoxins either block the receptor sites, block the synthesis of acetylcholine, or inhibit the enzyme cholinesterase.
28. Tabun and Sarin are anti-cholinesterase inhibitors; they prevent the breakdown of acetylcholine by inactivating the enzyme cholinesterase. They are absorbed through the skin, as well as the lungs. Victims lose muscle control and die within few minutes by suffocation.
30. Atropine occupies receptor sites on nerve endings that are also occupied by acetylcholine.
32. Ethyl alcohol is removed by being oxidized to acetaldehyde, which is then oxidized to acetic acid, which is subsequently oxidized to carbon dioxide and water.
34. They are found in the liver, where they help detoxify poisons.
36. The cytochrome P-450 enzymes in the liver act on fat-soluble substances to make them water-soluble, so they can be excreted.
38. Humans have 46 chromosomes (23 pairs).
40. Teratogens are chemical agents that cause birth defects. Mutagens are chemicals capable of altering genes and chromosomes, causing abnormalities in the offspring.
42. Benign tumors grow slowly, they often regress spontaneously, and they do not invade neighboring tissue. Malignant tumors are called cancers. Cancers invade and destroy neighboring tissue.
44. initiation, promotion
46. breast, colon
48. Compare the cancer rates of a large number of people who were exposed to a chemical with the cancer rates of a large number of people who weren't exposed.
50. No. About 90% of all chemicals that are carcinogens are also mutagens.
52. Almost all human cancers caused by chemicals have long induction periods, which means that cancers occur 20 or more years after exposure. Exposure to chemicals today may result in increased cancer rates 20–30 years from now.

Chapter 19

2. Resource Conservation and Recovery Act (RCRA)
4. Fly ash is the solid material that is blown upward in an incinerator smokestack with combustion gases. When discarded items such as cans, jar lids, and batteries are incinerated, many toxic metals, including beryllium, cadmium, chromium, lead, and mercury, are released and carried up the incinerator stack in fly ash.
6. More MSW will be recycled only when recycling becomes more cost-effective. Dealers and manufacturers will not be willing to accept items for recycling if they cannot count on steady markets for the recycled products.
8. **a.** *Hazardous waste* is waste material that poses a substantial hazard to human health or the environment if improperly treated, stored, transported, or disposed of.
   b. the producer of the waste
   c. the producer of the waste
10. Corrosive waste
12. Ignitable waste
14. Toxic waste
16. Neutralize it with a base to make water and a salt:
    $$HNO_3 + NaOH \longrightarrow H_2O + NaNO_3$$
18. Perpetual storage methods for hazardous wastes include (1) burial in a secured landfill, (2) deep-well injection, and (3) surface impoundment.
20. Holding in a pool with a nonreactive, plastic liner
22. A tax on chemical and petroleum industries
24. Not In My Back Yard!
26. State your own opinion.
28. U.S. Department of Energy
30. Very few radioactive wastes are regulated by RCRA. They are regulated by the Department of Energy.
32. Ten half-lives must elapse before radioactivity is reduced to negligible levels. Plutonium-239 (24,400-year half-life) requires 244,000 years; strontium-90 (28-year half-life), 280 years.
34. The waste is fused with sand at 2000° F, and the molten mixture is poured into stainless steel canisters and allowed to harden into a black glass.
36. The Low-Level Radioactivity Waste Policy Act set 1992 as the deadline for each state to take responsibility for disposal of its own low-level commercial wastes. If states prefer, they have the option of joining a multi-state compact in which one state provides a disposal site for the other states in the compact.
38. Spent fuel rods from nuclear reactors are stored under water in tanks at the reactor sites. The water shields the radiation and dissipates the heat that is generated. Many commercial plants are already running out of storage space. In 1982, Congress passed the Nuclear Waste Policy Act, which set up a timetable for building an underground repository for permanent storage of high-level waste.
40. Yucca Mountain, Nevada
42. Currently, the recommended procedure for disposing of high-level radioactive waste is to concentrate it, incorporate it into a glass or ceramic material, seal the glass or ceramic product in a corrosion-resistant steel canister, and then bury the canister in a concrete vault deep underground in a stable geologic formation composed of salt, tuff, or granite.
44. Entombment advantages: Workers not exposed to as much radiation as in dismantling; radioisotopes decay under concrete and become less radioactive; cheaper than dismantling. Entombment disadvantages: Site is ugly and has to be protected from tampering.
46. Option 1: Concentrate waste, incorporate in glass, seal in canister, and bury in concrete vault deep underground.
    Option 2: Concentrate waste, incorporate in glass, seal in canister, and bury in deep ocean trenches.
    Option 3: Transmutation of radioisotopes to less-harmful isotopes by means of neutron bombardment.
    Option 4: Shoot waste into the sun or into outer space.
48. (1) Very dry, (2) rock formation is tuff, (3) low probability of seismic activity
50. Not On Planet Earth

Appendix A

2. pH = 5

Appendix B

2. 0.153 kg
4. 2 oz
6. 1.6 kg
8. 29.6 mL
10. **a.** above by 5.4°F **b.** above by 3°C

Appendix C

2. 36,000 mg
4. 1420 mL
6. above by 34 $\mu g/L$
8. 670 nm and 370 nm
10. 250 mL
12. 7,700 g, 7.7 kg, 0.008 tons

Appendix D

2. 125
4. 252.10 mg
6. 700 m^3
8. 0.000508 g
10. 1.75 L

Appendix E

2. 3.6 kg
4. 0.80 g
6. 21.2 mol
8. 2.5 mol
10. 8.00 mol

Credits

p. iv, © 1999 S. Wanke/PhotoLink/PhotoDisc, Inc. **p. vi,** © 1996 National Archive/PhotoDisc, Inc. **p. vii,** © 1999 PhotoLink/PhotoDisc, Inc. **p. viii,** © 1995 CMCD/PhotoDisc, Inc. **p. ix,** © 1999 Russell Illig/PhotoDisc, Inc. **p. xi,** © 1995 CMCD/PhotoDisc, Inc. **p. xii,** © 1999 C. Lee/PhotoLink/PhotoDisc, Inc.

Chapter 1: p. 1, © Alan Fortune/Earth Scenes. **Fig. 1.2,** Tom Pantages. **Fig. 1.3a,** Digital Vision/PictureQuest; **b,** © 1999 Patricia Bryant/PhotoDisc, Inc. **Fig. 1.5,** Grant Heilman/Grant Heilman Photography, Inc. **Fig. 1.6a,** © PhotoDisc, Inc.; **b,** Edward R. Degginger/PictureQuest; **c,** © Gianni Dagli Orti/CORBIS. **Fig. 1.8a,** © 1997 Digital Stock Corp.; **b,** Tom McHugh/Photo Researchers, Inc.; **c,** Stephen Frisch. **Fig. 1.10,** © 1991 Richard Megna/Fundamental Photographs, NYC. **Fig. 1.11,** Adapted from *Chemistry: The Central Science,* 5th ed., by Theodore L. Brown, Eugene LeMay, and Bruce Bursten (Englewood Cliffs, NJ: Prentice Hall, 1991).

Chapter 2: p. 23, © Jeremy Woodhouse/PhotoDisc, Inc. **Fig. 2.1,** Courtesy of NASA. **Table 2.2,** Adapted from EARTH by Frank Press and Raymond Siever © 1974, 1978, 1982, 1986 by W. H. Freeman and Company. Used with permission. **Fig. 2.7a,** Breck P. Kent. **Fig. 2.8b,** © Paul Silverman/Fundamental Photographs, NYC. **Fig. 2.9a,** © J. & L. WEBER/Peter Arnold Inc. **Fig. 2.9,** © PhotoDisc, Inc. **Fig. 2.12a,** © 1999 Russell Illig/PhotoDisc, Inc.; **b,** © 1999 Ken Samuelson/PhotoDisc, Inc. **Fig. 2.13a,** © PhotoDisc, Inc.; **b,** Tom Pantages. **Fig. 2.15,** Adapted from *Environmental Science: The Way the World Works,* 3rd ed., by Bernard J. Nebel (Englewood Cliffs, NJ: Prentice Hall, 1990).

Chapter 3: p. 55, © Dr. Dennis Kunkel/Phototake NYC. **Fig. 3.1,** North Wind Picture Archives. **Fig. 3.2,** Culver Pictures, Inc. **Figs. 3.6, right, & 3.7c,** Richard Megna/Fundamental Photographs, NYC. **Fig. 3.14,** © IBM Research/Peter Arnold Inc. **Fig. 3.17,** Bettmann/CORBIS. **Fig. 3.20, left,** Lawrence Migdale/Photo Researchers, Inc.; **right,** Stephen Frisch. **p. 74,** North Wind Picture Archives.

Chapter 4: p. 82, © Pete Saloutos/Stock Market. **Fig. 4.1,** © American Institute of Physics/Emilio Segre Visual Archives. **Fig. 4.2b,** © 1996 L. Hobbs/PhotoLink/PhotoDisc, Inc. **Figs. 4.3b & 4.4,** Tom Pantages. **Fig. 4.5,** Adapted from *Modern Analytical Chemistry,* by J. N. Miller (Englewood Cliffs, NJ: Prentice Hall, 1992). **Fig. 4.7,** Science Photo Library/Photo Researchers, Inc. **Fig. 4.13a,** © Yoav Levy, Phototake NYC; **b,** © C. D. Winters, Photo Researchers, Inc. **Fig. 4.17a,** © 1988 Paul Silverman/Fundamental Photographs, NYC. **Fig. 4.18b,** © Steve Cole/PhotoDisc/PictureQuest. **p. 108,** © Michael Iscaro/Black Star/PNI.

Chapter 5: p. 116, © 1999 PhotoLink/PhotoDisc, Inc. **Fig. 5.1,** Adapted from *Chemistry: The Central Science,* 5th ed., by Theodore L. Brown, Eugene LeMay, and Bruce Bursten (Englewood Cliffs, NJ: Prentice Hall, 1991). **Fig. 5.2,** Adapted from *Introductory Chemistry: Investigating the Molecular Nature of Matter,* by J. P. Sevenair and A. R. Burkett (Dubuque, IA: Wm. C. Brown, 1997). **Fig. 5.3,** Adapted from *Chemistry: The Central Science,* 5th ed., by Brown, LeMay, and Bursten. **Fig. 5.5,** Argonne National Lab. **Fig. 5.6a,** Adapted from *Chemistry: The Central Science,* 5th ed., by Brown, LeMay, and Bursten; **b,** © Michael Collier/Stock Boston/PictureQuest. **Table 5.5,** Reprinted with permission from National Council on Radiation Protection and Measurement (NCRP87b). Copyright 1990 by the National Academy of Sciences. Courtesy of the National Academy Press. Washington, D. C. **Fig. 5.7,** © Yoav Levy/Phototake NYC. **Fig. 5.8,** From EPA survey on radon in residential dwellings, 1993. **Fig. 5.9,** Glen Allison/Getty Images/Stone. **Fig. 5.10,** Science VU/Visuals Unlimited, Inc. **Fig. 5.11a,** © Steve Callahan/Visuals Unlimited, Inc. **Fig. 5.11b,** © 1984 by Consumers Union of U.S., Inc. Yonkers, NY 10703-1057, a nonprofit organization. Reprinted with permission from the October 1984 issue of CONSUMER REPORTS for educational purposes only. No commercial use or photocopying permitted. To learn more about Consumers Union, log onto www.ConsumerReports.org. **Fig. 5.12a,** © Charles D. Winters/Photo Researchers, Inc.; **b,** © 1994 Jack Fields/Photo Researchers, Inc. **Fig. 5.13,** Adapted from *Chemistry: The Central Science,* 5th ed., by Brown, LeMay, and Bursten. **Fig. 5.16,** © 1996 U.S. Air Force/PhotoDisc, Inc. **p. 140,** Mary Evans Picture Library/Photo Researchers, Inc.

Chapter 6: p. 146, © 2000 CORBIS. **Fig. 6.1,** © Steve McCutcheon/Visuals Unlimited, Inc. **Fig. 6.2,** © Larry Lefever/Grant Heilman Photography, Inc. **Fig. 6.3,** © Bettmann/CORBIS. **Fig. 6.4a,** Adapted from *Chemistry, the Molecular Science,* 2nd ed., by John Olmsted III and Gregory M. Williams (Dubuque, IA: Wm. C. Brown, 1997); **b,** Joseph Sohm, ChromoSohm Inc./CORBIS. **Fig. 6.5,** Science Photo Library/Photo Researchers, Inc. **Fig. 6.6,** Grant Heilman/Grant Heilman Photography, Inc. **Fig. 6.8,** © 1990 Richard Megna/Fundamental Photographs, NYC. **Fig. 6.9,** Adapted from *Chemistry,* 2nd. ed., by Ronald Gillespie and David Humphreys (Englewood Cliffs, NJ: Prentice Hall, 1989). **Fig. 6.10,** Tom Pantages. **Fig. 6.12,** Photos courtesy of The Sherwin Williams Co. **Fig. 6.13,** Adapted from*Chemistry: The Central Science,* 5th ed., by Theodore L. Brown, Eugene LeMay, and Bruce Bursten (Englewood Cliffs, NJ: Prentice Hall, 1991). **Fig. 6.14,** © Science VU/Visuals Unlimited, Inc. **Fig. 6.15,** Adapted from *Chemistry,* 2nd. ed., by Gillespie and Humphreys. **p. 168,** Burton Silverman.

Chapter 7: p. 176, © CORBIS. **Figs. 7.5 & 7.8,** Tom Pantages. **Fig. 7.9,** Adapted from *Environmental Science: Systems and Solutions,* Web-Enhanced Edition, by Michael L. McKinney and Robert M. Schoch (Sudbury, MA: Jones & Barlett, 1998). **Fig. 7.10,** Tom Stack and Associates. **Fig. 7.11,** Thomas A. Schneider. **Fig. 7.12,** © Don and Pat Valenti. **Fig. 7.13,** Adapted from *Environmental Science: The Way the World Works,* 3rd ed., by Bernard J. Nebel (Englewood Cliffs, NJ: Prentice Hall, 1990). **Fig. 7.14,** Stephen Frisch. **Fig. 7.15,** Adapted from *Chemistry, the Molecular Science,* 2nd ed., by John Olmsted III and Gregory M. Williams (Dubuque, IA: Wm. C. Brown, 1997). **Fig. 7.16,** Tom Pantages. **Fig. 7.19,** Adapted from *Chemistry, the Molecular Science,* 2nd ed., by Olmsted and Williams.

Chapter 8: p. 213, John Warden/Getty Images/Stone. **Fig. 8.2a,** © J & L WEBER/Peter Arnold, Inc.; **b,** © Yoav Levy/Phototake. **Fig. 8.8,** © Charles D. Winters/Timeframe Photography/Photo Researchers, Inc. **Fig. 8.14b,** Tom Pantages. **Fig. 8.17,** North Wind Picture Archives. **Fig. 8.18b,** © 1998 PhotoDisc, Inc. **Fig. 8.19,** David Dennis/Tom Stack and Associates. **p. 242,** North Wind Picture Archives.

Chapter 9: p. 251, Bob Krist/Getty Images/Stone. **Fig. 9.2b,** © 2001 SuperStock, Inc. **Fig. 9.3b,** CORBIS. **Fig. 9.4,** Tom Pantages. **Fig. 9.5,** George Haling/Photo Researchers, Inc. **Figs. 9.6 & 9.7a,** Tom Pantages; **b,** John Coletti/Stock Boston. **Fig. 9.9,** Tom Pantages. **Figs. 9.10 & 9.12,** Patrick Watson. **Fig. 9.13a,** James Stevenson/Science Photo Library/Photo Researchers, Inc.; **b,** PictureQuest. **Fig. 9.14a,** Tom Pantages; **b,** Randy Faris/CORBIS. **Fig. 9.15,** Arthur Tilley/Getty Images/Stone. **p. 274,** Photo courtesy of Honda.

Chapter 10: p. 280, Kim Taylor/Bruce Coleman, Inc. **Fig. 10.1,** Data from D. Speidel and A. Agnew, "The World Water Budget," in D. Speidel et al., eds., *Perspectives on Water Uses and Abuses* (New York: Oxford University Press, 1988), p. 28. **Fig. 10.2,** Adapted from *Environmental Science: The Way the World Works,* 3rd ed., by Bernard J. Nebel (Englewood Cliffs, NJ: Prentice Hall, 1990). **Fig. 10.4b,** Richard C. Walters/Visuals Unlimited, Inc. **Fig. 10.7,** © Richard Megna/Fundamental Photographs, NYC. **Fig. 10.9,** PictureQuest. **Fig. 10.10,** Adapted from *Genetics: Analysis of Genes and Genomes,* 5th ed., by Daniel I. Hartl and Elizabeth W. Jones (Sudbury, MA: Jones and Bartlett, 2001). **Fig. 10.16,** © Doug Sokell/Visuals Unlimited, Inc. **Fig. 10.18,** G. L. Kiff/Visuals Unlimited, Inc. **Fig. 10.20,** Adapted from *Environmental Science: The Way the World Works,* 6th ed., by Bernard J. Nebel and Richard T. Wright (Englewood Cliffs, NJ: Prentice Hall, 1998). **p. 311,** © David Tumley/CORBIS.

Chapter 11: p. 316, Courtesy of NASA. **Fig. 11.2,** © Ray Nelson/Phototake NYC. **Fig. 11.5,** Stephen Frisch. **Fig. 11.6,** Adapted from *Environmental Science: Systems and Solutions,* Web-Enhanced Edition, by Michael L. McKinney and Robert M. Schoch (Sudbury, MA: Jones & Barlett, 1998). **Fig. 11.8,** Adapted from *Biology: Understanding Life,* 3rd ed., by Sandra Alters (Sudbury, MA: Jones and Bartlett, 2000). **Fig. 11.9,** © SUE FORD/SPL/Photo Researchers, Inc. **Fig. 11.10a,** Science Photo Library; **b,** Courtesy of Applied Ceramics, Inc., Atlanta, GA. **Fig. 11.11,**

© SYGMA. **Fig. 11.14,** © Malcom S. Kirk/Peter Arnold Inc. **Fig. 11.15,** © Frank Hanna/Visuals Unlimited, Inc. **Fig. 11.17,** Courtesy of NASA. **Fig. 11.18,** From D. H. Meadows, D. L. Meadows, and J. Randers, *Beyond the Limits* (White River Junction, VT: Chelsea Green Publishing, 1992), p.152. Used by permission of Chelsea Green Publishing Co. **Fig. 11.19,** From *Vital Signs,* 1998, p. 71. Used by permission of World Watch Institute, Washington, DC, www.worldwatch.org. **Fig. 11.21,** Adapted from *Modern Analytical Chemistry,* by J. N. Miller (Englewood Cliffs, NJ: Prentice Hall, 1992). **Fig. 11.23,** From *Vital Signs,* 1999. Used by permission of World Watch Institute, Washington, DC, www.worldwatch.org. **Fig. 11.24,** Adapted from *Environmental Science: Systems and Solutions,* Web-Enhanced Edition, by McKinney and Schoch. **Fig. 11.25,** Adapted from *Environmental Science: The Way the World Works,* 3rd ed., by Bernard J. Nebel (Englewood Cliffs, NJ: Prentice Hall, 1990). **p. 352,** © 1965 by California Institute of Technology and Carnegie Institution of Washington/Hale Observatories/Science Source/Photo Researchers, Inc.

Chapter 12: p. 358, © Lester Lefkowitz/CORBIS/Stock Market. **Fig. 12.1a,** © J. L. & F. Ziegler/Peter Arnold Inc.; **b,** Peter Arnold Inc.; **c,** Colin Garratt/CORBIS. **Fig. 12.2,** Adapted from *Environmental Science: The Way the World Works,* 6th ed., by Bernard J. Nebel and Richard T. Wright (Englewood Cliffs, NJ: Prentice Hall, 1998). **Fig. 12.3,** © A. J. Copley/Visuals Unlimited, Inc. **Fig. 12.4,** © Neil Beer/PhotoDisc, Inc. **Fig. 12.5,** Adapted from *Environmental Science: The Way the World Works,* 6th ed., by Nebel and Wright. **Fig. 12.7,** © ROSENFELD IMAGES LTD/SPL/Photo Researchers, Inc. **Fig. 12.8a,** © Kenneth J. Stein/Phototake NYC; **b,** © Phototake/PictureQuest. **Fig. 12.10,** Ben Osborne/Getty Images/Stone. **Fig. 12.11,** Ludek Pesek/Science Photo Library/Photo Researchers, Inc. **Fig. 12.12,** Adapted from *Environmental Science: The Way the World Works,* 6th ed., by Nebel and Wright. **Fig. 12.13a,** Grant Heilman/Grant Heilman Photography, Inc.; **b,** © C. P. Hickman/Visuals Unlimited, Inc. **p. 378,** Christopher Morris/Black Star. **Fig. 12.15,** U.S. Department of Energy, Energy Information Administration, 2000. **Fig. 12.16,** Adapted from *Environmental Science: The Way the World Works,* 6th ed., by Nebel and Wright.

Chapter 13: p. 385, © 1999 PhotoLink/PhotoDisc, Inc. **Fig. 13.5,** Based on information from the U.S. Department of Energy. **Fig. 13.6a,** Larry Lefever/Grant Heilman Photography; **b,** Science VU—API/Visuals Unlimited, Inc. **Fig. 13.8a** Photo Courtesy of Princeton University. **Fig. 13.9a,** Alexander Tsiara/Science Source/Photo Researchers, Inc. **Figs. 13.9b & 13.10,** Adapted from *Environmental Science: The Way the World Works,* 3rd ed., by Bernard J. Nebel (Englewood Cliffs, NJ: Prentice Hall, 1990). **Fig. 13.11,** Industrial Solar Technology: Parabolic Trough Solar Collector System at the California Correctional Institution, Tehachapi, CA. **Fig. 13.13b,** © PhotoDisc; **c,** National Renewable Energy Lab. **Fig. 13.14,** © 1998 EyeWire, Inc. **Fig. 13.15,** © David Halpern/Photo Researchers, Inc. **Fig. 13.16,** Lowell Georgia/Photo Researchers, Inc. **Fig. 13.17,** Grant Heilman Photography. **Fig. 13.18,** © Keven Schafer/Peter Arnold Inc. **Fig. 13.19,** Adapted from *Chemistry: The Central Science,* 5th ed., by Theodore L. Brown, Eugene LeMay, and Bruce Bursten (Englewood Cliffs, NJ: Prentice Hall, 1991). **Fig. 13.20,** Photo Courtesy of Honda. **p. 406,** © SYGMA.

Chapter 14: p. 411, Jim Cummins/Getty Images/FPG International. **Fig. 14.5,** Litvoy, Visuals Unlimited, Inc. **Fig. 14.10,** Harry Ransom/Humanities Research Center. **Fig. 14.11,** From *Vital Signs,* 1999, p. 148. Used by permission of World Watch Institute, Washington, DC, www.worldwatch.org. **Figs. 14.12 & 14.13,** Adapted from *Chemistry, the Molecular Science,* 2nd ed., by John Olmsted III and Gregory M. Williams (Dubuque, IA: Wm. C. Brown, 1997). **Fig. 14.14,** Adapted from *Genetics: Analysis of Genes and Genomes,* 5th ed., by Daniel I. Hartl and Elizabeth W. Jones (Sudbury, MA: Jones and Bartlett, 2001). **Fig. 14.15,** Patrick Watson. **Fig. 14.16,** Adapted from *Genetics: Analysis of Genes and Genomes,* 5th ed., by Hartl and Jones. **Fig. 14.17a,** Jackie Lewin, Royal Free Hospital/Science Photo Library/Photo Researchers, Inc.; **b,** CNRI/Science Photo Library/Photo Researchers, Inc. **Fig. 14.18,** Adapted from *Biology: Understanding Life,* 3rd ed., by Sandra Alters (Sudbury, MA: Jones and Bartlett, 2000). **Fig. 14.19b,** Dr. Thomas A. Steitz/Yale University. **Fig. 14.25,** Barrington Brown/Cold Springs Harbor Laboratory/Photo Researchers, Inc. **Fig. 14.26c,** Ken Eward/BioGrafx. **Figs. 14.27, 14.28, 14.29, 14.30,** Adapted from *Genetics: Analysis of Genes and Genomes,* 5th ed., by Hartl and Jones. **p. 446,** © Gerhard Stief/Photo Researchers, Inc. **Fig. 14.31,** Adapted from *Genetics: Analysis of Genes and Genomes,* 5th ed., by Hartl and Jones. **Fig. 14.32,** From "Lifecodes,"*Chemical and Engineering News,* October 1994, p. 9.

Chapter 15: p. 453, © 1997 Digital Vision. **Fig. 15.1,** Adapted from *Biology: Understanding Life,* 3rd ed., by Sandra Alters (Sudbury, MA: Jones and Bartlett, 2000). **Fig. 15.2,** © 1997 Digital Stock Corp. **Fig. 15.3,** Adapted from *Chemistry, the Molecular Science,* 2nd ed., by John Olmsted III and Gregory M. Williams (Dubuque, IA: Wm. C. Brown, 1997). **Fig. 15.6,** Charles Cecil/Visuals Unlimited, Inc. **Fig. 15.7,** Martin Rotker/Phototake. **Fig. 15.8,** Tom Pantages. **Fig. 15.10,** © David Toase/PhotoDisc, Inc. **p. 476,** © Steven King/Peter Arnold Inc. **Fig. 15.13,** USDA, 1992.

Chapter 16: p. 482, Isaac Geib/Grant Heilman Photography, Inc. **Fig. 16.1,** Adapted from *World of Chemistry,* Melvin D. Joesten, David O. Johnston, John T. Netterville, and James L. Wood (Philadelphia: Saunders College Publishing, 1991). **Fig. 16.3,** © Martin Wendler/Peter Arnold Inc. **Fig. 16.4,** Adapted from *Environmental Science: The Way the World Works,* 3rd ed., by Bernard J. Nebel (Englewood Cliffs, NJ: Prentice Hall, 1990). **Fig. 16.5,** Adapted from *Essentials of Physical Geography,* by Tom L. McKnight (Englewood Cliffs, NJ: Prentice Hall, 1992). **Fig. 16.6,** Garry D. McMichael/Photo Researchers, Inc. **Fig. 16.7,** Michael W. Nelson/Stock South/PNI. **Fig. 16.8,** Tom Pantages. **Fig. 16.9,** © Gianni Tortoli/Photo Researchers, Inc. **Fig. 16.14** From *Vital Signs,* 2000, p. 35. Used by permission of World Watch Institute, Washington, DC, www.worldwatch.org. **Fig. 16.15,** Adapted from *Environmental Science: The Way the World Works,* 3rd ed., by Nebel. **Fig. 16.17** From *Vital Signs,* 2000, p. 35. Used by permission of World Watch Institute, Washington, DC, www.worldwatch.org. **p. 504,** © Michael Andrews/Earth Scenes.

Chapter 17: p. 510, Chris Stutz/TWS Inc. **Figs. 17.1 & 17.3,** Tom Pantages. **Fig. 17.4,** © Betz/Visuals Unlimited, Inc. **Fig. 17.6,** Alan Pitcairn/Grant Heilman Photography, Inc. **Fig. 17.8,** Tom Pantages. **Fig. 17.9,** From Carl H. Snyder, *Extraordinary Chemistry of Ordinary Things,* © 1992, by John Wiley & Sons, Inc. Reprinted by permission of John Wiley & Sons, Inc. **Fig. 17.11,** © Jeff Greenberg/Visuals Unlimited, Inc. **Fig. 17.12,** Tom Pantages. **Fig. 17.13,** © Stanley Flegler/Visuals Unlimited, Inc. **Fig. 17.14,** Tom Pantages. **p. 528,** © BIOS/Peter Arnold Inc.

Chapter 18: p. 534, Gordon Roberts/Holt Studios/Photo Researchers. **Fig. 18.1,** © Terry G. Murphy/Animals Animals. **Fig. 18.4,** © Bettmann/CORBIS. **Fig. 18.5,** Chris Lowe/Index Stock Imagery/PictureQuest. **Fig. 18.7,** © David M. Phelps/Visuals Unlimited, Inc. **Fig. 18.8,** David Woodfall/Getty Images/Stone. **Fig. 18.9,** © David H. Wells/CORBIS. **Fig. 18.10,** CNRI/Science Photo/Photo Researchers, Inc. **Fig. 18.11,** Tom Pantages. **p. 553,** Table reprinted with permission from R. Wilson and E. A. C. Crouch, "Risk Assessment and Comparisons: An Introduction," *Science 236* (1987), p. 267. Copyright 1987, American Association for the Advancement of Science. Used by permission of Richard Wilson.

Chapter 19: p. 558, Sander/Gamma Liaison. **Fig. 19.1,** © Science VU/Visuals Unlimited, Inc. **Fig. 19.3,** Charles Rushing/Visuals Unlimited, Inc. **Fig. 19.4,** From EPA report, "Character of MSW in United States, 1997 Update." **Fig. 19.6,** D. Newman/Visuals Unlimited, Inc. **Fig. 19.7,** Adapted from *Sustaining the Earth,* 3rd ed., by G. Tyler Miller, Jr. (Belmont, CA: Wadsworth), Fig. 11-7. **Fig. 19.8,** © Tom Carroll/Phototake/PictureQuest. **Fig. 19.9,** National Academy of Sciences. **Fig. 19.10,** Courtesy of P.D.I. Inc., Mobile Incinerators. **Figs. 19.11 & 19.13,** Adapted from *Environmental Science: The Way the World Works,* 6th ed., by Bernard J. Nebel and Richard T. Wright (Englewood Cliffs, NJ: Prentice Hall, 1998). **Fig. 19.14,** Courtesy of Howard Gaskell, Baltimore County, MD. **Fig. 19.14a,** DOE Photo. **Fig. 19.15,** Holt Confer/The Image Works. **Fig. 19.16,** DOE Photo. **p. 579,** © Andy Levin/Photo Researchers, Inc. **Fig. 19.17,** Science VU/Visuals Unlimited, Inc.

Index

COMMON MONOATOMIC CATIONS

Cation	Name	Cation	Name
Al^{3+}	aluminum ion	Pb^{4+}	lead(IV) ion
Ba^{2+}	barium ion	Li^{+}	lithium ion
Bi^{3+}	bismuth ion	Mg^{2+}	magnesium ion
Cd^{2+}	cadmium ion	Mn^{2+}	manganese(II) ion
Ca^{2+}	calcium ion	Hg_2^{2+}	mercury(I) ion
Co^{2+}	cobalt(II) ion	Hg^{2+}	mercury(II) ion
Co^{3+}	cobalt(III) ion	Ni^{2+}	nickel(II) ion
Cu^{+}	copper(I) ion	K^{+}	potassium ion
Cu^{2+}	copper(II) ion	Ag^{+}	silver ion
Cr^{3+}	chromium(III) ion	Na^{+}	sodium ion
H^{+}	hydrogen ion	Sr^{2+}	strontium ion
Fe^{2+}	iron(II) ion	Sn^{2+}	tin(II) ion
Fe^{3+}	iron(III) ion	Sn^{4+}	tin(IV) ion
Pb^{2+}	lead(II) ion	Zn^{2+}	zinc ion

COMMON MONOATOMIC ANIONS

Anion	Name	Anion	Name
Br^{-}	bromide ion	N^{3-}	nitride ion
Cl^{-}	chloride ion	O^{2-}	oxide ion
F^{-}	fluoride ion	P^{3-}	phosphide ion
I^{-}	iodide ion	S^{2-}	sulfide ion

COMMON POLYATOMIC IONS

Cation	Name	Cation	Name
NH_4^{+}	ammonium ion	H_3O^{+}	hydronium ion

Anion	Name	Anion	Name
$C_2H_3O_2^{-}$	acetate ion	OH^{-}	hydroxide ion
HCO_3^{-}	bicarbonate ion	ClO^{-}	hypochlorite ion
CO_3^{2-}	carbonate ion	NO_3^{-}	nitrate ion
ClO_3^{-}	chlorate ion	NO_2^{-}	nitrite ion
ClO_2^{-}	chlorite ion	ClO_4^{-}	perchlorate ion
CrO_4^{2-}	chromate ion	MnO_4^{-}	permanganate ion
CN^{-}	cyanide ion	PO_4^{3-}	phosphate ion
$Cr_2O_7^{2-}$	dichromate ion	SO_4^{2-}	sulfate ion
HSO_4^{-}	hydrogen sulfate ion	SO_3^{2-}	sulfite ion
HSO_3^{-}	hydrogen sulfite ion	$S_2O_3^{2-}$	thiosulfate ion